SIGNAL TRANSDUCTION PATHWAYS, PART B
Stress Signaling and Transcriptional Control

ANNALS OF THE NEW YORK ACADEMY OF SCIENCES
Volume 1091

SIGNAL TRANSDUCTION PATHWAYS, PART B
Stress Signaling and Transcriptional Control

Edited by Marc Diederich

Published by Blackwell Publishing on behalf of the New York Academy of Sciences
Boston, Massachusetts
2006

Library of Congress Cataloging-in-Publication Data

Signal transduction pathways. Part B, Stress signaling and transcriptional control / edited by Marc Diederich.
  p. ; cm. – (Annals of the New York Academy of Sciences, ISSN 0077-8923 ; v. 1091)
  Includes bibliographical references and index.
  ISBN-13: 978-1-57331-647-7 (alk. paper)
  ISBN-10: 1-57331-647-4 (alk. paper)
  1. Pathology, Molecular–Congresses. 2. Cellular signal transduction–Congresses. 3. Chemoprevention–Congresses.
  I. Diederich, Marc. II. New York Academy of Sciences.
  III. Title: Stress signaling and transcriptional control. IV. Series.
  [DNLM: 1. Signal Transduction–physiology–Congresses.
  2. Chemoprevention–Congresses. 3. Neoplasms–physiopathology–Congresses. 4. Nervous System Diseases–pathology–Congresses. W1 AN626YL v. 1091 2006 / QU 375 S5786b 2006]

  RB113.S5263 2006
  500 s–dc22
  [616.07]
                                                              2006033064

The *Annals of the New York Academy of Sciences* (ISSN: 0077-8923 [print]; ISSN: 1749-6632 [online]) is published 28 times a year on behalf of the New York Academy of Sciences by Blackwell Publishing, with offices located at 350 Main Street, Malden, Massachusetts 02148 USA, PO Box 1354, Garsington Road, Oxford OX4 2DQ UK, and PO Box 378 Carlton South, 3053 Victoria Australia.

Information for subscribers: Subscription prices for 2006 are: Premium Institutional: $3850.00 (US) and £2139.00 (Europe and Rest of World).
Customers in the UK should add VAT at 5%. Customers in the EU should also add VAT at 5% or provide a VAT registration number or evidence of entitlement to exemption. Customers in Canada should add 7% GST or provide evidence of entitlement to exemption. The Premium Institutional price also includes online access to full text articles from 1997 to present, where available. For other pricing options or more information about online access to Blackwell Publishing journals, including access information and terms and conditions, please visit www.blackwellpublishing.com/nyas.

Membership information: Members may order copies of the *Annals* volumes directly from the Academy by visiting www.nyas.org/annals, emailing membership@nyas.org, faxing 212-298-3650, or calling 800-843-6927 (US only), or +1 212-298-8640, ext. 345 (International). For more information on becoming a member of the New York Academy of Sciences, please visit www.nyas.org/membership.

Journal Customer Services: For ordering information, claims, and any inquiry concerning your institutional subscription, please contact your nearest office:
UK: Email: customerservices@blackwellpublishing.com; Tel: +44 (0) 1865 778315; Fax +44 (0) 1865 471775
US: Email: customerservices@blackwellpublishing.com; Tel: +1 781 388 8599 or 1 800 835 6770 (Toll free in the USA); Fax: +1 781 388 8232
Asia: Email: customerservices@blackwellpublishing.com; Tel: +65 6511 8000; Fax: +61 3 8359 1120
Members: Claims and inquiries on member orders should be directed to the Academy at email: membership@nyas.org or Tel: +1 212 838 0230 (International) or 800-843-6927 (US only).

Printed in the USA.
Printed on acid-free paper.

Mailing: The *Annals of the New York Academy of Sciences* are mailed Standard Rate.
Postmaster: Send all address changes to *Annals of the New York Academy of Sciences*, Blackwell Publishing, Inc., Journals Subscription Department, 350 Main Street, Malden, MA 01248-5020. Mailing to rest of world by DHL Smart and Global Mail.

Copyright and Photocopying
© 2006 The New York Academy of Sciences. All rights reserved. No part of this publication may be reproduced, stored, or transmitted in any form or by any means without the prior permission in writing from the copyright holder. Authorization to photocopy items for internal and personal use is granted by the copyright holder for libraries and other users registered with their local Reproduction Rights Organization (RRO), e.g. Copyright Clearance Center (CCC), 222 Rosewood Drive, Danvers, MA 01923, USA (www.copyright.com), provided the appropriate fee is paid directly to the RRO. This consent does not extend to other kinds of copying such as copying for general distribution, for advertising or promotional purposes, for creating new collective works, or for resale. Special requests should be addressed to Blackwell Publishing at journalsrights@oxon.blackwellpublishing.com.

Disclaimer: The Publisher, the New York Academy of Sciences, and the Editors cannot be held responsible for errors or any consequences arising from the use of information contained in this publication; the views and opinions expressed do not necessarily reflect those of the Publisher, the New York Academy of Sciences, or the Editors.

Annals are available to subscribers online at the New York Academy of Sciences and also at Blackwell Synergy. Visit www.annalsnyas.org or www.blackwell-synergy.com to search the articles and register for table of contents e-mail alerts. Access to full text and PDF downloads of *Annals* articles are available to nonmembers and subscribers on a pay-per-view basis at www.annalsnyas.org.

The paper used in this publication meets the minimum requirements of the National Standard for Information Sciences Permanence of Paper for Printed Library Materials, ANSI Z39.48-1984.

ISSN: 0077-8923 (print); 1749-6632 (online)
ISBN-10: 1-57331-647-4 (paper); ISBN-13: 978-1-57331-647-7 (paper)

A catalogue record for this title is available from the British Library.

Digitization of the *Annals of the New York Academy of Sciences*

An agreement has recently been reached between Blackwell Publishing and the New York Academy of Sciences to digitize the entire run of the *Annals of the New York Academy of Sciences* back to volume one.

The back files, which have been defined as all of those issues published before 1997, will be sold to libraries as part of Blackwell Publishing's Legacy Sales Program and hosted on the Blackwell Synergy website.

Copyright of all material will remain with the rights holder. Contributors: Please contact Blackwell Publishing if you do not wish an article or picture from the *Annals of the New York Academy of Sciences* to be included in this digitization project.

ANNALS OF THE NEW YORK ACADEMY OF SCIENCES

Volume 1091
December 2006

SIGNAL TRANSDUCTION PATHWAYS, PART B
Stress Signaling and Transcriptional Control

Editor
MARC DIEDERICH

This volume is the result of a meeting entitled **Cell Signaling World: Signal Transduction Pathways as Therapeutic Targets**, held January 25–28, 2006, in Luxembourg. Parts A, C, and D can be found in Volumes 1090, 1095, and 1096, respectively.

CONTENTS

Preface. *By* MARC DIEDERICH xxvii

Part I. Oxidative Stress

Oxidative Upregulation of Bcl-2 in Healthy Lymphocytes. *By* SILVIA CRISTOFANON, SILVIA NUCCITELLI, MARIA D'ALESSIO, FLAVIA RADOGNA, MILENA DE NICOLA, ANTONIO BERGAMASCHI, CLAUDIA CERELLA, ANDREA MAGRINI, MARC DIEDERICH, AND LINA GHIBELLI 1

Intracellular Pro-oxidant Activity of Melatonin Deprives U937 Cells of Reduced Glutathione without Affecting Glutathione Peroxidase Activity. *By* MARIA CRISTINA ALBERTINI, FLAVIA RADOGNA, AUGUSTO ACCORSI, FRANCESCO UGUCCIONI, LAURA PATERNOSTER, CLAUDIA CERELLA, MILENA DE NICOLA, MARIA D'ALESSIO, ANTONIO BERGAMASCHI, ANDREA MAGRINI, AND LINA GHIBELLI 10

Mitochondrial "Movement" and Lens Optics following Oxidative Stress from UV-B Irradiation: Cultured Bovine Lenses and Human Retinal Pigment Epithelial Cells (ARPE-19) as Examples. *By* VLADIMIR BANTSEEV AND HYUN-YI YOUN 17

2-Methoxyestradiol Inhibits Superoxide Anion Generation while It Enhances Superoxide Dismutase Activity in Swine Granulosa Cells. *By* GIUSEPPINA BASINI, SUJEN ELEONORA SANTINI, AND FRANCESCA GRASSELLI 34

Role of Reactive Oxygen Species in Kv Channel Inhibition and Vasoconstriction Induced by TP Receptor Activation in Rat Pulmonary Arteries. *By* ANGEL COGOLLUDO, GIOVANNA FRAZZIANO, LAURA COBEÑO, LAURA MORENO, FEDERICA LODI, EDUARDO VILLAMOR, JUAN TAMARGO, AND FRANCISCO PEREZ-VIZCAINO 41

DNA Strand Breaks by Metal-Induced Oxygen Radicals in Purified *Salmonella typhimurium* DNA. *By* EZZATOLLAH KEYHANI, FATEMEH ABDI-OSKOUEI, FARNOOSH ATTAR, AND JACQUELINE KEYHANI 52

Antioxidant Enzymes during Hypoxia–Anoxia Signaling Events in *Crocus sativus* L. Corm. *By* EZZATOLLAH KEYHANI, LILA GHAMSARI, JACQUELINE KEYHANI, AND MAHNAZ HADIZADEH 65

Ataxia-Telangiectasia-Mutated-Dependent Activation of Ku in Human Fibroblasts Exposed to Hydrogen Peroxide. *By* JONG HWA LEE, JI HOON YU, KYUNG HWAN KIM, AND HYEYOUNG KIM 76

Regulation of 2-Deoxy-D-Glucose Transport, Lactate Metabolism, and MMP-2 Secretion by the Hypoxia Mimetic Cobalt Chloride in Articular Chondrocytes. *By* ALI MOBASHERI, NICOLA PLATT, COLIN THORPE, AND MEHDI SHAKIBAEI 83

Oxidative Stress Response in Telomerase-Immortalized Fibroblasts from a Centenarian. *By* CHIARA MONDELLO, MARIA GRAZIA BOTTONE, SAKON NORIKI, CRISTIANA SOLDANI, CARLO PELLICCIARI, AND ANNA IVANA SCOVASSI 94

Differential Modulation of AMPK Signaling Pathways by Low or High Levels of Exogenous Reactive Oxygen Species in Colon Cancer Cells. *By* IN-JA PARK, JIN-TAEK HWANG, YOUNG MIN KIM, JOOHUN HA, AND OCK JIN PARK 102

Alterations in Salivary Antioxidants, Nitric Oxide, and Transforming Growth Factor-β_1 in Relation to Disease Activity in Crohn's Disease Patients. *By* ALI REZAIE, FAKHTEH GHORBANI, AZADEH ESHGHTORK, MOHAMMAD J. ZAMANI, GHOLAMREZA DEHGHAN, BARDIA TAGHAVI, SHEKOUFEH NIKFAR, AZADEH MOHAMMADIRAD, NASSER E. DARYANI, AND MOHAMMAD ABDOLLAHI 110

Control of Bioamine Metabolism by 5-HT$_{2B}$ and α_{1D} Autoreceptors through Reactive Oxygen Species and Tumor Necrosis Factor-α Signaling in Neuronal Cells. *By* BENOIT SCHNEIDER, MATHÉA PIETRI, SOPHIE MOUILLET-RICHARD, MYRIAM ERMONVAL, VINCENT MUTEL, JEAN-MARIE LAUNAY, AND ODILE KELLERMANN 123

Determination of Oxidative Stress Status and Concentration of TGF-β1 in the Blood and Saliva of Osteoporotic Subjects. *By* GHOLAMREZA YOUSEFZADEH, BAGHER LARIJANI, AZADEH MOHAMMADIRAD, RAMIN HESHMAT, GHOLAMREZA DEHGHAN, ROJA RAHIMI, AND MOHAMMAD ABDOLLAHI 142

Part II. Transcriptional Control

Targeting Signal-Transducer-and-Activator-of-Transcription-3 for Prevention and Therapy of Cancer: Modern Target but Ancient Solution. *By* BHARAT B. AGGARWAL, GAUTAM SETHI, KWANG SEOK AHN, SANTOSH K. SANDUR, MANOJ K. PANDEY, AJAIKUMAR B. KUNNUMAKKARA, BOKYUNG SUNG, AND HARUYO ICHIKAWA 151

Gene Expression Modulation in A549 Human Lung Cells in Response to
Combustion-Generated Nano-Sized Particles. *By*
ANDREA ARENZ, CHRISTINE E. HELLWEG, NEVENA STOJICIC,
CHRISTA BAUMSTARK-KHAN, AND HORST-HENNING GROTHEER 170

Multiple Levels of Control of the Expression of the Human AβH-J-J Locus
Encoding Aspartyl-β-hydroxylase, Junctin, and Junctate. *By*
GIORDANA FERIOTTO, ALESSIA FINOTTI, GIULIA BREVEGLIERI,
SUSAN TREVES, FRANCESCO ZORZATO, AND ROBERTO GAMBARI 184

Activation of Nuclear Factor κB by Different Agents: Influence of Culture
Conditions in a Cell-Based Assay. *By* CHRISTINE E. HELLWEG,
ANDREA ARENZ, SUSANNE BOGNER, CLAUDIA SCHMITZ, AND
CHRISTA BAUMSTARK-KHAN 191

Atrial Appendage Transcriptional Profile in Patients with Atrial Fibrillation
with Structural Heart Diseases. *By* MARIA S. KHARLAP,
ANGELICA V. TIMOFEEVA, LUDMILA E. GORYUNOVA, GEORGE L. KHASPEKOV,
SERGEY L. DZEMESHKEVICH, VLADIMIR V. RUSKIN, RENAT S. AKCHURIN,
SERGEY P. GOLITSYN, AND ROBERT SH. BEABEALASHVILLI 205

DNA Hypomethylation of *CAGE* Promotors in Squamous Cell Carcinoma
of Uterine Cervix. *By* TAEK SANG LEE, JAE WEON KIM,
GYEONG HOON KANG, NOH HYUN PARK, YONG SANG SONG,
SOON BEOM KANG, AND HYO PYO LEE 218

The *MECP2* Gene Mutation Screening in Rett Syndrome Patients from
Croatia. *By* TANJA MATIJEVI, JELENA KNEŽEVIČ, INGEBORG BARIŠIĆ,
BISERKA REŠIĆ, VIDA ČULIĆ, AND JASMINKA PAVELIĆ 225

Prostaglandins Regulate Transcription by Means of Prostaglandin Response
Elements Located in the Promoters of Mammalian Na,K-ATPase β1
Subunit Genes. *By* KEIKANTSE MATLHAGELA AND MARY TAUB 233

Different Modulation of ER-Mediated Transactivation by Xenobiotic Nuclear
Receptors Depending on the Estrogen Response Elements and Estrogen
Target Cell Types. *By* GYESIK MIN 244

Effects of TK Promotor and Hepatocyte Nuclear Factor-4 in CAR-Mediated
Transcriptional Activity of Phenobarbital Responsive Unit of *CYP2B*
Gene in Monkey Kidney Epithelial-Derived Cell Line COS-7. *By*
GYESIK MIN 258

Expression of the E2F Family of Transcription Factors and Its Clinical
Relevance in Ovarian Cancer. *By* DANIEL REIMER, SUSANN SADR,
ANNEMARIE WIEDEMAIR, GEORG GOEBEL, NICOLE CONCIN,
GERDA HOFSTETTER, CHRISTIAN MARTH, AND ALAIN G. ZEIMET 270

Isolation and Characterization of the Rat SND p102 Gene Promoter:
Putative Role for Nuclear Factor-Y in Regulation of Transcription. *By*
LORENA RODRÍGUEZ, NEREA BARTOLOMÉ, BEGOÑA OCHOA, AND
MARÍA J. MARTÍNEZ 282

The cAMP-Responsive Unit of the Human Insulin-Like Growth Factor–Binding
Protein-1 Coinstitutes a Functional Insulin-Response Element. *By*
GHISLAINE SCHWEIZER-GROYER, GUILLAUME FALLOT,
FRANÇOISE CADEPOND, CHRISTELLE GIRARD, AND ANDRÉ GROYER 296

c-Jun and JunB Are Essential for Hypoglycemia-Mediated *VEGF* Induction. *By*
BJÖRN TEXTOR, MELANIE SATOR-SCHMITT, KARL HARTMUT RICHTER,
PETER ANGEL, AND MARINA SCHORPP-KISTNER 310

Altered Gene Expression Pattern in Peripheral Blood Leukocytes from Patients with Arterial Hypertension. *By* A.V. TIMOFEEVA, L.E. GORYUNOVA, G.L. KHASPEKOV, D.A. KOVALEVSKII, A.V. SCAMROV, O.S. BULKINA, YU.A. KARPOV, K.A. TALITSKII, V.V. BUZA, V.V. BRITAREVA, AND R.SH. BEABEALASHVILLI 319

Effects of AT_1 Receptor-Mediated Endocytosis of Extracellular Ang II on Activation of Nuclear Factor-κB in Proximal Tubule Cells. *By* JIA L. ZHUO, OSCAR A. CARRETERO, AND XIAO C. LI 336

Part III. Histone Deacetylase

Retinoic Acid and Histone Deacetylase Inhibitor BML-210 Inhibit Proliferation of Human Cervical Cancer HeLa Cells. *By* VERONIKA V. BORUTINSKAITE, RUTA NAVAKAUSKIENE, AND KARL-ERIC MAGNUSSON 346

Effects of Histone Deacetylase Inhibitors, Sodium Phenyl Butyrate, and Vitamin B3, in Combination with Retinoic Acid, on Granulocytic Differentiation of Human Promyelocytic Leukemia HL-60 Cells. *By* RASA MERZVINSKYTE, GRAZINA TREIGYTE, JURATE SAVICKIENE, KARL-ERIC MAGNUSSON, AND RUTA NAVAKAUSKIENE 356

The Histone Deacetylase Inhibitor FK228 Distinctly Sensitizes the Human Leukemia Cells to Retinoic Acid-Induced Differentiation. *By* JURATE SAVICKIENE, GRAZINA TREIGYTE, VERONIKA BORUTINSKAITE, RUTA NAVAKAUSKIENE, AND KARL-ERIC MAGNUSSON 368

Effect of Valproic Acid, a Histone Deacetylase Inhibitor, on Cell Death and Molecular Changes Caused by Low-Dose Irradiation. *By* DARINA ZÁŠKODOVÁ, MARTINA ŘEZÁČOVÁ, JIŘINA VÁVROVÁ, DORIS VOKURKOVÁ, AND ALEŠ TICHÝ 385

Part IV. Novel Technological and Therapeutical Approaches

Protein Folding Information in Nucleic Acids Which Is Not Present in the Genetic Code. *By* JAN C. BIRO 399

Antigens and Cytokine Genes in Antitumor Vaccines: The Importance of the Temporal Delivery Sequence in Antitumor Signals. *By* MARÍA JOSÉ HERRERO, RAFAEL BOTELLA, FRANCISCO DASÍ, ROSA ALGÁS, MARÍA SÁNCHEZ, AND SALVADOR F. ALIÑO 412

Arrest of Cancer Cell Proliferation by dsRNAs. *By* TATYANA O. KABILOVA, ALBINA V. VLADIMIROVA, ELENA L. CHERNOLOVSKAYA, AND VALENTIN V. VLASSOV 425

Design and Functional Activity of Phosphopeptides with Potential Immunomodulating Capacity, Based on the Sequence of Grb2-Associated Binder 1. *By* AKOS KERTESZ, BALAZS TAKACS, GYORGYI VARADI, GABOR K. TOTH, AND GABRIELLA SARMAY 437

Distinct Activity of Peptide Mimetic Intracellular Ligands (Pepducins) for Proteinase-Activated Receptor-1 in Multiple Cells/Tissues. *By* SATOKO KUBO, TSUYOSHI ISHIKI, ICHIKO DOE, FUMIKO SEKIGUCHI, HIROYUKI NISHIKAWA, KENZO KAWAI, HIROFUMI MATSUI, AND ATSUFUMI KAWABATA 445

A New Method to Assess Drug Sensitivity on Breast Tumor Acute Slices
  Preparation. *By* PEDRO MESTRES, ANDREA MORGUET, WERNER SCHMIDT,
  AXEL KOB, AND ELKE THEDINGA 460

Process Simulation in a Mechatronic Bioreactor Device with Speed-Regulated
  Motors for Growing of Three-Dimensional Cell Cultures. *By*
  MINA MIHAILOVA, VASSIL TRENEV, PENKA GENOVA, AND
  SPIRO KONSTANTINOV 470

Animal Model of Drug-Resistant Tumor Progression. *By*
  NADEZDA MIRONOVA, OLGA SHKLYAEVA, EKATERINA ANDREEVA,
  NELLY POPOVA, VASILYI KALEDIN, VALERYI NIKOLIN,
  VALENTIN VLASSOV, AND MARINA ZENKOVA 490

Preparation and Characterization of Recombinant Chicken Growth
  Hormone (chGH) and Its Putative Antagonist chGH G119R Mutein. *By*
  HELENA E. PACZOSKA-ELIASIEWICZ, GILI SALOMON, SHAY REICHER,
  EUGENE E. GUSSAKOWSKY, ANNA HRABIA, AND ARIEH GERTLER 501

Induction of Apoptosis of Osteoclasts by Targeting Transcription Factors
  with Decoy Molecules. *By* ROBERTA PIVA, LETIZIA PENOLAZZI,
  MARGHERITA ZENNARO, ERCOLINA BIANCHINI, EROS MAGRI,
  MONICA BORGATTI, ILARIA LAMPRONTI, ELISABETTA LAMBERTINI,
  ELISA TAVANTI, AND ROBERTO GAMBARI 509

Competition Effects Shape the Response Sensitivity and Kinetics of
  Phosphorylation Cycles in Cell Signaling. *By* CARLOS SALAZAR AND
  THOMAS HÖFER 517

Preparation of Leptin Antagonists by Site-Directed Mutagenesis of Human,
  Ovine, Rat, and Mouse Leptin's Site III: Implications on Blocking
  Undesired Leptin Action *In Vivo*. *By* GILI SOLOMON, LEONORA
  NIV-SPECTOR, DANA GONEN-BERGER, ISABELLE CALLEBAUT,
  JEAN DJIANE, AND ARIEH GERTLER 531

Dual Activity of Phosphorothioate CpG Oligodeoxynucleotides on HIV:
  Reactivation of Latent Provirus and Inhibition of Productive Infection
  in Human T Cells. *By* CARSTEN SCHELLER, ANETT ULLRICH, STEFAN
  LAMLA, ULF DITTMER, AXEL RETHWILM, AND ELENI KOUTSILIERI 540

Prostanoids with Cyclopentenone Structure as Tools for the Characterization
  of Electrophilic Lipid–Protein Interactomes. *By*
  KONSTANTINOS STAMATAKIS AND DOLORES PÉREZ-SALA 548

Index of Contributors 571

Contents of the Other Volumes

PART A: Apoptotic and Extracellular Signaling

Volume 1090, December 2006

CONTENTS

Preface. By MARC DIEDERICH xxvii

Part I. Apoptotic Cell Signaling Mechanisms

Pleiotropic Effects of PI-3′ Kinase/Akt Signaling in Human Hepatoma Cell Proliferation and Drug-Induced Apoptosis. By CATHERINE ALEXIA, MARLÈNE BRAS, GUILLAUME FALLOT, NATHALIE VADROT, FANNY DANIEL, MALIKA LASFER, HOUDA TAMOUZA, AND ANDRÉ GROYER 1

Expression of Bcl-xL, Bax, and p53 in Primary Tumors and Lymph Node Metastases in Oral Squamous Cell Carcinoma. By MAREK BALTAZIAK, EWA DURAJ, MARIUSZ KODA, ANDRZEJ WINCEWICZ, MARCIN MUSIATOWICZ, LUIZA KANCZUGA-KODA, MAGDALENA SZYMANSKA, TOMASZ LESNIEWICZ, AND BOGUSLAW MUSIATOWICZ 18

Reactive Structural Dynamics of Synaptic Mitochondria in Ischemic Delayed Neuronal Death. By CARLO BERTONI-FREDDARI, PATRIZIA FATTORETTI, TIZIANA CASOLI, GIUSEPPINA DI STEFANO, MORENO SOLAZZI, ELISA PERNA, AND CLARA DE ANGELIS 26

Selective Resistance of Tetraploid Cancer Cells against DNA Damage-Induced Apoptosis. By MARIA CASTEDO, ARNAUD COQUELLE, ILIO VITALE, SONIA VIVET, SHAHUL MOUHAMAD, SOPHIE VIAUD, LAURENCE ZITVOGEL, AND GUIDO KROEMER 35

Molecular Determinants Involved in the Increase of Damage-Induced Apoptosis and Delay of Secondary Necrosis due to Inhibition of Mono(ADP-Ribosyl)ation. By CLAUDIA CERELLA, CRISTINA MEARELLI, SERGIO AMMENDOLA, MILENA DE NICOLA, MARIA D'ALESSIO, ANDREA MAGRINI, ANTONIO BERGAMASCHI, AND LINA GHIBELLI 50

Magnetic Fields Protect from Apoptosis via Redox Alteration. By M. DE NICOLA, S. CORDISCO, C. CERELLA, M.C. ALBERTINI, M. D'ALESSIO, A. ACCORSI, A. BERGAMASCHI, A. MAGRINI, AND L. GHIBELLI 59

The Cleavage Mode of Apoptotic Nuclear Vesiculation Is Related to Plasma Membrane Blebbing and Depends on Actin Reorganization. By M. DE NICOLA, C. CERELLA, M. D'ALESSIO, S. COPPOLA, A. MAGRINI, A. BERGAMASCHI, AND L. GHIBELLI 69

Experimental Apoptosis Provides Clues about the Role of Mitochondrial Changes in Neuronal Death. *By* PATRIZIA FATTORETTI, CARLO BERTONI-FREDDARI, RINA RECCHIONI, BELINDA GIORGETTI, MARTA BALIETTI, YESSICA GROSSI, MORENO SOLAZZI, TIZIANA CASOLI, GIUSEPPINA DI STEFANO, AND FIORELLA MARCHESELLI 79

Alterations in mRNA Expression of Apoptosis-Related Genes *BCL2*, *BAX*, *FAS*, *Caspase-3*, and the Novel Member *BCL2L12* after Treatment of Human Leukemic Cell Line HL60 with the Antineoplastic Agent Etoposide. *By* KOSTAS V. FLOROS, HELLINIDA THOMADAKI, DIMITRA FLOROU, MAROULIO TALIERI, AND ANDREAS SCORILAS 89

Using Janus Green B to Study Paraquat Toxicity in Rat Liver Mitochondria: Role of ACE Inhibitors (Thiol and Nonthiol ACEi). *By* M. GHAZI-KHANSARI, A. MOHAMMADI-BARDBORI, AND M-J. HOSSEINI 98

Membrane Fluidity Changes Are Associated with Benzo[*a*]Pyrene-Induced Apoptosis in F258 Cells: Protection by Exogenous Cholesterol. *By* MORGANE GORRIA, XAVIER TEKPLI, ODILE SERGENT, LAURENCE HUC, FRANÇOIS GABORIAU, MARY RISSEL, MARTINE CHEVANNE, MARIE-THÉRÈSE DIMANCHE-BOITREL, AND DOMINIQUE LAGADIC-GOSSMANN 108

Metal-Containing Proteins in the Apoptosis and Redox Processes in the Rat Prostate and Human Prostate Cells. *By* I. GRBAVAC, C. WOLF, N. WENDA, D. ALBER, M. KÜHBACHER, D. BEHNE, AND A. KYRIAKOPOULOS 113

Does Transduced p27 Induce Apoptosis in Human Tumor Cell Lines? *By* MIRA GRDIŠA, ANA-MATEA MIKECIN, AND MIROSLAV POZNIC 120

Apoptotic Cell Signaling in Lymphocytes from HIV$^+$ Patients during Successful Therapy. *By* SANDRO GRELLI, EMANUELA BALESTRIERI, CLAUDIA MATTEUCCI, ANTONELLA MINUTOLO, GABRIELLA D'ETTORRE, FILIPPO LAURIA, FRANCESCO MONTELLA, VINCENZO VULLO, STEFANO VELLA, CARTESIO FAVALLI, ANTONIO MASTINO, AND BEATRICE MACCHI 130

Capacitation and Acrosome Reaction in Nonapoptotic Human Spermatozoa. *By* SONJA GRUNEWALD, THOMAS BAUMANN, UWE PAASCH, AND HANS-JUERGEN GLANDER 138

Caspase Activation and Extracellular Signal-Regulated Kinase/Akt Inhibition Were Involved in Luteolin-Induced Apoptosis in Lewis Lung Carcinoma Cells. *By* JIN-HYUNG KIM, EUN-OK LEE, HYO-JUNG LEE, JIN-SOOK KU, MIN-HO LEE, DEOK-CHUN YANG, AND SUNG-HOON KIM 147

Ceramide Modulation of Antigen-Triggered Ca^{2+} Signals and Cell Fate: Diversity in the Responses of Various Immunocytes. *By* ENDRE KISS, GABRIELLA SÁRMAY, AND JÁNOS MATKÓ 161

Caspase-3 Activation, Bcl-2 Contents, and Soluble FAS-Ligand Are Not Related to the Inflammatory Marker Profile in Patients with Sepsis and Septic Shock. *By* FABIAN KRIEBEL, SILKE WITTEMANN, HSIN-YUN HSU, THOMAS JOOS, MANFRED WEISS, AND E. MARION SCHNEIDER 168

Two Forms of the Nuclear Matrix–Bound p53 Protein in HEK293 Cells. *By* MARIA A. LAPSHINA, IGOR I. PARKHOMENKO, AND ALEXEI A. TERENTIEV 177

Jpk, a Novel Cell Death Inducer, Regulates the Expression of Hoxa7 in F9 Teratocarcinoma Cells, but not during Apoptosis. *By* EUN YOUNG LEE AND MYOUNG HEE KIM 182

HGF/SF Regulates Expression of Apoptotic Genes in MCF-10A Human Mammary Epithelial Cells. *By* CATHERINE LEROY, JULIEN DEHEUNINCK, SYLVIE REVENEAU, BÉNÉDICTE FOVEAU, ZONGLING JI, CÉLINE VILLENET, SABINE QUIEF, DAVID TULASNE, JEAN-PIERRE KERCKAERT, AND VÉRONIQUE FAFEUR 188

Arsenic Trioxide Represses NF-κB Activation and Increases Apoptosis in ATRA-Treated APL Cells. *By* JULIE MATHIEU AND FRANÇOISE BESANÇON 203

Cytotoxicity of TRAIL/Anticancer Drug Combinations in Human Normal Cells. *By* OLIVIER MEURETTE, ANNE FONTAINE, AMELIE REBILLARD, GWENAELLE LE MOIGNE, THIERRY LAMY, DOMINIQUE LAGADIC-GOSSMANN, AND MARIE-THERESE DIMANCHE-BOITREL 209

Hyperpolarization of Plasma Membrane of Tumor Cells Sensitive to Antiapoptotic Effects of Magnetic Fields. *By* S. NUCCITELLI, C. CERELLA, S. CORDISCO, M.C. ALBERTINI, A. ACCORSI, M. DE NICOLA, M. D'ALESSIO, F. RADOGNA, A. MAGRINI, A. BERGAMASCHI, AND L. GHIBELLI 217

Melatonin as an Apoptosis Antagonist. *By* FLAVIA RADOGNA, LAURA PATERNOSTER, MARIA CRISTINA ALBERTINI, AUGUSTO ACCORSI, CLAUDIA CERELLA, MARIA D'ALESSIO, MILENA DE NICOLA, SILVIA NUCCITELLI, ANDREA MAGRINI, ANTONIO BERGAMASCHI, AND LINA GHIBELLI 226

Prevention of p53 Degradation in Human MCF-7 Cells by Proteasome Inhibitors Does Not Mimic the Action of Roscovitine. *By* CARMEN RANFTLER, MARIETA GUEORGUIEVA, AND JÒZEFA WESIERSKA-GADEK 234

Role of ATP in Trauma-Associated Cytokine Release and Apoptosis by P2X7 Ion Channel Stimulation. *By* E. MARION SCHNEIDER, KATRIN VORLAENDER, XUELING MA, WEIDONG DU, AND MANFRED WEISS 245

Experimental Sepsis: Characteristics of Activated Macrophages and Apoptotic Cells in the Rat Spleen. *By* HELLE EVI SIMOVART, ANDRES AREND, HELLE TAPFER, KERSTI KOKK, MARINA AUNAPUU, ELLE POLDOJA, GUNNAR SELSTAM, AND AADE LIIGANT 253

Insulin-Like Growth Factor-I Receptor Correlates with Connexin 26 and Bcl-xL Expression in Human Colorectal Cancer. *By* STANISLAW SULKOWSKI, LUIZA KANCZUGA-KODA, MARIUSZ KODA, ANDRZEJ WINCEWICZ, AND MARIOLA SULKOWSKA 265

Characterization of the Proapoptotic Intracellular Mechanisms Induced by a Toxic Conformer of the Recombinant Human Prion Protein Fragment 90–231. *By* VALENTINA VILLA, ALESSANDRO CORSARO, STEFANO THELLUNG, DOMENICO PALUDI, KATIA CHIOVITTI, VALENTINA VENEZIA, MARIO NIZZARI, CLAUDIO RUSSO, GENNARO SCHETTINI, ANTONIO ACETO, AND TULLIO FLORIO 276

Role of NADPH Oxidase and Calcium in Cerulein-Induced Apoptosis: Involvement of Apoptosis-Inducing Factor. *By* JI HOON YU, KYUNG HWAN KIM, AND HYEYOUNG KIM 292

Part II. Extracellular Matrix Interactions

Signaling for Integrin $\alpha 5/\beta 1$ Expression in *Helicobacter pylori*–Infected Gastric Epithelial AGS Cells. *By* SOON OK CHO, KYUNG HWAN KIM, JOO-HEON YOON, AND HYEYOUNG KIM 298

Human Recombinant Vasostatin-1 May Interfere with Cell–Extracellular Matrix Interactions. *By* VALENTINA DI FELICE, FRANCESCO CAPPELLO, ANTONELLA MONTALBANO, NELLA ARDIZZONE, CLAUDIA CAMPANELLA, ANGELA DE LUCA, DANIELA AMELIO, BRUNO TOTA, ANGELO CORTI, AND GIOVANNI ZUMMO 305

Microgravity Signal Ensnarls Cell Adhesion, Cytoskeleton, and Matrix Proteins of Rat Osteoblasts: Osteopontin, CD44, Osteonectin, and α-Tubulin. *By* YASUHIRO KUMEI, SADAO MORITA, HISAKO KATANO, HIDEO AKIYAMA, MASAHIKO HIRANO, KEI'ICHI OYHA, AND HITOYATA SHIMOKAWA 311

14-3-3 Proteins Bind Both Filamin and $\alpha_L \beta_2$ Integrin in Activated T Cells. *By* SUSANNA M. NURMI, CARL G. GAHMBERG, AND SUSANNA C. FAGERHOLM 318

MAP Kinases

Grb2-Associated Binder 1 (Gab1) Adaptor/Scaffolding Protein Regulates Erk Signal in Human B Cells. *By* ADRIENN ANGYAL, DAVID MEDGYESI, AND GABRIELLA SARMAY 326

CXC Receptor and Chemokine Expression in Human Meningioma: SDF1/CXCR4 Signaling Activates ERK1/2 and Stimulates Meningioma Cell Proliferation. *By* FEDERICA BARBIERI, ADRIANA BAJETTO, CAROLA PORCILE, ALESSANDRA PATTAROZZI, ALESSANDRO MASSA, GIANLUIGI LUNARDI, GIANLUIGI ZONA, ALESSANDRA DORCARATTO, JEAN LOUIS RAVETTI, RENATO SPAZIANTE, GENNARO SCHETTINI, AND TULLIO FLORIO 332

Reduction of Bcr-Abl Function Leads to Erythroid Differentiation of K562 Cells via Downregulation of ERK. *By* A. BRÓZIK, N.P. CASEY, CS. HEGEDŰS, A. BORS, A. KOZMA, H. ANDRIKOVICS, M. GEISZT, K. NÉMET, AND M. MAGÓCSI 344

The MAPK Pathway and HIF-1 Are Involved in the Induction of the Human PAI-1 Gene Expression by Insulin in the Human Hepatoma Cell Line HepG2. *By* ELITSA Y. DIMOVA AND THOMAS KIETZMANN 355

Role of Mitogen-Activated Protein Kinases, NF-κB, and AP-1 on Cerulein-Induced IL-8 Expression in Pancreatic Acinar Cells. *By* KYUNG DON JU, JI HOON YU, HYEYOUNG KIM, AND KYUNG HWAN KIM 368

Upregulation of VEGF by 15-Deoxy-$\Delta^{12,14}$-Prostaglandin J_2 via Heme Oxygenase-1 and ERK1/2 Signaling in MCF-7 Cells. *By* EUN-HEE KIM, HYE-KYUNG NA, AND YOUNG-JOON SURH 375

SDF-1 Controls Pituitary Cell Proliferation through the Activation of ERK1/2 and the Ca^{2+}-Dependent, Cytosolic Tyrosine Kinase Pyk2. *By* ALESSANDRO MASSA, SILVIA CASAGRANDE, ADRIANA BAJETTO, CAROLA PORCILE, FEDERICA BARBIERI, STEFANO THELLUNG, SARA ARENA, ALESSANDRA PATTAROZZI, MONICA GATTI, ALESSANDRO CORSARO, MAURO ROBELLO, GENNARO SCHETTINI, AND TULLIO FLORIO 385

Insulin Primes Human Neutrophils for CCL3-Induced Migration: Crucial Role for JNK 1/2. *By* FABRIZIO MONTECUCCO, GIORDANO BIANCHI, MARIA BERTOLOTTO, GIORGIO VIVIANI, FRANCO DALLEGRI, AND LUCIANO OTTONELLO 399

Doxorubicin-Induced MAPK Activation in Hepatocyte Cultures Is Independent of Oxidant Damage. *By* ROSAURA NAVARRO, ROSA MARTÍNEZ, IDOIA BUSNADIEGO, M. BEGOÑA RUIZ-LARREA, AND JOSÉ IGNACIO RUIZ-SANZ 408

Superoxide Anions Are Involved in Doxorubicin-Induced ERK Activation in Hepatocyte Cultures. *By* ROSAURA NAVARRO, IDOIA BUSNADIEGO, M. BEGOÑA RUIZ-LARREA, AND JOSÉ IGNACIO RUIZ-SANZ 419

MAPKinase Gene Expression, as Determined by Microarray Analysis, Distinguishes Uncomplicated from Complicated Reconstitution after Major Surgical Trauma. *By* E. MARION SCHNEIDER, MANFRED WEISS, WEIDONG DU, GERHARD LEDER, KLAUS BUTTENSCHÖN, ULRICH C. LIENER, AND UWE B. BRÜCKNER 429

Effects of Chemical Ischemia on Cerebral Cortex Slices: Focus on Mitogen-Activated Protein Kinase Cascade. *By* ANNA SINISCALCHI, SABRINA CAVALLINI, SILVIA MARINO, SOFIA FALZARANO, LARA FRANCESCHETTI, AND RITA SELVATICI 445

Amyloid Precursor Protein Modulates ERK-1 and -2 Signaling. *By* VALENTINA VENEZIA, MARIO NIZZARI, EMANUELA REPETTO, ELISABETTA VIOLANI, ALESSANDRO CORSARO, STEFANO THELLUNG, VALENTINA VILLA, PIA CARLO, GENNARO SCHETTINI, TULLIO FLORIO, AND CLAUDIO RUSSO 455

Index of Contributors 467

PART C: Cell Signaling in Health and Disease

Volume 1095, January 2007

CONTENTS

Preface. By MARC DIEDERICH xxvii

Part I. Cancer

Cationic Surfactants Induce Apoptosis in Normal and Cancer Cells. By
RIYO ENOMOTO, CHIE SUZUKI, MASATAKA OHNO, TOSHINORI OHASI,
RYOKO FUTAGAMI, KEIKO ISHIKAWA, MIKA KOMAE, TAKAYUKI NISHINO,
YASUO KONISHI, AND EIBAI LEE 1

DMNQ S-64 Induces Apoptosis Via Caspase Activation and Cyclooxygenase-2
Inhibition in Human Nonsmall Lung Cancer Cells. By EU-SOO LIM,
YUN-HEE RHEE, MIN-KYU PARK, BEOM-SANG SHIM, KYOO-SEOK AHN,
HEE KANG, HWA-SEUNG YOO, AND SUNG-HOON KIM 7

Bcl-2 Expression in Oral Squamous Cell Carcinoma. By B. POPOVIĆ, B. JEKIĆ,
I. NOVAKOVIĆ, L.J. LUKOVIĆ, Z. TEPAVČEVIĆ, V. JURIŠIĆ,
M. VUKADINOVIĆ, AND J. MILAŠIN 19

Apoptotic Effect of Celecoxib Dependent Upon p53 Status in Human Ovarian
Cancer Cells. By YOO-CHEOL SONG, SU-HYEONG KIM,
YONG-SUNG JUHNN, AND YONG-SANG SONG 26

Breast Cancer Cells Response to the Antineoplastic Agents Cisplatin,
Carboplatin, and Doxorubicin at the mRNA Expression Levels of Distinct
Apoptosis-Related Genes, Including the New Member, *BCL2L12*. By
HELLINIDA THOMADAKI AND ANDREAS SCORILAS 35

Effect of Distinct Anticancer Drugs on the Phosphorylation of p53 Protein at
Serine 46 in Human MCF-7 Breast Cancer Cells. By
JÓZEFA WĘSIERSKA-GĄDEK, MARIETA GUEORGUIEVA,
IRENE HERBACEK, AND CARMEN RANFTLER 45

Significant Coexpression of GLUT-1, Bcl-xL, and Bax in Colorectal
Cancer. By ANDRZEJ WINCEWICZ, MARIOLA SULKOWSKA,
MARIUSZ KODA, LUIZA KANCZUGA-KODA, EWA WITKOWSKA, AND
STANISLAW SULKOWSKI 53

Combination of Doxorubicin and Sulforaphane for Reversing
Doxorubicin-Resistant Phenotype in Mouse Fibroblasts with p53^{Ser220}
Mutation. By CARMELA FIMOGNARI, MONIA LENZI, DAVIDE SCIUSCIO,
GIORGIO CANTELLI-FORTI, AND PATRIZIA HRELIA 62

Aromatase Expression Was Not Detected by Immunohistochemistry in
Endometrial Cancer. By YONG-TARK JEON, SO YEON PARK,
YONG-BEOM KIM, JAE WEON KIM, NOH-HYUN PARK, SOON-BEOM KANG,
HYO-PYO LEE, AND YONG-SANG SONG 70

ER Stress Induces the Expression of Jpk, which Inhibits Cell Cycle Progression in F9 Teratocarcinoma Cell. *By* HYE SUN KIM, KYOUNG-AH KONG, HYUNJOO CHUNG, SUNGDO PARK, AND MYOUNG HEE KIM 76

Akt Involvement in Paclitaxel Chemoresistance of Human Ovarian Cancer Cells. *By* SU-HYEONG KIM, YONG-SUNG JUHNN, AND YONG-SANG SONG 82

Expression of Leptin, Leptin Receptor, and Hypoxia-Inducible Factor 1α in Human Endometrial Cancer. *By* MARIUSZ KODA, MARIOLA SULKOWSKA, ANDRZEJ WINCEWICZ, LUIZA KANCZUGA-KODA, BOGUSLAW MUSIATOWICZ, MAGDALENA SZYMANSKA, AND STANISLAW SULKOWSKI 90

The Cyclooxygenase-2 Selective Inhibitor Celecoxib Suppresses Proliferation and Invasiveness in the Human Oral Squamous Carcinoma. *By* YOUNG EUN KWAK, NAM KYEOUNG JEON, JIN KIM, AND EUN JU LEE 99

The Epidermal Growth Factor Receptor Tyrosine Kinase Inhibitor ZD1839 (Iressa) Suppresses Proliferation and Invasion of Human Oral Squamous Carcinoma Cells via p53 Independent and MMP, uPAR Dependent Mechanism. *By* EUN JU LEE, JIN HA WHANG, NAM KYEONG JEON, AND JIN KIM 113

Role of Vascular Endothelial Growth Factor-D (VEGF-D) on IL-6 Expression in Cerulein-Stimulated Pancreatic Acinar Cells. *By* JANGWON LEE, KYUNG HWAN KIM, AND HYEYOUNG KIM 129

Lack of Association of the Cyclooxygenase-2 and Inducible Nitric Oxide Synthase Gene Polymorphism with Risk of Cervical Cancer in Korean Population. *By* TAEK SANG, LEE YONG TARK JEON, JAE WEON KIM, NOH HYUN PARK, SOON BEOM KANG, HYO PYO LEE, AND YONG SANG SONG 134

Increased Cyclooxygenase-2 Expression Associated with Inflammatory Cellular Infiltration in Elderly Patients with Vulvar Cancer. *By* TAEK SANG LEE, YONG TARK JEON, JAE WEON KIM, JAE KYUNG WON, NOH HYUN PARK, IN AE PARK, YONG SUNG JUHNN, SOON BEOM KANG, HYO PYO LEE, AND YONG SANG SONG 143

Viability of a Human Melanoma Cell after Single and Combined Treatment with Fotemustine, Dacarbazine, and Proton Irradiation. *By* IVAN M. PETROVIĆ, LELA B. KORIĆANAC, DANIJELA V. TODOROVIĆ, ALEKSANDRA M. RISTIĆ-FIRA, LUCIA M. VALASTRO, GIUSEPPE PRIVITERA, AND GIACOMO CUTTONE 154

Response of a Human Melanoma Cell Line to Low and High Ionizing Radiation. *By* ALEKSANDRA M. RISTIC-FIRA, DANIJELA V. TODOROVIC, LELA B. KORICANAC, IVAN M. PETROVIC, LUCIA M. VALASTRO, PABLO G.A. CIRRONE, LUIGI RAFFAELE, AND GIACOMO CUTTONE 165

Effect of Paclitaxel on Intracellular Localization of c-Myc and P-c-Myc in Prostate Carcinoma Cell Lines. *By* ROSANNA SUPINO, ENRICA FAVINI, GIUDITTA CUCCURU, FRANCO ZUNINO, AND A. IVANA SCOVASSI 175

Erufosine: A Membrane Targeting Antineoplastic Agent with Signal Transduction Modulating Effects. *By* M.M. ZAHARIEVA, S.M. KONSTANTINOV, B. PILICHEVA, M. KARAIVANOVA, AND M.R. BERGER 182

Part II. Cell Signaling in Health and Disease

SHP-1 Tyrosine Phosphatase in Human Erythrocytes. *By*
MARCANTONIO BRAGADIN, FLORINA ION-POPA, GIULIO CLARI, AND
LUCIANA BORDIN 193

Is the Distribution of Selenium and Zinc in the Sublocations of Spermatozoa
Regulated? *By* HOLGER BERTELSMANN, HARALD SIEME,
DIETRICH BEHNE, AND ANTONIOS KYRIAKOPOULOS 204

Immune Complexes Induce Monocyte Survival through Defined Intracellular
Pathways. *By* GIORDANO BIANCHI, FABRIZIO MONTECUCCO,
MARIA BERTOLOTTO, FRANCO DALLEGRI, AND LUCIANO OTTONELLO 209

Inhibition of Serine–Threonine Protein Phosphatases in Monocyte
Chemoattractant Protein-1 Expression in *Helicobacter pylori*-Stimulated
Gastric Epithelial Cells. *By* HAE-YUN CHUNG, BORAM CHA, AND
HYEYOUNG KIM 220

Cardioprotective Action of Urocortin in Early Pre- and Postconditioning. *By*
BARBARA CSEREPES, GABOR JANCSO, BALAZS GASZ, BOGLARKA RACZ,
ANDREA FERENC, LASZLO BENKO, BALAZS BORSICZKY, MARIA KURTHY,
SANDOR FERENCZ, JANOS LANTOS, JANOS GAL, ENDRE ARATO,
ATTILA MISETA, GYORGY WEBER, AND ELIZABETH ROTH 228

Novel Involvement of the Immunomodulator AS101 in IL-10 Signaling,
via the Tyrosine Kinase Fer. *By* RAMI HAYUN, SALI SHPUNGIN,
HANA MALOVANI, MICHAEL ALBECK, EITAN OKUN, URI NIR, AND
BENJAMIN SREDNI 240

Expression and Protective Role of Heme Oxygenase-1 in Delayed Myocardial
Preconditioning. *By* GÁBOR JANCSÓ, BARBARA CSEREPES, BALÁZS GASZ,
LÁSZLÓ BENKÓ, BALÁZS BORSICZKY, ANDREA FERENC, MÁRIA KÜRTHY,
BOGLÁRKA RÁCZ, JÁNOS LANTOS, JÁNOS GÁL, ENDRE ARATÓ,
LÓSZLÓ SÍNAYC, GYÖRGY WÉBER, AND ERZSÉBET RÓTH 251

Expression of Insulin Signaling Transmitters and Glucose Transporters at the
Protein Level in the Rat Testis. *By* KERSTI KOKK, ESKO VERÄJÄNKORVA,
XIAO-KE WU, HELLE TAPFER, ELLE PÕLDOJA, HELLE-EVI SIMOVART, AND
PASI PÖLLÄNEN 262

Effect of Endothelin on Sodium/Hydrogen Exchanger Activity of Human
Monocytes and Atherosclerosis-Related Functions. *By*
GEORGE KOLIAKOS, CHRISTINA BEFANI, KONSTANTINOS PALETAS, AND
MARTHA KALOYIANNI 274

Small GTPase Ras and Rho Expression in Rat Osteoblasts during
Spaceflight. *By* YASUHIRO KUMEI, HITOYATA SHIMOKAWA, KEI'ICHI OHYA,
HISAKO KATANO, HIDEO AKIYAMA, MASAHIKO HIRANO, AND
SADAO MORITA 292

Protein Expression in the Tissues of the Cardiovascular System of the Rat under
Selenium Deficiency and Adequate Conditions. *By* A. KYRIAKOPOULOS,
A. RICHTER, T. POHL, C. WOLF, I. GRBAVAC, A. PLOTNIKOV,
M. KÜHBACHER, H. BERTELSMANN, AND D. BEHNE 300

Study on the Correlations among Disease Activity Index and Salivary
Transforming Growth Factor-β1 and Nitric Oxide in Ulcerative Colitis
Patients. *By* ALI REZAIE, SARA KHALAJ, MARYAM SHABIHKHANI,
SHEKOUFEH NIKFAR, MOHAMMAD J. ZAMANI, AZADEH MOHAMMADIRAD,
NASER E. DARYANI, AND MOHAMMAD ABDOLLAHI 305

Generation of ΔTAp73 Proteins by Translation from a Putative Internal Ribosome Entry Site. *By* A. EMRE SAYAN, JEAN-PIERRE ROPERCH, BERNA S. SAYAN, MARIO ROSSI, M.J. PINKOSKI, RICHARD A. KNIGHT, ANNE E. WILLIS, AND GERRY MELINO 315

IRF-7: New Role in the Regulation of Genes Involved in Adaptive Immunity. *By* MARCO SGARBANTI, GIULIA MARSILI, ANNA LISA REMOLI, ROBERTO ORSATTI, AND ANGELA BATTISTINI 325

A New Transcript Splice Variant of the Human Glucocorticoid Receptor: Identification and Tissue Distribution of hGRΔ313–338, an Alternative Exon 2 Transactivation Domain Isoform. *By* JONATHAN D. TURNER, ANDREA B. SCHOTE, MARC KEIPES, AND CLAUDE P. MULLER 334

Exploitation of Host Signaling Pathways by B Cell Superantigens—Potential Strategies for Developing Targeted Therapies in Systemic Autoimmunity. *By* MONCEF ZOUALI 342

Part III. Chemoprevention

Antineoplastic and Anticlastogenic Properties of Curcumin. *By* TZVETAN ALAIKOV, SPIRO M. KONSTANTINOV, TZVETOMIRA TZANOVA, KYRIL DINEV, MARGARITA TOPASHKA-ANCHEVA, AND MARTIN R. BERGER 355

Sanguinarine Inhibits VEGF-Induced Akt Phosphorylation. *By* GIUSEPPINA BASINI, SUJEN ELEONORA SANTINI, SIMONA BUSSOLATI, AND FRANCESCA GRASSELLI 371

Effect of Curcumin Treatment on Protein Phosphorylation in K562 Cells. *By* ROMAIN BLASIUS, MARIO DICATO, AND MARC DIEDERICH 377

Dosage Effects of Ginkgolide B on Ethanol-Induced Cell Death in Human Hepatoma G2 Cells. *By* WEN-HSIUNG CHAN AND YAN-DER HSUUW 388

Attenuation of Aβ-Induced Apoptosis of Plant Extract (Saengshik) Mediated by the Inhibition of Mitochondrial Dysfunction and Antioxidative Effect. *By* CHU-YUE CHEN, JUNG-HEE JANG, MI HYUN PARK, SUNG JOO HWANG, YOUNG-JOON SURH, AND OCK JIN PARK 399

Wogonin Prevents Immunosuppressive Action but Not Anti-Inflammatory Effect Induced by Glucocorticoid. *By* RIYO ENOMOTO, CHIE SUZUKI, CHIKA KOSHIBA, TAKAYUKI NISHINO, MIKIKO NAKAYAMA, HIROYUKI HIRANO, TOSHIO YOKOI, AND EIBAI LEE 412

The Analgesic Effect of *Tribulus terrestris* Extract and Comparison of Gastric Ulcerogenicity of the Extract with Indomethacine in Animal Experiments. *By* M.R. HEIDARI, M. MEHRABANI, A. PARDAKHTY, P. KHAZAELI, M.J. ZAHEDI, M. YAKHCHALI, AND M. VAHEDIAN 418

Epigallocatechin Gallate Dose-Dependently Induces Apoptosis or Necrosis in Human MCF-7 Cells. *By* YAN-DER HSUUW AND WEN-HSIUNG CHAN 428

Resveratrol Induces Apoptosis in Chemoresistant Cancer Cells via Modulation of AMPK Signaling Pathway. *By* JIN-TAEK HWANG, DONG WOOK KWAK, SUN KYO LIN, HYE MIN KIM, YOUNG MIN KIM, AND OCK JIN PARK 441

Antioxidative Effects of Plant Polyphenols: From Protection of G Protein Signaling to Prevention of Age-Related Pathologies. *By* VIKTOR JEFREMOV, MIHKEL ZILMER, KERSTI ZILMER, NENAD BOGDANOVIC, AND ELLO KARELSON 449

Jaceosidin, a Pharmacologically Active Flavone Derived from *Artemisia argyi*, Inhibits Phorbol-Ester-Induced Upregulation of COX-2 and MMP-9 by Blocking Phosphorylation of ERK-1 and -2 in Cultured Human Mammary Epithelial Cells. *By* MIN A JEONG, KI WON LEE, DO-YOUNG YOON, AND HYONG JOO LEE 458

Effects of Selenium Diet on Expression of Selenoproteins in the Lung of the Rat. *By* KATARZYNA BUKALIS, DOROTHEA ALBER, GREGOR BUKALIS, DIETRICH BEHNE, AND ANTONIOS KYRIAKOPOULOS 467

Protective Effects of Piceatannol against Beta-Amyloid–Induced Neuronal Cell Death. *By* HYO JIN KIM, KI WON LEE, AND HYONG JOO LEE 473

Jaceosidin Induces Apoptosis in *ras*-Transformed Human Breast Epithelial Cells through Generation of Reactive Oxygen Species. *By* MIN-JUNG KIM, DO-HEE KIM, KI WON LEE, DO-YOUNG YOON, AND YOUNG-JOON SURH ... 483

Involvement of AMPK Signaling Cascade in Capsaicin-Induced Apoptosis of HT-29 Colon Cancer Cells. *By* YOUNG MIN KIM, JIN-TAEK HWANG, DONG WOOK KWAK, YUN KYUNG LEE, AND OCK JIN PARK 496

Epigallocatechin Gallate Inhibits Phorbol Ester-Induced Activation of NF-κB and CREB in Mouse Skin: Role of p38 MAPK. *By* JOYDEB KUMAR KUNDU AND YOUNG-JOON SURH 504

Peonidin Inhibits Phorbol-Ester–Induced COX-2 Expression and Transformation in JB6 P$^+$ Cells by Blocking Phosphorylation of ERK-1 and -2. *By* JUNG YEON KWON, KI WON LEE, HAENG JEON HUR, AND HYONG JOO LEE 513

Wogonin, a Plant Flavone, Potentiates Etoposide-Induced Apoptosis in Cancer Cells. *By* EIBAI LEE, RIYO ENOMOTO, CHIE SUZUKI, MASATAKA OHNO, TOSHINORI OHASHI, AZUSA MIYAUCHI, ERIKO TANIMOTO, KAORI MAEDA, HIROYUKI HIRANO, TOSHIO YOKOI, AND CHIYOKO SUGAHARA 521

Inhibitory Effects of 7-Carboxymethyloxy-3′,4′,5-Trimethoxy Flavone (DA-6034) on *Helicobacter pylori*-Induced NF-κB Activation and iNOS Expression in AGS Cells. *By* JEONG-SANG LEE, HYUN-SOO KIM, KI-BAIK HAHM, MI-WON SOHN, MOOHI YOO, JEFFREY A. JOHNSON, AND YOUNG-JOON SURH 527

Phenolic Phytochemicals Derived from Red Pine (*Pinus densiflora*) Inhibit the Invasion and Migration of SK-Hep-1 Human Hepatocellular Carcinoma Cells. *By* SANG JUN LEE, KI WON LEE, HAENG JEON HUR, JI YOUNG CHUN, SEO YOUNG KIM, AND HYONG JOO LEE 536

KG-135 Inhibits COX-2 Expression by Blocking the Activation of JNK and AP-1 in Phorbol Ester–Stimulated Human Breast Epithelial Cells. *By* SIN-AYE PARK, EUN-HEE KIM, HYE-KYUNG NA, AND YOUNG-JOON SURH 545

Resveratrol Inhibits IL-1β–Induced Stimulation of Caspase-3 and Cleavage of PARP in Human Articular Chondrocytes *in vitro*. *By* MEHDI SHAKIBAEI, THILO JOHN, CLAUDIA SEIFARTH, AND ALI MOBASHERI 554

Possible Link Between NO Concentrations and COX-2 Expression in Systems Treated with Soy-Isoflavones. *By* JANG-IN SHIN, YUN-KYUNG LEE, YOUNG MIN KIM, JIN-TAEK HWANG, AND OCK JIN PARK 564

Assessment of the Effect of *Echinacea purpurea* (L.) Moench on Apoptotic and Mitotic Activity of Liver Cells during Intoxication by Cadmium. *By* ALINA SMALINSKIENE, VAIVA LESAUSKAITE, STANISLOVAS RYSELIS, OLEG ABDRAKHMANOV, RIMA KREGZDYTE, ILONA SADAUSKIENE, LEONID IVANOV, NIJOLE SAVICKIENE, VIRGILIJUS ZITKEVIČIUS, AND ARUNAS SAVICKAS 574

Influence of *Echinacea purpurea* (L.) Moench Extract on the Toxicity of Cadmium. *By* VIRGILIJUS ZITKEVICIUS, ALINA SMALINSKIENE, VAIVA LESAUSKAITE, NIJOLE SAVICKIENE, ARUNAS SAVICKAS, STANISLOVAS RYSELIS, RIMA KREGZDYTE, OLEG ABDRAKHMANOV, ILONA SADAUSKIENE, AND LEONID IVANOV 585

Index of Contributors 593

PART D: Inflammatory Signaling Pathways and Neuropathology

Volume 1096, January 2007

CONTENTS

Preface. By MARC DIEDERICH xxvii

Part I. Inflammatory Signaling Pathways

Targeting Bacterial Endotoxin: Two Sides of a Coin. By HERBERT BOSSHART AND MICHAEL HEINZELMANN.. 1

Interaction between the *Helicobacter pylori* CagA and α-Pix in Gastric Epithelial AGS Cells. By HYE YEON BAEK, JOO WEON LIM, AND HYEYOUNG KIM 18

Expression of Suppressors of Cytokine Signaling-3 in *Helicobacter pylori*-Infected Rat Gastric Mucosal RGM-1Cells. By BORAM CHA, KYUNG HWAN KIM, HIROFUMI MATSUI, AND HYEYOUNG KIM............ 24

Role of Proteinase-Activated Receptor-2 on Cyclooxygenase-2 Expression in *H. pylori*–Infected Gastric Epithelial Cells. By JI HYE SEO, KYUNG HWAN KIM, AND HYEYOUNG KIM 29

Control of Human Herpes Virus Type 8-Associated Diseases by NK Cells. By MARIA C. SIRIANNI, MASSIMO CAMPAGNA, DONATO SCARAMUZZI, MAURIZIO CARBONARI, ELENA TOSCHI, ILARIA BACIGALUPO, PAOLO MONINI, AND BARBARA ENSOLI 37

Analysis of Tissue Distribution of TNF-α, TNF-α-Receptors, and the Activating TNF-α–Converting Enzyme Suggests Activation of the TNF-α System in the Aging Intervertebral Disc. By BEATRICE E. BACHMEIER, ANDREAS G. NERLICH, CHRISTOPH WEILER, GÜNTHER PAESOLD, MARIANNE JOCHUM, AND NORBERT BOOS 44

Upregulation of Apolipoprotein B Secretion, but Not Lipid, by Tumor Necrosis Factor-α in Rat Hepatocyte Cultures in the Absence of Extracellular Fatty Acids. By NEREA BARTOLOMÉ, LORENA RODRÍGUEZ, MARÍA J. MARTÍNEZ, BEGOÑA OCHOA, AND YOLANDA CHICO 55

Gene Expression Profiling of LPS-Stimulated Murine Macrophages and Role of the NF-κB and PI3K/mTOR Signaling Pathways. By S. DOS SANTOS, A.-I. DELATTRE, F. DE LONGUEVILLE, H. BULT, AND M. RAES 70

Modification of Proteins by Cyclopentenone Prostaglandins is Differentially Modulated by GSH *in vitro*. By JAVIER GAYARRE, M. ISABEL AVELLANO, FRANCISCO J. SÁNCHEZ-GÓMEZ, M. JESÚS CARRASCO, F. JAVIER CAÑADA, AND DOLORES PÉREZ-SALA 78

Signaling Pathways Involved in Proteinase-Activated Receptor$_1$-Induced Proinflammatory and Profibrotic Mediator Release Following Lung Injury. By PAUL F. MERCER, XIAOLING DENG, AND RACHEL C. CHAMBERS 86

Signaling Pathway Used by HSV-1 to Induce NF-κB Activation: Possible Role of Herpes Virus Entry Receptor A. *By* M. TERESA SCIORTINO, M. ANTONIETTA MEDICI, FRANCESCA MARINO-MERLO, DANIELA ZACCARIA, MARIA GIUFFRÈ, ASSUNTA VENUTI, SANDRO GRELLI, AND ANTONIO MASTINO 89

Melphalan Reduces the Severity of Experimental Colitis in Mice by Blocking Tumor Necrosis Factor-α Signaling Pathway. *By* GALINA SHMARINA, ALEXANDER PUKHALSKY, VLADIMIR ALIOSHKIN, AND ALEX SABELNIKOV 97

Part II. Neuropathology

Cellular Prion Protein Signaling in Serotonergic Neuronal Cells. *By* SOPHIE MOUILLET-RICHARD, BENOÎT SCHNEIDER, ELODIE PRADINES, MATHÉA PIETRI, MYRIAM ERMONVAL, JACQUES GRASSI, J. GRAYSON RICHARDS, VINCENT MUTEL, JEAN-MARIE LAUNAY, AND ODILE KELLERMANN 106

Strongly Reduced Number of Parvalbumin-Immunoreactive Projection Neurons in the Mammillary Bodies in Schizophrenia: Further Evidence for Limbic Neuropathology. *By* HANS-GERT BERNSTEIN, STEPHANIE KRAUSE, DIETER KRELL, HENRIK DOBROWOLNY, MARION WOLTER, RENATE STAUCH, KARIN RANFT, PETER DANOS, GUSTAV F. JIRIKOWSKI, AND BERNHARD BOGERTS 120

Alterations of Synaptic Turnover Rate in Aging May Trigger Senile Plaque Formation and Neurodegeneration. *By* CARLO BERTONI-FREDDARI, PATRIZIA FATTORETTI, BELINDA GIORGETTI, YESSICA GROSSI, MARTA BALIETTI, TIZIANA CASOLI, GIUSEPPINA DI STEFANO, AND GEMMA PERRETTA 128

Preservation of Mitochondrial Volume Homeostasis at the Early Stages of Age-Related Synaptic Deterioration. *By* CARLO BERTONI-FREDDARI, PATRIZIA FATTORETTI, BELINDA GIORGETTI, YESSICA GROSSI, MARTA BALIETTI, TIZIANA CASOLI, GIUSEPPINA DI STEFANO, AND GEMMA PERRETTA 138

Immunohistochemical Evidence for Impaired Neuregulin-1 Signaling in the Prefrontal Cortex in Schizophrenia and in Unipolar Depression. *By* IRIS BERTRAM, HANS-GERT BERNSTEIN, UWE LENDECKEL, ALICJA BUKOWSKA, HENRIK DOBROWOLNY, GERBURG KEILHOFF, DIMITRIOS KANAKIS, CHRISTIAN MAWRIN, HENDRIK BIELAU, PETER FALKAI, AND BERNHARD BOGERTS 147

Dysregulation of GABAergic Neurotransmission in Mood Disorders: A Postmortem Study. *By* HENDRIK BIELAU, JOHANN STEINER, CHRISTIAN MAWRIN, KURT TRÜBNER, RALF BRISCH, GABRIELA MEYER-LOTZ, MICHAEL BRODHUN, HENRIK DOBROWOLNY, BRUNO BAUMANN, TOMASZ GOS, HANS-GERT BERNSTEIN, AND BERNHARD BOGERTS 157

Release of β-Amyloid from High-Density Platelets: Implications for Alzheimer's Disease Pathology. *By* TIZIANA CASOLI, GIUSEPPINA DI STEFANO, BELINDA GIORGETTI, YESSICA GROSSI, MARTA BALIETTI, PATRIZIA FATTORETTI, AND CARLO BERTONI-FREDDARI 170

The Role of Selenite on Microglial Migration. *By* LISA DALLA PUPPA,
NICOLAI E. SAVASKAN, ANJA U. BRÄUER, DIETRICH BEHNE, AND
ANTONIOS KYRIAKOPOULOS 179

Altered Subcellular Distribution of the Alzheimer's Amyloid Precursor Protein
Under Stress Conditions. *By* SARA C.T.S. DOMINGUES, ANA GABRIELA
HENRIQUES, WENJUAN WU, EDGAR F. DA CRUZ E. SILVA, AND
ODETE A.B. DA CRUZ E. SILVA 184

Differential Distribution of Alzheimer's Amyloid Precursor Protein Family
Variants in Human Sperm. *By* MARGARIDA FARDILHA, SANDRA I. VIEIRA,
ALBERTO BARROS, MÁRIO SOUSA, ODETE A.B. DA CRUZ E. SILVA, AND
EDGAR F. DA CRUZ E. SILVA 196

The Effect of Repeated Physical Exercise on Hippocampus and Brain Cortex
in Stressed Rats. *By* DRAGANA FILIPOVIĆ, LJUBICA GAVRILOVIĆ,
SLADJANA DRONJAK, AND MARIJA B. RADOJČIĆ 207

Prion Protein Aggregation and Neurotoxicity in Cortical Neurons. *By*
JOANA BARBOSA MELO, PAULA AGOSTINHO, AND
CATARINA RESENDE OLIVEIRA 220

Intensive Remodeling of Purkinje Cell Spines after Climbing Fibers
Deafferentation Does Not Involve MAPK and Akt Activation. *By*
JELENA M. MILAŠIN, ANNALISA BUFFO, DANIELA CARULLI, AND
PIERGIORGIO STRATA 230

Immunomorphological Analysis of RAGE Receptor Expression and NF-κB
Activation in Tissue Samples from Normal and Degenerated Intervertebral
Discs of Various Ages. *By* ANDREAS G. NERLICH, BEATRICE E. BACHMEIER,
ERWIN SCHLEICHER, HELMUT ROHRBACH, GUENTHER PAESOLD, AND
NORBERT BOOS 239

Amyloid Precursor Protein and Presenilin 1 Interaction Studied by FRET in
Human H4 Cells. *By* MARIO NIZZARI, VALENTINA VENEZIA,
PAOLO BIANCHINI, VALENTINA CAORSI, ALBERTO DIASPRO,
EMANUELA REPETTO, STEFANO THELLUNG, ALESSANDRO CORSARO,
PIA CARLO, GENNARO SCHETTINI, TULLIO FLORIO, AND
CLAUDIO RUSSO 249

Amino-Terminally Truncated Prion Protein PrP90-231 Induces Microglial
Activation *in vitro*. *By* STEFANO THELLUNG, ALESSANDRO CORSARO,
VALENTINA VILLA, VALENTINA VENEZIA, MARIO NIZZARI,
MICHELA BISAGLIA, CLAUDIO RUSSO, GENNARO SCHETTINI,
ANTONIO ACETO, AND TULLIO FLORIO 258

Activation and Endocytic Internalization of Melanocortin 3 Receptor in
Neuronal Cells. *By* S.J.M. WACHIRA, B. GURUSWAMY, L. URADU,
C.A. HUGHES-DARDEN, AND F.J. DENARO 271

Index of Contributors 287

The New York Academy of Sciences believes it has a responsibility to provide an open forum for discussion of scientific questions. The positions taken by the participants in the reported conferences are their own and not necessarily those of the Academy. The Academy has no intent to influence legislation by providing such forums.

Preface

In 1998, we organized the first specialized meeting in the field of signal transduction and gene expression in Luxembourg. This type of meeting was originally intended to teach doctoral students attending the cellular and molecular biology program (DEA de Pharmacologie moléculaire) of the University of Nancy I (France).

From 1998 to 2004, this teaching program became a full-size international meeting, and more than 4,300 fundamental and clinical researchers have gathered in Luxembourg to discuss therapeutic applications in the field of signal transduction, transcription, and translation. Our meetings allow new insights into a rapidly moving field. Novel antibodies against receptors, protein kinase inhibitors, and siRNA targeting both signal transduction and gene expression will certainly lead to the therapeutic approaches that will be developed and used in this new century.

This is the second of four volumes forming the proceedings of our Cell Signaling World 2006 meeting. The contributions are divided according to areas of research. Part B focuses on basic research, and the chapters are divided into the following sections: oxidative stress, transcriptional control, HDAC, and novel technological and therapeutic approaches.

This field is moving forward so rapidly that another meeting has been set for January 23–25, 2008, at which fundamental and clinical researchers will gather again in Luxembourg for an eighth meeting, entitled Apoptosis World 2008: From Mechanisms to Applications. The details of this meeting, which will focus on the evolution of the therapeutic applications derived from the field of signal transduction, can be accessed at <http://www.transduction-meeting.lu>.

I would like to thank the editorial department of the *Annals of the New York Academy of Sciences* for its help in publishing these proceedings. I would also like to extend my special gratitude to the City of Luxembourg, as well as to the Fondation de Recherche Cancer et Sang (FRCS) for their generous contributions in support of the conference. I also thank Q8 Petroleum, Novartis, and Alexis for supporting our meetings, and I look forward to the next one in 2008.

—Marc Diederich
Laboratoire de Biologie Moléculaire et Cellulaire du Cancer
Hôpital Kirchberg
L-2540 Luxembourg

Oxidative Upregulation of Bcl-2 in Healthy Lymphocytes

SILVIA CRISTOFANON,[a,b] SILVIA NUCCITELLI,[b]
MARIA D'ALESSIO,[b] FLAVIA RADOGNA,[b] MILENA DE NICOLA,[b]
ANTONIO BERGAMASCHI,[c] CLAUDIA CERELLA,[b]
ANDREA MAGRINI,[c] MARC DIEDERICH,[a] AND LINA GHIBELLI[b]

[a]*Laboratoire de Biologie Moléculaire et Cellulaire du Cancer (LBMCC), Fondation Recherche sur le Cancer et les Maladies du Sang, Hôpital Kirchberg, L-2540 Luxembourg, Luxembourg*

[b]*Dipartimento di biologia, Università di Roma "Tor Vergata," 00133 Roma, Italy*

[c]*Cattedra di Medicina del Lavoro, Università di Roma "Tor Vergata," 00133 Roma, Italy*

ABSTRACT: In many cell systems, pharmacological glutathione (GSH) depletion with the GSH neosynthesis inhibitor buthionine sulfoximine (BSO) leads to cell death and highly sensitizes tumor cells to apoptosis induced by standard chemotherapeutic agents. However, some tumor cells upregulate Bcl-2 in response to BSO, thus surviving the treatment and failing to be chemosensitized. Cell lines of monocytic and lymphocytic origins respond to BSO treatment in an opposite way, lymphocytes being chemosensitized and unable to transactivate Bcl-2. In this article we investigate the response to BSO of lymphocytes freshly isolated from peripheral blood of healthy donors. After ensuring that standard separation procedures do not alter *per se* lymphocytes redox equilibrium nor Bcl-2 levels in the first 24 h of culture, we show that BSO treatment promotes the upregulation of Bcl-2, with a mechanism involving the increased radical production consequent to GSH depletion. Thus, BSO treatment may increase the differential cytocidal effect of cytotoxic drugs in tumor versus normal lymphocytes.

KEYWORDS: Bcl-2; buthionine sulfoximine; apoptosis; glutathione; lymphocyte

INTRODUCTION

Among the biochemical mechanisms responsible for apoptotic cell death of human tumor cell, particular attention has been devoted to those related to oxidative stress generated by oxygen-free radicals and peroxides.[1,2] The main features of oxidative stress are the loss of protective action of biological scavengers, such as superoxide dismutase and glutathione (GSH). This results in an increased level of peroxides, hydroperoxides, and free radical. The importance of intracellular GSH in the pathology of disease, particularly cancer, has long been appreciated.[3] However, the ubiquitous nature of GSH has made it difficult to ascribe to a specific molecular mechanism in disease accomplishment. In many cell systems, pharmacological GSH depletion with the GSH neosynthesis inhibitor buthionine sulfoximine (BSO) leads to cell death and highly sensitizes tumor cells to apoptosis induced by standard chemotherapeutic agents.[4] For these reasons, BSO is widely used in clinical practice as a chemosensitizing agent.[5]

We had shown that unlike most tumors, two human tumor cell lines (U937 monocyte, and Hep G2, hepatocyte) survive to BSO, not because BSO is unable to elicit an apoptotic response in these cells, but because the apoptotic process is stopped after cytochrome c release and before caspase activation, due to the development of an adaptive response.[6] We studied the mechanisms of such an adaptation and found that as a response to BSO, U937 upregulates Bcl-2 mRNA and protein levels by a mechanism possibly involving NF-κB transcription factor.[7] Moreover, BSO-dependent Bcl-2 upregulation is associated with the ability to survive the BSO; interestingly, GSH depletion upregulates Bcl-2 in BSO-resistant but not in BSO-sensitive cells.[7]

A fundamental problem in the field of lymphoid neoplasia concerns the relationship between normal immune elements in patients with lymphomas and lymphocytic leukemias, and the tumor cells, which themselves constitute a part of the host's immune system.[8] Bl-41, a cell line of lymphocytic origin deriving from an Epstein–Barr-negative Burkitt lymphoma, does not develop an adaptive response to BSO treatment, being (*a*) unable to transactivate Bcl-2, (*b*) induced to apoptosis, and (*c*) chemosensitized by BSO.[7] This suggests that BSO might act as a chemosensitizer and improve the treatment of some lymphomas. Here, we investigate the response to BSO of lymphocytes freshly isolated from peripheral blood of healthy donors, with the goal of understanding whether the use of BSO as an adjuvant may help to increase the differential cytocidal effect of antitumor therapy between normal and tumor counterparts.

MATERIALS AND METHODS

Mononuclear White Blood Cells Isolation

Peripheral blood mononucleated cells (PBMC) were isolated from the anticoagulated peripheral blood of 22 healthy adult human donors (13 men and 9 women) using the standard Ficoll–Hypaque (Sigma) density separation method. After isolation, PBMC were washed, counted, resuspended at cell density of 1×10^6 cells/mL in RPMI 1640 supplemented with 10% FCS, 2 mM L-glutamine, 100 IU/mL penicillin, and streptomycin and kept in a controlled atmosphere (5% CO_2) incubator at 37°C.

All the experiments were performed on mixed mononuclear cells, that is, lymphocytes+monocytes, distinguished by differential labeling upon flow cytometric analysis (see below).

Surface Staining

The expression of CD3-CD19 on lymphocyte was analyzed by adding 5 μL phycoerythrin (PE)-conjugated anti-CD3, 5 μL PE-conjugated anti-CD19 (Becton Dickinson) to leukocyte pellets. The suspensions were incubated in ice for 20 min in the dark and then washed twice in phosphate-buffered saline (PBS)-EDTA. The cells were finally resuspended in 0.5 mL of PBS and processed in a DAKO Galaxy flow cytometer. Statistics were elaborated in 50,000 events/sample by FlowMax software.

Flow Cytometric Analysis

FlowMax software permitted to study on the same dot plot FSC versus SSC quantization and counting of singular populations of monocytes and lymphocytes. Gating the area of labeled lymphocytes allowed to get information on parameters (abundance and fluorescence, corresponding to GSH, ROS, and Bcl-2 levels) of the specific area, as shown in FIGURE 1 A.

Cell Treatments

GSH depletion was performed by inhibiting GSH neosynthesis with 1 mM BSO (Sigma), which depletes PBMC in 18–24 h. Vitamin E analog Trolox C (Fluka) was used as a radical scavenger[9]; it was added to the cell culture at the final concentration of 500 μM (30 min prior to BSO when used together).

Bcl-2 Determination

Cells were fixed, permeabilized, and stained with anti-Bcl-2 monoclonal antibody (Calbiochem Novabiochem Corp., San Diego, CA) according to the manufacturer's instruction.[7] Detection was done with FITC-conjugated antibodies and processed in a DAKO Galaxy flow cytometer. Statistics were elaborated in 50,000 events/sample by FlowMax software. Mean values given

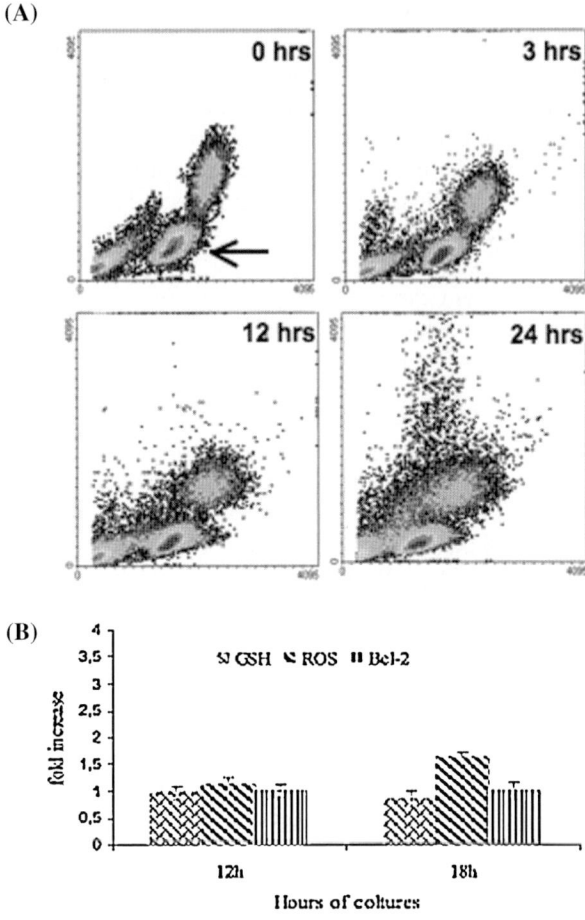

FIGURE 1. Identification of lymphocytes and quantification of basal parameters. (**A**) The area corresponding to CDX+ cells (B+T lymphocytes) is indicated; the population maintains a homogenous and compact distribution throughout the experimental time. (**B**) GSH, ROS, and Bcl-2 levels, measured as described in the section on "Materials and Methods," are shown at 0, 12, and 18 h of culture. The separation protocol does not change any oxidative characteristics in explanted lymphocytes. The weak changes are not statistically significant. All values are expressed as fold increase with respect to control cell sample at 0 h of culture, which was considered = 1. Results are the mean of 18 experiments performed on blood samples of different individuals ± SD.

by this analysis were used for further elaboration. For comparison between different experiments, the value of each treated cell sample was compared with the value of the control cell sample (untreated cells after separation protocol), which was considered equal to 1. The values were then given as a fold increase with the respect control.

GSH Determination

GSH intracellular levels were detected with Orto-ftaldeide (OPTA, Molecular Probes)[10] (ec.λ = 340 nm-em.λ = 445 nm). OPTA was added to the cell suspension at the final concentration of 50 µM. The suspensions were incubated at 37°C for 20 min in the dark and then washed. The cells were finally resuspended in PBS and processed in a DAKO Galaxy flow cytometer. Statistics were elaborated in 50,000 events/sample by FlowMax software. Mean values from this analysis were used for further elaboration. For comparison between different experiments, the value of each treated cell sample was compared with the value of the control cell sample (untreated cells after separation protocol), as fold increase (control = 1).

Determination of ROS Production

ROS levels were detected with 2-7-dichloro-fluorescein-diacetate (DHCFDA) (ec.λ = 490 nm-em.λ = 520 nm), which was added to the cells at the final concentration of 10 µM. DHCFDA fluoresces only when oxidized. The suspensions were incubated at 37°C in the dark for 20 min and then washed. Cells were finally resuspended in PBS and processed in a DAKO Galaxy flow cytometer. Statistics were elaborated in 50,000 events/sample by FlowMax software. Mean values given by this analysis were used for further elaboration. For comparison between different experiments, the value of each treated cell sample was compared with the value of the control cell sample (untreated cells after separation protocol), as fold increase (control = 1).

Statistical Analyses

Statistical analyses were performed using Student's t-test for unpaired data, and P values less than 0.05 were considered significant. Data are presented as fold increase versus 0 h of treatment cells ± SD.

RESULTS

Separation Procedures Do Not Stimulate a Significant Alteration of Oxidative Parameters and Bcl-2 Levels in Explanted Lymphocytes

The procedures required to purify the different types of white blood cells present in the regular buffy coats are quite harsh and may lead to cell alterations as it occurs, that is, for redox parameters in monocytes (Cristofanon et al., in preparation). To determine whether separation protocol and culture conditions could alter the redox state of lymphocytes, we analyzed the levels of GSH and radicals, with OPTA and DHCFDA, respectively, in the lymphocytes population for the first 24 h of culture.

No significant alterations in GSH levels were detectable in 24 h; intracellular-free radicals levels appeared slightly but not significantly increased at 18 h of culture (FIG. 1 B).

Since we have shown that Bcl-2 protein level may respond to redox alterations, we expect that in these quite stable conditions Bcl-2 levels remained unaltered. FIGURE 1 B shows that indeed Bcl-2 levels are stable in cultured lymphocytes.

BSO Induces Bcl-2 Upregulation in Lymphocytes as a Response to ROS Production

In order to explore the response of lymphocytes to redox disequilibrium, purified lymphocytes were treated with BSO. FIGURE 2 A shows that GSH is depleted by about 35% after 18 h of BSO in the treated lymphocytes. At the same time points, BSO treatment strongly stimulated ROS production (FIG. 2 A).

Thus, depleting the GSH we created an oxidative condition inside the cells; in these oxidative conditions, BSO induces a sudden Bcl-2 upregulation, as shown in FIGURE 2 A.

To investigate whether the strong BSO-dependent Bcl-2 upregulation was due to ROS production or to the decreased availability of GSH, we scavenged the ROS produced by BSO with Trolox C, which does not change the kinetic of depletion of GSH, but efficiently scavenges the BSO increased ROS levels (data not shown). FIGURE 2 B shows that radicals scavenging revert the BSO-dependent upregulation of Bcl-2, showing that in this cell system it is a response to radical production.

DISCUSSION

The theoretical basis for the use of BSO as a chemosensitizer stems from the general findings that cells deprived of GSH are less prone to extrude xenobiotics (via GSH-S-transferase), thus bypassing the problem of multi-drug-resistance, and to resist to apoptosis, due to reduced antioxidant ability and reduced levels of the antiapoptotic protein Bcl-2. Indeed, most cells respond to GSH depletion by downregulating Bcl-2: this is probably due to an increased breakdown of Bcl-2 protein in an oxidizing environment.[11] The different susceptibility to BSO treatment in different cellular lines, is probably linked to the intrinsic ability of some cells to adapt to GSH depletion by activating specific survival pathways.[7] We have shown that BL-41 (human B cell Burkitt lymphoma cell line), unlike U937(monocytic cell line), is induced to

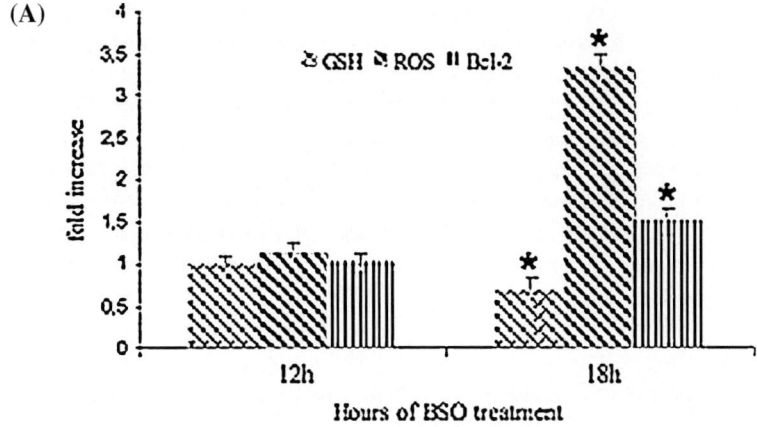

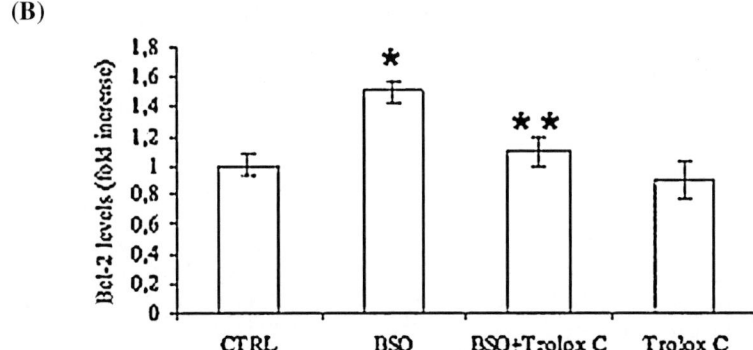

FIGURE 2. BSO upregulates Bcl-2 in healthy lymphocytes as a response to ROS production. (**A**) GSH, ROS, and Bcl-2 levels ROS levels in lymphocytes treated with BSO 1 mM. The increase of ROS and Bcl-2 levels at 18 h of cells treated with BSO is statistically significant ($P < 0.05$) with respect to 0 and 12 h. All values are expressed as fold increase with respect to control cell sample at 0 h of culture, which was considered = 1. Results are the mean of 18 experiments performed on blood samples of different individuals ± SD. (**B**) Levels of Bcl-2 in lymphocytes untreated and treated for 18 h with BSO, BSO+TroloxC, and Trolox C. Bcl-2 increase in cells treated with BSO with respect to untreated; and Bcl-2 decrease in the cells treated with BSO+Trolox C with respect to BSO alone are statistically significant ($P < 0.05$). All values are expressed as fold increase with respect to control cell sample at 0 h of cultures, which was considered = 1. Results are the mean of 18 experiments performed on blood samples of different individuals ± SD.

apoptosis by 1 mM BSO.[7] In this study we were interested to understand if the inability of BL-41 to respond to GSH depletion by upregulating Bcl-2 is a histotypic idiosyncrasy. Modulation of Bcl-2 protein levels as an adaptive response should conceivably be an important regulatory mechanism for cell survival in conditions of sudden environmental changes. Bcl-2 upregulation has been shown to occur in physiologically regulated situations, such as the rescue of target cells by prosurvival cytokines (i.e., NGF).[12,13] The upregulation of Bcl-2 as a response to damaging situations, such as stressing disulfide disturbances and oxidative stress, has been documented.[14]

Commonly used cells separation procedures do not induce activation and changes on oxidative profile (GSH extrusion and ROS production) of healthy lymphocytes, unlike monocytes (Cristofanon *et al.*, in preparation); redox challenges are among the most important trigger of leukocyte activation; indeed, ROS production and the ability to live in an oxidative environment is mandatory for an appropriate inflammatory response of monocyte; less for lymphocytes. It is important to stress that in the experiments here described the two types of mononuclear leukocytes were not physically separated: thus, any radical present in the extracellular medium, even though produced by monocytes, was able to cross lymphocytes plasma membrane. Thus, the different behavior toward radical challenging might be explained by hypothesizing that perhaps lymphocytes need more specific triggers than oxidative stress to be activated. The lack of proradical effect of the separation procedures *per se* allowed us to study the response of BSO-induced oxidative stress on the Bcl-2 levels without problems of overlapping events.

We found that BSO treatment induces the upregulation of Bcl-2 in healthy lymphocytes, unlike what occurs in their tumor counterpart, that is, BL-41. This means that either there is no such factor like an histotypic idiosyncrasy allowing to predict cell response to oxidative challenge, in terms of Bcl-2 upregulation; or, tumor transformation implies the loss of the ability to respond to an oxidative environment building up a defense that may allow longer survival time for the working lymphocyte in the inflammation site that conceivably is an oxidizing one.

As a last point, we want to stress that the mechanism at the basis of Bcl-2 upregulation as a response to GSH depletion is still unsolved. In this study, we show that it is not the decreased GSH concentration, but rather the increase in free radicals, which acts as the stimulus for Bcl-2 upregulation. Since it seems to involve NF-κB activation,[7] a transcription factor that is activated by redox imbalance,[15] it is conceivable that ROS may activate, via either the canonical[16] or the noncanonical[17] pathway, the activation of NF-κB, which in turn increases the transcription rate of the Bcl-2 gene.[18]

REFERENCES

1. LORENZI, M., D.F. MONTISANO, S. TOLEDO & H.C. WONG. 1987. Increased single strand breaks in DNA of lymphocytes from diabetic subjects. J. Clin. Invest. **79:** 653–656.
2. LORENZI, M., J.A. NORDBERG & S. TOLEDO. 1987. High glucose prolongs cell-cycle traversal of cultured human endothelial cells. Diabetes **36:** 1261–1267.

3. VOEHRINGER, D.W. 1999. BCL-2 and glutathione: alterations in cellular redox state that regulate apoptosis sensitivity. Free Radic. Biol. Med. **27:** 945–950.
4. REBER, U., U. WULLNER, M. TREPEL, *et al.* 1998. Potentiation of treosulfan toxicity by the glutathione-depleting agent buthionine sulfoximine in human malignant glioma cells: the role of bcl-2. Biochem. Pharmacol. **55:** 349–359.
5. RAPPA, G., M.P. GAMCSIK, R.L. MITINA, *et al.* 2003. Retroviral transfer of MRP1 and gamma-glutamyl cysteine synthetase modulates cell sensitivity to L-buthionine-S,R-sulphoximine (BSO): new rationale for the use of BSO in cancer therapy. Eur. J. Cancer **39:** 120–128.
6. GHIBELLI, L., C. FANELLI, G. ROTILIO, *et al.* 1998. Rescue of cells from apoptosis by inhibition of active GSH extrusion. FASEB J. **12:** 479–486.
7. D'ALESSIO, M., C. CERELLA, C. AMICI, *et al.* 2004. Glutathione depletion up-regulates Bcl-2 in BSO-resistant cells. FASEB J. **18:** 1609–1611.
8. SCHATTNER, E.J. 2002. Apoptosis in lymphocytic leukemias and lymphomas. Cancer Invest. **20:** 737–748.
9. CASINI, A.F., A. POMPELLA & M. COMPORTI. 1984. Glutathione depletion, lipid peroxidation, and liver necrosis following bromobenzene and iodobenzene intoxication. Toxicol. Pathol. **12:** 295–299.
10. TREUMER, J. & G. VALET. 1986. Flow-cytometric determination of glutathione alterations in vital cells by o-phthaldialdehyde (OPT) staining. Exp. Cell Res. **163:** 518–524.
11. CELLI, A., F.G. QUE, G.J. GORES & N.F. LARUSSO. 1998. Glutathione depletion is associated with decreased Bcl-2 expression and increased apoptosis in cholangiocytes. Am. J. Physiol. **275:** G749–G757.
12. AKBAR, A.N., N.J. BORTHWICK, R.G. WICKREMASINGHE, *et al.* 1996. Interleukin-2 receptor common gamma-chain signaling cytokines regulate activated T cell apoptosis in response to growth factor withdrawal: selective induction of anti-apoptotic (bcl-2, bcl-xL) but not pro-apoptotic (bax, bcl-xS) gene expression. Eur. J. Immunol. **26:** 294–299.
13. KATOH, S., Y. MITSUI, K. KITANI & T. SUZUKI. 1996. Nerve growth factor rescues PC12 cells from apoptosis by increasing amount of bcl-2. Biochem. Biophys. Res. Commun. **229:** 653–657.
14. SANDAU, K.B. & B. BRUNE. 2000. Up-regulation of Bcl-2 by redox signals in glomerular mesangial cells. Cell Death Differ. **7:** 118–125.
15. ZHOU, L.Z., A.P. JOHNSON & T.A. RANDO. 2001. NF kappa B and AP-1 mediate transcriptional responses to oxidative stress in skeletal muscle cells. Free Radic. Biol. Med. **31:** 1405–1416.
16. LI, X. & G.R. STARK. 2002. NFkappaB-dependent signaling pathways. Exp. Hematol. **30:** 285–296.
17. HEISSMEYER, V., D. KRAPPMANN, F.G. WULCZYN & C. SCHEIDEREIT. 1999. NF-kappaB p105 is a target of IkappaB kinases and controls signal induction of Bcl-3-p50 complexes. EMBO J. **18:** 4766–4778.
18. KURLAND, J.F., R. KODYM, M.D. STORY, *et al.* 2001. NF-kappaB1 (p50) homodimers contribute to transcription of the bcl-2 oncogene. J. Biol. Chem. **276:** 45380–45386.

Intracellular Pro-oxidant Activity of Melatonin Deprives U937 Cells of Reduced Glutathione without Affecting Glutathione Peroxidase Activity

MARIA CRISTINA ALBERTINI,[a] FLAVIA RADOGNA,[b] AUGUSTO ACCORSI,[a] FRANCESCO UGUCCIONI,[a] LAURA PATERNOSTER,[a] CLAUDIA CERELLA,[b] MILENA DE NICOLA,[b] MARIA D'ALESSIO,[b] ANTONIO BERGAMASCHI,[c] ANDREA MAGRINI,[c] AND LINA GHIBELLI[b]

[a]*Istituto di Chimica Biologica, Università di Urbino, Urbino, Italy*

[b]*Dipartimento di Biologia, Università di Roma "Tor Vergata," Roma, Italy*

[c]*Cattedra di Medicina del Lavoro, Università di Roma "Tor Vergata," Roma, Italy*

ABSTRACT: It was long believed that melatonin might counteract intracellular oxidative stress because it was shown to potentiate antioxidant endogenous defences, and to increase the activity of many antioxidant enzymes. However, it is now becoming evident that when radicals are measured within cells, melatonin increases, rather than decreasing, radical production. Herein we demonstrate a pro-oxidant effect of melatonin in U937 cells by showing an increase of intracellular oxidative species and a depletion of glutathione (GSH). The activity of glutathione peroxidase is not modified by melatonin treatment as it does occur in other experimental models.

KEYWORDS: melatonin; pro-oxidant; reduced glutathione; ROS; glutathione peroxidase.

INTRODUCTION

The pineal hormone melatonin has been reported to be an effective free radical scavenger and antioxidant *in vitro* as well as *in vivo*.[1-8] Melatonin is believed to scavenge the highly toxic hydroxyl radical, the peroxynitrite anion, and possibly the peroxyl radical; secondly, it reportedly scavenges the superoxide anion and also quenches singlet oxygen.[3] The neurotoxicity

Address for correspondence: Maria Cristina Albertini, Istituto di Chimica Biologica "G. Fornaini," Via Saffi, 2–61029 Urbino, Italy. Voice: +39-722-305288; fax: +39-722-320188.
e-mail: cemetbio@uniurb.it

of a number of compounds like kainic acid,[9] haloperidol,[10] and the cellular toxicity of hydrogen peroxide[11] is inhibited by melatonin administration. In some cases, the accompanying decrease in reduced glutathione (GSH) levels are also prevented by melatonin administration. On basis of these findings, melatonin was considered to exert its physiological and pharmacological effects, at least partly, via its antioxidant activity.[12] However, authors observed only a limited antioxidant activity of melatonin in several systems.[13–16]

Besides promoting deleterious effects at high concentrations, reactive oxygen species (ROS) function as intracellular downstream messengers targeting specific proteins and genes.[17–19] For example, programmed cell death in lymphocytes is known to be influenced by alterations of the cellular redox state as well as by intracellular ROS formation.[20–25] Because little is known about the effects of melatonin on the intracellular redox state, even though melatonin can easily cross cell membranes because of its amphiphilicity,[26] we examined whether melatonin interferes with intracellular ROS production, GSH intracellular concentration, and GSH peroxidase (GSH-Px) activity in U937 cells.

MATERIALS AND METHODS

Cell Cultures and Treatments

Melatonin was kindly provided by the Institute of Medicinal Chemistry, University of Urbino. L-buthionine-[S,R]-sulfoximine (BSO) was purchased from Sigma-Aldrich (Milan, Italy). Dichlorodihydrofluorescine diacetate (DHCFDA) was a product from Invitrogen (San Giuliano Milanese, Italy). U937 cells were cultured in RPMI 1640 medium (Invitrogen, Carlsbad, CA) supplemented with 10% fetal bovine serum (Biological Industries, Kibbutz Beit Haemek, Israel), penicillin (50 units/mL), and streptomycin (50 μg/mL) (Sera-Lab Ltd., Crawley Down, England), at 37°C in T-75 tissue culture flasks (Corning, Corning, NY) gassed with an atmosphere of 95% air to 5% CO_2. Treatments were performed in 2 mL of prewarmed saline A (8.182 g/L NaCl, 0.372 g/L KCl, 0.336 g/L $NaHCO_3$, and 0.9 g/L glucose) containing 1×10^6 cells.

Determination of ROS Formation in Cells

Levels of ROS were measured in U937 cells incubated for 2 h and 6 h in the presence and in the absence of 1 mM melatonin. The fluorescent probe dichlorodihydrofluorescine diacetate (DHCFDA) was used to detect intracellular ROS because it flouresces after oxidation by ROS and reactive nitrogen species (NOS). Fluorescence was measured with a Dako Galaxy flow cytometer (490-nm and 520-nm wavelegths for excitation and emission, respectively).

The protocol indicates the use of aliquots of 2×10^6 cells to be stained with 10 μM DHCFDA for 20 min in the dark. Samples are then centrifuged and the pellet resuspended in phosphate-buffered saline solution (PBS) at 2×10^6 cells/mL to be analyzed.

Determination of Intracellular GSH

Because GSH represents more than 90% of the nonprotein thiols, the latter will be referred to as GSH. After 2 h and 6 h incubation of U937 cells (1×10^6 cells/mL) with melatonin (1 mM) or BSO (1 mM), cells were washed in PBS and used for the determination of cellular GSH concentration by the method described by Beutler.[27] In brief, cells were centrifuged and the pellet was then resuspended in 150 μL of a solution containing 1.67% metaphosphoric acid, 0.2% EDTA, 30% NaCl, kept in ice for 5 min, and centrifuged at 10,000 $\times g$ for 5 min. The GSH content was measured spectrophotometrically in the supernatant, at 412 nm, using 5,5′-dithiobis(2-nitrobenzoic acid) ($\varepsilon_{412} = 13{,}600$ M^{-1} cm^{-1}).

Determination of Intracellular Glutathione Peroxidase (GSH-Px) Activity

For GSH-Px activity cells were washed twice with saline A, resuspended in the same medium at a density of 1×10^6 cells/mL, and finally lysed with a sonicator. The resulting lysates were centrifuged for 5 min at $18{,}000 \times g$ at 4 °C. GSH-Px activity was determined spectrophotometrically in the supernatant as already described.[28]

RESULTS

In contrast to other antioxidant molecules which decrease intracellular ROS concentration, melatonin significantly increases intracellular ROS formation following 2 h and 6 h of incubation with U937 cells (FIG. 1). The data also indicate that at 2 h of incubation intracellular ROS are strongly increased (more than sevenfold) while after 6 h the variation is less consistent (about threefold).

The increase of oxidative species is accompanied by a depletion of U937 intracellular GSH concentration during melatonin treatment. In fact, at 2 h and at 6 h of incubation, intracellular GSH concentration is 73.3% and 65.5% of the control, respectively (FIG. 2). In the presence of BSO, an inhibitor of γ-glutamylcysteine synthetase, the decrease of the thiol is similar to that induced by melatonin, being 66.6% and 62% of the control at 2 h and 6 h, respectively (FIG. 2).

To assess whether the treatment of U937 cells with melatonin may also cause an increase of cellular antioxidant enzymes, we measured the activity of GSH-Px, which has been found augmented in rat brain tissue following "*in vivo*"

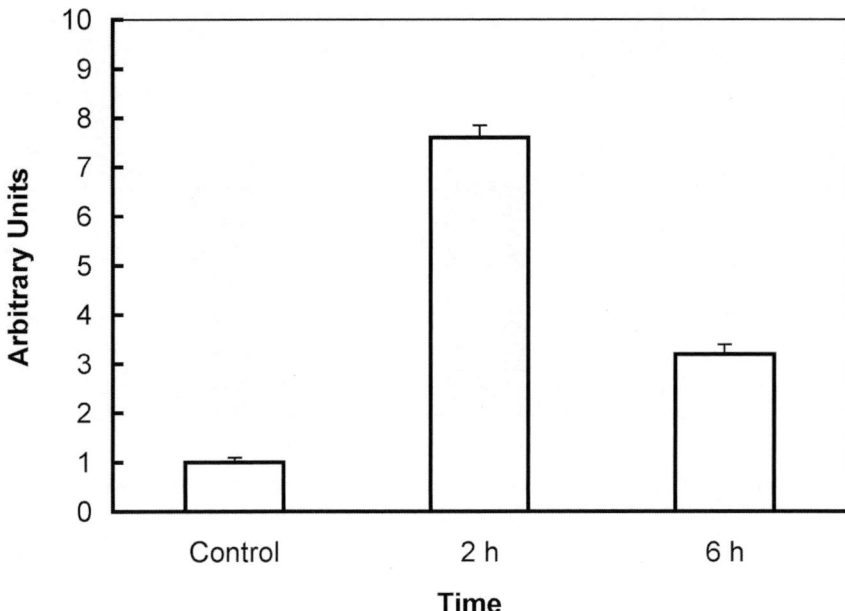

FIGURE 1. Evaluation of intracellular ROS concentration in U937 cells incubated with melatonin. Experimental conditions are reported in MATERIALS AND METHODS. ROS concentrations are expressed as arbitrary units of flourescence and are means ± SD ($n = 4$).

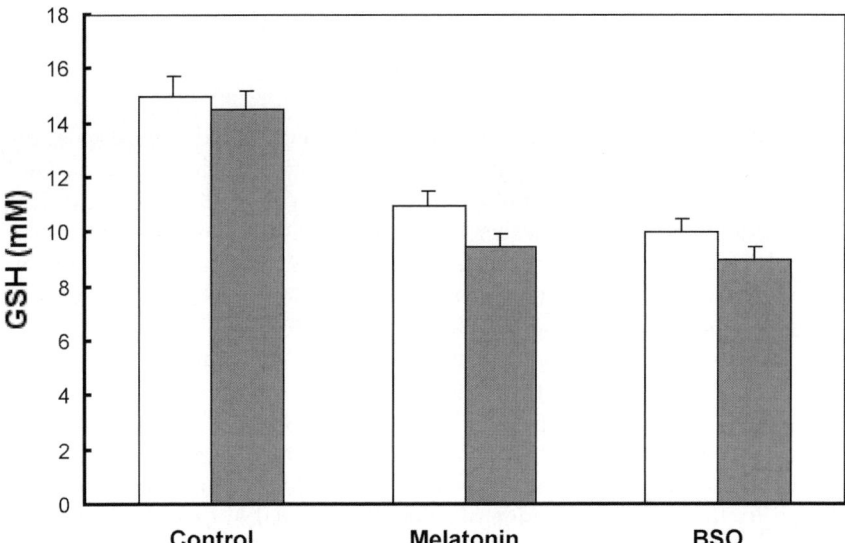

FIGURE 2. Measurement of intracellular GSH concentration in U937 cells incubated with melatonin and BSO for 2 h (white bars) and 6 h (dark bars). Experimental conditions are reported under "Materials and Methods." GSH concentrations are expressed as μmol/mL of cells and are means ± SD ($n = 4$).

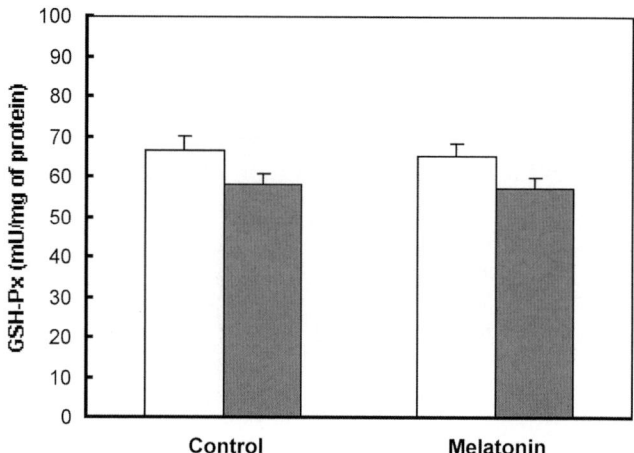

FIGURE 3. Evaluation of GSH-Px activity in U937 cells incubated with melatonin for 2 h (white bars) and 6 h (dark bars). Experimental conditions are reported under "Materials and Methods." GSH-Px activities are expressed as mU/mg of protein and are means ± SD ($n = 4$).

administration of melatonin.[28] Data reported in FIGURE 3 demonstrate that in our experimental conditions, U937 do not show any significantly increase of this enzyme at 2 h and 6 h of incubation.

DISCUSSION

The antioxidant role of melatonin has been demonstrated *in vitro* (possibly through interaction with iron?). Its potential use as an antioxidant *in vivo* comes from evidence showing its ability to reduce radical production in oxidant-related pathological situations, where radicals are measured in the extracellular fluids.[1-8] It was long believed that melatonin might also counteract intracellular oxidative stress because it was shown to potentiate antioxidant endogenous defences, and to increase the activity of many antioxidant enzymes.[29] However, it is now becoming evident that when radicals are measured within cells, melatonin increases, rather than decreases, radical production,[30] even though it is still unknown which types of radicals are produced. Also completely unknown is the mechanism at the basis of this phenomenon, which may not necessarily involve direct oxidant activity of melatonin (or one of its catabolites), but the ability to elicit the activation of some cellular radical producing enzymes (e.g., lipoxygenases, cyclooxygenases, NO-synthase, NADPH-oxidase). This problem is particularly timely nowadays, because it is emerging that oxidative radicals, considered so far as causative agents of important human diseases, may also mediate resistance to apoptosis by triggering survival pathways.[31] Our results demonstrate a pro-oxidant effect of melatonin by inducing the increase of intracellular ROS in U937. The increase of ROS is higher at 2 h than at 6 h, probably indicating that ROS production is transient. Intracellular GSH concentration decrease may reflect free radical formation in U937 cells treated with melatonin. Cell GSH-Px activity does not show any modification probably because, in our experimental conditions, U937 treated cells maintain sufficiently high levels of GSH to face the amounts of the formed radical species (note that after 6 h of incubation cells still contain more than 60% of control cell GSH, and the same is true for the positive control cells incubated with BSO). Experiments are in progress in our laboratory to investigate how melatonin may act on U937 cell metabolism and may directly or indirectly stimulate oxidative radical production.

REFERENCES

1. MONTILLA, P.L., I.F. TUNEZ, C. MUNOZ DE AGUEDA, et al. 1998. Protective role of melatonin and retinol palmitate in oxidative stress and hyperlipidemic nephropathy induced by adriamycin in rats. J. Pineal Res. **25:** 86–93.
2. PRINC, F.G., A.G. MAXIT, C. CARDALDA, et al. 1998. In vivo protection by melatonin against delta-aminolevulinic acid-induced oxidative damage and its antioxidant effect on the activity of haem enzymes. J. Pineal Res. **24:** 1–8.
3. REITER, R.J. 1998. Oxidative damage in the central nervous system: protection by melatonin. Prog. Neurobiol. **56:** 359–384.
4. BEYER, C.E., J.D. STEKETEE & D. SAPHIER. 1998. Antioxidant properties of melatonin: an emerging mystery. Biochem. Pharmacol. **56:** 1265–1272.
5. RODRIGUEZ, C., J.C. MAYO, R.M. SAINZ, et al. 2004. Regulation of antioxidant enzymes: a significant role for melatonin. J. Pineal Res. **36:** 1–9.
6. MAYO, J.C., D.X. TAN, R.M. SAINZ, et al. 2003. Protection against oxidative protein damage induced by metal-catalyzed reaction or alkylperoxyl radicals: comparative effects of melatonin and other antioxidants. Biochim. Biophys. Acta **1620:** 139–150.
7. ALLEGRA, M., R.J. REITER, D-X. TAN, et al. 2003. The chemistry of melatonin's interaction with reactive species. J. Pineal Res. **34:** 1–10.
8. ORTEGA-GUTIERREZ, S., J.J. GARCIA, E. MARTINEZ-BALLARIN, et al. 2002. Melatonin improves deferoxamine antioxidant activity in protecting against lipid peroxidation caused by hydrogen peroxide in rat brain homogenates. Neurosci. Lett. **323:** 55–59.
9. FLOREANI, M., S.D. SKAPER, L. FACCI, et al. 1997. Melatonin maintains glutathion homeostasis in kainic acid exposed rat brain tissues. FASEB J. **11:** 1309–1315.
10. POST, A., F. HOLSBOER & C. BEHL. 1998. Induction of NF-kappa B activity during haloperidol-induced oxidative toxicity in clonal hippocampal cells. J. Neurosci. **18:** 8236–8246.
11. BALDWIN, W.S. & J.C. BARRETT. 1998. Melatonin attenuates hydrogen peroxide toxicity in MCF7 cells only at pharmacological concentrations. Biochem. Biophys. Res. Commun. **250:** 602–605.
12. REITER, R.J., C.S. OH & O. FUJIMORI. 1996. Melatonin: its intracellular and genomic actions. Trends Endocr. Metab. **7:** 22–27.

13. MARSHALL, K.A., R.J. REITER, B. POEGGELER, *et al.* 1996. Evaluation of the antioxidant activity of melatonin *in vitro*. Free Radic. Biol. Med. **21:** 307–315.
14. ABUJA, P.M., P.M. LIEBMANN, M. HAYN, *et al.* 1997. Antioxidant role of melatonin in lipid peroxidation of human LDL. FEBS Lett. **413:** 289–293.
15. WÖLFLER, A., P.M. ABUJA, K. SCHAUENSTEIN & P.M. LIEBMANN. 1999. N-acetylserotonin is a better extra- and intracellular antioxidant than melatonin. FEBS Lett. **449:** 206–210.
16. MONTALDO, C., E. CANNAS, T. DETTORI, *et al.* 2000. Lack of melatonin effect on hydrogen peroxide induced bronchoconstriction in isolated and perfused rat lung. Life Sci. **66:** PL339–PL344.
17. FINKEL, T. 1998. Oxygen radicals and signaling. Curr. Opin. Cell. Biol. **10:** 248–253.
18. ALLEN, R.G. & M. TRESINI. 2000. Oxidative stress and gene regulation. Free Radic. Biol. Med. **28:** 463–499.
19. FINKEL, T. 2000. Redox-dependent signal transduction. FEBS Lett. **476:** 52–54.
20. SATO, N., S. IWATA, K. NAKAMURA, *et al.* 1995. Thiol-mediated redox regulation of apoptosis. Possible roles of cellular thiols other than glutathione in T cell apoptosis. J. Immunol. **154:** 3194–3203.
21. WILLIAMS, M.S. & P.A. HENKART. 1996. Role of reactive oxygen intermediates in TCR-induced death of T cell blasts and hybridomas. J. Immunol. **157:** 2395–2402.
22. CHIBA, T., S. TAKAHASHI, N. SATO & K. KIKUCHI. 1996. Fas-mediated apoptosis is modulated by intracellular glutathione in human T cells. Eur. J. Immunol. **26:** 1164–1169.
23. BAUER, M.K.A., M. VOGT, M. LOS, *et al.* 1998. Role of reactive oxygen intermediates in activation-induced CD95 (APO-1/Fas) ligand expression. J. Biol. Chem. **273:** 8048–8055.
24. BANKI, K., E. HUTTER, N.J. GONCHOROFF & A. PERL. 1999. Elevation of mitochondrial transmembrane potential and reactive oxygen intermediate levels are early events and occur independently from activation of caspases in Fas signaling. J. Immunol. **162:** 1466–1479.
25. CLEMENT, M.V. & J. PERVAIZ. 1999. Reactive oxygen intermediates regulate cellular response to apoptotic stimuli: an hypothesis. Free Radic. Res. **30:** 241–252.
26. MENENDEZ-PELAEZ, A., B. POEGGELER, R.J. REITER, *et al.* 1995. Nuclear localization of melatonin in different mammalian tissues: immunocytochemical and radioimmunoassay evidence. J. Cell. Biochem. **53:** 373–382.
27. BEUTLER, E. 1984. Red Cell Metabolism: A Manual of Biochemical Methods: 131–134. Grune & Stratton. New York.
28. BARLOW-WALDEN, L.R., R.J. REITER, M. ABE, *et al.* 1995. Melatonin stimulates brain glutathione peroxidase activity. Neurochem. Int. **26:** 497–502.
29. ANTOLIN, I., C. RODRIGUEZ, R.M. SAINZ, *et al.* 1996. Neurohormone melatonin prevents cell damage: effect on gene expression for antioxidant enzymes. FASEB J. **10:** 882–890.
30. WOLFLER, A., H.C. CALUBA, P.M. ABUJA, *et al.* 2001. Prooxidant activity of melatonin promotes fas-induced cell death in uman leukemic Jurkat cells. FEBS Lett. **502:** 127–131.
31. SCHOEMAKER, M., J.E. ROS, M. HOMAN, *et al.* 2002. Cytokine regulation of pro- and anti-apoptotic genes in rat hepatocytes:NF-kB-regulated inhibitor of apoptosis protein 2 (c-IAP2) prevents apoptosis. J. Hepatol. **36:** 742–750.

Mitochondrial "Movement" and Lens Optics following Oxidative Stress from UV-B Irradiation

Cultured Bovine Lenses and Human Retinal Pigment Epithelial Cells (ARPE-19) as Examples

VLADIMIR BANTSEEV AND HYUN-YI YOUN

School of Optometry, University of Waterloo, Waterloo, ON N2L 3G1 Ontario, Canada

ABSTRACT: Mitochondria provide energy generated by oxidative phosphorylation and at the same time play a central role in apoptosis and aging. As a byproduct of respiration, the electron transport chain is known to be the major intracellular site for the generation of reactive oxygen species (ROS). Exposure to solar and occupational ultraviolet (UV) radiation, and thus production of ROS and subsequent cell death, has been implicated in a large spectrum of skin and ocular pathologies, including cataract. Retinal pigment epithelial cell apoptosis generates photoreceptor dysfunction and ultimately visual impairment. The purpose of this article was to characterize *in vitro* changes following oxidative stress with UV-B radiation in (*a*) ocular lens optics and cellular function in terms of mitochondrial dynamics of bovine lens epithelium and superficial cortical fiber cells and (*b*) human retinal pigment epithelial (ARPE-19) cells. Cultured bovine lenses and confluent cultures of ARPE-19 cells were irradiated with broadband UV-B radiation at energy levels of 0.5 and 1.0 J/cm^2. Lens optical function (spherical aberration) was monitored daily up to 14 days using an automated laser scanning system that was developed at the University of Waterloo. This system consists of a single collimated scanning helium–neon laser source that projects a thin (0.05 mm) laser beam onto a plain mirror mounted at 45° on a carriage assembly. This mirror reflects the laser beam directly up through the scanner table surface and through the lens under examination. A digital camera captures the actual position and slope of the laser beam at each step. When all steps have been made, the captured data for each step position is used to calculate the back vertex distance for each position and the difference in that measurement between beams. To investigate mitochondrial

Address for correspondence: V. Bantseev, School of Optometry, University of Waterloo, Waterloo, ON N2L 3G1 Canada. Voice: +1-519-888-4567; ext.: 37091; fax: +1-519-725-0784.
e-mail: vbantsee@uwaterloo.ca

movement, the mitochondria-specific fluorescent dye Rhodamine 123 was used. Time series were acquired with a Zeiss 510 (configuration Meta 18) confocal laser scanning microscope equipped with an inverted Axiovert 200 M microscope and 40-× water-immersion C-Apochromat objective (NA 1.2). The optical analysis showed energy level-dependent increases in back vertex distance variability (loss of sharp focus) from 0.39 ± 0.04 mm (control, $n = 11$) to 1.63 ± 0.33 mm (1.0 J/cm^2, $n = 10$) and 0.63 ± 0.13 mm (0.5 J/cm^2, $n = 9$). Confocal laser scanning microscopy analysis of both bovine lenses and ARPE-19 cells showed that following treatment at 0.5 J/cm^2 the mitochondria stopped moving immediately whereas at 1.0 J/cm^2 not only did the mitochondria stop moving, but fragmentation and swelling was seen. Untreated control tissue exhibited up to 15 μm/min of movement of the mitochondria. This could represent normal morphological change, presumably allowing energy transmission across the cell from regions of low to regions of high ATP demand. Lack of mitochondrial movement, fragmentation, and swelling of mitochondria may represent early morphological changes following oxidative stress that may lead to activation of caspase-mediated apoptotic pathways.

KEYWORDS: lens optics; mitochondria; oxidative stress; confocal laser scanning microscopy; UV irradiation

INTRODUCTION

The key function of mitochondria is energy production through oxidative phosphorylation and lipid oxidation.[1] The process takes place within the mitochondrial inner membrane and includes five multi-sub-unit enzyme complexes. Several other metabolic functions are performed by mitochondria, including urea production and heme, nonheme iron and steroid biogenesis, and intracellular Ca^{2+} homeostasis as well as interaction with the endoplasmic reticulum. For many of these mitochondrial functions, there is only a partial understanding of the components involved, with even less information on mechanisms and regulation.[2]

The vertebrate lens is a cellular structure that is responsible for fine focusing of light on the retina. The lens consists of two types of cells organized in distinct spatial patterns: the epithelial monolayer that covers the anterior surface and fiber cells that comprise the bulk of the lens. Cell division in the lens is restricted to the equatorially located epithelial cells, which give rise to terminally differentiated fiber cells in a process that continues through life.[3] Early electron microscopy studies of mitochondria of vertebrate lenses showed an absence of mitochondria in lens fiber cells and the presence of very few, short mitochondria in the epithelial cells.[4] It became widely accepted that the lens epithelium plays the most important role in lens metabolism.[5] In recent years understanding of mitochondrial morphology and distribution has been enhanced by advances in confocal microscopy that permits imaging of living cells with the aid of fluorescent dye technology. Recent studies using specific fluorescent dyes and confocal microscopy of rat,[6] fish,[7] and bovine lenses[8,9] show that superficial cortical fiber cells contain numerous metabolically active mitochondria, suggesting that the superficial cortical fiber cells play a much more active role in lens metabolism than previously suspected.

Recognition of the role of mitochondria in processes, such as apoptosis and calcium homeostasis, has sparked a renewed interest in mitochondrial research. As a byproduct of respiration, the electron transport chain is known to be the major intracellular site for the generation of reactive oxygen species (ROS). Oxidative damage to the mitochondria has been experimentally demonstrated to cause an elevation in mitochondrially produced ROS.[10] Such mitochondrial changes have been manifested morphologically by the presence of short swollen mitochondria. Experimental lens damage has been found to correlate well with lens anatomy and with the integrity and activity of mitochondria found in the epithelium and superficial fiber layers of the lens.[8,9]

The lens represents a very useful model to study mitochondria. Even though the mitochondria are restricted to a minute portion of the lens, namely the epithelial and superficial cortical fiber cells, they play an important role in maintaining lens transparency. Our previous studies have showed that following treatment with the mitochondrial uncoupler carbonyl cyanide *m*-chlorophenylhydrazone (CCCP), rat[6] and bovine[8] lenses developed opacities. Similarly, bovine lenses treated with sodium dodecyl sulfate[9] showed a decrease in mitochondrial length and numbers and at the same time developed opacities, suggesting that mitochondria and lens optical quality are correlated. Our recent study, using confocal microscopy, was undertaken to image the movement of live mitochondria of bovine lens epithelial and superficial cortical fiber cells.[11] Using the mitochondria-specific dye tetra methyl rhodamine ethyl ester (TMRE), mitochondrial movement was acquired using confocal laser scanning microscopy by imaging the intact lens equatorial region for 3–5 min. Bidirectional dynamic movement of mitochondria with frequent reversals was observed in both epithelial and superficial cortical fiber cells of live bovine lenses. In the epithelium, this movement was up to 5 μm/min whereas in the superficial cortex the observed movement was up to 18.5 μm/min. The movement was abolished following treatment with CCCP. Whether the movement is true motion, or movement of TMRE across a mitochondrial network representing change in the distribution of potential across the inner membrane, presumably allowing energy transmission across the cell from regions of low to regions of high ATP demand, remains unclear.

Several epidemiological studies have showed a correlation between ultraviolet (UV) radiation and cataract development.[12] Once the lens is removed, the intraocular lens may not offer sufficient protection[13] and the retina may receive an increased amount of UV radiation. Earlier[14] and more recently[15] experimental investigations showed that UV irradiation affects the optics of bovine lenses. The retinal pigment epithelial cells, having the same embryonic origin as neurosensory retina have been recently used as an *in vitro* retinal model to study the molecular mechanisms following UV radiation.[16] Cellular and molecular mechanisms of UV damage on lens or retinal tissue are still not clear. Moreover, less is known as to how the mitochondria are affected. The purpose of this article was to characterize *in vitro* changes following oxidative stress with UV-B radiation in terms of mitochondrial dynamics of bovine lens epithelium and superficial cortical fiber cells and human retinal pigment epithelial (ARPE-19) cells.

MATERIALS AND METHODS

All chemicals were obtained from Sigma Chemical Co. (St. Louis, MO), unless indicated otherwise. Mitochondria-specific fluorescent dye Rhodamine-123 was obtained from Invitrogen Canada Inc. (Burlington, ON).

Bovine Eye Dissection and Lens Culture

Bovine eyes, obtained from a local abattoir, were dissected and the lenses were excised within 1–5 h post mortem under sterile conditions, as described previously.[9] Briefly, the lenses were suspended on a beveled washer (14 mm inner diameter) in a three-part chamber (made from glass, silicon rubber, and a metal base) filled with 21 mL of culture medium consisting of Medium 199 with Earle's salts, 100 mg/L L-glutamine, 3% dialyzed fetal bovine serum, 2.2 g/L sodium bicarbonate, 5.96 g/L HEPES, and 1% antibiotics (100 units/mL penicillin and 0.1 mg/mL streptomycin). Lenses were then preincubated at 37°C with 95% air and 5% CO_2 for 24 h prior to experimental use to ensure that lenses damaged during dissection were excluded from the study.

Human Retinal Pigment Epithelial Cell Culture

The human ARPE-19 cell line, originally obtained from American Type Culture Collection (ATCC), was kindly provided by Dr. R. Tchao, University of the Sciences in Philadelphia (Philadelphia, PA). ARPE-19 cells were plated in glass bottom culture dishes (MatTek Corporation, Ashland, MA) at low density and cultured in DMEM/Ham's F-12 50:50 mixture supplemented with L-glutamine and 15 mM HEPES (Mediatech, Herndon, VA), 10% fetal bovine serum (Hycolone, Logan, UT), and insulin-transferrin-sodium selenite media supplement (ITS supplement). The cell line was then grown to be at least 70% confluent in a humidified incubator with 95% air and 5% CO_2 at 37°C. To maintain the optimal growth conditions, the physiological solution was changed every 24 h.

UV-B Irradiation of Cultured Bovine Lenses and ARPE-19 Cells

Exposure to UV-B was produced by filtering banks of UV fluorescence tubes in a custom-designed UV irradiation unit with 95% air and 5% CO_2 as described previously.[15] Briefly, before irradiation, the UV source was calibrated with an Instaspec II diode-array spectroradiometer (Oriel Corporation, Stratford, CT). Cultured bovine lenses and at least 70% confluent ARPE-19 cell monolayer were then irradiated at 37°C with 1.0 and 0.5 J/cm^2 of broadband UV-B (290–320 nm), with the calculated biologically effective radiant energy levels of 0.445 and 0.223 J/cm^2, respectively. The UV source was positioned directly above the lenses or cells. In order to minimize absorption of the radiation by the medium, during irradiation the medium was removed except for a 1.0 mm layer that covered either the bovine lenses or the ARPE-19 cells. Following the radiation, the previously removed medium was replenished immediately with fresh warm stock.

Analysis of Lens Optical Properties—Spherical Aberration

Lens optical quality (spherical aberration) was assessed using the Scantox *in vitro* automated laser scanning system developed at the University of Waterloo before exposure and daily for 14 days after the irradiation at approximately the same time each day. In order to provide optimal conditions, the culture medium was changed every 48 h. The Scantox *in vitro* automated laser scanning system consists of a collimated laser source that projects a laser beam onto a plain mirror mounted at 45° on a carriage assembly. This mirror reflects the laser beam directly up through the scanner table surface and through the lens under examination. The mirror carriage is connected via a drive screw to a positioning motor. This positioning motor turns the drive screw and thereby moves the laser in user-defined steps across the lens in an automated fashion. A digital camera captures the actual position and slope of the laser beam at each step. When all steps have been made, the captured data for each step position is used to calculate the back vertex distance for each position and the difference in that measurement between beams. A series of 22 laser beams, passed at increments of 0.5 mm, for a total range of 11 mm are projected through the lens. The results for this part of the study involved 9900 objective optical measurements (30 lenses, 15 scan points, 22 beams).

Confocal Laser Scanning Microscopy

To investigate mitochondrial morphology and movement, the mitochondria-specific fluorescent dye Rhodamine 123 was used. Rhodamine 123 is a lipophilic cell-permeable, cationic nontoxic fluorescent dye that specifically stains live mitochondria. Rhodamine 123 is accumulated specifically by the mitochondria in proportion to membrane potential. At various time intervals the morphology, distribution, and movement of the mitochondria were analyzed immediately ($n = 3$ for each group), and after treatment at 7 ($n = 3$ for each group) and 14 days ($n = 3$ for each group) in bovine lenses. For ARPE-19 cells, the analysis was carried out immediately ($n = 2$ glass bottom culture dishes for each group) or 1 ($n = 2$ for each group) and 2 days ($n = 2$ for each group) after radiation. For confocal microscopy, samples were loaded with Rhodamine 123, by bathing them for 15 min at room temperature in either 10 mL serum-free M199 (bovine lenses) or 3 mL serum-free DMEM/Ham's F-12 50:50 mixture (ARPE-19 cells) containing 20 μm Rhodamine 123, and rinsed. For confocal microscopy imaging, bovine lenses were then immobilized in 1% agarose on glass bottom plates as described previously,[11] whereas ARPE-19 cells were imaged in glass bottom plates without agarose. Time series of bovine lens epithelial and superficial cortical fiber cells as well as ARPE-19 cells were acquired with a Zeiss 510 (configuration Meta 18) confocal laser scanning system (Carl Zeiss Inc., Toronto, Canada) equipped with an inverted Axiovert 200 M microscope and a high numerical aperture 40-× water-immersion C-Apochromat objective (NA 1.2). The combination of a 488 nm Argon laser and 505 long pass emission filters were used to visualize Rhodamine 123 fluorescence. Subsequent image analysis (such as the number and length of mitochondria and rate of movement) was achieved using commercial LSM510 VisArt and Physiology software packages (Carl Zeiss Inc., Jena, Germany) and Image Analysis toolbox of MatLab 7.1 software package (MathWorks Inc., Natick, MA). The representative length of mitochondria was measured from one end of the most visible structure to the other end using the measure function of the software. The rate of movement (expressed in μm/min) was determined by noticing the initial position (position zero) of moving mitochondria in the first X, Y gallery of time series, to the final movement observed during the time series by measuring the length in microns (μm).[11]

Statistical Analyses

Statistical analyses were carried out using either one- (number and length of the mitochondria) or two-way repeated (lens optics) measures of ANOVA, calculated by SAS® 9.1 statistical software (SAS Institute Inc., Cary, NC). Data were considered significantly different at probability levels equal to or less than 0.05. The difference was recorded numerically with all results expressed as mean ± standard error of the mean (SEM).

RESULTS

Bovine Lens Optical Function

In total, 30 bovine lenses were scanned, divided into three groups: controls and UV-B irradiated, at 1.0 and 0.5 J/cm^2. The lenses were scanned before radiation and daily for 14 days using the automated scanning laser system, Scantox. FIGURE 1 shows representative individual scans consisting of 22 laser beams scanned across that of a control (A), 1.0 (B) and 0.5 J/cm^2 (C) treated lenses at day 14 after irradiation. The control group showed little change in the amount of spherical aberration. By examining the data over time across all groups, a latent dose-dependent increase in the amount of spherical aberration (back vertex distance variability) over time was found (FIG. 2). A significant increase in the amount of spherical aberration became evident in the 1.0 J/cm^2 irradiated group lenses starting on day 11 ($P < 0.05$). By day 14 in both irradiated groups, the amount of spherical aberration was significantly higher than that of controls ($P < 0.05$). Moreover, at day 14 the 1.0 J/cm^2 treated group showed an exponential increase in the amount of spherical aberration (over 2.5 times higher value) in comparison to that of the 0.5 J/cm^2 group.

Mitochondrial Morphology and Movement of Controls

In this study the mitochondria of both bovine lenses and APRE-19 cells were stained with specific fluorescent dye, Rhodamine 123. In the epithelium of bovine lenses strikingly different mitochondrial morphologies could be seen ranging from short, dot-like structures less than 1 μm in length, to elongated

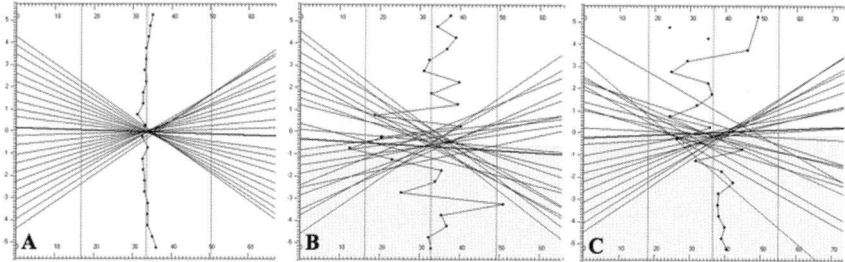

FIGURE 1. Representative individual scans comparing optics of control to that of 1.0 and 0.5 J/cm^2 14 days after UV-B irradiation. Lens optical quality was assessed using the automated laser scanning system, as described in the "Materials and Methods" section. UV-B-dependent increase in the amount of spherical aberration, as shown by a broken line, was seen in 1.0 (**B**) and 0.5 J/cm^2 (**C**) irradiated lenses as compared to control (**A**). The x axis (mm) indicates the back vertex distance and the y axis (mm) indicates the position of the laser beam across the lens.

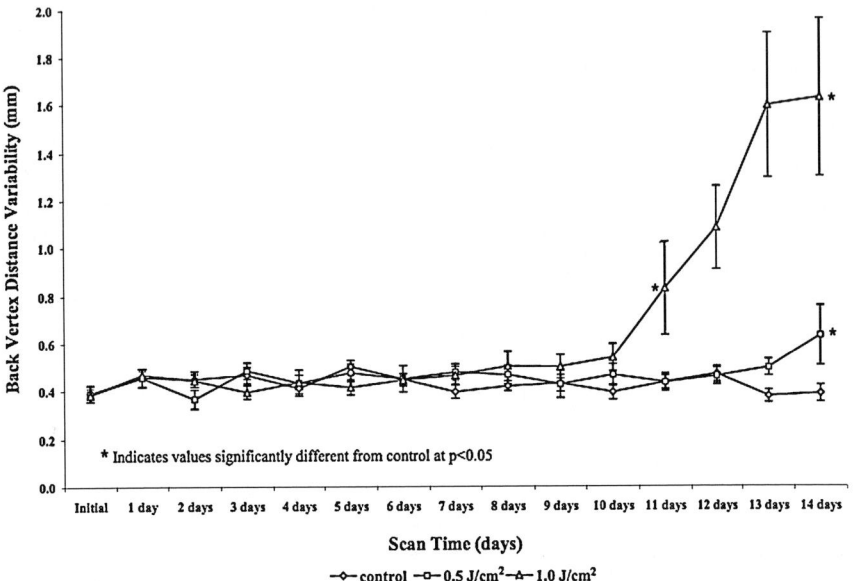

FIGURE 2. Comparative effect of UV-B irradiation over time on optics of bovine lenses. A latent dose-dependent increase in the amount of spherical aberration (back vertex distance variability, mm) was seen. Compared to controls, a significant increase ($P < 0.05$) in 1.0 J/cm² group lenses was noted on day 11. Spherical aberration continued to increase and by the end of the experiment the difference was significant in both group lenses.

threads, up to 10 μm. When the dot-like mitochondria could be found dispersed throughout the cytoplasm, the elongated dense threads appeared to surround the individual nuclei, filling the bulk of the epithelial cells (FIG. 3 A). Occasionally, branching could be seen on those elongated threads. In contrast, in the superficial cortex while short, dot-like mitochondria less than 1 μm in length were seen, more elongated and not as dense up to 74 μm in length, threads of mitochondria could be seen (FIG. 3 B). While occasional branching in these long threads could also be seen, as in the epithelium, the mitochondria were not as dense, and were roughly aligned along the long axis of the superficial cortical fiber cells. The morphology of the mitochondria of ARPE-19 cells was also heterogeneous, with some mitochondria measuring less than 1 μm while others up to 15 μm in length (FIG. 3 C). Frequently, branches in those thread-like structures could be seen. While in some cells dense threads of mitochondria could be seen arranged radially around the nuclei, in other cells the threads could be seen loosely arranged throughout the cytoplasm processes without particular reference to the nuclei. Using a MatLab semiautomated image analysis approach we were able to determine the number and the average length of the mitochondria, with values for controls and treated lenses and ARPE-19 cells listed in TABLE 1.

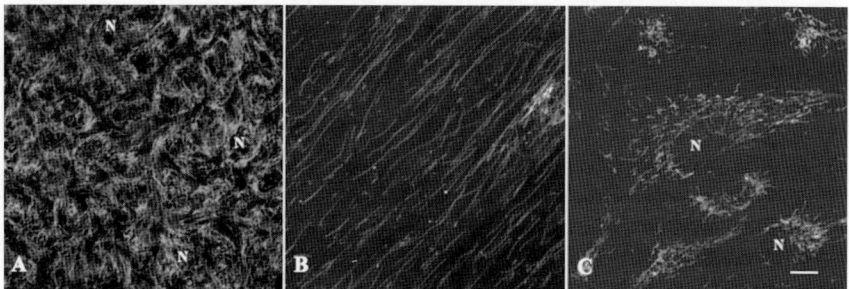

FIGURE 3. Representative confocal micrographs showing the distribution and morphology of the mitochondria of controls. Bovine lenses and ARPE-19 cells were loaded with Rhodamine 123 and viewed under confocal microscope as described in the "Materials and Methods" section. The mitochondria of bovine lens epithelium are shown in **A** and superficial cortex in **B**, whereas the mitochondria of ARPE-19 cells are shown in **C**. Bar = 10 μm.

In order to establish a control base line, prior to UV-B irradiation, the mitochondria of bovine lens epithelial and superficial cortical fiber cells and ARPE-19 cells were monitored by capturing Rhodamine 123 fluorescence over time using confocal laser scanning microscopy. The capturing of the time series was summed for 4 min of data acquisition 120 frames on average. In the epithelium of bovine lenses the most obvious and rapid movement was seen in the short, dot-like mitochondria. This movement, around the nuclei, was up to 8 μm/min (See Video Clip 1 in online version). In the superficial cortex apparent movement along the long axis of the cells could be seen in both dot-like short and elongated mitochondria with up to 5 μm/min and up to 25 μm/min, respectively (See Video Clip 2 in online version). In both the lens epithelium and superficial cortex movement of the mitochondria was bidirectional. Similarly, bidirectional movement of short and elongated mitochondria was seen in the ARPE-19 cells measuring up to 3 μm/min and 12 μm/min, respectively (See Video Clip 3 in online version). Moreover, as part of the movement branching was noted in many mitochondria.

UV-B Irradiation-Induced Morphological Changes and Alterations in Size and Movement of the Mitochondria

At both UV energy levels studied, morphological changes and alterations in movement of the mitochondria were monitored immediately, 7 and 14 days after irradiation of bovine lenses. Because of rapid cell division, changes in ARPE-19 cells were monitored immediately, 1 and 2 days after UV-B irradiation. Changes were compared to controls of each group. Immediately following 1.0 J/cm^2 irradiation of lenses perinuclear-arranged threads of mitochondria could be seen resembling that of controls in the epithelium, whereas in the

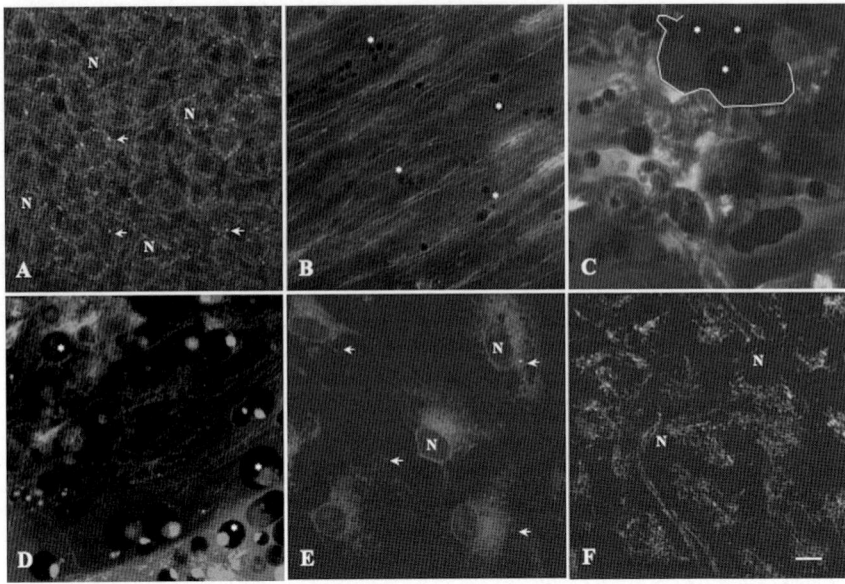

FIGURE 4. Representative confocal micrographs showing UV-B irradiation-induced morphological changes and alterations in size of the mitochondria. Compared to controls short swollen nodule-like mitochondria (indicated by *arrows*) were seen in the epithelium of bovine lenses immediately after 1.0 J/cm^2 irradiation (**A**). In the superficial cortex the immediate effect of 1.0 J/cm^2 was expressed by the presence of small numerous vacuoles (**B**, indicated by *asterisks*). Only few cells and large vacuoles (*solid line*) were seen in the epithelium of bovine lenses 7 days after 1.0 J/cm^2 (**C**). In the superficial cortex at the same time fewer and shorter mitochondria as well as vacuoles (*asterisks*) were seen (7 days after 1.0 J/cm^2, **D**). In the ARPE-19 cells immediately after 1.0 J/cm^2 only few dot-like short mitochondria (indicated by *arrows*) were seen (**E**). In the 0.5 J/cm^2 overall shorter mitochondria were seen (**F**). Bar = 10 μm.

periphery of the cell, numerous short and swollen, nodule-like mitochondria could be seen in contrast to the controls (FIG. 4 A, TABLE 1). Despite the close resemblance to controls, no movement of the mitochondria in the epithelial cells was seen. Moreover, Rhodamine 123 bleaching was observed 2 min after data acquisition (See Video Clip 4 in online version). In the superficial cortical fiber cells while the size of mitochondria was similar to that of controls, small numerous vacuoles scattered throughout cytoplasm could be seen (FIG. 4 B, TABLE 1). Movement of the mitochondria in the superficial cortical fiber cells was only seen occasionally in dot-like and the shorter mitochondria (See Video Clip 5 in online version). In the 0.5 J/cm^2 group, change in the lens epithelium was similar to that of 1.0 J/cm^2 group lenses, with perinuclear-arranged threads of mitochondria resembling that of controls (TABLE 1). In the periphery, numerous short and swollen, nodule-like mitochondria could be

TABLE 1. Comparative list of changes in the average mitochondrial length (μm $\pm$ SEM) over time in epithelium and superficial cortical fiber cells of control (untreated) and 1.0 or 0.5 J/cm^2 UV-B irradiated bovine lenses and ARPE-19 cells

Treatment	Control (range)	1.0 J/cm^2 (range)	0.5 J/cm^2 (range)
Immediate (epi)	2.54 $\pm$ 0.03 (1–10)	2.03 $\pm$ 0.02 (0.4–7)*	1.84 $\pm$ 0.02 (0.4–6)*
7 days (epi)	2.83 $\pm$ 0.05 (0.7–13)	N/A	2.0.4 $\pm$ 0.03 (0.4–7)*
Immediate (fib)	6.25 $\pm$ 0.26 (1–74)	3.38 $\pm$ 0.12 (0.4–29)*	3.38 $\pm$ 0.13 (0.4–44)*
7 days (fib)	4.63 $\pm$ 0.17 (0.4–53)	3.34 $\pm$ 0.14 (0.4–28)*	4.56 $\pm$ 0.18 (0.5–46)
Immediate ARPE-19	3.08 $\pm$ 0.095 (0.6–36)	N/A	2.77 $\pm$ 0.06 (0.7–13)*

*Indicates values significantly different from control at $P < 0.05$.

seen (data not shown). Occasional movement of dot-like mitochondria could be seen in the epithelial cells. In the superficial cortex of 0.5 J/cm^2 exposed lenses, the changes were similar to the 1.0 J/cm^2 group lenses and were limited to the presence of small numerous vacuoles (data not shown). While longer thread-like mitochondria showed occasional movement in some lenses, short dot-like mitochondria were seen moving readily in multidirectional, burst-like patterns.

Drastic morphological changes were observed in the lens epithelium 7 days after 1.0 J/cm^2 irradiation. Only a few cells and large vacuoles were seen (FIG. 4 C). In the remaining cells only diffuse Rhodamine 123 fluorescence was seen. In the superficial cortex, fewer and shorter mitochondria, as well as large vacuoles, could be seen (FIG. 4 D). However, the remaining mitochondria were not moving. In the 0.5 J/cm^2 group lenses 7 days after irradiation, while distinct epithelial cells could be seen, Rhodamine 123 fluorescence was very diffuse (data not shown). In the superficial cortical fiber cells numerous small vacuoles could be seen while the mitochondria resembled that of controls (data not shown). The major change was the lack of movement of the mitochondria. Moreover, complete Rhodamine 123 bleaching was evident immediately after data acquisition.

By 14 days after 1.0 J/cm^2 treatment few bovine epithelial cells could be seen. No mitochondria were seen. Large vacuoles scattered throughout the cytoplasm were seen in the superficial cortex. In the 0.5 J/cm^2 group lenses 14 days after irradiation, distinct epithelial cells could be captured. However, large prominent vacuoles were seen. While in some lenses only diffuse Rhodamine 123 fluorescence was seen, in other samples mitochondria, shorter than that of controls, could be seen. The remaining mitochondria exhibited quick burst-like movement over a very short distance (less than 1 μm). Moreover, Rhodamine 123 bleaching was evident 3 min after data acquisition. In the superficial cortex small vacuoles could be seen scattered throughout the cytoplasm. Mitochondria appeared shorter than that of controls. Occasionally, swollen, nodule-like mitochondria were seen. However the major change, as compared to 7 days after treatment, is the fact that some movement of the mitochondria over a short distance could be seen, especially in small, dot-like structures.

In the ARPE-19 cells immediately following 1.0 J/cm^2 irradiation very diffused Rhodamine 123 fluorescence could be captured and only a few dot-like, short mitochondria were seen (FIG. 4 E). No movement was seen and Rhodamine 123 photobleaching was observed immediately after the start of data acquisition (See Video Clip 6 in online version). Overall shortening of the mitochondria could be seen immediately after 0.5 J/cm^2 irradiation, with only few branched mitochondria seen (FIG. 4 F). Those shorter mitochondria were not moving and Rhodamine 123 bleaching was observed 3 min after data acquisition. In ARPE-19 cells for both 1 and 2 days after 1.0 and 0.5 J/cm^2 irradiation only a few cells lacking mitochondria remained, exhibiting only diffuse Rhodamine 123 fluorescence (data not shown).

DISCUSSION

Exposure to solar UV radiation has been implicated in a large spectrum of skin and ocular pathologies, including cataract.[12] The gradual depletion of stratospheric ozone due to chlorofluorocarbons (CFCs), aircraft pollutants, and other industrial pollutants substantially increases the levels of UV radiation, particularly UV-B, reaching the Earth's surface.[17] While the cornea looses its refractive function once removed from the eye, the lens, cultured in physiological solution, continues to maintain its cellular makeup and original refractive function. Similarly, cultured ARPE-19 cells represent an *in vitro* model of neurosensory retina.

The major and most novel consequence of the UV-B irradiation reported in this study is the cessation of mitochondrial movement. While our recent study showed the movement of mitochondria in bovine lenses,[11] to our knowledge this is the first study describing movement of mitochondria in ARPE-19 cells. The mechanism of organelle movement is not known in bovine lenses or ARPE-19 cells. In neurons, transport is thought to occur along microtubules, presumably involving kinesin-related proteins and cytoplasmic dynein.[18] This movement can be inhibited by stabilizing microtubules with drugs like nocodazole.[19–21] A recent study of rat lens showed the existence of a microtubule-based motor system containing both kinesin and dynein in the elongating fiber cells.[22] The presence of this system was attributed to the transportation of Golgi complexes to their target regions during increased cell growth accompanying elongation of the secondary fiber cells. A large number of microtubules were regularly arranged into bundles parallel to the long axis of fiber cells, a morphological observation similar to that of the distribution of the mitochondria seen in the superficial cortical fiber cells (FIG. 3 B) and this may represent the machinery responsible for the observed rapid mitochondrial movement in the superficial cortex of bovine lenses. Movement is also an ATP-dependent process, and the depolarization of mitochondria combined with the potential impairment of function associated with the calcium load could deplete local ATP concentrations quite rapidly. Thus, a number of different mechanisms could contribute to the cessation of movement. The mechanisms responsible for the alteration in movement may also underlie the shape change. If the rod-like structure associated with normal mitochondria is essentially a function of the organelle being stretched out on a cytoskeletal structure, the dissolution of that structure could result in the rounding that we report here.

The second finding of this study is related to the highly variable morphology and distribution of mitochondria seen in that of controls and also the observed morphological changes and alterations in size of the mitochondria following irradiation (TABLE 1). The observation in the controls may be related to the dynamic state of the mitochondria. In growing cells mitochondria are frequently found as extremely dynamic structures with tubular sections dividing in half, branching, and fusing presumably to form a complex network.[23] Moreover, there is an emerging appreciation that the decrease in the size of the mitochondria may be associated with early cell injury and that these changes are related to the offset of normal fusion and fission of the mitochondria.[24] In several *in vitro* models of apoptosis it has been reported that normal elongated mitochondria become fragmented.[25,26] Fragmentation of mitochondria following treatment may be a consequence of the association of proteins, such as dynamin-related protein 1 (DRP1) with mitochondria. Overexpression of this protein, responsible for mitochondrial fission in neurons, causes fragmentation,[26,27] whereas a dominant-negative form of DRP1 decreases sensitivity to mitochondrially mediated apoptosis.[26]

Alterations in mitochondrial morphology in neurons[28] and astrocytes[29] in response to calcium loading have also been reported. These authors found that mitochondria changed from elongated rod-like to short spherical morphology following treatment. In these studies the authors concluded that one of the possible mechanisms in the observed shape change was caused by mitochondrial permeability transition. In a previous study it was proposed that the elongated, thread-like or reticulated form of mitochondria is beneficial on the basis of studies in cardiac myocytes.[30] Thus, because of the cable properties of mitochondria, the proton motive force generated by the electron transport chain can be effectively distributed across the reticulum and thus facilitate ATP synthesis. This advantage is hypothetical at this point because whether the mitochondria are able to generate a spatially limited proton motive force is unclear. In this study, the consequence of the alteration in morphology for mitochondrial function appears to correlate with observed increase in the amount of spherical aberration in bovine lenses, whereas the connection is much less clear for ARPE-19 cells. Our previous studies using the mitochondrial uncoupler CCCP resulted in concentration-dependent fragmentation of mitochondria and an increase in the amount of spherical aberration of bovine lenses.[8] The current results and our previous studies[8,9] have suggested that modified mitochondrial function may have a direct consequence on lens optics.

A latent UV-B dose-dependent increase in the amount of spherical aberration seen in bovine lenses confirms earlier observations.[14,15] This latent effect of oxidative stress from UV-B irradiation on loss of sharp focus was first evident at the 11 days scan point, where an over twofold increase in back vertex distance variability in the 1.0 J/cm^2 irradiated group lenses was seen. By day 14 in both irradiated groups the amount of spherical aberration was significantly higher than that of controls ($P < 0.05$). While the difference in the amount of calculated biologically effective radiant energy levels was linear between the two groups (0.445 and 0.223 J/cm^2, respectively), at day 14 the 1.0 J/cm^2 treated group showed an exponential increase in the amount of spherical aberration (over 2.5 times higher value) in comparison to that of the 0.5 J/cm^2 group.

As mentioned in the introduction, the vertebrate lens is a cellular structure that consists of two types of cells organized in distinct spatial patterns: the epithelial monolayer that covers the anterior surface and fiber cells that comprise the bulk of the lens. Cell division in the lens is restricted to the equatorially located epithelial cells, which give rise to terminally differentiated fiber cells in a process that continues through life. The newly formed fiber cells overlie the preexisting cells in a process that continues throughout life, resulting in a steady increase in tissue volume. The oldest cells, which at one time were the embryonic lens, form the lens nucleus, or central zone. Loss of aging or damaged cells is impeded because of their central location and because of the complete enclosure of the lens by the acellular capsule. The observed latency in the optical response may be explained, at least partially, by the lens' unique morphology and its rate of growth as described above. Similar results were seen in the study of age and toxic treatment of rat lenses where delayed optical change was seen in older lenses in comparison to younger ones.[31] Moreover, the younger lenses showed the ability to recover from optical damage while the older ones did not.[31]

It is interesting to consider the role of oxidative stress from UV-B irradiation on lens optics and mitochondria and how these may be an expression of early cell injury. The effects are clearly produced by UV-B exposures that are above the normal threshold. However, the changes in shape and movement of the mitochondria that we observe are very similar to the effects of CCCP. This is notable, because we have demonstrated recently that the application of CCCP results in both mitochondrial rounding and movement cessation.[11] Based on the observation of a short dot-like mitochondria one can conclude that rounding of mitochondria or their temporary immobilization does not inevitably lead to injury. However, it is interesting to note that the lens is capable of recovery, both in terms of lens optics and the normal morphological phenotype of mitochondria, following low concentrations of CCCP treatment.[11] This suggests that the recovery of mitochondrial function aids in the restoration of lens optical function. Thus, in this experimental paradigm, the alteration of mitochondrial morphology would be considered a symptom rather than a cause of cell injury. Indeed, this raises the possibility that either shape or movement change could be an attempt to protect the cell from injury, although what exactly is accomplished by this maneuver remains to be established.

In summary, the main findings of this study are that UV-B irradiation, as a consequence of generation of ROS and subsequent oxidative stress, causes a cessation of mitochondrial movement in both bovine lenses and ARPE-19 cells. It also produces a latent change in lens optics and a rapid and substantial remodeling of mitochondrial morphology. Morphological changes and alterations in size and movement of the mitochondria could be detected immediately and over time following UV-B irradiation. These findings suggest that oxidative stress might alter the normal movement of mitochondria to cellular destinations where ATP synthesis is required and thus impair not only cellular function but also lens optics. More broadly, these experiments clearly establish an unappreciated dynamic equilibrium that exists between the distribution of mitochondria within cells and their biological function. Under conditions of oxidative stress or chemical treatment one might anticipate that impaired mitochondrial function would alter the ability of mitochondria to move and thereby prevent the normal distribution of mitochondria within either lens epithelium and superficial cortex or ARPE-19 cells. While cellular repair mechanisms may exist at or below threshold levels of UV-B radiation, under conditions of chronic stress over a life time, the oxidative stress could stop new mitochondria from being delivered to distal parts of the superficial cortical fiber cells and ARPE-19 cells, or possibly even prevent the retrieval of dysfunctional mitochondria to the cell body for degradation, either of which may result in cell damage. When one considers the number of endogenous toxins that the ocular surface is exposed to (such as oxidative stress) as well as accidental exposure to xenobiotics that may impair mitochondrial function and dissipate the mitochondrial membrane potential, it seems likely that the alteration of the movement and morphology of mitochondria is likely to be a broadly important phenomenon associated with lens and ARPE-19 cells injury.

SUPPLEMENTARY MATERIAL

The following supplementary material is available in QuickTime for this article:

Video Clip S1. Movement of mitochondria in the epithelium of control bovine lenses.
Video Clip S2. Movement of mitochondria in the superficial cortex of control bovine lenses.
Video Clip S3. Movement of mitochondria in control ARPE-19 cells.
Video Clip S4. The immediate effect of 1.0 J/cm^2 UV-B irradiation on movement of mitochondria in epithelium of bovine lenses.
Video Clip S5. The immediate effect of 1.0 J/cm^2 UV-B irradiation on movement of mitochondria in superficial cortex of bovine lenses.

Video Clip S6. Lack of movement of mitochondria in ARPE-19 cells immediately following 1.0 J/cm² UV-B irradiation.

This material is available as part of the online article from: http://www.blackwell-synergy.com/doi/abs/10.1196/annals.1378.051 (This link will take you to the article abstract).

Please note: Blackwell Publishing is not responsible for the content or functionality of any supplementary materials supplied by the authors. Any queries (other than missing material) should be directed to the corresponding author for the article.

ACKNOWLEDGMENTS

This work was supported by the E.A. Baker Foundation for the Prevention of Blindness and the Canadian Foundation for Innovation. The authors appreciate the assistance of Katelyn Wharram with dissection of bovine eyes and Daniel Lelli with MatLab programming.

REFERENCES

1. FREY, T.G. & C.A. MANNELLA. 2000. The internal structure of mitochondria. TIBS **25:** 319–324.
2. WALLACE, D.C., M.D. BROWN, S. MELOV, et al. 1998. Mitochondrial biology, degenerative diseases and aging. Biofactors **7:** 187–190.
3. MCAVOY, J.W. 1978. Cell division, cell elongation and the co-ordination of crystallin gene expression during lens morphogenesis in the rat. J. Embryol. Exp. Morphol. **45:** 271–281.
4. KUWABARA, T. 1975. The maturation of the lens cell: a morphologic study. Exp. Eye Res. **20:** 427–443.
5. KINSEY, V.A. & V.N. REDDY. 1965. Studies of the crystalline lens. XI. The relative role of the epithelium and capsule in transport. Invest. Ophthalmol. **4:** 104–116.
6. BANTSEEV, V., K.L. HERBERT, J.R. TREVITHICK & J.G. SIVAK. 1999. Mitochondria of rat lenses: distribution near and at the sutures. Curr. Eye Res. **19:** 506–516.
7. BANTSEEV, V., K.L. MORAN, D.G. DIXON, et al. 2004. Optical properties, mitochondria and sutures of lenses of fishes: a comparative study of nine species. Can. J. Zool. **82:** 86–93.
8. BANTSEEV, V., A.P. CULLEN, J.R. TREVITHICK & J.G. SIVAK. 2003. Optical function and mitochondrial metabolic properties in damage and recovery of bovine lens after in vitro carbonyl cyanide m-chlorophenylhydrazone treatment. Mitochondrion **3:** 1–11.
9. BANTSEEV, V., D. MCCANNA, A. BANH, et al. 2003. Mechanisms of ocular toxicity using the in vitro bovine lens and sodium dodecyl sulfate as a chemical model. Toxicol. Sci. **37:** 98–107.
10. SKULACHEV, V.P. 2001. Mitochondrial filaments and clusters as intracellular power-transmitting cables. Trends Biochem. Sci. **26:** 23–29.
11. BANTSEEV, V. & J.G. SIVAK. 2005. Confocal laser scanning microscopy imaging of dynamic TMRE movement in the mitochondria of epithelial and superficial cortical fiber cells of bovine lenses. Mol. Vis. **11:** 518–523.

12. ROBMAN, L. & H. TAYLOR. 2005. External factors in the development of cataract. Eye **19:** 1074–1082.
13. LIN, K., Y. LIN, J. LEE, *et al.* 2002. Spectral transmission characteristics of spectacle, contact, and intraocular lenses. Ann. Ophthalmol. **34:** 206–215.
14. STUART, D.D., A.P. CULLEN, J.G. SIVAK & M.J. DOUGHTY. 1994. Optical effects of UV-A and UV-B radiation on the cultured bovine lens. Curr. Eye Res. **13:** 371–376.
15. YOUN, H.Y., K.L. MORAN, O.M. ORIOWO, *et al.* 2004. Surfactant and UV-B-induced damage of the cultured bovine lens. Toxicol. *In Vitro* **18:** 841–852.
16. PATTON, W.P., U. CHAKRAVARTHY, R.J. DAVIES & D.B. ARCHER. 1999. Comet assay of UV-induced DNA damage in retinal pigment epithelial cells. Invest. Ophthalmol. Vis. Sci. **40:** 3268–3275.
17. JOHNSON, G.J. 2004. The environment and the eye. Eye **18:** 1235–1250.
18. LIGON, L.A. & O. STEWARD. 2000. Movement of mitochondria in the axons and dendrites of cultured hippocampal neurons. J. Comp. Neurol. **427:** 340–350.
19. NEUTZNER, A. & R.J. YOULE. 2005. Instability of the mitofusin Fzo1 regulates mitochondrial morphology during the mating response of the yeast Saccharomyces cerevisiae. J. Biol. Chem. **280:** 18598–18603.
20. KNOWLES, M.K., M.G. GUENZA, R.A. CAPALDI & A.H. MARCUS. 2002. Cytoskeletal-assisted dynamics of the mitochondrial reticulum in living cells. Proc. Natl. Acad. Sci. USA **99:** 14772–14777.
21. TSENG, C.J., Y.J. WANG, Y.C. LIANG, *et al.* 2002. Microtubule damaging agents induce apoptosis in HL 60 cells and G2/M cell cycle arrest in HT 29 cells. Toxicology **175:** 123–142.
22. LO, W.K., X.J. WEN & C.J. ZHOU. 2003. Microtubule configuration and membranous vesicle transport in elongating fiber cells of the rat lens. Exp. Eye Res. **77:** 615–626.
23. BEREITER-HAHN, J. 1990. Behavior of mitochondria in the living cell. Int. Rev. Cytol. **122:** 1–63.
24. RINTOUL, G.L., A.J. FILIANO, J.B. BROCARD, *et al.* 2003. Glutamate decreases mitochondrial size and movement in primary forebrain neurons. J. Neurosci. **23:** 7881–7888.
25. DESAGHER, S. & J.C. MARTINOU. 2000. Mitochondria as the central control point of apoptosis. Trends Cell Biol. **10:** 369–377.
26. FRANK, S., B. GAUME, E.S. BERGMANN-LEITNER, *et al.* 2001. The role of dynamin-related protein 1, a mediator of mitochondrial fission, in apoptosis. Dev. Cell **1:** 515–525.
27. FILIANO, J.J., M.J. GOLDENTHAL, A.C. MAMOURIAN, *et al.* 2002. Mitochondrial DNA depletion in Leigh syndrome. Pediatr. Neurol. **26:** 239–242.
28. DUBINSKY, J.M. & Y. LEVI. 1998. Calcium-induced activation of the mitochondrial permeability transition in hippocampal neurons. J. Neurosci. Res. **53:** 728–741.
29. KRISTAL, B.S. & J.M. DUBINSKY. 1997. Mitochondrial permeability transition in the central nervous system: induction by calcium cycling-dependent and -independent pathways. J. Neurochem. **69:** 524–538.
30. AMCHENKOVA, A.A., L.E. BAKEEVA, Y.S. CHENTSOV, *et al.* 1988. Coupling membranes as energy-transmitting cables. I. Filamentous mitochondria in fibroblasts and mitochondrial clusters in cardiomyocytes. J. Cell. Biol. **107:** 481–495.
31. HERBERT, K.L., D.G. DIXON & J.G. SIVAK. 1998. Effects of age on the sensitivity of the rat lens to hexanol *in vitro*. J. Toxicol. Cutaneous Ocul. Toxicol. **17:** 127–139.

2-Methoxyestradiol Inhibits Superoxide Anion Generation while It Enhances Superoxide Dismutase Activity in Swine Granulosa Cells

GIUSEPPINA BASINI, SUJEN ELEONORA SANTINI, AND FRANCESCA GRASSELLI

Dipartimento di Produzioni Animali, Biotecnologie Veterinarie, Qualità e Sicurezza degli Alimenti, Sezione di Fisiologia Veterinaria, Università di Parma, 43100 Parma, Italy

ABSTRACT: 2-Methoxyestradiol (2-ME) is an estradiol metabolite with antiangiogenic properties. It can be produced by granulosa cells and it is present in normal follicle at high concentrations. The identification of reactive oxygen species (ROS) role in the molecular pharmacology of 2-ME is an active area of research. The objective of this article was therefore to evaluate the effect of 2-ME on both superoxide anion (O_2^-) production and superoxide dismutase (SOD) activity in swine granulosa cells collected from follicles greater than 5 mm and treated for 48 h with 1 μM 2-ME. 2-ME inhibited ($P < 0.001$) O_2^- generation in swine granulosa cells, while it stimulated ($P < 0.05$) the SOD activity. We previously demonstrated that a stimulation of O_2^- generation triggers angiogenetic response in granulosa cells. Therefore, we argue that the inhibitory effect of 2-ME on O_2^- could be responsible for its antiangiogenetic effect.

KEYWORDS: 2-methoxyestradiol; granulosa cells; angiogenesis; O_2^-; SOD

INTRODUCTION

Angiogenesis is the generation of new blood vessels by capillary growth sprouting from preexisting ones. The adult vascular system is quiescent, except for pathological angiogenesis during wound healing and tumor growth. On the contrary, ovaries are distinctive in that the formation of new blood vessels plays an obligatory role in the successful growth and development of ovulatory follicles.[1] Therefore, in an angiogenesis perspective, the ovary offers

Address for correspondence: Giuseppina Basini, Dipartimento di Produzioni Animali, Biotecnologie Veterinarie, Qualità e Sicurezza degli Alimenti, Sezione di Fisiologia, Facoltà di Medicina Veterinaria, Via del Taglio 8, 43100 Parma, Italy. Voice: +39-0521-032775; fax: +39-0521-032770.
  e-mail: basini@unipr.it

a unique opportunity to unravel the molecular orchestration of blood vessels development and regression under normal conditions. The regulation of this process is very poorly understood. Among the various triggers of angiogenesis are reactive oxygen species (ROS), mainly in the form of superoxide anion (O_2^-). This molecule has long been regarded as a toxic product and much attention has been paid to its main detoxifying enzyme, superoxide dismutase (SOD).

Remarkably, both *in vitro* and *in vivo* studies indicate that angiogenic response is triggered by ROS signaling in a highly coordinated manner, pointing out a role for ROS as signal transducers.[2] In particular, our previous work[3] suggested that ROS may be the important intracellular messengers linking tissue hypoxia to the subsequent angiogenetic response.

Recently, antiangiogenesis has been the subject of intense interest because of its therapeutical implication for restricting tumor growth. 2-Methoxyestradiol (2-ME), an estradiol metabolite, is formed by granulosa cell catechol-O-methyltransferase activity and is present in the normal follicle at high concentrations.[4] In this unique microenvironment, it may regulate selected cell types via autocrine and/or paracrine action. Evidence is growing that 2-ME is a potent inhibitor of angiogenesis and tumor growth[5] but the identification of ROS role in the molecular pharmacology of 2-ME is a matter of concern. To this purpose, we evaluated the effect of 2-ME on both O_2^- and SOD activity in swine granulosa cells.

MATERIALS AND METHODS

All the reagents were obtained from Sigma (St. Louis, MO) unless otherwise specified.

Granulosa Cell Collection

Swine ovaries were collected at a local slaughterhouse, placed into cold phosphate-buffered saline (PBS) (4°C) supplemented with penicillin (500 IU/mL), streptomycin (500 μg/mL), and amphotericin B (3.75 μg/mL), and maintained in a freezer bag and transported to the laboratory within 1 h. After a series of washings with PBS and ethanol (70%), granulosa cells from follicles greater than 5 mm were aseptically harvested by aspiration with a 26-gauge needle and released in medium with heparin (50 IU/mL), centrifuged for pelleting, and then treated with 0.9% prewarmed ammonium chloride at 37°C for 1 min to remove the red blood cells. Cell number and viability were estimated using a hemocytometer under a phase contrast microscope after vital staining with trypan blue (0.4%) of an aliquot of cell suspension. Cells were seeded at different plating densities (see below) in culture medium (CM) composed by M199 supplemented with sodium bicarbonate (2.2 mg/mL), bovine serum albumin (BSA 0.1%), penicillin (100 IU/mL), streptomycin (100 μg/mL), amphotericin B (2.5 μg/mL), selenium (5 ng/mL), and transferrin (5 μg/mL).

WST-1 Assay for O_2^- Production

O_2^- production was evaluated by the cell proliferation WST-1 test (Roche, Mannheim, Germany) since evidence exists that tetrazolium salts can be used as a reliable measure of intracellular O_2^- production.[6,7] About 10^4 cells/200 μL CM were seeded in 96-well plates and treated for 48 h with 1 μM 2-ME. Twenty microliters of WST-1 were added to the cells during the last 4 h of incubation and absorbance was then determined using a Spectra Shell Microplate reader (SLT Spectra, Milan, Italy) at 450 nm against 620 nm.

SOD Activity

SOD activity was determined by a SOD Assay Kit (Dojindo Molecular Technologies, Japan). About 2×10^5 cells/200 μL CM were seeded in 96-well plates and treated for 48 h with 1 μM 2-ME. After centrifugation for 10 min at 400 g, the surnatants were discarded and cells were lysed adding cold Triton 1% in TRIS HCl (100 μL/10^5 cells) and incubating on ice for 30 min. Cell lysates were tested without dilution and a standard curve of SOD ranging from 0.156 to 20 U/mL was prepared. The colorimetric assay was performed measuring formazan produced by the reaction between a tetrazolium salt (WST-1) and O_2^-, produced by the reaction of an exogenous xantine oxidase. The remaining O_2^- is an indirect hint of the endogenous SOD activity. The absorbance was determined with a Spectra Shell Microplate Reader reading at 450 nm against 620 nm.

Statistical Analysis

Each parameter was examined at least four times (6 replicates/treatment). Experimental data are presented as mean ± SEM; statistical differences between treatments were calculated with analysis of variance (ANOVA) using Statgraphics package (STSC Inc., Rockville, MD). When significant differences were found, means were compared by Scheffè's F-test; P values < 0.05 were considered to be statistically significant.

RESULTS

O_2^- Production

As shown in FIGURE 1, a 48-h treatment with 1 μM 2-ME significantly inhibited ($P < 0.001$) O_2^- generation in swine granulosa cells.

SOD Activity

SOD activity (basal 280 ± 5 mU/mL mean $\pm$ SEM) in swine granulosa cells were significantly ($P < 0.05$) enhanced by a 48-h treatment with 1 μM 2-ME as evidenced in FIGURE 2.

DISCUSSION

2-ME is a naturally occurring compound that is formed during metabolism of estradiol by granulosa cell catechol-O-methyltransferase activity and it is present in the normal follicle at high concentrations.[4] Evidence is growing that 2-ME is a potent inhibitor of angiogenesis and tumor growth and that its ability to suppress tumorigenesis is due to its antiangiogenetic effect on vascular endothelial cell proliferation.[5] This study was designed to evaluate possible indirect antiangiogenetic effect of 2-ME mediated through ROS.

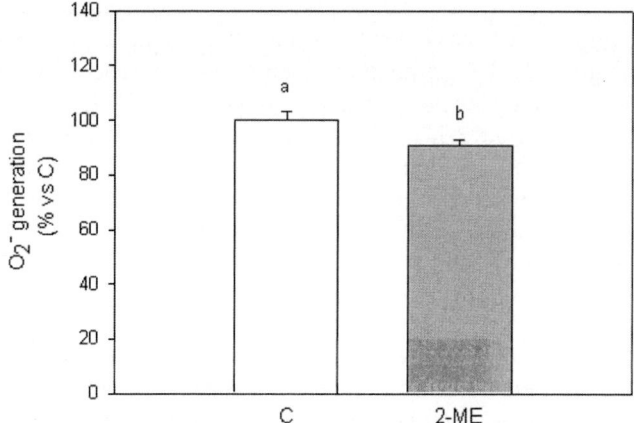

FIGURE 1. Effect of the treatment with 1 μM 2-ME for 48 h on O_2^- generation by 10^4 swine granulosa cells. Data represent the mean $\pm$ SEM of six replicates/treatment repeated in four different experiments. Different letters indicate significant differences ($P < 0.001$).

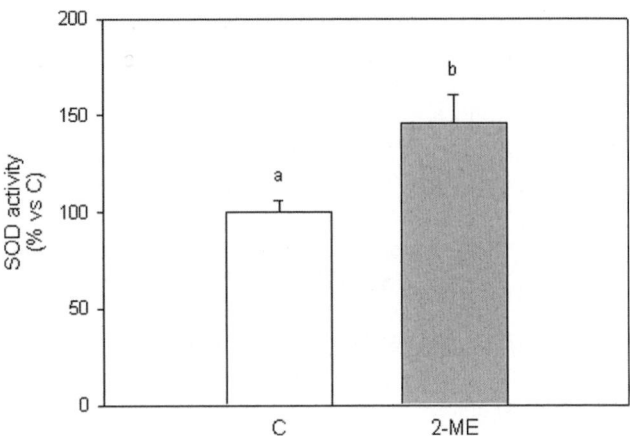

FIGURE 2. SOD activity (basal 280 ± 5 mU/mL mean ± SEM) evaluated spectrophotometrically in 2×10^5 swine granulosa cells treated with 1 μM 2-ME for 48 h. Data represent the mean ± SEM of six replicates/treatment repeated in four different experiments. Different letters indicate significant differences ($P < 0.05$).

In normal adult tissues, angiogenesis occurs conditionally. Wound healing is an example of angiogenesis wherein the sprouting of new vessels is regulated by growth factors released locally. The female reproductive system is a unique and exceptional organ system where new blood vessel formation and regression are orchestrated in each cycle through a complex process of hormonal actions. Development of ovarian follicles leading to ovulation is associated with changes in cellular components, tissue mass, fluid accumulation, and tissue remodeling. Tissue mass increase necessitates adequate blood supply to sustain perfusion. Fluid accumulation is contributed by vascular permeability that is often induced by the proangiogenic growth factor VEGF.[1] In a previous work,[8] we evidenced that 2-ME inhibits VEGF production by swine granulosa cells and therefore this hormone could act as a potential follicular antiangiogenic regulator. In general, antiangiogenesis has been the subject of intense interest because the potential to block tumor growth by inhibition of the neoangiogenic process represents an intriguing approach to the treatment of solid tumors. However, the molecular regulation of this process is very poorly understood and the ovary represents an outstanding model to study the regulatory events. Also, we[9] evidenced that during follicle growth swine granulosa cells are physiologically exposed to a progressive oxygen shortage. In addition, we have shown that hypoxia stimulates angiogenesis through an increase of VEGF production. However, despite considerable progress made in the understanding of the pathways, which are activated during cellular hypoxia, no consensus has been reached on the mechanism by which O_2 sensing is achieved.[10,11] Since the mitochondrion is the major oxygen-consuming organelle, it might be expected to play a central role in oxygen-sensitive processes by varying the production of ROS during hypoxia.[12,13] These molecules, mainly in the form of O_2^-, have long been regarded as toxic products. However, nowadays both *in vitro* and *in vivo* studies indicate that angiogenic response is triggered by ROS signaling in a highly coordinated manner,[14] pointing out a role for ROS as signal transducers.[2] In particular, in a previous study,[9] we evidenced that a stimulation of O_2^- generation triggers angiogenetic response in granulosa cells. In general, the identification of ROS role in the molecular pharmacology of 2-ME is an active area of research. In particular, the effect of the hormone on cellular superoxide radical appears to be cell type specific.[15] It has been reported[16] that 2-ME inhibits SOD activity and that this mechanism is responsible for the ability of 2-ME to kill cancer cells; this evidence, though, has been contrasted by Kachadourian *et al*.[17] Our study demonstrates that 2-ME stimulates SOD activity thus inhibiting O_2^- generation. Taken together, our present and previous results suggest that this effect could be responsible for the antiangiogenetic effect of 2-ME in swine granulosa cells.

ACKNOWLEDGMENTS

This work was supported by FIL COFIN grants.

REFERENCES

1. TAMANINI, C. & M. DE AMBROGI. 2004. Angiogenesis in developing follicle and corpus luteum. Reprod. Domest. Anim. **39:** 206–216.
2. SCHROEDL, C., D.S. MCCLINTOCK, G.R. SCOTT BUDINGER, *et al*. 2002. Hypoxic but not anoxic stabilization of HIF-1α requires mitochondrial reactive oxygen species. Am. J. Physiol. Lung. Cell. Mol. Physiol. **283:** 922–931.
3. BASINI, G., F. GRASSELLI, F. BIANCO, *et al*. 2004. Effect of reduced oxygen tension on reactive oxygen species production and activity of antioxidant enzymes in swine granulosa cells. Biofactors **20:** 61–69.
4. SPICER, L.J., M.A. WALEGA & J.M. HAMMOND. 1987. Metabolism of [3H] 2-hydroxyestradiol by cultured porcine granulosa cells: evidence for the presence of catechol-O-methyltransferase pathway and a direct stimulatory effect of 2-methoxyestradiol on progesterone production. Biol. Reprod. **36:** 562–571.
5. FOTSIS, T., Y. ZHANG, M.S. PEPPER, *et al*. 1994. The endogenous oestrogen metabolite 2-methoxyestradiol inhibits angiogenesis and suppresses tumour growth. Nature **368:** 237–239.
6. BENOV, L. & I. FRIDOVICH. 2002. Is reduction of the sulfonated tetrazolium 2,3-bis (2-methoxy-4-nitro-5-sulfophenyl)-2-tetrazolium 5-carboxanilide a reliable measure of intracellular superoxide production? Anal. Biochem. **310:** 186–190.
7. UKEDA, H., T. SHIMAMURA, M. TSUBOUCHI, *et al*. 2002. Spectrophotometric assay of superoxide anion formed in Maillard reaction based on highly water-soluble tetrazolium salt. Anal. Sci. **18:** 1151–1154.

8. BASINI, G., F. BIANCO & F. GRASSELLI. 2004. 2-methoxyestradiol, an endogenous estradiol metabolite, inhibits vascular endothelial growth factor production by swine granulosa cells: a potential follicular antiangiogenic regulator. 12th International Congress of Endocrinology. Lisbona. P499.
9. BASINI, G., F. BIANCO, F. GRASSELLI, *et al.* 2004. The effects of reduced oxygen tension on swine granulosa cell. Regul. Pept. **120:** 69–75.
10. CHANDEL, N.S. & P.T. SCHUMACKER. 2000. Cellular oxygen sensing by mitochondria: old questions, new insight. J. Appl. Physiol. **88:** 1880–1889.
11. MAXWELL, P.H. & P.J. RATCLIFFE. 2002. Oxygen sensors and angiogenesis. Semin. Cell. Dev. Biol. **13:** 29–37.
12. MAULIK, N. & D.K. DAS. 2002. Redox signaling in vascular angiogenesis. Free Radic. Biol. Med. **33:** 1047–1060.
13. KULISZ, A., N. CHEN, N.S. CHANDEL, *et al.* 2002. Mitochondrial ROS initiate phosphorylation of p38 MAP kinase during hypoxia in cardiomyocytes. Am. J. Physiol. Lung Cell Mol. Physiol. **282:** 1324–1329.
14. MAULIK, N. 2002. Redox signaling of angiogenesis. Antioxid. Redox Signal. **4:** 805–815.
15. MOOBERRY, S.L. 2003. Mechanism of action of 2-methoxyestradiol: new developments. Drug Resist. Updat. **6:** 355–361.
16. HUANG, P., L. FENG, E.A., M.J. OLDHAM, *et al.* 2000. Superoxide dismutase as a target for the selective killing of cancer cells. Nature **407:** 390–395.
17. KACHADOURIAN, R., S.I. LIOCHEV, D.E. CABELLI, *et al.* 2001. 2-methoxyestradiol does not inhibit superoxide dismutase. Arch. Biochem. Biophys. **392:** 349–353.

Role of Reactive Oxygen Species in Kv Channel Inhibition and Vasoconstriction Induced by TP Receptor Activation in Rat Pulmonary Arteries

ANGEL COGOLLUDO,[a] GIOVANNA FRAZZIANO,[a] LAURA COBEÑO,[a] LAURA MORENO,[a] FEDERICA LODI,[a] EDUARDO VILLAMOR,[b] JUAN TAMARGO,[a] AND FRANCISCO PEREZ-VIZCAINO[a]

[a]*Department of Pharmacology, School of Medicine, Universidad Complutense, 28040 Madrid, Spain*

[b]*Research Institute of Growth and Development (GROW), University of Maastricht, 6202 AZ Maastricht, The Netherlands*

ABSTRACT: Voltage-gated potassium channels (Kv) and thromboxane A_2 (TXA_2) have been involved in several forms of human and experimental pulmonary hypertension. We have reported that the TXA_2 analog U46619, via activation of TP receptors and PKCζ, inhibited Kv currents in rat pulmonary artery smooth muscle cells (PASMC), increased cytosolic calcium, and induced a contractile response in isolated rat and piglet pulmonary arteries (PA). Herein, we have analyzed the role of reactive oxygen species (ROS) in this signaling pathway. In rat PA, U46619 increased dichlorofluorescein fluorescence, an indicator of intracellular hydrogen peroxide, and this effect was prevented by the NADPH oxidase inhibitor apocynin and by polyethyleneglycol-catalase (PEG-catalase, a membrane-permeable form of catalase). U46619 inhibited Kv currents in native PASMC and these effects were strongly inhibited by apocynin. The contractile responses to U46619 in isolated PA were inhibited by PEG-catalase and the NADPH oxidase inhibitors diphenylene iodonium (DPI) and apocynin. A membrane permeable of hydrogen peroxide, t-butyl hydroperoxide, also inhibited Kv currents and induced a contractile response. Activation of NADPH oxidase and the subsequent production of hydrogen peroxide are involved in the Kv channel inhibition and the contractile response induced by TP receptor activation in rat PA.

KEYWORDS: voltage-gated potassium channels; thromboxane A_2; cyclooxygenases

Address for correspondence: F. Perez-Vizcaino, Department of Pharmacology, School of Medicine, Universidad Complutense, 28040 Madrid, Spain. Voice: 34-913941477; fax: 34-913941465.
e-mail: fperez@med.ucm.es

INTRODUCTION

Thromboxane A_2 (TXA_2), a major metabolite of arachidonic acid produced via cyclooxygenases (COX-1 and COX-2),[1] and isoprostanes, generated by oxygen-free radical-mediated peroxidation of arachidonic acid,[2] are potent pulmonary vasoconstrictors.[3,4] They have been involved in several forms of human and experimental pulmonary hypertension.[2,5,6] TXA_2 and several isoprostanes bind and activate thromboxane endoperoxide (TP) receptors, members of G protein–coupled membrane receptors.

Voltage-dependent potassium channels (Kv) make a substantial contribution to whole-cell K^+ conductance and membrane potential in pulmonary arteries (PA), and are a major mechanism for controlling PA tone.[7,8] Furthermore, changes in expression or function of Kv channels in PA smooth muscle cells (PASMC) have been involved in the pathogenesis of primary and anorexigen-induced pulmonary hypertension.[9–11] We have recently reported that inhibition of Kv channels is involved in the signaling of TP receptors in rat and porcine pulmonary arteries (PA), establishing a link between these two pathogenetic factors in pulmonary hypertension.[12,13] The atypical protein kinase C (PKCζ) plays a functional role in the Kv channel inhibition and the vasoconstriction induced by TP receptor activation.[12,13] Thus, the TP receptor agonist U46619, via PKCζ, inhibits voltage-gated K^+ channels (Kv) leading to membrane depolarization, activation of L-type Ca^{2+} channels, increase in cytosolic Ca^{2+} concentrations ($[Ca^{2+}]_i$), and a contractile response.

Reactive oxygen species (ROS), such as superoxide anion (O_2^-) and hydrogen peroxide (H_2O_2), are known to produce significant changes in vascular function.[14] Furthermore, ROS have also been proposed as intracellular signaling molecules in the transduction pathway for several vasoactive factors.[15–17] In fact, H_2O_2 modulates the activity of several K^+ channels, inhibiting Kv channels while activating large conductance calcium-dependent (BKCa) and ATP-dependent (K_{ATP}) K^+ channels.[18–20] Potential sources of vascular ROS production include NADPH-dependent oxidases, xanthine oxidase, cyclooxygenase, lipoxygenase, and endothelial NO synthases (eNOS).[14] Membrane NADPH oxidase is a multi-sub-unit enzymatic complex, which is considered to be the most important source of O_2^- in the vessel wall.[21–23]

In this article, we hypothesized that the Kv channel inhibition and the vasoconstriction following TP receptor activation might involve changes in ROS. We show that the TXA_2 analog U46619 increased intracellular ROS production and that the Kv channel inhibition and the subsequent vasoconstriction was sensitive to inhibitors of NADPH oxidase and to the H_2O_2 scavenger catalase.

MATERIALS AND METHODS

All experiments were carried out in accordance with the European Animals Act 1986 (Scientific Procedures) and approved by our institutional review board. Drugs and reagents were obtained from Sigma (Tres Cantos, Spain).

Tissue Preparation and Cell Isolation

Second- to third-order branches of the PA isolated from male Wistar rats (250–300 g) were dissected and endothelium was mechanically denuded. Cells were isolated in Ca^{2+}-free PSS containing (in mg/mL) papain 1, dithiothreitol 0.8, and albumin 0.7 and used within 8 h of isolation.[12,13]

Measurement of ROS

ROS generation in PA was assessed using 2,7-dichlorofluorescein (DCF).[24] PA were incubated with the membrane-permeable diacetate form of the dye (DCFH-DA, 10 μM for 60–90 min), which is cleaved and trapped intracellularly. ROS in the cells oxidize DCFH, yielding the fluorescent product DCF. PA were placed in the stage of a fluorescent inverted microscope (Leica DM IRB, Wetzlar, Germany), perfused with Krebs solution (2 mL min^{-1}) and illuminated through the luminal face using a 450–490 nm band-pass filter. Emitted fluorescence was filtered using a 515 nm long-pass emission filters. Images were taken with a Leica DC300F color digital camera and saved for offline analysis. Fluorescence was quantified using ImageJ (version 1.32j, NIH, http://rsb.info.nih/ij/). After subtracting background, intensity values are reported as a percentage of initial values. Preparations were equilibrated for 30 min before the application of drugs.

Electrophysiological Studies

Membrane currents were recorded using the whole-cell configuration of the patch clamp technique, normalized for cell capacitance and expressed in $pA \cdot pF^{-1}$ as previously described.[12,13] Spindle-shaped PASMC with an average capacitance of 20.2 ± 1.1 pF ($n = 17$) were used. Outward currents were recorded under essentially Ca^{2+}-free conditions, and EGTA and ATP were included in the pipette solution to minimize the component of ATP-dependent and BKCa K^+ currents. Under these conditions, the outward currents induced by depolarizing pulses to test potentials from −60 to +60 mV from a holding potential of −60 mV are essentially abolished by the Kv channel inhibitor 4-aminopyridine,[12] indicating that it was evoked by the activation of Kv channels.

Contractile Tension Recording

Contractile responses in isolated endothelium-denuded PA rings were recorded as previously reported.[12,13] Arteries were stimulated with U46619 (100 nM) and once a stable contraction was reached, were washed with Krebs solution. A second stimulation with U46619 was elicited after 60 min in the absence (controls) or presence with different drugs. The values of the second contraction were expressed as a percentage of the initial response to the agonist.

Statistical Analysis

Data are expressed as mean ± SEM; n indicates the number of arteries or cells tested from different animals. Statistical analysis was performed using Student's t-test for paired observations or one-way analysis of variance (ANOVA) followed by a Newman–Keul's test. Differences were considered statistically significant when $P < 0.05$.

RESULTS

TP Receptor Stimulation Increases ROS Production in PA

We analyzed whether the TP receptor agonist U46619 produces an increase in cytosolic H_2O_2 using DCFH-DA. In DCFH-loaded arteries, 100 nM U46619 produced a marked increase in fluorescence (FIG. 1). The membrane-permeable hydroperoxide, t-butyl-hydroperoxide (10 μM) also increased DCF fluorescence. The magnitude of the increase during the initial phase (first 4 min) was similar for both drugs. However, at longer periods, the increase induced by U46619 was only ~56% ($P < 0.01$) of that produced by t-butyl-hydroperoxide. U46619-induced increase in DCF fluorescence was abolished in PA that were previously incubated for 1 h with polyethyleneglycol-catalase (PEG-catalase) (100 U mL^{-1}), a membrane-permeable analog of the H_2O_2 scavenger catalase. Apocynin (300 μM), a specific inhibitor of NADPH oxidase, also abolished the U46619-induced increase in DCF fluorescence (FIG. 1).

Inhibition of Kv Currents by TP Receptor Activation Is Sensitive to Apocynin and Catalase

A family of Kv currents was obtained in PASMC when eliciting depolarizing steps from –60 to +60 mV (FIG. 2 A), which was usually reproducible for at least 1 h. As previously reported,[12] U46619 (100 nM) inhibited Kv currents present in PASMC (FIG. 2 A, B) in a voltage-independent manner (e.g., 36.8 ± 8.9% and 35.3 ± 2.1% inhibition at –20 and +60 mV, respectively). The inhibitory effect of U46619 occurred with no apparent changes in the activation and inactivation kinetics. Apocynin (300 μM) had no significant effect on Kv currents, but fully prevented the inhibitory effects of U46619 (FIG. 2 C, D).

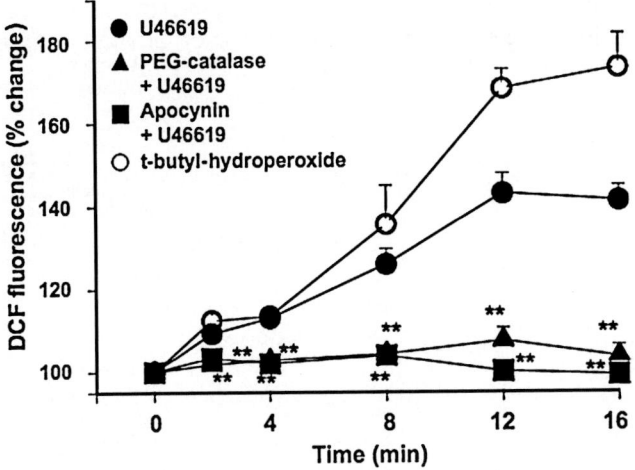

FIGURE 1. TP receptor activation increases intracellular H_2O_2 in PA. Time course of changes in fluorescent intensity in DCFH-loaded rat PA treated with U46619 (100 nM) or t-butyl-hydroperoxide (10 μM). Some arteries were preincubated for 60 min in the presence of PEG-catalase (100 U mL^{-1}) or for 30 min with apocynin (300 μM) before the addition of U46619. Results are mean ± SEM of 3 PA. **indicates $P < 0.01$ versus control, unpaired Student's t-test.

Inhibition of U46619-Induced Pulmonary Vasoconstriction by NADPH Oxidase Inhibitors and Catalase

Stimulation of endothelium-denuded PA rings with U46619 (100 nM, which produced ∼50% of its maximal response) induced a sustained contractile response of 102 ± 6 mg ($n = 15$), which was suitably reproduced after a 60-min washout (103 ± 7% of the first contraction). Pretreatment for 60 min with diphenylene iodonium (DPI, 10 μM), an inhibitor of flavin-containing enzymes, including NADPH oxidase, or with apocynin (300 μM) strongly inhibited the vasoconstrictor response (FIG. 3). Similarly, in the presence of PEG-catalase (100 U mL^{-1}), U46619-induced vasoconstriction was strongly inhibited.

T-Butyl-Hydroperoxide Inhibits Kv Currents and Induces a Vasoconstrictor Response

In PASMC, addition of t-butyl-hydroperoxide (5 μM) to the external solution inhibited Kv currents in a voltage-independent manner (e.g., 30.7 ± 1.8% and 27.4 ± 1.4 inhibition at –20 and +60 mV, respectively) and with no apparent changes in the activation and inactivation kinetics (FIG. 4 A and 4 B). A higher concentration of t-butyl-hydroperoxide (100 μM) produced a

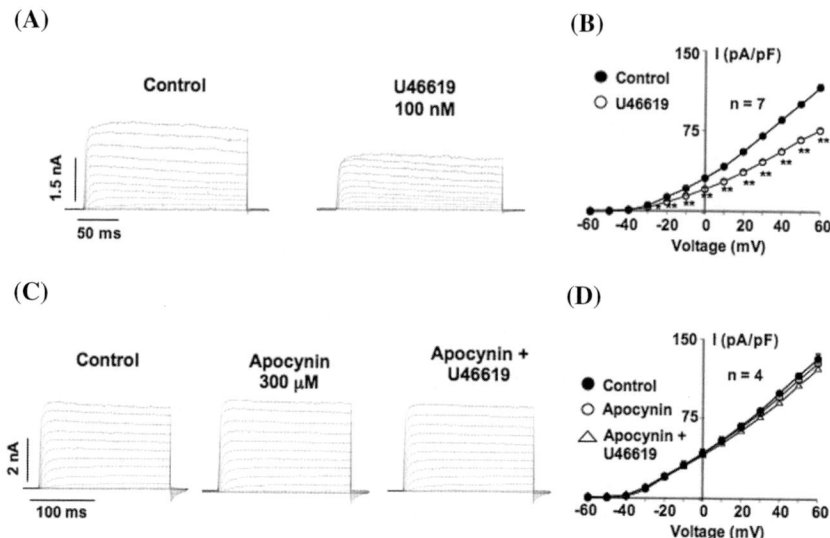

FIGURE 2. Inhibition of Kv currents by TP receptor activation is sensitive to apocynin in PASMC. Effects of U46619 (100 nM) on Kv currents in PASMC under control conditions (**A,B**) and in the presence of apocynin (300 μM, **B,C**): (**A,C**) current traces and (**B,D**) current–voltage relationships. Results are mean ± SEM. *and **indicate $P < 0.05$ and $P < 0.01$, respectively, versus control, paired Student's t-test.

similar inhibitory effect (28 ± 2% at +60 mV, FIG. 4 C) than that observed at the lower concentration.

In PA, addition of t-butyl-hydroperoxide (10 μM) induced a contractile response. However, although the response to U46619 was sustained, the response to t-butyl-hydroperoxide was biphasic (FIG. 5), a maximal response was achieved at about 2 min (50 ± 4%, of the response induced by 100 nM U46619) and was followed by a decay to a lower tone (23 ± 5%).

DISCUSSION

In a previous study,[12] we reported that, in intact PA and freshly isolated PASMCs, activation of TP receptors inhibits Kv channels, leading to membrane depolarization, activation of L-type Ca^{2+} channels, elevation of $[Ca^{2+}]_i$, and vasoconstriction. PKCζ played a role as a link between TP receptor activation and Kv channel inhibition. In this study, we have examined the role of ROS in this signaling pathway. Our results indicate that the TP receptor agonist U46619 increases intracellular ROS, inhibits Kv currents, and induces a vasoconstrictor response. Furthermore, all these effects were prevented by NADPH oxidase inhibitors or catalase. Addition of t-butyl-hydroperoxide, a membrane-permeable analog of H_2O_2, reproduced the effects of U46619.

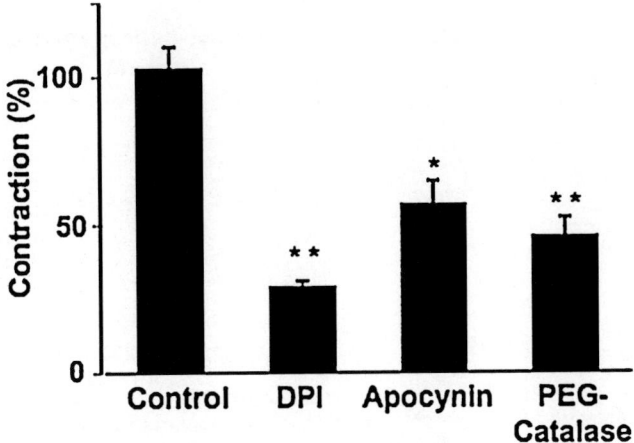

FIGURE 3. Inhibition of TP receptor activation-induced vasoconstriction in PA by NADPH oxidase inhibitors and catalase. Effects of DPI (10 μM), apocynin (300 μM), and PEG-catalase (100 U mL^{-1}) on 100 nM U46619-induced contractions in rat PA. Results are normalized to values obtained in a previous control U46619-induced contraction. Results are mean ± SEM of 4–15 PA. *and **indicate $P < 0.05$ and $P < 0.01$, respectively, versus control unpaired Student's t-test.

DCFH-AM is a cell-permeable dye, which is cleaved into DCFH and trapped intracellularly and fluoresces when oxidized by ROS, reflecting the overall oxidative status of the cell.[24] U46619 augmented DCF fluorescence indicating that it increased ROS intracellularly. However, DCFH is sensitive toward ONOO$^-$, H$_2$O$_2$, and hydroxyl radical.[24] The U46619-induced increase in DCF fluorescence was abolished by PEG-catalase, a membrane-permeable analog of the H$_2$O$_2$ scavenger catalase, indicating that it was mainly due to increased cytosolic concentrations of H$_2$O$_2$. This increase was also sensitive to the NADPH oxidase inhibitor apocynin, suggesting that TP receptor activation increases membrane NADPH oxidase activity leading to the production of O$_2^-$, which is then converted by endogenous superoxide dismutase into H$_2$O$_2$. Previous studies have shown that angiotensin II also increases ROS via NADPH oxidase, playing an important role in its signaling pathway.[15,16] In contrast, other vasoactive factors, such as endothelin-1 increase ROS via both NADPH oxidase-dependent and -independent pathways.[15,17] A previous study, showed that long-term treatment (16 h) with U46619 promoted the formation of O$_2^-$ in porcine PA segments, an effect inhibited by DPI and apocynin.[25] This effect was associated with an upregulation of the expression of gp91phox, the catalytic subunit of NADPH oxidase, suggesting that the effect of TP activation was due to changes in gene expression rather than an activation of the enzyme. In addition, the inhibitory effect of aspirin and the TXA$_2$ synthase inhibitor furegrelate on platelet-dependent augmentation of neutrophil ROS

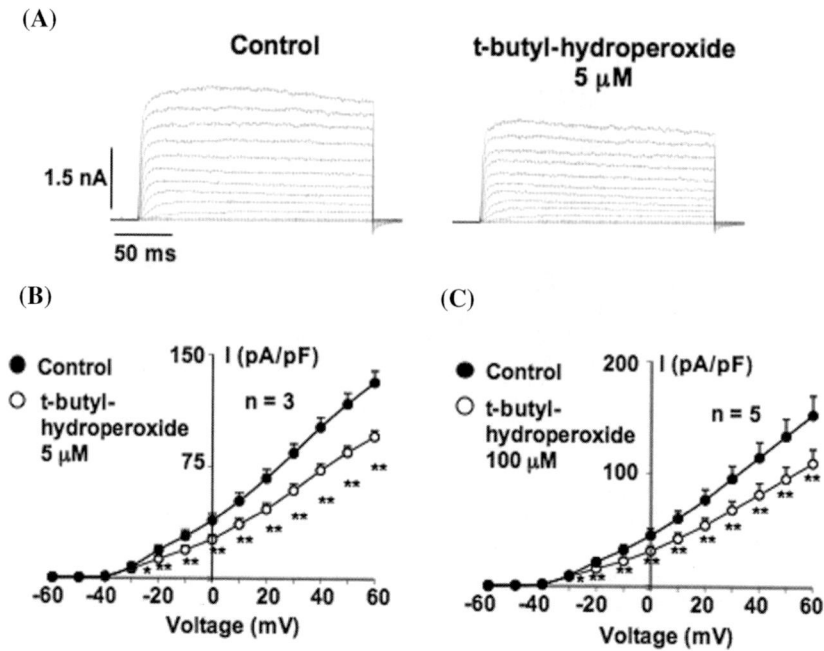

FIGURE 4. ROS inhibit Kv currents in PASMC. Effects of t-butyl-hydroperoxide (5 and 100 μM) on Kv currents in PASMC: **(A)** current traces and **(B,C)** current–voltage relationships. Results are mean ± SEM. * and ** indicate $P < 0.05$ and $P < 0.01$, respectively, versus control paired Student's t-test.

production provides indirect evidence that TP receptor activation can enhance ROS production.[26]

K^+ channel function is a main determinant of vascular smooth muscle tone.[7,8] Its activity is modulated by multiple vasoactive factors and intracellular signaling molecules.[7–13] In fact, ROS are known modulators of K^+ channel function. H_2O_2 enhances BKCa and K_{ATP} channel activity in several vascular beds[18] but inhibits Kv channels in the ductus arteriosus[19] and rat coronary arteries.[20] We therefore, hypothesized that increased ROS might mediate the Kv channel blockade induced by TP receptor activation. The prevention of the inhibitory effects of U46619 by apocynin strongly suggests that the activation of NADPH oxidase and the increase in intracellular H_2O_2 is involved in the signaling pathway leading to Kv channel blockade. This view is reinforced by the inhibitory effects of the membrane-permeable analog of H_2O_2 t-butyl-hydroperoxide in PASMCs. Moreover, the magnitude of this effect, its voltage-independency, and the lack of changes in the activation and inactivation kinetics were similar to those of U46619. In our previous study we reported the involvement of PKCζ in this same signaling pathway.[12,13] H_2O_2 has been reported to stimulate PKCζ activity in alveolar epithelial cells and COS-7

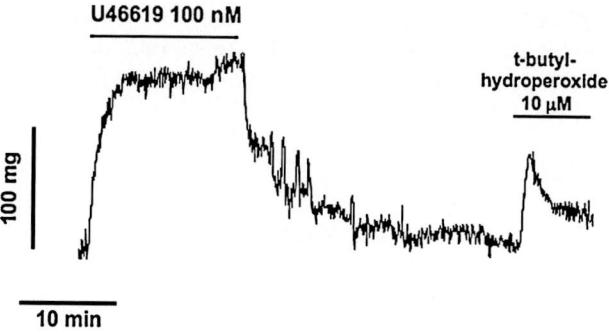

FIGURE 5. ROS induce a contractile response in endothelium-denuded PA. The trace shows the sustained contractile effect of U46619 (100 nM), followed by a washout period and the biphasic contractile response induced by t-butyl-hydroperoxide (10 μM).

cells.[27,28] Therefore, it might be suggested that NADPH oxidase-derived H_2O_2 might act as a link between TP receptor and PKCζ.

We also analyzed the role of ROS in the vasoconstrictor effects of U46619. ROS have been shown to modulate arterial tone via K^+ channel-dependent and -independent pathways.[19,29] Since Kv channel inhibition plays a role in vasoconstriction following TP receptor activation, it is expected that ROS were also involved in the vasoconstrictor effect of U46619. In fact, the contractile effect of U46619 in isolated PA was inhibited by the NADPH inhibitors DPI and apocynin and by PEG-catalase. These results indicate that activation of NADPH oxidase, the subsequent H_2O_2 production, and Kv channel inhibition are involved in the contractile effects of TP receptor activation. However, in contrast to the full inhibitory effect on U46619-induced increase in DCF fluorescence and Kv current inhibition, apocynin only partially inhibited the contractile responses. This indicates the existence of additional pathways for vascular smooth muscle contraction in PA stimulated by U46619. Accordingly, despite t-butyl-hydroperoxide perfectly reproduced the effect of U46619 on Kv currents, it induced a biphasic contractile response, which was only ∼50% of that induced by U46619.

To our knowledge, this study is the first one showing that TP receptor activation increases ROS via activation of NADPH oxidase. NADPH oxidase-derived superoxide and the subsequent production of H_2O_2 are involved in the Kv channel inhibition and the contractile response induced by TP receptor activation in rat PA.

ACKNOWLEDGMENTS

This work was supported by grants from the Comisión Interministerial de Ciencia y Tecnología (SAF SAF2005-03770, SAF2005-04609 and AGL2004-06685-C04-1) and from Comunidad Autónoma de Madrid (GR/SAL/0594/2004). A.C, L.M., and F.L. are supported by RECAVA (Red Temática de Investigación Cardiovascular), Ministerio de Educación y Ciencia (FPU) and CSIC (I3P grant), respectively.

REFERENCES

1. BABER, S.R. et al. 2003. Role of cyclooxygenase-2 in the generation of vasoactive prostanoids in the rat pulmonary and systemic vascular beds. Circulation **108:** 896–901.
2. JANSSEN, L.J. 2001. Isoprostanes: an overview and putative roles in pulmonary pathophysiology. Am. J. Physiol. **280:** L1067–L1082.
3. PEREZ-VIZCAINO, F. et al. 1997. Involvement of protein kinase C in reduced relaxant responses to the NO/cGMP pathway in piglet pulmonary arteries contracted by the thromboxane A_2 mimetic U46619. Br. J. Pharmacol. **121:** 1323–1333.
4. GONZALEZ-LUIS, G. et al 2005. Age-related differences in vasoconstrictor responses to isoprostanes in piglet pulmonary and mesenteric vascular smooth muscle. Pediatr. Res. **57:** 845–852.
5. CHRISTMAN, B.W. et al. 1992. An imbalance between the excretion of thromboxane and prostacyclin metabolites in pulmonary hypertension. N. Engl. J. Med. **327:** 70–75.
6. MONTALESCOT, G. et al. 1990. Thromboxane receptor blockade prevents pulmonary hypertension induced by heparin–protamine reactions in awake sheep. Circulation **82:** 1765–1777.
7. ARCHER, S. et al. 1998. Molecular identification of the role of voltage-gated K^+ channels, Kv1.5 and Kv1.2, in hypoxic pulmonary vasoconstriction and control of resting membrane potential in rat pulmonary artery myocytes. J. Clin. Invest. **101:** 2319–2330.
8. YUAN, X-J. 1995. Voltage-gated K^+ currents regulate resting membrane potential and $[Ca^{2+}]i$ in pulmonary arterial myocytes. Circ. Res. **77:** 370–378.
9. YUAN, X-J. et al. 1998. Attenuated K^+ channel gene transcription in primary pulmonary hypertension. Lancet **351:** 726–727.
10. ARCHER, S. & S. RICH. 2000. Primary pulmonary hypertension: a vascular biology and translational research "work in progress." Circulation **102:** 2781–2791.
11. WANG, J. et al. 1998. Action of fenfluramine on voltage-gated K^+ channels in human pulmonary artery smooth muscle cells. Lancet **352:** 290.
12. COGOLLUDO, A. et al. 2003. Thromboxane A_2-induced inhibition of voltage-gated K^+ channels and pulmonary vasoconstriction. Role of protein kinase Cζ. Circ. Res. **93:** 656–663.
13. COGOLLUDO, A. et al. 2005. Postnatal maturational shift from PKCζ and voltage-gated K^+ channels to RhoA/Rho kinase in pulmonary vasoconstriction. Cardiovasc. Res. **66:** 84–93.
14. CAI, H. & D.G. HARRISON. 2000. Endothelial dysfunction in cardiovascular diseases: the role of oxidant stress. Circ. Res. **87:** 840–844.
15. TOUYZ, R.M. et al. 2004. Angiotensin II and endothelin-1 regulate MAP kinases through different redox-dependent mechanisms in human vascular smooth muscle cells. J. Hypertens. **22:** 1141–1149.

16. GRIENDLING, K.K. & M. USHIO-FUKAI. 2000. Reactive oxygen species as mediators of angiotensin II signaling. Regul. Pept. **91:** 21–27.
17. LI, L. *et al.* 2003. Endothelin-1 increases vascular superoxide via endothelin(A)-NADPH oxidase pathway in low-renin hypertension. Circulation **107:** 1053–1058.
18. LIU, Y. & D.D. GUTTERMAN. 2002. Oxidative stress and potassium channel function. Clin. Exp. Pharmacol. Physiol. **29:** 305–311.
19. MICHELAKIS, E.D. *et al.* 2002. O_2 sensing in the human ductus arteriosus: regulation of voltage-gated K^+ channels in smooth muscle cells by a mitochondrial redox sensor. Circ. Res. **91:** 478–486.
20. LIU, Y. *et al.* 2001. High glucose impairs voltage-gated K^+ channel current in rat small coronary arteries. Circ. Res. **89:** 146–152.
21. GRIENDLING, K.K., D. SORESCU & M. USHIO-FUKAI. 2000. NAD(P)H oxidase: role in cardiovascular biology and disease. Circ. Res. **86:** 494–501.
22. SOUZA, H.P. *et al.* 2001. Vascular NAD(P)H oxidase is distinct from the phagocytic enzyme and modulates vascular reactivity control. Am. J. Physiol. Heart Circ. Physiol. **280:** H658–H667.
23. PEREZ-VIZCAINO, F. *et al.* 2002. Postnatal maturation in nitric oxide-induced pulmonary artery relaxation involving cyclooxygenase-1 activity. Am. J. Physiol. Lung Cell. Mol. Physiol. **283:** L839–L848.
24. MYHRE, O. *et al.* 2003. Evaluation of the probes 2,7-dichlorofluorescin diacetate, luminol, and lucigenin as indicators of reactive species formation. Biochem. Pharmacol. **65:** 1575–1582.
25. MUZAFFAR, S. *et al.* 2004. Iloprost inhibits superoxide formation and gp91phox expression induced by the thromboxane A_2 analogue U46619, 8-isoprostane $F_{2\alpha}$, prostaglandin $F_{2\alpha}$, cytokines and endotoxin in the pig pulmonary artery. Br. J. Pharmacol. **141:** 488–496.
26. CHLOPICKI, S. *et al.* 2004. Functional role of NADPH oxidase in activation of platelets. Antioxid. Redox Signal. **6:** 691–698.
27. DADA, L.A. *et al.* 2003. Hypoxia-induced endocytosis of Na,K-ATPase in alveolar epithelial cells is mediated by mitochondrial reactive oxygen species and PKC-zeta. J. Clin. Invest. **111:** 1057–1064.
28. KONISHI, H. *et al.* 1997. Activation of protein kinase C by tyrosine phosphorylation in response to H_2O_2. Proc. Natl. Acad. Sci. USA **94:** 11233–11237.
29. LUCCHESI, P.A. *et al.* 2005. Hydrogen peroxide acts as both vasodilator and vasoconstrictor in the control of perfused mouse mesenteric resistance arteries. J. Hypertens. **23:** 571–579.

DNA Strand Breaks by Metal-Induced Oxygen Radicals in Purified *Salmonella typhimurium* DNA

EZZATOLLAH KEYHANI,[a,b] FATEMEH ABDI-OSKOUEI,[b] FARNOOSH ATTAR,[b] AND JACQUELINE KEYHANI[a]

[a]*Laboratory for Life Sciences, 19979 Tehran, Iran*
[b]*Institute of Biochemistry and Biophysics, University of Tehran, 13145 Tehran, Iran*

ABSTRACT: Purified *Salmonella typhimurium* DNA was incubated for 1h at 37°C with various concentrations (10–100 μM) of transition metal ions (Fe^{2+}, Fe^{3+}, Cu^{2+}, Ni^{2+}, Cd^{2+}), with various concentrations (0.1–100 mM) of H_2O_2, and with various concentrations of each transition metal ion in the presence of various concentrations of H_2O_2. Damage to DNA was assessed by electrophoresis of the reaction mixtures in 1% agarose gel. Breakage of the DNA strands would produce a series of DNA fragments resulting in a smear in the gel, while intact DNA produced a single band. Results showed that no damage to the DNA was detectable after incubation with either H_2O_2 alone or either of the metal ions alone. However, all of the metal ions investigated triggered DNA breakage in the presence of H_2O_2. The extent of breakage depended on the metal ion and on its concentration, as well as on the H_2O_2 concentration. Addition of either EDTA or catalase to the reaction mixture completely inhibited the DNA degradation, confirming the involvement of both the metal ion and the H_2O_2 in the breakage of DNA strands. Production of the hydroxyl radical when H_2O_2 and a metal ion were both present in the reaction mixture was evidenced by the thiobarbituric acid method. The most extensive damage was caused by Cu^{2+} followed, in decreasing order, by Fe^{2+}, Fe^{3+}, Ni^{2+}, and Cd^{2+}.

KEYWORDS: transition metals; reactive oxygen species; DNA; oxidative stress; toxicity; *Salmonella typhimurium*

INTRODUCTION

Transition metals are essential for most living systems at trace levels, but they are usually identified as toxic, or even highly toxic, at elevated levels. Iron and

Address for correspondence: Dr. Ezzatollah Keyhani, Institute of Biochemistry and Biophysics, University of Tehran, P.O. Box 13145-1384, 13145 Tehran, Iran. Voice: +98-21-6695-6974; fax: +98-21-6640-4680.
e-mail: keyhanie@ibb.ut.ac.ir

copper are typical examples of essential transition metals that cause severe cell damage upon accumulation. Iron that is commonly found in one of two stable oxidation states in aqueous solutions, Fe^{2+} and Fe^{3+}, participates in a wealth of biochemical reactions, including, but not limited to, the control of electrons flow through bioenergetic pathways, the activation of molecular oxygen, nitrogen, and hydrogen, the decomposition of noxious derivatives of oxygen, such as peroxide and superoxide, and the binding of oxygen by hemoglobin, myoglobin, and hemerythrins.[1] Copper forms the essential redox-active center in a variety of metalloproteins, such as ceruloplasmin, Cu–Zn superoxide dismutase, cytochrome c oxidase, dopamine β-hydroxylase, tyrosinase, lysyl oxidase, and ascorbic oxidase.[2] Increased intake of iron has been associated with tissue damage observed in patients suffering from primary idiopathic hemochromatosis and those receiving blood transfusions for treatment of β-thalassemia;[3,4] excess copper has been identified as a cause of cancer.[5] Other transition metals, such as nickel and cadmium, have become the cause of great concern because of their increasing accumulation in the environment. Nickel, a metal abundantly used in modern industry, is known to be essential to most forms of life in trace amounts; but it is also known as one of the most powerful human metal carcinogens, as an embryotoxic and teratogenic agent, and as a leading cause of contact dermatitis.[6,7] Cadmium is a widely distributed metallic pollutant, which can be absorbed into biological systems through direct uptake as well as by accumulation in food chains. It has been identified as a potent animal and human carcinogen.[8,9]

Most often, transition metal toxicity stems from the metals' ability to promote or exacerbate oxidative cell damage.[7,10,11] This will occur, for example, when hydrogen peroxide (H_2O_2) accumulates in the cell. Because of aerobic respiration, oxidants such as H_2O_2 are generated in living organisms by normal cellular metabolism. H_2O_2 is normally disposed of by specialized enzymes, such as catalases and peroxidases, so that its concentration remains at a level beneficial to the cell. Indeed, in small concentrations (10^{-6} M), H_2O_2 is a signaling molecule capable of inducing chemotactic activity, stimulating the synthesis of cytoskeleton elements, and causing changes in cytosolic calcium concentrations.[12] However, physiologic perturbations of cellular homeostasis may lead to a dramatic increase in the amount of H_2O_2 within a cell. In the presence of transition metals, H_2O_2 is rapidly converted to the highly reactive and highly toxic hydroxyl radical (OH·).[13] The latter is responsible for lipid peroxidation, oxidative damage to proteins, and breakage of DNA strands.

In this study, the extent of DNA damage caused by H_2O_2 in the presence of Fe^{2+}, Fe^{3+}, Cu^{2+}, Ni^{2+}, and Cd^{2+} was evaluated *in vitro*, using purified *Salmonella typhymurium* DNA. *S. typhimurium* is a pathogen ubiquitously distributed worldwide and a major threat to human health. Whether considering the biological aspect or the medical aspect, this bacterium has been the subject of many investigations and is still being extensively studied.

Results showed that DNA degradation would occur only when both H_2O_2 and any one of the transition metal ions was present, and that it was caused by the production of the hydroxyl radical. Furthermore, a scale of metal toxicity was established, showing that, in this instance, Cu^{2+} was the most damaging transition metal ion and that Cd^{2+} was the least damaging one.

MATERIALS AND METHODS

Chemicals

Catalase (42,000 U/mg protein), $FeCl_3.6H_2O$ and boric acid were obtained from Sigma Chemical Co. (St. Louis, MO, USA). All other chemicals were from Merck Chemical Co. (Germany). All were of reagent grade.

DNA Preparation

S. typhimurium DNA was purified by phenol extraction and ethanol precipitation, essentially according to the method outlined by Sambrook *et al.*[14] The strain of *S. typhimurium* used was provided by Prof. N.O. Keyhani, Department of Microbiology and Cell Sciences, University of Florida at Gainesville, FL, USA. The purified DNA preparation was characterized by an absorbance$_{260nm}$/absorbance$_{280nm}$ ratio of 1.8. It was still treated with pronase E according to the method in Keyhani *et al.*[15] and repurified by phenol extraction and ethanol precipitation. The concentration, purity, and intactness of the isolated DNA was determined by spectrophotometry, diphenylamine assay,[16] orcinol assay,[17] and electrophoresis in agarose gel. The prepared DNA was found to be free of degradation fragments, RNA, and proteins.

Exposure of DNA to Metal Ions, H_2O_2, or Both

Damage to DNA caused by metal ions, H_2O_2 or a mixture of metal ions and H_2O_2, was assessed as follows:

1. *Metal ions*: Purified *S. typhimurium* DNA, at a concentration of 20 μg/mL, was incubated with increasing concentrations (2, 10, 20, 100, and 200 μM) of either Fe^{2+}, Fe^{3+}, Cu^{2+}, Ni^{2+}, or Cd^{2+}, at 37°C for 1 h, in a final volume of 10 μL (pH 7.4). Immediately following the incubation period, the samples were electrophoresed in a 1% agarose gel as described by Sambrook *et al.*,[14] except that ethidium bromide was added to the electrophoresis buffer only, at a final concentration of 0.5 μg/mL. The agarose was type II, low-endo-osmotic agarose from Merck. Electrophoresis was conducted for 12 h, at 25 volts.
2. *H_2O_2*: Purified *S. typhimurium* DNA, at a concentration of 20 μg/mL, was incubated with increasing concentrations (0.1, 0.3, 1, 3, 10, 30,

and 100 mM) of H_2O_2, at 37°C for 1 h, in a final volume of 10 μL (pH 7.4). Immediately following the incubation period, the samples were electrophoresed in a 1% agarose gel as described above. Alternatively, spectra of DNA preparations were recorded between 190 and 310 nm, before and after treatment with H_2O_2.
3. *Metal ions and H_2O_2*: Purified *S. typhimurium* DNA, at a concentration of 20 μg/mL, was incubated with increasing concentrations (10, 20, 50, 70, and 100 μM) of either Fe^{2+}, Fe^{3+}, Cu^{2+}, Ni^{2+}, or Cd^{2+}, in the presence of increasing concentrations (0.1, 0.3, 1, 3, 10, 30, and 100 mM) of H_2O_2, at 37°C for 1 h, in a final volume of 10 μL (pH 7.4). Immediately following the incubation period, the samples were electrophoresed in a 1% agarose gel as described above.

Exposure of DNA to Metal Ion and H_2O_2, in the Presence of EDTA or Catalase

Reaction mixtures consisting of *S. typhimurium* DNA (20 μg/mL), 50 μM metal ion, 100 mM H_2O_2, and 25 mM EDTA, in a final volume of 10 μL, were incubated at 37°C for 1 h (pH 7.4). Immediately following the incubation period, the samples were electrophoresed in a 1% agarose gel as described above.

To investigate the effect of catalase, reaction mixtures consisting of *S. typhimurium* DNA (20 μg/mL), 50 μM metal ion, 100 mM H_2O_2 and 3 u of catalase, in a final volume of 16 μL, were incubated at 37°C for 1 h (pH 7.4). Immediately following the incubation period, the samples were electrophoresed in a 1% agarose gel as described above.

Assay for Hydroxyl Radical Formation

The production of hydroxyl radical by various metal ions in the presence of H_2O_2 was assayed as described below.

1. *By measuring D(-)ribose degradation*: Increasing concentrations of D(-)ribose (1.4 – 8.4 mM) were incubated for 1 h, at 37°C, in the presence of 100 mM H_2O_2 and 100 μM of either Fe^{2+}, Fe^{3+}, Cu^{2+}, Ni^{2+} or Cd^{2+}, at pH 7.4. The extent of OH·-induced D(-)ribose degradation was determined by titrating the formation of thiobarbituric acid–reactive substances (TBARS) at 532 nm, using an extinction coefficient of 1.56×10^4 $M^{-1}.cm^{-1}$, according to the method described by Halliwell *et al.*[18]
2. *By measuring the loss of thymine absorbance at 265 nm*: Thymine, at a concentration of 1 A_{265}/mL, was incubated with 100 mM H_2O_2 and 100 μM of either Fe^{2+}, Fe^{3+}, Cu^{2+}, Ni^{2+} or Cd^{2+}, at pH 7.4, for 1 h at 37°C, in a final volume of 3 mL. The formation of OH· radical was assessed by following the decrease in absorbance at 265 nm as described by Lown *et al.*[19]

RESULTS

Effect of Transition Metal Ions on DNA

Exposure of purified *S. typhimurium* DNA to various concentrations of Fe^{2+}, Fe^{3+}, Cu^{2+}, Ni^{2+}, and Cd^{2+} in the range 2–200 μM for 1 h at 37°C resulted in no apparent DNA degradation as evaluated by electrophoresis in 1% agarose gels. As seen in FIGURE 1A (a–e), neither the mobility of the DNA, nor the fluorescence intensity after staining with ethidium bromide and examination under UV light, was altered. One exception was noticeable in the presence of 100 μM Fe^{3+}, where the sample did not migrate uniformly and some retardation was observed (FIG. 1A [b], lane 5) Results shown in FIGURE 1A were obtained with $FeSO_4$ (FIG. 1A [a]), $FeCl_3$ (FIG. 1A [b]), $CuSO_4$ (FIG. 1A [c]), $NiSO_4$ (FIG. 1A [d]) and $CdCl_2$ (FIG. 1A [e]); the control DNA was in lane 1 in all instances. The same results were obtained whether chloride, sulfate, or nitrate salts of any of the metals were used.

Effect of H_2O_2 on DNA

Exposure of *S. typhimurium* DNA to increasing H_2O_2 concentrations, from 0.1 to 100 mM, for 1 h at 37°C, resulted in no apparent alteration as evaluated by electrophoresis in 1% agarose gels. As seen in FIGURE 1B, neither the mobility of the DNA, nor the fluorescence intensity after staining with ethidium bromide and examination under UV light, was altered. Furthermore, no alterations in DNA spectral properties were observed when absorption spectra of DNA preparations, which had been exposed to H_2O_2, were recorded.

Effect of Transition Metal Ions on DNA in the Presence of H_2O_2

When *S. typhimurium* DNA was exposed to increasing concentrations of Fe^{2+}, Fe^{3+}, Cu^{2+}, Ni^{2+}, and Cd^{2+}, in the presence of various H_2O_2 concentrations, DNA degradation occurred as evidenced by the smears obtained after electrophoresis in 1% agarose gels. The extent of damage depended on the nature of the metal ion, its concentration, and the concentration of H_2O_2. Some representative gels are shown in FIGURE 1C, D. FIGURE 1C (a, b, and c) shows the effect, on *S. typhimurium* DNA, of increasing H_2O_2 concentrations in the presence of 20 μM Cd^{2+} (FIG. 1C [a], where H_2O_2 concentrations were, from lane 3 to 8, respectively 0.3, 1, 3, 10, 30, and 100 mM), 70 μM Cd^{2+} (FIG. 1C [b], where H_2O_2 concentrations were, from lane 3 to 8, respectively 0.3, 1, 3, 10, 30, and 100 mM), and 100 μM Cd^{2+} (FIG. 1C [c], where H_2O_2 concentrations were, from lane 3 to 8, respectively 0.1, 0.3, 1, 3, 10, and 30 mM). For a given H_2O_2 concentration, the extent of damage caused to DNA increased at higher Cd^{2+} concentrations. This can be seen, for example, by comparing the damage caused by 10 mM H_2O_2 in the presence of 20 μM Cd^{2+} (lane 6 in FIG. 1C [a]), 70 μM Cd^{2+} (lane 6 in FIG. 1C [b]), and 100 μM Cd^{2+} (lane 7 in FIG. 1C [c]); the control DNA was in lane 1 for all three figures.

FIGURE 1D (a) shows the effect, on *S. typhimurium* DNA, of 20 μM Fe^{2+} in the presence of 0.1, 0.3, 1, 3, 10, 30, and 100 mM H_2O_2. The extent of damage to the DNA observed in the presence of 20 μM Fe^{2+} and 10 mM H_2O_2 (lane 7, FIG. 1D [a]) was much greater than that observed in the presence of 20 μM Cd^{2+} and 10 mM H_2O_2 (lane 6, FIG. 1C [a]), where damage to DNA was barely detectable. The control DNA was in lane 1 in both figures.

FIGURE 1D (b and c) shows the effect, on *S. typhimurium* DNA, of 10 and 20 μM Cu^{2+} in the presence of 0.1, 0.3, 1, 3, 10, 30, and 100 mM H_2O_2. Here again, for a given H_2O_2 concentration, the extent of DNA damage would increase as the concentration of the metallic ion increased. This is seen, for example, by comparing lane 4 in FIGURE 1D (b) (0.3 mM H_2O_2 and 10 μM Cu^{2+}) and lane 4 in FIGURE 1D (c) (0.3 mM H_2O_2 and 20 μM Cu^{2+}). Furthermore, it is also observed that 20 μM Cu^{2+} caused as much damage to DNA in the presence of 1 mM H_2O_2 (lane 5, FIG. 1D [c]) as 20 μM Fe^{2+} in the presence of 10 mM H_2O_2 (lane 7, FIG. 1D [a]), or 20 μM Cd^{2+} in the presence of 100 mM H_2O_2 (lane 8, FIG. 1C [a]).

A comparison of the extent of DNA breakage caused by 20 μM Fe^{2+}, Fe^{3+}, Cu^{2+}, Ni^{2+}, Cd^{2+} in the presence of various concentrations of H_2O_2, as shown after electrophoresis in agarose gels, is presented in FIGURE 2A. The extent of DNA damage was evaluated from the size of DNA fragments produced, as shown by electrophoretic mobility. Complete DNA degradation, shown as 100 on the scale in FIGURE 2A, corresponded to the amount of damage that would break DNA into fragments small enough to be undetectable after electrophoresis in agarose gel under the conditions used in this work. Cu^{2+} was the most damaging metal ion, followed by Fe^{2+} and then, in decreasing order, Fe^{3+}, Ni^{2+}, and Cd^{2+}. Cu^{2+} (20 μM) caused complete DNA degradation in the presence of 3 mM H_2O_2, whereas 20 μM of either of the other metal ions would cause complete DNA degradation in the presence of 100 mM H_2O_2.

Protection by EDTA and Catalase

When *S. typhimurium* DNA was exposed to high concentrations of metal ions (50 μM) and H_2O_2 (100 mM), in the presence of either EDTA (25 mM) or catalase (3 u), DNA degradation was prevented (FIG. 1E). In FIGURE 1E (a), lanes 2, 4, and 6 show the results obtained after electrophoresis of *S. typhimurium* DNA that was incubated with 100 mM H_2O_2, 25 mM EDTA and, respectively, 50 μM Fe^{2+}, 50 μM Fe^{3+}, and 50 μM Cu^{2+}; lanes 3, 5, and 7 show the results obtained after electrophoresis of *S. typhimurium* DNA

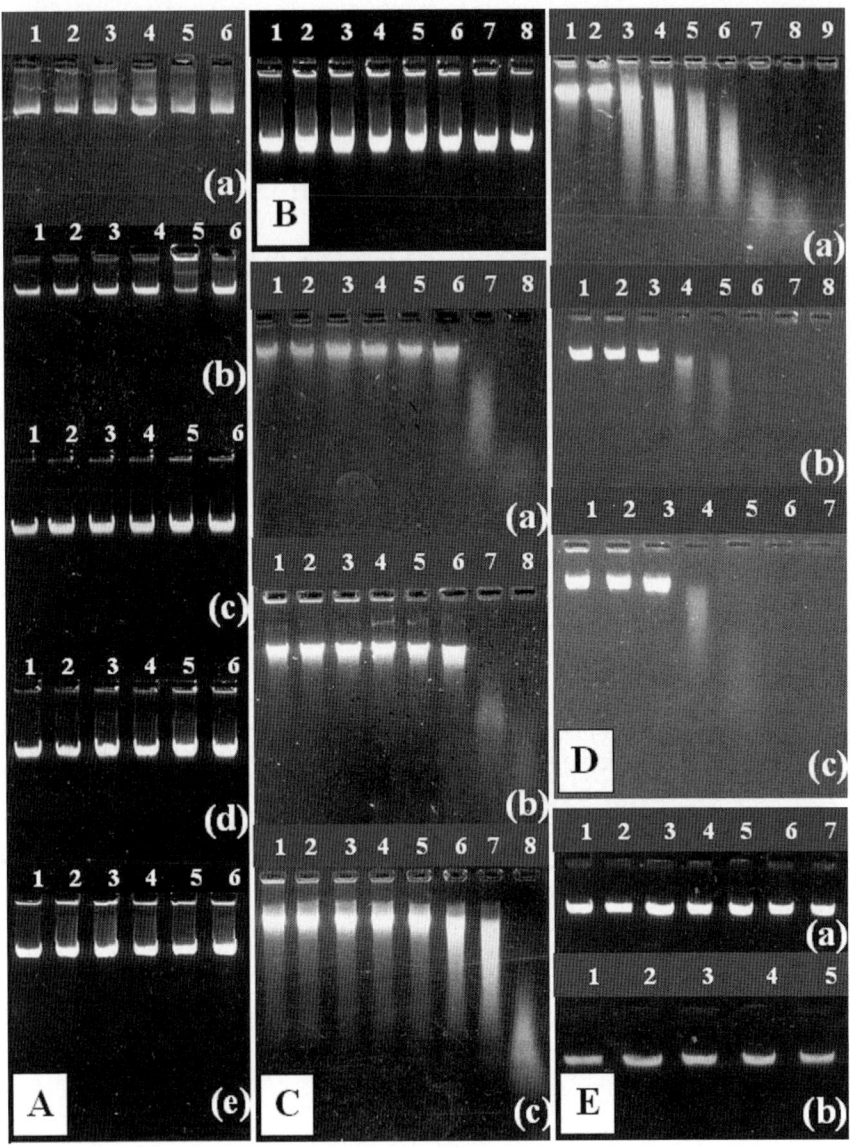

FIGURE 1. Electrophoresis in 1% agarose gel of *Salmonella typhimurium* DNA (0.2 μg per sample) incubated for 1 h at 37°C with increasing concentrations of metal ions (**A**), with increasing concentrations of H_2O_2 (**B**), with various metal ion concentrations in the presence of increasing H_2O_2 concentrations (**C** and **D**), with 50 μM metal ion, 100 mM H_2O_2 and either 25 mM EDTA, or 3 U catalase (**E**). (**A**) lane 1(a–e): *S. typhimurium* DNA; lanes 2–6: *S. typhimurium* DNA incubated with 2, 10, 20, 100, and 200 μM FeSO$_4$ (a), FeCl$_3$ (b), CuSO$_4$ (c), NiSO$_4$ (d), and CdCl$_2$ (e), respectively. (**B**) lane 1: *S. typhimurium* DNA; lanes 2–8: *S. typhimurium* DNA incubated with 0.1, 0.3, 1, 3, 10, 30, and 100 mM H_2O_2, respectively. (**C**) lane 1 (a, b, c): *S. typhimurium* DNA; lane 2: *S. typhimurium* DNA

that was incubated with 100 mM H_2O_2, 3 U catalase and, respectively, 50 μM Fe^{2+}, 50 μM Fe^{3+}, and 50 μM Cu^{2+}. Lane 1 shows control DNA. In FIGURE 1E (b), lanes 2 and 4 show the results obtained after electrophoresis of *S. typhimurium* DNA that was incubated with 100 mM H_2O_2, 25 mM EDTA and, respectively, 50 μM Ni^{2+} and 50 μM Cd^{2+}; lanes 3 and 5 show the results obtained after electrophoresis of *S. typhimurium* DNA that was incubated with 100 mM H_2O_2, 3 u catalase and, respectively, 50 μM Ni^{2+} and 50 μM Cd^{2+}. Lane 1 shows control DNA.

Hydroxyl Radical Formation

The formation of OH· by the various metal ions used in this study, in the presence of H_2O_2, is illustrated in FIGURE 2B, C. FIGURE 2B shows the formation of TBARS at 532 nm resulting from OH·-induced D(-)ribose degradation.[18] Under the testing conditions, Fe^{2+} generated the highest amount of OH·, followed closely by Fe^{3+}; Cu^{2+} was next, followed by Ni^{2+} and Cd^{2+}. Comparatively, the last two ions were poor generators of OH·. FIGURE 2C shows the decrease in absorbance at 265 nm due to thymine degradation, which was used as an assessment of OH· radical formation as described in Ref. 19. Optimum OH· production was observed in the presence of Cu^{2+}; it was still high in the presence of Fe^{2+} and Fe^{3+}; but it was considerably reduced in the presence of Ni^{2+} and Cd^{2+}. The results obtained closely matched those in FIGURE 2A, which reported the relative toxicity of the metal ions used in this study; the least toxic metal ions (Ni^{2+} and Cd^{2+}) were also those that produced the least OH•.

---

incubated with 20 μM (a), 70 μM (b), and 100 μM (c) Cd^{2+}; lanes 3–8 (a): *S. typhimurium* DNA, Cd^{2+} 20 μM and, respectively, 0.3, 1, 3, 10, 30, and 100 mM H_2O_2; lanes 3–8 (b): same as (a), but Cd^{2+} is 70 μM; lanes 3–8 (c): *S. typhimurium* DNA, Cd^{2+} 100 μM and, respectively, 0.1, 0.3, 1, 3, 10, and 30 mM H_2O_2. (**D**) lane 1 (a,b,c): *S. typhimurium* DNA; lane 2: *S. typhimurium* DNA incubated with 20 μM Fe^{2+} (a), 10 μM Cu^{2+} (b), and 20 μM Cu^{2+} (c); lanes 3–9 (a): *S. typhimurium* DNA, 20 μM Fe^{2+} and, respectively, 0.1, 0.3, 1, 3, 10, 30, and 100 mM H_2O_2; lanes 3–8 (b): *S. typhimurium* DNA, 10 μM Cu^{2+} and, respectively, 0.1, 0.3, 1, 3, 10, and 30 mM H_2O_2; lanes 3–7 (c): *S. typhimurium* DNA, 20 μM Cu^{2+} and, respectively, 0.1, 0.3, 1, 3, and 10 mM H_2O_2. (E) lane 1(a, b): *S. typhimurium* DNA; lanes 2, 4, 6 (a): *S. typhimurium* DNA, 25 mM EDTA, 100 mM H_2O_2 and either 50μM Fe^{2+} (lane 2), 50 μM Fe^{3+} (lane 4), or 50 μM Cu^{2+} (lane 6); lanes 3, 5, 7 (a): *S. typhimurium* DNA, 3 U catalase, 100 mM H_2O_2 and either 50 μM Fe^{2+} (lane 3), 50 μM Fe^{3+} (lane 5), or 50μM Cu^{2+} (lane 7); lanes 2, 4 (b): *S. typhimurium* DNA, 25 mM EDTA, 100 mM H_2O_2, and either 50 μM Ni^{2+} (lane 2), or 50 μM Cd^{2+} (lane 4); lanes 3, 5 (b): *S. typhimurium* DNA, 3 U catalase, 100 mM H_2O_2 and either 50 μM Ni^{2+} (lane 3), or 50 μM Cd^{2+} (lane 5).

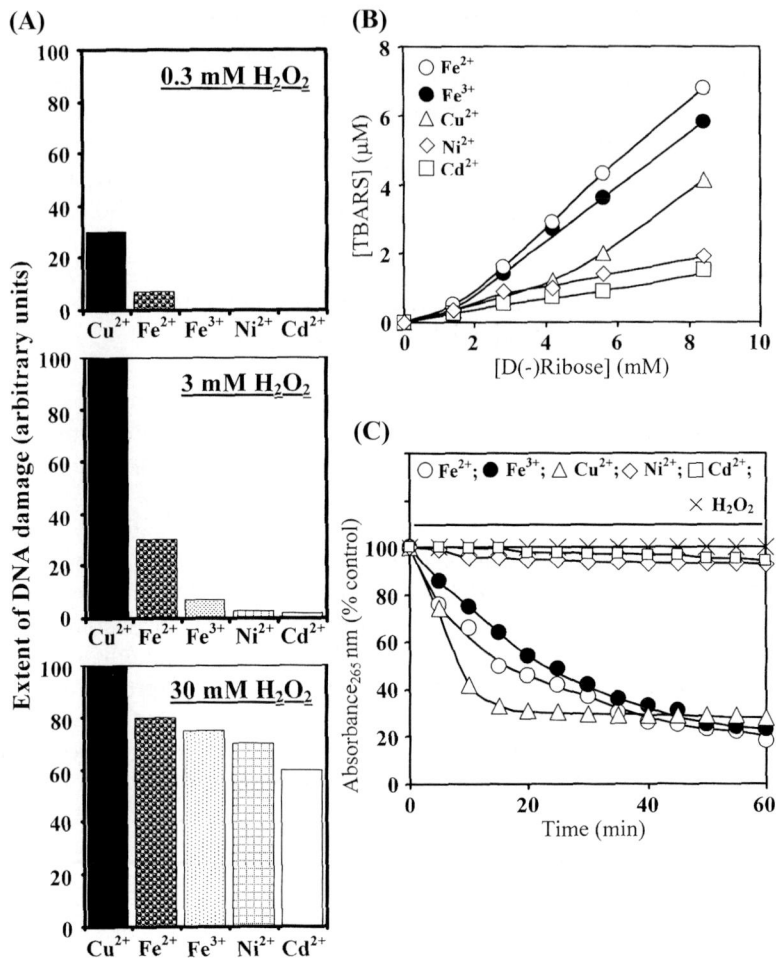

FIGURE 2. (**A**) Extent of DNA damage (as detected in 1% agarose gels) after incubation, for 1 h at 37°C, of 0.2 μg of purified *S. typhimurium* DNA with 20 μM of either Cu^{2+}, Fe^{2+}, Fe^{3+}, Ni^{2+}, or Cd^{2+} and 0.3 mM H_2O_2 (*top panel*), 3 mM H_2O_2 (*central panel*), or 30 mM H_2O_2 (*bottom panel*). Electrophoresis was conducted for 12 h at 25 volts. The extent of DNA damage was evaluated from the size of DNA fragments produced, as shown by electrophoretic mobility. Complete DNA degradation (100 on the scale) would produce DNA fragments small enough to be undetectable after electrophoresis under the conditions used in this work. (**B**) Formation of thiobarbituric acid substances (TBARS) when increasing D(-) ribose concentrations were incubated for 1 h at 37°C in the presence of 100 mM H_2O_2 and 100 μM of either Fe^{2+}, Fe^{3+}, Cu^{2+}, Ni^{2+}, or Cd^{2+}, at pH 7.4. (**C**) Decrease in the A_{265} of thymine during incubation with 100 mM H_2O_2 and 100 μM of either Fe^{2+}, Fe^{3+}, Cu^{2+}, Ni^{2+} or Cd^{2+}, at pH 7.4 and 37°C, in a final volume of 3 mL. Thymine was at a concentration of 1 A_{265}/mL, and the decrease in A_{265} was used to assess OH· radical formation (see text).

DISCUSSION

During the reduction of oxygen to water in aerobic life, superoxide radical ($O_2^{\cdot-}$), H_2O_2, and OH·, all known as reactive oxygen species (ROS), can be formed. $O_2^{\cdot-}$ is a moderately reactive ROS that is readily dismutated to H_2O_2. It is also able to reduce metal ions that are mainly present in the cell in the oxidized form [reaction (1)].

$$Me^{(n+1)+} + O_2^{\cdot-} \rightarrow Me^{n+} + O_2 \quad (1)$$

The reduced ions may, in turn, catalyze the conversion of H_2O_2 to OH· by the Fenton or Haber–Weiss reaction [reaction (2)].

$$Me^{n+} + H_2O_2 \rightarrow Me^{(n+1)+} + OH\cdot + OH^- \quad (2)$$

$O_2^{\cdot-}$, as well as the reduced metal ion, is also produced by reaction between the oxidized metal ions and H_2O_2 [reaction (3)].

$$Me^{(n+1)+} + H_2O_2 \rightarrow Me^{n+} + O_2^{\cdot-} + 2H^+ \quad (3)$$

Thus, even if metal ions are mainly present in the cell in the oxidized form, they will be readily reduced either by $O_2^{\cdot-}$ or by H_2O_2 itself; the reduced ion will then react with another H_2O_2 molecule to generate the hydroxyl radical. This is what is most likely happening in the *in vitro* system studied in this work, where only purified DNA, H_2O_2 and the metal ion (four of those tested here were in the oxidized form) were present in buffer.

As shown in the results, exposure of purified *S. typhimurium* DNA, either to metal ions alone in various concentrations, or to H_2O_2 alone in various concentrations, resulted in no apparent DNA damage as judged by electrophoresis in agarose gel. However, when the DNA was exposed to a metal ion and H_2O_2 simultaneously, DNA breakage occurred. Furthermore, when EDTA, a metal chelator, was added to the mixture of DNA, metal ion and H_2O_2, no damage to the DNA was observed. Similarly, when catalase, an H_2O_2 scavenger, was added to the mixture of DNA, metal ion and H_2O_2, no damage to the DNA was observed. Finally, the formation of OH• when the metal ions were in the presence of H_2O_2, was evidenced by two separate tests.

Although cadmium has been identified as a potent animal and human carcinogen, it appeared the least toxic among the metals tested in the system studied in this work. On the other hand, the high toxicity of copper illustrated here is not surprising in the light of its properties.[20] Copper is an essential transition metal with a key role in aerobic life on account of its function in cytochrome *c* oxidase and related enzymes. It is known to readily interact with radicals and with molecular oxygen, and it has been shown that its toxicity is based in part on the formation of hydroperoxide radicals.[21] As shown here, it would cause complete degradation of *S. typhimurium* DNA at a concentration of 20 µM in the presence of 3 mM H_2O_2, while the same concentration of the next most toxic metal in this work, Fe^{2+}, would cause complete DNA degradation in the presence of 100 mM H_2O_2. Copper was also among the metals that produced the most OH•.

Transition metals are known to cause mutations and have been demonstrated to cause cancer in a number of species, including the human. DNA breakage is among their reported harmful effects and, although metal ions have been shown occasionally to form complexes with DNA,[22–24] the strand breakage that they cause has been attributed to the generation of highly toxic ROS.[25–27] We did not evaluate the mutagenic effect of the metals ions studied in this work, but we showed that for each of the metals studied DNA breakage was caused by generation of the hydroxyl radical in the presence of H_2O_2.

The role of transition metals in increasing oxidative stress is not limited to the production of ROS. It also stems from their ability to bind to proteins and inhibit their enzymatic activity, in particular the activity of enzymes involved in the protection against oxidative stress, such as peroxidases.[10,28–30] H_2O_2 is constantly generated in living aerobic cells, and by impairing H_2O_2-scavenging enzymes, toxic concentrations of transition metals further increase the risk of oxidative stress and cell damage.

ACKNOWLEDGMENT

This work was supported in part by the J. and E. Research Foundation, Tehran, Iran.

REFERENCES

1. AISEN, P. 1980. Iron transport and storage proteins. Annu. Rev. Biochem. **49:** 357–393.
2. LINDER, M.C. 1991. Nutritional Biochemistry and Metabolism. Elsevier. New York.
3. BEAMISH, M.R., R. WALKER, F. MILLER, et al. 1974. Transferrin iron, chelatable iron and ferritin in idiopathic haemochromatosis. Br. J. Haematol. **27:** 219–228.
4. EDWARDS, C.Q., L.M. GRIFFEN, R.S. AJIOKA & J.P. KUSHNER. 1998. Screening for hemochromatosis phenotype versus genotype. Semin. Hematol. **35:** 72–76.
5. HARTWIG, A. 1995. Current aspects in metal genotoxicity. Biometals **8:** 3–11.
6. BARCELOUX, D.G. 1999. Nickel. J. Toxicol. Clin. Toxicol. **37:** 239–258.
7. MANINI, P., A. NAPOLITANO, E. CAMERA, et al. 2003. Ni^{2+} enhances Fe^{2+}/peroxide induced oxidation of arachidonic acid and formation of geno/cytotoxic 4-hydroxynonnal: a possible contributory mechanism in nickel toxicity and allergenicity. Biochim. Biophys. Acta **1621:** 9–16.
8. IARC. 1993. Cadmium and Cadmium Compounds. IARC Monographs on the Evaluation of Carcinogenic Risks to Human – Beryllium, Cadmium, Mercury and Exposure in the Glass Manufacturing Industry. **58:** 119–237. IARC. Lyon.

9. LIN, C.-J., K.-H. WU, F.-H. YEW & T.-C. LEE. 1995. Differential cytotoxicity of cadmium to rat embryonic fibroblasts and human skin fibroblasts. Toxicol. Appl. Pharmacol. **133:** 20–26.
10. BACCOUCH, S., A. CHAONI & E. EL FERJANI. 1998. Nickel-induced oxidative damage and antioxidant responses in *Zea mays* shoots. Plant Physiol. Biochem. **36:** 689–694.
11. HEGEDÜS, A., S. ERDEIS & G. HORVÁTH. 2001. Comparative studies of H_2O_2 detoxifying enzymes in green and greening barley seedlings under cadmium stress. Plant Sci. **160:** 1085–1093.
12. GECHEV, T.S. & J. HILLE. 2005. Hydrogen peroxidase a signal controlling plant programmed cell death. J. Cell Biol. **168:** 17–20.
13. VRANOVÁ, E., D. INZÉ & F. VAN BREUSEGEM. 2002. Signal transduction during oxidative stress. J. Exp. Bot. **53:** 1227–1236.
14. SAMBROOK, J., E.F. FRITSCH & T. MANIATIS. 1989. Molecular Cloning—A Laboratory Manual. 2nd ed. Cold Spring Harbor Laboratory Press. Cold Spring Harbor, NY.
15. KEYHANI, J., F. JAFARI-FAR, N. EINOLLAHI, *et al.* 1998. DNA-mobility shift assay and the detection of anti-DNA IgG in systemic lupus erythematosus patients. Immunol. Lett. **62:** 81–86.
16. ASHWELL, G. 1957. Colorimetric analysis of sugars. Methods Enzymol. **3:** 99–101.
17. LIN, R.I. & O.A. SCHJEIDE. 1969. Micro estimation of RNA by cupric ion catalyzed orcinol reaction. Anal. Biochem. **27:** 473–483.
18. HALLIWELL, B., M. GROOTVELD & J.M. GUTTERIDGE. 1988. Methods for the measurement of hydroxyl radicals in biomedical systems: deoxyribose degradation and aromatic hydroxylation. Methods Biochem. Anal. **33:** 59–90.
19. LOWN, J.W., S.K. SIM & H.H. CHEN. 1978. Hydroxyl radical production by free and DNA-bound aminoquinone antibiotics and its role in DNA degradation. Electron spin resonance detection of hydroxyl radicals by spin trapping. Can. J. Biochem. **56:** 1042–1047.
20. NIES, D.H. 1999. Microbial heavy-metal resistance. Appl. Microbiol. Biotechnol. **51:** 730–750.
21. RODRIGUEZ MONTELONGO, L., L.C. DE LA CRUZ RODRIGUEZ, R.N. FARIAS & E.M. MASSA. 1993. Membrane-associated redox cycling of copper mediates hydroperoxide toxicity in *Escherichia coli*. Biochim. Biophys. Acta **1144:** 77–84.
22. HOSSAIN, Z. & F. HUQ. 2002. Studies on the interaction between Cd^{2+} ions and DNA. J. Inorg. Biochem. **90:** 85–96.
23. CHIVERS, P.T. & R.T. SAUER. 2002. High-affinity nickel binding to the C-terminal domain regulates binding to operator DNA. Chem. Biol. **9:** 141–1148.
24. WAALKES, M.P. & L.A. POIRIER. 1984. *In vitro* cadmium-DNA interactions: cooperativity of cadmium binding and competitive antagonism by calcium, magnesium, and zinc. Toxicol. Appl. Pharmacol. **75:** 539–546.
25. SCHÜTZENDÜBEL, A. & A. POLLE. 2002. Plant responses to abiotic stresses: heavy metal–induced oxidative stress and protection by mycorrhization. J. Exp. Bot. **53:** 1351–1365.
26. DALLY, H. & A. HARTWIG. 1997. Induction and repair inhibition of oxidative DNA damage by nickel(II) and cadmium(II) in mammalian cells. Carcinogenesis **18:** 1021–1026.
27. SALMON, T.B., A.E. EVERT, B. SONG & P.W. DOETSCH. 2004. Biological consequences of oxidative stress-induced DNA damage in *Saccharomyces cerevisiae*. Nucleic Acids Res. **32:** 3712–3723.

28. KEYHANI, J., E. KEYHANI, N. EINOLLAHI, et al. 2003. Heterogeneous inhibition of horseradish peroxidase activity by cadmium. Biochim. Biophys. Acta **1621:** 140–148.
29. CONVERSO, D.A., M.E. FERNANDEZ & M.L. TOMARO. 2000. Cadmium inhibition of a structural wheat peroxidase. J. Enzyme Inhibition **15:** 171–183.
30. KEYHANI, J., E. KEYHANI, S. ZARCHIPOUR, et al. 2005. Stepwise binding of nickel to horseradish peroxidase and inhibition of the enzymatic activity. Biochim. Biophys. Acta **1722:** 312–323.

Antioxidant Enzymes during Hypoxia–Anoxia Signaling Events in *Crocus sativus* L. Corm

EZZATOLLAH KEYHANI,[a,b] LILA GHAMSARI,[b] JACQUELINE KEYHANI,[a] AND MAHNAZ HADIZADEH[b]

[a]*Laboratory for Life Sciences, 19979 Tehran, Iran*

[b]*Institute of Biochemistry and Biophysics, University of Tehran, 13145 Tehran, Iran*

ABSTRACT: The activity of reactive oxygen species (ROS)-scavenging enzymes, catalase, superoxide dismutase (SOD), glutathione peroxidase, o-dianisidine and ascorbate peroxidases, was investigated in *Crocus sativus* L. corms cultivated in normoxic and hypoxic–anoxic conditions. The activity of the ROS-scavenging enzymes studied varied during cultivation. However, the pattern of ROS-scavenging enzymes production was different in corms cultivated in normoxic and hypoxic–anoxic conditions. In normoxic conditions, only the activities of peroxidases and SOD were stimulated. In dormant corms placed under hypoxia–anoxia, the activities of catalase, SOD, and glutathione peroxidase were stimulated, with the highest stimulation observed for catalase, followed by SOD, and then glutathione peroxidase. In corms that had been rooted for 3 days before being placed in hypoxia–anoxia, the activities of all ROS-scavenging enzymes studied were stimulated with the highest stimulation still observed for catalase, followed by the peroxidases, and finally SOD. Thus catalase was the prevailing enzyme produced under hypoxia–anoxia.

KEYWORDS: saffron; hypoxia–anoxia; catalase; superoxide dismutase; o-dianisidine peroxidase; ascorbate peroxidase; glutathione peroxidase

INTRODUCTION

Aerobic organisms under hypoxic–anoxic conditions switch to glycolysis when oxidative phosphorylation is repressed.[1–3] A deficiency in oxygen is produced in plants by high rain precipitation and soil flooding and results in alterations in plant morphology, cell structure, and gene expression.[3–6] The signaling pathways that control the response to hypoxia–anoxia are complex,

Address for correspondence: Dr. Ezzatollah Keyhani, Institute of Biochemistry and Biophysics, University of Tehran, P.O. Box 13145-1384, 13145 Tehran, Iran. Voice: +98-21-6695-6974; fax: +98-21-6640-4680.

e-mail: keyhanie@ibb.ut.ac.ir

and the targets sensing oxygen deficiency are not definitely established and most probably multiple sensors exist in higher organisms.[7] Heme is believed to be involved in oxygen-sensing in the yeast *Saccharomyces cerevisiae*,[8,9] while cytochrome *c* oxidase is likely to be the hemoprotein sensor.[10] Mitochondria may function as O_2 sensors by increasing their generation of reactive oxygen species (ROS) during hypoxia.[11] Other oxygen-sensing molecules, such as the cytochrome b558/NADPH oxidase located in the plasma membrane and generating superoxide radical,[12] guanylate cyclase O_2-sensitive ion channel, and cytochrome P450,[13,14] have been proposed as oxygen sensors. In plants, nitric oxide (NO) levels increase in response to hypoxia. This increase appears to be modulated by levels of plant class I nonsymbiotic hemoglobin, which also increases in response to hypoxia.[15] The hypoxia-inducible factor 1 has been shown to play a key role in the transcriptional activation of gene by hypoxia,[16,17] including cancer and diabetes.[18,19] Alcohol dehydrogenase is one of the enzyme proteins synthesized under anaerobiosis and is considered essential for the preservation of plants during hypoxia.[20,21] Under hypoxia, or under any other biotic or abiotic stress, or even under physiological conditions (unstressed), ROS are produced. Plants have well-developed defense systems against ROS, involving both limiting the formation of ROS as well as their removal.[22] Thus two mechanisms are involved in the formation of ROS, one for signaling events (low level), and another mechanism to regulate their intracellular concentration by scavenging them.[23] Since the major ROS-scavenging mechanisms in plants include superoxide dismutase (SOD), ascorbate peroxidase, catalase, and glutathione peroxidase, the purpose of this research was to determine the amounts of these enzymes in corms cultivated under normoxic conditions as well as in dormant corms placed under hypoxia and in corms cultivated for 3 days and then subjected to hypoxia. The results showed that, in normoxic conditions, only the activities of peroxidases and SOD were stimulated. In dormant corms placed under hypoxia–anoxia, the activities of catalase, SOD, and glutathione peroxidase were stimulated, with the highest stimulation observed for catalase, followed by SOD, and then glutathione peroxidase. In corms that had been rooted for 3 days before being placed in hypoxia–anoxia, the activities of all ROS-scavenging enzymes studied were stimulated with the highest stimulation still observed for catalase, followed by the peroxidases, and finally SOD. Thus catalase was the prevailing enzyme produced under hypoxia–anoxia.

MATERIALS AND METHODS

Cultivation Under Normoxic Conditions

Dormant corms each weighing between 3 and 6 g were cultivated under normoxic conditions in distilled water for up to 17 days. All corms started to root within 24 h. Rooting corms were withdrawn daily and used immediately for extract preparation.

Hypoxia–Anoxia Induction by Flooding

Either dormant corms or corms that rooted for 3 days under normoxic conditions were immersed in distilled water and maintained under the water level. Dormant corms (3–6 g) were without roots and shoots; 3-day rooting corms exhibited 15 ± 2-mm roots (average length) and 40 ± 5-mm shoots (average length). Flooded corms (dormant or 3-day rooting) were withdrawn after 1, 2, 3, 4, 8, 10, and 14 days under water and used immediately for extract preparation. To assess the ability to resume root growth, dormant corms and 3-day rooting corms were transferred, after hypoxia–anoxia, to a glass jar with adequate amount of water and at room temperature (22–25°C). They were followed for up to 1 month for possible root growth.

Extract Preparation

Under our experimental conditions, dormant corms as well as flooded dormant corms did not exhibit any roots or shoots. Corms cultivated under normoxic conditions developed roots and shoots that were carefully removed before preparation of the extracts. Extracts were prepared according to a previously published method.[24] In brief: cleaned corms were homogenized in 0.1 M phosphate buffer, pH 7.0. After centrifugation at 3,000 g for 10 min and then at 35,000 g for 30 min, a clear, transparent supernatant, termed the *crude extract*, was obtained and used for our studies. Protein concentration in the extracts was determined by the Lowry method.[25]

Enzymatic Activity Assays

Catalase activity was assayed as described previously by following the dismutation of H_2O_2 spectrophotometrically using an extinction coefficient of 27 $M^{-1}cm^{-1}$ for H_2O_2 at 240 nm.[26] One unit was defined as the amount of enzyme decomposing 1 μmol H_2O_2 per minute.

Peroxidase activity was determined by following the H_2O_2-mediated oxidation of o-dianisidine at 460 nm, and ascorbate at 290 nm with extinction coefficients of 11.3 $mM^{-1}cm^{-1}$ and 2.8 $mM^{-1}cm^{-1}$, respectively, as previously described.[27] One unit was defined as the amount of enzyme needed for the oxidation of 1 μmol of substrate per minute.

SOD activity was assayed by a method based on the inhibition of pyrogallol autoxidation in alkaline solution as described elsewhere.[28] One unit of SOD activity corresponded to the amount required to inhibit pyrogallol autoxidation by 50%.

Glutathione peroxidase activity was assayed by following the oxidation of β-NADPH at 340 nm, using an extinction coefficient of 6.22 mM^{-1}cm^{-1}; the assay was conducted in the presence of glutathione, glutathione reductase, and β-NADPH, in 0.05 M phosphate buffer, pH 6.5, 1 mM azide. One unit was defined as the amount of enzyme that would oxidize 1 mol of β-NADPH per minute.

RESULTS

Catalase Activity

Corms cultivated under normoxic conditions showed up to 10% increase in the amount of catalase (expressed as units per milligram of protein in the extract) detectable during the first 3 days of development. This was followed, from the 4th day on, by a progressive decrease so that, at days 8 and 10, only 15% of the amount present in dormant corms was detectable (FIG. 1A). Thereafter, the amount of catalase increased again to reach 78% of the amount present in dormant corm, after 17 days of cultivation (FIG. 1 B).

Dormant corms placed in hypoxic–anoxic conditions showed a 10% increase in the amount of catalase at day 3 of hypoxia–anoxia; thereafter, the amount of catalase continued to increase so that, after 8 days in hypoxia–anoxia, it was 2.5 times the amount detected in dormant corms before hypoxia–anoxia. This high level of catalase was still detectable at day 14 of hypoxia–anoxia (FIG. 1A).

In contrast, in corms that were cultivated for 3 days in normoxic conditions before undergoing hypoxia–anoxia, the amount of catalase increased by a factor of 4 after only 1 day of hypoxia–anoxia, increased to 5 times its original value at days 4–10 in hypoxia–anoxia, and was still 4 times its original value after 14 days in hypoxia–anoxia (FIG. 1B). The original value was that after 3 days in normoxic conditions; it is shown as time 0 in FIGURE 1B.

Peroxidase Activity

Peroxidase activity was assayed using two different reducing substrates, o-dianisidine and ascorbate.

(a) o-Dianisidine Peroxidase Activity. The amount of o-dianisidine peroxidase detected in corms cultivated under normoxic conditions increased progressively during the first 3 days of cultivation, when it doubled. After 8 days of cultivation, the amount of o-dianisidine peroxidase decreased to 1.5 times the amount observed in dormant corms and remained at that level for the next 10 days (FIG. 1C, D).

In dormant corms placed in hypoxic–anoxic conditions, after an initial 20% increase at day 1, the amount of o-dianisidine peroxidase decreased by 40% at

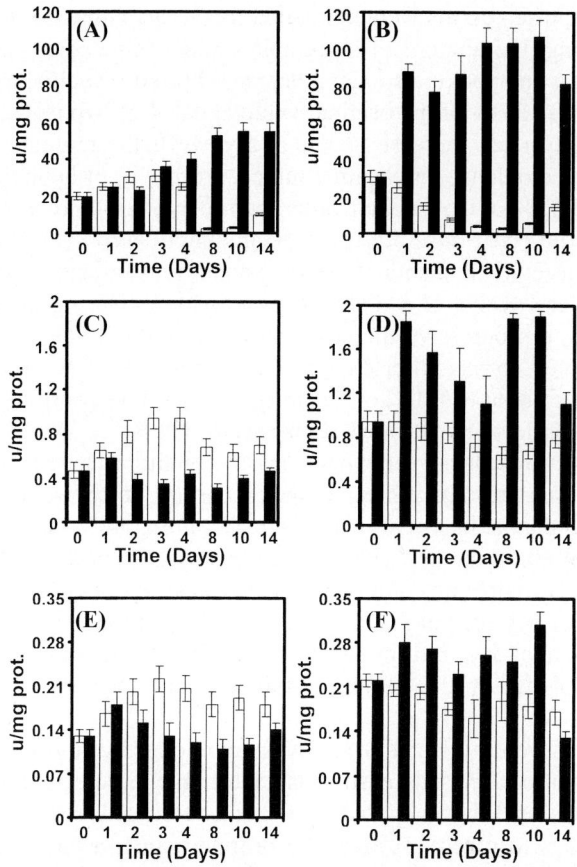

FIGURE 1. Effects of hypoxia–anoxia on the amounts of catalase, o-dianisidine peroxidase, and ascorbate peroxidase detectable in saffron corms. **A, C** and **E** report the results obtained when dormant corms were placed in hypoxia–anoxia; the *white bars* refer to the control situation where dormant corms were cultivated for up to 14 days in normoxic conditions and the *black bars* refer to dormant corms that were placed in hypoxia–anoxia for the same period. **B, D)** and **F** report the results obtained when corms were cultivated for 3 days in normoxic conditions prior to hypoxia–anoxia; the *white bars* refer to the control situation where corms were cultivated for up to 17 days in normoxic conditions (results are reported starting at day 3 of cultivation, which is indicated as time 0 on the graphs) and the *black bars* refer to corms cultivated for 3 days in normoxic conditions (time 0 on the graphs), then put in hypoxia–anoxia for up to 14 days. (**A** and **B**) catalase activity; (**C** and **D**) o-dianisidine peroxidase activity; (**E** and **F**) ascorbate peroxidase activity.

day 2 of hypoxia–anoxia and still decreased at day 8 of hypoxia–anoxia, when it was 65% of the original value in dormant corms before hypoxia–anoxia; thereafter, a progressive increase was observed so that at day 14 of hypoxia–anoxia, the amount of o-dianisidine peroxidase detectable was similar to that in dormant corms before hypoxia–anoxia (FIG. 1C).

In contrast, when corms were cultivated for 3 days in normoxic conditions before flooding, the amount of detectable o-dianisidine peroxidase doubled after only 1 day in hypoxia–anoxia; it decreased progressively during the next 3 days to reach 120 % of the original value at day 4 of hypoxia–anoxia. This was followed by a new increase, so that at days 8–10, the amount of detectable o-dianisidine peroxidase had again doubled. At day 14 of hypoxia–anoxia, it was back at 120 % of the original value (before hypoxia–anoxia) (FIG. 1 D).

(b) Ascorbate Peroxidase Activity. The amount of ascorbate peroxidase detected in corms cultivated under normoxic conditions followed approximately the same pattern as that of o-dianisidine peroxidase. During the first 3 days of cultivation, a progressive increase was observed to 170% of the original value in dormant corms. After 8 days of cultivation, the amount of ascorbate peroxidase decreased to 140% of the amount observed in dormant corms and remained at that level for the next 10 days (FIG. 1E, F).

In dormant corms placed in hypoxia–anoxia conditions, the amount of ascorbate peroxidase detectable changed also according to the same pattern observed with o-dianisidine peroxidase. After an initial 40% increase at day 1, the amount of ascorbate peroxidase decreased by 20% at day 2, and still decreased until day 8 of hypoxia–anoxia, when it was 80% of the original value in dormant corms before hypoxia–anoxia; thereafter, a progressive increase was observed so that at day 14 of hypoxia–anoxia, the amount of ascorbate peroxidase detectable was 110% of that in dormant corms before hypoxia–anoxia (FIG. 1 E).

When corms were cultivated for 3 days in normoxic conditions before flooding, the amount of ascorbate peroxidase detectable increased to 130% of the original value after 1 day in hypoxia–anoxia; it decreased progressively during the next 2 days to reach 110% of the original value at day 3 of hypoxia–anoxia. This was followed by a new increase so that at day 10, the amount of ascorbate peroxidase detectable reached 140% of the original value. At day 14 of hypoxia–anoxia, it had decreased to 60% of the original value (before hypoxia–anoxia) (FIG. 1F).

SOD Activity

The amount of SOD detectable in corms cultivated under normoxic conditions followed a cycle of alternate decreases and increases. An initial decrease brought the amount of enzyme detectable to 70% of the original value at day 3 of cultivation; this was followed by an increase to 120% of the original value at day 8 of cultivation. By day 14, the amount of SOD was similar to that in dormant corms and by day 17, it was 90% of that value (FIG. 2A, B).

In dormant corms placed in hypoxia–anoxia conditions, the amount of SOD doubled after 2 days, and then decreased progressively to 160% of the original value after 4 days in hypoxia–anoxia; it went back to the original value

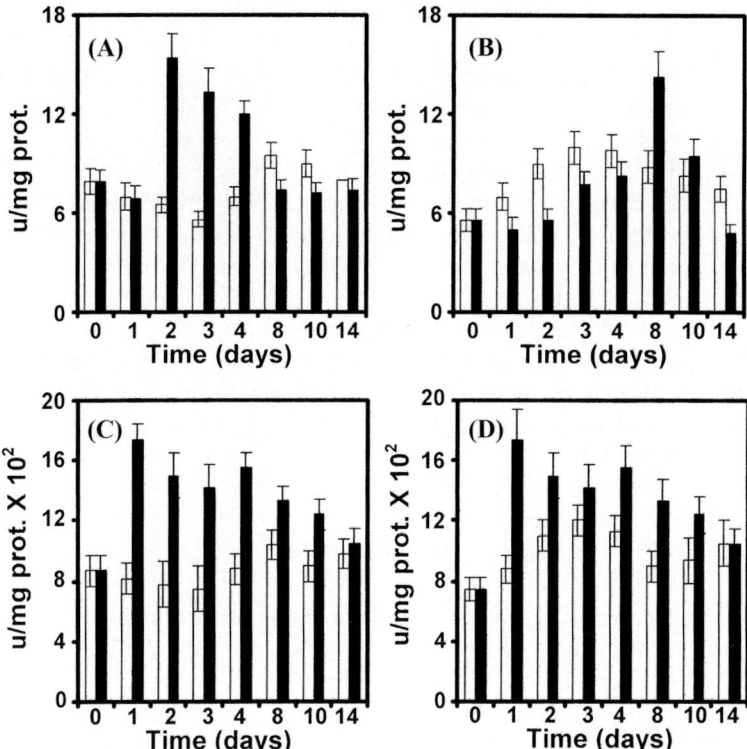

FIGURE 2. Effects of hypoxia–anoxia on the amounts of SOD and glutathione peroxidase detectable in saffron corms. **A** and **C** report the results obtained when dormant corms were placed in hypoxia–anoxia; the *white bars* refer to the control situation where dormant corms were cultivated for up to 14 days in normoxic conditions and the *black bars* refer to dormant corms that were placed in hypoxia–anoxia for the same period. **B** and **D** report the results obtained when corms were cultivated for 3 days in normoxic conditions prior to hypoxia–anoxia; the *white bars* refer to the control situation, in which corms were cultivated for up to 17 days in normoxic conditions (results are reported starting at day 3 of cultivation, which is indicated as time 0 on the graphs), and the *black bars* refer to corms cultivated for 3 days in normoxic conditions (time 0 on the graphs), then put in hypoxia–anoxia for up to 14 days. **A** and **B**: SOD activity; **C** and **D**: glutathione peroxidase activity.

in dormant corms after 8 days, and remained at that value until day 14 of hypoxia–anoxia (FIG. 2 A).

In contrast, when corms were cultivated for 3 days in normoxic conditions before flooding, the amount of SOD decreased slightly during the first 2 days in hypoxia–anoxia; it began to increase at day 3 and was 2.5 times the original value at day 8 of hypoxia–anoxia; it was back to the original value (before hypoxia–anoxia) after 14 days of hypoxia–anoxia (FIG. 2B).

Glutathione Peroxidase Activity

The amount of glutathione peroxidase detectable in corms cultivated under normoxic conditions increased progressively to 155% of the original value after 3 days of cultivation; it went down to 120% of the original value in dormant corms and remained at that level for the next 10 days (FIG. 2C, D).

In dormant corms placed in hypoxia–anoxia conditions, the amount of glutathione peroxidase doubled after 1 day of hypoxia–anoxia; decreased to 175% of the original value at day 2 and remained at that value for the next few days. It was at 150% of the original value at day 8 of hypoxia–anoxia and at 125% of the original value at day 14 of hypoxia–anoxia (FIG. 2C).

When corms were cultivated for 3 days in normoxic conditions before flooding, the amount of glutathione peroxidase evolved according to the same pattern as that observed for dormant corms placed in hypoxia–anoxia without prior cultivation (FIG. 2D).

DISCUSSION

In *Crocus sativus* L. corms subjected to hypoxia–anoxia, alcohol dehydrogenase, and lactate dehydrogenase increased in order to maintain adequate levels of ATP even though, ultimately, overloading of lactate and ensuing acidosis led to cell death and necrosis.[21] Evidence has accumulated on the involvement of ROS in the response to hypoxia–anoxia. After initial suggestion that ROS are detrimental to the metabolic process of living cells, it is now generally admitted that ROS participate in various cell signaling pathways and are necessary for many of the cells' biological activities. ROS allow for adaptation of the cell and, depending on the duration and intensity of the signal, activate the process responsible for cell damage or death. Hydrogen peroxide mediates cell growth[29] and may selectively activate signaling pathway[30] and angiogenesis.[31] H_2O_2 alone was sufficient to activate C-jun N-terminal kinase (JNK) and induce cell death. Moreover, oxidants have been demonstrated to induce cell proliferation,[29,31] various transcriptional factors,[32] changes in cellular shape, and cell death.[33] In all species so far studied, including plants, several enzymes scavenging ROS, such as SOD, catalase, glutathione peroxidase, and ascorbate peroxidase, have been identified.[33] Moreover, there is a balance between various ROS-scavenging enzymes. The balance between SOD and ascorbate peroxidase or catalase is important for determining the steady-state level of superoxide radicals and hydrogen peroxide.[23] Ascorbate peroxidase might be responsible for the fine modulation of ROS for signaling, whereas catalase might be responsible for the removal of excess ROS during stress.[23]

The relationship between ROS-producing and ROS-scavenging mechanisms was different in normoxic and dormant corms versus corms rooted for 3 days and then subjected to hypoxia. Neither ascorbate peroxidase, nor o-dianisidine peroxidase exhibited an increase in their level; actually their amounts decreased compared to the control values. In contrast, the amounts of catalase, SOD, and glutathione peroxidase increased compared to the control with the increase in catalase being the largest, followed by that of SOD, and finally that of glutathione peroxidase. In contrast, the amounts of all five enzymes studied increased compared to the control in rooted corms subjected to hypoxia, with the increase in catalase being the largest again, followed, in decreasing order, by those of o-dianisidine peroxidase, ascorbate peroxidase, glutathione peroxidase, and SOD. The mechanism that coordinates the relationship between different components of the ROS removal network in *Crocus sativus* L. corms is complex. Substrate affinity and concentration, enzyme concentration, and reaction rate are important parameters for assessing the relative contribution of the different enzymes to ROS detoxification. Catalase appears to be the enzyme implicated in the major scavenging mechanism under both hypoxic experimental conditions even though SOD has been considered the first line of defense against ROS.[22] Catalase regulates H_2O_2 homeostasis, as it functions as a cellular sink for H_2O_2.[34] Catalase deficiency leads to elevation of H_2O_2 levels and triggers programmed cell death.[34] During anaerobiosis, elevation in alcohol dehydrogenase expression correlated with increases in the H_2O_2 production.[35] The mechanism of ROS detoxification is considered to be multifactorial, depending on both the duration of stress and the concentration of ROS. At low concentrations, ROS are involved in cell signaling; at tolerable concentrations, ROS are scavenged by antioxidant enzymes; but when their concentrations increase too much, leading to cellular damage, cell death occurs.[34,36] Besides catalase, peroxidases can also mediate H_2O_2 scavenging, and their activities correlatively increased under our experimental conditions (FIG. 1D, F).

ACKNOWLEDGMENT

This work was supported in part by the J and E. Research Foundation, Tehran, Iran.

REFERENCES

1. SEMENZA, G.L., P.H. ROTH, H.-M. FANG & G.L. WANG. 1994. Transcriptional regulation of genes encoding glycolytic enzymes by hypoxia-inducible factor 1. J. Biol. Chem. **269:** 23757–23763.
2. DISCHER, D.J., N.H. BISHOPRIC, X. WU, *et al.* 1998. Hypoxia regulates β-enolase and pyruvate kinase-M promoters by modulating Sp1/Sp3 binding to a conserved GC element. J. Biol. Chem. **273:** 26087–26093.
3. BIEMELT, S., M.R. HAJIREZAI, M. MELZER, *et al.* 1999. Sucrose synthase activity does not restrict glycolysis in roots of transgenic potato plants under hypoxic conditions. Planta **210:** 41–49.

4. PERATA, P. & A. ALPI. 1993. Plant responses to anaerobiosis. Plant Sci. **93:** 1–17.
5. CRAWFORD, R.M.M. & R. BRÄNDLE. 1996. Oxygen deprivation stress in a changing environment. J. Exp. Bot. **47:** 145–159.
6. SACHS, M.M., C.C. SUBBAIAH & I.N. SAAB. 1996. Anaerobic gene expression and flooding tolerance in maize. J. Exp. Bot. **47:** 1–15.
7. FIRTH, J.D., B.L. EBERT, C.W. PUGH & P.J. RATCLIFFE. 1994. Oxygen-regulated control elements in the phosphoglycerate kinase 1 and lactate dehydrogenase A genes: Similarities with erythropoietin 3′ enhancer. Proc. Natl. Acad. Sci. USA **91:** 6496–6500.
8. IYER, N.V., L.E. KOTCH, F. AGANI, *et al.* 1998. Cellular and developmental control of O_2 homeostasis by hypoxia-inducible factor 1 α. Gene Dev. **12:** 149–162.
9. ZITOMER, R.S. & C.V. LOWRY. 1992. Regulation of gene expression by oxygen in *Saccharomyces cerevisiae*. Microbiol. Rev. **56:** 1–11.
10. KWAST, K.E., P.V. BURKE, B.T. STAAHL & R.O. POYTON. 1999. Oxygen sensing in yeast: evidence for the involvement of the respiratory chain in regulating the transcription of a subset of hypoxic genes. Proc. Natl. Acad. Sci. USA **96:** 5446–5451.
11. BELL, E.L., B.M. EMERLING & N.S. CHANDEL. 2005. Mitochondrial regulation of oxygen sensing. Mitochondrion **5:** 322–332.
12. WANG, D., C. YOUNGSON, V. WONG, *et al.* 1996. NADPH-oxidase and a hydrogen peroxide-sensitive K^+ channel may function as an oxygen sensor complex in airway chemoreceptors and small lung carcinoma cell lines. Proc. Natl. Acad. Sci. USA **93:** 13182–13187.
13. HARDER, D.R., J. NARAYANAN, E.K. BIRKS, *et al.* 1996. Identification of a putative microvascular oxygen sensor. Circ. Res. **79:** 54–61.
14. TYLOR, C.T., S.J. LISCO, C.S. AWTREY & S.P. COLGAN. 1998. Hypoxia inhibits cyclic nucleotide-stimulated epithelial ion transport: role of nucleotide cyclases as oxygen sensors. J. Pharmacol. Exp. Ther. **284:** 568–575.
15. HUNT, P.W., E.J. KLOK, B. TREVASKIS, *et al.* 2002. Increased level of hemoglobin 1 enhances survival of hypoxic stress and promotes early growth. Proc. Natl. Acad. Sci. USA **99:** 17197–17222.
16. BAILEY-SERRES, J. & R.K. DAWE. 1996. Both 5′ and 3′ sequences of maize adh1 mRNA are required for enhanced translation under low-oxygen conditions. Plant Physiol. **112:** 685–695.
17. DREW, M.C. 1997. Oxygen deficiency and root metabolism: injury and acclimation under hypoxia and anoxia. Annu. Rev. Plant Physiol. Plant Mol. Biol. **48:** 223–250.
18. SEMENZA, G.L. 2003. Targeting HIF-1 for cancer therapy. Nat. Rev. Cancer **10:** 721–732.
19. LASSO, J.N. & H.A. BAWADI. 2005. Hypoxia inducible factors as targets for functional food. J. Agric. Food Chem. **53:** 3751–3768.
20. JOHNSON, J.R., B.G. COBB & M.C. DREW. 1994. Hypoxic induction of anoxia tolerance in roots of *Adh1* null *Zea mays* L. Plant Physiol. **105:** 61–67.
21. KEYHANI, E. & J. KEYHANI. 2004. Hypoxia/anoxia as signaling for increased alcohol dehydrogenase activity in saffron (*Crocus sativus* L.) corm. Ann. N.Y. Acad. Sci. **1030:** 449–457.
22. ALSCHER, R.G., N. ERTURK & L.S. HEATH. 2002. Role of superoxide dismutases (SODs) in controlling oxidative stress in plants. J. Explt. Bot. **53:** 1331–1341.
23. MITTLER, R. 2002. Oxidative stress, antioxidants and stress tolerance. Trends Plant Sci. **7:** 405–410.

24. KEYHANI, E. & N. SATTARAHMADY. 2002. Catalytic properties of three L-lactate dehydrogenases from saffron corms (*Crocus sativus* L.). Mol. Biol. Rep. **29:** 163–166.
25. LOWRY, O.H., N.J. ROSEBROUGH, A.L. FARR & R.J. RANDALL. 1951. Protein measurement with the folin phenol reagent. J. Biol. Chem. **193:** 265–275.
26. KEYHANI, J., E. KEYHANI & J. KAMALI. 2002. Thermal stability of catalases active in dormant saffron (*Crocus sativus* L.) corms. Mol. Biol. Rep. **29:** 125–128.
27. KEYHANI, E., M. VEISSIZADEH & J. KEYHANI. 2000. Kinetics of isoperoxidases and their differential sensitivity to inhibitors in saffron (*Crocus sativus* L.) bulb. *In* BioThermoKinetics 2000: Animating the cellular map. J.-H. S. Hofmeyr, J.M. Rohwer & J.L. Snoep, Eds.: 37–42. Stellenbosch University Press. Stellenbosch.
28. ATTAR, F., E. KEYHANI & J. KEYHANI. 2006. Comparative study of superoxide dismutase activity assays in *Crocus sativus* L. corms. Appl. Biochem. Microbiol. **42:** 101–106.
29. ARNOLD, R.S., J. SHI, E. MURAD, *et al.* 2001. Hydrogen peroxide mediates the cell growth and transformation caused by the mitogenic oxidase Nox1. Proc. Natl. Acad. Sci. USA **98:** 5550–5555.
30. PANTANO, C., P. SHRIVASTAVA, B. MCELHINNEY & Y. JANSSEN-HEININGER. 2003. Hydrogen peroxide signaling through tumor necrosis factor receptor I leads to selective activation of C-jun N-terminal kinase. J. Biol. Chem. **278:** 44091–44096.
31. ARBISER, J.L., J. PETROS, R. KLAFTER, *et al.* 2002. Reactive oxygen generated by Nox1 triggers the angiogenic switch. Proc. Natl. Acad. Sci. USA **99:** 715–720.
32. JANSSEN-HEININGER, Y.M., M.E. POYNTER & P.A. BAEUERLE. 2000. Recent advances towards understanding redox mechanisms in the activation of nuclear factor kappaB. Free Radic. Biol. Med. **28:** 1317–1327.
33. DAT, J.F., R. PELLINEN, T. BEECKMAN, *et al.* 2003. Changes in hydrogen peroxide homeostasis trigger an active cell death process in tobacco. Plant J. **33:** 621–632.
34. GECHEV, T.S. & J. HILLE. 2005. Hydrogen peroxide as a signal controlling plant programmed cell death. J. Cell Biol. **168:** 17–20.
35. FUKDO, T. & J. BAILEY-SERRES. 2004. Plant responses to hypoxia—is survival a balancing act ? Trends Plant Sci. **9:** 449–456.
36. SOBERMAN, R.J. 2003. The expanding network of redox signaling: new observations, complexities and perspectives. J. Clin. Investig. **111:** 571–574.

Ataxia-Telangiectasia-Mutated-Dependent Activation of Ku in Human Fibroblasts Exposed to Hydrogen Peroxide

JONG HWA LEE,[a] KYUNG HWAN KIM,[a] TOMOHIRO MORIO,[b] AND HYEYOUNG KIM[c]

[a]*Department of Pharmacology and Institute of Gastroenterology, Brain Korea 21 Project for Medical Science, College of Medicine, Yonsei University, Seoul 120-752, Korea*

[b]*Department of Pediatrics and Developmental Biology, Tokyo Medical and Dental University Graduate School, Center for Cell Therapy, Tokyo Medical and Dental University Medical Hospital, Tokyo 113-8519, Japan*

[c]*Department of Food and Nutrition, Brain Korea 21 Project, College of Human Ecology, Biomolecule Secretion Research Center, Yonsei University, Seoul 120-749, Korea*

> ABSTRACT: DNA is damaged in cells during cell replication, by infection, or by various environmental stresses. The damaged cells stop cell cycle, repair damaged DNA, and when repaired progress into the next cell cycle stage. But when the attempt to repair the damage fails, the cells undergo apoptosis. The most deleterious damage of all is double-strand DNA breaks (DSBs), where ATM (ataxia-telangiectasia-mutated) serves as a sensor. The ATM pathway culminates in DNA repair through nonhomologous end-joining or through homologous recombination. Upon DNA damage, the DNA repair protein Ku70/80 translocates into the nucleus, which may be mediated by ATM. Previously, we found that pancreatic acinar cells undergo apoptosis upon oxidative stress, and the cell death stems from nuclear loss of Ku70/80. This study aims to investigate whether ATM has a role in Ku activation and prevention of cell death induced by oxidative stress (hydrogen peroxide) using A-T fibroblasts stably transfected with human full-length ATM cDNA or empty vector. As a result, hydrogen peroxide-induced cell death was augmented in A-T cells transfected with empty vector while cell death was prevented in A-T fibroblasts stably transfected with human full-length ATM cDNA. Ku DNA-binding activity induced by hydrogen peroxide treatment was increased in the A-T fibroblasts stably transfected with human full-length ATM cDNA compared to that in A-T

Address for correspondence: Hyeyoung Kim, Department of Food and Nutrition, Yonsei University, College of Human Ecology, Seoul 120-749, Korea. Voice: 82-2-2123-3125; fax : 82-2-364-5781.
e-mail: kim626@yonsei.ac.kr

cells transfected with empty vector. The results suggest that ATM may be essential for Ku activation to repair DNA damage from oxidative stress and prevent cell death caused by oxidative stress.

KEYWORDS: ataxia-telangiectasia-mutated; oxidative stress; A-T fibroblasts; Ku

INTRODUCTION

DNA double-strand breaks (DSB) in mammalian cells induce a series of responses that result in cell cycle arrest, DNA repair, gene transcription, and cell death.[1–5] Four members of the phosphatidylinositol 3-kinase-like protein kinase (PIKK) family—DNA-protein kinase catalytic subunit (DNA-PKcs), ataxia-telangiectasia-mutated (ATM), ataxia-telangiectasia-mutated and Rad3-related (ATR), and human suppressor with morphogenetic effect on genitalia (hSMG-1)—are critical regulators of this process. Among PIKK family proteins, DNA-PKcs is the best known member, and is critical for nonhomologous end-joining (NHEJ) pathway in double-strand break repair.[6] The DNA-PK holoenzyme contains a large (450-kDa) catalytic subunit (DNA-PKcs), and two accessory proteins, Ku70 and Ku80, which form a heterodimer. DNA-PKcs possesses an intrinsic DNA end-binding activity that is greatly stimulated by the Ku0/Ku80 heterodimer. Reduced activity of DNA-PKcs in mice leads to a profound defect in the adaptive immune system.[7] DNA-PKcs-deficient mice demonstrate extreme hypersensitivity to ionizing radiation (IR).[8] The mechanism has been unknown, but a defect in the DSB repair system may be implicated.

A part of the clinical phenotype of A-T might be associated with increased oxidative stress.[9] Support for this idea was provided by studies that showed compelling evidence for increased oxidative stress in neuronal tissues of ATM-deficient mice[10–12] and human A-T cell lines.[13] Furthermore, overexpression of superoxide dismutase 1, which elevates hydrogen peroxide levels in ATM-deficient mice exacerbated certain features of the A-T phenotype.[14] It is not clear whether ATM itself is directly involved in sensing the increase in reactive oxygen species (ROS) or whether oxidative stress in A-T cells is associated with unrepaired DSBs continuously present in the DNA. It is also possible that ATM regulates the expression of genes whose products are involved in oxidative stress responses.

We previously showed that oxidative stress induced nuclear loss of Ku proteins and thus induces apoptosis in pancreatic acinar AR42J cells.[15] Inhibition of Ku-DNA-binding activity by the transfection of the C-terminal Ku80 (427–732) expression gene resulted in an increase in apoptotic cell death in AR42J cells. In other reports, Ku70 or Ku80 inactivation diminished the expression of the other Ku subunit (Ku80 or Ku70) and inhibited Ku DNA binding as well as DNA-PK activity either Ku70- or Ku80-deficient cells.[16]

This study aims to investigate whether ATM has a role in Ku activation and prevention of cell death induced by oxidative stress (hydrogen peroxide) using A-T fibroblasts stably transfected with human full-length ATM cDNA or empty vector.

METHODS

Cell Line and Culture Condition

A-T fibroblasts GM05849C, stably transfected with human full-length ATM cDNA (pEBS-YZ5 [+ ATM]) or empty vector (pEBS7-MOCK [+vector]), were cultured in Dulbecco's modified Eagle's medium (Invitrogen, Carlsbad, CA, USA) supplemented with 10% fetal bovine serum (Invitrogen) and antibiotics (100 units/mL penicillin and 100 μg/mL streptomycin).

Experimental Protocol

To determine whether H_2O_2 induces cell death and the alteration in Ku-DNA-binding activity in A-T fibroblasts stably transfected with human full-length ATM cDNA (+ATM) or empty vector (+vector) (Calbiochem, San Diego, CA, USA), each AT cell was treated with hydrogen peroxide. To determine cell viability, the cells were treated with hydrogen peroxide (10 or 50 μM) and cultured for 12 h. Viable cell numbers were determined by trypan blue exclusion test. The cells were incubated at 37°C for 12 h. The Ku-DNA-binding activity was determined by electrophoretic mobility shift assay (EMSA). For determination of the specific Ku-DNA binding, the competition assay was performed using AT cells without hydrogen peroxide treatment. Nuclear extracts were prepared from both AT cells, either transfected with human full-length ATM cDNA (+ATM) or empty vector (+vector). Binding reactions were performed after incubation with an unlabeled wild-type (wt) Ku oligonucleotide.

Cell Counting

Cell viability was determined directly by counting the viable cells with hemocytometer by trypan blue exclusion test (0.2% trypan blue). For this test, the cells were plated at 2×10^5 cells/well in a 24-well culture plate and then incubated for 12 h.

EMSA

The nuclear extracts (3 μg of nuclear protein) of YZ5 (+ATM) and MOCK (+vector) cells were incubated with the ^{32}P-labeled double-stranded oligonucleotide, 5'-GGGCCAAGAATCTTAGCAGTTTCGGG-3', in a buffer containing 30% glycerol, 50 mM HEPES, pH 7.9, 5 mM EDTA, 1 mM DTT,

0.3 mM KCl, and 0.01 μg/mL poly(dI-dC) at room temperature for 30 min. The reaction mixtures were subjected to electrophoretic separation at room temperature on a nondenaturing 5% acrylamide gel at 30 mA using 0.25 × Tris-borate-EDTA buffer. The gels were dried at 80°C for 2 h and exposed to the radiography film for 6–18 h at −70 °C with intensifying screens.

Statistical Analysis

The viable cell numbers were expressed as mean ± SEM. The difference in the mean value was tested by one-way analysis of variance (ANOVA) followed by Newman–Keul's test. A P-value < 0.05 was considered significant.

RESULTS

Hydrogen peroxide was added to A-T fibroblasts stably transfected with human full-length ATM cDNA (+ATM) or empty vector (+vector) to determine whether H_2O_2 induces cell death (FIG. 1) and the alteration in Ku-DNA-binding activity (FIG. 2). For cell viability determination, the cells were treated with hydrogen peroxide (10, 50 μM) and cultured for 12 h. Hydrogen peroxide-induced cell death was augmented in A-T cells transfected with empty vector,

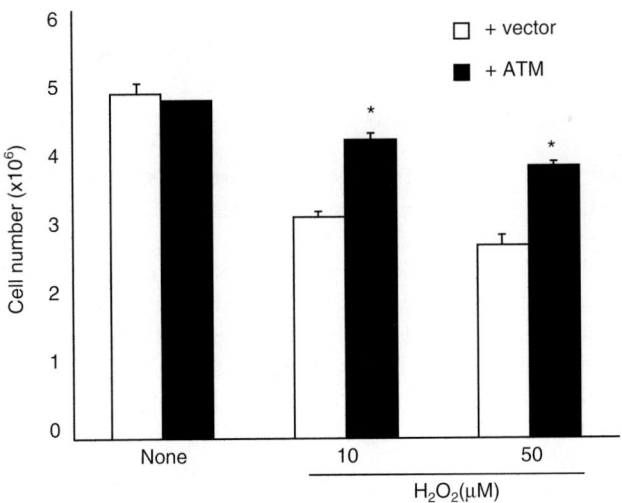

FIGURE 1. Hydrogen peroxide induced cell death in A-T fibroblasts stably transfected with human full-length ATM cDNA (+ATM) or empty vector (+vector). The cells were treated with hydrogen peroxide (10, 50 μM) and cultured for 12 h. Viable cell numbers, which were determined by trypan blue exclusion test, are expressed as mean ± SEM. *$P < 0.05$ versus the corresponding cells with empty vector (+vector).

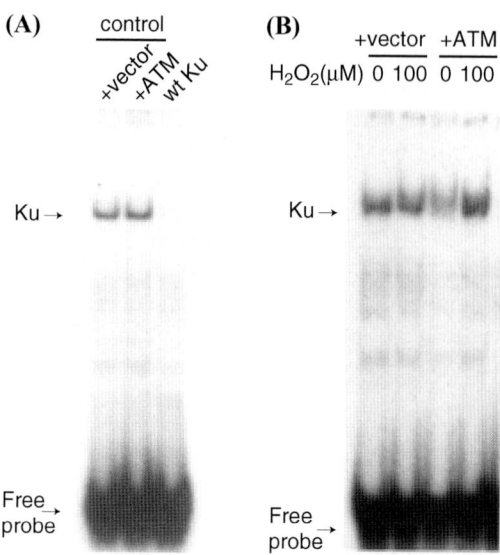

FIGURE 2. Hydrogen peroxide-induced alteration in the Ku-DNA-binding activity of A-T fibroblasts stably transfected with human full-length ATM cDNA (+ATM) or empty vector (+vector). For determination of the specific Ku-DNA binding, the competition assay was performed using A-T cells without hydrogen peroxide treatment of hydrogen peroxide (**A**). The cells were treated with hydrogen peroxide (100 μM) and cultured for 12 h. The Ku-DNA-binding activity was determined by EMSA (**B**). Nuclear extracts were prepared from both A-T cells, either transfected with human full-length ATM cDNA (+ATM) or empty vector (+vector). Binding reactions were performed after incubation with an unlabelled wt Ku oligonucleotide.

whereas cell death was prevented in A-T fibroblasts stably transfected with human full-length ATM cDNA. Hydrogen peroxide induced cell death in a dose-dependent manner. As shown in FIGURE 2, the cells were incubated at 37°C for 12 h for determination of the Ku-DNA-binding activity by EMSA. The Ku-DNA-binding activity induced by hydrogen peroxide treatment was increased in the A-T fibroblasts stably transfected with human full-length ATM cDNA compared to that in A-T cells transfected with empty vector. Ku activation in the cells with ATM gene in response to hydrogen peroxide may have a DNA repair function in the damaged cells. In contrast, A-T cells without ATM gene were sensitive to hydrogen peroxide treatment, which resulted in cell death. Prior to the experiment for the determination of the specific Ku-DNA binding, the competition assay was performed using A-T cells without hydrogen peroxide treatment of hydrogen peroxide (FIG. 2A). Nuclear extracts were prepared from both A-T cells, either transfected with human full-length ATM cDNA (+ATM) or empty vector (+vector). Binding reactions were performed after incubation with an unlabeled wt Ku oligonucleotide. Constitutive activity of Ku-DNA binding was not different between the cells transfected with human full-length ATM cDNA (+ATM) and those transfected with empty vector (+vector).

DISCUSSION

Of the many types of DNA damage that exist within the cell, probably the most dangerous are the double-strand breaks (DSBs). These result from exogenous agents, such as ionizing radiation (IR) and certain chemotherapeutic drugs, from endogenously generated ROS, and from mechanical stress on the chromosomes. They can also be produced when DNA replication forks encounter DNA single-strand breaks or other types of lesion. They can occur at the termini of chromosomes on account of the defective metabolism of chromosome ends (telomeres).[17] DNA repair proteins ATM and Ku have been linked to DNA repair pathways as cellular responses to DNA double-strand breaks.

It has been estimated that around 2×10^4 DNA damaging events occur in every cell of the human body every day.[18] A significant portion of the damage is caused by ROS. Excessive production of ROS could lead to the accumulation of chromosomal aberrations. A signaling pathway is induced by oxidative stress, which causes transfer of the repair enzymes to the nucleus, which is their site of action.[19] However, in this study, the Ku-DNA-binding activity of A-T cells was not induced in the response to oxidative stress and thus resulted in an increase in cell death. AT cells transfected with full-length ATM gene prevented cell death induced by oxidative stress, which may be contributed by the activation of Ku in response to oxidative stress. The results suggest that ATM may be essential for Ku activation to repair DNA damage upon oxidative stress and prevent cell death from exposure to oxidative stress.

ACKNOWLEDGMENT

This study was supported by a grant (F 01-2006-000-10063-0, Joint Research Project under the Korea-Japan Basic Scientific Cooperation Program) from the Korea Science and Engineering Foundation made in the program year of 2006 (to H. Kim). The study was supported by the Brain Korea 21 Project, Yonsei University College of Human Ecology.

REFERENCES

1. KASTAN, M.B. & D.S. LIM. 2000. The many substrates and functions of ATM. Nat. Rev. Mol. Cell Biol. **1:** 179–186.

2. KOBAYASHI, N., K. AGENMATSU, K. SUGITA, *et al*. 2003. Novel Atemis gene mutations of radiosensitive severe combined immunodeficiency in Japanese families. Hum. Genet. **112:** 348–352.
3. DUROCHER, D. & S.P. JACKSON. 2001. DNA-PK, ATM and ATR as sensors of DNA damage: variations on a theme? Curr. Opin. Cell Biol. **13:** 225–231.
4. LEVITT, N.C. & I.D. HICKSON. 2002. Caretaker tumor suppressor genes that defined genomic integrity. Trends Mol. Med. **8:** 179–186.
5. SHILOH, Y. 2003. ATM and related protein kinases: safeguarding genome integrity. Nat. Rev. Cancer **3:** 155–168.
6. SMITH, G.C. & S.P. JACLSON. 1999. The DNA-dependent protein kinase. Genes Dev. **13:** 916–934.
7. BLUNT, T., N.J. FINNIE, G.E. TACCIOLI & G. C. SMITH. 1995. Defective DNA-dependent protein kinase activity is linked to V(D)J recombination and DNA repair defects associated with the murine scid mutation. Cell **80:** 813–823.
8. BOSMAN, J. & A.M. CARROLL. 1991. The SCID mouse mutant: definition, characterization, and the potential uses. Annu. Rev. Immunol **9:** 323–350.
9. ROTMAN, G. & Y. SHILOH. 1997. Ataxia-telangiectasia: is ATM a sensor of oxidative damage and stress? Bioessays **19:** 911–917.
10. BARLOW, C., P.A. DENNERY, M.K. SHIEGENEGA & M.A. SMITH. 1999. Loss of the ataxia-telangiectasia gene product causes oxidative damage in target organs. Proc. Natl. Acad. Sci. USA **96:** 9915–9999.
11. KAMSLER, A., D. DAILY, A. HOCHMAN, *et al*. 2001. Increased oxidative stress in ataxia telangiectasia evidenced by alterations in redox state of brains from ATM-deficient mice. Cancer Res. **61:** 1849–1854.
12. QUICK, K.L. & L.L. DUGAN. 2001. Superoxide stress identifies neurons at risk in a model of ataxia-telangiectasia. Ann. Neurol. **49:** 627–635.
13. GATEI, M., D. SHKEDY & K.K. KHANNA. 2001. Ataxia-telangiectasia: chronic activation of damage-responsive functions is reduced by alpha-lipoic acid. Oncogene **20:** 289–294.
14. PETER, Y., G. ROTMAN, K. LOTEM & A. ELSON. 2001. Elevated Cu/Zn-SOD exacerbates radiation sensitivity and hematopoietic abnormalities of ATM-deficient mice. EMBO J. **20:** 1538–1546.
15. SONG, J.Y., J.W. LIM, H. KIM, *et al*. 2003. Oxidative stress induces nuclear loss of DNA repair proteins, Ku70 and Ku80, and apoptosis in pancreatic acinar AR42J cells. J. Biol. Chem. **278:** 36676–36687.
16. BARZILAI, A. & K. YAMAMOTO. 2004. DNA damage response to oxidative stress. DNA Repair **3:** 1109–1115.
17. KHANNA, K.K. & S.P. JACKSON. 2001. DNA double-strand breaks: signaling, repair and the cancer connection. Nat. Genet. **27:** 247–254.
18. MASER, R.S. 1997. hMere11 and hRad50 nuclear foci are induced during the normal cellular response to DNA double-strand breaks. Mol. Cell. Biol. **17:** 6087–6096.
19. ZHOU, B.B. & S.J. ELLEDGE. 2000. The DNA damage response: putting check-point in perspective. Nature **408:** 433–439.

Regulation of 2-Deoxy-D-Glucose Transport, Lactate Metabolism, and MMP-2 Secretion by the Hypoxia Mimetic Cobalt Chloride in Articular Chondrocytes

ALI MOBASHERI,[a,b] NICOLA PLATT,[a] COLIN THORPE,[a] AND MEHDI SHAKIBAEI[c]

[a]*Molecular Pathogenesis and Connective Tissue Research Groups, Department of Veterinary Preclinical Sciences, Faculty of Veterinary Science, University of Liverpool, Brownlow Hill and Crown Street, Liverpool, L69 7ZJ, United Kingdom*

[b]*Division of Veterinary Medicine, The School of Veterinary Medicine and Science, The University of Nottingham, Sutton Bonington Campus, Sutton Bonington, Leicestershire LE12 5RD, United Kingdom*

[c]*Musculoskeletal Research Group, Institute of Anatomy, Ludwig-Maximilian-University Munich, Pettenkoferstrasse 11, 80336 Munich, Germany*

ABSTRACT: Articular cartilage is an avascular tissue with significantly reduced levels of oxygen and nutrients compared to plasma and synovial fluid. Therefore, chondrocyte survival and cartilage homeostasis require effective mechanisms for oxygen and nutrient signaling. To gain a better understanding of the mechanisms responsible for oxygen and nutrient sensing in chondrocytes, we investigated the effects of hypoxic stimulation induced by cobalt chloride treatment (a hypoxiamimetic) on glucose uptake and lactate production in chondrocytes. We also studied the effects of cobalt chloride and glucose deprivation on the expression and secretion of active MMP-2. Primary cultures of articular chondrocytes were either maintained in 20% O_2 (normoxia) or exposed to the hypoxia-mimetic cobalt chloride for up to 24 h at the following concentrations: 15 μM, 37.5 μM, and 75 μM. Glucose transport was determined by measuring the net uptake of nonmetabolizable 2-deoxy-D-[2, 6-^{3}H] glucose into chondrocytes. Active MMP-2 secretion was assayed by gelatin zymography. Lactic acid production was assayed using a lactate kit. Exposure to cobalt chloride significantly increased the uptake of 2-deoxy-D-[2, 6-^{3}H] glucose and the production

Address for correspondence: Ali Mobasheri, B.Sc., A.R.C.S. (Hons), M.Sc., D.Phil. (Oxon), Division of Veterinary Medicine, The School of Veterinary Medicine and Science, The University of Nottingham, Sutton Bonington Campus, Sutton Bonington, Leicestershire LE12 5RD, United Kingdom. Voice: +44-0-115-951-6449; fax: +44-0-115-951-6415.
e-mail: ali.mobasheri@nottingham.ac.uk

Ann. N.Y. Acad. Sci. 1091: 83–93 (2006). © 2006 New York Academy of Sciences.
doi: 10.1196/annals.1378.057

of lactate. Glucose deprivation and cobalt chloride treatment increased levels of active MMP-2 in the culture medium. Our results suggest that these metabolic alterations are important events during adaptation to hypoxia. Upregulation of MMP-2 and the build-up of lactic acid will have detrimental effects on the extracellular matrix and may contribute to the pathogenesis and progression of osteoarthritis (OA).

KEYWORDS: articular cartilage; chondrocyte; hypoxia-inducible factor alpha (HIF-1α); GLUT1; GLUT3; matrix metalloproteinase-2 (MMP-2); glucose; lactate; osteoarthritis

INTRODUCTION

The maintenance of oxygen and glucose homeostasis is essential for many vital cellular functions including division, proliferation, differentiation, excitability, and secretion.[1] Oxygen and glucose homeostasis is also critical to cell fate, senescence, and apoptosis.[2,3] Cells can generally "sense" the levels of oxygen available for oxidative phosphorylation. However, only a limited number of specialized cell types can switch to anaerobic glycolysis when deprived of oxygen for extended periods of time. A significant advance in our understanding of the hypoxia response stems from the discovery of the hypoxia-inducible factor 1 (HIF-1) and related oxygen-regulated transcription factors. Depending on the duration and severity of hypoxia, cellular oxygen-sensor responses activate a variety of short- and long-term adaptations to save energy and protect cells. Transcriptional regulation of genes that maintain oxygen homeostasis is primarily mediated by HIF-1.[4] HIF-1 plays essential roles in mammalian development, physiology, and disease pathogenesis.[5] HIF-1 is a heterodimer whose α and β subunits are members of the PAS family of basic helix-loop-helix (bHLH) transcription factors.[6] HIF-1α primarily mediates responses to oxygen deprivation and prolonged hypoxia. The levels of HIF-1α increase during hypoxia, but the protein is unstable in the presence of oxygen because of the presence of an oxygen-dependent degradation domain that targets it for ubiquitination.[5,7] HIF-1α is involved in the transcriptional control of key genes involved in oxygen and glucose homeostasis including those encoding erythropoietin, glucose transporters, glycolytic enzymes, and vascular endothelial growth factor.[8]

Articular cartilage is a vascular tissue and hence the quantities of oxygen and glucose available to chondrocytes in the extracellular matrix can fluctuate considerably.[9–11] Therefore, chondrocytes must possess the ability to sense the amount of available oxygen and glucose in the matrix and respond appropriately to shortages or surpluses by altering metabolic rate. Chondrocytes consume less oxygen in comparison with most other cell types and consequently anaerobic glycolysis forms the principal source of cellular ATP in cartilage. Recent studies have clearly demonstrated the importance of HIF-1α and its target genes in the maintenance of anaerobic glycolysis[9] and cellular glucose levels[12,13] in response to hypoxia and nutrient stress.[10] The differential expression of the two HIF isoforms, HIF-1α and HIF-1β, GLUT-1 and MMP-2 in chondrocyte-like cells of the nucleus pulposus in the intervertebral disk have been used as a phenotypic and metabolic signature, allowing these cells to be distinguished from fibroblast-like cells in neighboring tissues.[10]

In light of these findings, in this study we designed and carried out a series of experiments aimed at testing the hypothesis that articular chondrocytes are also capable of adaptive metabolic responses to glucose deprivation and hypoxia mimetics, such as cobalt chloride. Other investigators have shown that this adaptation occurs by HIF-1α-mediated upregulation of genes encoding the hypoxia-responsive GLUT1 glucose transporter, which may promote glycolysis in chondrocytes from osteoarthritic cartilage.[14] Our results suggest that hypoxia mimetics (i.e., cobalt chloride) and glucose deprivation increase glucose transport, the production of lactic acid, and secretion of the active form of the matrix metalloproteinase MMP-2. We propose that in chronic hypoxia and in degenerate osteoarthritic joints these metabolic alterations will have important consequences for extracellular matrix turnover and may lead to the progression of arthritis.

MATERIALS AND METHODS

Chondrocyte Isolation and Culture

Normal equine articular cartilage was obtained from the stifle joints of horses euthanized for unrelated clinical reasons at the Philip Leverhulme Large Animal Hospital, University of Liverpool. The study was conducted with ethical approval in strict accordance with local guidelines (none of the horses were euthanized for the purpose of this study). Equine cartilage shavings were rinsed with phosphate-buffered saline (PBS), cut into small slices and incubated overnight with type I collagenase (EC 3.4.24.3 from *Clostridium histolyticum*) (approximately 100 collagen digestion units mL^{-1}) in Dulbecco's modified Eagle's medium (DMEM) supplemented with 1,000 mg L^{-1} glucose, 10% fetal calf serum, and 1% antibiotic/antimycotic solution. The filtered cell suspension was washed three times in fresh DMEM and the cells were counted on a hemocytometer; cell viability was determined by trypan blue dye exclusion and was usually 95% or higher. The cells (2×10^6 cells mL^{-1}) were cultured in monolayers for up to three passages as previously described.[13]

Monolayer cultured equine chondrocytes were used in this study because 2-deoxy-D-[2, 6-^{3}H] glucose uptake experiments can only be performed with monolayer cultured chondrocytes because the final washes in nonradioactive medium cannot be performed rapidly with cells in alginate beads. Chondrocytes used for glucose uptake, lactate production, and MMP-2 assays were not passaged more than three times to prevent chondrocyte dedifferentiation and phenotypic instability.

Experimental Design

Cobalt chloride was used as a hypoxia mimetic to simulate hypoxia by inducing HIF-1α in confluent monolayers of equine chondrocytes in 24-well plates and 75-cm^2 culture flasks. Net glucose transport was compared in normoxic and cobalt chloride–induced hypoxic conditions by measuring the uptake of nonmetabolizable 2-deoxy-D-[2, 6-^{3}H] glucose by chondrocytes and the production of lactic acid was assayed using a lactate kit. We also investigated the effects of cobalt chloride and glucose deprivation on active MMP-2 secretion by gelatin zymography. The MMP-2 assay was carried out with cells exposed to serum-free DMEM.

2-Deoxy-D-[2,6-^{3}H] Glucose Uptake

Glucose transport into equine chondrocytes was determined by measuring the uptake of nonmetabolizable 2-deoxy-D-[2, 6-^{3}H] glucose (Amersham/Pharmacia, Little Chalfont, UK) in the presence and absence of the glucose transport inhibitor cytochalasin B (1 μM mL^{-1}). Equine chondrocytes were cultured to 95% confluence in 24-well plates in DMEM containing 1,000 mg mL^{-1} glucose and supplemented with 4% fetal calf serum. The wells were rinsed three times with PBS before the assay was performed for 35 min at room temperature in glucose, pyruvate, and serum-free DMEM (Sigma Poole, Dorset, UK) containing 1 μCi mL^{-1} 2-deoxy-D-[2, 6-^{3}H] glucose. The wells were then washed three times with ice-cold PBS before the cells were lysed in 0.5 mL of a cell lysis solution consisting of 0.5% sodium dodecyl sulfate (SDS) and 0.5% Triton X-100 in PBS. Aliquots of cell lysates (0.45 mL) were transferred to 5-mL scintillation vials containing 3.55 mL of NACS104 scintillation cocktail for aqueous samples (Amersham/Pharmacia, Little Chalfont, UK). The uptake of the radiolabeled glucose was normalized to total cell protein content using a Bio-Rad detergent–compatible (DC) protein assay. All uptake experiments were carried out in triplicate and repeated under identical conditions at least three times. The data are presented as the percentage change in total 2-deoxy-D-[2, 6-^{3}H] glucose uptake.

Lactate Assays

Lactate assays were carried out on chondrocyte culture supernatants using a lactate kit from Sigma Diagnostics. The assay was performed using Sigma Lactate Reagent (Catalogue No. 735-10) following the manufacturer's instructions in the Sigma Lactate Assay Procedure No. 735. The optical density readings were obtained using a Novaspec II Visible Spectrophotometer at 540 nm. A 40 mgL^{-1} solution of lactic acid (Sigma) was used as a standard for calibration.

MMP-2 Zymography

The effects of cobalt chloride and glucose deprivation on the secretion of active MMP-2 was determined by gelatin zymography. Chondrocytes used in these experiments were incubated with serum-free DMEM culture medium in 24-well plates in the above conditions. At the end of the 24-h incubation period, aliquots of the culture medium supernatant were subjected to SDS-PAGE and zymography as recently described. The zymogram gels were scanned using a flatbed scanner and the densities of active MMP-2 bands quantified using Scion Image for Windows (version 4.0.2).

Statistical Analysis

Comparisons between experimental and control groups were made by means of the Student's *t*-test using GraphPad Instat software (version 3.05).

RESULTS

Cobalt Chloride Treatment Increases the Net Uptake of 2-Deoxy-D-[2,6-^{3}H] Glucose

The hypoxia-mimetic cobalt chloride is known to stimulate the expression of genes responsive to reduced oxygen tension. Concentrations ranging between 12 and 75 μM are tolerated by cells *in vitro* for up to 24 h without any cytotoxic effects, but concentrations higher than 250 μM are toxic.[15] Therefore, we used cobalt chloride at concentrations of 15, 37.5, and 75 μM and compared the uptake of 2-deoxy-D-[2, 6-^{3}H] glucose uptake by untreated, control equine chondrocytes to cells incubated for 24 h with increasing concentrations of cobalt chloride. The net uptake of 2-deoxy-D-[2, 6-^{3}H] glucose was significantly higher in chondrocytes incubated with cobalt chloride and the highest increase was seen with 37.5 μM cobalt chloride (FIG. 1A).

Cobalt Chloride Treatment Increases the Production of Lactate

Another cellular effect of cobalt chloride treatment is the stimulation of glycolysis. A byproduct of increased glycolysis is elevated production of lactate. We assayed the production of lactic acid by monolayer cultured equine chondrocytes in 24-well plates. The production of lactate in culture supernatants of control chondrocytes was compared with cells incubated for 24 h

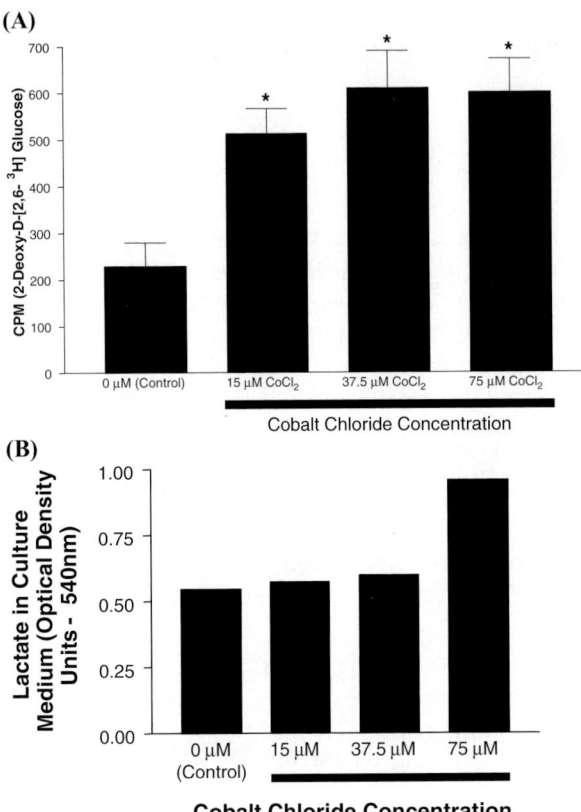

FIGURE 1. (A) Effects of the hypoxia-mimetic cobalt chloride on the uptake of 2-deoxy-D-[2,6-^{3}H] glucose by monolayer cultured equine chondrocytes in 24-well plates. The uptake of 2-deoxy-D-[2,6-3H] glucose in control chondrocytes was compared with cells incubated for 24 h with increasing concentrations of cobalt chloride (15, 37.5, and 75 μM). The net uptake of 2-deoxy-D-[2,6-^{3}H] glucose was significantly higher in chondrocytes incubated with cobalt chloride compared to control cells. The highest increase was seen with 37.5 μM cobalt chloride. *Error bars* indicate standard errors of the means ($n = 3$). In cases where a statistically significant difference between an experimental group and the control group was found, the bar is labeled with *. (B) Effects of cobalt chloride on the production of lactic acid by monolayer cultured equine chondrocytes in 24-well plates. The production of lactate in culture supernatants of control chondrocytes was compared with cells incubated for 24 h with increasing concentrations of cobalt chloride (15, 37.5, and 75 μM). Lactic acid production was higher in chondrocytes incubated with cobalt chloride compared to control cells and the effect seemed to plateau at a concentration of 75 μM cobalt chloride.

with increasing concentrations of cobalt chloride (15, 37.5, and 75 μM). Lactic acid production was higher in chondrocytes incubated with cobalt chloride compared to control cells and the effect was maximal at a concentration of 75 μM cobalt chloride (FIG. 1B).

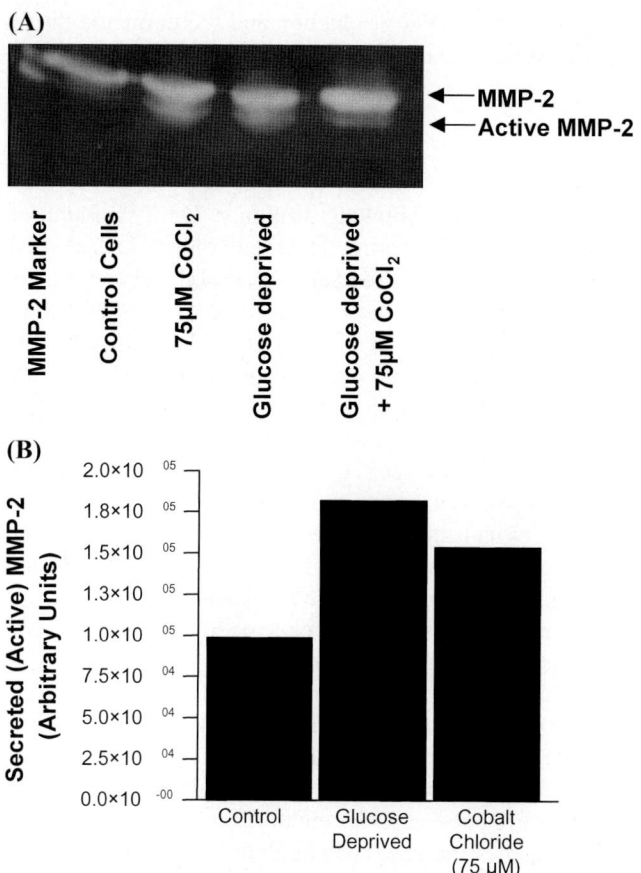

FIGURE 2. Effects of glucose deprivation and cobalt chloride on the expression of active MMP-2 secreted into the culture medium of chondrocytes. (**A**) A representative gelatin zymogram used for the quantitative analysis of active MMP-2 expression. (**B**) Glucose deprivation, exposure to 75 μM cobalt chloride, or a combination of both for periods of up to 24 h significantly increased MMP-2 production secretion by chondrocytes compared to the control group. MMP-2 is detected as two closely migrating bands on the zymogram; the lower molecular weight band corresponds to active MMP-2 and the higher molecular weight band represents inactive MMP-2.

Cobalt Chloride and Glucose Deprivation Increase Levels of Active MMP-2

We studied the effects of cobalt chloride and glucose deprivation on the expression of active MMP-2 secreted into the culture medium of chondrocytes. A typical gelatin zymogram is shown in FIGURE 2A. MMP-2 is seen as two closely migrating bands on the zymogram; the lower molecular weight band corresponds to active MMP-2. Exposure to 75 μM cobalt chloride for 24 h, glucose-free DMEM (glucose deprivation), or a combination of both significantly increased MMP-2 production and secretion and the levels of active MMP-2 compared to chondrocytes in the control cultures (FIG. 2 B).

DISCUSSION

This study leads to the following findings: (1) Treatment of chondrocyte cultures with the hypoxia mimetic cobalt chloride increases the glucose transporter-mediated uptake of 2-deoxy-D-[2, 6-^{3}H] glucose, which is suggestive of HIF-1α upregulation of GLUTs, presumably the hypoxia-responsive GLUT1 and GLUT3 proteins. (2) Cobalt chloride treatment also promotes glycolysis by increasing the production of lactic acid, which accumulates in the culture medium supernatant.

Recent research suggests that articular cartilage is a tissue routinely exposed to low oxygen tensions[9] and hypoxia.[16,17] Furthermore, the generation of reactive oxygen species (ROS) has been found to increase in joint diseases, such as OA.[18,19] Studies in other tissues have demonstrated that one of the hallmarks of hypoxia is functional activation of HIF-1α, an oxygen-sensitive transcriptional activator that mediates changes in gene expression in response to changes in oxygen levels. Under low oxygen tensions, cellular hypoxia results and ATP levels are diminished. Reduced degradation and stabilization of the HIF1-α protein increase its transcriptional activity as cellular oxygen concentrations decrease. The net effect of this is transactivation of several dozen target genes. These include the genes encoding erythropoietin (EPO), glucose transporters (GLUT1, GLUT3), glycolytic enzymes (lactate dehydrogenase, LDH; phosphofructokinase, PFK), and vascular endothelial growth factor (VEGF).[8] The products of these genes increase oxygen delivery and glycolysis to meet the metabolic demands of reduced oxygen availability.

Experimental studies from several laboratories have shown that chondrocytes express the oxygen-regulated transcription factor HIF-1α.[9,10,16] Chondrocytes also possess multiple isoforms of the GLUT/SLC2A family of glucose transporters.[11-13,20,21] Among these are the hypoxia-responsive GLUT1 and GLUT3 transporters.[12,13] We have recently proposed that the metabolic adjustment of chondrocytes to a low-oxygen, low-glucose environment in the extracellular matrix of articular cartilage involves HIF-1α as an "oxygen sensor" capable of activating hypoxia-responsive target genes involved in extracellular "glucose sensing" and uptake. GLUT1 and GLUT3 are therefore putative glucose sensors in chondrocytes as well as targets of HIF-1α.[22,23] It is likely that other HIF-1α gene targets concerned with anaerobic glycolysis (i.e., lactate dehydrogenase and phosphofructokinase)[24-26] may also be involved in glucose sensing in chondrocytes.

Inflammation is a process that is closely integrated in the pathogenesis and progression of osteoarthritis.[27] Notably, the synovial fluid of most OA patients has an increased number of mononuclear cells (mainly macrophages and T lymphocytes) and increased levels of immunoglobulins, complement proteins, and inflammatory cytokines. Synovial macrophages are known to accumulate in OA joints, especially when the synovium has been activated by proinflammatory cytokines and degraded components of the extracellular matrix of the damaged cartilage.[28] This activation of synovial macrophages also activates synovial fibroblasts resulting in the generation of a broad range of catabolic factors.[29] Recent studies suggest that hypoxia alters the phenotype of macrophages in a manner that promotes the progression of the inflammatory process. As commonly observed in other cell types, GLUT1 and matrix metalloproteinase-7 (MMP-7) are upregulated by hypoxia in macrophages.[29] Hypoxic upregulation of these proteins suggests that hypoxia may affect the functioning of synovial macrophages in OA joint tissues. The results of this study suggest that hypoxic conditions also result in overexpression of MMP-2 in chondrocytes, which may promote hypoxia-mediated degradation of the extracellular matrix in cartilage. Inflammation can stimulate angiogenesis, and angiogenesis can facilitate inflammation by facilitating innervation and altering tissue oxygenation.

In summary, our results suggest that chondrocyte adaptation to hypoxia occurs by metabolic alterations including enhancement of the glucose transporting capacity of the cells, increased glycolysis, and lactate production. These processes are likely to be HIF-1α-mediated and reversible. Furthermore, we have shown that hypoxia and glucose deprivation increase the production of lactic acid and production of the active form of the matrix metalloproteinase MMP-2. Upregulation of MMP-2 and the build-up of lactate will have detrimental effects on the extracellular matrix. We propose that chronic hypoxia may occur in structurally degenerated osteoarthritic joints and the consequent metabolic alterations may contribute to the pathogenesis and progression of osteoarthritis. Future studies will determine the effects of hypoxia on other key metabolites and biochemical pathways in chondrocytes in health and disease.

REFERENCES

1. ROLLAND, F., J. WINDERICKX & J.M. THEVELEIN. 2001. Glucose-sensing mechanisms in eukaryotic cells. Trends Biochem. Sci. **26:** 310–317.
2. NEMOTO, S., M.M. FERGUSSON & T. FINKEL. 2004. Nutrient availability regulates SIRT1 through a forkhead-dependent pathway. Science **306:** 2105–2108.
3. MARTENS, G. et al. 2005. Nutrient sensing in pancreatic beta cells suppresses mitochondrial superoxide generation and its contribution to apoptosis. Biochem. Soc. Trans. **33:** 300–301.
4. SEMENZA, G.L. 1998. Hypoxia-inducible factor 1 and the molecular physiology of oxygen homeostasis. J. Lab. Clin. Med. **131:** 207–214.
5. SEMENZA, G.L. 2001. HIF-1 and mechanisms of hypoxia sensing. Curr. Opin. Cell Biol. **13:** 167–171.

6. WANG, G.L. et al. 1995. Hypoxia-inducible factor 1 is a basic-helix-loop-helix-PAS heterodimer regulated by cellular O2 tension. Proc. Natl. Acad. Sci. USA **92:** 5510–5514.
7. SUTTER, C.H., E. LAUGHNER & G.L. SEMENZA. 2000. Hypoxia-inducible factor 1alpha protein expression is controlled by oxygen-regulated ubiquitination that is disrupted by deletions and missense mutations. Proc. Natl. Acad. Sci. USA **97:** 4748–4753.
8. SEMENZA, G.L. 1999. Regulation of mammalian O2 homeostasis by hypoxia-inducible factor 1. Annu. Rev. Cell. Dev. Biol. **15:** 551–578.
9. PFANDER, D. et al. 2003. HIF-1alpha controls extracellular matrix synthesis by epiphyseal chondrocytes. J. Cell Sci. **116:** 1819–1826.
10. RAJPUROHIT, R. et al. 2002. Phenotypic characteristics of the nucleus pulposus: expression of hypoxia inducing factor-1, glucose transporter-1 and MMP-2. Cell Tissue Res. **308:** 401–407.
11. MOBASHERI, A. et al. 2002. Glucose transport and metabolism in chondrocytes: a key to understanding chondrogenesis, skeletal development and cartilage degradation in osteoarthritis. Histol. Histopathol. **17:** 1239–1267.
12. MOBASHERI, A. et al. 2002. Human articular chondrocytes express three facilitative glucose transporter isoforms: GLUT1, GLUT3 and GLUT9. Cell Biol. Int. **26:** 297–300.
13. RICHARDSON, S. et al. 2003. Molecular characterization and partial cDNA cloning of facilitative glucose transporters expressed in human articular chondrocytes; stimulation of 2-deoxyglucose uptake by IGF-I and elevated MMP-2 secretion by glucose deprivation. Osteoarthritis Cartilage **11:** 92–101.
14. PFANDER, D., T. CRAMER & B. SWOBODA. 2005. Hypoxia and HIF-1alpha in osteoarthritis. Int. Orthop. **29:** 6–9.
15. BEHROOZ, A. & F. ISMAIL-BEIGI. 1997. Dual control of glut1 glucose transporter gene expression by hypoxia and by inhibition of oxidative phosphorylation. J. Biol. Chem. **272:** 5555–5562.
16. SCHIPANI, E. et al. 2001. Hypoxia in cartilage: HIF-1alpha is essential for chondrocyte growth arrest and survival. Genes Dev. **15:** 2865–2876.
17. LEE, R.B. & J.P. URBAN. 1997. Evidence for a negative Pasteur effect in articular cartilage. Biochem. J. **321**(Pt 1): 95–102.
18. HENROTIN, Y.E., P. BRUCKNER & J.P. PUJOL. 2003. The role of reactive oxygen species in homeostasis and degradation of cartilage. Osteoarthritis Cartilage **11:** 747–755.
19. HENROTIN, Y., B. KURZ & T. AIGNER. 2005. Oxygen and reactive oxygen species in cartilage degradation: friends or foes? Osteoarthritis Cartilage **13:** 643–654.
20. SHIKHMAN, A.R. et al. 2001. Cytokine regulation of facilitated glucose transport in human articular chondrocytes. J. Immunol. **167:** 7001–7008.
21. SHIKHMAN, A.R., D.C. BRINSON & M.K. LOTZ. 2004. Distinct pathways regulate facilitated glucose transport in human articular chondrocytes during anabolic and catabolic responses. Am. J. Physiol. Endocrinol. Metab. **286:** E980–E985.
22. VANNUCCI, S.J., L.B. SEAMAN & R.C. VANNUCCI. 1996. Effects of hypoxia-ischemia on GLUT1 and GLUT3 glucose transporters in immature rat brain. J. Cereb. Blood Flow Metab. **16:** 77–81.
23. BADR, G.A. et al. 1999. Glut1 and glut3 expression, but not capillary density, is increased by cobalt chloride in rat cerebrum and retina. Brain Res. Mol. Brain Res. **64:** 24–33.

24. SEMENZA, G.L. *et al.* 1997. Structural and functional analysis of hypoxia-inducible factor 1. Kidney Int. **51:** 553–555.
25. SEMENZA, G.L. *et al.* 1996. Hypoxia response elements in the aldolase A, enolase 1, and lactate dehydrogenase A gene promoters contain essential binding sites for hypoxia-inducible factor 1. J. Biol. Chem. **271:** 32529–32537.
26. SEMENZA, G.L. *et al.* 1994. Transcriptional regulation of genes encoding glycolytic enzymes by hypoxia-inducible factor 1. J. Biol. Chem. **269:** 23757–23763.
27. BONNET, C.S. & D.A. WALSH. 2005. Osteoarthritis, angiogenesis and inflammation. Rheumatology (Oxford). **44:** 7–16.
28. HENROTIN, Y., C. SANCHEZ & M. BALLIGAND. 2005. Pharmaceutical and nutraceutical management of canine osteoarthritis: present and future perspectives. Vet. J. **170:** 113–123.
29. BURKE, B. *et al.* 2003. Hypoxia-induced gene expression in human macrophages: implications for ischemic tissues and hypoxia-regulated gene therapy. Am. J. Pathol. **163:** 1233–1243.

Oxidative Stress Response in Telomerase-Immortalized Fibroblasts from a Centenarian

CHIARA MONDELLO,[a] MARIA GRAZIA BOTTONE,[a,b] SAKON NORIKI,[c] CRISTIANA SOLDANI,[b] CARLO PELLICCIARI,[b] AND ANNA IVANA SCOVASSI[a]

[a]*Istituto di Genetica Molecolare del CNR, Via Abbiategrasso 207, 27100 Pavia, Italy*

[b]*Dipartimento di Biologia Animale, Piazza Botta 10, 27100, Pavia, Italy*

[c]*Department of Oncological Pathology, Faculty of Medicine, Matsuoka, Yoshida-Gun, 910-1193, Fukui, Japan*

> ABSTRACT: It has been reported that cells with ectopic expression of telomerase are more resistant to apoptotic cell death than their normal counterpart. However, controversial results were obtained when the cellular response to oxidative stress was analyzed. The present research was therefore aimed at defining the effect of the oxidative stress induced by *tert*-butylhydroperoxide (tBOOH) and 2-deoxy-D-ribose (D-ribose) in human fibroblasts from a centenarian (cen3) and, in parallel, on the same cells after telomerase immortalization (cen3tel cells). By studying different parameters of apoptosis *in situ* (i.e., chromatin condensation, phosphatidylserine externalization, and DNA fragmentation), we found that both tBOOH and D-ribose induce apoptosis to a greater extent in cen3 than in cen3tel cells, suggesting a protective role of telomerase toward apoptotic death. However, monitoring the cell number during treatment with the drugs, we found a decrease in cell number; since this reduction was lower in cen3 fibroblasts compared to cen3tel cells, it is likely that telomerase does not fully protect cells from drug toxicity.
>
> KEYWORDS: apoptosis; oxidative stress; phagocytosis; telomerase; telomeres

INTRODUCTION

Telomeres, which constitute the ends of linear eukaryotic chromosomes, are formed by short tandemly repeated units (in humans, the hexamer TTAGGG),

Address for correspondence: Chiara Mondello, Istituto di Genetica Molecolare del CNR, Via Abbiategrasso 207, 27100 Pavia, Italy. Voice: +39-0382-546332; fax: +39-0382-422286.
 e-mail: mondello@igm.cnr.it

and are elongated by the specialized enzyme telomerase. This is a reverse transcriptase enzyme, with a protein catalytic subunit (TERT) and an RNA component containing the template sequence. Telomerase activity is almost undetectable in normal human somatic cells and its activity is not sufficient to maintain telomeres, which thus shorten at each cell division.[1–3]

Human somatic cells in culture have a limited life span, and after a finite number of cell divisions they enter a state of irreversible proliferative arrest, known as cellular senescence. It is widely accepted that shortening of telomeres plays a crucial role in triggering cellular senescence: telomeres falling below a threshold length are not recognized as chromosome ends anymore, but as DNA double-strand breaks, and activate cell cycle checkpoints. The causal relationship between telomere shortening and senescence has been demonstrated by showing that ectopic expression of the cDNA for human TERT (h*TERT*) allows several types of somatic cells, including fibroblasts, to maintain telomere functions and to escape replicative senescence, becoming immortal.[4–7] It has recently been shown that cells with ectopic expression of telomerase are more resistant to apoptotic cell death induced by different stimuli than their normal counterpart. However, controversial results were obtained when the response to oxidative stress was analyzed (reviewed in Ref. 8).

In this study, we investigated the effects of oxidative stress on fibroblasts from a centenarian individual (cen3) and on their counterpart immortalized by ectopic telomerase expression (cen3tel); two inducers of stress-mediated apoptosis have been used, that is, *tert*-butylhydroperoxide (tBOOH), a lipid hydroperoxide analogue that induces apoptosis *via* peroxynitrite-dependent and -independent DNA single-strand breakage,[9] and 2-deoxy-D-ribose (D-ribose), a reducing sugar that causes apoptosis by the disruption of actin filaments and GSH depletion.[10] We analyzed the effects of telomerase expression on cell survival and apoptosis.

MATERIALS AND METHODS

Cell Culture and Treatments

Cen3 fibroblasts and cen3tel cells, which were immortalized by ectopic telomerase expression, were previously described.[11–13] Cen3 fibroblasts were used around population doubling (PD) 18, whereas cen3tel cells around PD 30, that is, after about 13 PDs since the introduction of the cDNA for h*TERT*. Cells were grown either in petri dishes or on glass coverslips and maintained at 37°C in a humidified atmosphere containing 5% CO_2. Cells were treated with 25 mM D-ribose (Sigma-Aldrich, St. Louis, MO, USA) for 48 h, or with 0.1 mM tBOOH (Sigma) for 1 h. In cells treated with D-ribose, apoptosis was analyzed at the end of the treatment, whereas in cells treated with tBOOH, after 24 h of recovery in fresh medium. Experiments were carried out in triplicate. To determine the effect of the drug on cell survival, 10^5 cells of each cell line were seeded in 3-cm petri dishes and the cell number was counted at the beginning of the treatment and after different times; for each time point, the percentage of the cells recovered in treated samples compared to those found in controls was calculated.

Detection of Apoptosis

Dual staining by Annexin V and propidium iodide (PI). To detect early and late apoptotic cells, control and treated cells were detached by trypsinization. Then the samples were incubated for 15 min at r.t. with FITC-conjugated Annexin V in complete culture medium (3 μL/10^6 cells; Bender MedSystem, Prodotti Gianni, Italy) and 2 μg/mL PI according to the procedure reported by Pellicciari et al.[14] The cells were washed with Ca^{2+}-containing medium to preserve the Annexin binding, and observed by fluorescence microscope.

TUNEL procedure to detect DNA fragmentation. The cells grown on glass coverslips were fixed with 70% ethanol for 30 min. After rehydration with phosphate-buffered saline (PBS), the cells were incubated for 1 h at 37°C with 50 μL of the TUNEL mixture according to the manufacturer's instructions (Boehringer Mannheim, Germany) and following the procedure reported in Soldani et al.[15] After washing with PBS the cells were counterstained for cytoplasm with Evan's blue (0.1% in PBS) and for DNA with Hoechst 33258 (0.1 μg/mL in PBS for 10 min), washed with PBS and mounted on a nonfluorescent glass slide in a drop of Mowiol (Calbiochem, Inalco S.p.A., Italy).

Fluorescence microscopy and digital photomicrography. The stained slides were observed in fluorescence microscopy with an Olympus BX50 microscope equipped with a 100-W mercury lamp. The following conditions were used: 330- to 385-nm excitation filter (excf), 400-nm dichroic mirror (dm), and 420-nm barrier filter (bf), for Hoechst 33258; 450- to 480-nm excf, 500-nm dm, and 515-nm bf for FITC; and 540-nm excf, 580-nm dm, and 620-nm bf for Evan's blue and rhodamine. Images were recorded with an Olympus Camedia C-5050 digital camera and stored on a PC by the Olympus software, for processing and printing.

RESULTS AND DISCUSSION

Telomerase-immortalized cen3tel cells were studied around PD 30. At this stage of propagation, cen3tel behaved like the cen3 parental fibroblasts: they were characterized by a normal fibroblastic morphology, displayed contact growth inhibition, expressed p16^{INK4A}, and showed a functional p53.[13]

We analyzed apoptotic hallmarks, including phosphatidylserine (PS) externalization, loss of membrane integrity, and DNA degradation in cen3 and cen3tel cells treated with tBOOH or D-ribose. The double staining with Annexin V/PI made it possible to identify different cell types, that is, cells with early apoptotic signs (green fluorescence, corresponding to PS exposure), cells at an intermediate step of apoptosis (green/red fluorescence, accounting for cells which were also permeable to PI) and red, late apoptotic cells. The relatively late event of apoptotic DNA degradation was also monitored *in situ* by the TUNEL assay. Apoptotic cells were observed in both treated cell lines, thus suggesting that the drugs may activate a classical apoptotic pathway, in agreement with the data on the response to D-ribose of blood cells from centenarians.[16,17] Interestingly, we observed the removal of apoptotic cells by adjacent fibroblasts through phagocytosis (FIG. 1).

The apoptotic index (i.e., the percentage of apoptotic cells) is reported in TABLE 1, and indicates that D-ribose induces apoptosis at a higher extent than tBOOH. The values obtained with TUNEL assay were always lower that those calculated from Annexin V/PI staining; this discrepancy can be explained by the fact that DNA degradation occurs at a later time during apoptosis, whereas the positivity for Annexin V staining accounts for a much more precocious apoptotic event. The apoptotic index was lower in cen3tel cells compared to cen3 fibroblasts, suggesting that telomerase reactivation in cen3tel cells protects cells from oxidative stress-induced apoptosis.

These results suggest that induction of telomerase expression in normal fibroblasts is associated with a protection against apoptosis. Our conclusions are in agreement with the observation that telomerase-immortalized human lung fibroblasts treated with $FeSO_4$ and H_2O_2 under conditions that generate hydroxyl radicals show an increased resistance to apoptosis, compared to parental cells.[18] In contrast, Gorbunova *et al.*[19] found no protective effect against H_2O_2-induced apoptosis in telomerase-expressing human lung fibroblasts. The possible complexity of the apoptotic response mediated by ectopic telomerase expression is proven by the experiments carried out in human erythroleukemia K562 cells overexpressing telomerase, in which protection against apoptosis induced by serum deprivation, ionizing radiation, and etoposide was found, whereas no protection was observed after treatment with the tyrosine kinase inhibitor herbimycin A and DNA synthesis inhibitors, that is, Ara-C and hydroxyurea.[20] However, it has to be considered that parental K562 cells already express telomerase and the level of telomerase in parental cells could be sufficient to protect cells from oxidative stress-induced apoptosis.

A decrease in cell number was observed in parallel with the occurrence of apoptotic cell death; this decrease was greater during treatment with tBOOH than with D-ribose (FIG. 2). Upon treatment with D-ribose, a similar trend in cell number reduction was observed in cen3 and cen3tel cells; in contrast, after treatment with tBOOH, cell loss was greater in cen3tel than in cen3 cells.

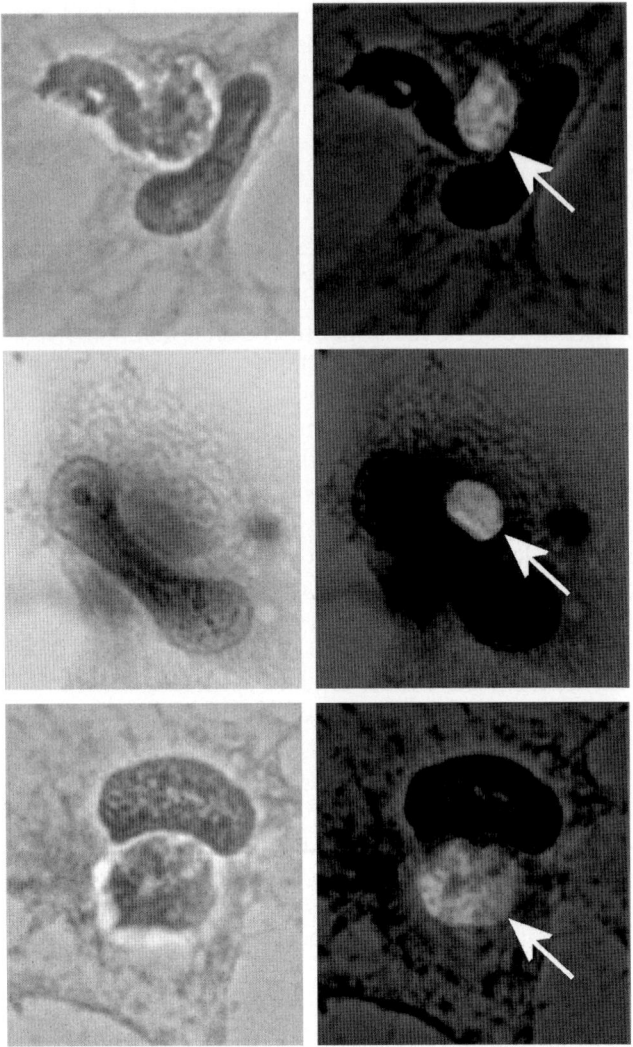

FIGURE 1. Phagocytosis of apoptotic cells as observed in cen3 and cen3tel cell cultures after treatment with D-ribose. *Left*: conventional morphological staining with hematoxylin and eosin of cell samples previously processed for TUNEL. *Right*: the nuclei of phagocytosed apoptotic cells (*arrows*) are TUNEL-positive (the micrographs were taken after setting the microscope for simultaneous observation in fluorescence and bright field at low illumination).

The greater effect of this drug on cen3tel cells might be related to the higher proliferation rate of these cells compared to cen3 fibroblasts. Phagocytosis, which we clearly detected in treated cells (FIG. 1), together with alternative pathways of cell death that might have taken place in addition to apoptosis,[21]

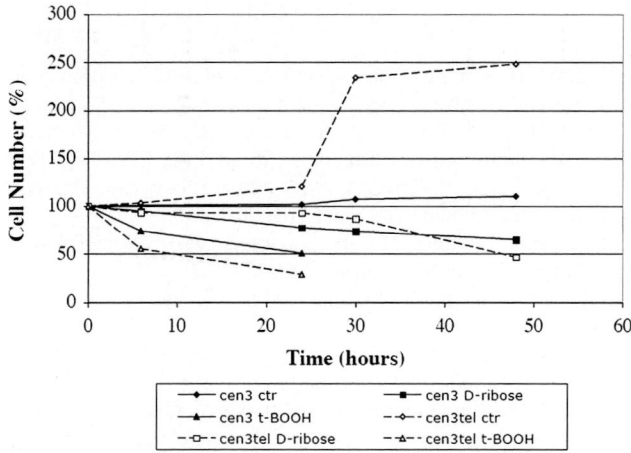

FIGURE 2. Effect of oxidant drugs on cen3 and cen3tel cells. Cells were treated with 0.1 mM tBOOH for 1 h and further recovered in fresh medium for 24 h; cells were counted before adding the drug (starting time: 0 h) and at 6 h and 24 h of recovery. For D-ribose (25 mM for 48 h)-treated cells, the analysis was performed at 0 h, 6 h, 24 h, 30 h, and 48 h of treatment. The analysis was carried out in parallel on untreated samples. For each time, the number of cells counted in each sample is expressed as a percentage of the number of cells at the starting time. Results of a typical experiment are shown.

could account for at least a part of the cell loss we observed. On the whole, our data suggest that telomerase expression does not fully protect cells against drug toxicity.

ACKNOWLEDGMENTS

The research at the laboratory of C.M. is supported by the EU grant FIGHT-CT 2002-217 and by the Ministero Italiano Istruzione, Università e Ricerca (FIRB projects RBNE01RNN7 and RBAU01ZB78), at the AIS laboratory by the FIRB project RBNE0132MY. This research was also supported by the MIUR PRIN Project 2005058254 (C.P.).

TABLE 1. Apoptotic index in control and treated cen3 and cen3tel cells

	Control		tBOOH		D-ribose	
Cell line	Annexin+	TUNEL+	Annexin+	TUNEL+	Annexin+	TUNEL+
cen3	2.2 ± 1.1	1.3 ± 0.06	37.4 ± 2.4	25.8 ± 0.6	51.1 ± 4.0	47.2 ± 5.8
cen3tel	1.5 ± 0.2	1.0 ± 0.05	15.3 ± 4.3	8.3 ± 0.8	35.1 ± 3.0	19.1 ± 1.2

Data are derived from three independent experiments (mean ± SD).

REFERENCES

1. CECH, T.R. 2004. Beginning to understand the end of the chromosome. Cell **116**: 273–279.
2. BLACKBURN, E.H. 2005. Telomeres and telomerase: their mechanisms of action and the effects of altering their functions. FEBS Lett. **579**: 859–862.
3. MASUTOMI, K., E.Y. YU, S. KHURTS, et al. 2003. Telomerase maintains telomere structure in normal human cells. Cell **114**: 241–253.
4. HARLEY, C.B., A.B. FUTCHER & C.W. GREIDER. 1990. Telomeres shorten during ageing of human fibroblasts. Nature **345**: 458–460.
5. DE LANGE, T. 1994. Activation of telomerase in a human tumor. Proc. Natl. Acad. Sci. USA **91**: 2882–2885.
6. BODNAR, A.G., M. OUELLETTE, M. FROLKIS, et al. 1998. Extension of life-span by introduction of telomerase into normal human cells. Science **279**: 349–352.
7. HARLEY, C.B. 2002. Telomerase is not an oncogene. Oncogene **21**: 494–502.
8. MONDELLO, C. & A.I. SCOVASSI. 2004. Telomeres, telomerase and apoptosis. Biochem. Cell Biol. **82**: 498–507.
9. PALOMBA, L., A. GUIDARELLI, A.I. SCOVASSI & O. CANTONI. 2001. Different effects of *tert*-butylhydroperoxide-induced peroxynitrite-dependent and -independent DNA single-strand breakage on PC12 cell poly(ADP-ribose) polymerase activity. Eur. J. Biochem. **268**: 5223–5228.
10. KLETSAS, D., D. BARBIERI, D. STATHAKOS, et al. 1998. The highly reducing sugar 2-deoxy-D-ribose induces apoptosis in human fibroblasts by reduced glutathione depletion and cytoskeletal disruption. Biochem. Biophys. Res. Commun. **243**: 416–425.
11. MONDELLO, C., C. PETROPOULOU, D. MONTI, et al. 1999. Telomere length in fibroblasts and blood cells from healthy centenarians. Exp. Cell Res. **248**: 234–242.
12. MONDELLO, C., M. CHIESA, P. REBUZZINI, et al. 2003. Karyotype instability and anchorage-independent growth in telomerase-immortalized fibroblasts from two centenarian individuals. Biochem. Biophys. Res. Commun. **308**: 914–921.
13. ZONGARO, S., E. DE STANCHINA, T. COLOMBO, et al. 2005. Stepwise neoplastic transformation of telomerase immortalized human fibroblasts. Cancer Res. **65**: 11411–11418.
14. PELLICCIARI, C., M.G. BOTTONE, A.I. SCOVASSI, et al. 2000. Rearrangement of nuclear ribonucleoproteins and extrusion of nucleolus-like bodies during apoptosis induced by hypertonic stress. Eur. J. Histochem. **44**: 247–254.
15. SOLDANI, C., A.I. SCOVASSI, U. CANOSI, et al. 2005. Multicolor fluorescence technique to detect apoptotic cells in advanced coronary atherosclerotic plaques. Eur. J. Histochem. **49**: 47–52.
16. MONTI, D., S. SALVIOLI, M. CAPRI, et al. 2000. Decreased susceptibility to oxidative stress-induced apoptosis of peripheral blood mononuclear cells from healthy elderly and centenarians. Mech. Aging Devel. **121**: 239–250.
17. BONAFÈ, M., S. SALVIOLI, C. BARBI, et al. 2004. The different apoptotic potential of the p53 codon 72 alleles increases with age and modulates *in vivo* ischaemia-induced cell death. Cell Death Differ. **11**: 962–973.
18. REN, J-G, XIA H-L, TIAN Y-M et al. 2001. Expression of telomerase inhibits hydroxyl radical-induced apoptosis in normal telomerase negative human lung fibroblasts. FEBS Lett. **488**: 133–138.

19. GORBUNOVA, V., A. SELUANOV & O.M. PEREIRA-SMITH. 2002. Expression of human telomerase (hTERT) does not prevent stress-induced senescence in normal human fibroblasts but protects the cells from stress-induced apoptosis and necrosis. J. Biol. Chem. **277:** 38540–38549.
20. AKIYAMA, M., O. YAMADA, N. KANDA, *et al.* 2002. Telomerase overexpression in K562 leukemia cells protects against apoptosis by serum deprivation and double-stranded DNA break inducing agents, but not against DNA synthesis inhibitors. Cancer Lett. **178:** 187–197.
21. GUIMARÃES, C.A. & R. LINDEN. 2004. Programmed cell death. Apoptosis and alternative cell death styles. Eur. J. Biochem. **271:** 1638–1650.

Differential Modulation of AMPK Signaling Pathways by Low or High Levels of Exogenous Reactive Oxygen Species in Colon Cancer Cells

IN-JA PARK,[a] JIN-TAEK HWANG,[b] YOUNG MIN KIM,[c] JOOHUN HA,[b] AND OCK JIN PARK[a]

[a]*Department of Food and Nutrition, Hannam University, 133 Ojeong-dong Daedeok-gu, Daejeon 306-791, Korea*

[b]*Department of Biochemistry and Molecular Biology, Medical Research Center for Bioreaction to Reactive Oxygen Species, Kyung Hee University College of Medicine, Seoul 130-701, Korea*

[c]*Department of Biological Sciences, Hannam University, 133 Ojeong-dong Daedeok-gu, Daejeon 306-791, Korea*

> ABSTRACT: This study was undertaken to examine the effect of low and high concentrations of H_2O_2 on cancer cell proliferation and apoptosis, and AMPK signaling pathways in HT-29 human colon cancer cells. Nontoxic doses of H_2O_2 (10 μM) induced cancer cell proliferation, whereas the toxic level of 1,000 μM H_2O_2 induced apoptosis. The stimulation of cell proliferation was accompanied with an increase in cyclooxygenase-2 (COX-2), and apoptosis induced by high-dose H_2O_2 was correlated with the activation of AMPK and negatively correlated with COX-2 expression. These results suggest that ROS at nontoxic levels can stimulate cancer cell growth by regulating AMP-activated protein kinase (AMPK) and/or COX-2, and the abundant exogenous ROS linked to the growth inhibition through modulating AMPK signaling pathways.
>
> KEYWORDS: reactive oxygen species; low-dose H_2O_2; high-dose H_2O_2; AMP-activated protein kinase; cyclooxygenase-2; HT-29 colon cancer cells

INTRODUCTION

Reactive oxygen species (ROS), oxygen-containing free radical molecules, are known as physiological byproducts of normal aerobic metabolism. The

---

Address for correspondence: Ock Jin Park, Department of Food and Nutrition, Hannam University, 133 Ojeong-dong Daedeok-gu, Daejeon 306-791, Korea. Voice: +82-42-629-7493; fax: +82-42-629-7490.
 e-mail: ojpark@hannam.ac.kr

most common ROS include superoxide anions (O^{2-}), hydrogen peroxide (H_2O_2), hydroxyl radicals ($HO°$), and oxides of nitrogen (NO_x). The predominant source of O^{2-} is from leakage through the mitochondrial electron transport chain,[1] but other enzymatic systems of NADPH oxidase or xanthine oxidase produce superoxide anions. Peroxisomes also possess several H_2O_2-generating enzymes, but most H_2O_2 is detoxified by peroxisomal catalase. The hydroxyl radicals are formed through the Fenton or Haber–Weiss reaction converting O^{2-} and H_2O_2 into $HO°$ in the presence of Fe^{2+} or Cu^+. Sporadic generation of ROS can be found in the host defense system of macrophages against invading microorganisms, where significant quantities of ROS are generated. ROS production can be also detected in non-phagocytic systems, such as human tumor cells. ROS have been regarded as undesirable metabolic byproducts often linked to macromolecule damage, aging, or degenerative diseases.[2–5] However, a physiological rate of ROS production activates cellular signaling pathways necessary for cell growth and proliferation. Some evidence suggests that ROS play signaling roles in physiologic or pathophysiologic processes of cell proliferation, adhesion, and atherosclerosis. Moreover, ROS emerge as important intracellular signaling molecules, which act as mediators or second messengers at nontoxic concentrations through receptor-transducing pathways. In particular, growth factor–induced ROS generation is considered to be important in the mitogenic process of cell cycle of neoplastic proliferation. An excessive production of ROS beyond the antioxidant capacities of the cell leads to an oxidative stress that results in metabolic disturbances and cell senescence. In the cancer cell system, ROS exert a paradoxical effect. ROS can promote tumor growth by transforming normal cells through activation of transcription factors or inhibition of tumor-suppressor genes, whereas its elevated levels would inhibit tumor cells through the stimulation of proapoptotic signals. The excess ROS generate cell cycle arrest and apoptotic cell death or even necrosis in severe cases.[6] Therefore, the maintenance of ROS homeostasis is extremely important to cell signaling and the regulation of cell death.

In this study we investigated the differential effect of tumor cell proliferation or apoptosis by low or high levels of exogenous H_2O_2 in HT-29 cancer cells. Also, the AMP-activated protein kinase (AMPK) signaling pathways, including cyclooxygenase-2 (COX-2), by H_2O_2 was examined. Cyclooxygenase (COX), the central enzyme in prostanoid biosynthesis, is involved in the first step of prostanoid synthesis from arachidonic acid. There are two distinct isoenzymes—COX-1 and COX-2—and COX-1 is constitutively expressed in almost every cell type, whereas COX-2 is induced by various stimuli such as cytokines and mitogens. The evidence indicates that overexpression of COX-2 is related to a number of solid malignancies including colon cancer and can influence apoptotic susceptibility.

MATERIALS AND METHODS

Cell Culture

HT-29 human colon cancer line was purchased from ATCC (Gaithersburg, MD, USA). Cells were cultured in RPMI1640 containing 10% FBS under normoxic conditions (20% O_2, 5% CO_2, 75% N_2).

Cell Proliferation by MTT Assay

Cells seeded on 96-well microplates at 4,000 cells/well were incubated with the test compounds for the indicated time period. Medium was removed and then incubated with 100 μL of MTT solution (2 mg/mL MTT in PBS) for 4 h. MTT is converted to a blue formazan. Absorbance was determined using an auto reader.

Chromatin Staining with Hoechst 33342

Apoptosis was observed by chromatin staining with Hoechst 33342, as previously described. Cells were incubated with each stimulus. After incubation the supernatant was discarded and cells were fixed with 3.5% formaldehyde in PBS for 30 min at room temperature, washed four times with PBS, and exposed to Hoechst 33342 at 10 μM for 30 min at room temperature. Cell preparations were examined under ultraviolet illumination with a fluorescence microscope (Olympus Optical Co., Tokyo, Japan).

Protein Extract and Western Blotting

Cells were rinsed twice with ice-cold PBS and scraped with lysis buffer (50 mM Tris-HCl, pH 7.4, 1% NP-40, 0.25% sodium deoxycholate, 150 mM NaCl, 1 mM EDTA, 1 mM PMSF, 1 mM sodium orthovanadate, 1mM NaF, 1 μg/mL aprotinin, 1 μg/mL leupeptin, 1 μg/mL pepstatin) and subjected to Western blot analysis. The anti-phospho-specific antibodies that recognize phosphorylated ACC-Ser[79] and PARP were obtained from Cell Signaling Technology. Antibodies for COX-2 and β-actin were purchased from Santa Cruz Biotechnology.

RESULTS

Mild Hydrogen Peroxide Promotes Cancer Cell Proliferation, while High-Dose Hydrogen Peroxide Reduces Cell Growth Markedly

Reactive oxygen species (ROS) might act as modulators of intracellular signaling cascades that use hydrogen peroxide as a principal second messenger

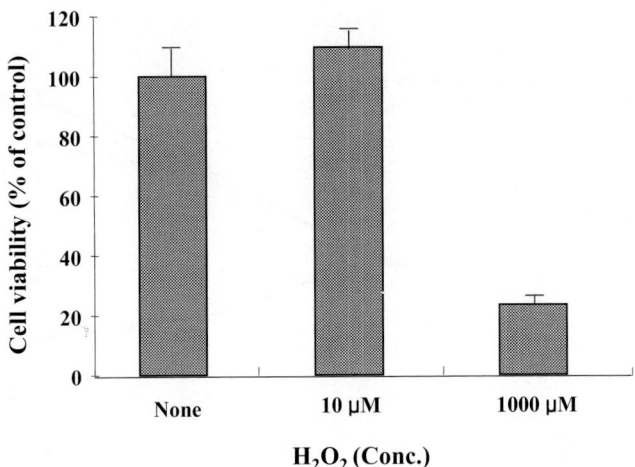

FIGURE 1. The cell growth differences of low-dose (10 μM) and high-dose H_2O_2 (1,000 μM) in HT-29 colon cancer cells. Cancer cell proliferation was increased at 10 μM concentration of H_2O_2, while the growth was decreased at 1,000 μM. These cells were exposed to H_2O_2 for 48 h, and the cell viability was measured by MTT assay.

molecule, and be thus involved in the modulation of cancer cell proliferation. Here we examined the proliferative effect of ROS in the form of hydrogen peroxide (H_2O_2). HT-29 colon cancer cells were treated with low- or high-dose H_2O_2. As shown FIGURE 1, treatment of HT-29 cells with 10 μM H_2O_2 increased cell proliferation, whereas the growth of HT-29 colon cancer cell at 1,000 μM H_2O_2 was greatly decreased. Throughout the 72-h incubation period (FIG. 2), the cell viability of 10 μM H_2O_2-treated cells showed stimulated growth compared to the control, and cancer cells treated with 1,000 μM H_2O_2 exhibited the accelerated loss of cell viability. These results indicated that cancer cell proliferation occurred at 10 μM H_2O_2, whereas 80% of growth inhibition was caused by high concentration of H_2O_2.

Only High Levels of Hydrogen Peroxide Induced Apoptosis of Cancer Cells

We determined whether an apoptosis motif exists in H_2O_2 treatment of HT-29 cells (FIG. 3). HT-29 cells were exposed to H_2O_2 10 μM and 1,000 μM, and cells were incubated with Hoechst 33342 dye, and apoptosis motif was confirmed by fluorescence microscopy. Increase in the nuclei cleavage was observed only in cells treated with 1,000 μM H_2O_2. There were no detectable apoptosis in 10 μM H_2O_2-treated HT-29 cells.

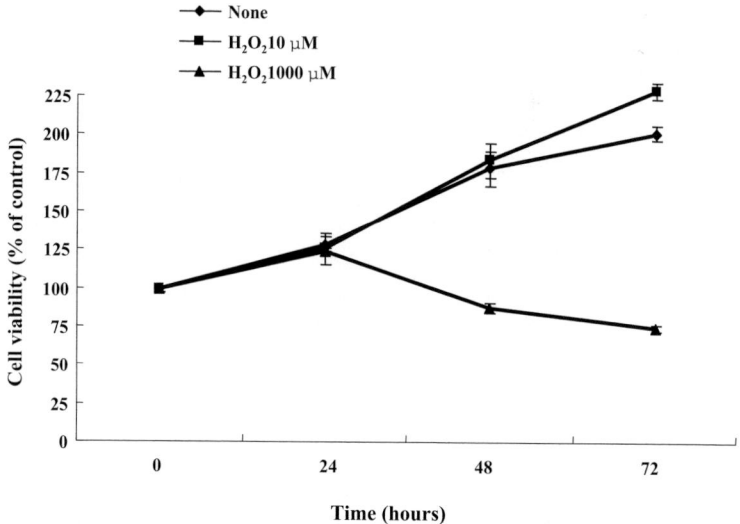

FIGURE 2. Cell viability of HT-29 cells exposed to low-dose (10 μM) and high-dose H_2O_2 (1,000 μM). Cells were treated with 10 μM or 1,000 μM H_2O_2 for 72h and the cell viability was observed.

Low Level of H_2O_2 Elevated COX-2 Expression with a Slight Change in AMPK Activation of HT-29 Cell Proliferation System

A number of studies have shown that cyclooxygenase-2 (COX-2) is a critical component of HT-29 colon cancer cell proliferation,[7] so we therefore tested

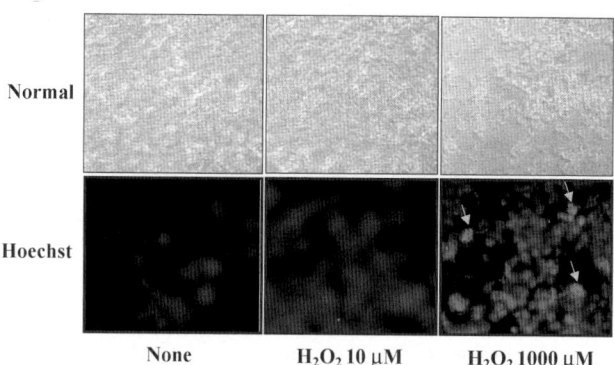

FIGURE 3. Effect of low-dose (10 μM) and high-dose H_2O_2 (1,000 μM) treatments on cell proliferation in HT-29 cancer cells. Cells were treated with H_2O_2 10 μM and 1,000 μM for 48 h. Cells were photographed using a Nikon digital camera and a fluoromicroscope after being treated with Hoechst 33342 staining. The *arrows* in panel indicate cleaved nuclei in the HT-29 cells.

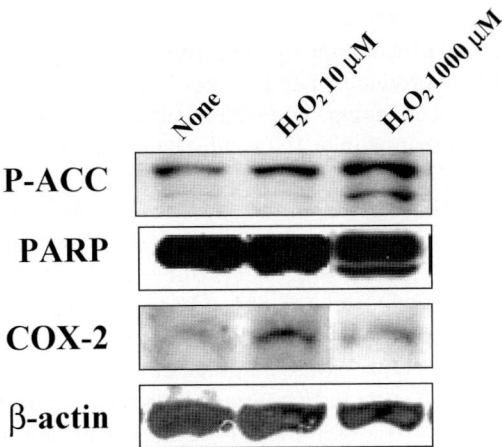

FIGURE 4. Changes in P-ACC expression, PARP cleavage, and COX-2 expression by different concentrations of H_2O_2. The cells were exposed to H_2O_2 (10 μM and 1,000 μM) for 24 h. The expression of P-ACC, PARP cleavage, and COX-2 was determined by Western blot as described in the section on "Materials and Methods."

the COX-2 expression in the H_2O_2-treated system. The low-dose H_2O_2 treatment increased COX-2 expression, whereas high doses of H_2O_2 significantly reduced COX-2 expression. Phospho-ACC was slightly increased in low-dose H_2O_2-treated cancer cells, with no change in PARP cleavage (FIG. 4). The phosphorylation level of ACC serine 79, the best-characterized phosphorylation site of AMPK, designates the status of AMPK activation.

High Level of H_2O_2 Decreased COX-2 Expression via Activated AMPK

The treatment of HT-29 cancer cells with high concentration of H_2O_2 (1,000 μM) resulted in the decreased expression of COX-2 via highly activated AMPK. Also, the evident cleavage of PARP was observed (FIG. 4). It was demonstrated that only high concentrations of H_2O_2 modulate the apoptotic marker PARP.

DISCUSSION

Reactive oxygen species (ROS), such as superoxide anion ($O \cdot_2^-$), hydrogen peroxide (H_2O_2), and hydroxyl radical ($\cdot OH$), are known to show indiscriminate effects on cellular targets such as DNA, favor cell transformation, and contribute to the development of a variety of malignant diseases.[8] ROS generated in the mitochondrial electron transport chain in the presence of oxygen are the primary source, and in other cellular compartments ROS generation can occur on account of the activities of oxidase, such as xanthine oxidase, NADPH oxidase, and monoamine oxidase. ROS have been classically considered as undesirable byproducts of cell metabolism and as factors contributing to inflammation, cancer, aging, and other degenerative diseases. However, from the observation that cellular H_2O_2 generation is rapidly induced following the production of growth factors, ROS emerged as an important factor in the regulation of neoplasia.[9] Also, the specific elevations of ROS in tumor cell lines has been noticed, and this can render increased resistance to ROS stress in cancer cells. The opposite effects of exogenous ROS on tumor cells have been observed, and the addition of H_2O_2 decreases cell proliferation and induces apoptotic death. It is known that in normal cells, the low levels of ROS originate from NADPH oxidase and the concentration of H_2O_2 is regulated by the glutathione system. In contrast, high levels of ROS are produced through the mitochondrial system, and H_2O_2 concentration is controlled by catalase. Several studies have suggested that ROS is an upstream signal of AMPK activation.[10] Recent investigations have shown that cellular signaling pathways are regulated by the redox state.[11] Generation of ROS leads to the activation of several kinases including MAP kinases, and it appears that MAP kinases are stimulated in the early stages of H_2O_2 treatment, leading to cell proliferation.[12–14]

Recent studies suggest that AMPK plays a critical role in the inhibition of cellular protein synthesis as well as stress-induced apoptosis.[15] AMPK cascades have been postulated to respond to the intracellular level of AMP or AMP:ATP ratio, and to be highly sensitive to oxidative stress.[16] ROS have been suggested to be upstream molecules of AMPK-activated signals.[17]

In conclusion, it was demonstrated that low-dose H_2O_2 induced cancer cell proliferation possibly via COX-2 elevation. High concentrations of H_2O_2 stimulated apoptotic cell death greatly, and this appeared to be accompanied by activated AMPK, indicating that ROS is upstream of AMPK activation. Our results indicated that ROS plays a significant role in cellular events of AMPK or COX-2 signaling and proliferation and apoptosis.

ACKNOWLEDGMENT

This research was supported by KOSEF Grant F01-2003-000-00081-0.

REFERENCES

1. AMES, B.N., M.K. SHIGENAGA & T.M. HAGEN. 1993. Oxidants, antioxidants, and the degenerative diseases of aging. Proc. Natl. Acad. Sci. USA **90:** 7915–7922.
2. SERACCI, R. 1987. The interactions of tobacco smoking and other agents in cancer etiology. Epidemiol. Rev. **9:** 175–193.

3. DELLINGER, B., W.A. PRYOR, R. CUETO, et al. 2001. Role of free radicals in the toxicity of airborne fine particulate matter. Chem. Res. Toxicol. **14:** 1371–1377.
4. CHURCH, D.F. & W.A. PRYOR. 1985. Free-radical chemistry of cigarette smoke and its toxicological implications. Environ. Health Perspect. **64:** 111–126.
5. KINNULA, V.L. & J.D. CRAPO. 2004. Superoxide dismutases in malignant cells and human tumors. Free Radic. Biol. Med. **36:** 718–744.
6. GONZALEZ, C., G. SANZ-ALFAYATE, M.T. AGAPITO, et al. 2002. Significance of ROS in oxygen sensing in cell systems with sensitivity to physiological hypoxia. Respir. Physiol. Neurobiol. **132:** 17–41.
7. LIU, E.S., V.Y. SHIN, Y.-N. YE, et al. 2005. Cyclooxygenase-2 in cancer cells and macrophages induces colon cancer cell growth by cigarette smoke extract. Eur. J. Pharmacol. **518:** 47–55.
8. PELICANO, H., D. CARNEY & P. HUANG. 2004. ROS stress in cancer cells and therapeutic implications. Drug Resist. Updat. **7:** 97–110.
9. BOONSTRA, J. & J.A. POST. 2004. Molecular events associated with reactive oxygen species and cell cycle progression in mammalian cells. Gene **337:** 1–13.
10. CHOI, S.-L., S.-J. KIM, K.-T. LEE, et al. 2001. The regulation of AMP-activated protein kinase by H_2O_2. Biochem. Biophys. Res. Commun. **287:** 92–97.
11. FINKEL, T. 2000. Redox-dependent signal transduction. FEBS Lett. **476:** 52–54.
12. LIU, S.L., X. LIN, D.Y. SHI, et al. 2002. Reactive oxygen species stimulated human hepatoma cell proliferation via cross-talk between PI3-K/PKB and JNK signaling pathways. Arch. Biochem. Biophys. **406:** 173–182.
13. SHACKELFORD, R.E., W.K. KAUFMANN & R.S. PAULES. 2000. Oxidative stress and cell cycle checkpoint function. Free Radic. Biol. Med. **28:** 1387–1404.
14. RUFFELS, J., M. GRIFFIN & J.M. DICKENSON. 2004. Activation of ERK1/2, JNK and PKB by hydrogen peroxide in human SH-SY5Y neuroblastoma cells: role of ERK1/2 in H_2O_2-induced cell death. Eur. J. Pharmacol. **483:** 163–173.
15. JIN, Q., B.S. JHUN, S.H. LEE, et al. 2004. Differential regulation of phosphatidylinositol 3-kinase/Akt, mitogen-activated protein kinase, and AMP-activated protein kinase pathways during menadione-induced oxidative stress in the kidney of young and old rats. Biochem. Biophys. Res. Commun. **315:** 555–561.
16. HARDIE, D.G. 2004. The AMP-activated protein kinase pathway — new players upstream and downstream. J. Cell Sci. **117:** 5479–**5**487.
17. HWANG, J., J. HA & O.J. PARK. 2005. Combination of 5-fluorouracil and genistein induces apoptosis synergistically in chemo-resistant cancer cells through the modulation of AMPK and Cox-2 signaling pathways. Biochem. Biophys. Res. Commun. **332:** 433–440.

Alterations in Salivary Antioxidants, Nitric Oxide, and Transforming Growth Factor-β_1 in Relation to Disease Activity in Crohn's Disease Patients

ALI REZAIE,[a] FAKHTEH GHORBANI,[a] AZADEH ESHGHTORK,[a] MOHAMMAD J. ZAMANI,[a] GHOLAMREZA DEHGHAN,[a] BARDIA TAGHAVI,[a] SHEKOUFEH NIKFAR,[a] AZADEH MOHAMMADIRAD, NASSER E. DARYANI,[b] AND MOHAMMAD ABDOLLAHI[a]

[a]*Department of Toxicology and Pharmacology, Faculty of Pharmacy, and Pharmaceutical Sciences Research Center, Tehran 14155-6451, Iran*

[b]*Department of Gastrointestinal Diseases, Faculty of Medicine, Tehran University of Medical Sciences, Tehran 14155-6451, Iran*

ABSTRACT: It has been postulated that oxidative stress, nitric oxide (NO), and transforming growth factor β_1 (TGF- β_1) have major roles in the pathophysiology of Crohn's disease (CD). The aim of this study was to determine the salivary levels of total antioxidant capacity (TAC), specific antioxidants (i.e., uric acid, albumin, transferrin, and thiol molecules), lipid peroxidation (LPO), NO, and TGF- β_1 in CD patients and control subjects and to also investigate their correlation with activity of the disease. Twenty-eight patients with confirmed diagnosis of CD were enrolled and whole saliva samples were obtained. Smokers, diabetics, those who suffered from periodontitis, and those who were consuming antioxidant supplements were excluded from the study. The Crohn's Disease Activity Index (CDAI) was used to determine the severity of the disease. Twenty healthy subjects were also recruited. In CD patients significant reductions in salivary levels of TAC (0.248 ± 0.145 vs. 0.342 ± 0.110 mmol/L), albumin (1.79 ± 0.42 vs. 2.3 ± 0.2 μg/mL), and uric acid (3.1 ± 1.4 vs. 4.1 ± 2.0 mg/dL) were found. TGF-β_1 was significantly increased in CD patients compared to healthy subjects (3.02 ± 1.54 vs. 2.36 ± 0.52 ng/mL). A fourfold increase in NO levels (198.8 ± 39.9 vs. 50.2 ± 21.3 μmol/L) along with a fivefold increase in LPO concentration (0.146 ± 0.064 vs. 0.027 ± 0.019 μmol/L) was documented in CD patients in comparison to the control group. CDAI significantly correlated with the TAC, LPO, and the interaction between TAC and LPO ($r^2 = 0.625$,

Address for correspondence: Prof. Mohammad Abdollahi, Department of Toxicology and Pharmacology, Faculty of Pharmacy, and Pharmaceutical Sciences Research Center, Tehran University of Medical Sciences, Tehran 14155-6451, Iran. Voice/fax: +98-21-66959104.
e-mail: mohammad.abdollahi@utoronto.ca

$r^2 = 0.8$, F-test's $P < 0.00005$). Saliva of CD patients exhibits an abnormal feature with respect to oxidative stress, NO, and TGF-β_1. TAC and LPO modify the effect of each other in determination of CD severity, which underlines the importance of oxidative stress in the pathogenesis of CD.

KEYWORDS: salivary antioxidants; nitric oxide; transforming growth factor; disease activity index; Crohn's disease; inflammatory bowel disease; saliva; ulcerative colitis

INTRODUCTION

Crohn's disease (CD) is an entity of inflammatory bowel diseases (IBD), which can affect any part of gastrointestinal tract from mouth to anus. The chronic and debilitating nature of the disease, lack of a cure, and the global increase in its incidence have drawn an enormous body of attention to the pathophysiology of CD. Although several hypotheses suggest a combination of environmental factors, genetic predisposition, and dysfunctional immunoregulation for the etiology of CD, the exact etiology remains uncertain.[1]

Over the past decade, there has been an extensive focus on cytokines,[2] nitric oxide (NO),[3] and oxidative stress[4-14] in the intestines and plasma of CD patients as potential factors involved in the initiation or propagation of this disease; however, saliva has received less interest and research remains limited.

Oxidative stress arises when there is a marked imbalance between the production of reactive oxygen species (ROS) and their removal by antioxidants. Detrimental effects of these highly reactive molecules have been well established in inflammation,[15] which could be measured indirectly by the concentration of the peroxidized cell wall lipids.[16] Increased levels of ROS and decreased total antioxidant capacity (TAC) in intestinal mucosa and plasma of CD patients have been reported in several studies.[4-14]

NO is produced via the action of nitric oxide synthase (NOS). Small (picomolar) amounts of NO are thought to be physiologic and protective, whereas large amounts of NO, produced by inducible NOS (iNOS), are proinflammatory and injurious.[3] Moreover, the NO interface with oxygen radicals, such as superoxide, leads to the formation of new compounds (e.g., peroxynitrite) with a higher potency to damage the cells.[17]

Among various cytokines, transforming growth factor-β_1 (TGF-β_1) seems to play a key anti-inflammatory role by counteracting tumor necrosis factor-α (TNF-α).[18-20] As TNF-α mediates most of its effects by ROS,[21] it is plausible that TGF-β_1 is capable of modulating oxidative stress. In addition, the tight interaction between TGF-β_1 and NO that has been demonstrated in studies concerning hypertension and kidney transplantation has not been studied in IBD.[22,23]

Recently, we reported novel findings about enhanced levels of lipid peroxidation (LPO) and NO along with decreased TAC, in saliva of CD patients.[24] To shed more light on dysregulation of salivary components in CD patients and the correlation of disease activity with these constituents, we recruited patients with different disease activity status and healthy controls and determined LPO, TAC, salivary main antioxidants (i.e., uric acid, albumin, transferrin, thiol groups), NO, and TGF-β_1 levels in their unstimulated whole saliva.

METHODS

Materials

Sodium acetate, 2,4,6-tripyridyl-s-triazine (TPTZ), 2-thiobarbituric acid (TBA), 1,1.3.3-tetramethoxypropan, trichloroacetic acid (TCA), dithionitrobenzoic acid (DTNB), sodium sulfate, $FeCl_3$ $6H_2O$, hydrochloric acid, distilled water, sulfuric acid, and n-butyl alcohol obtained from Merck Chemical Company (Darmstadt, Germany) were used in this study. Assay kits for uric acid, transferrin, and albumin were obtained from Pars-Azmoon Co. (Tehran, Iran).

Participants

Through a convenience sampling strategy and after considering our inclusion and exclusion criteria (see below) 28 patients with CD were enrolled in this study. The demographic and clinical characteristics of the patients are shown in TABLE 1. Diagnosis of CD was established according to clinical, endoscopic, and pathologic criteria. The disease activity was assessed according

TABLE 1. Clinical characteristics of the participants

	CD patients ($n = 28$)
Age (year)	$37.2 \pm 14.7,^a$ $(18-60)^b$
Male	11 (39.3%)
Female	17 (60.7%)
CDAI	$102.1 \pm 84.9,^a$ $(14-335)^a$
Disease	
Ileitis	12 (42.9%)
Ileocolitis	13 (46.4%)
Colitis	3 (10.7%)
Extraintestinal manifestations	2 (9.5%)
Current medications	
5-Aminosalicylates	28 (100%)
Corticosteroids	6 (21.4%)
Azathioprine	10 (35.7%)

[a]Standard Deviation.
[b]Range.

to Crohn's Disease Activity Index (CDAI).[28] CDAI is a composite scoring system based on the number of liquid stools, severity of abdominal pain, general well-being, extraintestinal manifestations, abdominal mass, use of antidiarrheal drugs, hematocrit, and body weight. CDAI is currently the most reliable activity index available for CD. The control group comprised 20 healthy individuals (8 women and 12 men, mean age 41 years). Smokers,[29] diabetics,[30] and those who consumed antioxidant supplements in recent months[31] were excluded from study on account of the potential confounding effects on oxidative stress. Participants were also seen to be free of periodontal inflammation of any grade. Patients receiving cyclosporine were excluded on the basis of a probable confounding effect on TGF-β_1 levels.[32] The Ethics Committee of Tehran University of Medical Sciences (TUMS) approved the study protocol and all the participants gave their informed consent before enrollment.

Sample Collection and Handling

Unstimulated whole saliva was collected during 5 min, allowed to drain into a plastic container, and was then centrifuged at 10,000 g for 5 min to remove bacterial and cellular debris and stored at $-80°C$ until analysis.

LPO Assay

Malonedialdehyde (MDA) level in the serum was determined using the thiobarbituric acid test. To precipitate the proteins of saliva, 2.5 mL of TCA 20% (w/v) was added into 0.5 mL of the sample, which was then centrifuged at 1500 g for 10 min. Then 2.5 mL of sulfuric acid (0.05 M) and 2 mL TBA 0.2% were added to the sediment, shaken, and incubated for 30 min in a boiling water bath. Then 4 mL n-butanol was added, and the solution was centrifuged and cooled, and emission of the supernatant was recorded by a spectrofluorimeter. The calibration curve was obtained using different concentrations of 1,1,3,3-tetramethoxypropane as a standard to determine the concentration of TBA–MDA adducts in samples.[33]

TAC Assay

The FRAP assay (ferric reducing ability of plasma), which depends upon the reducing of ferric tripyridyltriazine [Fe(III)–TPTZ] complex to the ferrous tripyridyltriazine [Fe(II)–TPTZ] was performed at low pH.[34] The Fe(II)–TPTZ complex gives a blue color with absorbance maximum at 593 nm.

NO and TGF-β_1 Assays

TGF-β_1 and NO concentrations were determined by enzyme-linked immunosorbent assay (ELISA) kits (Quntikine®) provided from R&D Systems GmbH, Germany. The assay employs the quantitative sandwich enzyme immunoassay technique. Total NO was assayed based on the enzymatic conversion of nitrate to nitrite by nitrate reductase as described in the previous report.[24]

Uric Acid Assay

The uric acid concentration was measured by the uricase/4-aminophenazone colorimetric method. In brief: uric acid present in the sample was converted by uricase into allantoin and hydrogen peroxide. The latter is then catalyzed by peroxidase present in the reagent mixture and oxidizes aminobenzene to produce red quinoneimine compound that absorbs at 520 nm.[35]

Transferrin Assay

Transferrin was determined by the immunoturbidimetric method. Transferrin in combination with its polyclonal antibody produces a complex that makes the solution turbid.[36]

Albumin Assay

Albumin concentration was measured by bromocresol purple colorimetric method. Albumin binds quantitatively to the indicator bromocresol purple (5,5 dibromo-cresol-sulfonephthalein) and the albumin complex absorbs at 600 nm.[37]

Assay of Thiol Groups

Thiol groups of plasma were measured spectrophotometrically at 412 nm using DTNB as the reagent.[38]

Statistical Analyses

Simple and multivariate linear regression analyses were used for determination of correlations and interactions after the assumptions for normality, constant variation, and linearity were met. Comparisons among patients and control groups were done by Student's t-test and in case of non-normal distribution logarithmic transformation or Welch's approximation was used. Level

TABLE 2. Salivary concentrations (mean ± SD) and the corresponding P-values for the difference between CD patients and controls

	CD	Controls	P-value
TGF-β_1 (ng/mL)	3.02 ± 1.54	2.36 ± 0.52	0.03
NO (μmol/L)	198.8 ± 39.9	50.2 ± 21.3	<0.00005
TAC (mmol/L)	0.248 ± 0.145	0.342 ± 0.110	0.02
LPO (μmol/L)	0.146 ± 0.064	0.027 ± 0.019	<0.00005
Albumin (μg/mL)	1.79 ± 0.42	2.3 ± 0.2	0.01
Uric acid (mg/dL)	3.1 ± 1.4	4.1 ± 2.0	0.03
Transferrin (mg/dL)	1.77 ± 0.42	1.91 ± 0.51	0.30
Total thiols (μmol/L)	251.6 ± 171.8	315.1 ± 103	0.11

of significance was set at 0.05. All statistical analyses were carried out by STATA 8.2. Graph is drawn by Graphis 2.6.2.

RESULTS

The salivary levels of NO, TGF-β_1, LPO, TAC, and specific antioxidants (i.e., uric acid, albumin, transferrin, and thiol groups) in CD patients and control group along with the corresponding P-values of the differences are shown in TABLE 2. In CD patients, significant reductions in salivary levels of TAC, albumin, and uric acid were found. Although CD patients had a lower level of transferrin and thiol, these reductions were not significant. TGF-β_1 was significantly increased in CD patients compared to healthy subjects. A fourfold increase in NO levels along with a fivefold increase in LPO concentration was documented in CD patients in comparison to the control group.

FIGURE 1 shows that CDAI significantly correlated with the TAC, LPO, and the interaction between TAC and LPO ($r^2 = 0.625$, $r^2 = 0.8$, F-test's $P < 0.00005$). A P-value of 0.001 for the coefficient of the interaction provides strong evidence that TAC and LPO modify the effect of each other. Centered model for the mean of TAC and LPO demonstrated a negative correlation between TAC and a positive correlation with LPO. Correlation of CDAI with specific antioxidants and their interaction with LPO was not significant.

Although NO and TGF-β_1 levels were significantly different between CD patients and controls, they had no correlation with CDAI or oxidative stress components. No significant correlation was found with respect to the basic characteristics of the patients (TABLE 1) and the laboratory results.

DISCUSSION

Whole saliva is a mixture of gingival crevicular fluid (GCF) and fluid secreted from salivary glands, of which the parotid, submandibular, and

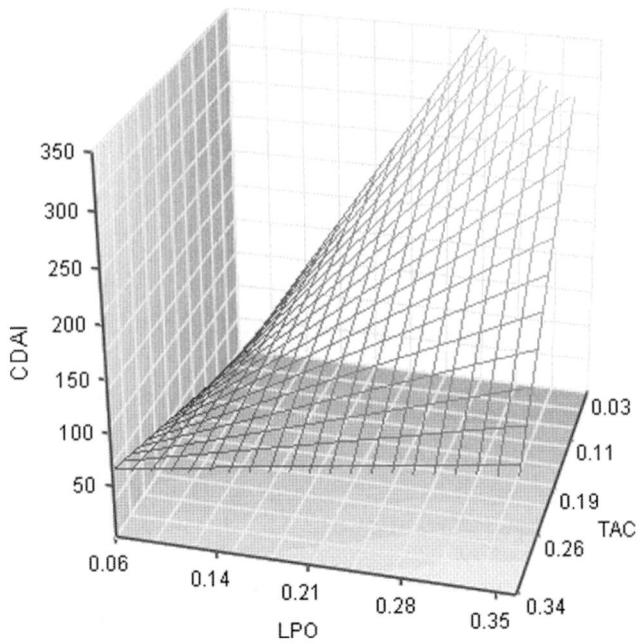

FIGURE 1. Correlation of CDAI with the salivary TAC (mmol/L) and LPO (μmol/L). The model yielded shows: CDAI = 534.7 × TAC + 1901 × LPO − 4981 × TAC × LPO −132. The t-test's P-values for each coefficient are 0.019, <0.0005, and 0.001, respectively.

sublingual are the major sources.[25] For analysis of saliva for immunoregulation and oxidative stress, whole saliva is more relevant as it contains GCF and immune-related cells.[25,26] Moreover, unstimulated whole saliva represents the major intraoral environment, except at the time of mastication.[27]

Salivary Oxidative Stress and CD

There is a great body of evidence that oxidative stress plays a role in the pathogenesis of CD, and several studies have reported increased ROS and altered antioxidant defense system in CD patients. Breath alkanes (i.e., ethane and pentane), as markers of LPO, have been reported to be significantly increased in nonsmoker CD patients.[4] 8-OhdG, a DNA oxidation product, is increased in CD patients.[5] There are also reports in the literature regarding the overproduction of oxygen radicals and reduction of antioxidant enzymes (e.g., glutathione and superoxide dismutase),[6–10] Vitamin C, α- and β-carotene,[4,5] and antioxidant micronutrients in the intestinal mucosa[11,12] and plasma[13,14] of patients suffering from CD. However, obvious correlations between these markers and disease activity have not been established.[16]

Following our recent observation[24] that saliva of CD patients contained higher levels of LPO and reduced TAC, we recruited more patients with more restricted inclusion criteria to show the possible correlation between CDAI and markers of oxidative stress.

Surprisingly, LPO, TAC and their interaction demonstrate a robust correlation with CDAI. The strong and statistically significant interaction found in the present data may explain the inability of the previous studies[16,39,40] to find a meaningful correlation between oxidative stress markers and disease activity, despite its theoretical plausibility. Another advantage is that we used TAC, which integrates the sum of endogenous and exogenous antioxidants that are present in saliva. It is superior in measuring individual antioxidants, since only limited number of radical scavengers could be measured and it is likely that there are compounds in our body and diet that antioxidant effects have not been revealed.

In reaction to mild oxidative stress, tissues often respond by producing more antioxidants and increasing TAC; however, severe persistent oxidative stress leads to cell injury, necrosis, or apoptosis, and consequently lower antioxidant capacity.[41] Therefore, normal values of TAC have two conflicting explanations: first, there is a balance between antioxidants and ROS and, second, in severe oxidative stress, excessive amounts of ROS consume the overproduced antioxidants and lower TAC to normal levels. The former situation is a sign of stability, while the latter is a sign of ongoing damage. As evident in our model (FIG. 1), those with low and normal levels of TAC have the lowest and also the highest disease activity, depending on their LPO status, while those with high TAC and high LPO have intermediate disease activity because their body is actively defending itself against oxygen radicals.

It should be considered that we excluded several known confounders of oxidative stress (i.e., smoking,[29] antioxidant supplements,[31] and periodontitis[42]) to yield more accurate results; however, there are definitely more confounders than we are aware of and that we did not control for. Moreover, antioxidant effects have been proposed for 5- aminosalicylic acid (5-ASA),[43] and as far as all the patients were receiving at least a type of this drug, we may have underestimated the true underlying oxidative stress in these patients.

Specific Salivary Antioxidants

The major antioxidants in saliva are uric acid and, to a lesser extent, albumin.[44] Uric acid displays a concentration similar to serum[44] and scavenges singlet O_2, peroxy and hydroxyl radicals and irreversibly degrades into several fragments.[45] Our data showed decreased levels of uric acid in saliva of CD patients when compared to control subjects. Iron and copper ions are powerful promoters of oxidative damage and accelerate LPO and formation of hydroxyl radicals.[46] Albumin and transferrin play a sacrificial antioxidant role by binding to copper and iron ions and consequently scavenging hydroxyl radical. This reaction will proteolytically degrade albumin and transferrin; however, this damage is biologically insignificant on account of their rapid turnover. In addition, albumin protects the targets from other potent oxidants, such as hypochlorous acid and peroxy radicals.[46] As we were expecting, we observed decreased levels of albumin in saliva of CD patients, but we could not show any significant difference in transferrin level between CD patients and controls.

Thiol molecules (e.g., gluthatione) are the most important intracellular antioxidants.[47,48] They are present in micromolar amounts in extracellular fluids and millimolar amounts inside the cells. Despite a trend toward higher levels of thiols in saliva of control subjects, the difference was not statistically significant. This may reflect the fact that a bigger sample size is needed to elucidate the effect of CD on salivary thiol concentration.

Generally, we were unable to show any significant relationship between uric acid, albumin, transferrin, and total thiol levels and CDAI with or without interaction with LPO levels. This might be due to the partial involvement of these antioxidants in TAC of saliva, which consequently needs a bigger sample size to yield a clear-cut conclusion.

Salivary TGF-β_1

Anti-inflammatory effects of TGF-β_1 are well established,[18–20] injection of cDNA vectors, encoding TGF-β_1, or topical administration of TGF-β_1 have ameliorated intestinal inflammation and accelerated wound/ulcer healing.[49] Increased serum, plasma, and colonic mucosal expression of TGF-β_1 is reported in several reports concerning IBD patients and animal colitis models.[50–53]

Similar to previous observations, our patients showed increased levels of TGF-β_1 in their saliva. This phenomenon has been explained as a defense mechanism to stimulate differentiation of epithelial cells to aid the repair process of the damaged mucosa.[54,55] In addition, it should be noted that TGF-β_1 as a growth factor tends to act locally on adjacent cells (paracrine or juxtacrine action) or on the same cell that has expressed the growth factor (autocrine action).[56] Increased salivary TGF-β_1 cannot be explained by its increased plasma levels and indicates an underlying pathophysiology originating from salivary glands. As in previous studies[55,57] we were unable to show any correlation between disease activity and TGF-β concentration.

Salivary NO

Generally, NO production and iNOS activity are increased in CD. Despite many studies in humans, animals, and cell lines, there is still substantial uncertainty regarding the role of NO in IBD. iNOS activity in intestinal mucosa of CD patients has demonstrated a 3.8-fold increase.[58] Using a chemiluminescence technique, NO level has been shown to be elevated in aspirated colonic gas and exhaled air.[59,60] Our patients showed a fourfold increase in salivary NO concentration. Similar to previous studies[17,61] we were unable to correlate NO to disease activity. No relationship was found between NO and oxidative stress markers and TGF-β_1. This may be due to the fact that the interaction between TGF-β_1 and NO is far more complex to be detected by our cross-sectional design, as TGF-β_1 enhances NO levels, while NO produces feedback inhibition of TGF-β_1.[22,23]

CONCLUSION

The prominent novel finding of our study is that TAC and ROS modify the effect of each other in the determination of CD severity. The aforenoted effect modification may emphasize that oxidative stress plays a role in pathophysiology of CD rather than a nonspecific marker of inflammation. Extrapolation of our results to other body regions (e.g., plasma or intestinal mucosa) or even to other oxidative stress-related diseases requires further investigation.

ACKNOWLEDGMENTS

This study was financially supported by a grant from TUMS.

REFERENCES

1. REZAIE, A., B.T. BAYAT & M. ABDOLLAHI. 2005. Biologic management of fistulizing Crohn's disease. Int. J. Pharmacol. **1:** 17–24.
2. PLAYFORD, R.J. & S. GHOSH. 2005. Cytokines and growth factor modulators in intestinal inflammation and repair. J. Pathol. **205:** 417–425.
3. CROSS, R.K. & K.T. WILSON. 2003. Nitric oxide in inflammatory bowel disease. Inflamm. Bowel Dis. **9:** 179–189.
4. WENDLAND, B.E., E. AGHDASSI, C. TAM, et al. 2001. Lipid peroxidation and plasma antioxidant micronutrients in Crohn disease. Am. J. Clin. Nutr. **74:** 259–264.
5. D'ODORICO, A., S. BORTOLAN & R. CARDIN. 2001. Reduced plasma antioxidant concentrations and increased oxidative DNA damage in inflammatory bowel disease. Scand. J. Gastroenterol. **36:** 1289–1294.
6. MIRALLES-BARRACHINA, O., G. SAVOYE, L. BELMONTE-ZALAR, et al. 1999. Low levels of glutathione in endoscopic biopsies of patients with Crohn's colitis: the role of malnutrition. Clin. Nutr. **18:** 313–317.
7. ARDITE, E., M. SANS, J. PANES, et al. 2000. Replenishment of glutathione levels improves mucosal function in experimental acute colitis. Lab Invest. **8:** 735–744.
8. SIDO, B., V. HACK, A. HOCHLEHNERT, et al. 1998. Impairment of intestinal glutathione synthesis in patients with inflammatory bowel disease. Gut **42:** 485–492.

9. HOLMES, E.W., S.L. YOUNG, D. EIZNHAMER & A. KESHAVARZIAN. 1998. Glutathione content of colonic mucosa: evidence for oxidative damage in active ulcerative colitis. Dig. Dis. Sci. **43:** 1088–1095.
10. STURNIOLO, G.C., C. MESTRINER, P.E. LECIS, et al. 1998. Altered plasma and mucosal concentrations of trace elements and antioxidants in active ulcerative colitis. Scand. J. Gastroenterol. **33:** 644–649.
11. GEERLING, B.J., A. BADART-SMOOK, C. VAN DEURSEN, et al. 2000. Nutritional supplementation with N-3 fatty acids and antioxidants in patients with Crohn's disease in remission: effects on antioxidant status and fatty acid profile. Inflamm. Bowel Dis. **6:** 77–84.
12. BUFFINGTON, G.D. & W.F. DOE. 1995. Altered ascorbic acid status in the mucosa from inflammatory bowel disease patients. Free Radic. Res. **22:** 131–143.
13. D'ODORICO, A., F. POZZATO, M. MINOTTO, et al. 1996. Plasma carotenoids and other antioxidant levels in Crohn's disease. Gastroenterology **110:** A897.
14. GEERLING, B.J., A. BADART-SMOOK, R.W. STOCKBRUGGER & R.J. BRUMMER. 1998. Comprehensive nutritional status in patients with long-standing Crohn disease currently in remission. Am. J. Clin. Nutr. **67:** 919–926.
15. SPITZ, D.R., E.I. AZZAM, J.J. LI & D. GIUS. 2004. Metabolic oxidation/reduction reactions and cellular responses to ionizing radiation: a unifying concept in stress response biology. Cancer Metastasis Rev. **23:** 311–322.
16. KRUIDENIER, L. & H.W. VERSPAGET. 2002. Review article: oxidative stress as a pathogenic factor in inflammatory bowel disease–radicals or ridiculous? Aliment Pharmacol. Ther. **16:** 1997–2015.
17. KIMURA, H., S. MIURA, T. SHIGEMATSU, et al. 1997. Increased nitric oxide production and inducible nitric oxide synthase activity in colonic mucosa of patients with active ulcerative colitis and Crohn's disease. Dig. Dis. Sci. **42:** 1047–1054.
18. LUETHVIKSSON, B.R. & B. GUNNLAUGSDOTTIR. 2003. Transforming growth factor-beta as a regulator of site-specific T-cell inflammatory response. Scand. J. Immunol. **58:** 129–138.
19. BARTOLOME, R.A., F. SANZ-RODRIGUEZ, M.M. ROBLEDO, et al. 2003. Rapid up-regulation of alpha4 integrin-mediated leukocyte adhesion by transforming growth factor-beta1. Mol. Biol. Cell **14:** 54–66.
20. GORELIK, L. & R.A. FLAVELL. 2002. Transforming growth factor-beta in T-cell biology. Nat. Rev. **2:** 46–53.
21. GARG, A.K. & B.B. AGGARWAL. 2002. Reactive oxygen intermediates in TNF signalling. Mol. Immunol. **39:** 509–517.
22. DAHLY, A.J., K.M. HOAGLAND, A.K. FLASCH, et al. 2002. Antihypertensive effects of chronic anti-TGF-beta antibody therapy in Dahl S rats. Am. J. Physiol. Regul. Integr. Comp. Physiol. **283:** R757–R767.
23. YING, W.Z. & P.W. SANDERS. 2003. The interrelationship between TGF-beta1 and nitric oxide is altered in salt-sensitive hypertension. Am. J. Physiol. Renal Physiol. **285:** F902–F908.
24. JAHANSHAHI, G., V. MOTAVASEL, A. REZAIE, et al. 2004. Alterations in antioxidant power and levels of epidermal growth factor and nitric oxide in saliva of patients with inflammatory bowel diseases. Dig. Dis. Sci. **49:** 1752–1757.
25. NAVAZESH, M. 1993. Methods for collecting saliva. Ann. N.Y. Acad. Sci. **694:** 72–77.
26. KAUFMAN, E. & I.B. LAMSTER. 2000. Analysis of saliva for periodontal diagnosis—a review. J. Clin. Periodontol. **27:** 453–465.

27. EDGAR, W.M. 1992. Saliva: its secretion, composition and functions. Br. Dent. J. **172:** 305–312.
28. BEST, W.R., J.M. BECKTEL, J.W. SINGLETON & F. KERN JR. 1976. Development of a Crohn's disease activity index. National Cooperative Crohn's Disease Study. Gastroenterology **70:** 439–444.
29. AGHDASSI, E. & J.P. ALLARD. 2000. Breath alkanes as a marker of oxidative stress in different clinical conditions. Free Radic. Biol. Med. **28:** 880–886.
30. AFSHARI, M., B. LARIJANI, A. REZAIE, et al. 2004. Ineffectiveness of allopurinol in reduction of oxidative stress in diabetic patients; a randomized, double-blind placebo-controlled clinical trial. Biomed. Pharmacother. **58:** 546–550.
31. AGHDASSI, E., B.E. WENDLAND & A.H. STEINHART, et al. 2003. Antioxidant vitamin supplementation in Crohn's disease decreases oxidative stress. A randomized controlled trial. Am. J. Gastroenterol. **98:** 348–353.
32. MOHAMED, M.A., H. ROBERTSON, T.A. BOOTH, et al. 2000. TGF-beta expression in renal transplant biopsies: a comparative study between cyclosporin-A and tacrolimus. Transplantation **69:** 1002–1005.
33. SATHO, K. 1978. Serum lipid peroxidation in cerebrovascular disorders determined by a new colorimetric method. Clin. Chim. Acta **90:** 37–43.
34. BENZIE, I.F.F. & J.J. STRAIN. 1996. Ferric reducing ability of plasma (FRAP) as a measure of antioxidant power: the FRAP assay. Anal. Biochem. **239:** 70–76.
35. DUNCAN, P.H., N. GOCHMAN, T. COOPER, et al. 1982. A candidate reference method for uric acid in serum. I. Optimization and evaluation. Clin. Chem. **28:** 284–290.
36. BUFFONE, G.J., S.A. LEWIS, M. IOSEFSOHN & J.M. HICKS. 1978. Chemical and immunochemical measurement of total iron-binding capacity compared. Clin. Chem. **24:** 1788–1791.
37. DOUMAS, B.T., W.A. WATSON & H.G. BIGGS. 1971. Albumin standards and the measurement of serum albumin with bromcresol green. Clin. Chim. Acta **31:** 87–96.
38. HU, M.L. & C.J. DILLARD. 1994. Plasma SH and GSH measurement. Methods Enzymol. **233:** 385–387.
39. SINGER, I.I., D.W. KAWKA, S. SCOTT, et al. 1996. Expression of inducible nitric oxide synthase and nitrotyrosine in colonic epithelium in inflammatory bowel disease. Gastroenterology **111:** 871–885.
40. DIJKSTRA, G., H. MOSHAGE, H.M. VAN DULLEMEN, et al. 1998. Expression of nitric oxide synthases and formation of nitrotyrosine and reactive oxygen species in inflammatory bowel disease. J. Pathol. **186:** 416–421.
41. SARAFIAN, T.A. & D.E. BREDESEN. 1994. Is apoptosis mediated by reactive oxygen species? Free Radic. Res. **21:** 1–8.
42. PANJAMURTHY, K., S. MANOHARAN & C.R. RAMACHANDRAN. 2005. Lipid peroxidation and antioxidant status in patients with periodontitis. Cell Mol. Biol. Lett. **10:** 255–264.
43. MILES, A.M. & M.B. GRISHAM. 1994. Antioxidant properties of aminosalicylates. Methods Enzymol. **234:** 555–572.
44. MOORE, S., K.A. CALDER, N.J. MILLER & C.A. RICE-EVANS. 1994. Antioxidant activity of saliva and periodontal disease. Free Radic. Res. **21:** 417–425.
45. GROOTVELD, M. & B. HALLIWELL. 1987. Measurement of allantoin and uric acid in human body fluids. A potential index of free-radical reactions *in vivo*? Biochem. J. **243:** 803–808.

46. HALLIWELL, B. 1994. Free radicals, antioxidants, and human disease: curiosity, cause, or consequence? Lancet **344:** 721–724.
47. GOULART, M., M.C. BATOREU, A.S. RODRIGUES, *et al.* 2005. Lipoperoxidation products and thiol antioxidants in chromium exposed workers. Mutagenesis **20:** 311–315.
48. BALCERCZYK, A. & G. BARTOSZ. 2003. Thiols are main determinants of total antioxidant capacity of cellular homogenates. Free Radic. Res. **37:** 537–541.
49. GILADI, E., E. RAZ, F. KARMELI, *et al.* 1995. Transforming growth factor-beta gene therapy ameliorates experimental colitis in rats. Eur. J. Gastroenterol. Hepatol. **7:** 341–347.
50. DIGNASS, A.U., J.L. STOW & M.W. BABYATSKY. 1996. Acute epithelial injury in the rat small intestine in vivo is associated with expanded expression of transforming growth factor alpha and beta. Gut **38:** 687–693.
51. BABYATSKY, M.W., G. ROSSITER & D.K. PODOLSKY. 1996. Expression of transforming growth factors alpha and beta in colonic mucosa in inflammatory bowel disease. Gastroenterology **110:** 975–984.
52. XIAN, C.J., X. XU, C.E. MARDELL, *et al.* 1999. Site-specific changes in transforming growth factor-alpha and beta1 expression in colonic mucosa of adolescents with inflammatory bowel disease. Scand. J. Gastroenterol. **34:** 591–600.
53. MAXWELL, L. & W. DOE. 2001. Inflammation location, but not type, determines the increase in TGF-beta1 and IGF-1 expression and collagen deposition in IBD intestine. Inflamm. Bowel Dis. **7:** 16–26.
54. BECK, P.L. & D.K. PODOLSKY. 1999. Growth factors in inflammatory bowel disease. Inflamm. Bowel Dis. **5:** 44–60.
55. KADER, H.A., V.T. TCHERNEV, E. SATYARAJ, *et al.* 2005. Protein microarray analysis of disease activity in pediatric inflammatory bowel disease demonstrates elevated serum PLGF, IL-7, TGF-beta1, and IL-12p40 levels in Crohn's disease and ulcerative colitis patients in remission versus active disease. Am. J. Gastroenterol. **100:** 414–423.
56. SPORN, M.B. & A.B. ROBERTS. 1992. Autocrine secretion–10 years later. Ann. Intern. Med. **117:** 408–414.
57. STURM, A., C. SCHULTE, R. SCHATTON, *et al.* 2000. Transforming growth factor-beta and hepatocyte growth factor plasma levels in patients with inflammatory bowel disease. Eur. J. Gastroenterol. Hepatol. **12:** 445–450.
58. RACHMILEWITZ, D., R. ELIAKIM, Z. ACKERMAN, *et al.* 1998. Direct determination of colonic nitric oxide level: a sensitive marker of disease activity in ulcerative colitis. Am. J. Gastroenterol. **93:** 409–412.
59. KOEK, G.H., G.M. VERLEDEN, P. EVENEPOEL & P. RUTGEERTS. 2002. Activity related increase of exhaled nitric oxide in Crohn's disease and ulcerative colitis: a manifestation of systemic involvement? Respir. Med. **96:** 530–535.
60. LJUNG, T., M. HERULF, E. BEIJER, *et al.* 2001. Rectal nitric oxide assessment in children with Crohn disease and ulcerative colitis. Indicator of ileocaecal and colorectal affection. Scand. J. Gastroenterol. **36:** 1073–1076.
61. GUIHOT, G., R. GUIMBAUD, V. BERTRAND, *et al.* 2000. Inducible nitric oxide synthase activity in colon biopsies from inflammatory areas: correlation with inflammation intensity in patients with ulcerative colitis but not with Crohn's disease. Amino Acids **18:** 229–237.

Control of Bioamine Metabolism by 5-HT$_{2B}$ and α_{1D} Autoreceptors through Reactive Oxygen Species and Tumor Necrosis Factor-α Signaling in Neuronal Cells

BENOIT SCHNEIDER,[a] MATHÉA PIETRI,[a]
SOPHIE MOUILLET-RICHARD,[a] MYRIAM ERMONVAL,[a]
VINCENT MUTEL,[b] JEAN-MARIE LAUNAY,[c]
AND ODILE KELLERMANN[a]

[a]*Institut André Lwoff-Institut Pasteur, CNRS FRE 2937, Laboratoire Différenciation Cellulaire et Prions, Villejuif Cedex, France*

[b]*Pharma Research Department, F. Hoffman-La-Roche Ltd., Basel, Switzerland*

[c]*Service de Biochimie, EA3621 Hôpital Lariboisière, Faculté de Pharmacie, Université Paris V, Paris, France*

ABSTRACT: Homeostasis of the central nervous system relies on the proper integration of cell-signaling pathways recruited by a variety of neuronal and non-neuronal factors, with the aim of tightly controlling neurotransmitter metabolism, storage, and transport. We took advantage of the 1C11 neuroectodermal cell line, endowed with the capacity to selectively differentiate into serotonergic (1C11^{5-HT}) or noradrenergic (1C11^{NE}) neurons, to identify functional targets of serotonin (5-hydroxytryptamine [5–HT]) and norepinephrine (NE) autoreceptors possibly involved in the control of neuronal functions. We demonstrate that 5-HT$_{2B}$ and adreno α_{1D} receptors are coupled to reactive oxygen species (ROS) production through NADPH oxidase activation in 1C11^{5-HT} and 1C11^{NE} neuronal cells, respectively. In the signaling cascade linking 5–HT$_{2B}$ receptors to NADPH oxidase, phospholipase A2-mediated arachidonic acid production is required for ROS synthesis. ROS, in turn, act as second message signals and control the activation of TACE (TNF-α converting enzyme), a member of a disintegrin and metalloproteinase family. 5–HT$_{2B}$ and α_{1D} receptor stimulation triggers TACE-dependent TNF-α shedding in the surrounding milieu of 1C11^{5-HT} and 1C11^{NE} cells. In these cells, shed TNF-α triggers degradation of 5-HT and NE into 5-HIAA and MHPG, respectively. Finally, we

Address for correspondence: Dr. Benoît Schneider, Institut André Lwoff, CNRS FRE 2937, Laboratoire Différenciation Cellulaire et Prions, 7, rue Guy Môquet, 94801 Villejuif Cedex, France. Voice: +33-1-49-58-33-31; fax: +33-1-49-58-33-29.
e-mail: bschneid@vjf.cnrs.fr

observe that 5-HT_{2B} and α_{1D} receptor couplings to the NADPH oxidase-TACE cascade are strictly restricted to 1C11-derived progenies that have implemented a complete neuronal phenotype. Altogether, our data indicate that couplings of 5-HT_{2B} and α_{1D} autoreceptors to ROS and TNF-α signaling control neurotransmitter metabolism in 1C11-derived neuronal cells. Eventually, we might explain the origin of oxidative stress and high level of TNF-α in neurodegenerative diseases as a consequence of deviation of normal signaling pathways coupled to neurotransmitters.

KEYWORDS: bioamine metabolism; reactive oxygen species; TNF-α; autoreceptor signaling; neuronal differentiation

INTRODUCTION

The vertebrate nervous system consists of many distinct neuronal cell types, each of which is engaged in multiple cellular interactions to generate functional circuits. The onset and maintenance of a defined neuronal phenotype rely on the coordinated expression of a whole set of specialized proteins involved in neurotransmitter synthesis, storage, and transport. Noradrenergic and serotonergic neurons produce their neurotransmitters through a series of enzymatic modifications of tyrosine and tryptophan, leading to norepinephrine (NE) and serotonin (5-HT [5-hydroxytryptamine]), respectively.[1] In the central nervous system, NE neurons are mainly concentrated within the locus coeruleus and their axons irrigate all brain regions. 5-HT neurons are essentially located within the raphe nuclei and also project to most parts of the brain.[1,2] By innervating many areas of the brain, the NE and 5-HT monoaminergic systems control a wide range of behavioral and physiological processes, including sleep, cognition, and mood. Homeostasis of central NE and 5-HT neurons necessitates a tight control of neuronal functions through NE and 5-HT autoreceptor signaling.

As many as 14 5-HT receptors[2,3] and 9 adrenoceptors,[4] most of which are G-protein-coupled receptors (GPCRs), transduce 5-HT- and NE-associated signals. This diversity, and the lack of specific pharmacological tools toward each receptor, however, hinders our understanding of how each receptor, alone or in combination with others, contributes to 5-HT or NE physiological functions. Dissecting 5-HT- and NE-coupled pathways with respect to neurotransmitter-associated functions cannot be achieved using heterogeneous brain cell populations. Clonal cell lines, such as 1C11, expressing a definite set of either 5-HT or NE receptors, may allow these limitations to be overcome.

The 1C11 clone has the properties of a neuroectodermal progenitor endowed with the capacity to differentiate into serotonergic ($1C11^{5-HT}$) or noradrenergic ($1C11^{NE}$) neuronal cells.[5] Upon induction by Bt_2cAMP, almost 100% of 1C11 cells acquire, within 4 days, a fully differentiated serotonergic phenotype

($1C11^{5-HT}$ d4), including 5-HT metabolism, storage, and transport. $1C11^{5-HT}$ cells also acquire $5-HT_{1B/1D}$, $5-HT_{2B}$, and $5-HT_{2A}$ receptors along neuronal differentiation.[6] Binding and transductional experiments excluded the functional presence of any other 5-HT receptor subtypes. By day 2 of the serotonergic differentiation, $5-HT_{1B/1D}$ and $5-HT_{2B}$ receptors become expressed and remain functional until day 4, at which time the $5-HT_{2A}$ receptor is induced. The cell surface expression of this latter receptor coincides with the onset of an active 5-HT transport system. Under combined addition of Bt_2cAMP and DMSO, 100% of 1C11 cells are converted within 12 days into fully functional noradrenergic neurons ($1C11^{NE}$ d12) able to synthesize, store, and take up NE. $1C11^{NE}$ cells respond to external NE from day 8 of noradrenergic differentiation via a single α_{1D} adrenoceptor.[5] This receptor plays a pivotal role in the onset of a functional NE transport system at day 12 of the neuronal program. Along either differentiation program, the bioaminergic receptors act as autoreceptors and mediate the effects of 5-HT or NE in the coordination and/or onset of all neurotransmitter-associated functions.[5] Such sequences of events in the inducible 1C11 cell line offer the opportunity to explore 5-HT and α_{1D} receptor couplings within a complete neuronal phenotype.

$5-HT_{1B/1D}$, $5-HT_{2B}$, $5-HT_{2A}$, and α_{1D} receptors expressed in the 1C11 cell system are G-protein-coupled receptors (GPCRs). The 1C11 cell line, as well as various cell systems, have allowed several of their coupling mechanisms to be characterized. In $1C11^{5-HT}$ serotonergic cells, $5-HT_{1B/1D}$ receptors elicit their response through Gi-mediated inhibition of adenylate cyclase. 5-HT binding to either $5-HT_2$ receptor subtypes activates Gq protein and stimulates phospholipase C (PLCβ), thereby initiating a release of IP_3 and a rise in intracellular Ca^{2+} level. $5-HT_{2B}$ receptors lose their ability to couple with the IP_3 pathway when $5-HT_{2A}$ receptors become induced at day 4 of differentiation. With respect to $5-HT_{2A}$ receptors, they are coupled to the PLA2 cascade.[7] We have also reported a coupling of $5-HT_{2B}$ receptors to the phospholipase A2 (PLA2)/arachidonic acid pathway in $1C11^{5-HT}$ cells.[7] In addition, the $5-HT_{2B}$ receptor recruits both cellular and inducible NO synthase transduction pathways through a type I PDZ domain.[8] Finally, with the help of the 1C11 cell system, we recently demonstrated that $5-HT_{2B}$ receptors are functionally coupled to NADPH oxidase–mediated reactive oxygen species (ROS) synthesis.[9] ROS act as "second message" signals and in turn control TNF-α shedding. By contrast with the $5-HT_{2B}/IP_3$ coupling, the $5-HT_{2B}$/ROS/TNF-α signaling cascade is specific to $1C11^{5-HT}$ cells that have implemented a complete serotonergic phenotype. In $1C11^{NE}$ noradrenergic cells, α_{1D} adrenoceptors are coupled to IP_3 production from day 8 of noradrenergic differentiation.[5] As observed for the $5-HT_{2B}$ receptor in $1C11^{5-HT}$ serotonergic cells, α_{1D} receptor stimulation triggers ROS and TNF-α productions in $1C11^{NE}$ cells. This coupling is also restricted to the terminal stage of differentiation (day 12).[9]

Because we suspect that the coupling of $5-HT_{2B}$ and α_{1D} autoreceptors to ROS and TNF-α signaling takes part in neuronal homeostasis, we further investigated these pathways in relation to neurotransmitter-associated functions. The present study substantiates the occurrence of functional links between $5-HT_{2B}$- or α_{1D} receptor-dependent ROS and TNF-α signaling and neurotransmitter metabolism.

MATERIALS AND METHODS

Materials

Dibutyryl cyclic AMP (Bt_2cAMP), cyclohexane carboxylic acid (CCA), and dimethyl sulfoxide (DMSO) were purchased from Sigma-Aldrich (St. Louis, MO, USA). Neurochemicals were obtained from RBI (Natick, MA, USA). All other chemicals, of the purest grade available, were from classical commercial sources. Agonists (MK212, Ro 60.0175, DOI, BW723C86) of 5-HT receptors were selected according to Refs. 10–12.

Cell Culture and Enzyme Inhibition

1C11 cells were grown and induced to differentiate along the serotonergic ($1C11^{5-HT}$) or catecholaminergic ($1C11^{NE}$) pathway as previously described.[5] NADPH oxidase activity was inhibited with diphenyleneiodonium (DPI; Sigma) by pretreating cells at 37°C for 45 min in culture medium containing the inhibitor. Phospholipase A2 (PLA2) activity was cancelled upon cell incubation with mepacrine (Sigma) at 37°C for 45 min. TNF-α-converting enzyme (TACE) was switched off by pretreating cells with TAPI-2 (TNF-α Processing Inhibitor-2, Peptides International, Louisville, KY, USA), following the same procedure as for DPI or mepacrine experiments.

Reactive Oxygen Species (ROS) Detection

ROS production with 1C11 precursor cells and their neuronal progenies ($1C11^{5-HT}$ and $1C11^{NE}$) was followed using the fluorogenic reagent OxyBurst^R Green H_2HFF BSA (Molecular Probes, Breda, the Netherlands).[13] In a typical assay, cells grown in a 96-well microculture plate were washed twice with PBS supplemented with 1 mM Ca^{2+} and Mg^{2+} and preincubated for 2 min at 37°C in the presence of the fluorogenic reagent (10 μg/mL) in a serotonin-free culture medium. For stimulation of $1C11^{5-HT}$ or $1C11^{NE}$ cells with $5-HT_2$ agonists or norepinephrine (NE), respectively, fluorescence was continuously recorded at $\lambda = 528$ nm (slit width = 10 nm) with excitation at $\lambda = 488$ nm (slit width = 10 nm) using a Cary Eclipse fluorometer (Varian Inc., Palo Alto, CA, USA).

Membrane Fraction Preparation and Immunoprecipitation of ^{32}P-labeled p47^{PHOX} NADPH Oxidase Subunits

1C11^{5-HT} or 1C11^{NE} cells were washed in PBS supplemented with 1 mM Ca^{2+} and Mg^{2+} and incubated in phosphate-free DMEM containing 50 μCi [^{32}P]PO$_4$$^{3-}$ (NEN–Life Science, Boston, MA, USA) per milliliter per 10^6 cells for 1 h at 37°C. After receptor stimulation with agonists or NE, ^{32}P-labeled cells were washed twice with ice-cold PBS, and then scraped off into PBS containing a cocktail of protease inhibitors. After centrifugation, the supernatant was removed and the pellet was frozen at –80°C. Plasma membranes were prepared according to Kellerman et al.[6] in the presence of 100 mM Na$_3$VO$_4$ to avoid protein dephosphorylation. p47^{PHOX} NADPH oxidase subunit was immunoprecipitated from the plasma membrane fraction overnight at 4°C under gentle mixing using G Sepharose beads (Amersham Pharmacia, Piscataway, NJ, USA) coupled to C-20 antibodies (Santa Cruz Biotechnology, Santa Cruz, CA, USA). Immunocomplexes were resolved by 10% SDS-PAGE and transferred to Immobilon membranes. ^{32}P-labeled p47^{PHOX} was detected using a phosphorImager (Molecular Dynamics, Sunnyvale, CA, USA).

TNF-α Quantification

TNF-α from culture media was measured with enzyme-linked immunosorbent assays (ELISA) according to the manufacturer's instructions (R&D System, Minneapolis, MN, USA). Cytosolic extracts of 1C11^{5-HT} and 1C11^{NE} neuronal cells were prepared by incubating cells for 30 min at 4°C in NET lysis buffer (50 mM Tris pH 7.4, 150 mM NaCl, 5 mM EDTA, 1% Triton X-100, 100 mM Na$_3$VO$_4$, and a cocktail of protease inhibitors). Extracts were centrifuged at 14,000 × g for 15 min. Protein concentrations in the supernatants were measured by using the bicinchoninic acid method (Pierce, Rockford, IL, USA). To standardize the results, induced TNF-α shedding was expressed in TNF-α units per milligram of cytosolic protein.

5-Hydroxy Indole Acetic Acid (5-HIAA) and 3-Methoxy-4-Hydroxyphenylglycol (MHPG) Quantifications

5-HIAA and MHPG, two degradative products of 5-HT and NE catabolism, were measured in cytosolic extracts of 1C11^{5-HT} and 1C11^{NE} neuronal cells by high-performance liquid chromatography as described by Bassant et al.[14] To standardize the results, 5-HIAA and MHPG accumulations were expressed in 5-HIAA and MHPG units per milligram of cytosolic protein.

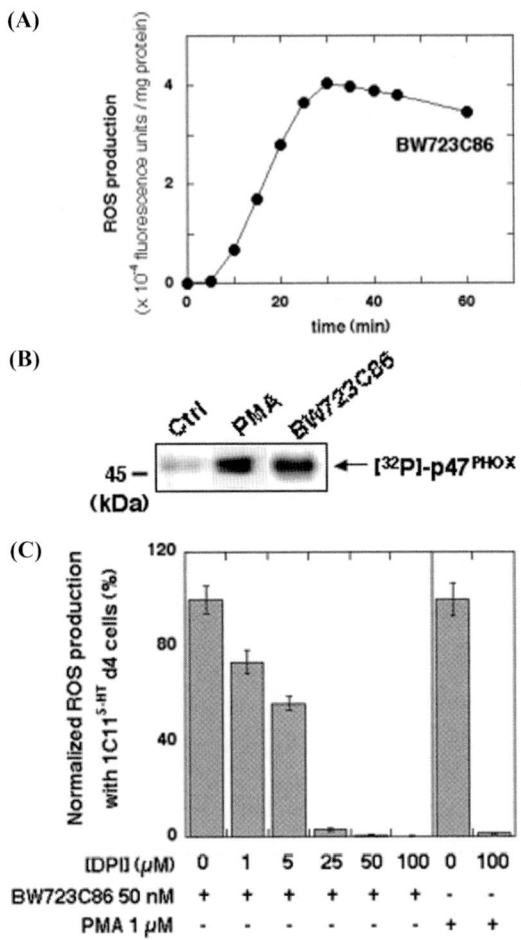

FIGURE 1. Stimulation of 5-HT$_{2B}$ receptors triggers NADPH oxidase-mediated ROS synthesis in 1C11^{5-HT} d4 cells. (**A**) Extracellular release of ROS was detected by using the fluorogenic reagent Oxyburst^R Green H2HFF-BSA (Molecular Probes). Time-dependent stimulation of ROS release was recorded in 1C11^{5-HT} d4 cells upon 5-HT$_{2B}$ receptor activation by 50 nM BW723C86. ROS release was completed within 30 min of 5-HT$_{2B}$ receptor stimulation. (**B**) Involvement of NADPH oxidase in 5-HT$_{2B}$ receptor-induced ROS production was assessed through activation of p47^{PHOX} NADPH oxidase subunit at the plasma membrane. After metabolic labeling with [^{32}P]PO$_4^{3-}$, 1C11^{5-HT} d4 cells were stimulated with 50 nM BW723C86 for 30 min. ^{32}P-labeled p47^{PHOX} (47 kDa) was immunoprecipitated from the cell membrane fraction and detected with a PhosphorImager after SDS-PAGE analysis. PMA (1 μM) was used as positive control. (**C**) The engagement of NADPH oxidase in 5-HT$_{2B}$ receptor-dependent signaling pathways was confirmed by using DPI, a specific inhibitor of NADPH oxidase activity. 1C11^{5-HT} d4 cells were preincubated with DPI concentrations ranging from 1 μM to 100 μM and exposed to 50 nM BW723C86 in combination with DPI. ROS release was followed over a period of 30 min. ROS production kinetics were normalized with respect to the velocity measured in the absence of DPI (100%).

RESULTS

Stimulation of 5-HT_{2B} Receptors Elicits NADPH Oxidase-Dependent ROS Production in Mature Serotonergic $1C11^{5-HT}$ d4 Cells

As underlined in a previous study,[9] 5-HT_{2B} receptors are functionally coupled to ROS production in fully differentiated $1C11^{5-HT}$ cells ($1C11^{5-HT}$ d4). A typical ROS response obtained upon exposure of these cells to 50 nM BW723C86, an agonist selective for 5-HT_{2B} receptors, is shown in FIGURE 1A. A salient observation was the lack of ROS production upon receptor stimulation with BW723C86 on $1C11^{5-HT}$ d2 cells, which already express 5-HT_{2B} receptors. We thus concluded that 5-HT_{2B} receptors are functionally coupled to ROS production only in $1C11^{5-HT}$ cells that have acquired a complete serotonergic phenotype.

NADPH oxidase constitutes a powerful superoxide anion generator and is a downstream effector in many signaling cascades.[15] We emphasized that this enzyme is the source of ROS in the 5-HT_{2B} receptor-induced ROS synthesis.[9] NADPH oxidase is a multicomponent enzyme and its activation relies in part on the phosphorylation steps of its cytosolic subunits ($p47^{PHOX}$, $p67^{PHOX}$) and translocation to the plasma membrane, where they associate with flavocytochrome b_{558} to form the catalytically active oxidase.[16,17] We observed that exposure of $1C11^{5-HT}$ d4 cells to 50 nM BW723C86 markedly raised the level of phosphorylation of the $p47^{PHOX}$ subunit associated with the plasma membrane fraction as early as 5 min following agonist addition (FIG.1B). This response resembled that observed on exposure of $1C11^{5-HT}$ d4 cells to the phorbol ester PMA, a potent activator of NADPH oxidase through $PKCs^{18}$ (FIG. 1B). The use of diphenyleneiodonium (DPI), a selective inhibitor of NADPH oxidase, allowed us to firmly establish the involvement of NADPH oxidase in 5-HT_{2B} receptor-coupled signaling cascades. Exposure of $1C11^{5-HT}$ cells to 100 μM DPI fully cancelled BW723C86- or PMA-induced ROS response (FIG.1C). Our data thus identified a NADPH oxidase-dependent ROS production pathway as a feature of 5-HT_{2B} receptor signaling activity in mature serotonergic neuronal cells.

The Coupling of 5-HT_{2B} Receptors to NADPH Oxidase Depends on Phospholipase A2 (PLA2)-Mediated Arachidonic Acid Production

The exact mechanisms sustaining NADPH oxidase activation upon 5-HT_{2B} receptor stimulation in $1C11^{5-HT}$ d4 cells remain elusive. Phospholipase A2 (PLA2)-mediated arachidonic acid production has been reported to modulate the plasma membrane translocation of $p47^{PHOX}$ and $p67^{PHOX}$ NADPH oxidase subunits in myeloid PLB-985 or monocytic U937 cells.[19,20] Because 5-HT_{2B} receptors are functionally linked to PLA2-mediated arachidonic acid production in $1C11^{5-HT}$ d4 cells,[7] we sought to determine whether arachidonic acid

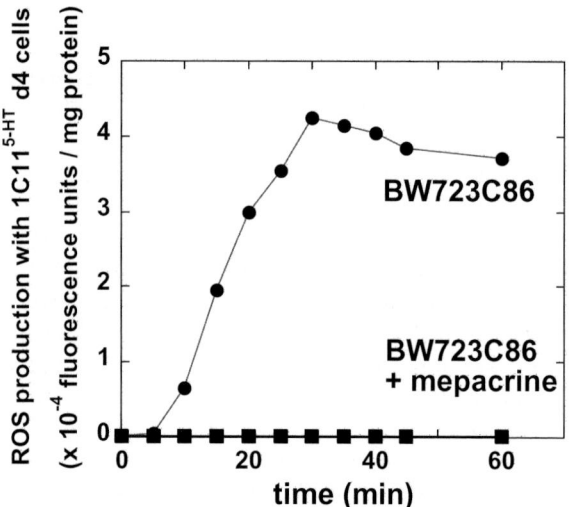

FIGURE 2. Phospholipase A2-mediated arachidonic acid production is necessary to 5-HT$_{2B}$ receptor-induced ROS synthesis. The engagement of phospholipase A2 (PLA2) in signaling pathways linking 5-HT$_{2B}$ receptors to NADPH oxidase was demonstrated by using mepacrine, a specific inhibitor of PLA2 activity. 1C11^{5-HT} d4 cells were preincubated with 100 μM mepacrine prior to 5-HT$_{2B}$ receptor stimulation with 50 nM BW723C86. ROS release was followed over a period of 30 min as in FIGURE 1. The 5-HT$_{2B}$ receptor-induced ROS production was fully cancelled with mepacrine.

could be involved in signaling cascades linking 5-HT$_{2B}$ receptors to NADPH oxidase. Treatment of 1C11^{5-HT} d4 cells with 100 μM mepacrine, a PLA2 inhibitor, prior to the addition of 50 nM BW723C86, fully abrogated the 5-HT$_{2B}$ receptor-induced ROS production (FIG. 2). The latter result demonstrates that PLA2 is a necessary intermediate in the pathway linking 5-HT$_{2B}$ receptor to NADPH oxidase activity in 1C11^{5-HT} serotonergic cells.

TNF-α Converting Enzyme is (TACE) a Downstream Target of the 5-HT$_{2B}$ Receptor-NADPH Oxidase Signaling Pathway

Beyond their cytotoxic role, ROS now emerge as signal transducers and have pleiotropic targets, including transcription factors, kinases, or metalloproteinases involved in ectodomain shedding of membrane-bound proteins.[21–23] We focused our search of downstream targets of the 5-HT$_{2B}$ receptors-NADPH oxidase cascade on members of the ADAM (a disintegrin and metalloproteinase) family, a class of proteases involved in the shedding of many cell surface proteins.[24] A well-known representative is the TACE (tumor necrosis factor-α converting enzyme) protein, which primarily governs the shedding of TNF-α.[25] TACE activation depends on a two-step mechanism, wherein the oxidative role of ROS is crucial.[23] We clearly established that the 5-HT$_{2B}$ receptor-NADPH oxidase signaling pathway targets TACE in 1C11^{5-HT} cells.[9] TACE enzymic activity was assessed through both the cleavage of the transmembrane pro-TNF-α into soluble TNF-α and its inhibition by TAPI-2. As depicted in FIGURE 3, BW723C86-mediated stimulation of 5-HT$_{2B}$ receptors triggered TNF-α shedding in the surrounding milieu of 1C11^{5-HT} d4 cells. TNF-α release became detectable in as early as 10 min and reached its maximum after 30 min of receptor stimulation. By contrast, with 1C11^{5-HT} d2 cells, agonist stimulation of 5-HT$_{2B}$ receptors failed to trigger any TNF-α response. In addition, 5-HT$_{2B}$ receptor-induced TNF-α shedding was cancelled upon 1C11^{5-HT} d4 cell exposure to 100 μM TAPI-2, prior to agonist stimulation (FIG. 3). Also, NADPH oxidase inhibition through DPI exposure switched off the agonist-induced TNF-α release (FIG. 3). These data exemplify that, in mature 1C11^{5-HT} serotonergic 5-HT$_{2B}$ receptors are functionally linked to TACE activation through a NADPH oxidase-mediated signaling cascade.

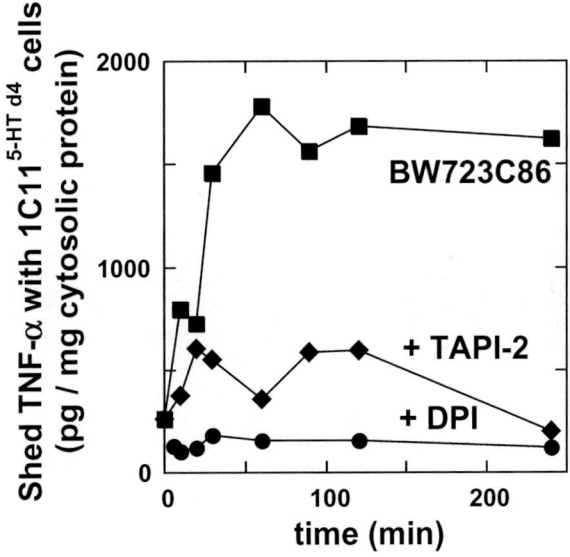

FIGURE 3. 5-HT$_{2B}$ receptor stimulation triggers TNF-α shedding in 1C11^{5-HT} d4 cells through ROS signaling. 5-HT$_{2B}$ receptor-mediated TNF-α shedding was followed for 240 min with 1C11^{5-HT} d4 cells stimulated with 50 nM BW723C86. The TNF-α response was completed within 30 min of stimulation. An involvement of TACE and NADPH oxidase in 5-HT$_{2B}$ receptor-induced TNF-α release was assessed by using TAPI-2, a specific inhibitor of TACE, or DPI. 1C11^{5-HT} d4 cells were preincubated for 45 min with 100 μM of the inhibitor prior to 50 nM BW723C86 exposure. TNF-α shedding was followed over a period of 240 min.

The 5-HT$_{2B}$ Receptor/NADPH Oxidase/TACE Cascade Controls 5-HT Metabolism

Next we examine, in fully differentiated 1C11^{5-HT} cells, the impact of 5-HT$_{2B}$ receptor-mediated ROS production and TNF-α shedding on 5-HT metabolism. Stimulation of 5-HT$_{2B}$ receptors with 50 nM BW723C86 in 1C11^{5-HT} d4 cells triggers an intracellular increase of 5-hydroxyindol acetic acid (5-HIAA), a degradative product of 5-HT. 5-HIAA accumulates as early as 20 min after agonist exposure (FIG. 4) and 5-HIAA production reaches a plateau at 90 min (FIG. 4). These results demonstrate that stimulation of 5-HT$_{2B}$ autoreceptors regulates the level of intraneuronal 5-HT by controlling 5-HT degradation into 5-HIAA.

We next wondered whether ROS could take part in the 5-HT$_{2B}$ receptor-mediated 5-HT catabolism. That ROS may take part in 5-HIAA generation is indeed supported by our observation that NADPH oxidase activation by 1 μM PMA triggers 5-HIAA accumulation in 1C11^{5-HT} d4 cells (data not shown). As shown in FIGURE 4, DPI addition quenches the BW723C86-induced degradation of 5-HT in these cells. Hence, the 5-HIAA response elicited by 5-HT$_{2B}$

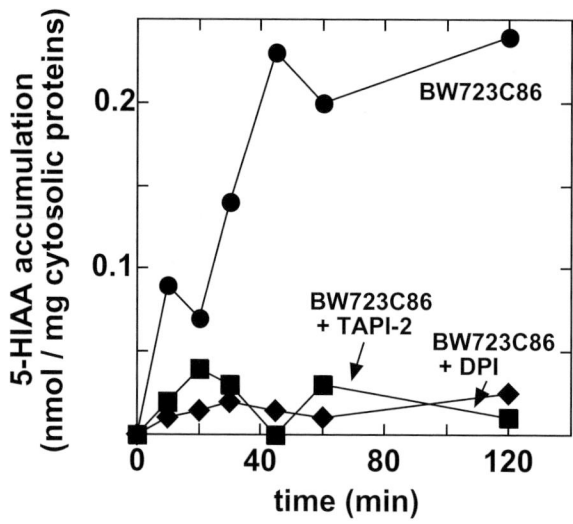

FIGURE 4. 5-HT$_{2B}$ receptors control 5-HT catabolism through ROS and TNF-α signaling. The accumulation of 5-HIAA, a degradative product of 5-HT, was followed during 120 min with 1C11^{5-HT} d4 cells exposed to 50 nM BW723C86. 5-HT$_{2B}$ receptor stimulation triggers 5-HT conversion into 5-HIAA within a 60-min time scale. An involvement of TNF-α and ROS signaling in 5-HT$_{2B}$ receptor-induced 5-HIAA accumulation was assessed by using TAPI-2 and DPI, respectively. 1C11^{5-HT} d4 cells were preincubated for 45 min with 100 μM of the inhibitor prior to 50 nM BW723C86 exposure. 5-HIAA production was followed over a period of 120 min.

receptor activation is under the control of NADPH oxidase-mediated ROS production. To further delineate the relation between 5-HT$_{2B}$ receptor-dependent 5-HIAA accumulation and the 5-HT$_{2B}$ receptor-NADPH oxidase-TACE cascade, we assessed the impact of TACE inhibition on agonist-induced 5-HT catabolism. Incubation of 1C11^{5-HT} d4 cells with 100 μM TAPI-2 prior to BW723C86-mediated 5-HT$_{2B}$ receptor stimulation fully cancels 5-HIAA accumulation (FIG. 4). As a whole, our data highlight that the 5-HT$_{2B}$ receptor control on 5-HT degradation is mediated by the ROS-TNF-α signaling pathway.

α$_{1D}$ Adrenoceptors Are Coupled to NADPH Oxidase-Mediated ROS Production and TACE-Dependent TNF-α Shedding in Fully Differentiated Noradrenergic 1C11^{NE} d12 Cells

An alternative fate of 1C11 precursor cells is to convert into fully functional noradrenergic neuronal cells. Along this differentiation pathway, 1C11^{NE} cells selectively implement a single α$_{1D}$ adrenoceptor at day 8 of the program.[5] No change in either the number of α$_{1D}$ binding sites or the pharmacological profile of the receptor has been observed from day 8 to day 12, when cells have acquired a mature noradrenergic phenotype.[5] In our previous report, we took advantage of this potential of the 1C11 cell line to probe for a NE-dependent ROS production.[9] FIGURE 5 recapitulates our main observations substantiating a coupling of α$_{1D}$ adrenoceptors to a NADPH oxidase-TACE cascade. First, ROS production is induced upon NE exposure in 1C11^{NE} d12 cells, which starts as early as 10 min after stimulation and reaches a plateau at 30 min (FIG. 5A). Because of the induction of a single adrenoceptor during noradrenergic differentiation, the NE-induced ROS response is transduced by α$_{1D}$ receptors. Second, the α$_{1D}$ receptor-associated ROS response is cancelled upon DPI treatment, which abrogates NADPH oxidase activity.[9] Third, activation of α$_{1D}$ receptors with NE on 1C11^{NE} d12 cells promotes p47^{PHOX} phosphorylation and translocation to the plasma membrane (FIG. 5B). This set of findings provides direct evidence for a functional coupling of α$_{1D}$ adrenoceptors to NADPH oxidase-dependent ROS production in mature 1C11^{NE} d12 cells.

Next we determined whether ROS produced upon stimulation of α$_{1D}$ receptors could serve as second messengers in 1C11^{NE} d12 cells and trigger TNF-α release. Stimulation of α$_{1D}$ adrenoceptors with 100 nM NE elicited TNF-α shedding in 1C11^{NE} d12 cells with a kinetics comparable to that observed upon 5-HT$_{2B}$ receptor activation with BW723C86 in 1C11^{5-HT} d4 cells.[9] NE-induced TNF-α shedding actually is a downstream event in the α$_{1D}$ receptor-NADPH oxidase-TACE cascade since it is cancelled under either NADPH oxidase inhibition with DPI or TACE inactivation using TAPI-2 (FIG. 5C). Altogether, our data highlight that, in mature 1C11^{NE} noradrenergic cells,

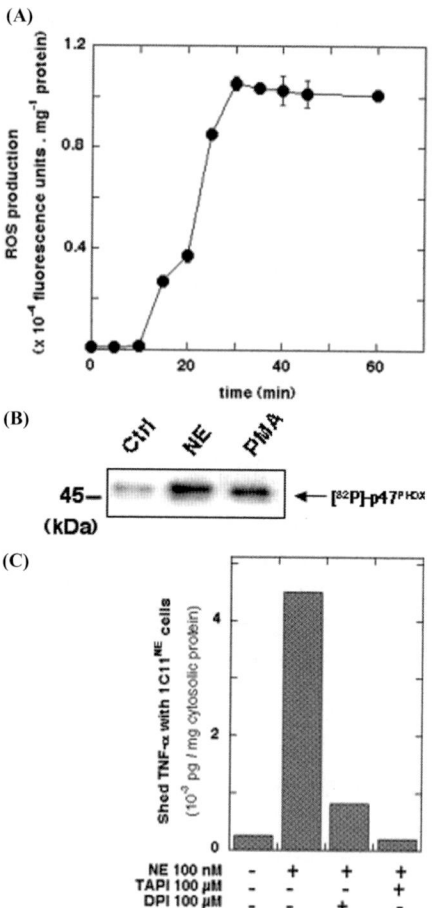

FIGURE 5. α_{1D} adrenoceptors are coupled to ROS production and TNF-α shedding in 1C11^{NE} d12 cells. (**A**) Stimulation of α_{1D} receptors induces ROS production in 1C11^{NE} d12 cells. Extracellular release of ROS was monitored as described in FIGURE 1. α_{1D} receptor-mediated ROS synthesis was measured over a period of 60 min in 1C11^{NE} d12 cells upon stimulation with 100 nM NE. Maximal ROS release was completed within 30-min stimulation of α_{1D} receptors. (**B**) Involvement of NADPH oxidase in NE-induced ROS production was monitored through activation of p47^{PHOX} NADPH oxidase subunit at the plasma membrane of 1C11^{NE} d12 cells following the same procedure as in FIGURE 1. PMA was used as positive control. (**C**) α_{1D} adrenoceptor stimulation induces TNF-α shedding through ROS signaling in 1C11^{NE} d12 cells. α_{1D} adrenoceptor-mediated TNF-α shedding was followed for 240 min with 1C11^{NE} d12 cells exposed to 100 nM NE. TNF-α response was completed within 30 min of stimulation. Quantities of shed TNF-α at 30 min in the culture medium of 1C11^{NE} cells are indicated. An involvement of TACE and NADPH oxidase activities in α_{1D} receptor-induced TNF-α release was assessed by using TAPI-2 and DPI, respectively. 1C11^{NE} d12 cells were pre-incubated for 45 min with 100 μM of the inhibitor prior to 100-nM NE exposure. TNF-α shedding was followed over a period of 240 min. Indicated levels correspond to 30-min NE stimulation of 1C11^{NE} d12 cells.

α_{1D} receptors are functionally linked to TACE activation through a NADPH oxidase-mediated signaling cascade.

TACE-Dependent TNF-α Release upon α_{1D} Adrenoceptor Stimulation Controls NE Metabolism

Following the example of the 5-HT$_{2B}$ receptor in 1C11^{5-HT} serotonergic cells, we examined the impact of the α_{1D} receptor-NADPH oxidase-TACE signaling cascade on NE metabolism with 1C11^{NE} d12 cells, by measuring the accumulation of 3- methoxy-4-hydroxyphenylglycol (MHPG), a physiological product of NE catabolism. Stimulation of 1C11^{NE} d12 cells with 100 nM NE triggers MHPG production in catecholaminergic cells (FIG. 6). MHPG accumulation becomes detectable at 20–30 min and reaches a maximum at 100–120 min (FIG. 6).

Quenching the α_{1D} receptor-induced ROS production with 100 μM DPI totally impairs MHPG accumulation in 1C11^{NE} d12 cells (FIG. 6). In addition, TACE inhibition with TAPI-2 fully cancels the α_{1D} receptor-related

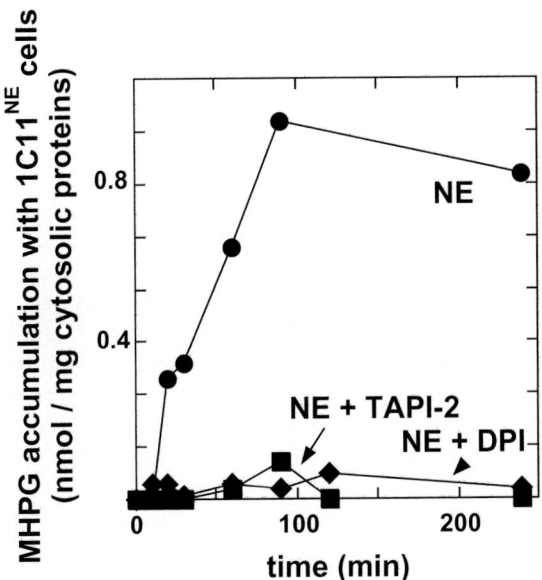

FIGURE 6. α_{1D} adrenoceptors control NE catabolism through ROS and TNF-α signaling. The accumulation of MHPG, a degradative product of NE, was followed for 240 min with 1C11^{NE} d12 cells exposed to 100 nM NE. α_{1D} receptor stimulation triggers NE conversion into MHPG within a 90-min time scale. An involvement of TNF-α and ROS signaling in α_{1D} receptor-induced MHPG accumulation was assessed by using TAPI-2 and DPI, respectively. 1C11^{NE} d12 cells were preincubated for 45 min with 100 μM of the inhibitor prior to 100 nM NE exposure. MHPG production was followed over a period of 240 min.

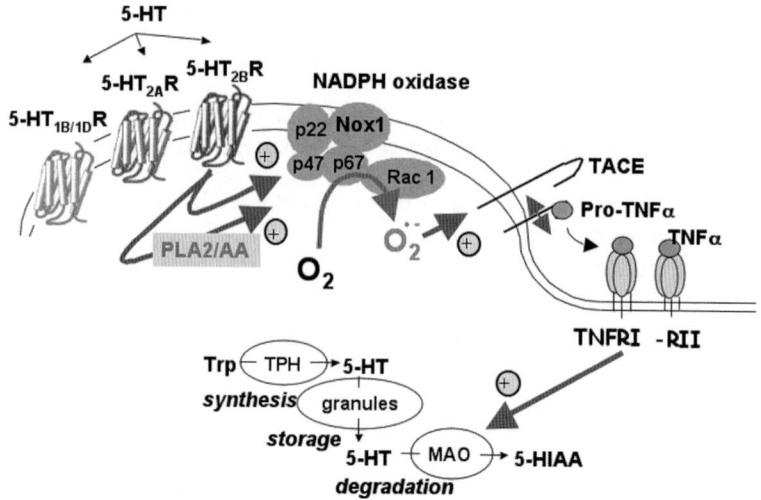

FIGURE 7. Model of 5-HT_{2B} receptor-dependent control of 5-HT metabolism through ROS and TNF-α signaling in serotonergic $1C11^{5\text{-HT}}$ cells. See text for further details.

NE degradation into MHPG (FIG. 6). We thus conclude that in noradrenergic $1C11^{NE}$ neuronal cells, α_{1D} adrenoceptor stimulation instructs NE degradation through a NADPH oxidase–TACE signaling cascade.

DISCUSSION

In vertebrates, homeostasis of the central nervous system relies on tight controls of neurotransmitter synthesis, storage, transport, and catabolism and depends on the signaling activity of many autoreceptors. *In vivo*, 5-HT_{1B} receptor signaling has been reported to potentiate 5-HT uptake by the serotonin transporter,[26] and 5-HT_{1A} receptors seem to control vesicular 5-HT storage,[27] and 5-HT_{2B} or α_{1D} receptors are also suspected to play the role of autoreceptors since transcripts encoding these receptors and related binding sites are found in brain serotonergic or noradrenergic neurons.[1–3,28–31] By establishing in 1C11-derived neuronal cells a functional link between 5-HT_{2B} and α_{1D} autoreceptor signaling and catabolism of 5-HT and NE neurotransmitters, our study sheds new light on how these bioamines, 5-HT or NE, take part in the fine-tuning of their own integrated metabolisms.

In a previous report, we established that 5-HT_{2B} or α_{1D} receptor stimulation elicits NADPH oxidase–dependent ROS synthesis in $1C11^{5\text{-HT}}$ and $1C11^{NE}$ cells.[9] This observation provided the prime evidence that bioaminergic autoreceptors take part in the control of the cellular redox equilibrium in a neuronal cell context. However, the capacity of 5-HT_{2B} or α_{1D} receptors to mobilize

NADPH oxidase is restricted to 1C11-derived neuronal cells that have acquired all the neuron- and neurotransmitter-associated functions, that is, at day 4 of the serotonergic program ($1C11^{5-HT}$ d4) or at day 12 of the noradrenergic differentiation ($1C11^{NE}$ d12). The molecular bases sustaining the delay in the coupling of these bioamine receptors to NADPH oxidase are difficult to dissect, since from day 2 of the serotonergic differentiation, or day 8 of the noradrenergic program, 5-HT_{2B} or α_{1D} receptors are functionally linked to other effectors including NOS and PLA2 for the 5-HT_{2B} receptor or PLC for the α_{1D} receptor.[5,7,8,32] Interestingly, the assembly of NADPH oxidase subunits into a functional complex appears to depend on sphingolipid- and cholesterol-enriched membrane microdomains ("lipid-rafts"), known as transduction platforms for many signaling processes.[33] We may speculate that the absence of 5-HT_{2B} or α_{1D} receptor couplings to NADPH oxidase before the terminal stage of neuronal differentiation reflects a localization of each protagonist in distinct raft microdomains. Along neuronal differentiation, neuritic extension and maturation of membrane properties would be associated with a spatial redistribution of signaling partners. This may allow $5\text{-HT}_{2B}/\alpha_{1D}$ receptors and NADPH oxidase subunits to concentrate in a proximate environment and to couple $5\text{-HT}_{2B}/\alpha_{1D}$ receptor signaling activity to ROS production. Alternatively, we cannot exclude that the coupling of 5-HT_{2B} or α_{1D} receptors to NADPH oxidase–mediated ROS production involves yet-to-be-identified intermediate molecules specifically expressed in fully differentiated cells. Nevertheless, it is very likely that the signaling intermediates between 5-HT_{2B} receptors and NADPH oxidase differ from those recruited by PMA, which mobilizes NADPH oxidase activity through PKCs.[18] Because, in $1C11^{5-HT}$ d4 cells, 5-HT_{2B} receptors are not coupled to the $PLC\beta/IP_3$ pathway,[34] we may exclude the participation of some classes of PKCs in 5-HT_{2B} receptor-mediated ROS response. These observations are in line with the occurrence of PKC-independent cascades in the phosphorylation steps of $p47^{PHOX}$ and $p67^{PHOX}$ NADPH oxidase subunits in monocytic cells.[35] On the other hand, in this study, we provide evidence that PLA2-mediated arachidonic acid production is involved in the 5-HT_{2B} receptor-mediated control of NADPH oxidase activity. Inhibition of PLA2 with mepacrine totally abrogates the 5-HT_{2B} receptor-induced ROS production in $1C11^{5-HT}$ d4 cells. This result supports the view that the control of NADPH oxidase activity operates through PLA2-dependent pathways in serotonergic neuronal cells (FIG. 7). With regard to noradrenergic $1C11^{NE}$ cells, α_{1D} adrenoceptors are coupled to $PLC\beta/IP_3$ through Gq.[5] A functional coupling of this adrenergic receptor to PLA2 has, however, never been reported in $1C11^{NE}$ cells. Investigations are in progress to examine whether the control of NADPH oxidase by α_{1D} receptors in $1C11^{NE}$ cells mobilizes PKC-dependent and/or PKC-independent pathways. In any case, restriction to mature bioaminergic neurons of the receptor couplings to NADPH oxidase favors the notion that such a signaling pathway contributes to the regulation and/or maintenance of neuron- and neurotransmitter-associated functions, rather than instructing early neuronal differentiation processes.

ROS produced upon 5-HT_{2B} or α_{1D} receptor stimulation play the role of "second message" signals and control the activation of TACE metalloproteinase, placing TACE as a novel downstream target of 5-HT_{2B}- and α_{1D}-associated signaling pathways in 1C11-derived neuronal cells[9] (FIG. 7). The control of TACE enzymic activity by ROS is in line with the previously reported two-step mechanisms of TACE activation, wherein the ROS-mediated oxidation of a crucial cysteine thiol group in the protease catalytic domain confers full activity.[23] Activated TACE, in turn, catalyzes the proteolytic cleavage of the transmembrane precursor pro-TNF-α and ensures the shedding of soluble TNF-α in the culture medium of 1C11$^{5\text{-HT}}$ and 1C11^{NE} cells. Again, because of the restriction of 5-HT_{2B}- or α_{1D} receptor-mediated TNF-α release to fully differentiated 1C11 neuronal cells, TNF-α likely plays a role in the maintenance of the neuronal phenotype. In line with this idea, we show in this study that 5-HT_{2B}- and α_{1D} receptor-associated TNF-α release has an impact on 5-HT and NE metabolism in 1C11$^{5\text{-HT}}$ and 1C11^{NE} cells, respectively. Upon agonist stimulation of these two bioaminergic receptors, TNF-α triggers the intraneuronal degradation of 5-HT and NE into 5-HIAA and MHPG, respectively. In agreement with this observation is the effect of brain administration of exogenous TNF-α, which enhances 5-HT catabolism.[36] In addition, TNF-α produced by neurons downregulates NE release from noradrenergic nerve terminals in the central nervous system.[37] In 1C11-derived neuronal progenies, by interacting with its two transmembrane receptors, TNF-R1 and TNF-R2,[38,39] TNF-α may mobilize signaling pathways targeting monoamine oxidases (MAOs), enzymes involved in 5-HT or NE catabolism (FIG. 7). The intermediate molecules in signaling cascades linking TNF receptors to MAOs are presently unknown. In any case, our overall data strongly suggest that TNF-α plays an important role in the regulatory processes that control bioamine metabolism in 1C11 neuronal cells.

Dysfunctions of the 5-HT or NE systems appear to be at the root of several neurological disorders, including Alzheimer's, Parkinson's, or prion diseases. In these neurodegenerative disorders, aberrant protein aggregation is widely suspected to interfere with bioaminergic signaling, which normally contributes to neuronal homeostasis.[40] Oxidative stress,[41,42] upregulated expression of proinflammatory cytokines such as TNF-α,[43–46] and reduction in neuronal bioamine metabolism[14,47] are also common hallmarks of these disorders. By linking 5-HT_{2B} and adreno α_{1D} receptors to ROS and TNF-α signaling, the present work gives some clues as to how deviation of 5-HT or NE normal signaling pathways, in relation to misfolded protein aggregation, may cause imbalances in ROS and TNF-α levels. Targeting these normal pathways to alleviate loss of neuronal homeostasis may represent new therapeutic potential in the field of neurodegenerative diseases.

REFERENCES

1. GORIDIS, C. & H. ROHRER. 2002. Specification of catecholaminergic and serotonergic neurons. Nat. Rev. Neurosci. **3:** 531–541.
2. GASPAR, P., O. CASES & L. MAROTEAUX. 2003. The developmental role of serotonin: news from mouse molecular genetics. Nat. Rev. Neurosci. **4:** 1002–1012.
3. HOYER, D., J.P. HANNON & G.R. MARTIN. 2002. Molecular, pharmacological and functional diversity of 5-HT receptors. Pharmacol. Biochem. Behav. **71:** 533–554.
4. PHILIPP, M. & L. HEIN. 2004. Adrenergic receptor knockout mice: distinct functions of 9 receptor subtypes. Pharmacol. Ther. **101:** 65–74.
5. MOUILLET-RICHARD, S., V. MUTEL, S. LORIC, et al. 2000. Regulation by neurotransmitter receptors of serotonergic or catecholaminergic neuronal cell differentiation. J. Biol. Chem. **275:** 9186–9192.
6. KELLERMANN, O., S. LORIC, L. MAROTEAUX & J.M. LAUNAY. 1996. Sequential onset of three 5-HT receptors during the 5-hydroxytryptaminergic differentiation of the murine 1C11 cell line. Br. J. Pharmacol. **118**(5): 1161–1170.
7. TOURNOIS, C., V. MUTEL, P. MANIVET, et al. 1998. Crosstalk between 5-HT receptors in a serotonergic cell line: involvement of arachidonic acid metabolism. J. Biol. Chem. **273:** 17498–17503.
8. MANIVET, P., S. MOUILLET-RICHARD, J. CALLEBERT, et al. 2000. PDZ-dependent activation of nitric-oxide synthases by the serotonin 2B receptor. J. Biol. Chem. **275:** 9324–9331.
9. PIETRI, M., B. SCHNEIDER, S. MOUILLET-RICHARD, et al. 2005. Reactive oxygen species-dependent TNF-alpha converting enzyme activation through stimulation of 5-HT2B and alpha1D autoreceptors in neuronal cells. FASEB J. **19:** 1078–1087.
10. PORTER, R.H., K.R. BENWELL, H. LAMB, et al. 1999. Functional characterization of agonists at recombinant human 5-HT2A, 5-HT2B and 5-HT2C receptors in CHO-K1 cells. Br. J. Pharmacol. **128:** 13–20.
11. CUSSAC, D., A. NEWMAN-TANCREDI, Y. QUENTRIC, et al. 2002. Characterization of phospholipase C activity at h5-HT2C compared with h5-HT2B receptors: influence of novel ligands upon membrane-bound levels of [3H]phosphatidylinositols. Naunyn Schmiedebergs Arch. Pharmacol. **365:** 242–252.
12. JERMAN, J.C., S.J. BROUGH, T. GAGER, et al. 2001. Pharmacological characterisation of human 5-HT2 receptor subtypes. Eur. J. Pharmacol. **414:** 23–30.
13. SCHNEIDER, B., V. MUTEL, M. PIETRI, et al. 2003. NADPH oxidase and extracellular regulated kinases 1/2 are targets of prion protein signaling in neuronal and nonneuronal cells. Proc. Natl. Acad. Sci. USA **100:** 13326–13331.
14. BASSANT, M.H., M. PICARD, D. OLICHON, et al. 1986. Changes in the serotonergic, noradrenergic and dopaminergic levels in the brain of scrapie-infected rats. Brain Res. **367:** 360–363.
15. SAUER, H., M. WARTENBERG & J. HESCHELER. 2001. Reactive oxygen species as intracellular messengers during cell growth and differentiation. Cell. Physiol. Biochem. **11:** 173–186.
16. QUINN, M.T., T. EVANS, L.R. LOETTERLE, et al. 1993. Translocation of Rac correlates with NADPH oxidase activation. Evidence for equimolar translocation of oxidase components. J. Biol. Chem. **268:** 20983–20987.

17. DUSI, S., V. DELLA BIANCA, M. GRZESKOWIAK & F. ROSSI. 1993. Relationship between phosphorylation and translocation to the plasma membrane of p47phox and p67phox and activation of the NADPH oxidase in normal and Ca(2+)-depleted human neutrophils. Biochem. J. **290**(Pt 1): 173–178.
18. KAWAKAMI, N., K. KITA, T. HAYAKAWA, et al. 2000. Phorbol myristate acetate induces NADPH oxidase activity of cytochalasin B-primed neutrophils through the protein kinase C-independent pathway. Biol. Pharm. Bull. **23:** 1100–1104.
19. DANA, R., T.L. LETO, H.L. MALECH & R. LEVY. 1998. Essential requirement of cytosolic phospholipase A2 for activation of the phagocyte NADPH oxidase. J. Biol. Chem. **273:** 441–445.
20. SELLMAYER, A., H. OBERMEIER, U. DANESCH, et al. 1996. Arachidonic acid increases activation of NADPH oxidase in monocytic U937 cells by accelerated translocation of p47-phox and co-stimulation of protein kinase C. Cell Signal. **8:** 397–402.
21. KAMATA, H. & H. HIRATA. 1999. Redox regulation of cellular signalling. Cell Signal. **11:** 1–14.
22. THANNICKAL, V.J. & B.L. FANBURG. 2000. Reactive oxygen species in cell signaling. Am. J. Physiol. Lung Cell Mol. Physiol. **279:** L1005–L1028.
23. ZHANG, Z., P. OLIVER, J.J. LANCASTER, et al. 2001. Reactive oxygen species mediate tumor necrosis factor alpha-converting, enzyme-dependent ectodomain shedding induced by phorbol myristate acetate. FASEB J. **15:** 303–305.
24. DUFFY, M.J., D.J. LYNN, A.T. LLOYD & C.M. O'SHEA. 2003. The ADAMs family of proteins: from basic studies to potential clinical applications. Thromb. Haemost. **89:** 622–631.
25. BLACK, R.A. 2002. Tumor necrosis factor-alpha converting enzyme. Int. J. Biochem. Cell. Biol. **34:** 1–5.
26. DAWS, L.C., G.G. GOULD, S.D. TEICHER, et al. 2000. 5-HT(1B) receptor-mediated regulation of serotonin clearance in rat hippocampus in vivo. J. Neurochem. **75:** 2113–2122.
27. PINEYRO, G. & P. BLIER. 1999. Autoregulation of serotonin neurons: role in antidepressant drug action. Pharmacol. Rev. **51:** 533–591.
28. CHOI, D.S. & L. MAROTEAUX. 1996. Immunohistochemical localisation of the serotonin 5-HT2B receptor in mouse gut, cardiovascular system, and brain. FEBS Lett. **391:** 45–51.
29. DUXON, M.S., T.P. FLANIGAN, A.C. REAVLEY, et al. 1997. Evidence for expression of the 5-hydroxytryptamine-2B receptor protein in the rat central nervous system. Neuroscience **76:** 323–329.
30. VOLGIN, D.V., M. MACKIEWICZ & L. KUBIN. 2001. Alpha(1B) receptors are the main postsynaptic mediators of adrenergic excitation in brainstem motoneurons, a single-cell RT-PCR study. J. Chem. Neuroanat. **22:** 157–166.
31. OSBORNE, P.B., M. VIDOVIC, B. CHIENG, et al. 2002. Expression of mRNA and functional alpha(1)-adrenoceptors that suppress the GIRK conductance in adult rat locus coeruleus neurons. Br. J. Pharmacol. **135:** 226–232.
32. LORIC, S., L. MAROTEAUX, O. KELLERMANN & J.M. LAUNAY. 1995. Functional serotonin-2B receptors are expressed by a teratocarcinoma-derived cell line during serotoninergic differentiation. Mol. Pharmacol. **47:** 458–466.
33. VILHARDT, F. & B. VAN DEURS. 2004. The phagocyte NADPH oxidase depends on cholesterol-enriched membrane microdomains for assembly. EMBO J. **23:** 739–748.

34. MOUILLET-RICHARD, S., M. PIETRI, B. SCHNEIDER, et al. 2005. Modulation of serotonergic receptor signaling and cross-talk by prion protein. J. Biol. Chem. **280**: 4592–4601.
35. BENNA, J.E., P.M. DANG, M. GAUDRY, et al. 1997. Phosphorylation of the respiratory burst oxidase subunit p67(phox) during human neutrophil activation. Regulation by protein kinase C-dependent and independent pathways. J. Biol. Chem. **272**: 17204–17208.
36. HAYLEY, S., K. BREBNER, S. LACOSTA, et al. 1999. Sensitization to the effects of tumor necrosis factor-alpha: neuroendocrine, central monoamine, and behavioral variations. J. Neurosci. **19**: 5654–5665.
37. IGNATOWSKI, T.A., B.K. NOBLE, J.R. WRIGHT, et al. 1997. Neuronal-associated tumor necrosis factor (TNF alpha): its role in noradrenergic functioning and modification of its expression following antidepressant drug administration. J. Neuroimmunol. **79**: 84–90.
38. BAUD, V. & M. KARIN. 2001. Signal transduction by tumor necrosis factor and its relatives. Trends Cell Biol. **11**: 372–377.
39. MACEWAN, D.J. 2002. TNF receptor subtype signalling: differences and cellular consequences. Cell Signal. **14**: 477–492.
40. MATTSON, M.P. & M. SHERMAN. 2003. Perturbed signal transduction in neurodegenerative disorders involving aberrant protein aggregation. Neuromol. Med. **4**: 109–132.
41. BARNHAM, K.J., C.L. MASTERS & A.I. BUSH. 2004. Neurodegenerative diseases and oxidative stress. Nat. Rev. Drug Discov. **3**: 205–214.
42. ANDERSEN, J.K. 2004. Oxidative stress in neurodegeneration: cause or consequence? Nat. Med. **10**(Suppl): S18–S25.
43. KORDEK, R., V.R. NERURKAR, P.P. LIBERSKI, et al. 1996. Heightened expression of tumor necrosis factor alpha, interleukin 1 alpha, and glial fibrillary acidic protein in experimental Creutzfeldt-Jakob disease in mice. Proc. Natl. Acad. Sci. USA **93**: 9754–9758.
44. WILLIAMS, A., A.M. VAN DAM, D. RITCHIE, et al. 1997. Immunocytochemical appearance of cytokines, prostaglandin E2 and lipocortin-1 in the CNS during the incubation period of murine scrapie correlates with progressive PrP accumulations. Brain Res. **754**: 171–180.
45. VIVIANI, B., S. BARTESAGHI, E. CORSINI, et al. 2004. Cytokines role in neurodegenerative events. Toxicol. Lett. **149**: 85–89.
46. MHATRE, M., R.A. FLOYD & K. HENSLEY. 2004. Oxidative stress and neuroinflammation in Alzheimer's disease and amyotrophic lateral sclerosis: common links and potential therapeutic targets. J. Alzheimer's Dis. **6**(2): 147–157.
47. GOTTFRIES, C.G. 1990. Neurochemical aspects on aging and diseases with cognitive impairment. J. Neurosci. Res. **27**: 541–547.

Determination of Oxidative Stress Status and Concentration of TGF-β1 in the Blood and Saliva of Osteoporotic Subjects

GHOLAMREZA YOUSEFZADEH,[a] BAGHER LARIJANI,[a]
AZADEH MOHAMMADIRAD,[b] RAMIN HESHMAT,[a]
GHOLAMREZA DEHGHAN,[b] ROJA RAHIMI,[b]
AND MOHAMMAD ABDOLLAHI[b]

[a]*Endocrinology and Metabolism Research Centre, Tehran University of Medical Sciences, Tehran, Iran*

[b]*Pharmaceutical Sciences Research Center, Tehran University of Medical Sciences, Tehran, Iran*

ABSTRACT: Preliminary reports indicate the influence of oxidative stress and interleukins, particularly TGF-β1, in maintenance of bone mass. This study was designed to determine any possible variations of cellular lipid peroxidation, the total antioxidant power, and concentration of TGF-β1 in blood and saliva of osteoporotic subjects in comparison to healthy people. Blood and saliva samples of 22 osteoporotic women and 22 age-matched healthy women were collected. Samples were analyzed for thiobarbituric acid-reactive substances (TBARS) as a marker of lipid peroxidation, ferric reducing ability (total antioxidant power, TAP), and concentration of TGF-β1. The blood and saliva TAP (mean ± SD) of osteoporotic subjects was significantly lower than that of healthy controls (606.65 ± 119.13 vs. 665.64 ± 63.73 mmol/L and 560.43 ± 84.70 vs. 612.05 ± 81.5, respectively). Blood and saliva TBARS (mean ± SD) of osteoporotic subjects were significantly higher than those of healthy controls (0.30 ± 0.04 vs. 0.26 ± 0.04 and 0.23 ± 0.03 vs. 0.16 ± 0.04 μmol/L, respectively). Concentrations of TGF-β1 (mean ± SD) in plasma and saliva of osteoporotic subjects were not different in comparison to healthy subjects. Results indicate that persons with osteoporosis have an increased oxidative stress that is not accompanied by changes in TGF-β1 levels. Use of supplementary antioxidants in osteoporotic patients may be helpful.

Address for correspondence: Prof. Mohammad Abdollahi, Department of Toxicology and Pharmacology, Faculty of Pharmacy, and Pharmaceutical Sciences Research Center, Tehran University of Medical Sciences, Tehran 14155-6451, Iran. Voice/fax: +98-21-66959104.
  e-mail: mohammad.abdollahi@utoronto.ca

KEYWORDS: oxidative stress; TGF-β; osteoporosis; blood; saliva

INTRODUCTION

Osteoporosis has been defined as a disease characterized by low bone mass and microarchitectural deterioration of bone tissue leading to increased bone fragility and a resulting increase in fracture risk.[1] Osteoporosis is one of the most common human disorders worldwide, especially in postmenopausal women. Osteoporosis results from a change in the balance between activities of osteoblasts and osteoclasts. These specialized cells are responsible for bone formation and resorption.[2] The activities of these bone cells can be influenced by a variety of nutritional and cellular factors, including the supply of oxygen, nutrients, endocrines, and growth factors.[3,4] In recent years, cytokines and free radicals have been viewed with interest as efficient factors in the pathophysiology of many chronic diseases[5–10] as well as the function of bone cells.[11]

Free radical products of oxygen metabolism, such as superoxide and hydrogen peroxide, are generated and released under environmental stimuli (e.g., cytokines, ultraviolet radiation), and pathologic circumstances. These reactive oxygen species are neutralized by the antioxidant system in the body. This system consists of agents, such as vitamins E and C, reduced glutathione (GSH), glutathione peroxidase, catalase, and superoxide dismutase (SOD).[12,13] Oxidative stress occurs when there is an imbalance between free radical production and antioxidant capacity. There is evidence of decreased antioxidant power in osteoporotic patients and a negative association between level of lipid peroxidation and bone mineral density.[14–16]

The cytokine TGF-β has been postulated to play a role in controlling bone density.[17] It promotes osteoblast proliferation[18] and its subsequent differentiation[19] and decreases osteoclast activity.[20,21] There are three closely related TGF-β isoforms (β1, β2, and β3).[22] Until now, only the β1 isoform has been detected at levels in excess of 1 ng/mL in human blood.[23,24] Several studies have investigated the concentration of bone TGF-β1 in osteoporotic patients to see whether there is any correlation between bone density and TGF-β 1. Nicolas *et al.* have shown that there is an indirect correlation between age and bone TGF-β and this can be a cause for bone loss with aging,[25] but Georgescu *et al.* have claimed that bone TGF-β1 is not an appropriate determinant of bone mass.[26] There is no study on the correlation of blood or saliva TGF-β concentration and bone loss.

This study was designed to examine the total antioxidant power (TAP), lipid peroxidation, and TGF-β1 concentration in the plasma and saliva of osteoporotic subjects in comparison to age- and sex-matched healthy people to determine whether there is any correlation between these parameters and bone loss.

MATERIALS AND METHODS

Subjects

Twenty-two postmenopausal women from the Shariati Hospital at the Tehran University of Medical Sciences (TUMS) with a history of osteoporosis were invited to participate in this study. Age, age at menopause, weight, lumbar spine bone mineral density (LS-BMD),[27] femoral neck bone mineral density (FN-BMD), LS Z-score, FN Z-score, LS T-score, and FN T-score were obtained from review of their medical records and personal interviews (TABLE 1). Patients were also shown to be free of periodontal inflammation, other oral pathologic conditions, and medical abnormalities that affect the salivary gland or salivary secretions.[10,28] Twenty-two healthy women free of medication and oral or systemic illness in the past 3 months, matched by age and smoking history, served as control.

The subjects were fully informed of the purpose and procedures of the study and written informed consent was obtained from all participants. The research protocol was approved by the Ethics Committee on Human Experimentation of TUMS.

Chemicals

Dithiobisnitrobenzoic acid (DTNB), Tris base, tetraethoxypropane (malonedialdehyde, MDA) were obtained from Sigma (Dorset, UK), 2-thiobarbituric (TBA), trichloroacetic acid (TCA), n-butanol from Merck (Germany), and 2,4,6-tripyridyl-s-triazine (TPTZ) from Fluka (Italy) were used in this study. TGF-β1 Determination Kit was provided by Bio Source Europe (Nivelles, Belgium).

TABLE 1. Baseline characteristics of patients at the time of entry into the study

Characteristic	Patients ($n = 22$)	Controls ($n = 22$)	P-value
Age (years)	59.27 ± 4.26	56.91 ± 6.23	0.149
Age at menopause (years)	48.50 ± 3.57	48.86 ± 3.81	0.753
BMI	25.39 ± 5.03	27.47 ± 3.88	0.33
LS-BMD (g/cm^2)	0.84 ± 0.05	1.132 ± 0.6	0.0001
LS Z-score	−2.06 ± 0.8	−0.02 ± 0.086	0.001
LS T-score	−3.10 ± 0.44	−0.482 ± 0.49	0.001
FN-BMD (g/cm^2)	0.778 ± 0.01	0.945 ± 0.09	0.0001
FN Z-score	−1.16 ± 0.58	0.11 ± 0.95	0.001
FN T-score	−1.69 ± 0.76	−0.13 ± 0.81	0.001

Values are mean ± SD. Weight is reported because height was not measured for the calculation of body mass index (BMI). LS-BMD is the lumber spine bone mineral density and FN-BMD is the femoral neck bone mineral density. Z-score and T-score are determined as described previously.[27]

Samples Collection and Handling

Blood and saliva samples were prepared at the same time. Unstimulated whole saliva produced in a 5-min period (about 3 mL) was collected, allowed to drain into a plastic container, and centrifuged at 3,000 × g, in 4°C for 5 min to remove bacterial and cellular debris. Saliva samples were stored at –80°C until analysis.

Blood samples were collected into Vacutainer tubes following a 12-h overnight fast. The blood was centrifuged at 1,700 g for 10 min and the plasma was separated. Plasma was stored at –80°C until analysis.

Measurement of Plasma TAP

Antioxidant power of plasma was determined by the FRAP test, based on measuring ability to reduce Fe^{3+} to Fe^{2+}.[29] In brief: the medium is exposed to Fe^{3+} and the antioxidants present in the medium start to produce Fe^{2+} as an antioxidant activity. The reagent included 300 mmol/L acetate buffer, pH 3.6, and 16 mL $C_2H_4O_2$ /L of buffer solution, 10 mmol/L 2,4,6-tripyridyl-s-triazine (TPTZ) in 40 mmol/L HCL, 20 mmol/L $FeCl_3$ $6H_2O$. Working FRAP reagent was prepared as required by mixing 25 mL acetate buffer, 2.5 mL TPTZ solution, and 2.5 mL $FeCl_3$ $6H_2O$ solution. Ten microliters of H_2O-diluted sample was then added to 300 μL freshly prepared reagent warmed at 37°C. The complex between Fe^{2+} and TPTZ gives a blue color with absorbance at 593 nm.

Measurement of Lipid Peroxidation

Malonedialdehyde (MDA) is the end product of the oxidation of polyunsaturated fatty acids and its concentration in the medium is an established measure of the extent of lipid peroxidation.[30] In this test the reaction of MDA with thiobarbituric acid (TBA) made a complex, which was determined spectrophotometrically, and lipid peroxidation in samples was assessed in terms of thiobarbituric acid reactive substances (TBARS) produced. The samples were diluted by buffered saline (1:5) and 800 μL of trichloroacetic acid (TCA, 28% w/v) was added to 400 μL of this mixture and centrifuged in 3,000 g for 30 min. Then, 600 μL of the supernatant was added to 150 μL of TBA (1% w/v). Then the mixture was incubated for 15 min in a boiling water bath and 4 mL n-butanol was added. The solution was centrifuged and cooled, and absorption of the supernatant was recorded at 532 nm by UV-160-A Shimadzu double-beam spectrophotometer. The calibration curve of 1,1,3,3-tetraethoxypropane standard solutions was used to determine the concentrations of TBA–MDA adducts in samples.

TGF-β1 Determination

Concentrations of TGF-β1 in saliva and plasma were measured using a solid-phase sandwich enzyme-linked immunosorbent assay (ELISA). A monoclonal antibody specific for TGF-β1 had been coated onto the wells of the microtiter strips provided. Samples, including standards of known TGF-β1 content, control specimen, and extracted unknowns, were pipetted into the well, followed by the addition of a biotinylated second antibody. During the first incubation, TGF-β1 antibody bound simultaneously to the immobilized antibody on one site, and to the solution-phase biotinylated antibody on a second site. After removal of excess detection antibody, stretavidin–peroxidase (enzyme) was added. This bound to the biotinylated antibody to complete the four-member sandwich. After a second incubation and washing to remove all unbound enzymes, a substrate solution was added, which was acted upon by the bound enzyme to produce color. The intensity of this colored product was directly proportional to the concentration of TGF-β1 present in the original specimen.[31]

Statistical Analysis

The statistical analysis was performed with the use of SPSS version 11.5. Continuous data were expressed as mean ± SD. General characteristics and laboratory findings were compared between groups using independent 2-sample Student's t-test for continuous data. Because the distribution of plasma and saliva concentration of TAP and TGF-β1 was not normal and so napierian logarithmic transformation was performed and results were expressed as geometric means. A P-value < 0.05 was considered to be statistically significant.

RESULTS

The characteristics of osteoporotic and healthy control subjects are shown in TABLE 1. There were no significant differences between osteoporotic and healthy subjects in the age, age at menopause, and weight.

TABLE 2 shows the results obtained from TBARS, TAP, and TGF-β1 measurement in blood and plasma of osteoporotic and healthy subjects. As observed, there were significant differences in plasma ($P = 0.002$) and saliva ($P < 0.001$) TBARS among osteoporotic and healthy subjects (0.30 ± 0.04 vs. 0.26 ± 0.04 and 0.23 ± 0.03 vs. 0.16 ± 0.04 μmol/L), respectively. Plasma ($P = 0.047$) and saliva ($P = 0.046$) TAP of osteoporotic subjects were significantly lower than those of healthy controls (606.65 ± 119.13 vs. 665.64 ± 63.73 mmol/L and 560.43 ± 84.70 vs. 612.05 ± 81.5), respectively. Concentration of TGF-β1 in blood and saliva of osteoporotic subjects was not different

TABLE 2. Levels of lipid peroxidation (TBARS), total antioxidant power (TAP), transforming growth factor-β1 (TGF-β1), and bone mineral density (BMD) in osteoporotic and control subjects

	Patients	Controls	P-value
Plasma TBARS (μmol/L)	0.30 ± 0.04	0.25 ± 0.04	0.002
Saliva TBARS (μmol/L)	0.22 ± 0.03	0.16 ± 0.04	0.000
Plasma TAP (mmol/L)	606.65 ± 119.13	665.64 ± 63.73	0.047
Saliva TAP (mmol/L)	560.43 ± 84.70	612.05 ± 81.50	0.046
Plasma TGF-β1 (pg/mL)	30.40 ± 6.62	33.07 ± 3.88	0.196
Saliva TGF-β1 (pg/mL)	23.79 ± 1.67	24.96 ± 2.38	0.154
BMD (g/cm^2)	0.84 ± 0.05	1.132 ± 0.6	0.001

from those of healthy controls ($P > 0.05$, 30.40 ± 6.62 vs. 33.07 ± 3.88 and 23.79 ± 1.67 vs. 24.96 ± 2.38 pg/mL).

DISCUSSION

This study shows that lipid peroxidation in plasma and saliva of osteoporotic subjects is higher than that of healthy subjects. Since lipid peroxidation is a determinant of oxidative stress, this result confirms that oxidative stress exists in osteoporotic subjects. It has been shown that oxidative stress enhances resorption by isolated osteoclasts and increases osteoclast formation in marrow culture.[32,33] Oxidative stress also inhibits differentiation and mineralization,[34] induces necrosis of osteoblasts,[35] and causes partial degradation and modification of fibronectin molecules.[36] Thus, it is confirmed that oxidative stress is involved in the pathophysiology of osteoporosis. The present results also demonstrate that TAP in plasma and saliva of osteoporotic subjects is significantly lower than that of healthy volunteers. This finding seems to be logical, because aging causes an increase in osteoclastic activity. Osteoclasts can generate reactive oxygen species, resulting in increased bone resorption.[37–39] Natural antioxidants, such as superoxide dismutase, catalase, GSH-peroxidase, and vitamins E and C, are consumed to neutralize these reactive oxygen species.[40] Thus, TAP gradually decreases. It appears that antioxidants may be beneficial in the prevention and treatment of osteoporosis. There are few studies that investigate the role of antioxidants in osteoporosis,[34,40] although further studies have been suggested to elucidate the role of antioxidants in this disorder.[35,41]

In this study, we also showed that the concentration of TGF-β1 in plasma and saliva of osteoporotic does not change significantly. This is in support of the findings of Georgescu et al., who report that bone TGF-β1 is not an appropriate determinant of bone mass;[26] it is also in contrast to the results of Nicolas et al., who reported that there is a correlation between age and bone TGF-β[25]

In conclusion, a positive association between the extent of oxidative stress in both blood and saliva and bone loss are confirmed in this study. This means that saliva, which is collected noninvasively, can be used as a good alternative sample to blood in these subjects. No significant change in TGF-β1 concentration of plasma and saliva in osteoporotic subjects was observed. Administration of antioxidant supplements may be helpful in the protection of bone mass density and treatment of osteoporosis.

ACKNOWLEDGMENT

This work was supported by a grant from Endocrinology and Metabolism Research Center of TUMS.

REFERENCES

1. 1991. Consensus development conference: prophylaxis and treatment of osteoporosis. Am J. Med. **90:** 107–110.
2. MONOLOGAS, S.C. & R.L. JILKA. 1995. Bone marrow, cytokines and remodeling: emerging insights into the pathophysiology of osteoporosis. N. Engl. J. Med. **322:** 305–311.
3. BOYLE, W.J., W.S. SIMONET & D.L. LACEY. 2003. Osteoclast differentiation and activation. Nature **423:** 337–342.
4. MONOLAGAS, S.C. 2000. Birth and death of bone cells. Basic regulatory mechanisms and implications for the pathogenesis and treatment of osteoporosis. Endocr. Rev. **21:** 115–117.
5. AFSHARI, M., B. LARIJANI, A. REZAIE, et al. 2004. Ineffectiveness of allopurinol in reduction of oxidative stress in diabetic patients; a randomized, double-blind placebo-controlled clinical trial. Biomed. Pharmacother. **58:** 546–550.
6. JAHANSHAHI, G., V. MOTAVASEL, A. REZAIE, et al. 2004. Alterations in antioxidant power and levels of epidermal growth factor and nitric oxide in saliva of patients with inflammatory bowel diseases. Dig. Dis. Sci. **49:** 1752–1757.
7. ASTANEIE, F., M. AFSHARI, A. MOJTAHEDI, et al. 2005. Total antioxidant capacity and levels of epidermal growth factor and nitric oxide in blood and saliva of insulin-dependent diabetic patients. Arch. Med. Res. **36:** 376–381.
8. RADFAR, M., B. LARIJANI, M. HADJIBABAIE, et al. 2005. Effects of pentoxifylline on oxidative stress and levels of EGF and NO in blood of diabetic type-2 patients; a randomized, double-blind placebo-controlled clinical trial. Biomed. Pharmacother. **59:** 302–306.
9. RAHIMI, R., S. NIKFAR, B. LARIJANI & M. ABDOLLAHI. 2005. A review on the role of antioxidants in the management of diabetes and its complications. Biomed. Pharmacother. **59:** 365–373.
10. MASHAYEKHI, F., F. AGHAHOSEINI, A. REZAIE, et al. 2005. Alteration of cyclic nucleotides levels and oxidative stress in saliva of human subjects with periodontitis. J. Contemp. Dent. Pract. **4:** 046–053.
11. ABDOLLAHI, M., B. LARIJANI, R. RAHIMI & P. SALARI. 2005. Role of oxidative stress in osteoporosis. Therapy **2:** 787–796.

12. ABDOLLAHI, M., A. RANJBAR, S. SHADNIA, et al. 2004. Pesticides and oxidative stress: a review. Med. Sci. Monit. **10:** RA144–RA147.
13. FINKEL, T. & N.J. HOLBROOK. 2000. Oxidants, oxidative stress and biology of aging. Nature **408:** 239–247.
14. YALIN, S., S. BAGIS, G. POLAT, et al. 2005. Is there a role of free oxygen radicals in primary male osteoporosis? Clin. Exp. Rheumatol. **23:** 689–692.
15. MAGGIO, D., M. BARABANI, M. PIERANDREI, et al. 2003. Marked decrease in plasma antioxidants in aged osteoporotic women: results of a cross-sectional study. J. Clin. Endocrinol. Metab. **88:** 1523–1527.
16. BASU, S., K. MICHAELSSON, H. OLOFSSON, et al. 2001. Association between oxidative stress and bone mineral density. Biochem. Biophys. Res. Commun. **288:** 275–279.
17. GRAINGER, D.J., J. PERCIVAL, M. CHIANO & T.D. SPECTOR. 1999. The role of serum TGF-beta isoforms as potential markers of osteoporosis. Osteoporos. Int. **9:** 398–404.
18. MOHAN, S. & D.J. BAYLINK. 1991. Bone growth factors. Clin. Orthop. **263:** 30–48.
19. CENTRELLA, M., S. CASINIGHINO, R. IGNOTZ & T.L. MCCARTHY. 1992. Multiple regulatory effects by TGF-β on type-I collagen levels in osteoblast-enriched cultures from fetal rat bone. Endocrinology **131:** 2863–2872.
20. HUGHES, D.E., A.H. DAI, J.C. TIFFEE, et al. 1996. Oestrogen promotes apoptosis of murine osteoclasts mediated by TGF-β. Nat. Med. **2:** 1132–1137.
21. BEAUDREUIL, J., G. MBALAVIELE, M. COHENSOLAL, et al. 1995. Short-term local injections of TGF-β1 decrease oviarectomy-stimulated osteoclastic resorption in vivo in rats. J. Bone Miner. Res. **10:** 971–977.
22. MASSAGUE, J. 1990. The transforming growth factor β family. Annu. Rev. Cell. Biol. **6:** 597–641.
23. GRAINGER, D.J., D.E. MOSEDALE, J.C. METCALFE, et al. 1995. Active and acid-activatable TGF-β in human sera, platelets and plasma. Clin. Chim. Acta **235:** 11–31.
24. DANIELPOUR, D., K.Y. KIM, L.L. DART, et al. 1990. Evidence for differential regulation of TGF-β1 and TGF-β2 expression in vivo by sandwich enzyme-linked immunosorbent assays. Ann. N.Y. Acad. Sci. **593:** 300–302.
25. NICOLAS, V., A. PREWETT, P. BETTICA, et al. 1994. Age-related decreases in insulin-like growth factor-I and transforming growth factor-beta in femoral cortical bone from both men and women: implications for bone loss with aging. J. Clin. Endocrinol. Metab. **78:** 1011–1016.
26. GEORGESCU, C., T. SECK, I. DIEL, et al. 2004. Bone matrix insulin-like growth factor (IGF)-I, IGF-II and transforming growth factor (TGF)-beta1 levels in men and postmenopausal women with osteoporosis: lack of association with circulating growth factors and bone mineral density. Rev. Med. Chir. Soc. Med. Nat. Iasi **108:** 281–286.
27. BARAN, D.T., K.G. FAULKNER, H.K. GENANT, et al. 1997. Diagnosis and management of osteoporosis: guidelines for the utilization of bone densitometry. Calcif. Tissue Int. **61:** 433–440.
28. ABDOLLAHI, M. & M. RADFAR. 2003. A review of drug-induced oral reactions. J. Contemp. Dent. Pract. **4:** 10–31.
29. BENSI, I.F. & S. STRAINS. 1999. Ferric reducing antioxidant assay. Methods Enzymol. **292:** 15–27.
30. SATHO, K. 1978. Serum lipid peroxidation in cerebrovascular disorders determined by a new colorimetric method. Clin. Chem. Acta **90:** 37–43.

31. KIM, S.J., D. ROMEO, Y.D. YOO & K. PARK. 1994. Transforming growth factor-beta: expression in normal and pathological conditions [review]. Horm. Res. **42:** 5–8.
32. BAX, B.E., A.S.T.M. ALAM, B. BANERJI, *et al.* 1992. Stimulation of osteoblastic bone resorption by hydrogen peroxide (H2O2). Biochem. Biophys. Res. Commun. **183:** 1153–1158.
33. SUDA, N., I. MORITA, T. KURODA & S. MUROTA. 1997. Participation of oxidative stress in the process of osteoclast differentiation. Biochim. Biophys. Acta **1157:** 318–323.
34. MODY, M., F. PARHAMI, T.A. SARAFIAN & L.L. DEMER. 2001. Oxidative stress modulates osteoblastic differentiation of vascular and bone cells. Free Radical Bio. Med. **31:** 509–519.
35. BRAYBOY, J.R., X.W. CHEN, Y.S. LEE & J.J.B. ANDERSON. 2001. The protective effects of Ginkgo biloba extract (EGb 761) against free radical damage to osteoblast-like bone cells (MC3T3-E1) and the proliferative effects of EGb 761 on these cells. Nutr. Res. **21:** 1275–1285.
36. SUZUKI, H., M. HAYAKAWA, K. KOBAYASHI, *et al.* 1997. H2O2-derived free radicals treated fibrinectin substratum reduces the bone nodule formation of rat calvarial osteoblast. Mech. Ageing Dev. **98:** 113–125.
37. DARDEN, A.G., W.L. RIES, W.C. WOLF, *et al.* 1996. Osteoclastic superoxide production and bone resorption: stimulation and inhibition by modulators of NADPH oxidase. J. Bone Miner. Res. **11:** 671–675.
38. RIES, W.L., L.L. KEY & R. RODRIGUIZ. 1992. Nitroblue tetrazolium reduction and bone resorption by osteoclasts *in vitro* inhibited by a manganese-based superoxide dismutase mimic. J. Bone Miner. Res. **7:** 931–939.
39. STEINBECK, M.J., W.H. APPEL, A.J. VERHOEVEN & M.J. KARNOVSKY. 1994. NADPH-oxidase expression and in situ production of superoxide byosteoclasts actively resorbing bone. J. Cell. Biol. **26:** 765–772.
40. PENCKOFER, S., D. SCHWERTZ & K. FLERCZAK. 2002. Oxidative stress and cardiovascular disease in type 2 diabetes: the role of antioxidants and pro-oxidants. J. Cardiovasc. Nurs. **16:** 68–85.
41. LEE, Y.S., X. CHEN & J.J.B. ANDERSON. 2001. Physiological concentrations of genistein stimulate the proliferation and protect against free radical induced oxidative damage of MC3T3-E1 osteoblast-like cells. Nutr. Res. **21:** 1287–1298.

Targeting Signal-Transducer-and-Activator-of-Transcription-3 for Prevention and Therapy of Cancer

Modern Target but Ancient Solution

BHARAT B. AGGARWAL, GAUTAM SETHI, KWANG SEOK AHN,
SANTOSH K. SANDUR, MANOJ K. PANDEY,
AJAIKUMAR B. KUNNUMAKKARA, BOKYUNG SUNG,
AND HARUYO ICHIKAWA

Department of Experimental Therapeutics, Cytokine Research Laboratory, The University of Texas M. D. Anderson Cancer Center, Houston, Texas 77030, USA

ABSTRACT: Recent evidence indicates a convergence of molecular targets for both prevention and therapy of cancer. Signal-transducer-and-activator-of-transcription-3 (STAT3), a member of a family of six different transcription factors, is closely linked with tumorigenesis. Its role in cancer is indicated by numerous avenues of evidence, including the following: STAT3 is constitutively active in tumor cells; STAT3 is activated by growth factors (e.g., EGF, TGF-α, IL-6, hepatocyte growth factor) and oncogenic kinases (e.g., Src); STAT3 regulates the expression of genes that mediate proliferation (e.g., c-myc and cyclin D1), suppress apoptosis (e.g., Bcl-x_L and survivin), or promote angiogenesis (e.g, VEGF); STAT3 activation has been linked with chemoresistance and radioresistance; and chemopreventive agents have been shown to suppress STAT3 activation. Thus inhibitors of STAT3 activation have potential for both prevention and therapy of cancer. Besides small peptides and oligonucleotides, numerous small molecules have been identified as blockers of STAT3 activation, including synthetic molecules (e.g., AG 490, decoy peptides, and oligonucleotides) and plant polyphenols (e.g., curcumin, resveratrol, flavopiridol, indirubin, magnolol, piceatannol, parthenolide, EGCG, and cucurbitacin). This article discusses these aspects of STAT3 in more detail.

KEYWORDS: STAT3; interleukin-6; cancer; chemoresistance

Address for correspondence: Dr. Bharat B. Aggarwal, Department of Experimental Therapeutics, The University of Texas M. D. Anderson Cancer Center, Box 143, 1515 Holcombe Boulevard, Houston TX 77030. Voice: 713-794-1817; fax: 713-794-1613.
  e-mail: aggarwal@mdanderson.org

Ann. N.Y. Acad. Sci. 1091: 151–169 (2006). © 2006 New York Academy of Sciences.
doi: 10.1196/annals.1378.063

INTRODUCTION

Signal-transducer-and-activator-of-transcription (STAT) is a family of six different transcription factors, first discovered in 1993 by James Darnell, which play major roles in cytokine signaling.[1,2] A typical STAT protein consists of a coiled-coil domain, a DNA-binding domain, a linker, an SH2 domain, and a transactivation domain (TAD) (FIG. 1). The TAD contains tyrosine and serine phosphorylation sites that are needed for the activation of STAT. Because it plays a major role in tumorigenesis, STAT3 is the focus of the remaining discussion.

STAT3 was discovered independently and simultaneously by two different groups in 1994.[3,4] It was initially regarded as an acute-phase response factor activated by interleukin-6 (IL-6), leukemia inhibitory factor, oncostatin M, and the ciliary neurotrophic factor (CNTF) family of cytokines, all known to mediate their signal through the gp130 protein.[4] Engagement of cell-surface cytokine receptors activates the janus kinase (JAK) family of protein kinases, which in turn phosphorylates and activates latent cytoplasmic STAT3 protein to an active dimer capable of translocating to the nucleus and inducing transcription of specific target genes (FIG. 1). Although four different members of

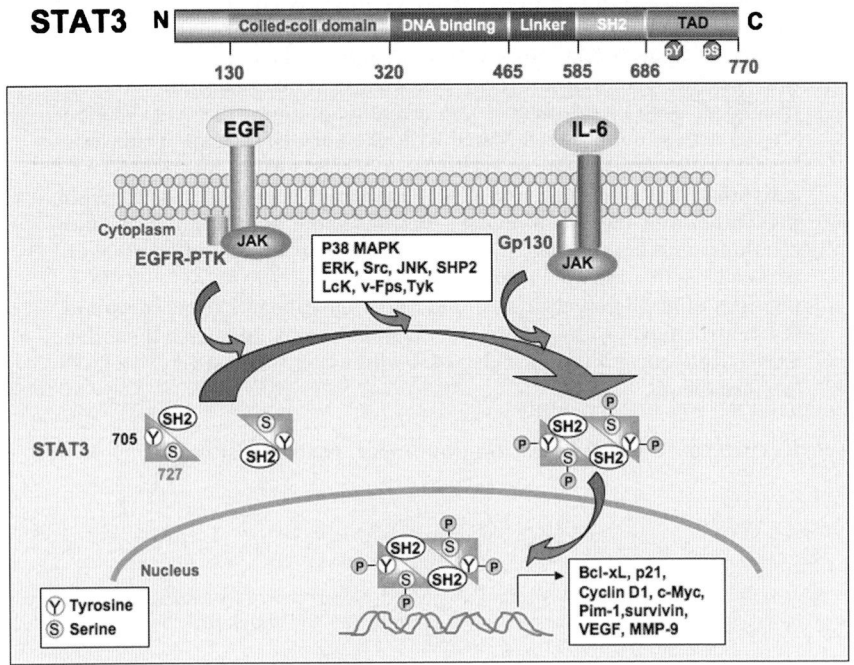

FIGURE 1. Signaling pathway leading to STAT3 activation.

the JAK family have been described (JAK1, JAK2, JAK3, and TYK2), JAK2 is one of the major mediators of STAT3 phosphorylation. Several other kinases have been implicated in the phosphorylation of STAT3, including members of the Src family (hck, src), Erb B1, Erb B2, anaplastic lymphoma kinase, protein kinase C (PKC)-δ, c-fes, gp130, and epithelial growth factor (EGF) receptor.[5–11] Although both serine and tyrosine phosphorylation of STAT3 are needed for full activation, evidence indicates that STAT3 contains a single serine phosphorylation site at position 727, which has no effect on its ability to bind to DNA.[12] Various studies also indicate that ERKs,[13] JNK,[14] p38 mitogen-activated protein kinase,[15] and PKC[9] participate in serine phosphorylation of STAT3.

While genetic deletion of the STAT3 gene is lethal to mouse embryos,[16] selective loss of STAT3 in keratinocytes results in impaired wound healing,[17] and skin-specific STAT3-transgenic mice develop psoriasis.[18] Increasing evidence also indicates that STAT3 mediates its tumorigenic effects not alone but through its cross-talk with various other transcription factors. These include PPAR-γ,[19] β-catenin,[20] nuclear factor-κB (NF-κB),[21,22] hypoxia-inducible factor 1-α,[23] c-myc,[24] c-fos,[25] c-jun,[26] glucocorticoid receptors,[27] and estrogen receptors.[28]

STAT3 IS ACTIVATED BY GROWTH FACTORS

Various growth factors that have been linked to tumor cell proliferation have been found to activate STAT3 (TABLE 1).[29–55] EGF signaling, which has been linked with proliferation of almost 30% of all tumor cells, has been shown to mediate its effect through activation of STAT3.[3] Similarly, IL-6, which has been linked with proliferation of multiple myeloma, renal cell carcinoma, prostate cancer, and other cancers, has been shown to mediate its effect through activation of STAT3.[3] The activation of STAT3 by both EGF and IL-6 involves tyrosine phosphorylation of STAT3 at position 705. Moreover, several other cytokines have been shown to activate STAT3. These include growth hormone,[56] transforming growth factor-α (TGF-α),[57] oncostatin M,[4] thrombopoietin,[58] platelet-derived growth factor (PDGF),[59] IL-5,[60] IL-6,[3] IL-9,[61] IL-10,[62,63] IL-12,[64] IL-22,[65] and leptin.[66] Certain chemokines, such as macrophage inflammatory protein 1α and RANTES (regulated upon activation of normal T cell expressed and secreted) also have been shown to activate STAT3.[67] Whether all these cytokines stimulate STAT3 through activation of JAK2 is not fully understood. Besides growth factors and cytokines, other factors that activate STAT3 include oxidative stress,[68] tobacco chewing,[69] hepatitis C virus,[70,71] ultraviolet B,[72] lipopolysaccharide,[73] osmotic shock,[74] and progestins.[75]

STAT3 IS CONSTITUTIVELY ACTIVE IN TUMOR CELLS

The role of STAT3 in cancer is supported through numerous lines of evidence. First, all Src-transformed cell lines have persistently activated STAT3,

TABLE 1. Tumors that express constitutively active STAT3, activators of STAT3, genes regulated by STAT3, and inhibitors of STAT3

Constitutive STAT3	Activators	Genes	Kinases	Inhibitors
Hematopoietic tumors	EGF[3]	**Antiapoptosis**	**Nonreceptor tyrosine kinases**	**Synthetic**
Multiple myeloma[37]	IL-6[3]	Bcl-x_L[95]	JAK[152]	AG490[32]
HTLV-1-dependent leukemia[29]	IL-5[60]	Bcl-2[110]	JAK2[6]	Sodium salicylate[117]
CLL[33]	IL-9[61]	Mcl-1[110]	JAK3[152]	Atiprimod[119]
CML[49]	IL-10[63]	cIAP-2[110]	TYK2[152]	BMS-354825[154]
AML[36]	IL-12[64]	Survivin[97]	Src[5]	Ethanol[116]
Large granular lymphocyte leukemia[40]	IL-22[65]			Nelfinavir[155]
Erythroleukemia[46]	TNF-α[104]	**Cell cycle progression**	**Receptor tyrosine kinases**	PS-341[120]
Polycythemia vera[39]	MCP-1[144]	Cyclin D1[142]	EGFR[57]	R115777[156]
EBV-related/Burkitt's[30]	GCSF[62]	c-Myc[148]	ErbB-2[6]	WP-1034[157]
Mycosis fungoides[32]	GMCSF[60]	c-Fos[25]	Gp130[153]	Platinum compound[115]
Cutaneous T cell lymphoma[35]	CSF[145]	p21[101]	Grb2[27]	15-Deoxy-delta 12, 14-PGJ2[158]
HSV saimiri-dependent (T cell)[34]	LIF[146]			UCN-01[159]
Hodgkin's disease[41]	OSM[4]	**Tumor invasion and metastasis**	**Serine kinases**	Statin[121]
Anaplastic lymphoma[48]	IFN-γ	MMP-2[149]	JNK[14]	
Hematopoietic tumors[37]	MIP-1α[67]	MMP-9[99]	P38MAPK[15]	**Peptides**
Multiple myeloma[37]	RANTES[67]	β-catenin[20]	ERK[13]	SOCS3[82]
HTLV-1-dependent leukemia[29]	SLF[147]	VEGF[100]		PIAS[160]
	UVB[72]	hTERT[150]	**Tyrosine**	GRIM-19[161]
Solid tumors	Osmotic shock[74]	IRF-1[106]	**Phosphatase**	Adiponectin[162]
Breast cancer[31]	Progestin[75]	NLK[151]	SHP2[77]	Duplin[163]
Brain tumor[47]	LPS[73]	MyD88[105]		SSI-1[102]
Colon carcinoma[54]	Tobacco[69]	RANKL[103]		α-Thrombin[164]
Ewig sarcoma[55]	HCV[70,71]	TNF[104]		Lipoxin A4[165]
Gastric carcinoma[52]	EGF[3]	β-macroglobulin[107]		DIF-1[166]
Lung cancer[51]		SOCS[102]		PTPεC[167]
Nasopharyngeal cancer[50]				STAT3-DN[77]
Ovarian carcinoma[38]				

Continued

TABLE 1. Continued

Constitutive STAT3	Activators	Genes	Kinases	Inhibitors
Pancreatic adenocarinoma[45]		Angiotensinogen[109]		Decoy peptide[168]
Prostate carcinoma[44]		Antichymotrypsin[108]		
Renal cell carcinoma[43]				**Naturals**
SCCHN cancer[57]				Flavopiridol[129]
				Indirubin[126]
				Magnolol[130]
				Resveratrol[124]
				Magnolol[130]
				Piceatannol[127]
				Parthenolide[128]
				EGCG[131]
				Curcumin[122,123]
				Cucurbitacin[125]
				Resveratrol[124]
				Magnolol[130]
				Others
				Rituximab[169]
				GQ-ODN[170]
				Retinoic acid[118]
				STA-21[171]
				EKB569[172]

STAT, signal-transducer-and-activator-of-transcription; CLL, chronic lymphocytic leukemia; CML, chronic myeloid leukemia; AML, acute myelogenous leukemia; MCL, mantle cell lymphoma; SCCHN, squamous cell carcinoma of the head and neck; HTLV, human T cell lymphotropic virus; EBV, Epstein–Barr virus; Nelfinavir, HIV-1 protease inhibitor; R115777, farnesyl transferase inhibitor; AG490 and piceatannol, tyrosine kinase inhibitors; PIAS, protein inhibitor of activated STAT3; GQ-ODN, G-quartet oligonucleotides; SOCS, suppressor of cytokine signaling; GRIM, gene associated with retinoid-IFN-induced mortality; EGCG, epigallocatechin-3-gallate; SSI, STAT-induced STAT inhibitor; PTPeC, protein tyrosine phosphatase εC; DN, dominant negative; EKb-569, EGF-R inhibitor; DIF-1, differentiation-inducing factor-1; JAB, SH2-domain-containing protein; IL, interleukin; TNF, tumor necrosis factor; MDA, melanoma differentiation antigen; MCP, monocyte chemoattractant protein; GCSF, granulocyte colony-stimulating factor; LIF, leukemia inhibitory factor; OSM, oncostatin M; IFN, interferon; MIP, macrophage inflammatory protein; RANTES, regulated upon activation, normal T cell expressed and secreted; HB-EGF, heparin-binding epidermal growth factor; LPS, lipopolysaccharide; VEGF, vascular endothelial growth factor; MMP, matrix metalloproteinase; HSP, heat shock protein, hTERT, human telomerase reverse transcriptase; trh, thyrotropin-releasing hormone; ATL, adult T cell leukemia/lymphoma., SLF, steel factor, HCV, hepatitis C virus.

and dominant-negative STAT3 blocks transformation.[76–79] Second, STAT3-C, a constitutively active mutant dimerized by cysteine–cysteine bridges instead of pTyr-SH2 interaction, can transform cultured cells so that they form tumors when injected into mice.[77] Indeed, STAT3 functions in fibroblast development to resist apoptosis.[80] Third, constitutive activation of STAT3 has been reported in large number of tumors, including breast cancer,[31] prostate cancer,[44] head and neck squamous cell carcinoma,[42] multiple myeloma,[37] lymphomas and leukemia,[81] brain tumor,[47] colon cancer,[54] Ewing sarcoma,[55] gastric cancer,[52] esophageal cancer,[51] ovarian cancer,[38] nasopharyngeal cancer,[50] and pancreatic cancer.[45]

Why STAT3 is constitutively active in tumor cells is not fully understood. Because no mutations in STAT3 have been reported that results in persistent activation, the only putative mechanisms to account for the constitutive activity of STAT3 are dysregulation of signaling molecules or mutation or deletions in the protein that negatively regulates STAT3 (e.g., protein inhibitor of activated STAT3 or suppressor of cytokine signaling [SOCS]).[82] For instance SOCS-1, a negative regulator of cytokine signaling is frequently silenced by methylation in various tumors.[82–86] Both receptor tyrosine kinases and nonreceptor tyrosine kinases have been linked with activation of STAT3. Besser et al. found that a single amino acid substitution in the v-Eyk intracellular domain results in activation of STAT3 and enhances cellular transformation.[87] Yu et al. showed that enhanced cells transformed by the Src oncoprotein have constitutively active STAT3.[78] Besides Src, inducers of STAT3 activation include a nonreceptor tyrosine kinase, v-Fps; polyoma virus middle T antigen, which activates Src family kinases; and v-Sis, which acts as a ligand for the PDGF receptor.[88] Additional nonreceptor tyrosine kinases include c-Fes,[11] Lck,[89] Ras/Rac1-mediated p38, and c-Jun N-terminal kinase.[15] STAT3 activation also is regulated by protein tyrosine phosphatases.[90]

STAT3 REGULATES EXPRESSION OF GENES INVOLVED IN TUMORIGENESIS

STAT3 is one of the major mediators of tumorigenesis.[91,92] The oncogenic significance of activated STAT3 molecules is due to their effects on numerous parameters of the development and progression of malignancy, such as apoptosis, cell proliferation, angiogenesis, and immune system evasion.[79,93,94] Constitutively active STAT3 has been implicated in the induction of resistance to apoptosis,[37] possibly through the expression of Bcl-x_L[95] and cyclin D1.[42] Its role in tumorigenesis is mediated through the expression of various genes that suppress apoptosis, mediate proliferation, invasion, and angiogenesis. These include Mcl-1,[40,96] Bcl-x_L,[95] and survivin,[97] all of which suppress apoptosis; c-myc[98] and cyclin D1,[42] which mediate cell proliferation; matrix metalloproteinase-9,[99] which mediates cellular invasion; and vascular endothelial growth factor (VEGF), which mediates angiogenesis.[100] Other genes that have been shown to be regulated by STAT3 include p21,[101] SOCS-3,[102] receptor activator of NF-κB ligand (RANKL),[103] tumor necrosis factor (TNF),[104] MyD 88,[105] interferon-regulatory factor 1,[106] c-fos,[25] β-macroglobulin,[107] antichymotrypsin,[108] and angiotensinogen,[109] which also have been linked with tumorigenesis.

STAT3 ACTIVATION INHIBITS APOPTOSIS

Numerous reports suggest that activation of STAT3 suppresses apoptosis.[80,95,97,110] For instance, Shen et al. showed that constitutively activated STAT3 protects fibroblasts from serum withdrawal and ultraviolet-induced apoptosis and antagonizes the proapoptotic effects of activated STAT1.[80] Thus constitutively active STAT3 can contribute to oncogenesis by protecting cancer cells from apoptosis. This implies that suppression of STAT3 activation could facilitate apoptosis.

INHIBITORS OF STAT3 HAVE POTENTIAL IN PREVENTION AND THERAPY OF CANCER

Because of the critical role of STAT3 in tumorigenesis as reviewed here, inhibitors of STAT3 have potential in both prevention and treatment of cancer. Perhaps one of the best-known inhibitors of STAT3 activation is AG490, which inhibits the activation of JAK2.[32] Other blockers of STAT3 include small peptides, oligonucleotides,[44,111,112] and small molecules. Turkson et al. identified phosphotyrosyl peptides that block STAT3-mediated DNA-binding activity, gene regulation, and cell transformation.[113] Various small molecules that block STAT3 include 15-PGJ2,[114] platinum complex,[115] ethanol,[116] sodium salicylate,[117] retinoic acid,[118] atiprimod,[119] PS-341,[120] and statins.[121]

Several plant polyphenols have been identified that can suppress STAT3 activation. These include curcumin,[122,123] resveratrol,[124] curcurbitacin,[125] indirubin,[126] piceatannol,[127] parthenolide,[128] flavopiridol,[129] magnolol,[130] and epigallocatechin-3-gallate.[131] How these agents suppress STAT3 activation is not fully understood. Curcumin, a well-established chemopreventive agent, has been shown to inhibit JAK2,[132] Src,[133] Erb2,[134] and EGFR,[135] all of which are implicated in STAT3 activation. Furthermore, curcumin has been shown to downregulate the expression of Bcl-x_L, cyclin D1, VEGF, and TNF[136] all of which are known to be regulated by STAT3. A recent study by Kim et al. has shown that curcumin phosphorylates SHP-2, which in turn associates with JAK1 and JAK2, thus inhibiting initiation of the JAK–STAT pathway.[137] Thus, pharmacologically safe and effective therapeutic agents that can block constitutive or inducible activation of STAT3 have potential for efficacy in treatment of cancer. Given that growing evidence implicates a number of important STAT3 target genes in the formation of tumors,[138] it seems logical to conclude that inhibition of STAT3 through pharmacological blockage of upstream molecules, such as Src and JAK may reduce tumor formation.

CONCLUSION

The major hallmarks of cancer include deregulated cell growth, tumor cell invasion, angiogenesis, and metastasis.[139] This review clearly shows that STAT3 can regulate all these different phases of tumorigenesis. Suppression of STAT3 in certain tumor cell models has led to expression of proinflammatory cytokines and chemokines by tumor cells that are needed for the innate immune response against the tumor cells.[140] Because molecular targets in the prevention of cancer do not differ from those in the treatment of cancer,[141–143] STAT3 is an ideal target for both prevention and treatment of cancer. Thus small molecules that can suppress STAT3 activation and are pharmacologically safe have potential for suppression of tumorigenesis.

ACKNOWLEDGMENT

This work was supported by a grant from the Clayton Foundation for Research (to B.B.A.), a lung cancer chemoprevention grant from the National Institutes of Health (PO1 Grant CA91844; to B.B.A.), and a P50 Head and Neck SPORE grant from the National Institutes of Health (P50CA97007; to B.B.A.). We would like to thank Kathryn Hale for critically reading this manuscript.

REFERENCES

1. SHUAI, K., G.R. STARK, I.M. KERR, *et al*. 1993. A single phosphotyrosine residue of Stat91 required for gene activation by interferon-gamma. Science **261:** 1744–1746.
2. LEVY, D.E. & J.E. DARNELL, JR. 2002. Stats: transcriptional control and biological impact. Nat. Rev. Mol. Cell Biol. **3:** 651–662.
3. ZHONG, Z., Z. WEN & J.E. DARNELL, JR. 1994. Stat3: a STAT family member activated by tyrosine phosphorylation in response to epidermal growth factor and interleukin-6. Science **264:** 95–98.
4. AKIRA, S., Y. NISHIO, M. INOUE, *et al*. 1994. Molecular cloning of APRF, a novel IFN-stimulated gene factor 3 p91-related transcription factor involved in the gp130-mediated signaling pathway. Cell **77:** 63–71.
5. SCHREINER, S.J., A.P. SCHIAVONE & T.E. SMITHGALL. 2002. Activation of STAT3 by the Src family kinase Hck requires a functional SH3 domain. J. Biol. Chem. **277:** 45680–45687.

6. REN, Z. & T.S. SCHAEFER. 2002. ErbB-2 activates Stat3 alpha in a Src- and JAK2-dependent manner. J. Biol. Chem.; **277:** 38486–38493.
7. GARCIA, R., T.L. BOWMAN, G. NIU, et al. 2001. Constitutive activation of Stat3 by the Src and JAK tyrosine kinases participates in growth regulation of human breast carcinoma cells. Oncogene **20:** 2499–2513.
8. ZHANG, Y., J. TURKSON, C. CARTER-SU, et al. 2000. Activation of Stat3 in v-Src-transformed fibroblasts requires cooperation of Jak1 kinase activity. J. Biol. Chem. **275:** 24935–24944.
9. JAIN, N., T. ZHANG, W.H. KEE, et al. 1999. Protein kinase C delta associates with and phosphorylates Stat3 in an interleukin-6-dependent manner. J. Biol. Chem. **274:** 24392–24400.
10. SELLERS, L.A., W. FENIUK, P.P. HUMPHREY, et al. 1999. Activated G protein-coupled receptor induces tyrosine phosphorylation of STAT3 and agonist-selective serine phosphorylation via sustained stimulation of mitogen-activated protein kinase. Resultant effects on cell proliferation. J. Biol. Chem. **274:** 16423–16430.
11. NELSON, K.L., J.A. ROGERS, T.L. BOWMAN, et al. 1998. Activation of STAT3 by the c-Fes protein-tyrosine kinase. J. Biol. Chem. **273:** 7072–7077.
12. WEN, Z. & J.E. DARNELL, JR. 1997. Mapping of Stat3 serine phosphorylation to a single residue (727) and evidence that serine phosphorylation has no influence on DNA binding of Stat1 and Stat3. Nucleic Acids Res. **25:** 2062–2067.
13. CHUNG, J., E. UCHIDA, T.C. GRAMMER, et al. 1997. STAT3 serine phosphorylation by ERK-dependent and -independent pathways negatively modulates its tyrosine phosphorylation. Mol. Cell Biol. **17:** 6508–6516.
14. LIM, C.P. & X. CAO. 1999. Serine phosphorylation and negative regulation of Stat3 by JNK. J. Biol. Chem. **274:** 31055–31061.
15. TURKSON, J., T. BOWMAN, J. ADNANE, et al. 1999. Requirement for Ras/Rac1-mediated p38 and c-Jun N-terminal kinase signaling in Stat3 transcriptional activity induced by the Src oncoprotein. Mol. Cell Biol. **19:** 7519–7528.
16. TAKEDA, K., K. NOGUCHI, W. SHI, et al. 1997. Targeted disruption of the mouse Stat3 gene leads to early embryonic lethality. Proc. Natl. Acad. Sci. USA **94:** 3801–3804.
17. SANO, S., S. ITAMI, K. TAKEDA, et al. 1999. Keratinocyte-specific ablation of Stat3 exhibits impaired skin remodeling, but does not affect skin morphogenesis. EMBO J. **18:** 4657–4668.
18. SANO, S., K.S. CHAN, S. CARBAJAL, et al. 2005. Stat3 links activated keratinocytes and immunocytes required for development of psoriasis in a novel transgenic mouse model. Nat. Med. **11:** 43–49.
19. WANG, L.H., X.Y. YANG, X. ZHANG, et al. 2004. Transcriptional inactivation of STAT3 by PPARgamma suppresses IL-6-responsive multiple myeloma cells. Immunity **20:** 205–218.
20. HAO, J., T.G. LI, X. QI, et al. 2006. WNT/beta-catenin pathway up-regulates Stat3 and converges on LIF to prevent differentiation of mouse embryonic stem cells. Dev. Biol. **290:** 81–91.
21. YU, Z., W. ZHANG & B.C. KONE. 2002. Signal transducers and activators of transcription 3 (STAT3) inhibits transcription of the inducible nitric oxide synthase gene by interacting with nuclear factor kappaB. Biochem. J. **367:** 97–105.
22. JANG, H.D., K. YOON, Y.J. SHIN, et al. 2004. PIAS3 suppresses NF-kappaB-mediated transcription by interacting with the p65/RelA subunit. J. Biol. Chem. **279:** 24873–24880.

23. JUNG, J.E., H.G. LEE, I.H. CHO, et al. 2005. STAT3 is a potential modulator of HIF-1-mediated VEGF expression in human renal carcinoma cells. FASEB J. **19:** 1296–1298.
24. SHIROGANE, T., T. FUKADA, J.M. MULLER, et al. 1999. Synergistic roles for Pim-1 and c-Myc in STAT3-mediated cell cycle progression and antiapoptosis. Immunity **11:** 709–719.
25. YANG, E., L. LERNER, D. BESSER, et al. 2003. Independent and cooperative activation of chromosomal c-fos promoter by STAT3. J. Biol. Chem. **278:** 15794–15799.
26. YOO, J.Y., W. WANG, S. DESIDERIO, et al. 2001. Synergistic activity of STAT3 and c-Jun at a specific array of DNA elements in the alpha 2-macroglobulin promoter. J. Biol. Chem. **276:** 26421–26429.
27. ZHANG, Z., S. JONES, J.S. HAGOOD, et al. 1997. STAT3 acts as a co-activator of glucocorticoid receptor signaling. J. Biol. Chem. **272:** 30607–30610.
28. WANG, L.H., X.Y. YANG, K. MIHALIC, et al. 2001. Activation of estrogen receptor blocks interleukin-6-inducible cell growth of human multiple myeloma involving molecular cross-talk between estrogen receptor and STAT3 mediated by co-regulator PIAS3. J. Biol. Chem. **276:** 31839–31844.
29. MIGONE, T.S., J.X. LIN, A. CERESETO, et al. 1995. Constitutively activated Jak-STAT pathway in T cells transformed with HTLV-I. Science **269:** 79–81.
30. WEBER-NORDT, R.M., C. EGEN, J. WEHINGER, et al. 1996. Constitutive activation of STAT proteins in primary lymphoid and myeloid leukemia cells and in Epstein-Barr virus (EBV)-related lymphoma cell lines. Blood **88:** 809–816.
31. SARTOR, C.I., M.L. DZIUBINSKI, C.L. YU, et al. 1997. Role of epidermal growth factor receptor and STAT-3 activation in autonomous proliferation of SUM-102PT human breast cancer cells. Cancer Res. **57:** 978–987.
32. NIELSEN, M., K. KALTOFT, M. NORDAHL, et al. 1997. Constitutive activation of a slowly migrating isoform of Stat3 in mycosis fungoides: tyrphostin AG490 inhibits Stat3 activation and growth of mycosis fungoides tumor cell lines. Proc. Natl. Acad. Sci. USA **94:** 6764–6769.
33. FRANK, D.A., S. MAHAJAN & J. RITZ. 1997. B lymphocytes from patients with chronic lymphocytic leukemia contain signal transducer and activator of transcription (STAT) 1 and STAT3 constitutively phosphorylated on serine residues. J. Clin. Invest. **100:** 3140–3148.
34. LUND, T.C., R. GARCIA, M.M. MEDVECZKY, et al. 1997. Activation of STAT transcription factors by herpesvirus Saimiri Tip-484 requires p56lck. J. Virol. **71:** 6677–6682.
35. SUN, W.H., C. PABON, Y. ALSAYED, et al. 1998. Interferon-alpha resistance in a cutaneous T-cell lymphoma cell line is associated with lack of STAT1 expression. Blood **91:** 570–576.
36. XIA, Z., M.R. BAER, A.W. BLOCK, et al. 1998. Expression of signal transducers and activators of transcription proteins in acute myeloid leukemia blasts. Cancer Res. **58:** 3173–3180.
37. CATLETT-FALCONE, R., T.H. LANDOWSKI, M.M. OSHIRO, et al. 1999. Constitutive activation of Stat3 signaling confers resistance to apoptosis in human U266 myeloma cells. Immunity **10:** 105–115.
38. HUANG, M., C. PAGE, R.K. REYNOLDS, et al. 2000. Constitutive activation of stat 3 oncogene product in human ovarian carcinoma cells. Gynecol. Oncol. **79:** 67–73.

39. RODER, S., C. STEIMLE, G. MEINHARDT, et al. 2001. STAT3 is constitutively active in some patients with polycythemia rubra vera. Exp. Hematol. **29:** 694–702.
40. EPLING-BURNETTE, P.K., B. ZHONG, F. BAI, et al. 2001. Cooperative regulation of Mcl-1 by Janus kinase/stat and phosphatidylinositol 3-kinase contribute to granulocyte-macrophage colony-stimulating factor-delayed apoptosis in human neutrophils. J. Immunol. **166:** 7486–7495.
41. KUBE, D., U. HOLTICK, M. VOCKERODT, et al. 2001. STAT3 is constitutively activated in Hodgkin cell lines. Blood **98:** 762–770.
42. MASUDA, M., M. SUZUI, R. YASUMATU, et al. 2002. Constitutive activation of signal transducers and activators of transcription 3 correlates with cyclin D1 overexpression and may provide a novel prognostic marker in head and neck squamous cell carcinoma. Cancer Res. **62:** 3351–3355.
43. HORIGUCHI, A., M. OYA, T. SHIMADA, et al. 2002. Activation of signal transducer and activator of transcription 3 in renal cell carcinoma: a study of incidence and its association with pathological features and clinical outcome. J. Urol. **168:** 762–765.
44. MORA, L.B., R. BUETTNER, J. SEIGNE, et al. 2002. Constitutive activation of Stat3 in human prostate tumors and cell lines: direct inhibition of Stat3 signaling induces apoptosis of prostate cancer cells. Cancer Res. **62:** 6659–6666.
45. GRETEN, F.R., C.K. WEBER, T.F. GRETEN, et al. 2002. Stat3 and NF-kappaB activation prevents apoptosis in pancreatic carcinogenesis. Gastroenterology **123:** 2052–2063.
46. KIRITO, K., T. NAGASHIMA, K. OZAWA, et al. 2002. Constitutive activation of Stat1 and Stat3 in primary erythroleukemia cells. Int. J. Hematol. **75:** 51–54.
47. SCHAEFER, L.K., Z. REN, G.N. FULLER, et al. 2002. Constitutive activation of Stat3alpha in brain tumors: localization to tumor endothelial cells and activation by the endothelial tyrosine kinase receptor (VEGFR-2). Oncogene **21:** 2058–2065.
48. ZAMO, A., R. CHIARLE, R. PIVA, et al. 2002. Anaplastic lymphoma kinase (ALK) activates Stat3 and protects hematopoietic cells from cell death. Oncogene **21:** 1038–1047.
49. COPPO, P., I. DUSANTER-FOURT, G. MILLOT, et al. 2003. Constitutive and specific activation of STAT3 by BCR-ABL in embryonic stem cells. Oncogene **22:** 4102–4110.
50. HSIAO, J.R., Y.T. JIN, S.T. TSAI, et al. 2003. Constitutive activation of STAT3 and STAT5 is present in the majority of nasopharyngeal carcinoma and correlates with better prognosis. Br. J. Cancer **89:** 344–349.
51. SONG, L., J. TURKSON, J.G. KARRAS, et al. 2003. Activation of Stat3 by receptor tyrosine kinases and cytokines regulates survival in human non-small cell carcinoma cells. Oncogene **22:** 4150–4165.
52. TO, K.F., M.W. CHAN, W.K. LEUNG, et al. 2004. Constitutional activation of IL-6-mediated JAK/STAT pathway through hypermethylation of SOCS-1 in human gastric cancer cell line. Br. J. Cancer **91:** 1335–1341.
53. WARIS, G., J. TURKSON, T. HASSANEIN, et al. 2005. Hepatitis C virus (HCV) constitutively activates STAT-3 via oxidative stress: role of STAT-3 in HCV replication. J. Virol. **79:** 1569–1580.
54. LIN, Q., R. LAI, L.R. CHIRIEAC, et al. 2005. Constitutive activation of JAK3/STAT3 in colon carcinoma tumors and cell lines: inhibition of JAK3/STAT3 signaling induces apoptosis and cell cycle arrest of colon carcinoma cells. Am. J. Pathol. **167:** 969–980.

55. LAI, R., F. NAVID, C. RODRIGUEZ-GALINDO, et al. 2006. STAT3 is activated in a subset of the Ewing sarcoma family of tumours. J. Pathol. **208:** 624–632.
56. GRONOWSKI, A.M., Z. ZHONG, Z. WEN, et al. 1995. In vivo growth hormone treatment rapidly stimulates the tyrosine phosphorylation and activation of Stat3. Mol. Endocrinol. **9:** 171–177.
57. GRANDIS, J.R., S.D. DRENNING, A. CHAKRABORTY, et al. 1998. Requirement of Stat3 but not Stat1 activation for epidermal growth factor receptor-mediated cell growth In vitro. J. Clin. Invest. **102:** 1385–1392.
58. GURNEY, A.L., S.C. WONG, W.J. HENZEL, et al. 1995. Distinct regions of c-Mpl cytoplasmic domain are coupled to the JAK-STAT signal transduction pathway and Shc phosphorylation. Proc. Natl. Acad. Sci. USA **92:** 5292–5296.
59. VIGNAIS, M.L., H.B. SADOWSKI, D. WATLING, et al. 1996. Platelet-derived growth factor induces phosphorylation of multiple JAK family kinases and STAT proteins. Mol. Cell Biol. **16:** 1759–1769.
60. STOUT, B.A., M.E. BATES, L.Y. LIU, et al. 2004. IL-5 and granulocyte-macrophage colony-stimulating factor activate STAT3 and STAT5 and promote Pim-1 and cyclin D3 protein expression in human eosinophils. J. Immunol. **173:** 6409–6417.
61. DEMOULIN, J.B., C. UYTTENHOVE, E. VAN ROOST, et al. 1996. A single tyrosine of the interleukin-9 (IL-9) receptor is required for STAT activation, antiapoptotic activity, and growth regulation by IL-9. Mol. Cell Biol. **16:** 4710–4716.
62. NISHIKI, S., F. HATO, N. KAMATA, et al. 2004. Selective activation of STAT3 in human monocytes stimulated by G-CSF: implication in inhibition of LPS-induced TNF-alpha production. Am. J. Physiol. Cell Physiol. **286:** C1302–C1311.
63. WILLIAMS, L., L. BRADLEY, A. SMITH, et al. 2004. Signal transducer and activator of transcription 3 is the dominant mediator of the anti-inflammatory effects of IL-10 in human macrophages. J. Immunol. **172:** 567–576.
64. JACOBSON, N.G., S.J. SZABO, R.M. WEBER-NORDT, et al. 1995. Interleukin 12 signaling in T helper type 1 (Th1) cells involves tyrosine phosphorylation of signal transducer and activator of transcription (Stat)3 and Stat4. J. Exp. Med. **181:** 1755–1762.
65. RADAEVA, S., R. SUN, H.N. PAN, et al. 2004. Interleukin 22 (IL-22) plays a protective role in T cell-mediated murine hepatitis: IL-22 is a survival factor for hepatocytes via STAT3 activation. Hepatology **39:** 1332–1342.
66. VAISSE, C., J.L. HALAAS, C.M. HORVATH, et al. 1996. Leptin activation of Stat3 in the hypothalamus of wild-type and ob/ob mice but not db/db mice. Nat. Genet. **14:** 95–97.
67. WONG, M. & E.N. FISH. 1998. RANTES and MIP-1alpha activate stats in T cells. J. Biol. Chem. **273:** 309–314.
68. CARBALLO, M., M. CONDE, R. EL BEKAY, et al. 1999. Oxidative stress triggers STAT3 tyrosine phosphorylation and nuclear translocation in human lymphocytes. J. Biol. Chem. **274:** 17580–17586.
69. NAGPAL, J.K., R. MISHRA & B.R. DAS. 2002. Activation of Stat-3 as one of the early events in tobacco chewing-mediated oral carcinogenesis. Cancer **94:** 2393–2400.
70. YOSHIDA, T., T. HANADA, T. TOKUHISA, et al. 2002. Activation of STAT3 by the hepatitis C virus core protein leads to cellular transformation. J. Exp. Med. **196:** 641–653.
71. SARCAR, B., A.K. GHOSH, R. STEELE, et al. 2004. Hepatitis C virus NS5A mediated STAT3 activation requires co-operation of Jak1 kinase. Virology **322:** 51–60.

72. AHSAN, H., M.H. AZIZ & N. AHMAD. 2005. Ultraviolet B exposure activates Stat3 signaling via phosphorylation at tyrosine705 in skin of SKH1 hairless mouse: a target for the management of skin cancer? Biochem. Biophys. Res. Commun. **333:** 241–246.
73. CARL, V.S., J.K. GAUTAM, L.D. COMEAU, *et al*. 2004. Role of endogenous IL-10 in LPS-induced STAT3 activation and IL-1 receptor antagonist gene expression. J. Leukoc. Biol. **76:** 735–742.
74. GATSIOS, P., L. TERSTEGEN, F. SCHLIESS, *et al*. 1998. Activation of the Janus kinase/signal transducer and activator of transcription pathway by osmotic shock. J. Biol. Chem. **273:** 22962–22968.
75. PROIETTI, C., M. SALATINO, C. ROSEMBLIT, *et al*. 2005. Progestins induce transcriptional activation of signal transducer and activator of transcription 3 (Stat3) via a Jak- and Src-dependent mechanism in breast cancer cells. Mol. Cell Biol. **25:** 4826–4840.
76. BROMBERG, J.F., C.M. HORVATH, D. BESSER, *et al*. 1998. Stat3 activation is required for cellular transformation by v-src. Mol. Cell Biol. **18:** 2553–2558.
77. BROMBERG, J.F., M.H. WRZESZCZYNSKA, G. DEVGAN, *et al*. 1999. Stat3 as an oncogene. Cell **98:** 295–303.
78. YU, C.L., D.J. MEYER, G.S. CAMPBELL, *et al*. 1995. Enhanced DNA-binding activity of a Stat3-related protein in cells transformed by the Src oncoprotein. Science **269:** 81–83.
79. BOWMAN, T., R. GARCIA, J. TURKSON, *et al*. 2000. STATs in oncogenesis. Oncogene **19:** 2474–2488.
80. SHEN, Y., G. DEVGAN, J.E. DARNELL, JR., *et al*. 2001. Constitutively activated Stat3 protects fibroblasts from serum withdrawal and UV-induced apoptosis and antagonizes the proapoptotic effects of activated Stat1. Proc. Natl. Acad. Sci. USA **98:** 1543–1548.
81. ZHANG, Q., P.N. RAGHUNATH, L. XUE, *et al*. 2002. Multilevel dysregulation of STAT3 activation in anaplastic lymphoma kinase-positive T/null-cell lymphoma. J. Immunol. **168:** 466–474.
82. YOSHIKAWA, H., K. MATSUBARA, G.S. QIAN, *et al*. 2001. SOCS-1, a negative regulator of the JAK/STAT pathway, is silenced by methylation in human hepatocellular carcinoma and shows growth-suppression activity. Nat. Genet. **28:** 29–35.
83. GALM, O., H. YOSHIKAWA, M. ESTELLER, *et al*. 2003. SOCS-1, a negative regulator of cytokine signaling, is frequently silenced by methylation in multiple myeloma. Blood **101:** 2784–2788.
84. OSHIMO, Y., K. KURAOKA, H. NAKAYAMA, *et al*. 2004. Epigenetic inactivation of SOCS-1 by CpG island hypermethylation in human gastric carcinoma. Int. J. Cancer **112:** 1003–1009.
85. KOMAZAKI, T., H. NAGAI, M. EMI, *et al*. 2004. Hypermethylation-associated inactivation of the SOCS-1 gene, a JAK/STAT inhibitor, in human pancreatic cancers. Jpn. J. Clin. Oncol. **34:** 191–194.
86. NAGAI, H., T. NAKA, Y. TERADA, *et al*. 2003. Hypermethylation associated with inactivation of the SOCS-1 gene, a JAK/STAT inhibitor, in human hepatoblastomas. J. Hum. Genet. **48:** 65–69.
87. BESSER, D., J.F. BROMBERG, J.E. DARNELL, JR., *et al*. 1999. A single amino acid substitution in the v-Eyk intracellular domain results in activation of Stat3 and enhances cellular transformation. Mol. Cell Biol. **19:** 1401–1409.

88. GARCIA, R., C.L. YU, A. HUDNALL, *et al.* 1997. Constitutive activation of Stat3 in fibroblasts transformed by diverse oncoproteins and in breast carcinoma cells. Cell Growth Differ. **8:** 1267–1276.
89. YU, C.L., R. JOVE & S.J. BURAKOFF. 1997. Constitutive activation of the Janus kinase-STAT pathway in T lymphoma overexpressing the Lck protein tyrosine kinase. J. Immunol. **159:** 5206–5210.
90. TANUMA, N., H. SHIMA, K. NAKAMURA, *et al.* 2001. Protein tyrosine phosphatase epsilonC selectively inhibits interleukin-6- and interleukin- 10-induced JAK-STAT signaling. Blood **98:** 3030–3034.
91. TURKSON, J., T. BOWMAN, R. GARCIA, *et al.* 1998. Stat3 activation by Src induces specific gene regulation and is required for cell transformation. Mol. Cell Biol. **18:** 2545–2552.
92. SCHLESSINGER, K. & D.E. LEVY. 2005. Malignant transformation but not normal cell growth depends on signal transducer and activator of transcription 3. Cancer Res. **65:** 5828–5834.
93. GAMERO, A.M., H.A. YOUNG & R.H. WILTROUT. 2004. Inactivation of Stat3 in tumor cells: releasing a brake on immune responses against cancer? Cancer Cell **5:** 111–112.
94. KORTYLEWSKI, M., M. KUJAWSKI, T. WANG, *et al.* 2005. Inhibiting Stat3 signaling in the hematopoietic system elicits multicomponent antitumor immunity. Nat. Med. **11:** 1314–1321.
95. ZUSHI, S., Y. SHINOMURA, T. KIYOHARA, *et al.* 1998. STAT3 mediates the survival signal in oncogenic ras-transfected intestinal epithelial cells. Int. J. Cancer **78:** 326–330.
96. PUTHIER, D., R. BATAILLE & M. AMIOT. 1999. IL-6 up-regulates mcl-1 in human myeloma cells through JAK / STAT rather than ras / MAP kinase pathway. Eur. J. Immunol. **29:** 3945–3950.
97. AOKI, Y., G.M. FELDMAN & G. TOSATO. 2003. Inhibition of STAT3 signaling induces apoptosis and decreases survivin expression in primary effusion lymphoma. Blood **101:** 1535–1542.
98. KIUCHI, N., K. NAKAJIMA, M. ICHIBA, *et al.* 1999. STAT3 is required for the gp130-mediated full activation of the c-myc gene. J. Exp. Med. **189:** 63–73.
99. DECHOW, T.N., L. PEDRANZINI, A. LEITCH, *et al.* 2004. Requirement of matrix metalloproteinase-9 for the transformation of human mammary epithelial cells by Stat3-C. Proc. Natl. Acad. Sci. USA **101:** 10602–10607.
100. NIU, G., K.L. WRIGHT, M. HUANG, *et al.* 2002. Constitutive Stat3 activity upregulates VEGF expression and tumor angiogenesis. Oncogene **21:** 2000–2008.
101. SINIBALDI, D., W. WHARTON, J. TURKSON, *et al.* 2000. Induction of p21WAF1/CIP1 and cyclin D1 expression by the Src oncoprotein in mouse fibroblasts: role of activated STAT3 signaling. Oncogene **19:** 5419–5427.
102. NAKA, T., M. NARAZAKI, M. HIRATA, *et al.* 1997. Structure and function of a new STAT-induced STAT inhibitor. Nature **387:** 924–929.
103. O'BRIEN, C.A., I. GUBRIJ, S.C. LIN, *et al.* 1999. STAT3 activation in stromal/osteoblastic cells is required for induction of the receptor activator of NF-kappaB ligand and stimulation of osteoclastogenesis by gp130-utilizing cytokines or interleukin-1 but not 1,25-dihydroxyvitamin D3 or parathyroid hormone. J. Biol. Chem. **274:** 19301–19308.
104. MISCIA, S., M. MARCHISIO, A. GRILLI, *et al.* 2002. Tumor necrosis factor alpha (TNF-alpha) activates Jak1/Stat3-Stat5B signaling through TNFR-1 in human B cells. Cell Growth Differ. **13:** 13–18.

105. STRENGELL, M., T. SARENEVA, D. FOSTER, et al. 2002. IL-21 up-regulates the expression of genes associated with innate immunity and Th1 response. J. Immunol. **169:** 3600–3605.
106. SATO, T., C. SELLERI, N.S. YOUNG, et al. 1997. Inhibition of interferon regulatory factor-1 expression results in predominance of cell growth stimulatory effects of interferon-gamma due to phosphorylation of Stat1 and Stat3. Blood **90:** 4749–4758.
107. ZHANG, X. & J.E. DARNELL, JR. 2001. Functional importance of Stat3 tetramerization in activation of the alpha 2-macroglobulin gene. J. Biol. Chem. **276:** 33576–33581.
108. KORDULA, T., R.E. RYDEL, E.F. BRIGHAM, et al. 1998. Oncostatin M and the interleukin-6 and soluble interleukin-6 receptor complex regulate alpha1-antichymotrypsin expression in human cortical astrocytes. J. Biol. Chem. **273:** 4112–4118.
109. MASCARENO, E., M. DHAR & M.A. SIDDIQUI. 1998. Signal transduction and activator of transcription (STAT) protein-dependent activation of angiotensinogen promoter: a cellular signal for hypertrophy in cardiac muscle. Proc. Natl. Acad. Sci. USA **95:** 5590–5594.
110. BHATTACHARYA, S., R.M. RAY & L.R. JOHNSON. 2005. STAT3-mediated transcription of Bcl-2, Mcl-1 and c-IAP2 prevents apoptosis in polyamine-depleted cells. Biochem. J. **392:** 335–344.
111. NAGEL-WOLFRUM, K., C. BUERGER, I. WITTIG, et al. 2004. The interaction of specific peptide aptamers with the DNA binding domain and the dimerization domain of the transcription factor Stat3 inhibits transactivation and induces apoptosis in tumor cells. Mol. Cancer Res. **2:** 170–182.
112. FLOWERS, L.O., P.S. SUBRAMANIAM & H.M. JOHNSON. 2005. A SOCS-1 peptide mimetic inhibits both constitutive and IL-6 induced activation of STAT3 in prostate cancer cells. Oncogene **24:** 2114–2120.
113. TURKSON, J., D. RYAN, J.S. KIM, et al. 2001. Phosphotyrosyl peptides block Stat3-mediated DNA binding activity, gene regulation, and cell transformation. J. Biol. Chem. **276:** 45443–45455.
114. NIKITAKIS, N.G., H. SIAVASH, C. HEBERT, et al. 2002. 15-PGJ2, but not thiazolidinediones, inhibits cell growth, induces apoptosis, and causes downregulation of Stat3 in human oral SCCa cells. Br. J. Cancer **87:** 1396–1403.
115. TURKSON, J., S. ZHANG, L.B. MORA, et al. 2005. A novel platinum compound inhibits constitutive Stat3 signaling and induces cell cycle arrest and apoptosis of malignant cells. J. Biol. Chem. **280:** 32979–32988.
116. CHEN, J., G. KUNOS & B. GAO. 1999. Ethanol rapidly inhibits IL-6-activated STAT3 and C/EBP mRNA expression in freshly isolated rat hepatocytes. FEBS Lett. **457:** 162–168.
117. WANG, Z., B. JIANG & P. BRECHER. 2002. Selective inhibition of STAT3 phosphorylation by sodium salicylate in cardiac fibroblasts. Biochem. Pharmacol. **63:** 1197–1207.
118. ZANCAI, P., R. CARIATI, M. QUAIA, et al. 2004. Retinoic acid inhibits IL-6-dependent but not constitutive STAT3 activation in Epstein-Barr virus-immortalized B lymphocytes. Int. J. Oncol. **25:** 345–355.
119. AMIT-VAZINA, M., S. SHISHODIA, D. HARRIS, et al. 2005. Atiprimod blocks STAT3 phosphorylation and induces apoptosis in multiple myeloma cells. Br. J. Cancer **93:** 70–80.

120. HIDESHIMA, T., D. CHAUHAN, T. HAYASHI, *et al*. 2003. Proteasome inhibitor PS-341 abrogates IL-6 triggered signaling cascades via caspase-dependent downregulation of gp130 in multiple myeloma. Oncogene **22:** 8386–8393.
121. ARNAUD, C., F. BURGER, S. STEFFENS, *et al*. 2005. Statins reduce interleukin-6-induced C-reactive protein in human hepatocytes: new evidence for direct antiinflammatory effects of statins. Arterioscler. Thromb. Vasc. Biol. **25:** 1231–1236.
122. BHARTI, A.C., N. DONATO & B.B. AGGARWAL. 2003. Curcumin (diferuloylmethane) inhibits constitutive and IL-6-inducible STAT3 phosphorylation in human multiple myeloma cells. J. Immunol. **171:** 3863–3871.
123. CHAKRAVARTI, N., J.N. MYERS & B.B. AGGARWAL. 2006. Targeting constitutive and interleukin-6-inducible signal transducers and activators of transcription 3 pathway in head and neck squamous cell carcinoma cells by curcumin (diferuloylmethane). Int. J. Cancer **119**(6): 1268–1275.
124. WUNG, B.S., M.C. HSU, C.C. WU, *et al*. 2005. Resveratrol suppresses IL-6-induced ICAM-1 gene expression in endothelial cells: effects on the inhibition of STAT3 phosphorylation. Life Sci. **78:** 389–397.
125. BLASKOVICH, M.A., J. SUN, A. CANTOR, *et al*. 2003. Discovery of JSI-124 (cucurbitacin I), a selective Janus kinase/signal transducer and activator of transcription 3 signaling pathway inhibitor with potent antitumor activity against human and murine cancer cells in mice. Cancer Res. **63:** 1270–1279.
126. NAM, S., R. BUETTNER, J. TURKSON, *et al*. 2005. Indirubin derivatives inhibit Stat3 signaling and induce apoptosis in human cancer cells. Proc. Natl. Acad. Sci. USA **102:** 5998–6003.
127. SU, L. & M. DAVID. 2000. Distinct mechanisms of STAT phosphorylation via the interferon-alpha/beta receptor. Selective inhibition of STAT3 and STAT5 by piceatannol. J. Biol. Chem. **275:** 12661–12666.
128. SOBOTA, R., M. SZWED, A. KASZA, *et al*. 2000. Parthenolide inhibits activation of signal transducers and activators of transcription (STATs) induced by cytokines of the IL-6 family. Biochem. Biophys. Res. Commun. **267:** 329–333.
129. LEE, Y.K., C.R. ISHAM, S.H. KAUFMAN, *et al*. 2006. Flavopiridol disrupts STAT3/DNA interactions, attenuates STAT3-directed transcription, and combines with the Jak kinase inhibitor AG490 to achieve cytotoxic synergy. Mol. Cancer Ther. **5:** 138–148.
130. CHEN, S.C., Y.L. CHANG, D.L. WANG, *et al*. 2006. Herbal remedy magnolol suppresses IL-6-induced STAT3 activation and gene expression in endothelial cells. Br. J. Pharmacol. **148:** 226–232.
131. MASUDA, M., M. SUZUI & I.B. WEINSTEIN. 2001. Effects of epigallocatechin-3-gallate on growth, epidermal growth factor receptor signaling pathways, gene expression, and chemosensitivity in human head and neck squamous cell carcinoma cell lines. Clin. Cancer Res. **7:** 4220–4229.
132. NATARAJAN, C. & J.J. BRIGHT. 2002. Curcumin inhibits experimental allergic encephalomyelitis by blocking IL-12 signaling through Janus kinase-STAT pathway in T lymphocytes. J. Immunol. **168:** 6506–6513.
133. REDDY, S. & B.B. AGGARWAL. 1994. Curcumin is a non-competitive and selective inhibitor of phosphorylase kinase. FEBS Lett. **341:** 19–22.
134. HONG, R.L., W.H. SPOHN & M.C. HUNG. 1999. Curcumin inhibits tyrosine kinase activity of p185neu and also depletes p185neu. Clin. Cancer Res. **5:** 1884–1891.

135. KORUTLA, L. & R. KUMAR. 1994. Inhibitory effect of curcumin on epidermal growth factor receptor kinase activity in A431 cells. Biochim. Biophys. Acta **1224:** 597–600.
136. SHISHODIA, S., H.M. AMIN, R. LAI, et al. 2005. Curcumin (diferuloylmethane) inhibits constitutive NF-kappaB activation, induces G1/S arrest, suppresses proliferation, and induces apoptosis in mantle cell lymphoma. Biochem. Pharmacol. **70:** 700–713.
137. KIM, H.Y., E.J. PARK, E.H. JOE, et al. 2003. Curcumin suppresses Janus kinase-STAT inflammatory signaling through activation of Src homology 2 domain-containing tyrosine phosphatase 2 in brain microglia. J. Immunol. **171:** 6072–6079.
138. BROMBERG, J. 2002. Stat proteins and oncogenesis. J. Clin. Invest. **109:** 1139–1142.
139. HANAHAN, D. & R.A. WEINBERG. 2000. The hallmarks of cancer. Cell **100:** 57–70.
140. WANG, T., G. NIU, M. KORTYLEWSKI, et al. 2004. Regulation of the innate and adaptive immune responses by Stat-3 signaling in tumor cells. Nat. Med. **10:** 48–54.
141. AGGARWAL, B.B., Y. TAKADA & O.V. OOMMEN. 2004. From chemoprevention to chemotherapy: common targets and common goals. Expert Opin. Investig. Drugs **13:** 1327–1338.
142. ABBRUZZESE, J.L. & S.M. LIPPMAN. 2004. The convergence of cancer prevention and therapy in early-phase clinical drug development. Cancer Cell **6:** 321–326.
143. DORAI, T. & B.B. AGGARWAL. 2004. Role of chemopreventive agents in cancer therapy. Cancer Lett. **215:** 129–140.
144. MELLADO, M., J.M. RODRIGUEZ-FRADE, A. ARAGAY, et al. 1998. The chemokine monocyte chemotactic protein 1 triggers Janus kinase 2 activation and tyrosine phosphorylation of the CCR2B receptor. J. Immunol. **161:** 805–813.
145. NOVAK, U., A.G. HARPUR, L. PARADISO, et al. 1995. Colony-stimulating factor 1-induced STAT1 and STAT3 activation is accompanied by phosphorylation of Tyk2 in macrophages and Tyk2 and JAK1 in fibroblasts. Blood **86:** 2948–2956.
146. MEGENEY, L.A., R.L. PERRY, J.E. LECOUTER, et al. 1996. bFGF and LIF signaling activates STAT3 in proliferating myoblasts. Dev. Genet. **19:** 139–145.
147. GOTOH, A., H. TAKAHIRA, C. MANTEL, et al. 1996. Steel factor induces serine phosphorylation of Stat3 in human growth factor-dependent myeloid cell lines. Blood **88:** 138–145.
148. BOWMAN, T., M.A. BROOME, D. SINIBALDI, et al. 2001. Stat3-mediated Myc expression is required for Src transformation and PDGF-induced mitogenesis. Proc. Natl. Acad. Sci. USA **98:** 7319–7324.
149. XIE, T.X., D. WEI, M. LIU, et al. 2004. Stat3 activation regulates the expression of matrix metalloproteinase-2 and tumor invasion and metastasis. Oncogene **23:** 3550–3560.
150. KONNIKOVA, L., M.C. SIMEONE, M.M. KRUGER, et al. 2005. Signal transducer and activator of transcription 3 (STAT3) regulates human telomerase reverse transcriptase (hTERT) expression in human cancer and primary cells. Cancer Res. **65:** 6516–6520.
151. KOJIMA, H., T. SASAKI, T. ISHITANI, et al. 2005. STAT3 regulates Nemo-like kinase by mediating its interaction with IL-6-stimulated TGFbeta-activated kinase 1 for STAT3 Ser-727 phosphorylation. Proc. Natl. Acad. Sci. USA **102:** 4524–4529.

152. IHLE, J.N. 1996. STATs: signal transducers and activators of transcription. Cell **84:** 331–334.
153. FUJIO, Y., K. KUNISADA, H. HIROTA, *et al.* 1997. Signals through gp130 upregulate bcl-x gene expression via STAT1-binding cis-element in cardiac myocytes. J. Clin. Invest. **99:** 2898–2905.
154. CHEN, Z., F.Y. LEE, K.N. BHALLA, *et al.* 2006. Potent inhibition of platelet-derived growth factor-induced responses in vascular smooth muscle cells by BMS-354825 (Dasatinib). Mol. Pharmacol. **69:** 1527–1533.
155. YANG, Y., T. IKEZOE, T. TAKEUCHI, *et al.* 2005. HIV-1 protease inhibitor induces growth arrest and apoptosis of human prostate cancer LNCaP cells in vitro and in vivo in conjunction with blockade of androgen receptor STAT3 and AKT signaling. Cancer Sci. **96:** 425–433.
156. VENKATASUBBARAO, K., A. CHOUDARY & J.W. FREEMAN. 2005. Farnesyl transferase inhibitor (R115777)-induced inhibition of STAT3(Tyr705) phosphorylation in human pancreatic cancer cell lines require extracellular signal-regulated kinases. Cancer Res. **65:** 2861–2871.
157. FADERL, S., A. FERRAJOLI, D. HARRIS, *et al.* 2005. WP-1034, a novel JAK-STAT inhibitor, with proapoptotic and antileukemic activity in acute myeloid leukemia (AML). Anticancer Res. **25:** 1841–1850.
158. KIM, H.J., Y.H. RHO, S.J. CHOI, *et al.* 2005. 15-Deoxy-delta12,14-PGJ2 inhibits IL-6-induced Stat3 phosphorylation in lymphocytes. Exp. Mol. Med. **37:** 179–185.
159. BHONDE, M.R., M.L. HANSKI, R. MAGRINI, *et al.* 2005. The broad-range cyclin-dependent kinase inhibitor UCN-01 induces apoptosis in colon carcinoma cells through transcriptional suppression of the Bcl-x(L) protein. Oncogene **24:** 148–156.
160. CHUNG, C.D., J. LIAO, B. LIU, *et al.* 1997. Specific inhibition of Stat3 signal transduction by PIAS3. Science **278:** 1803–1805.
161. LUFEI, C., J. MA, G. HUANG, *et al.* 2003. GRIM-19, a death-regulatory gene product, suppresses Stat3 activity via functional interaction. EMBO J. **22:** 1325–1335.
162. MIYAZAKI, T., J.D. BUB, M. UZUKI, *et al.* 2005. Adiponectin activates c-Jun NH2-terminal kinase and inhibits signal transducer and activator of transcription 3. Biochem. Biophys. Res. Commun. **333:** 79–87.
163. YAMASHINA, K., H. YAMAMOTO, K. CHAYAMA, *et al.* 2006. Suppression of STAT3 activity by Duplin, which is a negative regulator of the Wnt signal. J. Biochem. (Tokyo) **139:** 305–314.
164. BHAT, G.J., R.A. HUNT & K.M. BAKER. 1998. alpha-Thrombin inhibits signal transducers and activators of transcription 3 signaling by interleukin-6, leukemia inhibitory factor, and ciliary neurotrophic factor in CCL39 cells. Arch. Biochem. Biophys. **350:** 307–314.
165. WU, S.H., X.H. WU, C. LU, *et al.* 2006. Lipoxin A4 inhibits proliferation of human lung fibroblasts induced by connective tissue growth factor. Am. J. Respir. Cell Mol. Biol. **34:** 65–72.
166. KANAI, M., Y. KONDA, T. NAKAJIMA, *et al.* 2003. Differentiation-inducing factor-1 (DIF-1) inhibits STAT3 activity involved in gastric cancer cell proliferation via MEK-ERK-dependent pathway. Oncogene **22:** 548–554.
167. TANUMA, N., K. NAKAMURA, H. SHIMA, *et al.* 2000. Protein-tyrosine phosphatase PTPepsilon C inhibits Jak-STAT signaling and differentiation induced by interleukin-6 and leukemia inhibitory factor in M1 leukemia cells. J. Biol. Chem. **275:** 28216–28221.

168. COLEMAN, D.R. 4th, Z. REN, P.K. MANDAL, *et al.* 2005. Investigation of the binding determinants of phosphopeptides targeted to the SRC homology 2 domain of the signal transducer and activator of transcription 3. Development of a high-affinity peptide inhibitor. J. Med. Chem. **48:** 6661–6670.
169. ALAS, S. & B. BONAVIDA. 2001. Rituximab inactivates signal transducer and activation of transcription 3 (STAT3) activity in B-non-Hodgkin's lymphoma through inhibition of the interleukin 10 autocrine/paracrine loop and results in down-regulation of Bcl-2 and sensitization to cytotoxic drugs. Cancer Res. **61:** 5137–5144.
170. JING, N., Q. ZHU, P. YUAN, *et al.* 2006. Targeting signal transducer and activator of transcription 3 with G-quartet oligonucleotides: a potential novel therapy for head and neck cancer. Mol. Cancer Ther. **5:** 279–286.
171. SONG, H., R. WANG, S. WANG, *et al.* 2005. A low-molecular-weight compound discovered through virtual database screening inhibits Stat3 function in breast cancer cells. Proc. Natl. Acad. Sci. USA **102:** 4700–4705.
172. NUNES, M., C. SHI & GREENBERGER, L.M. 2004. Phosphorylation of extracellular signal-regulated kinase 1 and 2, protein kinase B, and signal transducer and activator of transcription 3 are differently inhibited by an epidermal growth factor receptor inhibitor, EKB-569, in tumor cells and normal human keratinocytes. Mol. Cancer Ther. **3:** 21–27.

Gene Expression Modulation in A549 Human Lung Cells in Response to Combustion-Generated Nano-Sized Particles

ANDREA ARENZ,[a] CHRISTINE E. HELLWEG,[a] NEVENA STOJICIC,[a] CHRISTA BAUMSTARK-KHAN,[a] AND HORST-HENNING GROTHEER[b]

[a]*Cellular Biodiagnostics, Department of Radiobiology, Institute of Aerospace Medicine, German Aerospace Center, Linder Höhe, D-51170 Köln, Germany*

[b]*Institute of Combustion Technology, German Aerospace Center, Pfaffenwaldring 38, D-70569 Stuttgart, Germany*

ABSTRACT: High levels of ambient air pollution are associated in humans with aggravation of asthma and of respiratory and cardiopulmonary morbidity; long-term exposures to particulate matter (PM) have been linked to possible increases in lung cancer risk, chronic respiratory disease, and increased death rates. The Biodiagnostics Group of the DLR Institute of Aerospace Medicine develops cellular test systems capable of monitoring the biological consequences of environmental conditions on humans already on cellular and molecular level. Such bioassays rely on the receptor–reporter principle, where cell lines are transfected with plasmids carrying a reporter gene under control of environment-dependent promoters (receptor), which play a key role in regulating gene expressions in response to extracellular signals. We developed the recombinant human lung epithelial cell line A549-NF-κB-EGFP/Neo carrying a genetically encoded fluorescent indicator for monitoring activation of the NF-κB signaling pathway in living cells in response to genotoxic and cytotoxic environmental influences. With this cell line we screened several candidate human radiation-responsive genes (GADD45β, CDKN1A) and NF-κB-dependent genes (IL-6, NFκBIA, and pNF-κB-EGFP) for gene expression changes by quantitative reverse transcriptase polymerase chain reaction (qRT-PCR) assay, using cDNA obtained from total RNA isolated at various time points after exposure to combustion generated nano-sized particle samples.

KEYWORDS: A549-NF-κB-EGFP/Neo; NF-κB signaling pathway; ambient particulate matter

Address for correspondence: Christa Baumstark-Khan, Cellular Biodiagnostics, Department of Radiobiology, Institute of Aerospace Medicine, German Aerospace Center, Linder Höhe, D-51170 Köln, Germany. Voice: +49-2203-6013140; fax: +49-2203-61970.
  e-mail: christa.baumstark-khan@dlr.de

Ann. N.Y. Acad. Sci. 1091: 170–183 (2006). © 2006 New York Academy of Sciences.
doi: 10.1196/annals.1378.064

INTRODUCTION

Cell response to different kinds of environmental pollution is complex and involves the participation of different classes of genes for different cellular outcomes (DNA repair, cell cycle control, signal transduction, inflammation, apoptosis, and oncogenesis). Ambient particulate matter (PM) is a complex mixture of chemicals and particles that may be compositionally diverse depending on geography and season. High levels of ambient air pollution are associated with aggravation of asthma and of respiratory and cardiopulmonary morbidity; long-term exposures to PM have been linked to possible increases in lung cancer risk, chronic respiratory disease, and increased death rates.[1] The generation of ultra-fine and nano-sized particles (<10 nm) in flames has been known for three decades, when extremely small particles (diameters down to 1.5 nm) as ions in low-pressure acetylene/oxygen flames were discovered.[2] Nano-sized particles are not only restricted to the flame itself, where they act as more or less transient precursors of soot, but are also emitted in remarkable concentrations in flame exhaust gases.[3] Although their concentration in flame exhaust gases is relatively small on a mass basis (mass numbers ranging from 1,000 to 40,000 amu), their numeric densities exceed those of soot by orders of magnitude.[4] In contrast to the better known soot particles, these extremely small soot precursors show surprising features, such as water solubility and transparency. The damaging effects of nano-sized particles are attributed to the fact that they are respirable and water-soluble and can penetrate lung mucosa. They are less readily cleared than fine particles, thereby prolonging their interaction with the lung epithelium and facilitating induction of cellular damage. Particulate matter has been shown to invoke inflammatory responses after exposure in animal models.

A family of transcription factors involved in proliferation, inflammation and apoptosis[5–8] that can be activated by various intra- and extracellular stimuli such as cytokines, oxidant-free radicals, ultraviolet irradiation and bacterial or viral products is the NF-κB/Rel family, with its members NF-κB1 (p50/p105), NF-κB2 (p52/p100), RelA (p65), RelB, and c-Rel. In the inactive state, NF-κB is bound to its inhibitor IκB,[9] mainly to IκBα, which controls nuclear uptake of NF-κB by masking the nuclear localization sequence of p65 and p50. Upon activation, IκBα can be degraded by several proteases[10] and the released NF-κB translocates into the nucleus and stimulates the expression of genes involved in a wide variety of biological functions by binding to κB or κB-like DNA motifs (NREs). NF-κB sites have been identified in the promoter or enhancer regions of a number of growth factors, cytokines, and adhesion molecules involved in fibrosis and inflammation.[11] In addition, NF-κB also regulates the expression of many genes whose products are involved in the control of cell proliferation and cell death.[12] Inappropriate activation of NF-κB has been associated with a number of inflammatory diseases, while persistent inhibition of NF-κB leads to inappropriate immune cell development or delayed cell growth.

In vitro experiments on diesel exhaust and PM_{10} have revealed that these particles are capable of inducing the release of proinflammatory cytokines[13] such as IL-6 and IL-8 from bronchial epithelial cells mediated by the transcription factor NF-κB through a mechanism partially involving tumor necrosis factor (TNF)-α.

Recently, monitoring of NF-κB activation in living cells after exposure to ionizing radiation was shown by the recombinant human embryonic kidney cell line HEK-pNF-κB-d2EGFP/Neo,[14] which was generated as a stably transfected clone carrying the NF-κB-responsive plasmid pNF-κB-d2EGFP/Neo that controls expression of the destabilized variant of enhanced green fluorescent protein (d2EGFP) by a synthetic promoter containing four copies of the NF-κB response element.[15] Cells which activated NF-κB in response to treatment with TNF-α or radiation showed an increase in the expression of d2EGFP which was quantified by flow cytometry.[14] In the current study the plasmid pNF-κB-EGFP/Neo carrying stable EGFP was used to generate the recombinant human epithelial cell line A549-pNF-κB-EGFP/Neo for investigating changing expression levels of several candidate human radiation-responsive genes (GADD45β, CDKN1A) and NF-κB-dependent genes (IL-6, NFκBIA, and pNF-κB-EGFP) by quantitative reverse transcriptase polymerase chain-reaction (qRT-PCR), using cDNA obtained from total RNA isolated at various time points after exposure to combustion-generated nano-sized particle samples.

MATERIALS AND METHODS

Generation of the Recombinant Cell Line A549-pNF-κB-EGFP/Neo and Growth Conditions

A549 parental cells, which are derived from a human bronchus alveolar carcinoma[16] and show some features characteristic of alveolar epithelial type II cells,[17] were obtained from the American Type Culture Collection (CCL-185, Rockville, MD). Cells were routinely maintained in α-MEM-medium (modified MEM, Biochrom KG, Berlin, Germany) supplemented with 10% fetal bovine serum (FBS) + 1 mM sodium pyruvate in a humidified 5% CO_2 incubator at 37°C.

Stable cell lines were generated by transfection of A549 cells with the pNF-κB-EGFP/Neo construct using cationic lipids (FuGENE, Roche Applied Science, Mannheim, Germany). pNF-κB-EGFP/Neo contains the EGFP gene sequences driven by an artificial promoter element with four NF-κB binding sites.[15] Cells with genomic incorporation were selected on the basis of antibiotic resistance (G418-sulfate, Calbiochem-Novabiochem, Schwalbach, Germany) and clonal populations were expanded and screened for TNF-α inducible EGFP expression as previously described for HEK/293 cells.[15] The newly generated cell line A549-pNF-κB-EGFP/Neo was maintained in medium supplemented with G418-sulfate (1.5 mg/mL) and split (1:5) every 7 days. Medium was changed after a growth period of 4 days. Gene expression experiments after exposure with nano-sized particles were performed in the absence of G418.

Preparation of Water-Soluble Samples of Combustion-Generated Nano-Sized Particles

The preparation of water-soluble nano-sized particles from combustion processes is described in detail.[4] A hybrid setup consisting of a low-pressure burner, a flow reactor and a photo-ionization mass spectrometer was used for the simultaneous detection of primary soot and of flame-generated nanoparticles as soot precursors. Low-pressure model flames were produced by combustion of premixed gases consisting of different oxygen and fuel (in this study propane was used as fuel) ratios. Masses of the generated particles and other species were determined by measurements with a time-of-flight mass spectrometer (TOF-MS). For the preparation of samples for toxicity analysis, an extra quartz probe was inserted laterally into a lower portion of the flame and exhaust gases containing freshly-formed nano-sized particles and soot-precursors were collected and condensation water of the exhaust gases was trapped by a liquid nitrogen trap. Samples were taken for 20 min at a flow rate of 2.3–2.5 L/min of the burned propane gas (CAS No. 74985) at a C/O ratio of 0.847. For biological experiments the frozen condensate samples containing the combustion water and dissolved material were thawed and used in appropriate dilutions.

Exposure with Water-Soluble Preparations of Nano-Sized Particles

A549-pNF-κB-EGFP/Neo cells were seeded at a density of 5×10^4 cells/cm^2 either in 96 strip-well plates (for growth curves) or in 35-mm petri dishes (for RNA isolation); A total of 48 h after seeding, cells were incubated with different dilutions of condensation water samples over indicated time spans. Control cells were treated similarly but without addition of nano-sized particles. TNF-α-treated cells (10 ng/mL) were used for comparison.

Determination of Cytotoxic Effects for Exposure to Nano-Sized Particles

Growth curves for nano-particles treated and untreated A549-pNF-κB-EGFP/Neo cells were obtained by combined fixation and staining of cells grown in one 8-well strip of a 96-strip-well plate per defined time point using crystal violet (0.1 mg mL^{-1}) solution with paraformaldehyde (0.35 mg mL^{-1}). The absorbance at 590 nm of the cell layers washed three times in PBS was measured in a microplate reader (Lambda Fluoro 320 plus, MWG Biotech, Ebersberg, Germany). The dose responsiveness was calculated as "relative cell growth" (A590$_{treated}$/A590$_{untreated}$) for 48-h incubation after start of treatment.

Total RNA Extraction and cDNA Synthesis

Total RNA of treated and nontreated cells was isolated as described earlier[18] using RNeasy Mini Kit (Qiagen, Hilden, Germany) including on-column DNase-digestion according to the manufacturer's recommendations. The integrity of the total RNA was qualitatively assessed by electrophoresis under denaturing conditions in an agarose-formaldehyde gel (1.2% [w/v]). Nucleic acid concentration was measured at 260 nm (Genequant *pro*, Amersham Biosciences, Freiburg, Germany). Purity of the RNA preparation was determined as the 260/280 nm ratio with expected values between 1.8 and 2. The RNA was first denatured at 65°C for 5 min. For the subsequent RT reaction, constant amounts of 500 ng total RNA were converted to double-stranded cDNA using the iScript cDNA synthesis kit, containing a blend of oligo(dT) and random primers, according to the manufacturer's instructions (Bio-Rad Ltd., Munich, Germany).

Oligonucleotide Primers

Sets of primer pairs were designed using published human nucleic acid cDNA/mRNA sequences (NCBI) or evaluated primer sequences from a primer database (RTPrimerDB; available at http://medgen.ugent.be/rtprimerdb) evaluated with regard to primer dimer formation, self-priming formation and primer melting temperature (Netprimer, PREMIER Biosoft International Palo Alto, CA). Primer sequences were designed to span at least one intron and validated using BLAST searches. PCR amplicons were characterized using Mfold[19] in order to predict the nature of any secondary structures that might interfere with primer access during the reverse transcription step. After optimization of the amplification conditions, gene-specific amplification was confirmed by a single peak in melting-curve analysis and a single band with the expected size in agarose gel electrophoresis. No primer-dimer formation was detected and identities of the PCR products were confirmed by restriction mapping. Primer information of the commercially synthesized primers (Invitrogen, Karlsruhe, Germany) for genes of interest and reference genes are listed in TABLE 1.

TABLE 1. Primer sequences used to amplify target genes and housekeeping genes

Primer	NCBI #	Sequence 5′– 3′	Annealing temperature [°C]	Amplicon length [bp]	PCR Efficiency
Reference genes					
B2M	NM_004048	fwd GGCTATCCAGCGTACTCCAAA rev CCAGTCCTTGCTGAAAGACAA	59.1	184	1.70
GAPDH	NM_002046	fwd CAATGACCCCTTCATTGACC rev GATCTCGCTCCTGGAAGATG	60	146	1.61
HPRT	NM_000194	fwd TGACACTGGCAAAACAATGCA rev GGTCCTTTTCACCAGCAAGCT	62	120	2.02
Target genes					
IL-6	NM_000600	fwd CTCAGCCCTGAGAAAGGAGA rev CCAGGCAAGTCTCCTCATTG	61	146	2.03
GADD45β	NM_015675	fwd ACAGTGGGGGTGTACGAGTC rev TTGATGTCGTTGTCACAGCA	61.9	155	1.87
NFκBIA	NM_020529	fwd AACCTGCAGCAGACTCCAC rev TGCTCACAGGCAAGGTGTAG	61	137	2.01
CDKN1A (p21/WAF)	NM_000389	fwd: GGAAGACCATGTGGACCTGT rev: AAATCTGTCATGCTGGTCTGC	61.9	129	1.97
EGFP	U55761	fwd GACGTAAACGGCCACAAGTT rev TAGGTCAGGGTGGTCACGAG	60	137	2.12

Quantification by Real-Time PCR

Real-time PCR using SYBR Green I technology was performed in the Opticon 2 with constant amounts of 20 ng reversely transcribed total RNA. All reactions were run in triplicates and nontemplate and RT minus controls were used as recommended. The final PCR reaction volume was 25 µL and consisted of 2× qPCR Mastermix for SYBR Green I (Eurogentec, Inc., Cologne, Germany) used according to the manufacturer's directions and 200 nM of each specific primer. The following amplification protocol was applied: After 10 min of denaturation at 95°C to activate the Hot Start DNA polymerase, 40 cycles of amplification were accomplished with: (a) 15 sec at 95°C for denaturation, (b) 30 sec at respective annealing temperature (TABLE 1), and (c) 30 sec at 60°C for elongation. Subsequently, a melting step was performed with slow heating starting at 60°C with a rate of 0.3°C/sec up to 95°C with continuous measurement of fluorescence. The same gene was always quantified in each run to prevent any inter-run variation. To generate the data basis for determination of PCR efficiency of each transcript, dilution series in triplets of a pool of all available cDNA was performed.[20]

Data Acquisition

Expression level data of the studied genes were obtained by measuring Ct values within the exponential part of the amplification curve. The Ct-value is defined as the number of cycles needed for the fluorescence signal to reach a specific threshold level of detection and is inversely correlated to the amount of starting molecules in the reaction. The real-time PCR efficiencies (E) were calculated from the slope, according to the equation $E = 10^{[-1/\text{slope}]}$, where E is in the range from 1 (minimal value) to 2 (theoretical maximum and optimum). In relative quantification the expression of a target gene is standardized by a reference gene, whose expression is considered to be constant.[21–22] The stability of standard gene expression is an elementary prerequisite for internal standardization of target gene expression. In order to determine the best suited housekeeping genes under the chosen experimental conditions, the geNorm Visual Application[22] was used. The significance of differences between expression ratios within the different treatment groups versus the mock-treated control was calculated by the REST-mcs software.[24] The software compares groups and test differences for significance with the pair wise fixed reallocation randomization test, where the Ct values for reference and target genes are jointly reallocated to control and sample groups and the expression ratios are calculated on the basis of the mean values. The error estimation of the calculated ratio uses a Taylor's series.

RESULTS

Cytotoxic Effects After Exposure to Nano-Sized Particles

A549-pNF-κB-EGFP/Neo cells showed reduced growth after 48-h treatment with condensation water samples obtained from combustion of propane at a C/O ratio of 0.85. The survival curve (FIG. 1) clearly shows the effective concentration for killing 50% of the cells to lie in the dilution range of about 1:400. The sample diluted by only slightly more than a factor of 2 (1:1,000) allows survival of nearly all of the treated cells, while incubation of cells with the sample diluted 1:200 reduced cell survival to about 10%.

Real-Time PCR Amplification Efficiencies and Linearity

Real-time PCR efficiencies were calculated using standard curves from fivefold dilutions of pooled cDNA[19] from the given slopes in the Opticon Monitor software. Investigated transcripts showed real-time PCR efficiency rates (TABLE 1) from 1.61 (GAPDH) to 2.12 (EGFP) in the investigated range from 120 pg to 75 ng cDNA input with high linearity (Pearson coefficient $r > 0.98$). To prove sensitivity and accuracy of the real-time PCR, intra- and interassay variations were determined for different template concentrations. Intraassay variation was determined in three repeats within one qPCR run. Interassay variations were determined from three runs on three different days using the same cDNA sample. The sensitivity of the assay was evaluated using

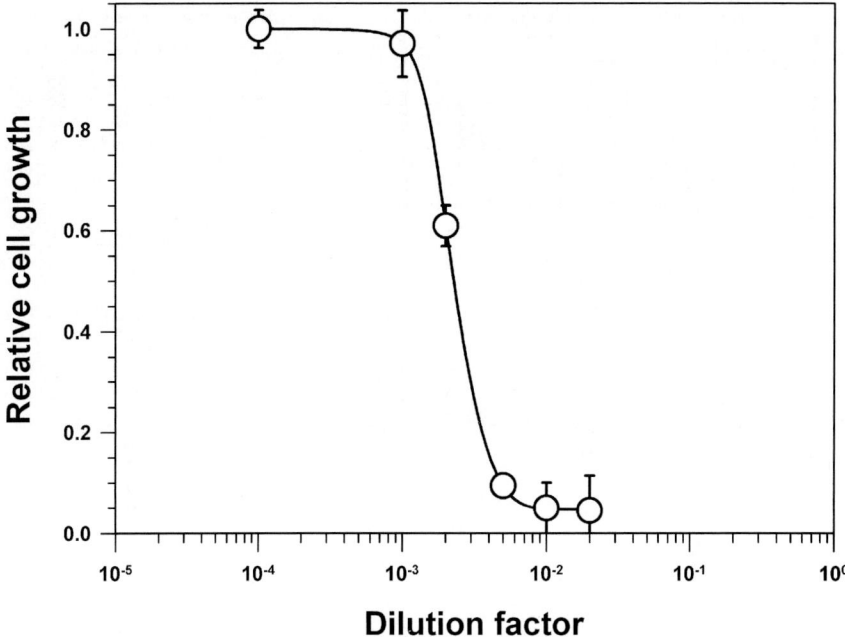

FIGURE 1. Survival of cells exposed to condensation water containing nano-sized particles obtained from combustion of propane (C_3H_8, CAS# 74985) at a C/O ratio of 0.85. Relative cell growth was calculated from the growth curves after 48 h of incubation.

different starting amounts of cDNA in a standard curve. SYBR green I fluorescence determination resulted in a reliable and sensitive cDNA quantification with high linearity ($r = 0.98$) over six orders of magnitude for RNA starting molecules.

Relative Quantification of Target Gene Expression

We have studied the time course of gene expression modulation after treatment with nano-sized particles generated by combustion processes in different dilutions in comparison to the gene expression response after treatment with TNF-α (10 ng/mL). Data are presented on a 2-based log scale.

Incubation with 10 ng/mL of TNF-α did not reduce cell survival (results not shown). At this concentration, gene expression (FIG. 2A) of the cell cycle arrest inducing gene CDKN1A is shown to be significantly downregulated. Under these conditions, the DNA damage-dependent gene GADD45β is induced shortly after treatment, followed by a downregulation 1 h later. Genes known to respond to inflammatory stress, such as IL-6 and NFκBIA, are upregulated as early as 30 min after treatment. Expression of the reporter gene EGFP, which is controlled in our system by activated NF-κB via NF-κB binding sequences

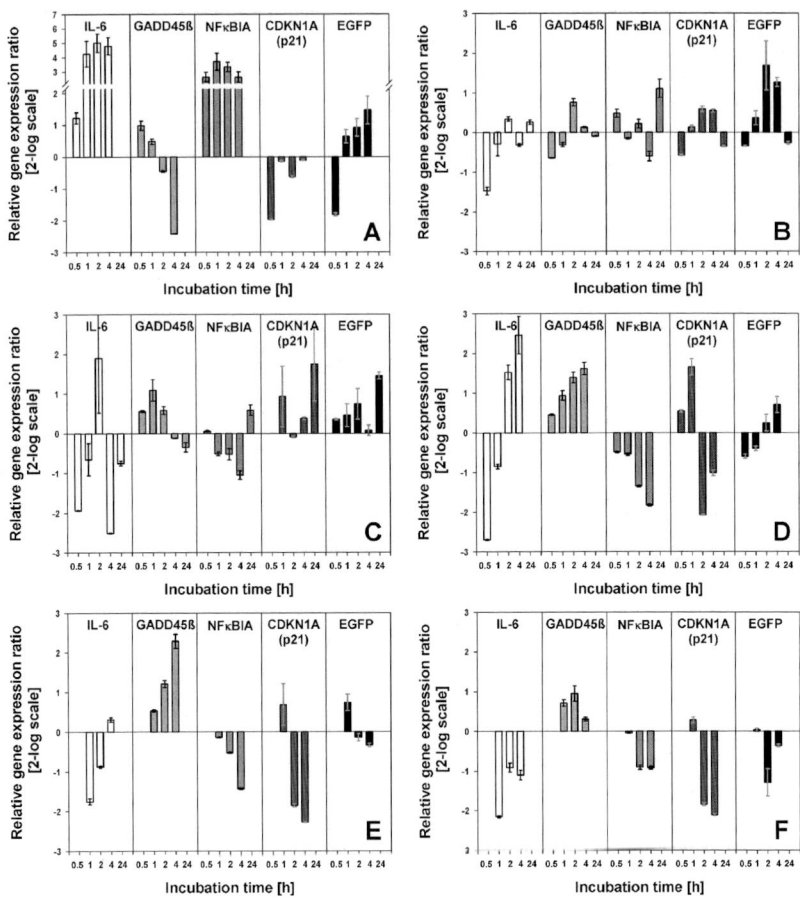

FIGURE 2. Gene expression profiles after exposure to 10 ng/mL TNF-α (**A**) and condensation water samples containing nano-sized particles obtained from combustion of propane (C_3H_8, CAS# 74985) at a C/O ratio of 0.85 (**B–F**). Changes in mRNA expression of the indicated genes were detected by quantitative real-time PCR. Relative expression ratios were calculated on the basis of group means for target genes versus a series of housekeeping genes (GAPDH, HPRT, and B2M). Gene expression patterns are shown for cells incubated with the 1:1,000 diluted combustion water sample (**B**) resulting in a survival level of 0.97 (see FIG. 1), with the 1:500 diluted sample (**C**) resulting in a survival level of 0.61, the 1:200 diluted sample (**D**) resulting in a survival level of 0.10, the 1:100 diluted sample (**E**) resulting in a survival level of 0.05, and the 1:50 diluted sample (**F**) resulting in a survival level of 0.04. "

in its artificial promoter, is continuously upregulated by TNF-α from 1 h of incubation onward.

For cells treated with nano-sized particles from propane combustion in a concentration range, where nearly all cells are able to survive the treatment (FIG. 2B), damage-dependent genes are only slightly upregulated. While EGFP expression depending on activation of NF-κB is clearly upregulated after 2 h, expression of NFκBIA is still not induced. A total of 24 h of incubation resulted in a significant upregulation of NFκBIA. At that time EGFP is not further expressed. IL-6 shows no upregulation for these conditions. Treatment with the 1:500 diluted sample of nano-sized particles (FIG. 2C), which allows only about half of the population to survive, shows an induction of GADD45β for short-term exposure (<4 h) and of CDKN1A for long-term exposure (24 h), while NF-κB-dependent EGFP expression is continuously upregulated. For exposure conditions with nano-sized particles which allow only for 10% of survivors (1:200 diluted sample), gene expression (FIG. 2D) for GADD45β increases continuously. CDKN1A is upregulated for the first posttreatment hour, while for later time points the picture changes and CDKN1A is going to be downregulated. Activation of NF-κB, as shown by EGFP expression, is delayed with moderate upregulation at 4 h. For NFκBIA, mRNA expression is downregulated with increasing exposure times. The proinflammatory cytokine IL-6 was upregulated at later time points in the investigated time range up to 5.5-fold upregulation ($\log_2 x = 2.46$) at 4 h after treatment. Exposure of cells to conditions that allow less than 5% of the population to survive (FIGS. 2E and 2F) show all genes to be downregulated with the exception of GADD45β.

DISCUSSION

The complex molecular responses to genotoxic stress are mediated by a diversity of regulatory pathways. Two pathways controlled by the ATM protein kinase, the master regulator of the cellular network induced by DNA double-strand breaks, have been reported to be of special interest for transduction of environmental signals. It has been shown that pro- and antiapoptotic signals were simultaneously induced, with the proapoptotic pathway mediated by p53 targets, and the prosurvival pathway by NF-κB targets.[25]

Thus, an important aspect of our work was to determine whether combustion-generated nano-sized particles cause DNA damage and/or activation of NF-κB and expression of DNA damage-dependent and/or NF-κB-dependent genes in pulmonary epithelial cells. Recombinant A549 lung cells (A549-NF-κB-EGFP/Neo) were incubated with different concentrations of condensation water samples collected from model flames (combustion of the gaseous fuel propane under laboratory conditions). TNF-α treated cells were used for comparison. RNA was extracted from exposed cells after various recovery times and a real-time qRT-PCR assay was applied, which employs relative quantification of candidate mRNA biomarkers. The expressions of the DNA p53-dependent gene CDKN1A and different NF-κB-dependent genes (GADD45β, NFκBIA, IL-6, "κB-EGFP") were analyzed.

The transcription factor p53 plays a central role as a principal regulator of the G_1 cell cycle checkpoint in maintaining the integrity of genome after exposure to DNA-damaging agents, thereby acting as a tumor suppressor. p53 protein accumulates within 1–3 h after a DNA-damaging event by a posttranscriptional mechanism involving phosphorylation.[26] Increased levels of p53 regulate the expression of specific genes involved in growth regulation and apoptosis.[27] p53 mediates cell cycle arrest and apoptosis by binding to DNA and activating the transcription of specific genes. One of the major targets of p53 involved in cell cycle control is the cyclin-dependent kinase inhibitor 1A ($p21^{CIP1/WAF1}$), encoded by the gene CDKN1A. Elevated levels of p21 protein can prevent G_1/S progression by binding and inhibiting the activities of the G_1 and S phase kinases, cyclin E-Cdk2, cyclin D-Cdk4-6, and cyclin A-Cdk2.[28] Its expression is tightly controlled in response to a variety of stress stimuli. In this study, we showed that low-to-moderate concentrations of combustion-generated nano-sized particles induce expression of CDKN1A in A549 lung cells within short exposure times with a possible G1 arrest and enhanced DNA repair as a consequence. For higher particle concentrations the downregulation of CDKN1A may be correlated with increased apoptotic cell death.

The pathway that antagonizes apoptosis and hence favors cell survival upon exposure to extracellular stressors is controlled by the transcription factor NF-κB. Members of the NF-κB family share a Rel homology domain (RHD), which mediates DNA binding, dimerization and interactions with specific inhibitory factors, the IκBs, which retain NF-κB dimers in the cytoplasm. Many stimuli activate NF-κB, mostly through IκB kinase-dependent (IKK-dependent) phosphorylation and subsequent degradation of IκB proteins. The liberated NF-κB dimers enter the nucleus, where they regulate transcription of diverse genes encoding cytokines, growth factors, cell adhesion molecules, and pro- and antiapoptotic proteins.[29] In addition, p53 protein stabilization was reported to be reduced upon NF-κB activation.[30] Recently, we developed a receptor–reporter assay for visualizing NF-κB activation in human embryonic kidney cells,[15] which we adapted in human lung epithelial cells for the present study. With this system EGFP synthesis is controlled by activated NF-κB, which is able to bind to its consensus sequences in the artificial promoter. Hence, EGFP mRNA expression can be used in our recombinant cell system to measure NF-κB activation. This could well be demonstrated for treatment with 10 ng/mL of TNF-α (FIG. 2A). For exposure of A549 lung cells to low-to-moderate concentrations of combustion-generated nano-sized particles, expression of EGFP mRNA was upregulated. This upregulation was most pronounced for exposure conditions that allowed better cell survival. For nano-particle samples, which downregulate EGFP expression, fewer than 5% of the cells survived the treatment.

Activated NF-κB regulates a large group of genes that have different functions. This control is crucial to a wide range of biological processes, including B and T cell development, adaptive immunity, oncogenesis, and cancer chemoresistance. During an inflammatory response, NF-κB activation protects cells from apoptosis by suppression of the Jun N-terminal kinase (JNK) cascade.[31] This suppression can involve upregulation of the growth arrest and DNA damage (GADD45) -family member GADD45β/MyD118, which are associated with the JNK kinase MKK7/JNKK2 and block its catalytic activity. Comparison of putative transcription factor-binding sites found in sequences of members of this gene family provides evidence that p53 and NF-κB may be involved in regulating their expression.[32] It was shown that GADD45β and GADD45α differentially accumulated upon induction of distinct pathways of growth arrest and apoptosis; notably, GADD45β was induced by transforming growth factor (TGF)-β, whereas GADD45α was induced by activating wild-type (wt) p53 function.[33] Our experiments clearly show that expression of GADD45β was induced also by exposure to TNF-α. For all particle concentrations used in the current study, we could demonstrate a significant time-dependent upregulation of GADD45β.

Interleukin-6 (IL-6) is a proinflammatory cytokine that also plays an important role in immunity, autoimmune diseases, and tumor development. It is produced by many different cells in response to stimuli such as bacteria, viruses and other cytokines, particularly IL-1 and TNF-α in a NF-κB0-dependent manner.[34] Elevated levels of IL-6 may be associated with an increased risk of heart attack, stroke, and chronic pulmonary disease,[35] health effects seen as a consequence of air pollution. Gene expression after treatment with 10 ng/mL TNF-α was upregulated in a time-dependent manner, reaching its maximum at 2 h of incubation. For treatment with combustion water, a clear upregulation for IL-6 was prominent for incubation times of 2 and 4 h for moderately concentrated particle samples, allowing a cell survival rate of 10–50%. At lower survival levels, IL-6 expression was clearly reduced.

Other factors regulated transcriptionally by NF-κB are the inhibitory IκB proteins (NFκBIA or NFκBIB), which inactivate NF-κB by trapping it in the cytoplasm. Upon activation, NFκBIA expression is upregulated and is postulated to be part of a negative feedback loop. For nano-particle-treated cells, expression of NFκBIA was demonstrated to be downregulated, while for exposure with 10 ng/mL TNF-α, a significant upregulation occurred. Such an upregulation could only be seen for late exposure times (24 h) for samples of nano-sized particles.

Comparing the tendency of gene expression pattern, we observed differences in the magnitude of induction of the investigated genes after exposure to nano-sized particles generated by combustion of propane. The results strongly suggest that the response of lung epithelial cells towards combustion-generated nano-sized particles is mediated in great part by activation of the transcription factor NF-κB. Even though the data reported above cannot be considered complete and/or definitive, nevertheless, on the whole, they confirm that the chosen method may constitute a suitable model system to study—at the molecular level—the effects of PM.

REFERENCES

1. ENGLERT, N. 2004. Fine particles and human health—a review of epidemiological studies. Toxicol. Lett. **149:** 235–242.
2. WERSBORG, B.L., L.K. FOX & J.B. HOWARD. 1975. Concentration and absorption coefficient in a low-pressure flame. Combust. Flame **24:** 1–10.
3. SGRO, L.A., G. BASILE, A.C. BARONE, et al. 2003. Detection of combustion formed nanoparticles. Chemosphere **51:** 1079–1090.
4. GROTHEER, H.H., H. POKORNY, K.L. BARTH, et al. 2004. Mass spectrometry up to 1 million mass units for the simultaneous detection of primary soot and of soot precursors (nanoparticles) in flames. Chemosphere **57:** 1335–1342.
5. SCHRECK, R., P. RIEBER & P.A. BAEUERLE. 1991. Reactive oxygen intermediates as apparently widely used messengers in the activation of the NF-κB transcription factor and HIV-1. EMBO J. **10:** 2247–2258.
6. BARNES, P.J. & M. KARIN. 1997. Nuclear factor kappa B: a pivotal transcription factor in chronic inflammatory diseases. New Engl. J. Med. **336:** 1066–1071.
7. WU, M., H. LEE, R.E. BELLAS, et al. 1996. Inhibition of NF-κB/Rel induces apoptosis of murine B cells. EMBO J. **15:** 4682–4690.
8. SONNENSHEIN, G.E. 1997. Rel/NF-κB transcription factors and the control of apoptosis. Semin. Cancer Biol. **8:** 113–119.
9. ZABEL, U., T. HENKEL, M.S. SILVA & P.A. BAEUERLE. 1993. Nuclear uptake control of NF-κB by MAD-3, an I kappa B protein present in the nucleus. EMBO J. **12:** 201–211.
10. ALKALAY, I., A. YARON, A. HATZUBAI, et al. 1995. Stimulation-dependent IκBα phosphorylation marks the NF-κB inhibitor for degradation via the ubiquitin-proteasome pathway. Proc. Natl. Acad. Sci. USA **92:** 10599–10603.
11. GHOSH, S., M.J. MAY & E.B. KOPP. 1998. NF-κB and Rel proteins: evolutionarily conserved mediators of immune responses. Annu. Rev. Immunol. **16:** 225–260.
12. BAICHWAL, V.R. & P.A. BAEUERLE. 1997. Activate NF-κB or die. Curr. Biol. **7:** R94–R96.
13. LEE, C.C., Y.W. CHENG & J.J. KANG. 2005. Motorcycle exhaust particles induce IL-8 production through NF-kappaB activation in human airway epithelial cells. J. Toxicol. Environ. Health A **68:** 1537–1555.
14. BAUMSTARK-KHAN, C., C.E. HELLWEG, A. ARENZ & M.M. MEIER. 2005. Cellular monitoring of the nuclear factor kappaB pathway for the assessment of space environmental radiation. Radiat. Res. **164:** 527–530.
15. HELLWEG, C.E., C. BAUMSTARK-KHAN & G. HORNECK. 2003. Generation of stably transfected mammalian cell lines as fluorescent NF-κB activation reporter assay. J. Biomol. Screen. **8:** 511–521.
16. GIARD, D.J., S.A. AARONSON, G.J. TODARO, et al. 1973. In vitro cultivation of human tumors: establishment of cell lines derived from a series of solid tumors. J. Natl. Cancer Inst. **51:** 1417–1423.
17. LIEBER, M., B. SMITH, A. SZAKAL, et al. 1976. A continuous tumor-cell line from a human lung carcinoma with properties of type II alveolar epithelial cells. Int. J. Cancer **17:** 62–70.
18. ARENZ, A., C.E. HELLWEG & C. BAUMSTARK-KHAN. 2005. Gene expression in mammalian cells after exposure to 95 MeV/amu argon ions. Adv. Space Res. **36:** 1680–1688.
19. ZUKER, M. 2003. Mfold web server for nucleic acid folding and hybridization prediction. Nucleic Acids Res. **31:** 3406–3415.

20. PFAFFL, M.W. 2001. A new mathematical model for relative quantification in real-time RT-PCR. Nucleic Acids Res. **29:** 2002–2007.
21. SERAZIN-LEROY, V., D. DENIS-HENRIOT, M. MOROT, et al. 1998. Semi-quantitative RT-PCR for comparison of mRNAs in cells with different amounts of housekeeping gene transcripts. Mol. Cell Probes **12:** 283–291.
22. SUZUKI, T., P.J. HIGGINS & D.R. CRAWFORD. 2000. Control selection for RNA quantitation. Biotechniques **29:** 332–337.
23. VANDESOMPELE, J., K. DE PRETER, F. PATTYN, et al. 2002. Accurate normalization of real-time quantitative RT-PCR data by geometric averaging of multiple internal control genes. Genome Biol. **3:** 1–12.
24. PFAFFL, M.W., G.W. HORGAN & L. DEMPFLE. 2002. Relative expression software tool (REST) for group-wise comparison and statistical analysis of relative expression results in real-time PCR. Nucleic Acids Res. **30:** e36.1–e36.10.
25. RASHI-ELKELES, S., R. ELKON, N. WEIZMAN, et al. 2006. Parallel induction of ATM-dependent pro- and antiapoptotic signals in response to ionizing radiation in murine lymphoid tissue. Oncogene **25:** 1584–1952.
26. MEEK, D.W. 1994. Posttranslational modification of p53. Semin. Cancer Biol. **5:** 203–210.
27. GOTTLIEB, T.M. & M. OREN. 1996. p53 in growth control and neoplasia. Biochim. Biophys. Acta **1287:** 77–102.
28. DULIC, V., K.W. KAUFMANN, S.J. WILSON, et al. 1994. p53-dependent inhibition of cyclin-dependent kinase activities in human fibroblasts during radiation-induced G_1 arrest. Cell **76:** 1013–1023.
29. KARIN, M. & A. LIN. 2002. NF-kappaB at the crossroads of life and death. Nat. Immunol. **3:** 221–227.
30. TERGAONKAR, V., M. PANDO, O. VAFA, et al. 2002. p53 stabilization is decreased upon NFkappaB activation: a role for NFkappaB in acquisition of resistance to chemotherapy. Cancer Cell **1:** 493–503.
31. PAPA, S., F. ZAZZERONI, C. BUBICI, et al. 2004. Gadd45 beta mediates the NF-kappa B suppression of JNK signalling by targeting MKK7/JNKK2. Nat. Cell Biol. **6:** 146–153.
32. BALLIET, A.G., K.S. HATTON, B. HOFFMAN & D.A. LIEBERMANN. 2001. Comparative analysis of the genetic structure and chromosomal location of the murine MyD118 (Gadd45beta) gene. DNA Cell Biol. **20:** 239–347.
33. VAIRAPANDI, M., A.G. BALLIET, A.J. JR. FORNACE, et al. 1996. The differentiation primary response gene MyD118, related to GADD45, encodes for a nuclear protein which interacts with PCNA and p21WAF1/CIP1. Oncogene **12:** 2579–2594.
34. AHN, K.S. & B.B. AGGARWAL. 2005. Transcription factor NF-κB: a sensor for smoke and stress signals. Ann. N. Y. Acad. Sci. **1056:** 218–233.
35. YENDE, S., G.W. WATERER, E.A. TOLLEY, et al. 2006. Inflammatory markers are associated with ventilatory limitation and muscle dysfunction in obstructive lung disease in well functioning elderly subjects. Thorax **61:** 10–16.

Multiple Levels of Control of the Expression of the Human AβH-J-J Locus Encoding Aspartyl-β-hydroxylase, Junctin, and Junctate

GIORDANA FERIOTTO,[a,b] ALESSIA FINOTTI,[b] GIULIA BREVEGLIERI,[a] SUSAN TREVES,[c] FRANCESCO ZORZATO,[d] AND ROBERTO GAMBARI[a,b]

[a]*Biotechnology Center, Università degli Studi di Ferrara, Ferrara, Italy*

[b]*Department of Biochemistry and Molecular Biology, Section of Molecular Biology, Università degli Studi di Ferrara, Ferrara, Italy*

[c]*Departments of Anaesthesia and Research, Hebelstrasse 20, Basel University Hospital, Basel, Switzerland*

[d]*Department of Experimental and Diagnostic Medicine, Section of General Pathology, Università degli Studi di Ferrara, Ferrara, Italy*

ABSTRACT: The human AβH-J-J locus is a genomic sequence which generates three functionally distinct proteins, the enzyme aspartyl-β-hydroxylase (AβH), the structural protein of sarcoplasmic reticulum junctin, and the membrane-bound calcium binding protein junctate. The first and second exons are mutually exclusive when mature mRNAs are produced. Moreover, the use of different splice donors has been shown to be involved in the generation of protein diversity by alternative splicing. As to transcriptional regulation, two promoters (P1 and P2) were identified. When the P1 and P2 promoter sequences are compared, important differences are clearly detectable. The most interesting result emerging from studies focused on the P2 promoter is that the calcium-dependent transcriptional factor MEF-2 activates the transcription of junctin, junctate, and AβH in excitable tissues and, to a lesser extent, in kidney. No Sp1 binding sites are present in the P2 promoter. In contrast, P1 promoter contains GC-rich sequences, which have homologies with the Sp1 consensus binding site.

KEYWORDS: transcription; aspartyl-β-hydroxylase; junctin; junctate; MEF-2; Sp1

Address for correspondence: Prof. Roberto Gambari, Department of Biochemistry and Molecular Biology, Section of Molecular Biology, Via Fossato di Mortara 74, 44100 Ferrara, Italy. Voice: +39-0532-424443; fax: +39-0532-424450.
e-mail: gam@unife.it

INTRODUCTION

The human AβH-J-J locus is a genomic sequence which generates three functionally distinct proteins, the enzyme aspartyl-β-hydroxylase (AβH), the structural protein of sarcoplasmic reticulum junctin, and the membrane-bound calcium binding protein junctate.[1] AβH catalyzes posttranslational hydroxylation of aspartate and asparagine residues in certain epidermal growth factor-like domains present in a number of proteins, including receptors and receptor ligands, involved in cell growth and differentiation, as well as extracellular matrix molecules.[2] Junctate is a novel integral calcium-binding protein of sarco(endo)plasmic reticulum membrane, which forms a supramolecular complex with the inositol 1,4,5 trisphosphate receptor (InsP$_3$R) and modulates calcium entry through receptor- and store-activated channels.[1,3] In addition, junctate induces and/or stabilizes peripheral couplings between the endoplasmic reticulum and the plasma membrane.[1,3] Junctin is a structural protein of sarcoplasmic reticulum.

The structural organization of the human AβH-J-J locus is shown in FIGURE 1. The scheme presented is based both on results previously reported in detail elsewhere by Treves et al.[1] and Feriotto et al.[4] The combination of data obtained by polymerase chain reaction (PCR) amplification and sequencing allowed us to define the splicing events (FIG. 1) as well as the structure of the 5′ region of this locus. The data obtained indicate that the first and second exons are mutually exclusive when mature mRNAs are produced. Moreover, the use of different splice donors has been shown to be involved in the generation of protein diversity by alternative splicing (see the lower part of FIGURE 1). Our results are in agreement with data from Dinchuk et al.[5] obtained on the homologous mouse system and suggest that at least two promoter sequences (tentatively identified as P1 and P2) are present in the 5′ region of the AβH-J-J locus.

DIFFERENTIAL RNA TRANSCRIPTION DIRECTED BY THE P1 AND P2 PROMOTERS

We employed a reverse transcription (RT)-PCR approach to examine total RNA samples extracted from several human adult tissues and confirmed by DNA sequencing the fidelity of the PCR products. Total RNA from human adult normal tissue was purchased from BD Biosciences Clontech (Palo Alto, CA). The results obtained from pancreas, brain, adrenal gland, liver, heart, and skeletal muscle show that transcription directed from the P1 promoter occurs in all the tissues analyzed (Feriotto et al., manuscript in preparation). On the contrary, the expression directed by the P2 promoter is tissue-specific, with high levels of transcription occurring particularly in skeletal muscle, cardiac muscle, and brain and lower level of transcription in kidney.[4] These results

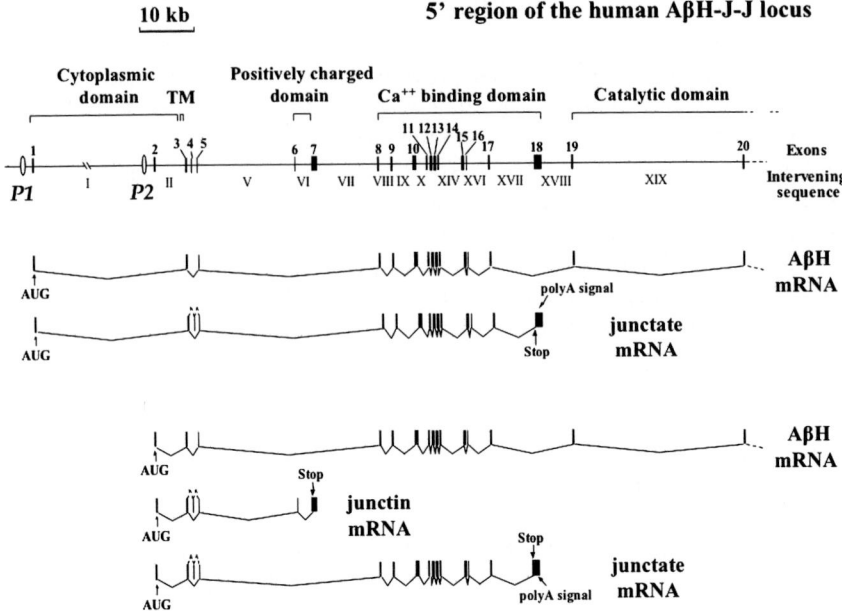

FIGURE 1. Structure of the 5' end of the human locus for aspartyl-β-hydroxylase, junctin, and junctate. Arabic numbers over black boxes indicate exons. Intervening sequences are indicated in Roman numbers. The two putative promoters P1 and P2 are indicated. A schematic representation of aspartyl-β-hydroxylase, junctin, and junctate exon splicing is shown at the bottom portion of the figure. The cytoplasmic, TM, positively charged, calcium binding, and catalytic domains are indicated. The locations of AUG, stop codons, and poly(A) signals are shown.

sustain the concept that the roles of the P1 and P2 promoters are sharply different.

We, therefore, have mapped the transcription initiation starting from these two promoters by RNA ligase-mediated 5' rapid amplification of cDNA ends (5'-RACE) with total RNA, isolated from skeletal muscle and human embryonic rhabdomyosarcoma (RD) cells. The identification of transcription initiation sites allowed us to map the putative P1 and P2 promoters [see Ref. 4 and Feriotto *et al.*, manuscript in preparation].

THE P1 PROMOTER SEQUENCES OF THE HUMAN AβH-J-J LOCUS

Similar to many housekeeping gene promoters, the P1 promoter sequence of the AβH-J-J locus lacks a TATA box and an initiator element. In contrast, this sequence is GC-rich and presents homologies with the Sp1 consensus binding

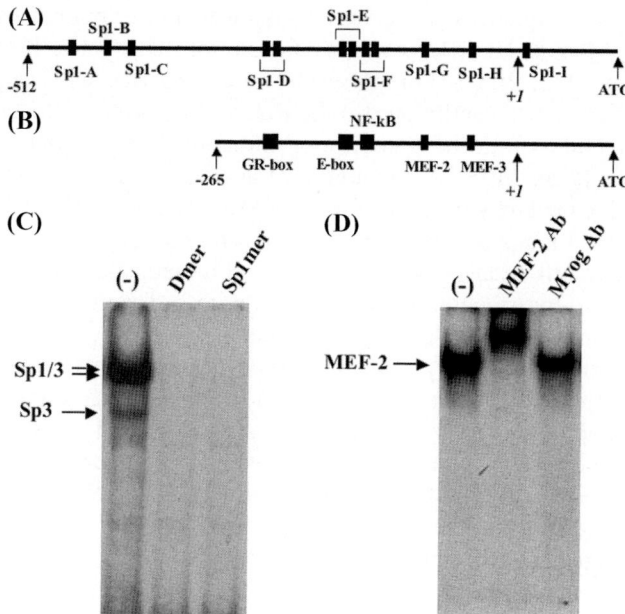

FIGURE 2. (**A**) Identification of multiple GC-rich elements, homologous to Sp1 binding site, within the AβH-J-J P1 promoter. (**B**) Location of MEF-3, MEF-2, E-box, NF-kB, GR binding sites within the AβH-J-J P2 promoter. (**C**) Binding of the AβH-J-J P1 promoter D-element, homologous to Sp1 box, to HeLa nuclear factors. Competition experiments with cold D-element and consensus Sp1 oligonucleotide demonstrate specificity of the binding (specific complexes are shown by *arrows*). (**D**) MEF-2 factors, present in C2C12 nuclear extracts, bind the P2 promoter MEF-2 box. In this case, specificity is demonstrated by supershift experiments performed using monoclonal antibodies anti-MEF-2 and anti-myogenin, used as negative control.

site, according with previous studies showing that the transcription of other TATA-less promoters frequently involves the action of proximal Sp1 sites.[6,7] Computer analysis indicates the presence of at least 12 sites that match the structural determinants of Sp1-binding specificity (FIG. 2A). A representative example of the migration profiles of the complexes produced by nuclear extracts binding to our GC-rich elements resembles the well-known electrophoresis pattern obtained with the consensus binding site for Sp1 transcription factor (FIG. 2C).[8–12]

ROLE OF MEF-2 IN THE ACTIVATION OF P2 PROMOTER SEQUENCES OF THE HUMAN AβH-J-J LOCUS

We have recently published a paper on the characterization of the P2 promoter, with the following conclusions: (*a*) the P2 promoter of the AβH-J-J locus contains sequences recognized by known transcription factors, such as MEF-2, MEF-3, NF-kB, as well as signals for other yet to be characterized proteins (FIG. 2B); (b) cis-elements were identified with negative and positive effects on transcription; (c) the minimal promoter is located within −159 nucleotides from the transcription initiation site; and (d) the nuclear transcription factor MEF-2 is required for transcriptional activation.

The involvement of MEF-2 in the transcriptional activity of the P2 promoter of the AβH-J-J locus is demonstrated by several convergent approaches, including supershift assays (FIG. 2D), *in vivo* chromatin immunoprecipitation (ChIP), and co-transfection of C2C12 cells with a plasmid carrying the P2 AβH-J-J minimal promoter and an expression plasmid carrying MEF-2 cDNA under the control of CMV promoter.[4] Moreover, the involvement of MEF-2 in transcriptional regulation of the AβH-J-J locus is further sustained by experiments employing constructs carrying mutated MEF-2 DNA binding sites. These promoters were found to be less active in sustaining transcription of the luciferase reporter gene both *in vitro* and *in vivo*. The results obtained support the hypothesis that the muscle-specific MEF-2 DNA binding activity is necessary to reach maximum levels of transcription.

CONCLUSION

When the P1 and P2 promoter sequences are compared, important differences are clearly detectable. The most interesting result emerging from studies focused on the P2 promoter is that the calcium-dependent transcriptional factor MEF-2[13,14] activates the transcription of junctin, junctate, and AβH in excitable tissues and to a lesser extent in kidney.[4] No Sp1 binding sites are present in the P2 promoter.[4] In contrast, the P1 promoter contains elements similar to Sp1 consensus binding sequence.

In addition, our data do not exclude a concerted regulation of the two promoter sequences based on interactions between different transcription factors. Strong experimental evidence demonstrates that transcription factors belonging to the Sp1 family interact with other transcription factors, including some proteins binding to the P2 promoter.[15,16] For instance, physical interactions between Sp1 and MEF-2 have been demonstrated in DNA binding complexes formed *in vitro* by nuclear extracts.[17] An intriguing possibility that should be analyzed in the future is the generation of a looping structures directed by the physical interactions between P1 and P2 promoters driven by transcription factors able to form heterodimers, such as Sp1 and MEF-2.

ACKNOWLEDGMENTS

This work was supported in part by grants from FIRB, Ministero Università e Ricerca Scientifica e Tecnologica 60 and 40%, the Department of Anesthesia Kantosspital Basel, HPRN-CT-2002-00331 from the European Union, the Italian Space Agency, and the Swiss Muscle Foundation. Roberto Gambari is supported by grants from AIRC.

REFERENCES

1. TREVES, S. et al. 2000. Molecular cloning, expression, functional characterization, chromosomal localization, and gene structure of junctate, a novel integral calcium binding protein of sarco(endo)plasmic reticulum membrane. J. Biol. Chem. **275:** 39555–39568.
2. MAEDA, T. et al. 2003. Antisense oligodeoxynucleotides directed against aspartyl (asparaginyl) beta-hydroxylase suppress migration of cholangiocarcinoma cells. J. Hepatol. **38:** 615–622.
3. TREVES, S. et al. 2004. Junctate is a key element in calcium entry induced by activation of InsP3 receptors and/or calcium store depletion. J. Cell. Biol. **166:** 537–548.
4. FERIOTTO, G. et al. 2005. Myocyte enhancer factor 2 activates promoter sequences of the human AbetaH-J-J locus, encoding aspartyl-beta-hydroxylase, junctin, and junctate. Mol. Cell. Biol. **25:** 3261–3275.
5. DINCHUK, J.E. et al. 2000. Aspartyl β-hydroxylase (Asph) and an evolutionarily conserved isoform of Asph missing the catalytic domain share exons with junctin. J. Biol. Chem. **275:** 39543–39554.
6. SUSKE, G. 1999. The Sp-family of transcription factors. Gene **238:** 291–300.
7. SUSKE, G., E. BRUFORD & S. PHILIPSEN. 2005. Mammalian SP/KLF transcription factors: bring in the family. Genomics **85:** 551–556.
8. KIELA, P.R. et al. 2003. Transcriptional regulation of the rat NHE3 gene. Functional interactions between GATA-5 and Sp family transcription factors. J. Biol. Chem. **278:** 5659–5668.
9. JEONG, J.H. et al. 2004. Regulation and autoregulation of the promoter for the latency-associated nuclear antigen of Kaposi's sarcoma-associated herpesvirus. J. Biol. Chem. **279:** 16822–16831.
10. LI, M. & R.E. KELLEMS. 2003. Sp1 and Sp3 are important regulators of AP-2 gamma gene transcription. Biol. Reprod. **69:** 1220–1230.
11. WONG, C.C. & W.M. LEE. 2002. The proximal cis-acting elements Sp1, Sp3 and E2F regulate mouse mer gene transcription in Sertoli cells. Eur. J. Biochem. **269:** 3789–3800.
12. TAPIAS, A. et al. 2005. Characterization of the 5'-flanking region of the human transcription factor Sp3 gene. Biochim. Biophys. Acta **1730:** 126–136.
13. GOSSETT, L.A. et al. 1989. A new myocyte-specific enhancer-binding factor that recognizes a conserved element associated with multiple muscle-specific genes. Mol. Cell. Biol. **9:** 5022–5033.
14. ESSER, K. et al. 1999. The CACC box and myocyte enhancer factor-2 sites within the myosin light chain 2 slow promoter cooperate in regulating nerve-specific transcription in skeletal muscle. J. Biol. Chem. **274:** 12095–12102.
15. FLUCK, C.E. & W.L. MILLER. 2004. GATA-4 and GATA-6 modulate tissue-specific transcription of the human gene for P450c17 by direct interaction with Sp1. Mol. Endocrinol. **18:** 1144–1157.

16. LOEFFLER, S. *et al.* 2005. Interleukin-6 induces transcriptional activation of vascular endothelial growth factor (VEGF) in astrocytes *in vivo* and regulates VEGF promoter activity in glioblastoma cells *via* direct interaction between STAT3 and Sp1. Int. J. Cancer **115:** 202–213.
17. GRAYSON, J., R. BASSEL-DUBY & R.S. WILLIAMS. 1998. Collaborative interactions between MEF-2 and Sp1 in muscle-specific gene regulation. J. Cell. Biochem. **70:** 366–375.

Activation of Nuclear Factor κB by Different Agents

Influence of Culture Conditions in a Cell-Based Assay

CHRISTINE E. HELLWEG, ANDREA ARENZ, SUSANNE BOGNER, CLAUDIA SCHMITZ, AND CHRISTA BAUMSTARK-KHAN

DLR, Institut für Luft- und Raumfahrtmedizin, Strahlenbiologie, 51147 Köln, Germany

ABSTRACT: The transcription factor nuclear factor κB (NF-κB) or other components of this pathway have been identified as possible therapeutic targets in inflammatory processes, cancer, and autoimmune diseases. In order to clarify the role of NF-κB in epithelial cells in response to different stresses, a cell-based screening assay for activation of NF-κB-dependent gene transcription in human embryonic kidney cells (HEK/293) was developed. This assay allows detection of NF-κB activation by measurement of the fluorescence of the reporter protein destabilized enhanced green fluorescent protein (d2EGFP). For characterization of the cell-based assay, activation of the pathway by several agents, for example, tumor necrosis factor α (TNF-α), interleukin-1β (IL-1β), lipopolysaccharide (LPS), camptothecin and phorbol ester (PMA), and the influence of the culture conditions on NF-κB activation by TNF-α were examined. NF-κB was activated by TNF-α, IL-1β, PMA, and camptothecin in a dose-dependent manner, but not by LPS. TNF-α results in the strongest induction of NF-κB-dependent gene expression. However, this response fluctuated from 30 to 90% of the cell population showing d2EGFP expression. This variation can be explained by differences in growth duration and cell density at the time of treatment. With increasing confluence of the cells, the activation potential decreased. In a confluent cell layer, only 20–35% of the cell population showed d2EGFP expression. The underlying mechanism of this phenomenon can be the production of soluble factors by the cells inhibiting the NF-κB activation or direct communication via gap junctions in the cell layer diminishing the TNF-α response.

Address for correspondence: Christine E. Hellweg, Radiobiology Division, Institute of Aerospace Medicine, DLR, Linder Höhe, 51147 Köln, Germany. Voice: +49-2203-601-3243; fax: +49-2203-61970.
e-mail: christine.hellweg@dlr.de

KEYWORDS: nuclear factor kappa B; human embryonic kidney cells; green fluorescent protein; gene expression; bioassay; tumor necrosis factor alpha; interleukin; phorbol ester; camptothecin; lipopolysaccharide

INTRODUCTION

Cellular stress protection responses lead to increased transcription of several genes via modulation of transcription factors. The transcription factor nuclear factor κB (NF-κB) or other components of this pathway represent essential components of the basic immune response in mammals and have been identified as possible therapeutic targets in inflammatory processes, cancer, and autoimmune diseases. Activation of the NF-κB pathway in response to cellular stress can have antiapoptotic effects in many cell types.[1–4] The pathway is induced by lipopolysaccharide (LPS), by phorbol esters, or by stimulation of the tumor necrosis factor α (TNF-α), interleukin-1β (IL-1β), or other proinflammatory lymphokine receptors, as well as by chemotherapeutics, oxidant stresses, and radiation. The ubiquitously expressed transcription factor is regulated by the cytoplasmic inhibitor I-κBα, which is phosphorylated and subsequently degraded upon activation of the pathway, followed by translocation of the liberated NF-κB to the nucleus and binding to κ-enhancer elements in promoters of several genes.

In order to clarify the role of NF-κB in epithelial cells in the response to different stressors, a cell-based screening assay for activation of NF-κB-dependent gene transcription in human embryonic kidney cells (HEK/293) was developed.[5] This assay allows detection of NF-κB activation by measurement of the fluorescence of the reporter protein destabilized enhanced green fluorescent protein (d2EGFP). Expression of the EGFP, originally isolated from the bioluminescent jellyfish *Aequorea victoria*,[6] allows fluorescent labeling of viable cells in gene expression studies. Unlike other reporter systems which require incubation of the cells with specific substrates or cofactors to produce a signal, EGFP fluorescence only requires exposure to blue light for visualization. The destabilized EGFP (d2EGFP) represents an EGFP variant with a reduced half-life in mammalian cells.[7] The quantification of EGFP fluorescence in cells by different means was described previously.[8] In brief, a quick measurement of fluorescent light output is achieved using a microplate reader. If single-cell data and an extremely sensitive measurement are required, a fluorescence-activated cell scanner (FACS) analysis can be performed.

The cell-based assay was established by stable transfection of HEK/293 cells with a vector carrying the d2EGFP gene under control of a synthetic promoter consisting of four κB elements and the minimal thymidine kinase promoter. NF-κB-dependent gene expression was quantified by flow cytometry as green fluorescence of d2EGFP per cell. Treatment of the recombinant HEK-pNF-κB-d2EGFP/Neo cells with the known NF-κB activator TNF-α results in strong induction of NF-κB-dependent gene expression in up to 90% of the cells 20 h after treatment. Involvement of the p65 subunit in this response was shown using an oligonucleotide-assay.[5] The HEK-pNF-κB-d2EGFP/Neo cells showed increased d2EGFP expression in response to X rays and heavy ions of different particle parameters.[9,10]

For a better characterization of the cell-based assay, activation of the pathway by other agents, and the influence of the culture conditions on NF-κB activation by TNF-α were examined. As possible NF-κB-activating agents, the cytokine IL-1β, the bacterial cell wall component LPS, the topoisomerase inhibitor camptothecin, and the phorbol ester PMA were selected. Important culture conditions that are shown to influence the NF-κB activation by TNF-α in this work are the growth duration before treatment and the cell density at the time of treatment.

MATERIALS AND METHODS

The development of the cell-based screening assay for detection of activation of NF-κB-dependent gene transcription in human embryonic kidney cells (HEK/293) was described before.[5] In brief: HEK/293 cells were stably transfected with a vector carrying the d2EGFP gene under control of a synthetic promoter consisting of four κB elements and the minimal thymidine kinase promoter (pNF-κB-d2EGFP/Neo). The plasmid contains a G418 resistance gene for stable transfection.

Chemicals

Chemicals were obtained from Sigma (Taufkirchen, Germany) if not stated otherwise. Recombinant human tumor necrosis factor α (hTNF-α) was dissolved in phosphate-buffered saline (PBS) containing 1% bovine serum albumin (BSA), recombinant human interleukin 1β (hIL-1β) (PeproTech, Inc., Rocky Hill, NJ) in PBS, the phorbol ester phorbol 12-myristate 13-acetate (PMA) in dimethylsufoxide (DMSO), all at a concentration of 10 μg/mL. Camptothecin (MP Biochemicals, Inc., Aurora, OH) was dissolved in DMSO at a concentration of 10 mmol/L. Lipopolysaccharide (LPS) from *Escherichia coli* O111:B4 (Sigma L-4391) was dissolved in PBS at a concentration of 1 mg/mL.

Cell Line and Culture Conditions

Human embryonic kidney (293/HEK) cells (CRL-1573) were obtained from the American Type Culture Collection, Manassas, VA. HEK and HEK-pNF-κB-d2EGFP/Neo cells were grown under standard conditions in 80-cm^2 flasks

(Nunc, Wiesbaden, Germany) in α-MEM-medium (modified MEM, Biochrom KG, Berlin, Germany), with 10% fetal bovine serum (FBS) at 37°C, saturated humidity and in a 5% CO_2/95% air atmosphere. The stably transfected clone (HEK-pNF-κB-d2EGFP/Neo L2) was maintained in medium containing 0.75 mg/mL of the amino glycoside antibiotic G418 (Calbiochem-Novabiochem, Schwalbach, Germany). Cultures were split (1:20) every 7 days using standard detachment procedures (0.05% trypsin containing 0.02% EDTA solution, Biochrom KG, Berlin, Germany). Medium was changed after 4 days.

Exposure of Cells to Different Agents

HEK-pNF-κB-d2EGFP/Neo cells were seeded into polylysin-coated 24-well-plates (Nunc) at a density of 3×10^4 cells/cm^2. After 2 days, hTNF-α, hIL-1β, PMA, camptothecin or LPS were added with fresh medium. After 20 h, the cells were harvested by trypsination for FACS-analysis (see *Fluorescence Microscopy* below).

Growth Kinetics

HEK-pNF-κB-d2EGFP/Neo cells were seeded into polylysin-coated petri dishes (Nunc) at a density of 3×10^4 cells/cm^2. Twice a day, the cells were harvested by trypsination and counted in a Fuchs-Rosenthal counting chamber.

Kinetics of the NF-κB Activation Potential

HEK-pNF-κB-d2EGFP/Neo cells were seeded into polylysin-coated petri dishes (Nunc) at a density of 3×10^4 cells/cm^2. Three times a day, 10 ng/mL hTNF-α was added in a set of the petri dishes and these cells were harvested for FACS-analysis (see **Fluorescence Measurements**) after 20 h incubation with hTNF-α.

Influence of Cell Density on NF-κB Activation Potential

HEK-pNF-κB-d2EGFP/Neo cells were seeded into polylysin-coated 24-well-plates (Nunc) in cell density from 10^2 to 2×10^5 cells/cm^2. After 2 days, 10 ng/mL hTNF-α was added and the cells were harvested for FACS-analysis (see **Fluorescence Measurements**) after 20-h incubation with hTNF-α.

Fluorescence Measurements

Fluorescence Microscopy

HEK-pNF-κB-d2EGFP/Neo cells treated with different agents were screened for d2EGFP expression in an inverted fluorescence microscope (Axiovert 135, Carl Zeiss, Oberkochen, Germany).

Flow Cytometry

Distribution patterns of fluorescent protein-expressing cells were analyzed using a FACScan (Becton Dickinson; San Jose, CA) equipped with an argon laser (488 nm) as excitation source for d2EGFP. The samples' forward and sideward scatter and d2EGFP fluorescence in fluorescence channel FL1 were determined and analyzed using the CellQuest software (version 1.2, Becton Dickinson). Samples of untransfected, stably transfected clones and cells treated with different agents were prepared as follows: cells were detached from the growth surface using trypsin and fixed with cold 3.5% formaldehyde in PBS for 30 min at 4°C. The fixative was diluted with PBS (1:3) and cells were stored at 4°C. Before FACS analysis, cells were centrifuged and suspended in PBS. With untransfected HEK cells, stably transfected untreated and TNF-α-treated cells were used for instrument adjustment, and the optimal compensation levels were determined and regions for the different populations were defined. A total of 2×10^4 cells were analyzed at a rate of 200–600 cells per second.

RESULTS

The stably transfected cell line HEK-pNF-κB-d2EGFP/Neo L2 was generated in order to create a fast and reliable reporter assay for the activation of NF-κB-dependent gene expression by different agents and conditions.[5] Activation of the pathway leads to binding of the endogenous NF-κB to the κB enhancer element, and thereby it increases the transcription of d2EGFP. The use of d2EGFP as reporter protein allows monitoring of the gene expression level in living cells or in formaldehyde-fixed cells by flow cytometry.[8,5] The cell line shows d2EGFP expression in response to treatment with the known NF-κB activator TNF-α within 3 h after addition, reaching a maximum after 20–24 h.

Dose-Response of NF-κB-Dependent Gene Expression after Treatment of Stably Transfected HEK Cells with Different Agents

In this work, the response of this cell line to treatment with several known NF-κB activators (cytokines, chemicals, and bacterial cell wall lipopolysaccharide) was determined. Cells were exposed to the agents for 20 h and d2EGFP fluorescence as indicator of NF-κB activation was analyzed by flow cytometry. The argon laser (488 nm) of the FACScan (Becton Dickinson) allows an optimal excitation of d2EGFP. The number of cells falling in the range defined as d2EGFP$^{(+)}$ by calibration with untransfected HEK cells, untreated and TNF-α-treated HEK-pNF-κB-d2EGFP/Neo L2 cells, was used as indicator of NF-κB-dependent d2EGFP expression.

The dose-effect curves for the d2EGFP expression after treatment with TNF-α, IL-1β, PMA, camptothecin, and LPS are shown in FIGURE 1. TNF-α elicits the highest response at a concentration of 5 ng/mL, resulting d2EGFP fluorescence in about 70% of the population (FIG. 1A). An increase in fluorescence with dose can be observed till the maximum is reached, and a decrease of fluorescence thereafter. At 0.1 ng/mL, an increase from 5 (vehicle-treated) to 10% EGFP$^{(+)}$ cells is observed.

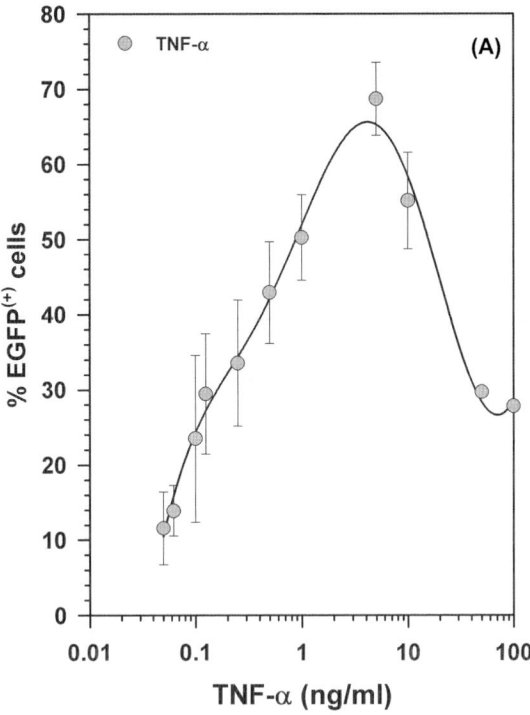

FIGURE 1. d2EGFP expression 20 h after treatment of stably transfected HEK-pNF-κB-d2EGFP/Neo cells with TNF-α (**A**), IL-1β (**B**), PMA (**C**), campothecin, (**D**) or LPS (**E**). Cells were grown for 48 h before treatment. d2EGFP was quantified by FACS analysis of formaldehyde-fixed cells. Mean and standard error of up to 10 independent experiments are shown.

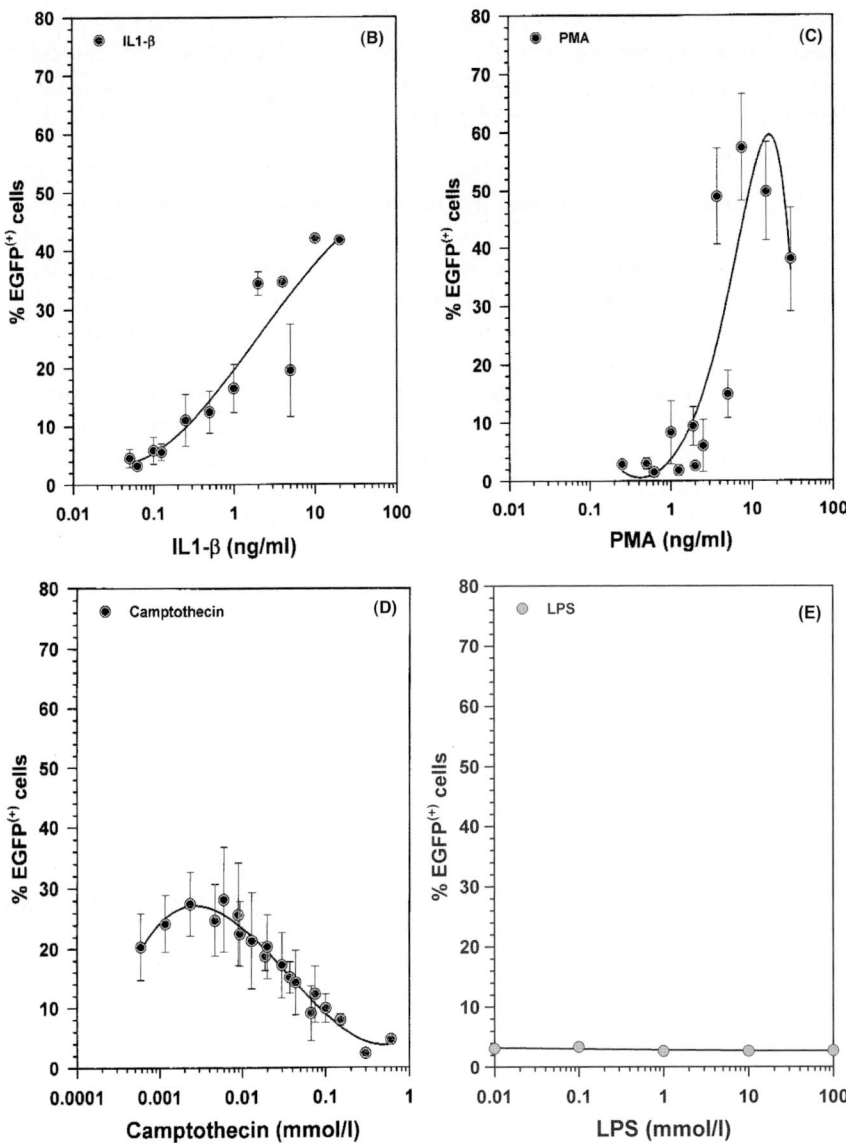

FIGURE 1. *Continued.*

The response to IL-1β is less pronounced, encompassing fewer 50% EGFP$^{(+)}$ cells at 10 ng/mL (FIG. 1B). NF-κB activation by PMA reaches its maximum at 7.5 ng/mL, with nearly 60% of the population expressing d2EGFP. Camptothecin is also an activator of NF-κB-dependent gene expression in HEK cells, reaching a maximum at 0.006 mmol/L, with a gradual decrease in d2EGFP expression with increasing concentrations. LPS had no effect on d2EGFP expression in the dose range 0.01–100 ng/mL. TNF-α results in the strongest induction of NF-κB-dependent gene expression. However, this response fluctuated from 30 to 90% EGFP$^{(+)}$ cells.

Kinetics of the NF-κB Activation Potential in HEK Cells

In order to discover factors influencing the NF-κB activation by TNF-α in stably transfected HEK cells, the dependence of d2EGFP expression after TNF-α treatment on growth duration before treatment and on cell density at the moment of treatment were examined.

First, several batches of cells seeded at the same time point were treated sequentially with 10 ng/mL TNF-α and harvested for FACS analysis 20 h later. FIGURE 2 shows the resulting kinetics of the NF-κB activation potential in stably transfected HEK cells. The background fluorescence in vehicle-treated cells is slightly higher after treatments during the first 60 h of growth. TNF-α treatment simultaneously with seeding of cells results in d2EGFP expression in about 50% of the cells 20 h later. The maximal response is reached when incubation with TNF-α starts around 20–40 h after seeding (corresponding to approximately one to two doubling times, which is 21.4 h for the transfected

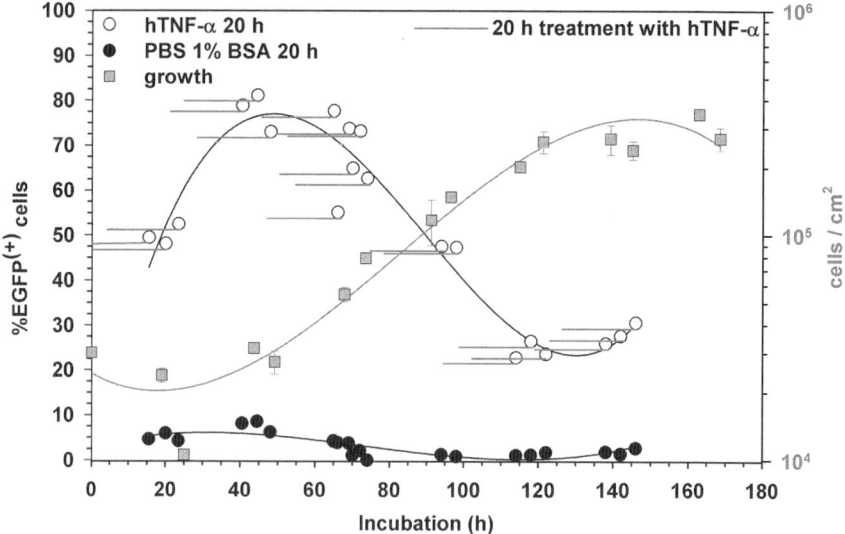

FIGURE 2. NF-κB-dependent gene expression in HEK-pNF-κB-d2EGFP/Neo L2 cells, 20 h after treatment with TNF-α (*open circles*), and growth as determined by cell counting (*gray squares*). The cells were seeded and treated with 10 ng/mL TNF-α for 20 h (*gray bars*) at several time points after seeding. d2EGFP expression was quantified by FACS analysis. Bars show the standard deviation of two independent experiments.

HEK cells). Activation of NF-κB by TNF-α was maximal within 48 h after seeding of cells. With increasing confluence of the cells, the activation potential decreased. If cells are treated 80 h after plating, only 50% react as measured 20 h later. Less than 30% of the cells show NF-κB-dependent gene expression if cells are treated after a growth period of 100 h. In a confluent cell layer, only 20–35% of the cell population showed d2EGFP expression.

The growth curve of stably transfected HEK cells, seeded at a density of 3×10^4 cells/cm^2 is shown in the same graph (FIG. 2). During the lag phase, NF-κB activation potential is moderate. The maximum is reached during the first third of the exponential phase, with cell numbers of 2×10^4 to 8×10^4 cells/cm^2. The minimum is reached during the stationary phase, with cell densities of 2×10^5 to 3.5×10^5 cells/cm^2.

Influence of Cell Density on NF-κB Activation Potential

Regarding the dependence of NF-κB activation potential on the growth period preceding the TNF-α treatment, the question was whether variation in cell density can explain this phenomenon. The possible connection of cell density and NF-κB activation was examined in cells that were seeded in different cell densities and treated at the same time point. The d2EGFP expression 20 h after treatment with TNF-α in relation to the cell density is shown in FIGURE 3. A total of 70–85% of the cells express d2EGFP if seeded in a density of 5×10^2 to 3×10^4 cells/cm^2 2 days prior to treatment, reaching a cell density of 1.5×10^3 to 6×10^4 cells/cm^2 at the time of treatment. Beyond the threshold of approximately 2×10^4 cells/cm^2, the higher the cell density, the lower was the d2EGFP expression after 20 h incubation with TNF-α. Cell densities higher than 3×10^4/cm^2 (standard seeding density for HEK cells) resulted in reduced NF-κB activation by TNF-α. At 3×10^5 cells/cm^2, only 37% of the cells responded to TNF-α treatment, compared to 84% at 1.5×10^4 cells/cm^2. This range represents the cell number span in a standard growth protocol for HEK cells, starting in the lag phase with around 1.5×10^4 cells/cm^2 (seeding of 3×10^4 cells/cm^2) and reaching up to 3.5×10^5 cells/cm^2 in the confluent state. The background fluorescence observed in vehicle-treated cells was higher if cells were seeded in density under 4×10^3 cells/cm^2. Analysis of very low cell numbers by flow cytometry resulted in high variation and high standard deviations.

DISCUSSION

In order to clarify the role of NF-κB in epithelial cells in the response to different stressors, a cell-based screening assay for activation of NF-κB-dependent gene transcription in HEK cells was developed.[5] The destabilized

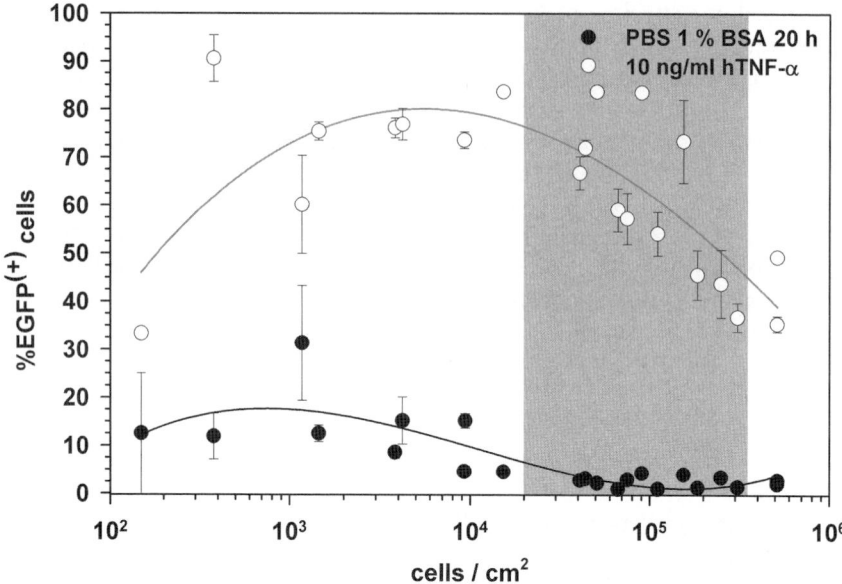

FIGURE 3. NF-κB-dependent gene expression in HEK-pNF-κB-d2EGFP/Neo L2 cells, 20 h after treatment with 10 ng/mL TNF-α. The cells were seeded in increasing densities and treated with TNF-α 48 h later. d2EGFP expression and cell number was quantified by FACS analysis. Data of four experiments were combined and regression analysis was performed, *bars* show standard errors of 8 wells (2 per independent experiment). The *gray shaded area* represents the cell densities spanned by the growth curve from FIGURE 2 (starts with a cell density of $2 \times 10^4/cm^2$ in the lag phase and reaches up to 3.5×10^5 cells/cm² in the confluent state during stationary phase).

d2EGFP carries a PEST sequence that targets the protein for degradation.[7,11] d2EGFP halves every 5.3 h in stably transfected HEK cells.[12] This assay allows detection of NF-κB activation by measurement of the fluorescence of the reporter protein d2EGFP. For a better characterization of this assay, activation of the pathway by IL-1β, LPS, camptothecin, and PMA, and the influence of the culture conditions on NF-κB activation by TNF-α were examined.

NF-κB-Dependent Gene Expression after Treatment with Different Agents

In human embryonic kidney cells, TNF-α, PMA, IL-1β, and camptothecin activate the NF-κB pathway in a dose-dependent manner, whereas LPS has no effect. The response to TNF-α was strong, but displayed high variability between experiments. IL-1 has been described as NF-κB activator in epithelial cells by Ray and Kennard.[13] The phorbol ester PMA is known to activate NF-κB[14] with a slightly slower kinetics compared to TNF-α.[15,16] The lack of response to LPS can be explained by the lack of the toll-like receptors on HEK cells' surface.[17]

Camptothecin binds to and stabilizes the topoisomerase-DNA covalent complex, which reversibly inhibits nuclear topoisomerase I. Presumably by blocking this enzyme's function, camptothecin generates double-strand breaks (DSB).[18–20] Camptothecin has been described to be a slow and weak NF-κB activator, involving NEMO (IKKγ).[21–23] Cells from ataxia telangiectasia (AT) patients exhibit a defect in NF-κB activation in response to treatment with camptothecin.[24] These findings are in line with this work, where it has been shown that camptothecin is a weaker activator of NF-κB-driven gene expression in human embryonic kidney cells than the cytokines TNF-α and IL-1β.

Factors Modulating the NF-κB Activation Potential in HEK Cells

Activation of NF-κB by TNF-α was maximal within 48 h after seeding of cells. With increasing confluence of the cells, the activation potential decreased. In a confluent cell layer, only 20–35% of the cell population showed d2EGFP expression. This effect was also observed when cells were seeded in different cell densities and treated at the same time point; the higher the cell density, the lower the d2EGFP expression was after 20 h incubation with TNF-α, but to a lower extent. The experiments have shown that the NF-κB activation by TNF-α depends on the growth duration after seeding of cells until treatment with the cytokine. The cell density reached during these growth periods cannot explain the full effect observed in growth experiments, as shown by the experiments with cells seeded in different densities.

The mechanism underlying this phenomenon could be the production of soluble factors by the cells inhibiting the NF-κB activation or the direct communication via gap junctions in the cell layer diminishing the TNF-α response or up- and downregulation of the TNF receptors during the growth period. During the exponential growth phase, activation of cell-cycle-dependent genes might influence the NF-κB activation potential. Directly after passage, the receptor number might be reduced by the superficial digestion with trypsin which is used for routine subculturing of adherent cells. With increasing cell density, the production of growth factors or the direct communication via gap junctions may modulate the number of TNF receptors of the cell surface. Chan and Aggarwal have shown that only a small fraction of the TNF receptors is needed for maximal NF-κB activation, but comparison of different cell lines indicated a correlation between receptor expression and activation of the transcription factor.[15]

Both TNF receptor types appear to be expressed on all nucleated cells in various numbers and distributions.[25] Cells of lymphoid and monocytic lineages contain high levels of TNF-R1, whereas epithelial cells express mostly the TNF-R2 receptor.[26,27] TNF receptors can be modulated at the protein or mRNA levels by a number of agents. The agents that increase the cell surface expression of TNF receptors include interferons, dibutyryl CAMP, retinal, butyrate, concanavalin A, wheat-germ agglutinin, and thyroid-stimulating hormone, whereas those that are known to downregulate these receptors are TNF, phorbol esters, interleukin-1, lipopolysaccharide, okadaic acid, cycloheximide, and actinomycin D.[15]

In a confluent state, most or all of the cells in the population are in contact. These cell contacts can contain gap junctions which allow exchange of small molecules, for example, second messenger. Gap junctions allow cell-to-cell electrical communication as well as passage of small molecular weight substances and are critical for synchronizing cellular activity in certain tissues. Hek/293 cells are known to express gap junction connexins which form functional gap junctions.[28,29] Communication via these gap junctions may modulate the response of HEK cells to TNF-α.

Detailed descriptions of comparable effects in cell culture experiments are rare. In breast cancer cell lines, the cell density was shown to influence the expression level of matrix metalloproteases (MMPs), and it was seen that transcription factors such as NF-κB, activated protein 1 (AP-1), and CRE were involved in this process.[30] Erl *et al.* describe a cell density-dependent sensitivity to apoptosis in rat aortic smooth muscle cells (rSMCs) by radical scavenger pyrrolidinedithiocarbamate PDTC, an inhibitor of NF-κB activation.[31]

The experiment conditions of the cell-based assay have to be very well controlled because cell density and growth duration influence the NF-κB activation. In order to elucidate whether this effect is stimulus-independent, experiments on NF-κB activation by PMA and camptothecin at different time points after seeding are in progress. Furthermore, possible receptor downregulation of the TNFR will be assessed during a standard culture period.

ACKNOWLEDGMENTS

The authors would like to thank Nevena Stoijcic for technical assistance.

REFERENCES

1. BOURS, V., M. BENTIRES-ALJ, A.C. HELLIN, *et al.* 2000. Nuclear factor-kappa B, cancer, and apoptosis. Biochem. Pharmacol. **60:** 1085–1089.
2. BURSTEIN, E. & C.S. DUCKETT. 2003. Dying for NF-kappaB? Control of cell death by transcriptional regulation of the apoptotic machinery. Curr. Opin. Cell Biol. **15:** 732–737.
3. CAMPBELL, K.J., S. ROCHA & N.D. PERKINS. 2004. Active repression of antiapoptotic gene expression by RelA(p65) NF-kappa. B. Mol. Cell **13:** 853–865.
4. CATALDI, A., M. RAPINO, L. CENTURIONE, *et al.* 2003. NF-kappaB activation plays an antiapoptotic role in human leukemic K562 cells exposed to ionizing radiation. J. Cell Biochem. **89:** 956–963.

5. HELLWEG, C.E., C. BAUMSTARK-KHAN & G. HORNECK. 2003. Generation of stably transfected Mammalian cell lines as fluorescent screening assay for NF-kappaB activation-dependent gene expression. J. Biomol. Screen **8:** 511–521.
6. CHALFIE, M., Y. TU, G. EUSKIRCHEN, et al. 1994. Green fluorescent protein as a marker for gene expression. Science **263:** 802–805.
7. LI, X., X. ZHAO, Y. FANG, et al. 1998. Generation of destabilized green fluorescent protein as a transcription reporter. J. Biol. Chem. **273:** 34970–34975.
8. HELLWEG, C.E., C. BAUMSTARK-KHAN, P. RETTBERG & G. HORNECK. 2001. The suitability of enhanced green fluorescent protein (EGFP) as a reporter component for bioassays. Anal. Chim. Acta. **426:** 175–184.
9. HELLWEG, C.E., A. ARENZ, M.M. MEIER & C. BAUMSTARK-KHAN. 2005. Cellular monitoring systems for the assessment of space environmental factors. Adv. Space Res. **3:** 1673–1679.
10. BAUMSTARK-KHAN, C., C.E. HELLWEG, A. ARENZ & M.M. MEIER. 2005. Cellular monitoring of the nuclear factor kappaB pathway for the assessment of space environmental radiation. Radiat. Res. **164:** 527–530.
11. RECHSTEINER, M. 1990. PEST sequences are signals for rapid intracellular proteolysis. Semin. Cell Biol. **1:** 433–440.
12. HELLWEG, C.E., A. ARENZ & C. BAUMSTARK-KHAN. Assessment of space environmental factors by cellular bioassays. Acta Astronaut. In press.
13. RAY, K.P. & N. KENNARD. 1993. Interleukin-1 induces a nuclear form of transcription factor NF kappa B in human lung epithelial cells. Agents Actions **38** Spec No: C61–C63.
14. BAEUERLE, P.A., M. LENARDO, J.W. PIERCE & D. BALTIMORE. 1988. Phorbol-ester-induced activation of the NF-kappa B transcription factor involves dissociation of an apparently cytoplasmic NF-kappa B/inhibitor complex. Cold Spring Harbor Symp. Quant. Biol. **53:** 789–798.
15. CHAN, H. & B.B. AGGARWAL. 1994. Role of tumor necrosis factor receptors in the activation of nuclear factor kappa B in human histiocytic lymphoma U-937 cells. J. Biol. Chem. **269:** 31424–31429.
16. MULLER, J.M., H.W. ZIEGLER-HEITBROCK & P.A. BAEUERLE. 1993. Nuclear factor kappa B, a mediator of lipopolysaccharide effects. Immunobiology **187:** 233–256.
17. KIRSCHNING, C.J., H. WESCHE, A.T. MERRILL & M. ROTHE. 1998. Human toll-like receptor 2 confers responsiveness to bacterial lipopolysaccharide. J. Exp. Med. **188:** 2091–2097.
18. BEIDLER, D.R. & Y.C. CHENG. 1995. Camptothecin induction of a time- and concentration-dependent decrease of topoisomerase I and its implication in camptothecin activity. Mol. Pharmacol. **47:** 907–914.
19. BOROVITSKAYA, A.E. & P. D'ARPA. 1998. Replication-dependent and -independent camptothecin cytotoxicity of seven human colon tumor cell lines. Oncol. Res **10:** 271–276.
20. DESAI, S.D., L.F. LIU, D. VAZQUEZ-ABAD & P. D'ARPA. 1997. Ubiquitin-dependent destruction of topoisomerase I is stimulated by the antitumor drug camptothecin. J. Biol. Chem. **272:** 24159–24164.
21. PIRET, B. & J. PIETTE. 1996. Topoisomerase poisons activate the transcription factor NF- kappaB in ACH-2 and CEM cells. Nucleic Acids Res. **24:** 4242–4248.
22. HABRAKEN, Y., O. JOLOIS & J. PIETTE. 2003. Differential involvement of the hMRE11/hRAD50/NBS1 complex, BRCA1 and MLH1 in NF-kappaB activation by camptothecin and X-ray. Oncogene **22:** 6090–6099.

23. HUANG, T.T., S.L. FEINBERG, S. SURYANARAYANAN & S. MIYAMOTO. 2002. The zinc finger domain of NEMO is selectively required for NF-kappa B activation by UV radiation and topoisomerase inhibitors. Mol. Cell Biol. **22:** 5813–5825.
24. PIRET, B., S. SCHOONBROODT & J. PIETTE. 1999. The ATM protein is required for sustained activation of NF-kappaB following DNA damage. Oncogene **18:** 2261–2271.
25. LEWIS, M., L.A. TARTAGLIA, A. LEE, et al. 1991. Cloning and expression of cDNAs for two distinct murine tumor necrosis factor receptors demonstrate one receptor is species specific. Proc. Natl. Acad. Sci. USA **88:** 2830–2834.
26. HOHMANN, H.P., M. BROCKHAUS, P.A. BAEUERLE, et al. 1990. Expression of the types A and B tumor necrosis factor (TNF) receptors is independently regulated, and both receptors mediate activation of the transcription factor NF-kappa B. TNF alpha is not needed for induction of a biological effect via TNF receptors. J. Biol. Chem. **265:** 22409–22417.
27. BROCKHAUS, M., H.J. SCHOENFELD, E.J. SCHLAEGER, et al. 1990. Identification of two types of tumor necrosis factor receptors on human cell lines by monoclonal antibodies. Proc. Natl. Acad. Sci USA **87:** 3127–3131.
28. BUTTERWECK, A., U. GERGS, C. ELFGANG, et al. 1994. Immunochemical characterization of the gap junction protein connexin45 in mouse kidney and transfected human HeLa cells. J. Membr. Biol **141:** 247–256.
29. DEL RE, A.M. & J.J. WOODWARD. 2005. Inhibition of gap junction currents by the abused solvent toluene. Drug Alcohol Depend. **78:** 221–224.
30. BACHMEIER, B.E., R. VENE, C.M. IANCU, et al. 2005. Transcriptional control of cell density dependent regulation of matrix metalloproteinase and TIMP expression in breast cancer cell lines. Thromb. Haemost. **93:** 761–769.
31. ERL, W., G.K. HANSSON, M.R. DE, et al. 1999. Nuclear factor-kappa B regulates induction of apoptosis and inhibitor of apoptosis protein-1 expression in vascular smooth muscle cells. Circ. Res. **84:** 668–677.

Atrial Appendage Transcriptional Profile in Patients with Atrial Fibrillation with Structural Heart Diseases

MARIA S. KHARLAP, ANGELICA V. TIMOFEEVA,
LUDMILA E. GORYUNOVA, GEORGE L. KHASPEKOV,
SERGEY L. DZEMESHKEVICH, VLADIMIR V. RUSKIN,
RENAT S. AKCHURIN, SERGEY P. GOLITSYN,
AND ROBERT SH. BEABEALASHVILLI

Clinical Electrophysiology Department, and the Laboratory of Genetic Engineering, The A.L. Myasnikor Institute of Clinical Cardiology, Russian Cardiology Research and Production Center, Moscow, Russia

ABSTRACT: During the last few years DNA microarray studies of gene expression changes in human atrial tissues from patients with and without atrial fibrillation (AF) have been performed. For this purpose, tissue samples are usually collected from AF patients undergoing open heart surgery. These investigations have limitations associated with the unavoidable heterogeneity of compared groups which is due to the presence of various structural changes accompanying different sets of underlying heart diseases in both groups. It is thus reasonable to compare the atrial tissue samples from AF patients with those from individuals without signs of cardiovascular disease. To address this, we selected the atrial tissue samples from 12 AF patients (who underwent open heart surgery) and compared them with control atrial tissue samples from 10 individuals with no signs of cardiovascular diseases (those who died due to street accident). cDNA microarray method and reverse transcription–polymerase chain reaction (RT-PCR) analysis were used to identify genes which can discriminate between control and pathologically altered atrial tissues. Thirty-nine genes were found to be differentially expressed in pathologically altered tissues samples independently of the type of the underlying structural heart disease. These genes are involved in signal transduction, gene transcription regulation, cell proliferation, and apoptosis. The greatest alterations were observed for NOR1, DEC1, MSF, and Bcl2A1 genes (5 to 28-fold decrease, $P < 0.05$). Additional studies are needed to determine the specific role of each selected gene in pathophysiological changes leading to AF.

Address for correspondence: Maria S. Kharlap, Research Scientist, Clinical Electrophysiology Department, The A.L. Myasnikov Institute of Clinical Cardiology, Cardiology Research and Production Center, 121 552, 3-d Cherepkovskaya Street 15-A, Moscow, Russia. Voice: +7-095-414-66-33; fax: +7-095-414-69-98.
e-mail: shadow1@mail.cnt.ru

KEYWORDS: cDNA microarray; RT-PCR; atrial fibrillation

INTRODUCTION

Atrial fibrillation (AF) itself and underlying heart diseases lead to changes in atrial function and structure. Often it is difficult to distinguish the contribution of AF to these multiple processes known as atrial remodeling.[1,2]

Molecular research of structurally changed human atria has been preferably focused on various ion channels[3,4] and proteins involved in calcium homeostasis,[5,6] while the links between different pathways involved in the pathology of AF have so far not been investigated.[7] This has stimulated the search for novel approaches that could help to find the relationship between the underlying conditions.

New molecular biology technology, cDNA, and oligonucleotide microarrays have been used in several studies to generate the expression data of human fibrillating atria.[8–13] The present investigation is aimed at finding new information about gene expression changes in human pathologically altered atrial tissue from AF patients with structural heart diseases in comparison with control atrial tissue from individuals without the signs of cardiovascular disease and disease-related gene identification.

MATERIALS AND METHODS

Patients and Tissue Samples

Patients were enrolled in the Clinical Departments of the Moscow Cardiology Research Center. After written informed consent, samples of right atrial appendages (RAA) were obtained from 12 AF patients who underwent cardiac bypass surgery, mitral valve replacement, or left atrial myxoma excision.

The cardiovascular medication of all patients was long-term, and the last administration was on the day before the surgical procedure. No patients received amiodarone. All patients were euthyroid. Clinical characteristics of the patients are summarized in TABLE 1.

RAA tissue samples from 10 individuals (7 male and 3 female; mean age 33 ± 15 years) without signs of cardiovascular diseases according to the postmortem examination protocol, who died on account of street accidents were used as a comparison group. The samples were obtained within 4 h after death. The procedures for collecting tissue samples for the study were approved by the Russian Cardiology Research Center's Human Ethics Experimentation Committee.

RNA Isolation

Immediately after excision, RAA samples were frozen in liquid nitrogen and stored at $-80°C$ until RNA isolation. Total RNA was isolated by acid guanidine

TABLE 1. Clinical characteristics of study patients

						Medication			
Patient No.	Age (yr)/Sex	AF	Underlying heart disease	LVEF (%)	LA diameter (mm)	Beta-blockers	Calcium antagonists	Digitalis	ACE inhibitor
1	56/M	PAF	MVD	60	46	−	−	−	−
2	62/M	PAF	MVD	60	41	−	−	−	−
3	51/F	PAF	LA myxoma	60	32	−	−	−	−
4	50/F	PAF	LA myxoma	60	45	+	+	−	−
5	58/F	CAF	LA myxoma	60	52	+	−	+	−
6	60/M	PAF	LA myxoma	60	45	+	−	−	−
7	48/F	CAF	MVD	60	47	+	−	−	+
8	43/M	PAF	MVD	55	53	+	−	+	+
9	68/M	CAF	MVD	53	52	+	−	−	−
10	65/M	PAF	CAD	55	37	+	+	−	+
11	64/M	PAF	CAD	60	37	+	−	+	+
12	67/M	CAF	CAD	50	47	+	−	+	+

NOTE: ACE = angiotensin converting enzyme; AF = atrial fibrillation; CAD = coronary artery disease; CAF = constant form of AF; LA = left atria; LVEF = left ventricular ejection fraction; M/F = male/female; MVD = mitral valve disease; PAF = paroxysmal form of AF.

thiocyanate/phenol/chloroform method according to the prescriptions of Clontech kit, Cat.# K1038-1 and purified by 2 M LiCl precipitation. The 28S and 18S rRNAs band ratio was 2:1. DNA contamination level did not exceed 0.1%.

Probe Preparation and Microarray Hybridization

Total RNA (1.5 μg) was reverse-transcribed in the presence of 35 μCi of [α-^{32}P]dATP (10mCi/mL, 3,000 Ci/mmol, Obninsk, Russia) according to the manufacturer's instructions (Clontech). The resulting ^{32}P-cDNA probes were purified from unincorporated nucleotides and hybridized with Atlas™ Human cDNA Expression Arrays (Clontech) membranes: Cardiovascular, Human 1.2, Human 1.2 II, and Human 1.2 III (4100 genes in total).

After hybridization and washing, the membranes were exposed to screens for 3–5 days and scanned in a PhosphorImager SI system (Molecular Dynamics). Images were analyzed using AtlasImage 2.01 software (Clontech) followed by statistical analyses.

RT-PCR Analysis

A total of 1.5 μg total RNA was reverse-transcribed in a 20-μL reaction mixture containing 50 mM Tris-HCl (pH 8.3), 75 mM KCl, 3 mM $MgCl_2$, 10 mM DTT, 1 mM each of the deoxynucleosidetriphosphates (USB), 2.5 μM random nanomers and 200 U of MMLV reverse transcriptase (Promega). The reactions were performed at 37°C for 60 min and stopped by incubation at 95°C for 5 min with subsequent addition of four volumes (80 μL) of the stop solution (1 mM EDTA, 10 mM Tris-HCl, pH 8.0). PCR amplification was carried with 1 μL diluted cDNA in 15 μL reaction mixture containing 10 mM Tris-HCl (pH 8.5), 50 mM KCl, 2 mM $MgCl_2$, 200 μM each dNTP, 1 μM each of gene-specific forward and reverse primers, 2.5 U of Taq polymerase (Biomaster, Moscow, Russia), "TaqStart" antibodies (Clontech) and 60 mM tetramethylammonium chloride (Fluka Chemie) in a DNA Thermal Cycler (Hybaid). Gene-specific primers for PCR analysis were designed using Oli99 software (Technogene, Russia). PCR primers for glyceraldehyde-3-phosphate dehydrogenase (GAPDH) were obtained from Clontech. The sequences of PCR-primers, number of PCR cycles used to obtain gene-specific PCR-product in exponential reaction phase, gene names and Genbank accession numbers are indicated in TABLE 2. Thermal profile for each gene was 94°C for 5 min followed by desired number of cycles at 94°C—40 sec, gene-specific annealing temperature (57–60°C)—40 sec, 72°C—60 sec. A total of 5 μL of each PCR mixture was run on 1.5% agarose gel in Tris-borate-EDTA electrophoresis buffer with ethidium bromide (1 μg/mL). Fluorescence-stained bands corresponding to PCR-products were quantified using an AlphaImager™ System

(Alpha Innotech Corporation). In semiquantitative reverse-transcription polymerase chain reaction (RT-PCR) analysis, serial dilutions of external standard (the product of amplification of analyzed cDNA, amplicon) were simultaneously amplified in different tubes. Expression levels of experimental samples were interpolated from standard curves of amplicon dilutions. PCR product amount of each studied gene was referred to that of GAPDH. Each PCR experiment was performed in double series.

Statistical Analysis

All cDNA-array and RT-PCR data were normalized to the set of stable expressed genes including housekeeping genes and statistically analyzed using the Mann-Whitney U test to evaluate the significance of differences between gene expression mean values for control and AF groups. All values were expressed as mean $\pm$ SD and a P value <0.05 was considered to be statistically significant.

RESULTS

Hybridization experiments were performed for 12 RAA tissue samples obtained from AF patients with valve disease, myxoma, and coronary artery disease who underwent open heart surgery. Eight patients had paroxysmal AF and four patients had constant AF. Ten RAA tissue samples from individuals with no signs of cardiovascular diseases were used as a group for comparison. Thirty-nine genes with altered expression in AF group (P value < 0.05) were revealed. Of these, 15 genes showing the most pronounced changes between AF and the control group were analyzed by RT-PCR (FIG. 1). The ubiquitously and stable expressed GAPDH gene was used to normalize RT-PCR data. The histogram in FIGURE 2 represents the mean relative gene expression (mRNA) level for control and AF groups. Standard deviation was calculated for each category. In agreement with microarray results, RT-PCR demonstrates a significant (5–28-fold) decrease in mRNA level for neuron-derived orphan receptor 1 (NOR1), basic helix-loop-helix transcription factor (DEC1), BCL-2-related protein A1 (BCL2A1), and megakaryocyte-stimulating factor (MSF). These changes were independent of the underlying heart disease. Gene expression levels of monocyte chemotactic protein 1 (MCP1), stratifin (SFN), stanniocalcin (STC1), zinc-finger protein tristetraprolin (TTP), vascular endothelial growth factor (VEGF) and transcription factor AP-1 were decreased in AF to a lesser extent (1.9-to 2.8-fold). For MMP14, YY1, HSPA1, MAPKK1, and HPI31 genes, difference in their expression between compared groups did not exceed 1.5-fold (P for all genes was <0.05).

TABLE 2.

Gene name	Abbreviation	GenBank mRNA accession number	Primer sequence[a]	PCR cycle number	PCR product size (bp)
Neuron-derived orphan receptor 1	NR4A3 (NOR1)	NM_006981 NM_173198 NM_173200	F5′-GAGCCTGAACCTTGATATCCAAGCC R5′-CAGTTCTACCAGGGCAACCAGGAC	32	196
Basic helix-loop-helix protein (differentially expressed in chondrocytes gene 1)	BHLHB2 (DEC1)	NM_003670	F5′-AAATTGCCGCACCGGCTCATCGAG R5′-TCGTGTCTTGGCCAGATACTGAAGC	24	323
70-kDa heat-shock protein 1 (intronless gene)	HSPA1A (HSP72)	NM_005345	F5′-GGGCAAGCGGTCCGGATAACGG R5′-TGTCGGCTCGGCTCTGAGATTGG	35	214
Zinc finger protein 36 (tristetraproline)	ZFP36 (TTP)	NM_003407	F5′-CATGGATCTGACTGCCATCTACGAG R5′-CTGCGGCCCTCCACTAGGCTG	30	198
Chemokine (C–C motif) ligand 2 (monocyte chemotactic protein 1)	CCL2 (MCP1)	NM_002982	F5′-CAGTCACCTGCTGTTATAACTTCACC R5′-GGGGTAGAACTGTGGTTCAAGAGG	29	448
Proteoglycan 4 megakaryocyte-stimulating factor	PRG4 (MSF)	NM_005807	F5′-GAGGTCATTATTTCTGATGCTAAGTC R5′-GAAAGCGCTGCCACTATTTGTCCAG	26	247

Continued

TABLE 2. Continued

Gene name	Abbreviation	GenBank mRNA accession number	Primer sequence[a]	PCR cycle number	PCR product size (bp)
Vascular endothelial growth factor	VEGF	NM_003376	F5′-GTGTGCCCACTGAGGAGTCCAAC R5′-GTCTGCGGATCTTGTACAAACAAATGC	28	271 199
BCL-2-related protein A1	BCL2A1/Bfl1	NM_004049	F5′-CCCGGATGTGGATACCTATAAGGAG R5′-GGCCGGTTTCACAATATGGAGTGTC	32	235
Stratifin/14-3-3σ protein	14-3-3σ(SFN)	NM_006142	F5′-GAGACACAGAGTCCGGCATTGGTC R5′-ACCAGTTCTTATAGGCTACTGAGAG	36	224
Stanniocalcin 1	STC1	NM_003155	F5′-CCATTCGGAGGTGCTCCACTTTCC R5′-TCTCTGATTGTGCTGACTGTGTCTTC	30	211
c-jun proto oncogene	JUN(AP-1)	NM_002228	F5′-CTGCAGGGTCCGCACTGATCCG R5′-GGGAGCGCAGGGTTAATTAAGATGC	29	190
Matrix metalloproteinase 14, membrane associated	MMP14	NM_004995	F5′-GAGTCTCCCAGAGGGTCATTCATG R5′-TCTCTACCCTCAACAAGATTAGATTCC	33	889
Transcriptional repressor protein yin & yang 1	YY1	NM_003403	F5′-GACCCTGGAGGGCGAGTTCTC R5′-CAGTTGTTTGGGATCTGAGAGGTC	29	190
Mitogen-activated protein kinase kinase 1	MAP2K1 MEK, MAPKK1	NM_002755	F5′-GGAGTGTTCAGTCTGGAATTTCAAG R5′-ATCCTTTGTCACAGGTGAAATGCAC	29	317
Proteasome inhibitor hPI31 subunit	PSMF1 (hPI31)	NM_006814	F5′-GCAGGACGCGCTCGTCTGCTTC R5′-TCCAAGTTCAGGGTCAAGTCTGCC	28	273
Glyceraldehyde-3-phosphate dehydrogenase/Housekeeping gene	GAPD (GAPDH)	NM_002046	F5′-ACCACAGTCCATGCCATCAC R5′-TCCACCACCCTGTTGCTGTA	22	452

[a]F indicates forward primer, R indicates reverse primer.

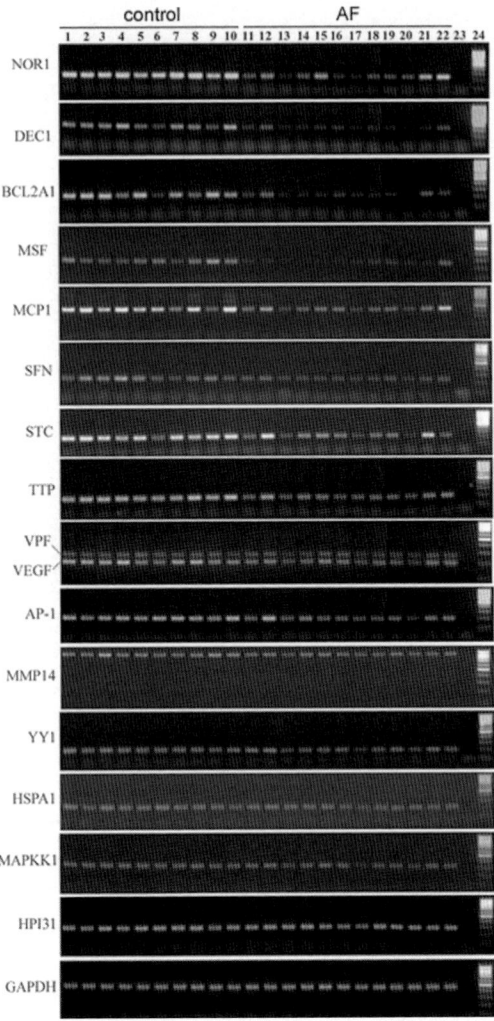

FIGURE 1. Semiquantitative RT PCR analysis of gene expression. Electrophoregrams of PCR products of differentially expressed genes in AF and ubiquitously expressed GAPDH gene used for data normalization. *Lanes* 1–10: RT-PCR products synthesized on mRNA from RAA control subjects (N1–N10, respectively). *Lanes* 11–22: RT-PCR products synthesized on mRNA from RAA of AF patients (P1–P12, respectively). *Lane* 23 is negative PCR-control; *lane* 24 is 100-bp DNA ladder. Gene names are indicated.

DISCUSSION

Using cDNA microarray technique and RT-PCR analysis, we have identified 15 genes discriminating between pathologically altered and control atrial

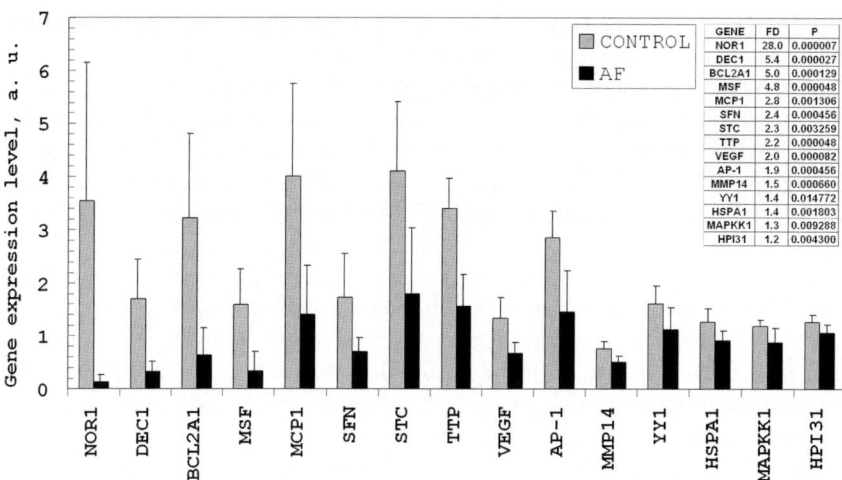

FIGURE 2. Histogram of averaged gene expression level in control and AF groups. The data are plotted after densitometrical analysis of the electrophoregrams in FIGURE 1. Standard deviation was calculated for each group and represented as whiskers on the plot. The significance value (*P*) is indicated for each gene.

tissues. All these genes were downregulated in AF patients compared with the control group. Predominant downregulation of atrial-specific transcription has been recently reported in permanent AF[13] and in persistent AF.[10,14] The majority of genes identified relate to regulatory genes acting at the level of transcription, translation and protein–protein communication. This is not surprising because regulatory genes, being often early-response genes, are destined for preservation of cell viability in new physiological or pathophysiological conditions.

The greatest alteration was observed in the expression of NOR1 gene (NR4A3, also know as NOT), a member of the NF4A subfamily of orphan nuclear hormone receptors including Nurr1 (NF4A2) and Nurr77 (NF4A1). NF4A receptors were originally identified as early-response genes induced by nerve growth factor or membrane depolarization in PC12 cells differentiating into cells with a neuronal phenotype.[15,16] Expression of NF4A receptors is highly inducible in macrophages by diverse inflammatory stimuli.[17] In HUVEC, all three members of NR4A receptors can be induced by VEGF via intracellular Ca^{2+}, protein kinase C, and calcineurin-dependent pathways.[18] The function of these receptors in the heart is, however, poorly understood. In mouse myogenic C2C12 cells, regulatory cross-talk between nuclear hormone receptor and beta-adrenergic receptor signaling pathways was demonstrated.[19,20] Nurr77 was induced by isoprenaline, beta-adrenergic agonist, while the repression of endogenous level of Nurr77 mRNA by small interfering RNA (siRNA)-mediated knock-out of Nurr77 suppressed genes involved in the regulation of lipid and energy homeostasis.[19] Sustained dysregulation of energy metabolism may lead to contractile dysfunction of the heart.[21]

Apoptotic alterations in cardiomyocytes (DNA damage and repression of antiapoptotic protein Bcl2) were observed in atrial tissue from chronic AF subjects.[12,13] In this study, we observed downregulation of Bcl2A1 (Bcl2-related protein A1, also known as Bfl-1, GRS, Bcl2L5), a member of the Bcl2 family including both antiapoptotic and proapoptotic proteins. Bcl2A1 inhibits apoptosis induced by a variety of stimuli in a variety of cells and tissues.[19] He et al.[22] reported that Bfl-1, the human homologue of Bcl2A1, is an important regulator of oxidant-induced cell death responses and a critical mediator of IL-11 and VEGF-induced cytoprotection. Little is known about the mechanism of this protective response in the heart, but downregulation of the Bcl2A1 gene may make a negative contribution to preservation of cardiomyocytes from apoptosis induced by reactive oxygen species.

Another putative protective gene repressed in AF patients in our study is stratifin (STF, 14-3-3σ), one of seven isotypes of the 14-3-3 family of acid proteins. All 14-3-3 proteins bind to common phosphoserine/phosphothreonine-containing peptide motifs and modulate a wide variety of cellular processes. The 14-3-3σ is unique among 14-3-3 isotypes.[23] This isotype is evolutionarily distant from other 14-3-3 proteins, preferentially undergoes homodimerization, and exhibits a binding partner specifically distinct from that of the other isotypes. In colon cancer cells 14-3-3σ binds and sequesters the proteins initiating mitosis (cyclin B1 and cdc2) and prevents them from entering the nucleus after DNA damage. The $14\text{-}3\text{-}3\sigma^{-/-}$ cells (cells in which both alleles are inactivated) could not maintain cell-cycle arrest and died (mitotic catastrophe) as they entered mitosis. Thus, 14-3-3σ may be an important regulator of G2/M checkpoint.[24]

Potential role in cardiac function was reported for 14-3-3ε protein.[25,26] The interaction between 14-3-3 and the human HERG K^+ channel pore-forming subunit leads to amplification and a prolongation of the effect of adrenergic stimulation upon HERG activity and may provide a mechanism for plasticity in the control of membrane excitability and cardiac rhythm. The association of 14-3-3η expression with alterations in the susceptibility toward apoptosis was reported in animal models.[27] In cell cultures, 14-3-3 protein regulates stability and translation of mRNA released from polyribosomes in response to environmental stress.[28] 14-3-3 binds to tristetraprolin (TTP) after MAPKAP kinase-2 (MK-2)-induced phosphorylation TTP at serines 52 and 178, which promotes the assembly of TTP:14-3-3 complexes. TTP is a tandem CCCH zinc finger protein which degrades mRNA through binding to an AU-rich element (ARE) located within 3′-untranslated region of mRNA.[29] TTP has been identified as a component of stress granules (SG) in cells subjected to environmental stress. The 14-3-3 binding excludes TTP from SG and inhibits TTP-dependent degradation of ARE-containing mRNA.[28]

Downregulation of transcription factor participating in the mammalian circadian clock system was detected first in our study. Differentially expressed in chondrocytes, gene 1 (DEC1, also known as Stra13/Sharp2/BHLHB2b) belongs to basic helix-loop-helix (bHLH) transcription factor proteins. DEC1 is a negative regulator of the mammalian circadian system, which represses Clock/Bmal1-induced transactivation of the Per1 promotor through direct protein–protein interaction with Bmal1 and/or competition for E-box elements.[30] The circadian clock has been demonstrated to exist and operate within cardiomyocytes of the heart, to be under the control of the sympathetic neurotransmitters, and to be associated with diurnal variation of genes involved in energy metabolism of the myocardium.[31,32] Whether decreased DEC1 expression in AF patients is related to the development of heart failure and/or subsequent contractile dysfunction development is not known and requires further investigation.

We observed downregulation of calcium-regulating hormone stanniocalcin 1 (STC1) in AF patients. STC1, known to regulate calcium and phosphate homeostasis in fish, causes reversible inhibition of transmembrane calcium currents through L-channels in the human heart.[33] In addition, STC1 protects brain neurons from hypoxia or ischemia[34] and attenuates chemokinesis and diminishes the chemotactic response to MCP1.[35]

Recognition of the function of genes investigated in the development and maintenance of AF itself and accompanied structural heart diseases will contribute to the understanding of molecular mechanisms of corresponding phenotypes and may lead to the identification of new therapeutic strategies.

Study Limitations

Every study with tissue samples obtained from patients undergoing cardiac surgery has potential limitations. All studied patients had underlying heart diseases and received different medications, which could influence gene expression and have a negative impact on the values of standard deviations. In the present study we analyzed gene expression only in RAA because it is the most accessible heart tissue in patients withAF. It is unconditionally of interest to describe gene expression in different parts of the atria in patients with structural heart diseases and AF. In addition, changes in gene expression are not always directly associated with changes in protein synthesis on account of existing regulatory mechanisms at the posttranscriptional and posttranslational levels.

REFERENCES

1. TIELEMAN, R.G. 2003. The pathophysiology of maintenance of atrial fibrillation. Pacing. Clin. Electrophysiol. **26:** 1569–1571.

2. SCHOONDERWOERD, B.A., I.C. VAN GELDER, D.J. VAN VELDHUISEN, et al. 2005. Electrical and structural remodeling: role in the genesis and maintenance of atrial fibrillation. Prog. Cardiovasc. Dis. **48:** 153–168.
3. VAN WAGONER, D.R., A.L. POND, P.M. MC CARTHY, et al. 1997. Outward K current densities and Kv1.5 expression are reduced in chronic human atrial fibrillation. Circ. Res. **80:** 772–781.
4. BRUNDEL, B.J., I.C. VAN GELDER, R.H. HENNING, et al. 2001. Alterations in potassium channel gene expression in atria of patients with persistent and paroxysmal atrial fibrillation: differential regulation of protein and mRNA levels for K channels. J. Am. Coll. Cardiol. **37:** 926–932.
5. BRUNDEL, B.J., I.C. VAN GELDER, R.H. HENNING, et al. 1999. Gene expression of proteins influencing the calcium homeostasis in patients with persistent and paroxysmal atrial fibrillation. Cardiovasc. Res. **42:** 443–454.
6. LAI, L.P., M.J. SU, J.L. LIN, et al. 1999. Down-regulation of L-type calcium channel and sarcoplasmic reticular Ca-ATPase mRNA in human atrial fibrillation without significant change in the mRNA of ryanodine receptor, calsequestrin and phospholamban: an insight into mechanism of atrial electrical remodeling. J. Am. Coll. Cardiol. **33:** 1231–1237.
7. BRUNDEL, B.J., R.H. HENNING, H.H. KAMPINGA, et al. 2002. Molecular mechanisms of remodeling in human atrial fibrillation. Cardiovasc. Res. **54:** 315–324.
8. LAMIRAULT, G., N. GABORIT, N. LE MEUR, et al. 2006. Gene expression profile associated with chronic atrial fibrillation and underlying valvular heart disease in man. J. Mol. Cell. Cardiol. **40:** 173–184.
9. LAI, L.P., J.L. LIN & C.S. LIN. 2004. Functional genomic study on atrial fibrillation using cDNA microarray and two-dimensional protein electrophoresis techniques and identification of the myosin regulatory light chain isoform reprogramming in atrial fibrillation. J. Cardiovasc. Electrophysiol. **15:** 214–223.
10. OHKI, R., K. YAMAMOTO, S. UENO, et al. 2005. Gene expression profiling of human atrial myocardium with atrial fibrillation by DNA microarray analysis. Int. J. Cardiol. **102:** 233–238.
11. KIM, Y.H., S. LIM DO, J.H. LEE, et al. 2003. Gene expression profiling of oxidative stress on atrial fibrillation in humans. Exp. Mol. Med. **35:** 336–349.
12. KIM, N.-H., Y. AHN, S.K. OH, et al. 2005. Altered patterns of gene expression in response to chronic atrial fibrillation. Int. Heart J. **46:** 383–395.
13. BARTH, A.S., S. MERK, E. ARNOLDI, et al. 2005. Reprogramming of the human atrial transcriptome in permanent atrial fibrillation: expression of a ventricular-like genomic signature. Circ. Res. **96:** 1022–1029.
14. OHKI-KANEDA, R., J. OHASHI, K. YAMAMOTO, et al. 2004. Cardiac function-related gene expression profiles in human atrial myocytes. Biochem. Biophys. Res. Commun. **320:** 1328–1336.
15. MILBRANDT, J. 1988. Nerve growth factor induces a gene homologous to the glucocorticoid receptor gene. Neuron **1:** 183–188.
16. YOON, J.K. & L.F. LAU. 1993. Transcriptional activation of the inducible nuclear receptor gene nur77 by nerve growth factor and membrane depolarization in PC12 cells. J. Biol. Chem. **268:** 9148–9155.
17. PEI, L. & A. CASTRILLO. 2005. Regulation of macrophage inflammatory gene expression by the orphan nuclear receptor Nur77. Mol. Endocrin. **8**: Epub.
18. LIU, D., H. JIA, D.I.R. HOLMES, et al. 2003. Vascular endothelial growth factor-regulated gene expression in endothelial cells. Arterioscler. Thromb. Vasc. Biol. **23:** 2002–2007.

19. MAXWELL, M.A., M.E. CLEASBY, A. HARDING, et al. 2005. Nurr77 regulates lipolysis in skeletal muscle cells. J. Biol. Chem. **280:** 12573–12584.
20. LIN, E.Y., A. ORLOFSKY, H-G. WANG, et al. 1996. A1, a Bcl-2 family member, prolongs cell survival and permits myeloid differentiation. Blood **87:** 983–992.
21. TAEGTMEYER, H., C.R. WILSON, P. RAZEGHI, et al. 2005. Metabolic energetics and genetics in the heart. Ann. N. Y. Acad. Sci. **1047:** 208–218.
22. HE, C-H., A.B. WAXMAN, C.G. LEE, et al. 2005. Bcl-2-related protein A1 is an endogenous and cytokine-stimulated mediator of cytoprotaction in hyperoxic acute lung injury. J. Clin. Invest. **115:** 1039–1048.
23. WILKER, E.W., R.A. GRANT, S.C. ARTIM, et al. 2005. A structural basis for 14-3-3σ functional specificity. J. Biol. Chem. **280:** 18891–18898.
24. CHAN, T.A., H. HERMEKING, C. LENGAUER, et al. 1999. 14-3-3σ is required to prevent mitotic catastrophe after DNA damage. Nature **7:** 535–537.
25. KAGAN, A., Y.F. MELMAN, A. KRUMERMAN, et al. 2002. 14-3-3 amplifies and prolongs adrenergic stimulation of HERG K channel activity. EMBO J. **21:** 1889–1898.
26. KAGAN, A. & T.V. MCDONALD. 2005. Dynamic control of HERG/I(Kr) by PKA-mediated interactions with 14-3-3. Novartis Found. Symp. **266:** 75–89 [discussion 89–99].
27. XING, H.S. ZHANG, C. WEINHEIMER, et al. 2000. 14-3-3 Proteins block apoptosis and differentially regulate MAPK cascades. EMBO J. **19:** 349–358.
28. STOECKLIN, G., T. STUBBS, N. KEDERSHA, et al. 2004. MK2-induced tristetraprolin: 14-3-3 complexes prevent stress granule association and ARE-mRNA. EMBO J. **23:** 1313–1324.
29. BLACKSHEAR, P.J., W.S. LAI, E.A. KENNINGTON, et al. 2003. Characteristics of the interaction of a synthetic human tristetraprolin tandem zinc finger peptide with AU-rich element-containing RNA substrates. J. Biol. Chem. **278:** 19947–19955.
30. HONMA, S., T. KAWAMOTO, Y. TAKAGI, et al. 2002. Dec1 and Dec2 are regulators of the mammalian molecular clock. Nature **419:** 841–844.
31. YOUNG, M.E., P. RAZEGHI, A.M. CEDARS, et al. 2001. Intrinsic diurnal variations in cardiac metabolism and contractile function. Circ. Res. **89:** 1199–1208.
32. DURGAN, D.J., M.A. HOTZE & T.M. TOMLIN. 2005. The intrinsic circadian clock within the cardiomyocyte. Am. J. Physiol. Heart Circ. Physiol. **289:** H1530–H1541. Epub.
33. SHEIKH-HAMAD, D., R. BICK, GANG-YI WU, et al. 2003. Stanniocalcin-1 is a naturally occuring L-channel inhibitor in cardiomyocytes: relevance to human heart failure. Am. J. Physiol. Heart Circ. Physiol. **285:** H442–H448.
34. ZHANG, K., P.J. LINDSBERG, T. TATLISUMAK, et al. 2000. Stanniocalcin: a molecular guard of neurons during cerebral ishemia. Proc. Nat. Acad. Sci. USA **97:** 3637–3642.
35. KANELLIS, J., R. BICK, G. GARCIA, et al. 2003. Stanniocalcin-1, an inhibitor of macrophage chemotaxis and chemokinesis. Am. J. Physiol. **286:** F356–F362.

DNA Hypomethylation of *CAGE* Promotors in Squamous Cell Carcinoma of Uterine Cervix

TAEK SANG LEE,[a] JAE WEON KIM,[b,c] GYEONG HOON KANG,[c,d] NOH HYUN PARK,[b,c] YONG SANG SONG,[b,c] SOON BEOM KANG,[b,c] AND HYO PYO LEE[b,c]

[a]*Department of Obstetrics and Gynecology, Seoul Metropolitan Boramae Hospital, Seoul 156-707, Korea*

[b]*Department of Obstetrics and Gynecology, College of Medicine, Seoul National University, Seoul 110-744, South Korea*

[c]*Cancer Research Institute, College of Medicine, Seoul National University, Seoul 110-744, South Korea*

[d]*Department of Pathology, College of Medicine, Seoul National University, Seoul 110-744, South Korea*

> ABSTRACT: This study was performed to determine whether promotor hypomethylation of *CAGE* is involved in cervical carcinogenesis. The surgical specimens of 40 cervical squamous cell carcinoma patients treated at Seoul National University Hospital and those of 48 healthy controls were used, with informed consent. We investigated the promotor hypomethylation status of *CAGE* by methylation-specific polymerase chain reaction (MSP) using primers specific for unmethylated sequences, and found hypomethylation of *CAGE* promotor at a frequency approaching 90% in cervical squamous cell carcinomas (35/40, 87.5%), but at less than 4% in controls ($P < 0.001$). This finding provides experimental evidence of the frequent hypomethylation of normally methylated *CAGE* promotor CpG islands in cervical cancer, and indicates that this hypomethylation is likely to be a valuable surrogate marker for the expression of *CAGE*. It also provides a clue concerning the molecular mechanisms of carcinogenesis in cervical squamous cell carcinoma.
>
> KEYWORDS: hypomethylation; *CAGE*; cervical cancer

INTRODUCTION

Cervical cancer remains the most common gynecologic malignancy in Korea, although the widespread application of cytologic screening has reduced

Address for correspondence: Jae Weon Kim, M.D., Department of Obstetrics and Gynecology and Cancer Research Institute, College of Medicine, Seoul National University, 28 Yungun-Dong, Chongno-Ku, Seoul 110-744, South Korea. Voice: +82-2-2072-3511; fax: +82-2-762-3599.
e-mail: kjwksh@snu.ac.kr

its incidence.[1] It is generally considered that human papillomavirus (HPV) infection is the major causative agent. However, in addition to HPV infection, it is clear that other factors are also involved in cervical carcinogenesis because the majority of patients with HPV-associated lesions do not progress to invasive cancer. Therefore, it is of some importance that other genetic and epigenetic events in cervical carcinogenesis be identified to facilitate the development of new strategies and the identification of molecular biomarkers for the early detection and prevention of cervical cancer. Widespread epigenetic abnormalities have been identified in various human cancers and studied as potential molecular biomarkers,[2] and abnormal DNA methylation is recognized as a molecular basis for the genesis of various types of cancer.[3–11] The hypermethylation of normally unmethylated CpG islands in promotor regions is well known, and this can silence tumor suppressor genes and promote cancer development and progression.[12–14] On the other hand, hypomethylation is also observed in normally methylated CpG islands in promotor regions, and also induces aberrant downstream gene expression when transcription factors are available. Few CpG islands are known that are normally methylated, examples are provided by some cancer/testis antigen genes, such as, the *MAGE* genes.[15,16] The expressions of these genes are restricted to the testis in normal tissues. Because germ cell line cells are devoid of major histocompatibility complex (MHC) molecules, they do not display antigenic peptides at their surface, and therefore, peptides encoded by cancer germline genes are strictly tumor specific.[17] Recently, another novel cancer/testis antigen *CAGE* was described,[18] and it has been reported that the methylation status of its CpG site determines its expression, and that its promotor hypomethylation precedes the developments of several cancer types.[19]

These methylation changes can be found in virtually every type of human neoplasm, and a correlation between DNA hypomethylation and cervical carcinogenesis has been strongly suggested.[20] Based on these observations, we analyzed the promotor methylation status of *CAGE* in uterine cervical carcinoma and normal cervical tissues by using a methylation-specific polymerase chain reaction (MSP) approach to clarify the relationship between *CAGE* promotor hypomethylation and the presence of uterine cervical carcinoma.

MATERIALS AND METHODS

Sample Collection and DNA Extraction

Forty tumor samples were collected from cases of invasive cervical squamous cell carcinoma resected at Seoul National University Hospital, and 48 normal cervical tissues were obtained from the hysterectomy specimens of patients who had myoma uteri with negative Pap smear results within 3 months of operation. All samples were analyzed after obtaining informed consent. Samples were digested with proteinase K and DNA was isolated using conventional phenol/chloroform/isoamyl alcohol extraction and ethanol precipitation.

Bisulfite Modification

Tumor DNAs were subjected to sodium bisulfite modification, as described previously.[21] Briefly, 2 μg of DNA was denatured with 2 M NaOH and then treated with 1 mM hydroquinone and 3.5 M sodium bisulfite for 16 h at 55°C. After purification using the Wizard DNA Clean-Up System (Promega, Madison, WI), DNA was treated with 3 M NaOH and subsequently precipitated with three volumes of 100% ethanol and one-third volume of 7.5 M NH4Ac at −20°C. The precipitated DNA was then washed with 70% ethanol and dissolved in distilled water.

MSP Analysis

After sodium bisulfite modification, MSP was performed to examine the methylation status of the CpG islands of *CAGE*. The sequences of the primer used for the amplification of unmethylated alleles were 5′-GTTTT TTATATGATTTGGAATTTGAT-3′(forward primer), 5′-AATTCAAATCTA CAACCTATTTCCCA-3′ (reverse primer).

Amplification was performed over 33 cycles at 95°C for 30 s, 53°C for 30 s, and 72°C for 1 min. The PCR products obtained were separated by electrophoresis through a 1.0% low melting agarose-gel containing ethidium bromide.

RESULTS

FIGURE 1 A shows the results of the MSP analysis of the DNA of 40 cancer patients and 48 healthy controls using the above-described primers specific for unmethylated sequences. Without regarding weakly positive cases, hypomethylation of the *CAGE* promotor was found in 35 (87.5%) of the 40 tumor tissues and in 2 (4%) of the 48 nonneoplastic cervical tissues, and this difference was highly significant ($P < 0.001$; χ^2 test).

Cases found to be negative for the unmethylated alleles may have been caused by adjacent normal cervical mucosa contamination, and a proportion of the positive broad band in normal cervical tissues may have been caused by the presence of dysplastic tissue.

DISCUSSION

We evaluated the methylation status of the CpG islands in *CAGE*, a novel cancer/testis antigen, to determine whether its hypomethylation is involved

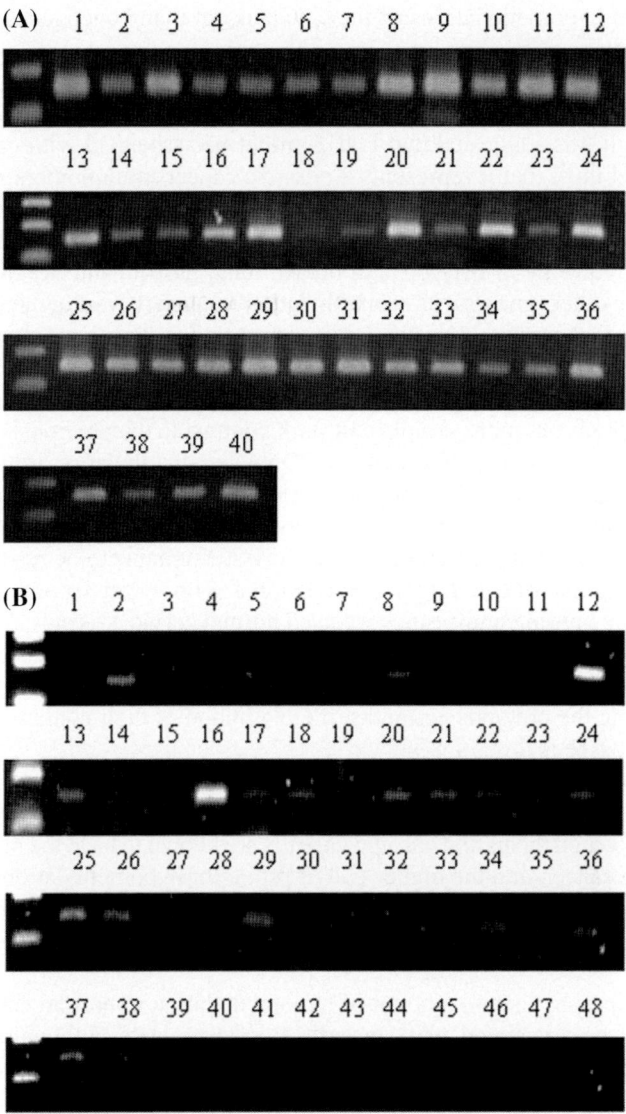

FIGURE 1. MSP analysis of DNA from 40 cancer patients (**A**), and 48 healthy controls (**B**) using primers specific for unmethylated sequences. (**A**) Squamous cell carcinoma of the uterine cervix. (**B**) Normal cervical tissues.

in cervical carcinogenesis. It has been determined that the expressions of tissue-specific genes are regulated by methylation, e.g., the hypermethylation of *MAGE*, a well-known cancer/testis antigen, led to its transcription silencing.[22–24] Moreover, several reports have demonstrated associations between the aberrant hypomethylations of the CpG sites of many oncogenes and their overexpressions.[25–27]

The *CAGE* gene was originally identified by the c-DNA library screening of sera from gastric cancer patients, and because of its cell cycle-dependent expression it was suggested that *CAGE* might be associated with cancer cell growth, and thus, that it represents a possible cancer immunotherapeutic target.[18] Subsequently, Cho et al.[19] revealed that the *CAGE* promotor is frequently hypomethylated in different cancer types, and reported that it is hypomethylated at a frequency of over 60% in breast, lung, gastric, and hepatic cancer. However, in other cancers its hypomethylation frequency was low, and of these uterine cervical cancer, in particular, was reported to have a much lower frequency (9%) of hypomethylation than other cancer types.[19] For this reason, we re-explored the hypomethylation status of *CAGE* in cervical cancer using similar methods but more samples. In stark contrast to the previously reported results, we found that the frequency of DNA hypomethylation in the *CAGE* promotor region was significantly increased in cervical cancer. In view of the different sample work-ups used, and the possibilities of cell contamination, we cannot account for this difference. Thus the issue remains to be resolved.

A small amount of *CAGE* hypomethylation was observed in normal cervical tissues in the present study. Since we used normal cervical tissues as controls, i.e., not from tissues adjacent to tumors, but from hysterectomy specimens of healthy individuals who had only myoma uteri and a history of a negative Pap smear during the previous 3 months, the likelihood of their contamination by cancerous tissue is low.

Promotor hypomethylation of *CAGE* was found in some nonneoplastic tissues, i.e., in liver cirrhosis and chronic gastritis tissues in a previous study.[19] However, liver cirrhosis and chronic gastritis are known to have the potential to progress to cancer, and the higher *CAGE* promotor hypomethylation frequencies observed in hepatic and gastric cancer than in liver cirrhosis and gastritis, respectively, in the previous report, support the precancerous statuses of these lesions.

For this reason, a small amount of hypomethylation in normal cervical tissues might be associated with an early dysplastic state, although this was not detected by conventional cytologic testing. Therefore, we believe that the relationship between cervical carcinogenesis stage and the degree of *CAGE* expression induced by hypomethylation warrants investigation.

The present study is limited because we did not include an examination of the relationship between *CAGE* promotor methylation status and gene expression. However, the relation between *CAGE* expression and hypomethylation was demonstrated by a previous study, which showed that *CAGE* is expressed in cancer cell lines not expressing *CAGE* when treated with 5′-aza-2′-deoxycytidine.[19]

However, despite the limitations mentioned above, the result of our study, namely that the *CAGE* promotor is hypomethylated in cervical squamous cell carcinoma more so than in any other cancer reported to date, reemphasizes the importance of promotor hypomethylation. Moreover, we believe that hypomethylation of *CAGE* promotor could be used as a valuable biochemical marker in cervical cancer, and that further research should be undertaken to clarify the potential usefulness of CAGE protein is this context.

REFERENCES

1. KOREAN SOCIETY OF OBSTETRICS AND GYNECOLOGY. 2003. Annual Report of Gynecologic Cancer Registry Program in Korean for 2001 (Jan. 1, 2001–Dec. 31, 2001). Korean J. Obstet. Gynecol. **46:** 1849–1887.
2. LAIRD, P.W. 2003. The power and the promise of DNA methylation markers. Nat Rev Cancer **3:** 253–266.
3. BAYLIN, S.B. & J.G. HERMAN. 2000. DNA hypermethylation in tumorigenesis: epigenetics joins genetics. Trends Genet. **16:** 168–174.
4. ESTELLER, M., O.G. CORN, S.B. BAYLIN, *et al.* 2001. A gene hypermethylation profile of human cancers. Cancer Res. **61:** 3225–3229.
5. BAYLIN, S.B., M. ESTELLER, M.R. ROUNTREE, *et al.* 2001. Aberrant patterns of DNA methylation, chromatin formation and gene expression in cancer. Hum. Mol. Genet. **10:** 687–692.
6. FEINBERG, A.P. & B. VOGELSTEIN. 1983. Hypomethylation distinguishes genes of some human cancers from their normal counterparts. Nature **301:** 89–92.
7. DIALA, E.S., M.S.C. CHEAH, D. ROTWICH, *et al.* 1983. Extent of DNA methylation in tumor cells. J. Natl. Cancer Inst. **71:** 755–764.
8. FEINBERG, A.P., C.W. GEHRKE, K.C. KUO, *et al.* 1988. Reduced genomic 5-methylcytosine in human colonic neoplasia. Cancer Res. **48:** 1159–1161.
9. LAIRD, P.W. & R. JAENISCH. 1994. DNA methylation and cancer. Hum. Mol. Genet. **3:** 1487–1495.
10. LAIRD, P.W., L. JACKSON-GRUSBY, A. FAZELI, *et al.* 1995. Suppression of intestinal neoplasia by DNA hypomethylation. Cell **81:** 197–205.
11. COUNTS, J.L., J.I. GOODMAN. 1995. Alterations in DNA methylation may play a variety of roles in carcinogenesis. Cell **83:** 13–15.
12. JONES, P.A. & S.B. BAYLIN. 2002. The fundamental role of epigenetic events in cancer. Nat. Rev. Genet. **3:** 415–428.
13. SHIM, Y.H., G.H. KANG & J.Y. RO. 2000. Correlation of p16 hypermethylation with p16 protein loss in sporadic gastric carcinomas. Lab. Invest. **80:** 689–695.
14. FLEISHER, A.S., M. ESTELLER, S. WANG, *et al.* 1999. Hypermethylation of the hMLH1 gene promotor in human gastric cancers with microsatellite instability. Cancer Res. **59:** 1090–1095.
15. DE SMET, C., C. LURQUIN, B. LETHE, *et al.* 1999. DNA methylation is the primary silencing mechanism for a set of germ line- and tumor-specific genes with a CpG-rich promotor. Mol. Cell Biol. **19:** 7327–7335.
16. TAKAHASHI, K., S. SHICHIJO, M. NOGUCHI, *et al.* 1995. Identification of MAGE-1 and MAGE-4 proteins in spermatogonia and primary spermatocytes of testis. Cancer Res. **55:** 3478–3482.
17. HAAS, G.G. JR., O.J. D'CRUZ & L.E. DE BAULT. 1988. Distribution of human leukocyte antigen-ABC and -D/DR antigens in the unfixed human testis. Am. J. Reprod. Immunol. Microbiol. **18:** 47–51.

18. PARK, S., Y. LIM, D. LEE, et al. 2003. Identification and characterization of a novel cancer/testis antigen gene CAGE-1. Biochim. Biophys. Acta **1625:** 173–182.
19. CHO, B., H. LEE, S. JEONG, et al. 2003. Promotor hypomethylation of a novel cancer/testis antigen gene CAGE is correlated with its aberrant expression and is seen in premalignant stage of gastric carcinoma. Biochem. Biophys. Res. Commun. **307:** 52–63.
20. KIM, Y.I., A. GIULIANO, K.D. HATCH, et al. 1994. Global DNA hypomethylation increases progressively in cervical dysplasia and carcinoma. Cancer **74:** 893–899.
21. HERMAN, J.G., J.R. GRAFF, S. MYOHANEN, et al. 1996. Methylation-specific PCR: a novel PCR assay for methylation status of CpG islands. Proc. Natl. Acad. Sci. USA **93:** 9821–9826.
22. SIGALOTTI, L., S. CORAL, M. ALTOMONTE, et al. 2002. Cancer testis antigens expression in mesothelioma: role of DNA methylation and bioimmunotherapeutic implications. Br. J. Cancer **86:** 979–982.
23. JANG, S.J., J.C. SORIA, L. WANG, et al. 2001. Activation of melanoma tumor antigens occurs early in lung cancer carcinogenesis. Cancer Res. **61:** 7559–7563.
24. SIGALOTTI, L., S. CORAL, G. NARDI, et al. 2002. Promotor methylation controls the expression of MAGE 2, 3 and 4 genes in human cutaneous melanoma. J. Immunother. **25:** 16–26.
25. CHO, M., H. UEMURA, S.C. KIM, et al. 2001. Hypomethylation of the MN/CA9 promotor and upregulated MN/CA9 expression in human renal cell carcinoma. Br. J. Cancer **85:** 563–567.
26. TAO, L., S. YANG, M. XIE, et al. 2000. Hypomethylation and overexpression of c-jun and c-myc protooncogenes and increased DNA methyltransferase activity in dichloroacetic and trichloroacetic acid-promoted mouse liver tumors. Cancer Lett. **158:** 185–193.
27. ROSTY, C., T. UEKI, P. ARGANI, et al. 2002. Overexpression of S100A in pancreatic ductal adenocarcinomas is associated with poor differentiation and DNA hypomethylation. Am. J. Pathol. **160:** 45–50.

The *MECP2* Gene Mutation Screening in Rett Syndrome Patients from Croatia

TANJA MATIJEVIĆ,[a] JELENA KNEŽEVIĆ,[a] INGEBORG BARIŠIĆ,[b] BISERKA REŠIĆ,[c] VIDA ČULIĆ,[c] AND JASMINKA PAVELIĆ[a]

[a]*Laboratory of Molecular Oncology, Division of Molecular Medicine, Rudjer Bošković Institute, 10002 Zagreb, Croatia*

[b]*Department of Pediatrics, Children's Hospital Zagreb, University of Zagreb, Medical School, 10000 Zagreb, Croatia*

[c]*Department of Medical Genetics, Pediatric Clinic, Clinical Hospital Split, 21000 Split, Croatia*

ABSTRACT: Rett syndrome (RTT) is an X-linked dominant neurodevelopmental disorder almost exclusively affecting females and is usually sporadic. Mutations in *MECP2* gene have been found in more than 80% of females with typical features of RTT. In this study, we analyzed 15 sporadic cases of RTT. In 7 of 15 patients (47%), we detected pathogenic mutations in the coding parts of *MECP2* fourth exon. We found two missense (T158M, R133C), two nonsense (R168X, R270X), two frameshift mutations (P217fs and a double deletion of 28-bp at 1132–1159 and 10-bp at 1167–1176), and one in-frame deletion (L383_E392del10). To our knowledge, the last two mutations have not been reported yet. We also detected one previously described polymorphism (S194S). In conclusion, these results show that the fourth exon should be the first one analyzed because it harbors most of the known mutations. Moreover, mutation-negative cases should be further analyzed for gross rearrangements. This is the first study of its kind in Croatia and it enabled us to give the patients an early confirmation of RTT diagnosis.

KEYWORDS: Rett syndrome; Croatia; mutation analysis

INTRODUCTION

Rett syndrome (RTT, MIM 312750) is an X-linked dominant progressive neurodevelopmental disorder and one of the most common causes of mental retardation in females. RTT usually occurs sporadically and is characterized by apparently normal development for 6–18 months, followed by a period of

Address for correspondence: Prof. Jasminka Pavelić, Rudjer Bošković Institute, Division of Molecular Medicine, Laboratory of Molecular Oncology, Bijenička 54, 10002 Zagreb, Croatia. Voice: +385-1-4560-926; fax: +385-1-4561-010.
e-mail: jpavelic@irb.hr

regression in language and motor skills. The patients lose purposeful hand use, which is replaced by repetitive stereotyped hand movements. Additional characteristics include autistic features, panic-like attacks, respiratory dysfunction (episodic apnea and/or hyperpnea), microcephaly, and decreased somatic growth.

MeCP2 (methyl-CpG-binding protein type 2) belongs to a family of methyl-binding proteins involved in transcriptional repression. Its methyl-binding domain (MBD) binds specifically to methylated DNA, and the transcriptional repression domain (TRD) interacts with histone deacetylase and transcriptional corepressor. Interactions between this transcription repressor complex and chromatin-bound MeCP2 lead to deacetylation of core histones H3 and H4 by histone deacetylases (HDACs) resulting in compaction of the chromatin, making it inaccessible to components of transcriptional machinery. Two additional domains include nuclear localization signal (NLS), which may be responsible for the transport of MeCP2 into the nucleus, and the C-terminal segment, which facilitates its binding to the nucleosome core.

Mutations in *MECP2*, encoding methyl-CpG-binding protein type 2 (MeCP2), cause approximately 60–80% of Rett cases.[1] Almost all mutations in *MECP2* occur *de novo*. About 65% of all *MECP2* mutations are caused by C > T transitions at eight CpG dinucleotides (R106, R133, T158, R168, R255, R270, R294, and R306), which are located in the third and fourth exon. The most common mutation is R168X. Although mutations are dispersed throughout the gene, a clustering of missense mutations occurs 5' of TRD, mostly in methyl-CpG-binding domain (MBD); they all involve evolutionarily conserved amino acids in functional domains of the protein. Nonsense, frameshift and splicing mutations appear distal to the MBD and result in premature termination of the protein. Larger multinucleotide deletions occur in the C-terminal domain.

In this study we established that mutation frequency in Croatian patients with RTT in 47% (7 of 15 patients). This conclusion was obtained by analyzing the coding region of *MECP2* by polymerase chain reaction (PCR) amplification and subsequent analysis by direct sequencing in 15 Croatian girls with RTT.

MATERIALS AND METHODS

DNA Analysis

We have analyzed the samples of 15 patients from Croatia; 7 of them were diagnosed with RTT at Children's Hospital Zagreb and 8 of them in Clinical Hospital Split.

Exons 2 and 3 were amplified and sequenced in one fragment and exon 4 was divided in three overlapping fragments for amplification and sequencing.

Genomic DNA was extracted from peripheral blood leukocytes from 15 RTT patients by the standard salting-out method. PCR amplification of the MECP2 coding exons[2,3,4] was performed by using primers in part previously described[1,2] and in part redesigned (TABLE 1). It was performed in a final volume of 20 μL using 0.4 μL DNA (100 ng), 0.6 μL of each primer (100 μM), 2 μL PCR buffer, 1.2 μL dNTPs (25 mM), and 0.1 μL (5U μL^{-1}) Taq polymerase (Applied Science, Roche Diagnostics, Almere, the Netherlands). The touchdown thermal protocol included denaturation at 95°C for 30 sec, annealing at 68–55°C (decreasing 1°C/cycle for 13 cycles, followed by 30 cycles at 55°C) for 1 min, and extension at 72°C for 1 min. Final elongation was performed at 72°C for 7 min.

DNA Sequencing

For direct sequencing, PCR products were treated with 3 μL ExoSAPIT (USB) at 37°C for 15 min and 80°C for another 15 min. A total of 2 μL of the treated product was used in sequencing reaction with BigDye terminator chemistry on ABI310 DNA sequencer (Applied Biosystems, Foster City, CA, USA). Sequencing analysis was performed by using primers partially previously described[1,2] and partially redesigned (TABLE 1).

RESULTS

We have used direct sequencing to analyze the coding region of *MECP2*. First we sequenced the fourth exon, because it is the largest, and two main domains, MBD and TRD, which harbor most of the known mutations, are encoded by it. All mutations found in this study are located in the fourth exon.

In this study, we analyzed 15 sporadic cases of RTT and in 7 of 15 patients (47%) we detected pathogenic mutations in the coding parts of *MECP2*. The mutations previously described which we detected in this study are presented

TABLE 1. PCR primer sets for amplifying and sequencing exons of *MECP2*

Exon	Forward/reverse primer	Sequence
2	F	5'-TCAATGGGGGCTTTCAACTTAC-3'
	R	5'-GTTATGTCTTTAGTCTTTGG-3'
3	F	5'-CCTGCCTCTGCTCACTTGTT-3'
	R	5'-GTTCCCCCCGACCCCACCCT-3'
4	F	5'-AAAACATCCCCAATGCTCCA-3'
	R	5'-GAAGAGCGGGAAAGGACTGAA-3'
	F	5'-TTGCTTTTCCGCCCAGG-3'
	R	5'-AAAGTCCTGGGAAGCTCCTTGT-3'
	F	5'-GCCCCCTGGCGAAGTTT-3'
	R	5'-TCAATAGTAACGTTTGTCAGAGCGT-3'

NOTE: F, forward primer; R, reverse primer.
DNA sequence cited from GenBank accession number AF030876.

TABLE 2. Mutations in *MECP2* found in Croatian patients with RTT

Exon	Domain	Nucleotide change[a]	Amino acid change[a]	Patients (N)[b]	Reference
4	MBD	397C > T	R133C	1	1
4	MBD	473C > T	T158M	1	1
4	Interdomain region	502C > T	R168X	1	17
4	TRD	651_652delTG	P217fs	1	19
4	TRD	808C > T	R270X	1	18
4	C-term	(1132_1159del28 + 1167_1176del10)	A378fs	1	Our study
4	C-term	1146_1177del30	L383_E392del10	1	Our study

[a] Nucleotide and amino acid position of the mutation, numbered from the ATG initiator codon.
[b] Number of patients with certain mutation in this study.

in TABLE 2. Two novel deletions (1146_1177del30 and (1132_1159del28 + 1167_1176del10)) were also detected. In a patient with pathogenic mutation R168X, we have also found one polymorphism (S194S), previously reported.[5]

DISCUSSION

In this study we detected pathogenic mutations in 47% of the patients. This is a rather low percentage of found mutations compared to the other studies where the range is between approximately 60–80%.[3,4,6] However, several studies have also reported quite low mutation occurrence, between 35% and 55%.[7,8] The discrepancy in our and in former studies is most probably a result of the use of different diagnostic procedures in making a diagnosis. Recently it was proved that diagnostic procedures strongly influence mutation detection rates: The detection rate in groups of patients in which organized diagnostic criteria were not utilized was 20%. To the contrary, in a group that included patients diagnosed with a strict adherence to the RTT criteria,[9] mutation detection rate was 72%.[10] The difference in results when comparing these two diagnostic methods is enormous, most likely accounting for the discrepancy in published data. Another recently proposed possibility is another gene involvement in disease development.[11–13] Mutations in another X-linked gene, cyclin-dependent kinase-like 5 (*CDKL5*), located in Xp22, have been identified in patients with an RTT-like phenotype or the early-onset seizures variant of RTT (Hanefeld variant). CDKL5 was shown to interact with MeCP2 *in vitro* and *in vivo*, and it was suggested that they belong to the same molecular pathway.[14] For example, it was demonstrated that the repression of BDNF (brain-derived neurotrophic factor) is dependent on MeCP2 phosphorylation[15] and CDKL5 is a putative kinase. However, *CDKL5* is primary connected with an early-onset seizure variant of RTT.

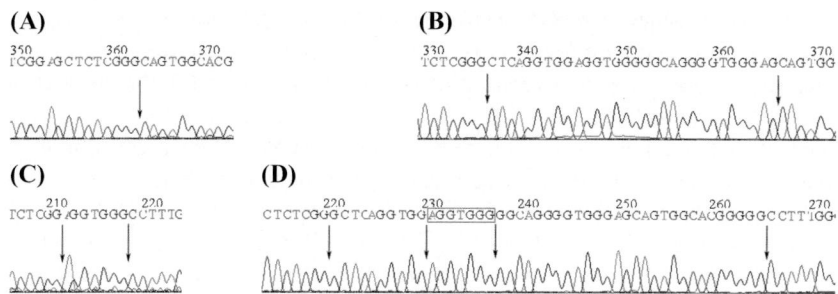

FIGURE 1. Sequence tracings of the two novel mutations described in this article. (**A**) DNA sequence of a patient with 1146_1177del30 mutation; (**B**) DNA sequence of a healthy control; (**C**) DNA sequence of a patient with (1132_1159del28 + 1167_1176del10) mutation; (**D**) DNA sequence of a healthy control. *Arrows* indicate the deletion breakpoints. Frame in (**D**) indicates the original sequence between two deletions. All DNA sequences are shown in the antisense direction.

In addition, gross rearrangements cannot be detected by standard DNA analysis procedures that we used in this study. Recently, in a study in which 110 patients who fulfilled the diagnostic criteria for RTT but had no detectable mutations in *MECP2* were analyzed, large deletions were identified in 37.8% of classic and in 7.5% of atypical RTT patients.[16] The authors concluded that quantitative methods should be included in standard diagnostic procedures because they showed that large deletions of *MECP2* are an important and recurrent cause of atypical RTT.

Most of the mutations reported in this study have already been reported,[1,17,18] but we also found two mutations which are, according to our knowledge, not yet described.[19] Novel mutations are deletions located in the deletion-prone region (DPR) of *MECP2*. The first one is an in-frame homozygous deletion 1146_1177del30 (L383_E392del10; FIGS. 1A–1D). Deletions of this kind that occur in a position near this one have already been reported by other researchers.[20,21]

This deletion was found in an 8-year-old girl born at term after an uneventful pregnancy. The birth weight was 2,920 g, the length was 47 cm, and the head circumference was 33 cm. The Apgar score was 7/9. There were problems during the delivery due to a cephalopelvic disproportion. The patient was treated with antibiotics for the perinatal infection. On brain ultrasound signs of perinatal hypoxia were noted.

Psychomotor development in the first months was normal. She sat independently at the age of 8–9 months and was able to stand by objects and to walk when held by both hands, but she could never walk independently. She spoke 2–3 meaningful words.

At the age of 12 months the parents observed psychomotor decline. The movement stereotypes of rocking and washing hands appeared, and she lost the ability to hold objects. At the age of 20 months, her head circumference (OFC) was at the third percentile (45 cm). Her psychomotor development was at the level of moderate-to-severe retardation (IQ = 40). The brain CT showed diffuse atrophic changes. The EEG was markedly epileptically changed, but at that time she experienced no seizures. Her hearing was normal, and the ophthalmologic examination revealed no pathology. The karyotype was normal.

Now, at the age of 8 years, her height is 115 cm (<3c), weight 18 kg (<3c), and OFC 48 cm (<3c). She does not speak, but only vocalizes. She sits independently, but she is unable to stand. Deformity of the spine (scoliosis/kyphosis) as well as contractures of both feet developed. Because of short generalized epileptic seizures that appear once every few months, antiepileptic therapy was introduced.

The other novel mutation is a complex frameshift homozygous deletion of 28 bp at 1132–1159 and 10 bp at 1167–1176, resulting in frameshift at A378. Complex deletions of this kind have also been reported by other researchers.[7,22] These findings are consistent with previous reports that this region is a hotspot for deletion because it contains tandem repeats and DNA polymerase slippage is frequent.[7,18]

The patient with (1132_1159del28 + 1167_1176del10) mutation is currently 13 years old; she was born after an uneventful pregnancy and no birth complications. Initially development was normal, but later her motor and language development slowed down (after speaking her first word at 9 months, it regressed). At 15 months she developed aggressive and autoaggressive behavior. At this point, the ability to understand was rather weak. She also had an impaired sleeping pattern. Further diagnosis showed disturbance in communication, severe mental retardation, epilepsy, and scoliosis. She is unable to walk independently. Stereotypic hand movements (hand-washing), characteristic of RTT patients, are also present. The genotyping of parents was not accomplished because DNA samples were not taken, but we are planning to complete this inquiry.

It is noteworthy that all, except one (P217fs), of the previously described mutations that we found in this study (4/7, 57%) are located in one of the eight CpG hotspot dinucleotides. Our results are therefore in concordance with other studies which show that approximately 65% of the mutations in *MECP2* are located in these positions.

We conclude here that the fourth exon should be the first one analyzed because all mutations found in this study are located in it. Two of the mutations found are novel; these are deletions characteristic for the C-terminal part of the *MECP2* gene. Our results indicate that an early diagnosis could enable families to maximize utilization of existing treatments. It would also significantly improve genetic counseling and possible prenatal diagnosis. This is the first study of this kind in Croatia. Therefore, mutation analyses of additional patients are necessary to determine the spectrum of *MECP2* mutations in the Croatian population.

ACKNOWLEDGMENT

This work was supported by a grant number 0098092 from the Ministry of Science, Education and Sport, Republic of Croatia.

REFERENCES

1. AMIR, R.E., I.B. VAN DEN VEYVER, M. WAN, et al. 1999. Rett syndrome is caused by mutations in X-linked MECP2, encoding methyl-CpG-binding protein 2. Nat. Genet. **23**: 185–188.
2. GIUNTI, L., S. PELAGATTI, V. LAZZERINI, et al. 2001. Spectrum and distribution of MECP2 mutations in 64 Italian Rett syndrome girls: tentative genotype/phenotype correlation. Brain Dev. **23**: 242–245.
3. BIENVENU, T., L. VILLARD, N. DE ROUX, et al. 2002. Spectrum of MECP2 mutations in Rett syndrome. Genet. Test. **6**: 1–6.
4. VACCA, M., F. FILIPPINI, A. BUDILLON, et al. 2001. MECP2 gene mutation analysis in the British and Italian Rett syndrome patients: hot spot map of the most recurrent mutations and bioinformatic analysis of a new MECP2 conserved region. Brain Dev. **23**: 246–250.
5. AMIR, R.E., I.B. VAN DEN VEYVER, R. SCHULTZ, et al. 2000. Influence of mutation type and X chromosome inactivation on Rett syndrome phenotypes. Ann. Neurol. **47**: 670–679.
6. YARON, Y., B. BEN ZEEV, R. SHOMRAT, et al. 2002. MECP2 mutations in Israel: implications for molecular analysis, genetic counseling, and prenatal diagnosis in Rett syndrome. Hum. Mutat. **20**: 323–324.
7. XIANG, F., S. BUERVENCIH, P. NICOLAO, et al. 2000. Mutation screening in Rett syndrome patients. J. Med. Genet. **37**: 250–255.
8. PAN, H., Y.P. WANG, X.H. BAO, et al. 2002. MECP2 gene mutation analysis in Chinese patients with Rett syndrome. Eur. J. Hum. Genet. **10**: 484–486.
9. HAGBERG, B., F. HANEFELD, A. PERCY, et al. 2002. An update on clinically applicable diagnostic criteria in Rett syndrome. Comments to Rett Syndrome Clinical Criteria Consensus Panel Satellite to European Paediatric Neurology Society Meeting, Baden Baden, Germany, 11 September 2001. Eur. J. Paediatr. Neurol. **6**: 293–297.
10. GAUTHIER, J., G. DE AMORIM, G.N. MNATZAKANIAN, et al. 2005. Clinical stringency greatly improves mutation detection in Rett syndrome. Can. J. Neurol. Sci. **32**: 321–326.
11. SCALA, E., F. ARIANI, F. MARI, et al. 2005. CDKL5/STK9 is mutated in Rett syndrome variant with infantile spasms. J. Med. Genet. **42**: 103–107.
12. TAO, J., H. VAN ESCH, M. HAGEDORN-GREIWE, et al. 2004. Mutations in the X-linked cyclin-dependent kinase-like 5 (CDKL5/STK9) gene are associated with severe neurodevelopmental retardation. Am. J. Hum. Genet. **75**: 1149–1154.
13. WEAVING, L.S., J. CHRISTODOULOU, S.L. WILLIAMSON, et al. 2004. Mutations of CDKL5 cause a severe neurodevelopmental disorder with infantile spasms and mental retardation. Am. J. Hum. Genet. **75**: 1079–1093.
14. MARI, F., S. AZIMONTI, I. BERTANI, et al. 2005. CDKL5 belongs to the same molecular pathway of MeCP2 and it is responsible for the early-onset seizure variant of Rett syndrome. Hum. Mol. Genet. **14**: 1935–1946.

15. CHEN, W.G., Q. CHANG, Y. LIN, et al. 2003. Derepression of BDNF transcription involves calcium-dependent phosphorylation of MeCP2. Science **302:** 885–889.
16. ARCHER, H.L., S.D. WHATLEY, J.C. EVANS, et al. 2006. Gross rearrangements of the MECP2 gene are found in both classical and atypical Rett syndrome. J. Med. Genet. **43:** 451–456.
17. WAN, M., S.S. LEE, X. ZHANG, et al. 1999. Rett syndrome and beyond: recurrent spontaneous and familial MECP2 mutations at CpG hotspots. Am. J. Hum. Genet. **65:** 1520–1529.
18. HUPPKE, P., F. LACCONE, N. KRAMER, et al. 2000. Rett syndrome: analysis of MECP2 and clinical characterization of 31 patients. Hum. Mol. Genet. **9:** 1369–1375.
19. Rett BASE: IRSA MECP2 Variation Database. Available at (http://mecp2.chw.edu.au//).
20. LACCONE, F., P. HUPPKE, F. HANEFELD, et al. 2001. Mutation spectrum in patients with Rett syndrome in the German population: evidence of hot spot regions. Hum. Mutat. **17:** 183–190.
21. TRAPPE, R., F. LACCONE, J. COBILANSCHI, et al. 2001. MECP2 mutations in sporadic cases of Rett syndrome are almost exclusively of paternal origin. Am. J. Hum. Genet. **68:** 1093–1101.
22. BUYSE, I.M., P. FANG, K.T. HOON, et al. 2000. Diagnostic testing for Rett syndrome by DHPLC and direct sequencing analysis of the MECP2 gene: identification of several novel mutations and polymorphisms. Am. J. Hum. Genet. **67:** 1428–1436.

Prostaglandins Regulate Transcription by Means of Prostaglandin Response Elements Located in the Promoters of Mammalian Na,K-ATPase β1 Subunit Genes

KEIKANTSE MATLHAGELA AND MARY TAUB

Biochemistry Department, School of Medicine and Biomedical Sciences, State University of New York at Buffalo, Buffalo, New York 14214, USA

ABSTRACT: Prostaglandins are potent products of arachidonic acid metabolism that play significant roles in regulating ion transport in the kidney. In the Madin Darby canine kidney (MDCK) cell line prostaglandin E_1 (PGE_1) stimulates the activity of the Na,K-ATPase and regulates transcription. Transient transfection studies conducted in MDCK cells with a human Na,K-ATPase β1 subunit promoter/luciferase construct, pHβ1-1141 Luc, showed a PGE_1 stimulation. The PGE_1 stimulation was inhibited by the PGE receptor antagonists SC19220 and AH6809, indicating the involvement of EP1 receptors (coupled to phospholipase C) and EP2 receptors (coupled to adenylate cyclase), respectively. A prostaglandin-regulatory element (PGRE) within the β1 subunit promoter (−110 to −92, AGTCCCTGC) is sufficient to elicit a PGE_1 stimulation in a heterologous promoter (in pLUC-MCS). Studies with promoter mutants indicated that in addition to the PGRE, an adjacent Sp1 site was also essential for regulation by PGE_1. Consistent with the involvement of Sp1 are the results of DNA affinity precipitation studies, which indicate that Sp1 as well as CREB, and Sp3 all bind to the PGRE. The involvement of this PGRE in transcriptional regulation of the Na,K-ATPase β1 gene was examined in a number of species. Only human and chimpanzee promoters possessed an identical PGRE site, unlike dog, rat, and mouse, which possessed Sp1 sites in similar locations. Two alternative PGREs were subsequently identified. The sequence of the one of these PGREs (TGACCTTC, −445 to −438) was conserved throughout all species examined, suggesting its physiologic significance.

KEYWORDS: prostaglandins; MDCK cells; Na,K-ATPase; transcription; Sp1; CREB

Address for correspondence: Dr. Mary Taub, Biochemistry Department, 140 Farber Hall, State University of New York at Buffalo, 3435 Main Street, Buffalo, New York 14214, USA. Voice: 716-829-3300; fax: 716-829-2725.
e-mail: biochtau@buffalo.edu

INTRODUCTION

Prostaglandins are biologically potent products of arachidonic acid metabolism by cyclooxygenase, and they play significant roles in blood pressure regulation.[1] The kidney is a major site of prostaglandin production, and in addition is a site where these renal prostaglandins act so as to affect blood pressure.[1] Included amongst the renal processes modulated by prostaglandins are ion transport systems, blood flow, glomerular filtration, and renin release.

Of particular interest to this report are the effects of prostaglandins on renal Na,K-ATPase activity. Prostaglandins have both acute and chronic effects on renal Na,K-ATPase activity. Acute prostaglandin treatments affect the integration of the Na,K-ATPase into the plasma membrane, whereas chronic treatments affect Na,K-ATPase gene expression. An *in vitro* model system that is particularly amenable for studies concerning such prostaglandin effects is the Madin Darby canine kidney (MDCK) cell line. The MDCK cell line is a well-characterized distal tubule model system in which the role of the Na,K-ATPase in transepithelial solute has been clearly defined. In addition, MDCK cells are particularly amenable for studies concerning prostaglandin effects due to the availability of a hormonally defined serum-free medium.[2]

Previously, we reported that prostaglandins E_1 (PGE_1) and 8-bromocyclic adenosine monophosphate (8B-cAMP) stimulate the activity of the Na,K-ATPase in MDCK cells in such a serum-free medium.[3] The stimulatory effects of PGE_1 and 8-bromocyclic AMP on Na,K-ATPase activity were explained by an increase in the level of expression of the Na,K-ATPase.[3] The Na,K-ATPase consists of both an α subunit, responsible for the catalytic activity, and a β subunit, required for the integration of the Na,K-ATPase into the plasma membrane.[4] PGE_1 was observed to increase β subunit mRNA levels up to tenfold, and to affect α subunit mRNA levels to a smaller extent. As β subunit levels limit α/β heterodimer formation,[4] regulation of $\beta1$ subunit transcription was examined in greater detail.[3–5] Transient transfection studies were conducted with a human Na,K-ATPase $\beta1$ promoter/luciferase construct, pHβ1-1141.[6] The results indicated that PGE_1 stimulates Na,K-ATPase $\beta1$ subunit gene transcription by both the cyclic adenosine 3′,5′-phsophate (cAMP) and calcium signaling pathways. Studies with 5′ deletion mutants led to the identification of a region in the human Na,K-ATPase $\beta1$ subunit promoter (-167 to -72) that was required in order to elicit a PGE_1 response.[5] A prostaglandin-responsive element (PGRE) was identified in this region.[5] In this report we further further analyze the receptor signaling pathways involved in mediating the prostaglandin response, as well as the role of PGREs in regulating the transcription of the Na,K-ATPase $\beta1$ subunit gene.

MATERIALS AND METHODS

Materials

Hormones and chemicals were from Sigma Aldrich Co. (St.Louis, MO), and prostaglandin agonists and antagonists were from Cayman Chemicals (Ann Arbor, MI). Powdered medium, lipofectamine and synthetic oligonucleotides were from Invitrogen Corp. (Carlsbad, CA). Galacto-Star reagent was from Applied Biosystems (Bedford, MA) and reporter lysis buffer was from Promega (Madison, WI). Supplies for electrophoresis and blotting were from Bio-Rad (Hercules, CA). Rabbit polyclonal antibodies against CREB (SC-186X), Sp1 (sc-1402X), and Sp3 (sc-644) were from Santa Cruz Biotech (Santa Cruz, CA), and streptavidin agarose was from Pharmingen (San Diego, CA).

Expression Vectors

The human Na,K-ATPase $\beta 1$ subunit/luciferase construct pHβ1-1141 Luc[6] was from Dr. Jerry Lingrel (University of Cincinnati). The vector pSVβgal was from Promega (Madison, WI) and pLUC MCS was from Stratagene Corp. (La Jolla, CA). Synthetic oligonucleotides were inserted into the Hind III/Xho I sites of pLUC MCS.

Cell Culture

Stock cultures of MDCK cells were routinely cultured in a 50:50 mixture of Dulbecco's modified Eagle's medium and Ham's F12 supplemented with 15 mM HEPES (pH 7.4) 20 mM sodium bicarbonate, 92 U/mL penicillin, 200 µg/mL streptomycin (DME/F12).The DME/F12 was supplemented with 5 µg/mL bovine insulin, 5 µg/mL human transferrin, 5×10^{-12} M triiodothyronine, 5×10^{-8} M hydrocortisone and 25 ng/mL PGE_1 at the time of use. The cells were grown at 37°C in a humidified 5% CO_2/95% air environment.

Transient Transfection Studies

MDCK cells were plated at 10^5 cells/35-mm dish into DME/F12 supplemented with 5 µg/mL insulin and 5 µg/mL transferrin. The next day, the monolayers were transiently transfected by lipofectamine with 1 µg of promoter/luciferase construct and 0.2 µg of pSVβgal.[7] Following an overnight incubation, the medium was first changed to DME/F12 with 5 µg/mL insulin and 5 µg/mL transferrin, followed 2 h later by the addition of effector molecules.

A total of 4 h later, the monolayers were solublized in reporter lysis buffer, and centrifuged 1 min at 14,000 rpm. The luciferase and β-galactosidase activity of the cell lysates was then determined as previously described.[7] The luciferase activity of each culture was normalized with respect to β-galactosidase activity, and compared with the control value. For each experimental condition, the normalized luciferase value was the mean ± SEM of quadruplicate determinations. The significance of the effect was determined by ANOVA ($P < 0.05$).

DNA Affinity Precipitation

Nuclear extracts were prepared from confluent monolayers of MDCK monolayers in 100-mm dishes containing DME/F12 supplemented with 5 μg/mL insulin and 5 μg/L transferrin, by a modification of the method of Dignam et al.[5,8] The nuclear extracts were incubated for 16 h either in the presence or the absence of a biotinylated double-stranded oligonucleotide. Control experiments conducted with a control oligo (5′-CTACTGCTATTCTAGTAACTGAC-3′) did not account for the observed binding with the experimental oligo. The biotinylated DNA-protein complexes were precipitated with streptavidin agarose beads (1 h; 4°C), washed three times with HKMG (10 mM HEPES, pH 7.9, 100 mM KCl, 5 mM $MgCl_2$, 10% glycerol, 1 mM EDTA), followed by separation on SDS-PAGE and Western blotting.[7]

Blasts of 5′ Flanking Sequences

The Na,K-ATPase β1 subunit genes for human, chimpanzee, dog, mouse, rat, and *Drosophila* were identified through Homologene on the National Center for Biotechnology (NCBI) web site. After identifying the location of the Na,K-ATPase β1 subunit gene in a particular species (e.g., chromosome 1, region 165807640 to 165833615 for human ATP1B1), the sequence of the 5′ flanking region was located on the appropriate chromosome contig (e.g., NT_00487 for human ATP1B1). The 5′ flanking regions of the Na,K-ATPase β1 subunit genes from different species were compared using the Blast 2 Sequences program (bl2seq), also located on the NCBI web site.

RESULTS

PGE_1 Stimulation of the β1 promoter: Signaling through EP1 and EP2 Receptors

To determine the effects of PGE_1 and prostaglandin E_2 (PGE_2) on transcription of the human Na,K-ATPase β1-subunit gene expression in MDCK cells,

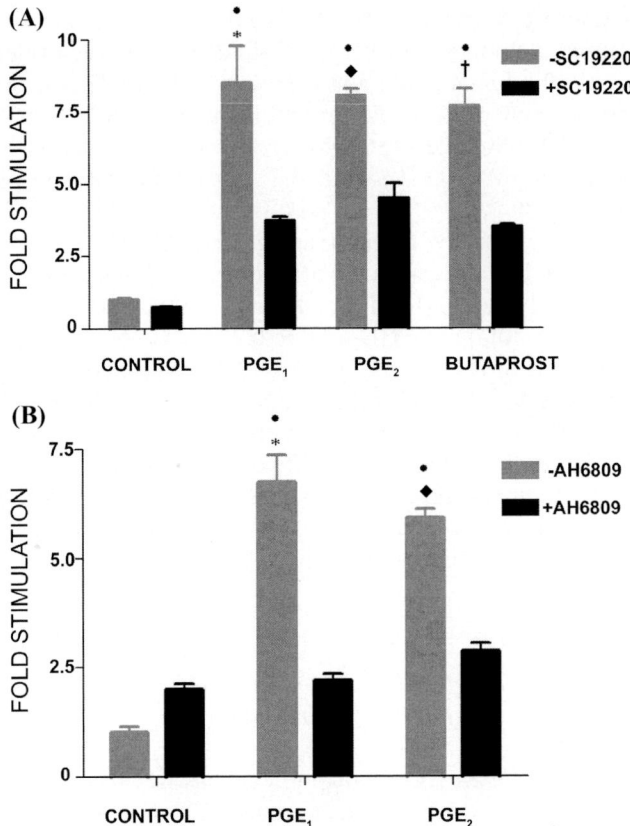

FIGURE 1. Stimulatory effects of EP agonists on Na^+,K^+-ATPase $\beta 1$ subunit gene expression. MDCK cells were transiently transfected with pHβ1-1141Luc. A total of 24 h later, a portion of the cultures were treated with either (**A**) 70.5 nM PGE_1, 86 nM PGE_2 or 8.1 μM butaprost or (**B**) 70.5 nM PGE_1 or 86 nM PGE_2. A portion of the cultures in (**A**) were treated with 30 μM SC-19220 for 30 min prior to, and during the 4-h incubation with agonist. A portion of the cultures in (**B**) were treated with 10 μM AH6809 for 30 min prior to, and during the 4-h incubation with agonist. The luciferase activity of cell lysates was determined in quadruplicate after the 4-h incubation period, and compared with untreated controls. At least three experiments were performed. Values are means ± SEM. *,♦, † $P < 0.01$ control and antagonist-treated cells compared to agonist-treated cells; •, $P > 0.05$ agonist-treated cells in comparison to each other.

transient transfection studies were conducted with pHβ1-1141Luc.[6] FIGURE 1A shows that PGE_1 (70.5 μM) and PGE_2 86 μM had equivalent stimulatory effects (8.5 ± 1.3–fold and 8.0 ± 0.3–fold, respectively). The possibility was evaluated that the PGE_1 and PGE_2 stimulation was mediated by EP1 and/or EP2 receptors. In order to determine whether the PGE_1 and PGE_2 stimulation of Na,K-ATPase β1 transcription was mediated by EP2 receptors, the effect of the EP2 agonist butaprost[9–11] was examined. FIGURE 1A shows butaprost (8.1 μM) caused a stimulation (8.5 ± 1.3–fold) that was equivalent to the stimulation caused by both PGE_1 and PGE_2. In order to determine whether EP1 receptors were involved, the EP1 receptor antagonist SC19220[9,12,13] was employed (FIG. 1A). SC19220 (30 μM) reduced the stimulation by PGE_1, PGE_2 and butaprost by 6.2 ± 1.3–fold, 6.2 ± 0.3–fold, and 6.3 ± 0.6–fold, respectively.

The effect of another EP receptor antagonist, AH6809, was examined (FIG. 1B). AH6809 has a nearly equal affinity for the human EP1, EP2, and EP3 III receptors.[9,14] FIGURE 1B shows that AH 6809 caused the stimulation by PGE_1 (6.7 ± 0.6–fold) and PGE_2 (5.9 ± 0.2–fold) to be reduced by 4.5 ± 0.1–fold (in the case of PGE_1), and by 3.8 ± 0.2–fold (in the case of PGE_2). AH-6809 alone caused 2.0 ± 0.1–fold stimulation, possibly because AH 6809 also acts as an EP3 antagonist,[14] and prevents the inhibition of adenylate cyclase by EP3. Thus, in the presence of AH 6809, the PGE_1 stimulation was not significant in comparison to cultures incubated with AH-6809 alone. In contrast, AH 6809 had little effect on the stimulation caused by butaprost (the butaprost stimulation was reduced by only 26 ± 0.1%) (Matlhagela and Taub, unpublished).

Role of a Prostaglandin-Response Element in Mediating the PGE_1 Stimulation

The human Na,K-ATPase β1 subunit promoter has previously been shown to consist of a 1,141-basepair region which contains a thyroid hormone response element (−459 to −438)[6] as well as a mineralocorticoid/glucocorticoid-response element (−650 to −630).[15] We have identified three putative prostaglandin-response elements (PGREs) that have some homology to a cyclic AMP response element (CRE). In order to determine whether any of these putative PGREs were required in order to observe the PGE_1 response, transient transfection studies have been conducted with 5′ deletion mutants (at the positions indicated on FIG. 2A).[5,6] The results indicated that the region between −182 to −83 was sufficient in order to obtain a PGE_1 stimulation. PGRE3 is located within this region of the Na,K-ATPase β1 subunit promoter (FIG. 2A).

The region in the Na,K-ATPase β1 subunit promoter that contains PGRE3 (−167 to −82) was introduced into a vector (pLUC MCS) adjacent to a TATA box. A PGE_1 stimulation of transcription was observed in MDCK cells transiently transfected with this construct (pLUC MCS PGRE) (FIG. 2B).[5] However, the PGE_1 stimulation was lost when the GC boxes immediately downstream of PGRE3 were translocated further downstream (FIGS. 2B and 2D).[5] Possibly, the PGE_1 stimulation is simply dependent upon the binding of Sp1 to the GC box immediately downstream from PGRE3. Indeed FIGURE 2B shows that the PGE_1 stimulation is lost by mutation of the GC box (−118 to −112) immediately downstream of PGRE3 (−100 to −92) from CCCGCCC

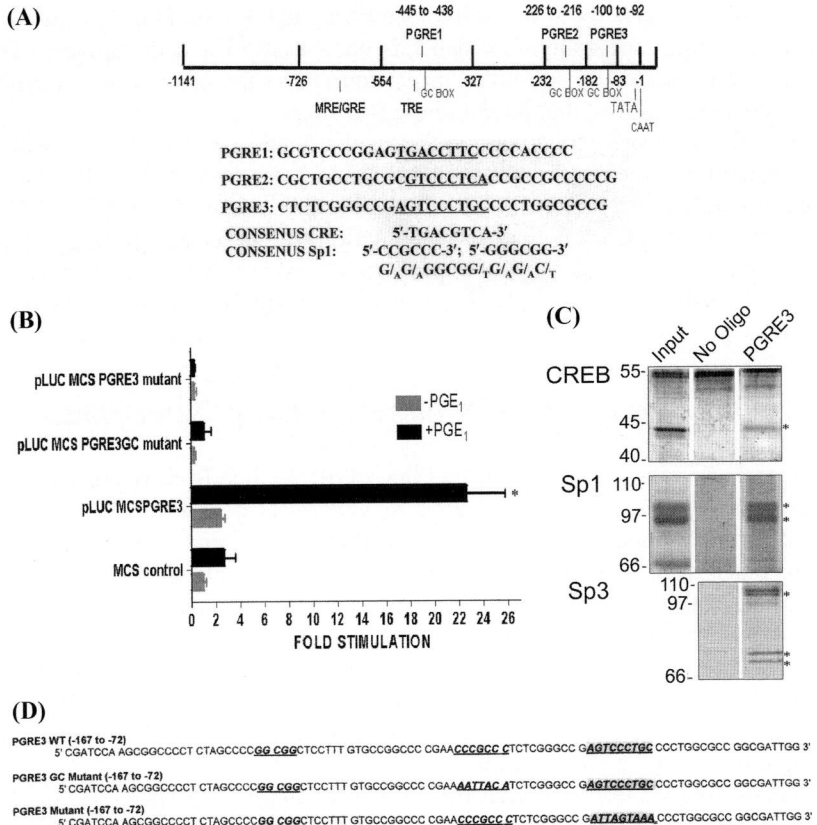

FIGURE 2. Stimulatory effect of PGE$_1$ on transcription by a PGRE. (**A**) Human Na,K-ATPase β1 subunit promoter contains three putative PGREs. The function of PGRE3 is dependent upon an adjacent GC box. GC boxes are also adjacent to PGRE1 and PGRE2. An MRE/GRE is located at −650 to −630 and a TRE is located at −459 to −438. (**B**) Effect of PGE$_1$ on transcription by pLUC MCSPGRE3. MDCK cells were transiently transfected with either the pLUC MCS vector containing the region between −167 to −72 in the human Na,K-ATPase β1 promoter (pLUC MCSPGRE3), pLuc MCSPGRE3mutant (containing a mutation in GC adjacent to PGRE3), pLUC MCSPGRE3muant (containing a mutation in PGRE3), or with the empty vector pLUC MCS. (**C**) DNA affinity precipitation of nuclear proteins. Nuclear extracts from MDCK cells were incubated with the biotinylated oligo 5′-CTCTCGGGCCG**AGTCCCTGC**CCCTGG-3′. Nuclear protein/oligo complexes were purified with streptavidin-agarose beads, separated on SDDS/PAGE, and subjected to a Western blot analysis using either anti-CREB, anti-Sp1, or anti-Sp3 polyclonal antibody. * indicates the band of interest. (**D**) The sequences of the oligos inserted into pLUCMCS in the vectors employed in (**B**) above.

to AATTACA. The dependence of the PGE$_1$ stimulation upon PGRE3 was similarly examined taking the genetic approach. FIGURE 2B shows that the PGE$_1$ stimulation was similarly lost by mutation of PGRE3 from AGTCCCTGC to

ATAGTAAA. Thus these observations indicate that (i) the PGE_1 stimulation of β1 subunit gene expression depends upon PGRE3 and the adjacent GC box, and (ii) the PGE_1 stimulation depends upon interactions between the transcription factor(s) that bind to these two sites.

In order to identify transcription factors that bind to PGRE3, DNA affinity precipitation studies were conducted. Nuclear extracts of MDCK cells were incubated with the biotinylated PGRE3 oligo CTCTCGGGCCGAGTCCCTGC, followed by streptavidin-agarose precipitation of the nuclear proteins which were binding to this oligo. A Western blot analysis of these nuclear proteins (FIG. 2C) indicated that CREB, Sp1, and Sp3 were all associated with PGRE3.

PGREs Present within Different Na,K-ATPase β1 Subunit Genes

Our 5′ deletion analysis indicated that PGRE3 alone is sufficient in order to obtain regulation of the human Na,K-ATPase β1 gene by PGE_1. In order to determine whether the Na,K-ATPase β1 subunit gene in canine kidney cells (such as MDCK) is regulated by a similar mechanism, the sequence of "PGRE3" in the canine Na,K-ATPase β1 subunit gene was examined. TABLE 1 shows that in the canine promoter, the sequence present at the location equivalent to human "PGRE3" consists of a consensus Sp1 site, rather than a "PGRE." Similar observations were made with the rat and mouse 5′ flanking regions. Only chimpanzee genomic DNA contained a "PGRE3" sequence that was identical to the human sequence.

Previously, we observed that the level of expression of Na,K-ATPase β1 subunit mRNA increases up to tenfold in MDCK cells treated with PGE_1, and that the level of the Na,K-ATPase itself was similarly elevated.[1] In MDCK cells this cAMP-mediated regulation may possibly occur via another mechanism of transcriptional control that involves another regulatory element. Indeed TABLE 1 shows that the PGRE1 sequence is identical in humans, chimpanzees, dogs, rats, and mice. Only *Drosophila* differed with regards to its PGRE1 sequence.

DISCUSSION

The results of our studies indicate that PGE_1 stimulates transcription of the human Na,K-ATPase β1 subunit gene by interacting with EP1 receptors (coupled to phosphdipase), as well as EP2 receptors (coupled to adenylate cyclase). As a consequence, both protein kinase C (PKC) and cyclic AMP-dependent protein kinase (PKA) are activated. A consequence is CREB phosphorylation and subsequent effects on gene regulation.

Although three possible PGREs are present in the human Na,K-ATPase β1 subunit promoter, our 5′ deletion analysis indicates PGRE3 alone is sufficient

TABLE 1. Regulatory elements in the 5′ flanking regions Na,K-ATPase β1 subunit genes

Na,K-ATPase β1 subunit gene	5′ flanking region: % identity to human	PGRE1 −438 →	PGRE2 −217 →	PGRE3 −92 →
NM_001677 ATP1B1_Man	100	CCCGGAGTGACCTTCCCCCA	GCGTCCCTCA	GGGCCGAGTCCCTGCCCCT
XM_513981 ATP1B1_Chimp	97	CCCGGAGTGACCTTCCCCCC	NS	GGGCCGAGTCCCTGCCCCT
NM_001003283 ATP1B1_Dog	32	CCTCGAGTGACCTTCCCCCA	—	GACTGGAGGCCCGCCCCG
NM_013113 Atp1b1_Rat	50	CCTGGAGTGACCTTCCCCCA	GTCCA	GGGCTCAGGCCCGCT
NM_009721 Atp1b1_Mouse	36	CCTGGAGTGACCTTCCCCCA	CGGGGCA	GGGCCCCGCGCCGCCCCT
NM_57819 Nrv1_Drosophila	31	AGTCTGCTGAATCATTCGCA	TAACCGTA	—

to elicit a PGE_1 response. Our DNA affinity precipitation studies indicate that both CREB and Sp1 bind to this PGRE3. In addition Sp1 binds to an adjacent Sp1 site. Thus, PGE_1 may be regulate human Na,K-ATPase β1 gene expression through the phosphorylation of CREB, resulting in an interaction between these two transcription factors. Indeed the studies presented here indicate that the regulation of gene expression by PGE_1 was lost as a consequence of a mutation in either PGRE3 or the adjacent Sp1 site.

Although the level of Na,K-ATPase β1 subunit mRNA increases in PGE_1-treated MDCK cells, the Na,K-ATPase β1 subunit promoter of canine cells does not contain a regulatory element identical to human PGRE3. Instead, the homologous sequence in the canine promoter is a consensus Sp1 site. Similar observations were made with respect to the mouse and rat Na,K-ATPase β1 promoters. Only the chimpanzee Na, K-ATPase β1 promoter contains a sequence that is identical to human PGRE3. Thus, another PGRE may be involved in mediating the regulation by the cAMP and calcium pathways in addition to PGRE3. It is unlikely that PGRE2 is such a regulatory element, as the PGRE2 sequence is not conserved in any species other than human.

However, the PGRE1 sequence is identical in the human, chimpanzee, dog, mouse, and rat Na,K-ATPase β1 subunit promoters. The conservation of the PGRE1 sequence suggests that PGRE1 is of physiologic importance in Na,K-ATPase β1 subunit gene regulation. The PGRE1 sequence TGACCTTC contains a CRE half site, which also indicates that CREB has a binding affinity for PGRE1 (as in the case of PGRE3). Also, like PGRE1, PGRE3 is adjacent to a GC box. ThusSp1 may also be involved in mediating the regulation through the PGRE1 site, as well as the PGRE3 site. Further studies are in progress to evaluate these possibilities.

ACKNOWLEDGMENTS

We thank Dr. Jerry Lingrel for pHβ1-1141 Luc. This work was supported by NHLBI Grant 1RO1HL69676-01 to Mary Taub.

REFERENCES

1. BONVALET, J.P., P. PRADELLES & N. FARMAN. 1987. Segmental synthesis and actions of prostaglandins along the nephron. Am. J. Physiol. **253:** F377–F387.
2. TAUB, M., et al. 1979. Growth of Madin-Darby canine kidney epithelial cell (MDCK) line in hormone-supplemented, serum-free medium. Proc. Natl. Acad. Sci. USA **76:** 3338–3342.
3. TAUB, M.L., et al. 1992. Regulation of the Na,K-ATPase activity of Madin-Darby canine kidney cells in defined medium by prostaglandin E1 and 8-bromocyclic AMP. J. Cell. Physiol. **151:** 337–346.

4. GEERING, K. 1991. The functional role of the beta-subunit in the maturation and intracellular transport of Na,K-ATPase. FEBS Lett. **285:** 189–193.
5. MATLHAGELA, K. *et al.* 2005. Identification of a prostaglandin-responsive element in the Na,K-ATPase {beta}1 promoter that is regulated by cAMP and Ca2+: evidence for an interactive role of cAMP regulatory element-binding protein and Sp1. J. Biol. Chem. **280:** 334–346.
6. FENG, J., J. ORLOWSKI & J.B. LINGREL. 1993. Identification of a functional thyroid hormone response element in the upstream flanking region of the human Na,K-ATPase beta 1 gene. Nucleic Acids Res. **21:** 2619–2626.
7. TAUB, M. *et al.* 2004. Regulation of the Na,K-ATPase in MDCK cells by prostaglandin E1: a role for calcium as well as cAMP. Exp. Cell Res. **299:** 1–14.
8. DIGNAM, J.D., R.M. LEBOVITZ & R.G. ROEDER. 1983. Accurate transcription initiation by RNA polymerase II in a soluble extract from isolated mammalian nuclei. Nucleic Acids Res. **11:** 1475–1489.
9. KIRIYAMA, M. *et al.* 1997. Ligand binding specificities of the eight types and subtypes of the mouse prostanoid receptors expressed in Chinese hamster ovary cells. Br. J. Pharmacol. **122:** 217–224.
10. COLEMAN, R.A., W.L. SMITH & S. NARUMIYA. 1994. International Union of Pharmacology classification of prostanoid receptors: properties, distribution, and structure of the receptors and their subtypes. Pharmacol. Rev. **46:** 205–229.
11. REGAN, J.W. 2003. EP2 and EP4 prostanoid receptor signaling. Life Sci. **74:** 143–153.
12. COLEMAN, R.A., I. KENNEDY & R.L. SHELDRICK. 1987. New evidence with selective agonists and antagonists for the subclassification of PGE2-sensitive (EP) receptors. Adv. Prostaglandin Thromboxane Leukot Res. **17A:** 467–470.
13. ZENG, L., S. AN & E.J. GOETZL. 1996. Selective regulation of RNK-16 cell matrix metalloproteinases by the EP4 subtype of prostaglandin E2 receptor. Biochemistry **35:** 7159–7164.
14. ABRAMOVITZ, M. *et al.* 2000. The utilization of recombinant prostanoid receptors to determine the affinities and selectivities of prostaglandins and related analogs. Biochim. Biophys. Acta **1483:** 285–293.
15. DERFOUL, A. *et al.* 1998. Regulation of the human Na/K-ATPase beta1 gene promoter by mineralocorticoid and glucocorticoid receptors. J. Biol. Chem. **273:** 20702–20711.

Different Modulation of ER-Mediated Transactivation by Xenobiotic Nuclear Receptors Depending on the Estrogen Response Elements and Estrogen Target Cell Types

GYESIK MIN

Department of Microbiological Engineering, Jinju National University, Jinju, Gyeongsangnam-Do, 660-758 Korea

ABSTRACT: Recent studies demonstrated that constitutive androstane receptor (CAR) inhibits ER-mediated transactivation of both endogenous and synthetic estrogen responsive promotor in Hep G2. Whereas steroid and xenobiotic receptor (SXR) but not peroxisome proliferator-activated receptor-γ (PPAR-γ) was also reported to repress estrogen receptor (ER) transactivation of the synthetic 4ERE in Hep G2, the effects of these xenobiotic nuclear receptors (XNRs) on the endogenous estrogen responsive promotor remain to be determined. Effects of CAR, SXR, and PPAR-γ on ER transactivation were also examined in three different kinds of breast cancer cell lines. However, except in MCF-7, studies were limited either in single dose response (MDA-MB-231) or with CAR only (MCF-7-K3). And there is presently no report on the effects of CAR, SXR, and PPAR-γ on ER-mediated transactivation in ovarian-derived CHO-S cells. Accordingly, this article further examined the effects of the endogenous vitellogenin B1 estrogen responsive promotor on the SXR- and PPAR-γ-modulated ER transactivation in Hep G2, and either dose-dependent or single dose effects of SXR, PPAR-γ, and CAR in two different breast cancer cell lines and the ovarian-derived cell line respectively, on the ER-mediated transactivation of the synthetic (4ERE)-tk-luciferase reporter. Consistent with the previous report, CAR significantly repressed ER-mediated transactivation of the endogenous vitellogenin B1 promotor in Hep G2 cells. However, contrary to the effects on the synthetic promotor, PPAR-γ potentiated whereas SXR did not have any effects on the ER transactivation of the vitellogenin promotor in Hep G2. In the breast cancer cell line of MDA-MB-231 in which endogenous ER is known not to be expressed, CAR modestly stimulated ER transactivation of the synthetic 4ERE in a low dose whereas both

Address for correspondence: Dr. Gyesik Min, Department of Microbiological Engineering, Jinju National University, Jinju, Gyeongsangnam-Do, 660-758 Korea. Voice: 055-751-3396; fax: 055-751-3399.
e-mail: g-min@jinju.ac.kr

SXR and PPAR-γ did not have any effects in all doses examined (20–500 ng). And in both CHO-S and estrogen-independent breast cancer cell line, MCF-7-K3, none of the three xenobiotic receptors significantly influenced the ER-mediated 4ERE transactivation in all doses examined. XNRs modulate ER-mediated transactivation depending on the estrogen response elements (EREs) and estrogen target cell types.

KEYWORDS: estrogen receptor; transactivation; xenobiotic nuclear receptor; estrogen response element; HepG2; breast cancer cell; CHO-S cell

INTRODUCTION

The hormone estrogen plays diverse physiological roles not only in the development and maintenance of female reproductive organs but also in the control of metabolic functions in other tissues including liver, brain, and cardiovascular system.[1] For example, estrogen has been implicated in the control of gene expression unrelated to cell growth and reproduction, such as lipid metabolism in the liver.[2,3] Estrogen is also involved in the initiation and progression of tumors in these organs including mammary and ovarian tumors.[4,5] Abnormal uncontrolled cellular proliferation can be caused by excessive secretion of the hormone, hyperresponsiveness of the estrogen target cells, or abnormalities in signal transduction pathways. Estrogens regulate the transcription of target genes by binding to estrogen receptors (ERs) that function as ligand-dependent transcription factors. ERs consist of several distinct functional modular domains and transcriptional activation of ER-regulated genes is mediated by two activation domains, AF-1 at the N terminus and AF-2 in the ligand-binding domain.[6,7] Ligand-bound ERs dissociate from heat shock proteins, translocate into the nucleus, homodimerize, and bind to DNA at an estrogen response element (ERE), which then cause interaction with basal transcriptional machinery in the promotor to initiate gene transcription.[8,9] Alternatively, ER can also interact with other transcription factors and steroid receptor coregulators stimulating or inhibiting activation of estrogen responsive target genes.[8] Some of these regulatory factors and other members of nuclear receptor superfamily have been implicated to modulate ER activity in human hepatic cell line and the development of breast cancer.[9–11] Moreover, recent studies suggest that ERs may have functional interactions with a wide range of other signaling molecules. Epidermal growth factor, transforming growth factor, and dopamine promote ER-dependent gene transcription of estrogen responsive genes.[10,12] In addition, ligand-bound ER can interact with other transcription factors, such as AP1 and Sp1.[10] These complexes bind to DNA through direct interaction of the ER-associated transcription factors, thus influencing the transcription activity of genes that do not contain EREs but rather have recognition sites specific for AP1 or Sp1.[10]

Recently, three different kinds of xenobiotic nuclear receptors (XNRs), steroid and xenobiotic receptor (SXR), peroxisome proliferator-activated receptor-γ (PPAR-γ), and constitutive androstane receptor (CAR) have been identified and drawn to our attention as possible candidates for regulating estrogen-mediated signal transduction pathways. These xenobiotic receptors play important biological roles in modulating steroid hormone homeostasis, lipid metabolism, and drug metabolism in the body by mediating cellular responses not only to the endogenous compounds, such as hormones, but also to the exogenous chemical compounds, such as drugs and environmental pollutants.[13–18]

SXR is activated by a variety of xenobiotic compounds, including drugs, such as rifampicin, steroid receptor agonists, and antagonists, such as estrogen and tamoxifen, and bioactive dietary compounds, such as phytoestrogens.[14,19] SXR is highly expressed in liver, the major expression site of steroid and xenobiotic-metabolizing enzymes.[14] SXR binds as a heterodimer with the retinoid X receptor (RXR) to the putative SXR response elements that are found in the promotor regions of these enzymes including *CYP3A4*, *CYP2A*, and *CYP2C*, suggesting that expression of these enzymes is induced by SXR.[14] SXR is also expressed in breast carcinoma tissues and its expression is correlated with cellular proliferation thus suggested to play important roles in the biology of human breast cancers.[20] The structural diversity of the compounds that activate SXR and induce transcription of genes related to xenobiotic metabolism including *CYP3A4* and the fact that some of these compounds were known to interact with classic steroid hormone receptors suggested that multiple signal transduction cross-talk might be involved.[21]

PPARs regulate the expression of genes involved in lipid and glucose metabolism, adipogenesis, inflammation, and cellular proliferation.[22] This orphan nuclear receptor family includes three kinds of closely related gene products, α, β, and γ and they show different expression patterns among tissues suggesting distinct biological roles.[18] PPARs respond to a broad class of structurally diverse ligands, including xenobiotic chemicals called peroxisome proliferators and both naturally occurring and synthetic fatty acids.[22] Ligand-bound PPARs heterodimerize with RXR and bind to specific peroxisome proliferator response elements (PPRE) in the upstream of a number of target genes.[22] In addition, PPAR-α binds to ERE with high affinity to inhibit expression of estrogen responsive genes that affect lipoprotein and fatty acid metabolism, whereas PPAR-γ represses cellular proliferation and promotes differentiation by its agonist, thiozolidinedione (TZD), or endogenous ligand, prostaglandin, in both breast cancer and liposarcoma cells.[18,23] PPAR ligands have also been reported to inhibit proliferation and stimulate apoptosis of breast cancer cells when they are treated with retinoids of either 9-*cis* retinoic acid or *all-trans* retinoic acid.[16]

CAR influences steroid homeostasis through transcriptional regulation of CYP2B genes, which are steroid hydroxylases.[24] CAR is sequestered in the cytoplasm, and treatment with agonists, such as phenobarbital and TCPOBOP, results in the translocation of CAR into the nucleus, where it binds to its cognate recognition sites as a heterodimer with RXR.[25,26] The transcriptional regulation by CAR is not limited to cytochrome P450 genes and influences on a broad spectrum of genes with diverse cellular functions, such as PPRE-regulated genes involved in peroxisomal oxidation and genes regulated by retinoic acid.[27] Thus, CAR may involve cross-regulation with other cellular signaling pathways. Previous studies also demonstrated that CAR/RXR bound to the ERE and CAR significantly inhibits ER-mediated transcriptional activity of the vitellogenin B1 promotor as well as a synthetic ERE-driven promotor by a mechanism in which CAR squelches limiting amounts of p160 coactivators, such as steroid receptor coactivator-1 (SRC-1) and glucocorticoid receptor interacting protein-1 (GRIP-1) which are essential for ER action.[28] These studies suggest that xenobiotics may influence ER function of lipid metabolism, steroid homeostasis, and tumorigenesis in both liver and female reproductive tissues.

Whereas recent studies demonstrated that CAR inhibits ER-mediated transactivation of both endogenous and synthetic estrogen responsive promotor in Hep G2, the effects of SXR and PPAR-γ on the ER transactivation have not been examined in detail in Hep G2. Although the effects of SXR and PPAR-γ on the ER transactivation were examined in the synthetic 4ERE in which SXR but not PPAR-γ represses ER transactivation in Hep G2,[28] the effects of these XNRs on the endogenous estrogen responsive promotor remain to be determined. Effects of CAR, SXR, and PPAR-γ on ER transactivation were also examined in three different kinds of breast cancer cell lines.[29,30] However, except in MCF-7, studies for the effects of these xenobiotic receptors on ER transactivation were limited either in single dose response (MDA-MB-231) or with CAR only (MCF-7-K3). And there is presently no report on the effects of CAR, SXR, and PPAR-γ on ER-mediated transactivation in ovarian-derived CHO-S cells.

Accordingly, the purpose of this article was to further examine the effects of the endogenous vitellogenin estrogen responsive promotor on the SXR- and PPAR-γ-modulated ER transactivation in human hepatoma cells, Hep G2, and either dose-dependent or single dose effects of SXR, PPAR-γ, and CAR in two different breast cancer cell lines, MDA-MB-231 and MCF-7-K3, and ovarian-derived cell line CHO-S, respectively, on the ER-mediated transactivation of the synthetic (4ERE)-tk-luciferase reporter.

MATERIALS AND METHODS

Ligand and Plasmid Construction

The ligands for ER, moxestrol and 17β-estradiol, for SXR, rifampicin, and for PPAR-γ, prostaglandin were purchased from Sigma Chemical Company (St. Louis, MO). The synthetic ligand for PPAR-γ, rosiglitazone (BRL49653), was obtained from Glaxo-Smith-Kline (Philadelphia, PA). The ligand for CAR, TCPOBOP, was obtained from B. Kemper (University of Illinois,

Urbana-Champaign, IL). Mouse CAR1 cDNA was isolated from a mouse liver cDNA library by polymerase chain reaction and verified by sequencing. For expression of CAR in mammalian cells, a *Bam*H1/*Eco*R1 fragment containing the CAR1 cDNA isolated from pGEX2TK-CAR was inserted into pcDNA3 (Invitrogen, Carlsbad, CA) digested with the same enzymes to produce pcDNA3-CAR. The mammalian expression plasmids, pCMX-SXR and pSG5-PPAR-γ, were from B. Kemper and plasmids pCMV-hERα, 4ERE-tk-luceferase, and vitellogenin-luciferase were from J. K. Kemper (University of Illinois).

Cell Culture

The cell lines used in this study were a human hepatoma cell line, HepG2, two types of human breast cancer cell lines, MDA-MB-231 and MCF-7-K3, and an ovarian cell line, CHO-S cells. Each cell type was maintained in an appropriate culture medium selected for optimal growth and transfection. HepG2 cells were maintained in Dulbecco's modified Eagle's medium (DMEM; Gibco, Grand Island, NY) supplemented with 10% charcoal dextran-striped calf serum (CDCS), 100 units/mL penicillin, and 100 μg/mL streptomycin. The MDA-MB-231 cell line that does not express ER was maintained in Leibovitz's L-15 medium (Gibco) supplemented with 5% CDCS, 10 μg/mL phenol red, 100 units/mL penicillin, 100 μg/mL streptomycin, 25 μg/mL gentamicin, 3.75 μg/mL hydrocortisone, 6 ng/mL bovine insulin, and 16 μg/mL glutathione. MCF-7-K3 cells were maintained in Eagle's minimum essential medium (MEM) supplemented with 5% CDCS, 100 units/mL penicillin, 100 μg/mL streptomycin, 25 μg/mL gentamicin, 4 ng/mL hydrocortisone, and 6 ng/mL bovine insulin. And CHO-S cells were maintained in DME/F-12 medium (Gibco) supplemented with 5% CDCS, 100 units/mL penicillin, and 100 μg/mL streptomycin.

Transfection

For transfection, the human breast cancer cells were cultured until about 80% confluency and weaned for 3 days in the absence of phenol red, hydrocortisone, and insulin. Cells were transfected using LipofectAMINE 2000 (Invitrogen) as described.[31] Briefly, cells were seeded in 24-well plates up to about 80% confluency, and at 12 to 16 h post seeding 250 ng of ATC4 (4ERE-tk-luciferase) or 500 ng of vitellogenin-luciferase vector, 10 ng of pRL-SV40 for measuring the transfection efficiency, and varying amounts of expression plasmids for ER, SXR, PPAR-γ, and CAR were added to each well. In order to examine the effects of ligand for each receptor, 10 nM of moxestrol or 17β-estradiol, 10 μM of rifampicin, 20 μM of prostaglandin or 1 μM of BRL49653, and

5 μM of TCPOBOP, which are ligands for the receptors, ER, SXR, PPAR-γ, and CAR, respectively, were added. The cells were incubated for 16–24 h after transfection, fresh medium containing the ligands was added, and the cells were incubated for an additional 24 h.

Dual Luciferase Assay

The cells were lysed and luciferase activities were determined by the Dual Luciferase Reporter assay system (Promega Biotech, Madison, WI). For each sample, the background of extracts from untreated cells was subtracted and the firefly luciferase values were normalized by dividing by *Renilla* luciferase values.

RESULTS

Effects of the Endogenous Vitellogenin Promoter on the SXR- and PPAR-γ-Modulated ER Transactivation in Hep G2

To investigate whether the endogenous vitellogenin estrogen responsive promotor can affect the SXR- and PPAR-γ-modulated ER transactivation, transient transfection experiments with a reporter containing the *Xenopus* vitellogenin B1 promotor were carried out in human hepatoma cells. The vitellogenin-luciferase reporter contains a 610-bp fragment from the 5′-flanking region of the the *Xenopus* vitellogenin B1 genomic clone and contains a functional ERE in the fragment.[28] HepG2 cells were cotransfected with vitellogenin-luciferase reporter and ER expression plasmid in the presence of SXR (100 ng or 500 ng), PPAR-γ (100 ng or 500 ng), or CAR (20 ng or 100 ng). Moxestrol was used as a ligand for the ER in Hep G2 cells because moxestrol is resistant to metabolism in hepatic cells. As shown in FIGURE 1, treatment with 10 nM of moxestrol induced the ER-mediated vitellogenin–luciferase activity over 100-fold when ER and expression plasmid of SXR, PPAR-γ, or CAR were cotransfected. Ligands for the xenobiotic receptors, rifampicin, BRL49653, and TCPOBOP, did not increase luciferase activity and thus are not ligands for ER-α in this assay system, nor do xenobiotic receptors activate the vitellogenin B1 promotor. Consistent with the previous report,[28] ligand-activated CAR significantly repressed the moxestrol-induced ER-mediated transactivation of the vitellogenin B1 promotor regardless of the dose of exogenously transfected CAR (FIG. 1). In contrast, increasing amounts of transfected SXR expression plasmid did not have any effects on the moxestrol-induced ER-mediated transactivation. In addition, it is interesting to note that treatment with 500 ng of PPAR-γ potentiated the ER-mediated transactivation of the vitellogenin B1 promotor (FIG. 1).

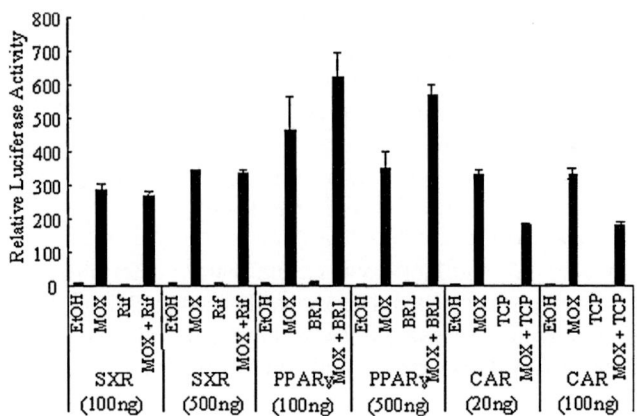

FIGURE 1. ER-mediated transcriptional activity of the vitellogenin promotor can be modulated differently by xenobiotic orphan nuclear receptors, SXR, PPAR-γ, and CAR in Hep G2 cells. Hep G2 cells were transfected with 500 ng of a vector containing vitellogenin promotor fused to the luciferase reporter gene (vitellogenin–luciferase), 10 ng of pRLSV40, 5 ng of CMV-hER-α in the presence of pCMX-SXR (100 or 500 ng), pSG5-PPAR-γ (100 or 500 ng), or pcDNA3-CAR (20 or 100 ng). The ligands were added for 24 h after transfection as indicated: 10 nM of moxestrol (MOX) for ER, 10 μM of rifampicin (Rif) for SXR, 1 μM of BRL49653 (BRL) for PPAR-γ, or 5 μM of TCPOBOP (TCP) for CAR. The cells were harvested for dual luciferase assays. The values for firefly luciferase were normalized by dividing by *Renilla* luciferase values. The standard errors of the mean were calculated from three independent transfection experiments.

Dose-Dependent Effects of XNRs, SXR, PPAR-γ, and CAR on the ER Transactivation in Human Breast Cancer Cells

Two different human breast cancer cell lines, MDA-MB-231 and MCF-7-K3, were used in the reporter assay in order to determine modulatory effects of SXR, PPAR-γ, and CAR on the ER-mediated transactivation of the synthetic 4ERE-tk-luciferase. Breast cancer cells were weaned for 3 days before transfection experiments were performed in order to eliminate the estrogenic effects of phenol red and other exogenous growth promoting hormones that can possibly affect ER signal transduction pathways, thus maximizing transactivation effects of ER by estrogen treatment. Human breast cancer cells were transfected with 4ERE-tk-luciferase reporter plasmid in the absence or presence of increasing amounts (20–500 ng) of the xenobiotic receptor expression plasmids, pCMX-SXR, pSG5-PPAR-γ, or pcDNA3-CAR. Because MDA-MB-231 cells do not express endogenous ER, these cells were also cotransfected with the expression plasmid for hER-α, pCMV-hER-α. Whereas, MCF-7-K3 cells that express endogenous ER were not transfected with the ER expression plasmid.

In MDA-MB-231 cells, treatment with 10 nM of 17β-estradiol induced the ER-mediated 4ERE-tk-luciferase activity over 10-fold (FIG. 2). Treatment with ligands alone for the xenobiotic receptors, rifampicin, prostaglandin, and TCPOBOP, did not increase luciferase activity, and thus they are not the ligands for the ER-α in this cell type. Whereas increasing doses of both SXR and PPAR-γ did not have any effects on the 17β-estradiol-induced ER-mediated transcriptional activity in the presence of the corresponding xenobiotic receptor ligands, a low dose of CAR (20 ng) modestly enhanced the 17β-estradiol-induced ER transactivation (FIG. 2).

As in the MDA-MB-231, treatment in MCF-7-K3 cell with 10 nM of 17β-estradiol also induced the endogenous ER-mediated 4ERE-tk-luciferase activity over 10-fold (FIG. 3). And treatment with ligands alone for the xenobiotic receptors, rifampicin, prostaglandin, and TCPOBOP, did not increase luciferase activity. In MCF-7-K3 cell line, increasing doses of SXR, PPAR-γ, and CAR did not have any significant effects on the 17β-estradiol-induced ER-mediated transcriptional activity in the presence of the corresponding xenobiotic receptor ligands (FIG. 3).

Effects of XNRs, SXR, PPAR-γ, and CAR on the ER Transactivation in CHO-S Cell

To investigate whether functional cross-talk occurs between the ER and the XNRs, SXR, PPAR-γ, and CAR, in an ovarian-derived cell line, transient

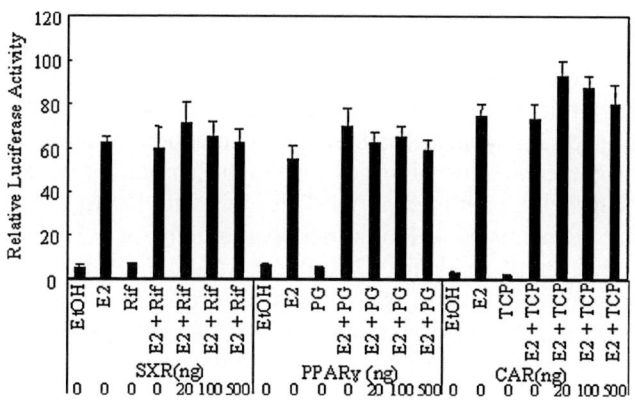

FIGURE 2. Effects on the ER transactivation by XNRs, SXR, PPAR-γ, and CAR in MDA-MB-231 cells. MDA-MB-231 cells were transfected with 250 ng of 4ERE-tk-luciferase vector, 10 ng of pRLSV40 for the internal control for transfection efficiency, and 50 ng of pCMV-hER-α in the absence or presence (20–500 ng) of pCMX-SXR, pSG5-PPAR-γ, or pcDNA3-CAR. The ligands were added for 24 h after transfection as indicated: 10 nM of 17β-estradiol (E2) for ER, 10 μM of rifampicin (Rif) for SXR, 20 μM of prostaglandin (PG) for PPAR-γ, or 5 μM of TCPOBOP (TCP) for CAR. The cells were harvested for dual luciferase assays. The values for firefly luciferase were normalized by dividing by *Renilla* luciferase values. The standard errors of the mean were calculated from three independent transfection experiments.

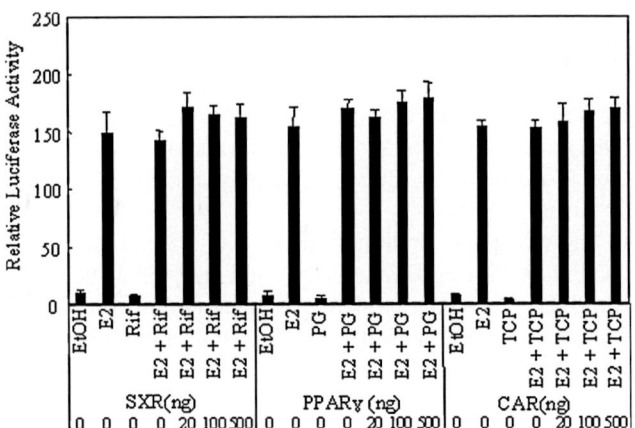

FIGURE 3. Effects on the ER transactivation by XNRs, SXR, PPAR-γ, and CAR in MCF-7-K3 cells. MCF-7-K3 cells were transfected with 250 ng of 4ERE-tk-luciferase vector and 10 ng of pRLSV40 for the internal control for transfection efficiency in the absence or presence (20–500 ng) of pCMX-SXR, pSG5-PPAR-γ, or pcDNA3-CAR. The ligands were added for 24 h after transfection as indicated: 10 nM of 17β-estradiol (E2) for ER, 10 μM of rifampicin (Rif) for SXR, 20 μM of prostaglandin (PG) for PPAR-γ, or 5 μM of TCPOBOP (TCP) for CAR. The cells were harvested for dual luciferase assays. The values for firefly luciferase were normalized by dividing by *Renilla* luciferase values. The standard errors of the mean were calculated from three independent transfection experiments.

transfection experiments were performed in CHO-S cells and the modulatory effects on the ER-mediated transactivation by the xenobiotic receptors were examined. A human ER mammalian expression plasmid was cotransfected with the expression plasmids for the orphan nuclear receptors and a synthetic 4ERE-tk-luciferase reporter. Treatment with 10 nM of 17β-estradiol induced the ER-mediated 4ERE-tk-luciferase activity over 10-fold when ER and the expression plasmids for SXR, PPAR-γ, and CAR were cotransfected (FIG. 4). And treatment with ligands, rifampicin, prostaglandin, and TCPOBOP, in the presence of the corresponding xenobiotic receptors did not induce the luciferase activity thus indicating that ligand-bound xenobiotic receptor did not transactivate 4ERE. In addition, the xenobiotic receptors, SXR, PPAR-γ, and CAR did not have any effects on the 17β-estradiol-induced ER-mediated transcriptional activity of the synthetic 4ERE-tk-luciferase (FIG. 4).

DISCUSSION

Whereas the results from this study are in consistence with the previous reports in the respect that CAR suppresses the ER-mediated transcriptional activity in both the endogenous ERE-containing vitellogenin B1 promotor and

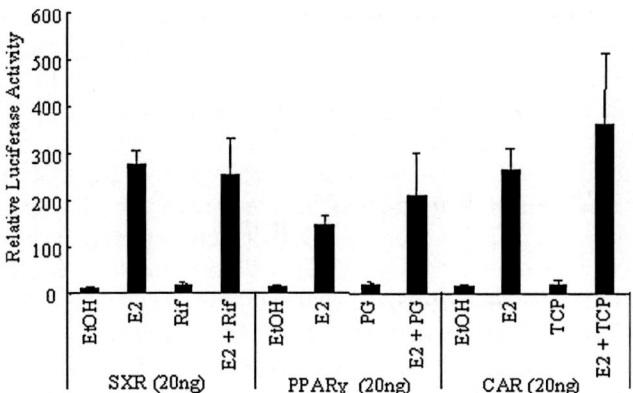

FIGURE 4. Effects on the ER transactivation by XNRs, SXR, PPAR-γ, and CAR in CHO-S cells. CHO-S cells were transfected with 250 ng of 4ERE-tk-luciferase vector, 10 ng of pRLSV40 for the internal control for transfection efficiency, and 5 ng of pCMV-hER-α in the presence of 20 ng of pCMX-SXR, pSG5-PPAR-γ, or pcDNA3-CAR. Ligands were added for 24 h after transfection as indicated: 10 nM of moxestrol (MOX) for ER, 10 μM of rifampicin (Rif) for SXR, 20 μM of prostaglandin (PG) for PPAR-γ, or 5 μM of TCPOBOP (TCP) for CAR. The cells were harvested for dual luciferase assays. The values for firefly luciferase were normalized by dividing by *Renilla* luciferase values. The standard errors of the mean were calculated from three independent transfection experiments.

the synthetic 4ERE promotor in HepG2 cells,[28,29] this study also suggests an important role of the estrogen responsive *cis*-element in modulating ER-mediated transactivation by XNRs. In contrast to the effects on the synthetic promotor as reported in the previous studies in which SXR repressed but PPAR-γ did not have any effects on the ER-mediated transactivation of 4ERE,[28] this study demonstrates that PPAR-γ potentiates but SXR does not have any effects on the ER transactivation of the endogenous estrogen responsive vitellogenin B1 promoter in Hep G2. These results indicate that in Hep G2 cells whereas CAR has inhibitory functional cross-talk with ER, PPAR-γ may have stimulatory cross-talk with ER at least on the endogenous estrogen responsive vitellogenin B1 promoter in regulating ER-mediated gene transcription. The molecular mechanisms by which CAR and PPAR-γ modulate ER-mediated transactivation are not known, but recent studies provide evidence that estrogen action can be modulated by functional cross-talk between ER and these XNRs. The xenobiotic orphan receptor CAR, which is activated by diverse xenobiotic chemicals, was demonstrated to antagonize ER-mediated transcriptional activity by squelching limiting amounts of p160 coactivator.[28] Alternatively, CAR may also regulate estrogen action by its induction of the cytochrome P450 xenobiotic-metabolizing enzymes that alter hepatic metabolism of estrogens. PPAR/RXR heterodimer was reported to bind to the vitellogenin A2 ERE and furthermore, natural ERE-containing promotors, including the pS2, very-low-density apolipoprotein II and vitellogenin A2 genes, revealed considerable differences in the binding of PPAR/RXR heterodimers to these EREs.[32] Thus, the promotor structure of the EREs may play an important role(s) in modulating ER-mediated transactivation by XNRs.

Although there is extensive evidence that estrogen plays a crucial role in the initiation and progression of breast cancer and in ovarian physiology, the molecular mechanism(s) of estrogen-mediated breast tumorigenesis and ovarian function is not clearly understood. This study examined the effects of the XNRs, SXR, PPAR-γ, and CAR, in different doses on the transcriptional activity of ER in two different kinds of human breast cancer cells, MDA-MB-231 and MCF-7-K3, and an ovarian-derived cell line, CHO-S. SXR and PPAR-γ did not have any effects on the estrogen-dependent ER transactivation of the synthetic 4ERE-tk-luciferase in both types of breast cancer cells as well as CHO-S cell. Whereas CAR did not influence the ER-mediated transactivation in MCF-7-K3 and CHO-S cells, a low dose of CAR modestly stimulated ER transactivation of the synthetic 4ERE in the MDA-MB-231 breast cancer cell line. This is in contrast to the previous report that CAR and SXR repress ER-mediated transactivation of the synthetic 4ERE in Hep G2 cells.[28] These results suggest different cross-talk interactions between ER and xenobiotic orphan nuclear receptors, SXR, PPAR-γ, and CAR, depending on the estrogen target cell types, such as liver, mammary gland, and ovary. Differentiated cells respond variably to the same chemical stimulation and this may be caused by receptor specificity and distinct expression profiles of different receptors and/or transcription coregulators, which then lead to receptor cross-talk mechanisms unique to the target cell. For example, different breast cancer cell lines have distinct characteristics in the level of ER expression and estrogen dependence for tumor cell growth: MCF-7 expresses ER and is estrogen dependent for cellular proliferation. Whereas, MCF-7-K3 expresses very high level of ER and estrogen independent for its growth, and MDA-MB-231 does not express ER.[30] In addition, XNRs, CAR, SXR, PPAR-γ, and their heterodimeric partner RXR, are differentially expressed in five human breast cancer cell lines.[28]

It is postulated that differential expression of ER, XNRs, CAR, SXR, and PPAR-γ, and/or other transcriptional coregulators in different breast cancer cell lines and CHO-S cell from HepG2 may contribute to different transcriptional regulatory mechanism(s) mediated by ER. However, it cannot still be ruled out that ERE of the endogenous estrogen target genes may be influenced by activation of the XNRs in breast cancer cells and/or CHO-S cells. One of the reasons for this possibility is that chromatin structures of the endogenous gene promotors are different from the exogenously transfected plasmid DNAs and the accessibility of transcription factors to these response elements may be influenced by the structural status of the chromatins. Further studies for the identification of nuclear receptors that can regulate estrogen action in the signal transduction pathways involved in growth and proliferation of breast cancer cells and ovarian function will provide important information toward understanding the molecular mechanism(s) of cross-talk interactions between ER and other XNRs in these processes. In conclusion, this study demonstrates that XNRs modulate ER-mediated transactivation depending on the EREs and estrogen target cell types.

ACKNOWLEDGMENTS

The author gratefully acknowledges Drs. B. Kemper and J. K. Kemper for providing materials for this study.

REFERENCES

1. KULA, K., J. SLOWIKOWSKA-HILCZER, R. WALCZAK-JEDRZEJOWSKA, et al. 2005. Physiological significance of estrogens in men-breakthrough in endocrinology. Endokrynol. Pol. **56**: 314–321.
2. ARCHER, T.K., S.P. TAM & R.G. DEELEY. 1986. Kinetics of estrogen-dependent modulation of apolipoprotein A-I synthesis in human hepatoma cells. J. Biol. Chem. **261**: 5067–5074.
3. CROSTON, G.E., L.B. MILAN, K.B. MARCHKE, et al. 1997. Androgen receptor-mediated antagonism of estrogen-dependent low density lipoprotein receptor transcription in cultured hepatocytes. Endocrinology **138**: 3779–3786.
4. BERNSTEIN, L. 2006. The risk of breast, endometrial and ovarian cancer in users of hormonal preparations. Basic Clin. Pharmacol. Toxicol. **98**: 288–296.
5. SEOL, W., B. HANSTEIN, M. BROWN, et al. 1998. Inhibition of estrogen receptor action by the orphan receptor SHP (short heterodimer partner). Mol. Endocrinol. **12**: 1551–1557.
6. EVANS, R.M. 1988. The steroid and thyroid hormone receptor superfamily. Science **240**: 889–895.
7. BEATO, M. 1989. Gene regulation by steroid hormones. Cell **56**: 335–344.
8. PAECH, K., P. WEBB, G.G. KUIPER, et al. 1997. Differential ligand activation of estrogen receptors ERα and ERβ at AP1 sites. Science **277**: 1508–1510.
9. JOHANNA, M., B.F. GEORGE, A.Y. SOPHIA, et al. 1993. Transcriptional activation by the estrogen receptor requires a conformational change in the ligand binding domain. Mol. Endocrinol. **7**: 1266–1274.
10. SOMMER, S. & S.A.W. FUQUA. 2001. Estrogen receptor and breast cancer. Cancer Biol. **11**: 339–352.
11. MAO, C. & D.J. SHAPIRO. 2000. A histone deacetylase inhibitor potentiates estrogen receptor activation of a stably integrated vitellogenin promotor in HepG2 cells. Endocrinology **141**: 2361–2369.
12. HARRINGTON, W.R., S. SHENG, D.H. BARNETT, et al. 2003. Activation of estrogen receptor alpha- and beta-selective ligands at diverse estrogen responsive gene sites mediating transactivation or transrepression. Mol. Cell Endocrinol. **206**: 13–22.
13. BLUMBERG, B. & R.M. EVANS. 1998. Orphan nuclear receptors-new ligands and new possibilities. Genes Dev. **15**: 3149–3155.

14. BLUMBERG, B., W.J. SABBAGH, H. JUGUILON, et al. 1998. SXR, a novel steroid and xenobiotic-sensing nuclear receptor. Genes Dev. **12:** 3195–3205.
15. CLAY, C.E., A.M. NAMEN, G. ATSUMI, et al. 1999. Influence of J series prostaglandins on apoptosis and tumorigenesis of breast cancer cells. Carcinogenesis **20:** 1905–1911.
16. ELSTNER, E.C., K. MULLER, K. KOSHIZUKA, et al. 1998. Ligands for peroxisome proliferator-activated receptor γ and retinoic acid receptor inhibit growth and induce apoptosis of human breast cancer cells in vitro and in BNX mice. Proc. Natl. Acad. Sci. USA **21:** 8806–8811.
17. FORMAN, B.M., I. TZAMELI, H.S. CHOI, et al. 1998. Androstane metabolites bind to and deactivate the nuclear receptor CAR-β. Nature **395:** 612–615.
18. KLIEWER, S.A., J.M. LEHMANN & T.M. WILLSON. 1999. Orphan nuclear receptors: shifting endocrinology into reverse. Science **284:** 757–760.
19. TAKESHITA, A., N. KOIBUCHI, J. OKA, et al. 2001. Bisphenol-A, an environmental estrogen, activates the human orphan nuclear receptor, steroid and xenobiotic receptor-mediated transcription. Eur. J. Endocrinol. **145:** 513–517.
20. MIKI, Y., T. SUZUKI, K. KITADA, et al. 2006. Expression of the steroid and xenobiotic receptor and its possible target gene, organic anion transporting polypeptide-A, in human breast carcinoma. Cancer Res. **66:** 535–542.
21. LEHMANN, J.M., D.D. MCKEE, M.A. WATSON, et al. 1998. The human orphan nuclear receptor PXR is activated by compounds that regulate *CYP3A4* gene expression and cause drug interactions. J. Clin. Invest. **102:** 1016–1023.
22. KASSAM, A., C.J. WINROW, F. FERNANDEZ-RACHUBINSKI, et al. 2000. The peroxisome proliferator response element of the gene encoding the peroxisomal ß-oxidation enzyme enoyl-CoA hydratase/3-hydroxyacyl-CoA dehydrogenase is a target for constitutive androstane receptor ß/9-*cis*-retinoic acid receptor-mediated transactivation. J. Biol. Chem. **275:** 4345–4350.
23. YANG, W., C. RACHEZ & L.P. FREEDMAN. 2000. Discrete roles for peroxisome proliferator-activated receptor γ and retinoid X receptor in recruiting nuclear receptor coactivators. Mol. Cell. Biol. **20:** 8008–8017.
24. WEI, P., J. ZHANG, M. EGAN-HAFLEY, et al. 2000. The nuclear receptor CAR mediates specific xenobiotic induction of drug metabolism. Nature **407:** 920–923.
25. TZAMELI, I., P. PISSIOS, E.G. SCHUETZ, et al. 2000. The xenobiotic compound 1,4-bis[2-(3,5-dichloropyridoxyl)]benzene is an agonist ligand for the nuclear receptor CAR. Mol. Cell Biol. **20:** 2951–2958.
26. KAWAMOTO, T., T. SUEYOSHI, I. ZELKO, et al. 1999. phenobarbital-responsive nuclear translocation of the receptor CAR in induction of the CYP2B gene. Mol. Cell Biol. **19:** 6318–6322.
27. SUGATANI, J., H. KOJIMA, A. UEDA, et al. 2001. The phenobarbital response enhancer module in the human bilirubin UDP-glucuronosyltransferase UGT1A1 gene and regulation by the nuclear receptor CAR. Hepatology **33:** 1232–1238.
28. MIN, G., H. KIM, Y. BAE, et al. 2002. Inhibitory cross-talk between estrogen receptor (ER) and constitutively activated androstane receptor (CAR). J. Biol. Chem. **277:** 34626–34633.
29. MIN, G. 2003. Comparison and analysis between human breast cancer cells and hepatoma cells for the effects of xenobiotic nuclear receptors (constitutive androstane receptor, steroid and xenobiotic receptor, and peroxisome-proliferator-activated receptor γ) on the transcriptional activity of estrogen receptor. Korean J. Life Sci. **13:** 314–323.

30. MIN, G., J. KIM, Y. CHO, *et al.* 2003. Effect of constitutive androstane receptor on transcriptional activity of estrogen receptor in estrogen independent breast cancer cell line MCF-7-K3. J. Indust. Tech. Res. Inst. **10:** 259–267.
31. MIN, G., J.K. KEMPER & B. KEMPER. 2002. Glucocorticoid receptor-interacting protein I mediates ligand-independent nuclear translocation and activation of constitutive androstane receptor *in vivo*. J. Biol. Chem. **277:** 25356–26363.
32. KELLER, H., F. GIVEL, M. PERROUD, *et al.* 1995. Signaling cross-talk between peroxisome proliferator-activated receptor/retinoid X receptor and estrogen receptor through estrogen response elements. Mol. Endocrinol. **9:** 794–804.

Effects of TK Promotor and Hepatocyte Nuclear Factor-4 in CAR-Mediated Transcriptional Activity of Phenobarbital Responsive Unit of *CYP2B* Gene in Monkey Kidney Epithelial-Derived Cell Line COS-7

GYESIK MIN

Department of Microbiological Engineering, Jinju National University, Jinju, Gyeongsangnam-Do, 660-758 Korea

ABSTRACT: Previous studies reported that constitutive androstane receptor (CAR) does not transactivate phenobarbital responsive unit (PBRU)2C1luciferase reporter gene in COS cells in which endogenous *CYP2B1* gene is not induced with PB. In order to understand molecular mechanism(s) whereby PBRU is transactivated, this article determined if the use of strong thymidine kinase (TK) promotor rather than the minimal *CYP2C1* promotor, and hepatocyte nuclear factor-4 (HNF-4) can affect CAR-mediated transactivation of PBRU in the monkey kidney epithelial-derived COS-7 cells. To examine CAR-mediated transactivation, cultured COS-7 cells were transfected with CAR expression plasmid, pEGFP-mCAR1, and confirmed for high level of the protein expression. In COS-7 cells, TK promotor induced CAR-mediated PBRU transactivation in a dose-dependent manner. Whereas expression of HNF-4 slightly promoted PBRU transactivation with low amount of CAR transfected, it repressed PBRU transactivation in a dose-dependent manner with high amount of CAR. Consistent with the previous reports in Hep G2 cells, CAR transactivated PBRU2C1luciferase in a dose-dependent manner and this CAR-mediated transactivation required functional NR-1 and NF-1 sites. However, HNF-4 did not affect CAR-mediated PBRU transactivation in Hep G2 cells. These results suggest that proximal promotor and a *trans*-acting factor, HNF-4, can affect CAR-mediated transactivation of PBRU in COS-7 cells.

KEYWORDS: transactivation; TK promotor; HNF-4; constitutive androstane receptor; PBRU; *CYP2B*; COS-7

Address for correspondence: Dr. Gyesik Min, Department of Microbiological Engineering, Jinju National University, Jinju, Gyeongsangnam-Do, 660-758 Korea. Voice: 055-751-3396; fax: 055-751-3399.
 e-mail: g-min@jinju.ac.kr

Ann. N.Y. Acad. Sci. 1091: 258–269 (2006). © 2006 New York Academy of Sciences.
doi: 10.1196/annals.1378.072

INTRODUCTION

Cytochrome P450 (P450s), a family of heme-containing monooxygenase proteins, are present in representatives of all classes of living organisms, from bacteria to humans[1] and play an important role(s) in the metabolism of xenobiotics and in the biosynthesis and catabolism of endogenous compounds.[2–7] The expression of these P450 genes shows tissue and developmental specificity and is regulated by diverse chemicals, such as endogenous hormones, growth factors, cytokines, and structurally different foreign compounds.[8] A characteristic of the xenobiotic-metabolizing enzymes is that subsets of the P450s are induced by a variety of chemicals.[5,7] This induction response has a major impact on P450-dependent drug metabolism and drug interactions, on the toxicity and carcinogenicity of xenobiotics, and on the activity and metabolism of endogenous hormones.[9] Phenobarbital (PB) is a representative prototype for a large number of structurally and functionally diverse xenochemicals that induce *CYP2B* genes.[10] *CYP2B* is a large gene subfamily that encodes versatile enzymes for the metabolism of xenobiotics and steroid hydroxylation. Even closely related isoforms display distinct sex- and tissue-specific regulation.[11] PB can upregulate several other hepatic enzymes involved in xenobiotic metabolism, such as glutathione *S*-transferase and UDP-glucuronosyltransferases.[12]

Considerable progress has been made in understanding the molecular mechanisms of PB induction of *CYP2B* genes. Earlier studies demonstrated that a sequence at about –2.3 kb in the *CYP2B2* gene contained the properties of a PB-responsive enhancer either in its normal context or fused to a heterologous gene in transient transfection assays.[13] Mutation analysis established that the PB-responsive enhancer was a complex enhancer that contained multiple redundant regulatory binding sites, including two nuclear receptor binding sites, NR-1 and NR-2, and an NF-1 site.[14–16] The nuclear receptor, constitutive androstane receptor (CAR), has been identified as the mediator of induction of *CYP2B* genes in the liver by PB. CAR is translocated to the nucleus of the liver cell from PB-treated animals, and binds to both NR-1 and NR-2 sites as a CAR/RXR heterodimer to transactivate PB-responsive unit (PBRU) of *CYP2B* gene.[17–19] CAR transactivation may also require coactivator proteins. The p160 steroid receptor coactivator-1 (SRC-1) has been shown to bind to CAR, and the binding was increased by the agonist, TCPOBOP, and decreased by antagonists, androstanes.[20] In addition, glucocorticoid receptor interacting protein-1 (GRIP-1) was demonstrated to interact with CAR and mediate ligand-independent nuclear translocation and activation of CAR *in vivo*.[19] Thus, PB induction may involve translocation of CAR to the nucleus and activation of CAR.

The mechanism by which CAR transactivates the PBRU still remains to be established. Previous studies reported that transient transfection of CAR does not transactivate *CYP2B1* PBRU fused to the minimal *CYP2C1* promotor/firefly luciferase reporter gene in COS cells in which the endogenous

$CYP2B1$ is not induced by PB.[21] This suggests that factors other than CAR are required in inducing PBRU activation and the expression of these factors may be different among different tissues. In order to identify those factors that may be involved in CAR-mediated PBRU transactivation, kidney epithelial-derived COS-7 cells can be served as a useful cellular model for introducing and screening different candidate factors from exogenous expression plasmids. Hepatocyte nuclear factor-4 (HNF-4) is a liver-enriched orphan receptor and has been suggested to serve as a common regulator for the liver-specific transcriptional activation of many CYP2 genes.[22] The $CYP2B1/B2$ sequence, ATCAAAAGCTGGAGG, has some similarity to the HNF-4 site in CYP genes (matching nucleotides underlined) and has been implicated in PB action by *in vitro* transcription assay and changes in protein binding in response to PB.[22] On the basis of these studies and the presence of this sequence in other PB-responsive genes, it was proposed as a potential PB-responsive element.[22,23]

Accordingly, the purpose of this article was to determine if the use of strong thymidine kinase (TK) promotor and HNF-4 can affect CAR-mediated transactivation of PBRU in monkey kidney epithelial-derived COS-7 cells.

MATERIALS AND METHODS

Plasmid Construction

Mouse CAR1 and CAR2 cDNAs were isolated from a mouse liver cDNA library by polymerase chain reaction and verified by sequencing. For bacterial expression of CAR1 with a His-tag at the N terminus, CAR1 cDNA digested with *Bam*HI and *Eco*RI was inserted into pET28a+ (Novagen Corp., Madison, WI) to produce pETCAR. For expression of CAR in mammalian cells, a *Bam*HI/*Eco*RI fragment containing the CAR1 cDNA isolated from pGEX2TK-CAR was inserted into pcDNA3 (Invitrogen, Carlsbad, CA) digested with the same enzymes to produce pcDNA3-CAR. For expression of GFP-mCAR1 in mammalian cells, a *Bam*HI/*Eco*RI fragment containing the CAR1 cDNA isolated from pcDNA3-CAR was inserted into pEGFP-C2 (BD Biosciences Clontech, Palo Alto, CA) digested with *Bgl*II/*Eco*RI to produce pEGFP-mCAR1. The mammalian expression vector pCMV5-HNF-4 and the reporter plasmid pPBRUTK-luc were obtained from B. Kemper (University of Illinois, Urbana-Champaign, IL). The reporter plasmids, pPBRU2C1-luc, pPBRU(NR-1m)2C1-luc, and pPBRU(NF-1m)2C1-luc have been described previously[16] and were also obtained from B. Kemper.

Bacterial Expression and Purification of CAR

For expression of CAR protein in bacteria, 1 L of *Escherichia coli* (*E. coli*) BL21(DE3)pLysS in LB broth were inoculated with 1/20 volume of overnight culture and incubated at 37°C for about 1 h to an $A_{600} = 0.6$. Expression was induced by addition of 0.5 mM isopropyl β-D-thioglucopyranoside, and the samples were incubated for 4 h at 37°C. The bacteria were pelleted by centrifugation, and the pellet was resuspended in 20 mL of 20 mM Tris-HCL, pH 7.9, 500 mM NaCl, 20% glycerol, 0.2 mM EDTA, 0.1% Nonidet-40, 4 mM DTT, and protease inhibitors. The cells were lysed by sonication and centrifuged at 22,000 g for 20 min. The supernatant was mixed with nickel-NTA slurry (Qiagen, Valencia, CA) for 6His-CAR and incubated at 4°C overnight. After washing by centrifugation and resuspension five times with 20 mM Tris-Hcl, pH 7.9, 300 mM NaCl, 20% glycerol, and 0.2 mM EDTA, and protease inhibitors, proteins were eluted at 4°C by resuspension and incubation for 20 min in 0.2–0.5 mL of 20 mM Tris-HCl, pH 7.9, 100 mM NaCl, 20% glycerol, and 0.2 mM EDTA, which contained 150 mM imidazol. The identity of the protein was established by Western analysis, and the purity and concentration of the protein were estimated by Coomassie Blue staining of SDS-polyacrylamide gels using bovine serum albumin as standard (Sigma Chemical Co., St. Louis, MO).

Antisera

For production of CAR antisera, CAR2 cDNA was inserted into pET28a+, and *E.coli* BL21(DE3)pLysS was transformed with the resulting plasmid. Expression of CAR2 was induced by incubation with 1 mM isopropyl β-D-thioglucopyranoside for 4 h at 37°C. Bacteria were lysed in Ni-NTA equilibrium buffer (20 mM Tris-HCl, pH 8.0; 500 mM NaCl; 10 mM imidazole; 0.2 mM phenylmethylsulfonyl fluoride; 1 mM DTT; 2 μg/mL leupeptin, pepstatin, and aprotinin; and 10 μg/mL benzamidine) by passing through a French press. CAR 2, which was in inclusion bodies, was pelleted by centrifugation for 15 min at 15,000 g. The pellet was washed twice with 8 M urea and once with 6 M guanidine-HCl in Ni-NTA equilibrium buffer plus 1% Nonidet P-40 minus the protease inhibitors. The protein was solubilized in 6 M guanidine-HCl and 15 mM β-mercaptoethanol in Ni-NTA equilibrium buffer, and the solubilized protein was purified by affinity chromatography on a nickel-NTA column. The sample was dialyzed against phosphate-buffered saline, which resulted in precipitation of the proteins. The precipitated protein was resuspended at a concentration of 5 mg/mL in phosphate-buffered saline for injection into a rabbit for antibody production.

Protein Expression Analysis by Western Blot

In order to determine expression of CAR in Hep G2 and COS-7 cells, cultured cells were either nontransfected or transfected with 0.5 μg of pEGFP-mCAR1 plasmid. Total cell extracts were prepared from both nontransfected control and transfected cells. And the cellular proteins were separated by 8.5% SDS-polyacrylamide gel electrophoresis and transferred onto nitrocellulose membrane. The membrane was probed with rabbit anti-mCAR2 antibody (1:1000 dilution) and horseradish peroxidase-conjugated goat anti-rabbit IgG. Proteins were then visualized using ECL detection system (Amersham Life Science Corp., Arlington Heights, IL).

Cell Culture and Transfection

Human Hep G2 and COS-7 cells were maintained in Dulbecco's modified Eagle's medium supplemented with 10% charcoal dextran-stripped fetal calf serum, 100 units/mL penicillin, and 0.01% streptomycin. For transfections, Hep G2 cells were transfected with 1 μg of plasmids containing either the wild-type PBRU, the PBRU with mutation of the NR-1 site (NR-1m), or the PBRU with mutation of the NF-1 site (NF-1m),[16] fused to the minimal *CYP2C1* promotor/firefly luciferase reporter,[24] 3 ng of pRLSV40 containing the SV40 promotor and *Renilla* luciferase reporter as an internal standard, and 2 uL of LipofectAMINE 2000 (Invitrogen) as described by the manufacturer. COS-7 cells were transfected with 1 μg of plasmid containing wild-type PBRU fused to the strong TK promotor/firefly luciferase reporter, 3 ng of pRLSV40 as an internal standard, and 2 uL of LipofectAMINE 2000. Expression plasmids for CAR, pcDNA3-CAR1, and HNF-4, pCMV5-HNF-4, were cotransfected as indicated in the figures. After 24–36 h of transfection, cells were lysed and luciferase activities were determined by the Dual Luciferase Reporter assay system (Promega Biotech, Madison, WI). For each sample, the background of extracts from untransfected cells was subtracted and the firefly luciferase values were normalized by dividing by *Renilla* luciferase values.

RESULTS

Expression of CAR Protein in Hep G2 and COS-7 Cells

Consistent with the previous report,[21] in nontransfected cells of Hep G2 and COS-7 either low level (Hep G2; FIG. 1 A) or no expression of endogenous CAR (COS-7; FIG. 1 B) was observed. When cells were transfected with the CAR expression plasmid, pEGFP-mCAR1, high levels of the GFP-CAR fusion protein (GFP-mCAR1) were expressed in both cell types (FIG. 1 A, B). These results allowed to examine CAR-mediated PBRU transactivation in both Hep G2 and COS-7 cells.

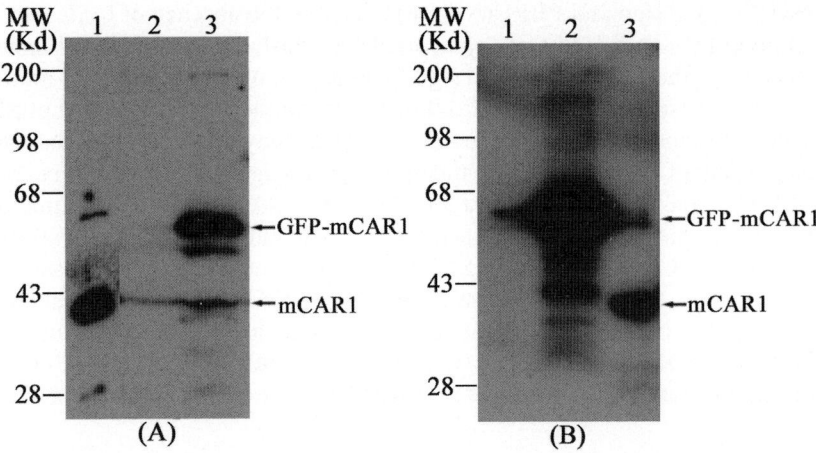

FIGURE 1. Expression of CAR protein in Hep G2 and COS-7 cells. Total cell extracts were prepared from either nontransfected control or cells transfected with 0.5 μg of pEGFP-mCAR1 plasmid. Cellular proteins were separated by 8.5% SDS-polyacrylamide gel electrophoresis and transferred onto nitrocellulose membrane. The membrane was probed with rabbit anti-mCAR2 antibody (1:1000 dilution) and horseradish peroxidase-conjugated goat anti-rabbit IgG. Proteins were visualized using ECL detection system. (**A**) CAR expression in Hep G2 cells. MW; molecular weight marker in kDa, Lane 1; Ni-NTA-purified 6His-mCAR1 protein from *Escherichia coli* BL21(DE3)pLysS transformed with pET28+ 6His-mCAR1 plasmid and induced with 1 mM IPTG, Lane 2; nontransfected Hep G2 cells, Lane 3; Hep G2 cells transfected with 0.5 μg of pEGFP-mCAR1 plasmid. The lower band (about 39 kDa) in lane 3 is a cleaved mCAR1 protein from the conjugated GFP-mCAR1 protein and the upper band in lane 1 is presumed to be a doublet of purified 6His-mCAR1 protein. (**B**) CAR expression in COS-7 cells. MW; molecular weight marker in kDa, Lane 1; nontransfected COS-7 cells, Lane 2; COS-7 cells transfected with 0.5 μg of pEGFP-mCAR1 plasmid, Lane 3; Ni-NTA-purified 6His-mCAR1 protein from *E. coli* BL21(DE3)pLysS transformed with pET28+6His-mCAR1 plasmid and induced with 1 mM IPTG. The lower band (about 39 kDa) in lane 2 is a cleaved mCAR1 protein from the conjugated GFP-mCAR1 protein and the upper band in lane 3 is presumed to be a doublet of purified 6His-mCAR1 protein.

CAR-Dependent Transactivation of the CYP2B1 PBRU Requires Functional NR-1 and NF-1 Sites and HNF-4 Does Not Affect CAR Transactivation in Hep G2 Cells

Mutations in the 5′ part of the bipartite NF-1 motif (NF-1m) of the *CYP2B1* PBRU have been shown to eliminate NF-1 binding and reduce PB induction in hepatocytes transfected *in situ*.[16] Consistent with the previous studies,[12] mutation of the NF-1 motif had little effect on basal transcriptional activity of the PBRU in Hep G2 cells that were not transfected with the CAR expression vector. Compare relative luciferase activity between reporters of

PBRU(WT)2C1luc and PBRU(NF-1m)2C1luc in the absence of CAR transfection in FIGURE 2. However, as reported previously[12] CAR-mediated transactivation of the PBRU decreased significantly by mutation of the NF-1 motif. These results suggest that NF-1 itself had little transactivating activity but enhances substantially the response to CAR. This study also demonstrated that independent of NF-1 site, NR-1 mutation caused a significant inhibition to the CAR-mediated PBRU transactivation. At 5 ng of CAR, the transactivation of the NR-1 mutant PBRU reporter reached a plateau indicating that the available NR sites for CAR/RXR binding, such as NR-1 and NR-2 reached saturation. Together, these results suggest that not only CAR but also NF-1 binding to its cognate binding sites at the *cis*-acting element play important roles in PBRU transactivation in hepatic cells. In Hep G2 cells, however, HNF-4 did not influence either CAR-mediated PBRU transactivation or 2C1luc activity (FIG. 2).

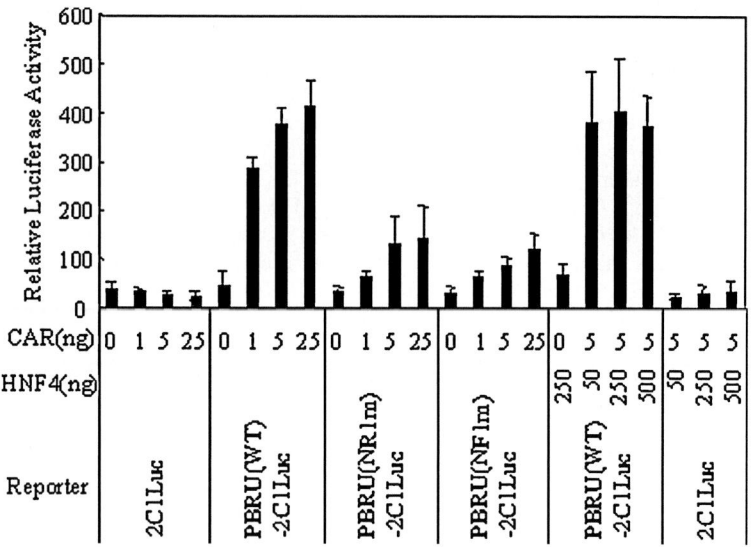

FIGURE 2. CAR-dependent transactivation of the *CYP2B1* PBRU requires functional NR-1 and NF-1 sites, and HNF-4 does not affect CAR transactivation in Hep G2 cells. Hep G2 cells were transfected with 1 μg of 2C1luciferase, PBRU(WT)2C1luciferase (wild type), PBRU(NR-1m)2C1luciferase (NR-1 site mutation), or PBRU(NF-1m)2C1luciferase (NF-1 site mutation) reporter plasmid, 3 ng of pRLSV40 for the internal control for transfection efficiency, and increasing amounts (1–25 ng) of pcDNA3-CAR expression plasmid as indicated. In some experiments, Hep G2 cells were cotransfected with increasing amounts (50–500 ng) of the mammalian expression plasmid, pCMV5-HNF-4. The cells were harvested for dual luciferase assays. The values for firefly luciferase were normalized by dividing by *Renilla* luciferase values. The standard errors of the mean were calculated from three independent transfection experiments.

TK Promotor Induces CAR-Mediated PBRU Transactivation in a Dose-Dependent Manner in COS-7 Cells

When COS-7 cells were transiently transfected with PBRU(WT)tk-luciferase reporter plasmid, the strong TK promotor induced CAR-mediated PBRU transactivation in a dose-dependent manner (FIG. 3). The CAR-mediated transactivation of PBRU was increased up to about 10-fold when 125 ng of CAR expression vector was transfected. This is in contrast to the earlier report that CAR does not transactivate PBRU2C1luciferase reporter gene in COS cells.[21] Thus, the proximal promotor in context to the PBRU enhancer region may also play a role in responding to CAR for PBRU transactivation at least in COS-7 cells.

HNF-4 Has Biphasic Effects on CAR-Mediated PBRU Transactivation in COS-7 Cells

In an attempt to identify other *trans*-acting factor(s) that may be required to stimulate transactivation of PBRU, COS-7 cells were cotransfected with PBRU(WT)tk-luciferase reporter plasmid and different amounts CAR

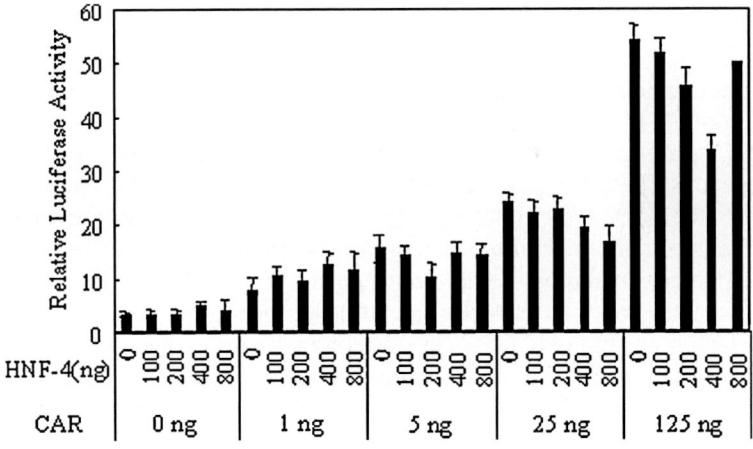

FIGURE 3. Effects of TK promotor and HNF-4 expression in CAR-dependent transactivation of the *CYP2B1* PBRU in COS-7 cells. COS-7 cells were transfected with 1 μg of PBRU(WT)tk-luciferase (wild type) reporter plasmid, 3 ng of pRLSV40 for the internal control for transfection efficiency, and increasing amounts (1–125 ng) of pcDNA3-CAR expression plasmid either in the absence or presence of increasing amounts (100–800 ng) of pCMV5-HNF-4 expression plasmid as indicated. The cells were harvested for dual luciferase assays. The values for firefly luciferase were normalized by dividing by *Renilla* luciferase values. The standard errors of the mean were calculated from three independent transfection experiments.

expression plasmid, pcDNA3-CAR1, in the absence or presence of increasing amounts of HNF-4 expression plasmid, pCMV5-HNF-4. Whereas expression of HNF-4 slightly promoted CAR-mediated PBRU transactivation with low amount of CAR (1 ng) transfected, it repressed the CAR-mediated transactivation in a dose-dependent manner with high amount of transfected CAR (5–125 ng) (FIG. 3). This is in contrast to the effects of HNF-4 in Hep G2 cells where HNF-4 did not influence CAR-mediated PBRU transactivation (FIG. 2).

DISCUSSION

The *CYP2* genes exhibit tissue-, development-, and gender-specific expression, as well as diverse responsiveness to a variety of inducers including PB.[22] Previous studies show that the mammalian *CYP2B* genes are predominantly expressed in the liver by transcriptional activation and further provided strong evidence that the PBRU in *CYP2B* genes is the principal mediator of PB induction in these genes.[4,22] For PB induction, the PB treatment most likely causes the translocation of CAR from the cytoplasm into the nucleus, where this constitutively active receptor binds to and transactivates the PBRU. In addition, transactivation of PBRU requires other accessory factors and coactivators, such as NF-1 and GRIP-1 for maximal activation by PB in both transient transfection and *in situ* analysis.[12,19]

The mechanism by which CAR transactivates the PBRU still remains to be established. Whereas most of the studies on the role of CAR in PBRU induction have been focused on the liver cells, earlier studies indicate that transient transfection of CAR does not transactivate *CYP2B1* PBRU fused to the minimal *CYP2C1* promotor/firefly luciferase reporter gene in COS cells in which the endogenous *CYP2B1* is not induced by PB.[21] This suggests that factors other than CAR are involved in inducing PBRU activation. Transcription of genes is influenced not only by *trans*-acting factors but also by *cis*-acting elements within the proximal promotor as well as enhancer regions. This study demonstrates that the proximal promotor can influence the responsiveness to CAR for PBRU transactivation in COS-7 cells. In contrast to the earlier report that CAR does not induce PBRU transactivation in COS cells when the enhancer was fused to the minimal 2C1 proximal promoter, the strong TK promoter induced CAR-mediated transactivation of PBRU in the same kidney-derived COS-7 cells. Similar to the liver-derived hepatoma cells, the PBRU transactivation in COS-7 cells was dose-dependent on the transfected CAR expression plasmid. Thus, although CAR transactivation is dependent on the PBRU sequence, the proximal promotor in context to the PBRU is an important factor in PBRU induction depending on different cell types. In addition, kidney epithelial-derived COS-7 cells can be used as an experimental cellular model for introducing and screening other *trans*-acting factors that may be involved in CAR-mediated PBRU transactivation.

The expression of *trans*-acting factors may be different among different tissues and may be introduced from exogenous expression plasmids in transient transfection system. HNF-4 is a liver-enriched orphan receptor and has been suggested to serve as a common regulator for the liver-specific transcriptional activation of many *CYP2* genes.[22] In this study, HNF-4 did not influence either CAR-mediated PBRU transactivation or 2C1luc activity in Hep G2 cells. This result is consistent with the previous reports that HNF-4 binding to *CYP2C* genes was not changed by PB treatment or inhibited by the presence of the17-bp oligonucleotide containing the proposed *CYP2B2* PB response element, and many liver-specific genes with functional HNF-4 motifs are not induced by PB.[22] In contrast, however, HNF-4 has biphasic effects on CAR-mediated PBRU transactivation in COS-7 cells. Whereas expression of HNF-4 slightly promoted CAR-mediated PBRU transactivation with low amount of CAR transfected, it repressed the CAR-mediated transactivation in a dose-dependent manner with high amount of transfected CAR. Thus, although HNF-4 does not appear to be involved in CAR-mediated PBRU induction in hepatoma cells, it may play a role in interacting with a PB-responsive factor(s) in COS-7 cells.

CAR needs to form a heterodimer with RXR in order to bind NR sites of PBRU for its transactivation.[4] In order to allow for studying CAR-mediated transactivation, expression of the endogenous CAR at the protein level needs to be examined in cultured cell types. In agreement with the earlier reports,[21] expression of the endogenous CAR was either low in Hep G2 or not detectable in COS-7 cells, and transient transfection with GFP–CAR expression plasmid caused high expression of CAR protein in both cell types. Since previous studies also indicate that both cell types express high levels of the endogenous RXR,[21] transfection with only CAR expression plasmid used in this study allowed to examine the transcriptional effects of CAR on PBRU transactivation in both Hep G2 and COS-7 cells.

Consistent with the previous reports in Hep G2 cells,[12] CAR transactivated PBRU2C1luciferase in a dose-dependent manner and this CAR-mediated transactivation required functional NR-1 and NF-1 sites. The mutation of the NF-1 motif had little effect on the basal transcriptional activity in Hep G2 cells that were not transfected with the CAR expression vector. However, CAR-mediated PBRU transactivation was decreased significantly by mutation of the NF-1 motif. These results suggest that NF-1 itself had little transactivating activity, but enhances substantially the response to CAR. This study also demonstrated that independent of NF-1 site, NR-1 mutation also caused a significant inhibition to the CAR-mediated PBRU transactivation. This result is in agreement with previous reports that mutation of the NR-1 and NR-2 sites decreases the response to PB in transient transfection studies.[15–17] Together, these results suggest that not only CAR but also NF-1 binding to its cognate-binding sites at the *cis*-acting element play important roles in PBRU transactivation in hepatic cells.

In conclusion, this study suggests that proximal promotor and a *trans*-acting factor, HNF-4, can affect CAR-mediated PBRU transactivation in COS-7 cells.

ACKNOWLEDGMENT

The author thanks Dr. B. Kemper for providing pCMV5-HNF-4, pPBRUTK-luc, pPBRU2C1-luc, pPBRU(NR-1m)2C1-luc, and pPBRU(NF-1m)2C1-luc vectors.

REFERENCES

1. NELSON, D.R., L. KOYMANS, T. KAMATAKI, et al. 1996. P450 superfamily: update on new sequences, gene mapping, accession numbers and nomenclature. Pharmacogenetics **6:** 1–42.
2. AXELROD, J. 1955. The enzymatic deamination of amphetamine (benzedrine). J. Biol. Chem. **214:** 753–763.
3. BHAGWAT, S.V., B.C. LEELAVATHI & S.K. SHANKAR. 1995. Cytochrome P450 and associated monooxygenase activities in the rat and human spinal cord-induction, immunological characterization, and immunocytochemical localization. Neuroscience **68:** 593–601.
4. HONKAKOSKI, P. & M. NEGISHI. 2000. Regulation of cytochrome P450 (CYP) genes by nuclear receptors. Biochem. J. **347:** 321–337.
5. SAVAS, U., K.J. GRIFFIN & E.F. JOHNSON. 1999. Molecular mechanisms of cytochrome P-450 induction by xenobiotics: an expanded role for nuclear hormone receptors. Mol. Pharmacol. **56:** 851–857.
6. SUEYOSHI, T. & M. NEGISHI. 2001. Phenobarbital response elements of cytochrome P450 genes and nuclear receptors. Annu. Rev. Pharmacol. Toxicol. **41:** 123–143.
7. WAXMAN, D.J. 1999. P450 gene induction by structurally diverse xenochemicals: central role of nuclear receptors CAR, PXR, and PPAR. Arch. Biochem. Biophys. **369:** 11–23.
8. WAXMAN, D.J. & T.K.H. CHANG. 1995. Hormonal regulation of liver cytochrome P450 enzymes. *In* Cytochrome P450: Structure, Mechanism, and Biochemistry, 2nd ed. P.R. Ortiz de Montellano, Ed.: 391–417. Plenum Press. New York.
9. CONNEY, A.H. 1982. Induction of microsomal enzymes by foreign chemicals and carcinogenesis by polycyclic aromatic hydrocarbons: G. H. A. Clowes Memorial Lecture. Cancer Res. **42:** 4875–4917.
10. WAXMAN, D.J. & L. AZAROFF. 1992. Phenobarbital induction of cytochrome P-450 gene expression. Biochem. J. **281:** 577–592.
11. HONKAKOSKI, P., A. KOJO & M.A. LANG. 1992. Regulation of mouse liver P450 *2B* subfamily by sex hormones and phenobarbital. Biochem. J. **285:** 979–983.
12. KIM, J., G. MIN & B. KEMPER. 2001. Chromatin assembly enhances binding to the CYP2B1 phenobarbital–responsive unit (PBRU) of nuclear factor-1, which binds simultaneously with constitutive androstane receptor (CAR/retinoid X receptor (RXR) and enhances CAR/RXR-mediated activation of the PBRU. J. Biol. Chem. **276:** 7559–7567.

13. TROTTIER, E., A. BELZIL & C. STOLTZ. 1995. Localization of a phenobarbital-responsive element (PBRE) in the 5'-flanking region of the rat CYP2B2 gene. Gene **158:** 263–268.
14. HONKAKOSKI, P. & M. NIGISHI. 1997. Characterization of a phenobarbital-responsive enhancer module in mouse P450 Cyp2b10 gene. J. Biol. Chem. **272:** 14943–14949.
15. STOLTZ, C., M.H. VACHON, E. TROTTIER, et al. 1998. The CYP2B2 phenobarbital response unit contains an accessory factor element and a putative glucocorticoid response element essential for conferring maximal phenobarbital responsiveness. J. Biol. Chem. **273:** 8528–8536.
16. LIU, S., Y. PARK, I. RIVERA-RIVERA, et al. 1998. Nuclear factor-1 motif and redundant regulatory elements comprise phenobarbital-responsive enhancer in CYP2B1/2. DNA Cell Biol. **17:** 461–470.
17. HONKAKOSKI, P., I. ZELKO & T. SUEYOSHI. 1998. The nuclear orphan receptor CAR-retinoid X receptor heterodimer activates the phenobarbital-responsive enhancer module of the CYP2B gene. Mol. Cell Biol. **18:** 5652–5658.
18. KAWAMOTO, T., T. SUEYOSHI & I. ZELKO. 1999. Phenobarbital-responsive nuclear translocation of the receptor CAR in induction of the CYP2B gene. Mol. Cell Biol. **19:** 6318–6322.
19. MIN, G., J.K. KEMPER & B. KEMPER. 2002. Glucocorticoid receptor-interacting protein 1 mediates ligand-independent nuclear translocation and activation of constitutive androstane receptor *in vivo*. J. Biol. Chem. **277:** 26356–26363.
20. TZAMELI, I., P. PISSIOS, E.G. SCHUETZ, et al. 2000. The xenobiotic compound 1,4-bis[2-(3,5-dichloropyridyloxy)] benzene is an agonist ligand for the nuclear receptor CAR. Mol. Cell Biol. **20:** 2951–2958.
21. MIN, G. 2003. Effects of constitutive androstane receptor (CAR) on PBRU transactivation of CYP2B gene in different culture cell types: comparison between HepG2 and COS-cells. Korean J. Life Sci. **13:** 324–332.
22. CHEN, D., G. LEPAR & B. KEMPER. 1994. A transcriptional regulatory element common to a large family of hepatic cytochrome P450 genes is a functional binding site of the orphan receptor HNF-4. J. Biol. Chem. **269:** 5420–5427.
23. SHAW, G.C. & A.J. FULCO. 1993. Inhibition by barbiturates of the binding of Bm3R1 repressor to its operator site on the barbiturate-inducible cytochrome P450BM-3 gene of *Bacillus megaterium*. J. Biol. Chem. **268:** 2997–3004.
24. PARK, Y., H. LI & B. KEMPER. 1996. Phenobarbital induction mediated by a distal CYP2B2 sequence in rat liver transiently transfected *in situ*. J. Biol. Chem. **271:** 23725–23728.

Expression of the E2F Family of Transcription Factors and Its Clinical Relevance in Ovarian Cancer

DANIEL REIMER,[a] SUSANN SADR,[a] ANNEMARIE WIEDEMAIR,[a] GEORG GOEBEL,[b] NICOLE CONCIN,[a] GERDA HOFSTETTER,[a] CHRISTIAN MARTH,[a] AND ALAIN G. ZEIMET[a]

[a]*Department of Obstetrics and Gynecology, Medical University Innsbruck, Innsbruck, Austria*

[b]*Department of Medical Statistics, Informatics and Health Economics (MSIG), Medical University Innsbruck, Innsbruck, Austria*

ABSTRACT: The E2F family of transcription factors plays a pivotal role in the regulation of cellular proliferation. On the basis of sequence homology and function, eight distinct members of E2F transcription factors (E2F-1 to E2F-8) have been distinguished to date. The regulation of E2F transcription factors is closely associated with the function of the retinoblastoma family of tumor suppressors (RB pathway). In the last decade various alterations of distinct components of the RB-E2F pathway were found to be associated with tumor progression. However, no data on the role of E2F family members are available in tumor biology of ovarian cancer. Here we describe an expression study of E2F transcription factors in various human ovarian cancer cell lines; its clinical relevance was examined in a training set of 77 ovarian cancer patients. Expression levels of E2F-1, E2F-2, and E2F-8 were elevated in all the ovarian cancer cell lines studied when compared with human peritoneal mesothelial cells (HPMCs). Interestingly, EGF treatment showed a time-dependent upregulation of the activating transcription factor E2F-3 and a simultaneous increase of DP-1, the heterodimeric partner of E2F-3. High expression of E2F-1, E2F-2, and E2F-8 was found to be associated with histopathologic grade 3 tumors and residual tumor over 2 cm in diameter after primary debulking surgery in ovarian cancer patients. Taken together, these data suggest that the proliferation-promoting E2F transcription factors E2F-1 and especially E2F-2 play a pivotal role in tumor biology of ovarian cancer and may be candidates for specific therapeutic targets.

KEYWORDS: E2F; transcription factor; tumor progression; EGF; ovarian cancer

Address for correspondence: Daniel U. Reimer, M.D., Department of Obstetrics and Gynecology, Medical University Innsbruck, Anichstrasse 35, A-6020 Innsbruck. Voice: +43-0-512-504-23051; fax: +43-0-512-504-23055.
e-mail: daniel.reimer@uibk.ac.at

INTRODUCTION

The E2F family of transcription factors is the downstream effector of the retinoblastoma (RB) pathway and is involved in the control of cell division via regulation of cell proliferation and apoptosis. E2F was first identified as a cellular activity which binds to and activates the adenovirus E2 promoter.[1] These transcription factors act by forming heterodimeric protein complexes with one member of the DP family of proteins (DP-1 and DP-2). The regulation of the E2F transcription factors is mediated by "pocket proteins," namely the retinoblastoma protein (RB), p107, and p130. In nonproliferating cells inactivation of E2F function is mediated by binding to nonphosphorylated RB. Moreover, RB acts as a transcriptional repressor complex by recruiting histone deacetylase (HDAC) and remodeling chromatin. When cells move in a transcriptionally active status, RB is phosphorylated by G1 cyclin-dependent kinase complexes (cyclinD/cdk4 and cyclinE/cdk2), leading to an inability to bind E2F, which is released and becomes transcriptionally active. Interaction of p107 and p130 with E2F is regulated similarly.[2–5]

Based upon sequence homology the E2F family of transcription factors includes eight members (E2F-1 to -8), which can be subdivided into different functional groups. Moreover, alternative splicing in E2F-3 (E2F-3a and E2F-3b) and in DP-2 contributes to the complexity of the system.[2–5] E2F-1 to -6 share an N-terminal DNA-binding domain, followed by a conserved region, termed the *marked box*, which is possibly involved in dimerization and DNA binding.[2,4] E2F-4 to -6 differ from E2F-1 to -3 by a truncated N terminus lacking a cyclinA-binding domain. The C-terminal transactivation domain is existent in E2F-1 to -5 and is responsible for the interaction with the "pocket proteins." E2F-7 and E2F-8 are structurally and functionally unique by lacking the dimerization and the transactivation domain and containing two DNA-binding domains[6] (FIG. 1).

Functionally, E2F-1, E2F-2, and E2F-3a represent the group of proliferation-promoting transcription factors which are required for transactivation of target genes in G1/S phase of the cell cycle. E2F-3b, E2F-4, and E2F-5 share in contrary repressive functions mainly promoter-associated with G0/G1 phase. E2F-6 and E2F-7 are considered to act as transcriptional repressors.[6] The recently identified E2F-8 share homology with E2F-7. In mouse and human fibroblasts these transcription factors seem to act synergistically in a repressive way.[7,8] However, the role of these transcription factors is not fully understood.

Various alterations, such as point mutations, deletions, or promoter methylation in components of the RB pathway were found to be associated with tumor formation. Moreover, deregulated E2F activity was often considered to be associated with poor survival in a variety of different tumor entities.[9–12] Although most alterations were found in upstream components of the RB pathway,

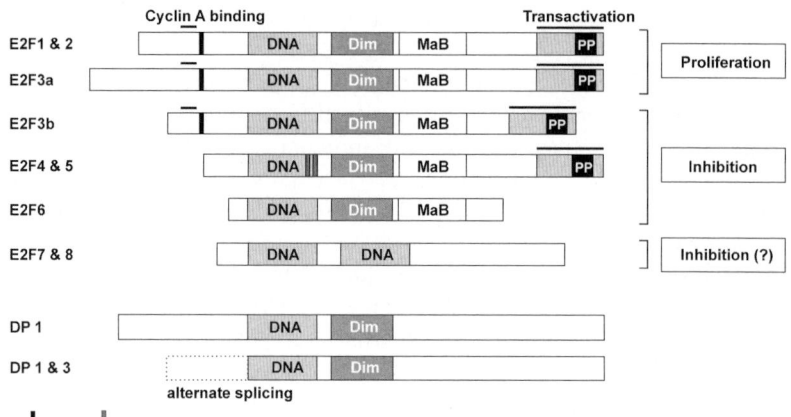

FIGURE 1. Schematic alignment of the E2F family of transcription factors and their coactivators DP-1 and DP-2. DNA, DNA binding domain; Dim, dimerization domain; MaB, marked box; PP, pocket protein-binding domain; NLS, nuclear localization signal; NES, nuclear export signal.

two recent studies report an overexpression of E2F-3 in human bladder cancer, underscoring the importance of deregulated E2F function as a causative role in human tumors.[13,14] However, no data are available concerning the putative role of E2F in the biology of ovarian cancer.

Motivated by these recent results we investigated the role of E2F transcription factors in various ovarian cancer cell lines. In this context the constitutive degree of expression of the E2F factors was determined by polymerase chain reaction (RT-PCR) and the possible role of epidermal growth factor (EGF) in governing the expression of these transcription factors was assessed. Furthermore, the clinical relevance of deregulated E2F transcription factors was examined in a training set of 77 ovarian cancer patients.

MATERIALS AND METHODS

Cell Culture Experiments

Various established ovarian cancer cell lines, HTB-77, OVCAR-3, SKOV-6, A2780, AG6000 (a dCK-deficient A2780 cell line, kindly provided by Prof. Peters, Department of Molecular Biosciences, Swedish University of Agricultural Sciences, Uppsala, Sweden), and normal human peritoneal mesothelial cells (HPMCs) derived from patients during surgery for other than inflammatory or malignant disease, were cultured and passaged in Dulbecco's modified minimum essential medium (Biochrom KGF, Schöller Pharma, Vienna) containing 10% fetal bovine serum, 1% L-glutamine, 0.2% penicillin,

0.2% streptomycin, and 1% nonessential amino acids (all from PAA Laboratories GmbH, Linz, Austria). For EGF treatment, cells were plated into 25-cm^2 culture flasks (Falkon; Becton Dickinson, Franklin Lakes, NJ, USA) and allowed to attach overnight. About 75–80% confluent cells were washed with phosphate-buffered saline solution and treated with EGF (Sigma, St. Louis, MO, USA) over various time periods (3 h, 6 h, 12 h, 24 h, and 72 h). The final concentration of the reagents was 10 ng/mL. At the end, incubation cells were counted with an electronic particle counter (Coulter, Dunstable, UK) and collected for subsequent total RNA isolation.

RNA Extraction and RT Reaction

Total RNA was isolated from different cancer cell lines and patient samples using the guanidium thiocyanate-phenol-chloroform method according to the manufacturer's protocol (RNAgents® Total RNA Isolation System, Promega, Madison, WI, USA). Integrity was evaluated by assessing the 18S- and 28-S-ribosomal RNA bands in 1% ethidium bromide–stained agarose gels. To remove any contaminating genomic DNA, DNAse treatment of typically 4 μg total RNA was performed in a final volume of 30 μL containing 40 U of rRNAsin® RNAse Inhibitor (Promega, Madison, WI, USA) and 2 U/μg RNA DNAse I® (Roche, Basel, Switzerland) according to the manufacturer's protocol (Roche). Incubation periods were 25°C for 30 min followed by heating at 70°C for 10 min.

Reverse transcription of typically 2 μg total RNA was performed in a final volume of 25 μL containing 1× RT-buffer (50 mM Tris HCl, pH 8.3, 75 mM KCl, 5 mM MgCl$_2$), 40 U of rRNAsin® RNAse Inhibitor (Promega, Madison, WI, USA), 10 mM dithiothreitol, 250 nM random hexamers, and 200 U of M-MLV reverse transcriptase according to the manufacturer's protocol (Invitrogen, Carlsbad, CA, USA). Incubation periods were 25°C for 10 min, and 37°C for 50 min, followed by heating at 70°C for 15 min to inactivate the reverse transcriptase enzyme.

Primers and Probes

Specific primers and probes for E2F-1 to E2F-8, DP-1 and DP-2, and for the TATA box-binding protein (TBP; a component of the DNA-binding protein complex TFIID as an endogenous RNA control) were determined with the assistance of the computer program "Primer Express" (Applied Biosystems, Foster City, CA, USA). To prevent amplification of contaminating genomic DNA, the probe was placed at a junction between two exons. Sequences of primers and probes can be obtained on request.

RT-PCR Amplification

RT-PCRs were performed using an ABI Prism 7900 Detection System (Applied Biosystems, Foster City, CA, USA) in a total volume of 25 μL reaction mixture containing 5 μL of each appropriately diluted RT sample (standard curve points and test samples), 12.5 μL TaqMan Universal PCR Master Mix (Applied Biosystems, Foster City, CA, USA), 900 nM of each primer and 250 nM of the probe. Cycling conditions were an initial step at 50°C for 2 min, a denaturing step at 95°C for 10 min, and 45 cycles at 95°C for 15 sec and 65°C for 1 min. Each experiment contained a standard curve, a control sample, 25 test samples (cell line–derived cDNA or patient sample) and a no-template control. The standard curves were generated with serially diluted solutions of standard cDNA derived from the HTB-77 cell line. RT-PCR assays were conducted as triplicates and the mean value was used for calculation. The gene expression levels were determined with the Comparative C_T method according to User Bulletin 2 (Applied Biosystems, Foster City, CA, USA).

Patients

Tissue samples from patients with invasive ovarian cancer were collected during primary debulking surgery at the Department of Obstetrics and Gynecology at Medical College Innsbruck between 1992 and 1999 ($n = 77$). Ovaries removed for reasons other than ovary-related disease serve as control ($n = 8$). All patients were monitored at the outpatient clinic of the Department of Obstetrics and Gynecology and the median observation period of surviving ovarian cancer patients was 5.0 years.

Statistical Analysis

Differences in expression levels in cell culture experiments were evaluated by comparison of mean values of three independently performed experiments by the Student's t-test. Differences in expression levels between normal and malignant tissues and between groups characterized by different clinicopathologic features were evaluated by the Mann–Whitney U test.

RESULTS

RT-PCR was used to determine constitutive expression levels of all members of the E2F family of transcription factors and their coactivators DP-1 and DP-2 in various ovarian cancer cell lines and the breast cancer cell line T47D. HPMCs derived from patients during surgery for other than inflammatory or malignant disease, were grown under the same conditions and served as control. Expression levels of E2F transcripts determined in tumor cells were normalized to the expression measured in HPMCs. As expected, we determined a significant overexpression of the proliferation-promoting transcription factors E2F-1 and E2F-2 in all tumor cell lines compared to HPMCs. Whereas expression level of E2F-1 mRNA was elevated three- to fivefold in the SKOV-6, A2780, and AG6000 cell line (2.762-fold, 3.696-fold, and 5.134-fold, respectively), a six- to eightfold overexpression of E2F-1 mRNA expression level was determined in the OVCAR-3- and T47D-cell line (6.434-fold, and 7.920-fold, respectively). Levels of E2F-2 transcripts were up to 30-fold overexpressed in the OVCAR-3-, A2780-, AG6000-, and the T47D cell line compared to HPMC (22.278-fold, 28.38-fold, 30.330-fold, and 17.920-fold, respectively), pointing to its pivotal role in the biology of these investigated tumor cell lines (FIG. 2A, B). No significant change in mRNA expression level was detected in E2F-3 (data not shown). Although we assume that uncontrolled proliferation could possibly be associated with a downregulation of putative inhibiting transcription factors, we could not detect downregulation of the transcript levels of the growth-inhibiting transcription factors E2F-4, E2F-5, and E2F-6. On the contrary, E2F-4 expression levels were slightly upregulated in the tumor cell lines and E2F-5 expression levels were estimated to be enhanced 4- to 13-fold in tumor cell lines compared with the very low expression level

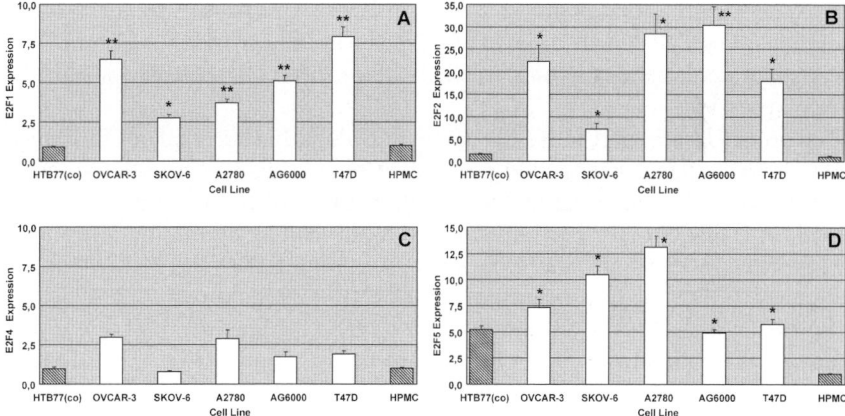

FIGURE 2. Constitutive expression of mRNA levels for E2F-1, E2F-2, E2F-4, and E2F-5 in tumor cell lines and in normal HPMCs. Results are given as means of three independently performed experiments. The levels of E2F transcripts were normalized to TBP and all data were further referred to E2F expression determined in HPMCs (set to 1). Differences in expression levels in cell culture experiments were assessed by the Student's t-test. Asterisks indicate statistical significance, (*) $P < 0.05$; (**) $P < 0.01$. (**A**) E2F-1 (**B**) E2F-2; (**C**) E2F-4; (**D**) E2F-5 expression.

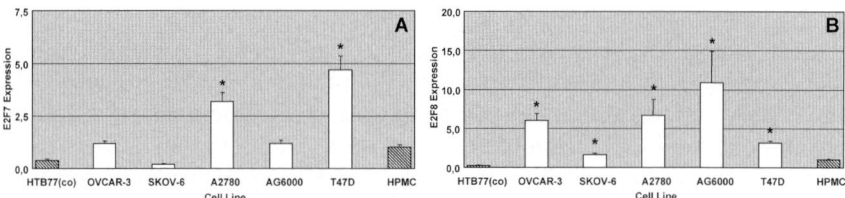

FIGURE 3. Constitutive expression of mRNA levels for E2F-7, and E2F-8 in tumor cell lines and normal HPMC. Results are given as means of three independently performed experiments. The levels of E2F transcripts were normalized to TBP and all data were further referred to E2F levels determined in HPMC (set to 1). Differences in expression levels in cell culture experiments were assessed by the Student's t-test. Asterisks indicate statistical significance, (*) $P < 0.05$. (**A**) E2F-7 and (**B**) E2F-8 expression.

detected in HPMCs (FIG. 2C, D). Furthermore, expression of the recently discovered E2F transcription factors E2F-7 and E2F-8, which are considered to act predominantly as inhibitors of E2F-target genes, were found to be upregulated in proliferating tumor cell lines. Whereas this effect did not reach statistical significance for E2F-7 in all the tumor cell lines investigated, we detected a 6- to 11-fold upregulation of E2F-8 mRNA levels in the OVCAR-3, A2780, and the AG6000 cell line (6.063-fold, 6.635-fold, and 10.843-fold, respectively). Expression level of E2F-8 was 3.157-fold upregulated in the breast cancer cell line and 1.649-fold upregulated in SKOV-6 cells (FIG. 3A, B). No statistically significant changes in expression levels were revealed for DP-2 expression, but DP-1 mRNA expression was shown to be slightly upregulated in all tumor cell lines (data not shown). However, for DP-1, statistical significance was only reached in the A2780 cell line with a 6.088 fold overexpression.

In order to learn more about the pathophysiologic background of E2F deregulation in ovarian cancer cell lines, we investigated changes in E2F transcription factor expression after treatment with the proliferation-stimulating agent EGF. In HTB-77 cells, EGF treatment did not lead to significant alterations in expression levels of the proliferation-promoting E2F-1 and E2F-2. On the contrary, E2F-3 mRNA increased significantly to a maximum of 2.473-fold expression after 12 h of treatment followed by a decline in expression levels measured in untreated cells (FIG. 4). Interestingly, a similar EGF-mediated effect was observed in DP-1 expression, whereas EGF had no effect on DP-2. DP-1 mRNA levels rise after a 12-h incubation with EGF to a 1.644-fold overexpression, followed also by a decline after a time period of 24 h (FIG. 4).

Preliminary data suggest opposite effects of interferon-γ treatment on E2F expression. Whereas expression of E2F-1 and E2F-2 declined in a time-dependent manner, we detected upregulation of the inhibiting transcription factors E2F-4 and E2F-5 (data not shown).

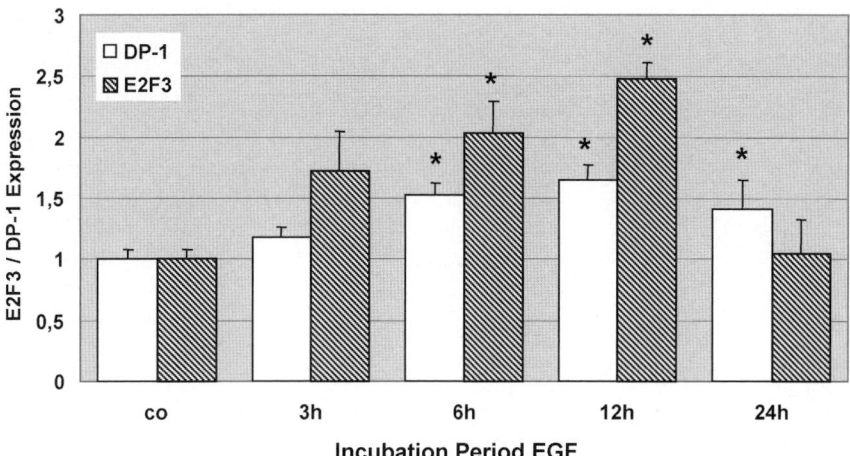

FIGURE 4. Expression patterns of E2F-3 and DP-1 in HTB-77 cells treated with 10 ng/mL EGF. About 75–80% confluence cells were treated with EGF over various time periods and mRNA levels were measured by RT-PCR. Results are given as means of three independently performed experiments. The level of expression was normalized to TBP and all data were further referred to expression levels determined in simultaneously cultured untreated HTB-77 cells (set to 1). Differences in expression levels in cell culture experiments were assessed by the Student's t-test. Asterisks indicate statistical significance, (*) $P < 0.05$. DP-1-(*white*), E2F-3 expression (*stippled*).

Motivated by these results we examined the influence of E2F deregulation on the clinical outcome in a training set of 77 ovarian cancer patients. In accordance with our cell culture data, we found a significant overexpression of all members of the E2F family of transcription factors except E2F-6 in the investigated ovarian cancer samples as compared with eight control samples from ovaries removed for reasons other than ovarian-related disease. Highest levels of mRNA were observed for E2F-2 and E2F-8. Interestingly, E2F-8 mRNA could not be detected in our control samples ($n = 8$), whereas in the ovarian cancer samples, the transcript level was ninefold upregulated compared to the control samples.

High E2F-1, E2F-2, and E2F-8 expression was significantly associated with histopathologic grade 3 tumors and residual disease greater than 2 cm in diameter after primary debulking surgery (FIG. 5A–F; TABLE 1).

DISCUSSION

Mutations in the RB pathway are found in a wide range of tumors.[15] Whereas most of these alterations occur in upstream regulators of RB, there is growing evidence in the literature that deregulated E2F activity plays a pivotal

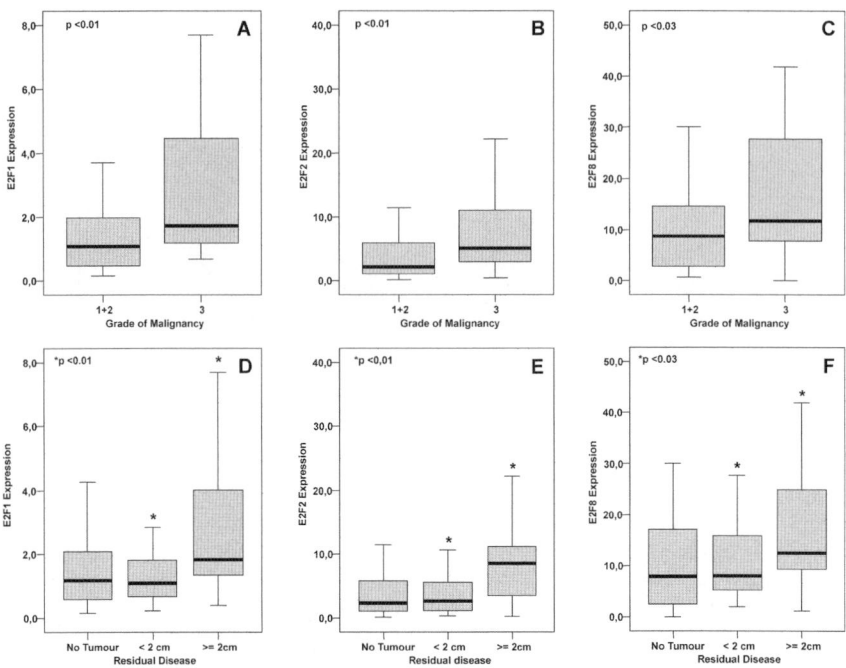

FIGURE 5. Correlation of E2F-1, E2F-2, and E2F-8 transcript levels with histopathologic grade and residual disease after primary debulking surgery. Seventy-seven ovarian cancer patients were stratified into different groups according to histopathologic grade and residual disease. Differences in E2F expression were evaluated by the Mann–Whitney U test. (**A** and **D**) E2F-1; (**B** and **E**) E2F-2; (**C** and **F**) E2F-8 expression.

role in tumorigenesis and correlates with poor prognosis.[9–12] Most studies indicate that E2F deregulation in tumors is based upon an overexpression of the proliferation-promoting transcription factors, rather than deregulation of mainly inhibiting transcription factors. In non–small lung cancer and in esophageal squamous cell carcinoma, tumor progression and poor outcome were correlated with overexpression of the proliferation-promoting transcription factor E2F-1.[9,10] The high E2F-1 transcription levels were claimed to be based upon deregulation at the transcriptional level, rather than gene amplification, which was seen in a small subset of the investigated tumor samples.[9] Concerning E2F-1 mRNA expression, similar results were obtained in our cell line experiments, whereas this effect was by far exceeded by the overexpression of E2F-2, indicating that in ovarian cancer cell lines E2F-1 and mainly E2F-2 upregulation contributes to uncontrolled proliferation. These findings are corroborated by our clinical data, suggesting high expression levels of E2F-1 and E2F-2 to be associated with highly malignant and fast growing tumors *in vivo*.

TABLE 1. Results of RT–PCR of E2F-1, E2F-2, and E2F-8 transcript levels in ovarian cancer patients ($n = 77$)

	Histopathologic Grade				
	1 + 2		3		P-value
E2F-1	1.0624	(0.2961–1.8284)	1.7227	(0.0592–3.3874)	0.01
E2F-2	2.1190	(0.0000–4.6463)	5.0718	(0.8884–9.2560)	0.01
E2F-8	8.6643	(2.5861–14.7421)	11.6971	(10.0632–21.7602)	0.03
	Residual Disease				
	No tumor + Residual < 2 cm		Residual > 2 cm		P-value
E2F-1	1.1721	(0.4212–1.9234)	1.8271	(0.1373–3.5172)	0.01
E2F-2	1.9956	(0.0000–4.5650)	8.4690	(4.4144–12.5242)	0.01
E2F-8	7.2879	(0.6441–13.9322)	12.5083	(4.1260–20.8901)	0.02

Patients were stratified into two groups according to difference in the grade of malignancy and residual disease after primary debulking operation. Data are given as median and interquartile range. Differences in expression levels between the groups were evaluated by the Mann–Whitney U test.

However, E2F-1 and E2F-3 are known to act both as oncogenes and as tumor suppressors. The latter property is mainly mediated by targeting genes involved in the p53- and p73-pathway. Assuming that among the group of proliferation-promoting transcription factors, proapoptotic activity is thought to be unique for E2F-1 and E2F-3,[3,4,16,17] an overexpression of E2F-2 can probably overwhelm the E2F-1-mediated proapoptotic effects and cause uncontrolled proliferation in ovarian cancer. Indeed, preliminary survival data of herein investigated ovarian cancer patients indicate high E2F-2 expression to be the most powerful predictor of poor outcome among the proliferation-promoting transcription factors.

Three recent studies report that E2F-3 deregulation is associated with growth advantage of tumors and poor survival in human bladder and prostate cancer.[12–14] However, we could not detect a significant alteration of E2F-3 expression in ovarian cancer cell lines. These differences are probably due to known tissue-specific functions of various E2F transcription factors. Nonetheless, E2F-3 was the only E2F transcription factor influenced by EGF treatment, pointing to its pivotal role in cell proliferation under growth stimulatory conditions. Possibly, EGF treatment leads to enhanced E2F-3 function via an upregulation of its mRNA level and a simultaneous increase of DP-1, the heterodimeric partner of E2F-3.

As mentioned above, deregulation of the inhibiting transcription factors is believed to be involved only to a minor extent in tumorigenesis. In fact, E2F-4 and E2F-5 transcripts were not downregulated in ovarian cancer cell lines, but, on the contrary, we detected a significant overexpression of E2F-5 in ovarian cancer cell lines and in the analyzed tumor samples. This effect could possibly be explained as a reactive upregulation of inhibiting transcription factors in response to E2F-1 and E2F-2 deregulation in ovarian cancer. In addition, the recently described transcription factors E2F-7 and E2F-8 were also shown to be overexpressed in the cell lines and in the ovarian cancer samples. The *in vivo* function of both E2F-7 and E2F-8 is not fully understood. In mouse and human fibroblasts they have been shown to act predominantly as inhibitors of E2F-targeted genes.[7,8] Again, an autoregulative enhanced expression, as mentioned for E2F-5, could explain this upregulation in malignant cell lines. The role of inhibiting members of the E2F family of transcription factors is, however, yet to be fully elucidated in malignancies. Therefore analysis of patient survival in ovarian cancers with high expression of inhibiting E2F factors could provide a more detailed insight into their role in that disease.

In conclusion, our data underline the pivotal role of the proliferation-promoting E2F transcription factors E2F-1 and especially E2F-2 in the biology of ovarian cancer. However, further investigations in the field of E2F transcription factors are needed to uncover the exact cross-talk between the members of the E2F family in malignant disease and to evaluate their potential usefulness as targets in anticancer strategies.

REFERENCES

1. KOVESDI, I. et al. 1986. Identification of a cellular transcription factor involved in E1A trans-activation. Cell **45:** 219–228. review.
2. BLACK, A.R. & J. AZIZKHAN-CLIFFORD. 1999. Regulation of E2F: a family of transcription factors involved in proliferation control. Gene **237:** 281–302. review.
3. DEGREGORI, J. 2002. The genetics of the E2F family of transcription factors: shared functions and unique roles. Biochim. Biophys. Acta 21; **1602:** 131–150. Review.
4. STEVENS, C. & N.B. LA THANGUE. 2003. E2F and cell cycle control: a double-edged sword. Arch. Biochem. Biophys. 15; **412:** 157–169. Review.
5. DIMOVA, D.K. & N.J. DYSON. 2005. The E2F transcriptional network: old acquaintances with new faces [review]. Oncogene **24:** 2810–2826.
6. ATTWOOLL, C. et al. 2004. The E2F family: specific functions and overlapping interests [Review]. EMBO J. **23:** 4709–4716.
7. MAITI, B. et al. 2005. Cloning and characterization of mouse E2F8, a novel mammalian E2F family member capable of blocking cellular proliferation. J. Biol. Chem. **280:** 18211–18220.
8. CHRISTENSEN, J. et al. 2005. Characterization of E2F8, a novel E2F-like cell-cycle regulated repressor of E2F-activated transcription. Nucleic Acids Res. **33:** 5458–5470.
9. GORGOULIS, V.G. et al. 2002. Transcription factor E2F-1 acts as a growth-promoting factor and is associated with adverse prognosis in non-small cell lung carcinomas. J. Pathol. **198:** 142–156.
10. EBIHARA, Y. et al. 2004. Over-expression of E2F-1 in esophageal squamous cell carcinoma correlates with tumor progression. Dis. Esophagus **17:** 150–154.
11. NEVINS, J.R. 2001. The Rb/E2F pathway and cancer. Hum. Mol. Genet. **10:** 699–703. Review.
12. FOSTER, C.S. et al. 2004. Transcription factor E2F3 overexpressed in prostate cancer independently predicts clinical outcome. Oncogene **23:** 5871–5879.
13. FEBER, A. et al. 2004. Amplification and overexpression of E2F3 in human bladder cancer. Oncogene **23:** 1627–1630.
14. OEGGERLI, M. et al. 2004. E2F3 amplification and overexpression is associated with invasive tumor growth and rapid tumor cell proliferation in urinary bladder cancer. Oncogene **23:** 5616–5623.
15. SHERR, C.J. 1996. Cancer cell cycles [review]. Science 6; **274:** 1672–1677.
16. DENCHI, E.L. & K. HELIN. 2005. E2F1 is crucial for E2F-dependent apoptosis. EMBO Rep. **6:** 661–668.
17. PAULSON, Q.X. et al. 2006. E2F3a stimulates proliferation, p53-independent apoptosis and carcinogenesis in a transgenic mouse model. Cell Cycle **5:** 184–190.

Isolation and Characterization of the Rat SND p102 Gene Promoter

Putative Role for Nuclear Factor-Y in Regulation of Transcription

LORENA RODRÍGUEZ, NEREA BARTOLOMÉ, BEGOÑA OCHOA, AND MARÍA J. MARTÍNEZ

Department of Physiology, Faculty of Medicine and Dentistry, University of the Basque Country, Sarriena s/n, 48940-Leioa, Bizkaia, Spain

ABSTRACT: In this work, we report the isolation and characterization of a 1,688-bp sequence corresponding to the promoter region of the rat endoplasmic reticulum (ER) cholesterol ester hydrolase gene, renamed as staphylococcal nuclease domain–containing protein of 102 kDa (SND p102) in GenBank database according to the structural properties and molecular weight of the protein. The transcription start site was located 216 bases upstream of the ATG start codon by RNA ligase mediated-rapid amplification of cDNA ends (RLM-RACE). Bioinformatic analysis of the isolated sequence revealed a lack of typical promoter TATA box and the presence of GC-rich motifs and CCAAT boxes recognized by Sp 1 and nuclear factor-Y among other putative binding sites for a number of transcription factors implicated in both basal and regulated processes. Electrophoretic mobility shift and supershift assays using nuclear extracts from human (HepG2) and rat (McA-RH7777) hepatoma cells demonstrated that nuclear factor-Y (NF-Y) transcription factor bound to the core sequences at ($-257, -253$), ($-290, -286$), and ($-370, -366$) upstream translation initiation site. The absence of TATA box and the location and reverse orientation of the CCAAT boxes in the promoter region strongly suggest a role for NF-Y in the regulation of transcription of SND p102 gene.

KEYWORDS: SND p102; gene promoter; transcription; nuclear factor-Y

INTRODUCTION

The staphylococcal nuclease domain–containing protein of 102 kDa (SND p102) is a rat protein localized in the liver endoplasmic reticulum (ER) whose

Address for correspondence: María José Martínez, Department of Physiology, Faculty of Medicine and Dentistry, University of the Basque Country, Sarriena s/n, 48940-Leioa, Bizkaia, Spain. Voice: 34-946-012-832; fax: 34-946-015-662.
 e-mail: mariajose.martinez@ehu.es

Ann. N.Y. Acad. Sci. 1091: 282–295 (2006). © 2006 New York Academy of Sciences.
doi: 10.1196/annals.1378.074

function and biological role is still unclear. The purified protein was previously demonstrated to be responsible for more than 90% of the hydrolysis of cholesteryl esters in rat hepatocyte microsomes,[1] but very recent evidence points to a role for the protein in the secretion of phospholipids into lipoproteins of very low density and high density.[2] Denomination of the protein as SND p102 (GenBank accession number AAU05374) was based on the presence of four repeated staphylococcal nuclease domains and a single Tudor domain in the amino acid sequence, and on a polypeptide molecular weight of 102 kDa. The cDNA (GenBank accession number AY697864) and amino acid sequence of SND p102 showed 98–99% homology with the sequences of rat p105 and human p100, two proteins that are members of a highly conserved family of nuclear transcriptional coactivators. Finding such extensive homology in the sequences of SND p102 and nuclear coactivators was unexpected, since SND p102 was exclusively immunolocalized in the ER of the liver of rat, mouse, and human as well as in HepG2 cells.[3] However, this extensive homology and localization is consistent with a number of studies reporting the presence of homologues of p100 protein in extranuclear compartments of mammalian cells and *Drosophila*,[4-6] which indicated additional functions of the protein rather than that of a transcriptional coactivator. Thus, a p100-like protein has been found in the ER and cytosolic lipid droplets from cow and mouse lactating mammary gland, in lipid droplets from mouse adipocytes, and in the ER of mouse liver,[4] suggesting a role for the protein in the formation or secretion of milk lipid droplets and/or in the storage of triacylglycerol lipid bodies in adipocytes.[4] The presence of p100 and SND p102 in the ER of mouse and rat liver makes likely their participation in the formation of lipoproteins or lipid bodies within hepatocytes. It is well known that the liver is a critical organ for the maintenance of whole-body lipid balance and plasma lipoprotein homeostasis. Hepatocytes produce and secrete very-low-density lipoproteins (VLDL) in a highly complex process. The ER is crucial for the proper recruitment of triacylglycerol, cholesteryl ester, phospholipid and cholesterol, and assembly with the apolipoprotein B into the lipoprotein particles to be released. In a very recent work using recombinant adenoviruses encoding the cDNA of SND p102 or a specific antisense sequence, we have demonstrated that the level of expression of SND p102 protein is directly related to the amount of phospholipid released into lipoprotein particles from infected hepatocytes.[2]

Based on the current knowledge of SND p102 protein, it is relevant to characterize the mechanisms of its transcriptional regulation. As an initial step toward understanding the factors that control SND p102 gene expression, we have isolated a 1,688-bp region corresponding to the gene promoter (GenBank accession number AY957585) and studied its characteristics and the putative transcription factors involved in the expression of the gene. Nuclear factor-Y (NF-Y) is an ubiquitous heterotrimeric transcriptional activator composed of three subunits NF-YA, B, and C, all necessary for DNA binding. NF-Y complexes with CCAAT boxes and has a role in the initiation and regulation of transcription in TATA-containing and TATA-less promoters.[7] We have found within the SND p102 promoter region three binding sites for NF-Y that might be involved in the regulation of gene transcription.

MATERIALS AND METHODS

Materials

TaKaRa LA Taq™ polymerase was obtained from TaKaRa Bio, Inc. (Shiga, Japan). Oligonucleotides were synthesized by Isogen Life Science (Ijsselstein, the Netherlands). HepG2 and McA-RH7777 cell lines were purchased from the American Type Culture Collection (ATCC, Manassas, VA, USA), as well as Eagle's minimum essential medium (EMEM), Dulbecco's modified Eagle's medium (DMEM), horse and fetal bovine sera. L-glutamine, penicillin and streptomycin were obtained from Sigma Aldrich (St. Louis, MO, USA).

Methods

Isolation of the Rat SND p102 Gene Promoter

A double polymerase chain reaction (PCR)-based method using the Rat Genome Walker kit (Clontech, Mountain View, CA, USA) was performed. Libraries of rat genomic DNA (*Eco* RV, *Dra* I, *Pvu* II, and *Ssp* I) were obtained by digestion of rat genomic DNA and binding to an adaptor sequence, complementary to sense oligonucleotides (AP). Antisense oligonucleotides specific to the cDNA sequence of the rat SND p102 gene (GSP) were designed with the Primer 3 program[8] (available at http://frodo.wi.mit.edu/cgi-bin/primer3/primer3_www.cgi). First PCR was performed with oligonucleotides AP2 (5' ACTATAGGGCACGCGTGGT 3') and GSP54-87 (5' CCCAGAGAGGACCATCTTGACGATGCCCCGTTG 3') and genomic DNA libraries as template. PCR products were diluted and used as templates for nested PCR with oligonucleotides AP2 and GSP85-56 (5' GAGGGTGC-CGGTGCTGGTGAGAGGTTGCAG 3'). Nested PCR products were electrophoresed, purified from the agarose gel with the High Pure PCR Product Purification Kit (Roche, Mannheim, Germany) and sequenced with an ABI Prism 3100 Genetic Analyzer (Applied Biosystems, Foster City, CA, USA) by the BigDye Terminator method (Faculty of Science and Technology, University of the Basque Country). New antisense oligonucleotides were used in order to elongate the isolated promoter sequence. First PCR was performed with oligonucleotides AP2 and PRMTER785 (5' GGAATGTG-GCCGCGAATCTTCCTGCGGTCTGG 3') and DNA libraries as template. Nested PCR was performed with oligonucleotides AP2 and PRMTER935 (5' CCTGCTGTTCCAGCTGTGGTTAGGAAAAGCCTTGG 3').

Bioinformatic Analysis of the Sequence

Identification of putative binding sites for transcription factors implicated in both basal and regulated transcription was done by bioinformatic analysis using TESS[9] (Transcription Element Search System, available at http://www.cbil.upenn.edu/tess), MatInspector[10] (Genomatix Software GmbH, Munich, Germany, available at http://www.genomatix.de/cgi-bin/matinspector/matinspector.pl), and Jaspar[11,12] (available at http://jaspar.cgb.ki.se/cgi-bin/jaspar_db.pl).

Identification of Transcription Initiation Site by RLM-RACE

Total RNA was extracted from isolated rat hepatocytes[13] with Trizol (Invitrogen Life Technologies, Carlsbad, CA, USA) using 1 mL of reagent per 10×10^6 cells. Ten μg RNA were treated with calf intestine alkaline phosphatase (CIP) to remove 5′-phosphates from fragmented RNA, and with tobacco acid pyrophosphatase (TAP) to remove the cap structure from full-length mRNA. An adapter oligonucleotide was ligated to RNA by T4 RNA ligase. Both CIP and TAP treatments and adapter ligation were performed using First-Choice RNA Ligase Mediated-Rapid Amplification of cDNA Ends (RLM-RACE) (Ambion, TX). RNA was reverse-transcribed using random decamers. cDNA was used as template for PCR with oligonucleotides 5′ RACE Outer Primer (5′ GCTGATGGCGATGAATGAACACTG 3′), which binds to the adaptor sequence previously ligated to RNA, and GSP55-77 (5′ ACCATCTTGACGATGCCCCGTTG 3′) complementary to region (+55, +77) of the rat SND p102 cDNA (A of ATG start codon is considered as +1). A nested PCR was performed with oligonucleotides 5′ RACE Inner Primer (5′ CGCGGATCCGAACACTGCGTTTGCTGGCTTTGATG 3′) and GSP5-17 (5′ CTTTGCGCGAATTCCATGTGTG 3′), complementary to the region (−5, +17) of the cDNA. The PCR product was digested with *Eco* RI and *Bam* HI (Roche, Germany) and cloned into pUC19 for sequencing as mentioned above.

Preparation of Nuclear Extracts

Extracts were prepared from HepG2 or McA-RH7777 cell cultures using the Nuclear Extraction kit (Panomics, Fremont, CA, USA). Human hepatoblastoma HepG2 cells were cultured in EMEM containing 2 mM L-glutamine, 100 U/mL penicillin, 100 mg/mL streptomycin and 10% fetal bovine serum for 96 h before nuclear extracts were obtained. Rat hepatoma McA-RH777 cells were grown in DMEM supplemented with 20% horse serum and 5% fetal bovine serum for 96 h prior to obtaining nuclear extracts. Protein concentration was determined by the Bradford method,[14] using bovine seroalbumin as standard.

Electrophoretic Mobility-Shift Assay (EMSA)

Probes were obtained by annealing complementary single-stranded oligonucleotides corresponding to three different putative NF-Y binding sites present in the SND p102 gene promoter. Bicatenary probes were labeled at the 3' end with digoxigenin using the DIG Gel shift kit second generation (Roche, Germany). For mobility shift assays, 3 µg of HepG2 nuclear extracts or 6 µg of McA-RH7777 nuclear extracts were incubated at room temperature for 15 min with 160 fmol of the 3'-DIG-labeled dimerized probe, 1 µg poly[d(I-C)], 100 ng poly-L-lysine in 20 mM HEPES (pH 7,6), 1 mM EDTA, 10 mM $(NH_4)_2SO_4$, 1 mM DTT, 0,2% (w/v) Tween 20, 30 mM KCl. For specific or nonspecific competition assays (sequences of the probes are shown below), 25- to 125-fold molar excess of unlabeled probe was used.

NFY (−271, −239).wt 5' GGTTCTGGACGCTA*ATTGG*CCAGAGGGAGGTGG 3'
NFY (−303, −271).wt 5' AAACAGAGTGCT*ATTGG*CTTGAGTTCTAGCAG 3'
NFY (−384, −352).wt 5' TTTGCCCGTCACCG*CAAAT*CAGCTCTTTCAGCG 3'
NFY (−270, −241).mut 5' GTTCTGGACGCTA*CCTGT*CCAGAGGG-AGG 3'
NFY (−303, −275).mut 5' AAACAGAGTGCCT*CCTGT*CTTGAGTTCTA 3'
NFY (−381, −353).mut 5' GCCCGTCACCG*AAAGG*CAGCTCTTTCAGC 3'

For supershift experiments, 1.5 µg of anti-NF-Y antibody (BD Biosciences Pharmingen; Franklin Lakes, NJ, USA) was added, and incubated for 15 additional minutes at room temperature. Electrophoresis was performed at 4°C in 6% nondenaturing polyacrylamide gels in 45 mM Tris-borate, 1 mM EDTA (pH 8) at 110V for 70 min. Complexes were transferred to a HyBond N+ nylon membrane (Amersham Biosciences, Buckinghamshire, UK) at 300 mA for 30 min. Oligonucleotides were cross-linked at 120 mJ in a Stratalinker 1800 (Stratagene, LaJolla, CA, USA) and visualized by an enzyme immunoassay using anti-digoxigenin-AP. Chemilumiscence signal detection was performed in Hyperfilm™ ECL (Amersham Biosciences).

RESULTS

Isolation of the Rat SND p102 Gene Promoter Region

A genome walking method based on two consecutive PCRs was used to isolate the promoter region of the rat SND p102 gene. Rat genomic DNA libraries, obtained by digestion of genomic DNA with different endonucleases (*Eco* RV, *Dra* I, *Pvu* II, and *Ssp* I) and ligation to an adaptor sequence, were used as template for the first PCR. Both DNA libraries and sense oligonucleotides were part of the kit used. Antisense oligonucleotides complementary to the 5' end of the rat SND p102 cDNA were designed. For the first PCR, oligonucleotides AP2, complementary to the adaptor sequence of genomic DNA, and GSP54-87, complementary to region (+54, +87) of the SND p102 cDNA

(A from ATG translation start codon is considered as +1) were used. PCR products of different sizes were amplified from each library, and they were used as templates for nested PCR with oligonucleotides AP2 and GSP85-56, complementary to region (−85, −56) of the SND p102 cDNA. Fragments of 1.27 kb (*Eco* RV), 1.29 kb (*Dra* I), 0.89 kb (*Pvu* II), and 1.20 kb (*Ssp* I) were amplified and purified. Sequence analysis showed that shorter fragments were included in the 1.29-kb amplicon. A region of 1,256 pairs of bases upstream of the ATG start codon was obtained, whose last 155 bases corresponded to the 5′ region of the cDNA sequence reported for the rat SND p102 gene and 1,101 bases corresponded to a new sequence upstream of the cDNA sequence.

A second double-PCR run was performed in order to elongate the isolated sequence. Oligonucleotides AP2 and PRMTER785, complementary to the region (−785, −754) of the new isolated sequence, were used for the first PCR, and AP2 and PRMTER935, complementary to the region (−935, −901) for the second PCR. Fragments of 0.58 kb (*Eco* RV), 1.07 kb (*Dra* I), and 0.50 kb (*Pvu* II) were amplified. Sequence analysis showed again that the shorter fragments were included in the 1.07-kb amplicon. A new fragment of 989 bases was obtained whose last 411 bases corresponded to the 5′ end of the previously isolated promoter sequence.

In this way, a region of 1,899 bases was isolated (FIG. 1) whose last 211 bases corresponded to the cDNA sequence for SND p102 gene (GenBank accession number AY697864) and 1,688 bases corresponded to a new sequence upstream the cDNA. Sequence analysis using the Blast Rat Genome tool from NCBI established the chromosomal location of the new isolated sequence on chromosome 4, coinciding with the previously reported location for the SND p102 gene. Considering all these data, the new isolated sequence was considered as part of the promoter region for the SND p102 gene.

Determination of Transcription Initiation Site of Rat SND p102 Gene in Hepatocytes

RLM-RACE using RNA isolated from rat hepatocytes was performed. By this method, retrotranscription to cDNA was selectively achieved from full-length capped mRNA, as explained in the MATERIALS AND METHODS section, and was used to determine the transcription initiation site of the rat SND p102 gene. Two specific antisense oligonucleotides complementary to regions (+55, +77) and (−5, +17) of the 5′ end of the SND p102 cDNA were used for amplification from cDNA. Two products, differing only in 30 bases, were amplified in the nested PCR (FIG. 2) and cloned into pUC19 vector for sequencing. Sequences from seven clones of RLM-RACE products were analyzed. The most upstream 5′ ends for these fragments were −177, −179, −182 for the shorter fragment, and −208, −212 (2), and −216 for the longer fragment. These multiple 5′ ends may reflect incomplete reverse transcription. The existence of two

```
AATTACATT TTTAATATAT ATGTTATATA TAATGTATAT CTGTGTGTGT -1851
ATGCAAGCAT CACGTCACCG TTTGCATAAG CAGTCAGAAG ACGACTTGCA -1801
GGAGTTGGTT TTCTCCTTCC ACCATGTAGG TCCCAGGGAT TGAACTCGGG -1751
TCAACAGTCT TGCATGCAAA TAACTTTACC TCTTGGACCA TTCTGAGCTT -1701
CTTGTTCACT TAAAAATAAA CAGATGAAAA ATAATACATC AATGCCCGCC -1651
TTTGAAGAAA CAGATTGTAT GAACTTTGTA GACTTGTCCT GAGGACTAAA -1601
TTATATTCTG CAGGGAAAAA GTATTTAACA GGTAGTATAT TCTGTGCATG -1551
TTCATGCTCC ACATGATTTG GTTGTTAGGA TTGACTAGTA GGATCAATGG -1501
TTGTGTAGAA ATAGAAAATT AGTACAAGAC AAAAGAGAAC TGGCTGAGTA -1451
GGGGAAAAGG ACAGTAATTA TGATCAGAAA ATGAGATGTT TGGAATTACA -1401
                                              HNF-4
ATGTCAACTG CTTCAAATAT CCTGGAGAGT GTGGTTAAAT TGTTGCAAAG -1351
GTCAAATTCA TTAAAGACAA GGTCCTGGAG CTGAGATTTC AGGGTACTGG -1301
GCAAACCAGA ATTTATATAT ATTCGAATGT ATCCATAAAG ACCCACTTTC -1251
TTTTTGAATA GGGTGAATTC ATGGTATGGA CTTGCCTTTG GAGATATAGG -1201
AAGATGCCAA CCCCAAAACT TGACCCTGAT TTGAGACATG GGTGACTACT -1151
CAAACTTAAG CTATAGTGGA AGTGGTGTCT TTAGCATAGG GCCAGGTTTC -1101
ATTATTAAGT AATGCATACT AGAACCTATG CTCTAAAAAG GGGCACTCCA -1051
AGGACATAAA TTCTCCCTAA ACATCCAATT GAATATGGCT TTCCTGTCCC -1001
                           C/EBP
ATCTGGAGTT GAGGTTAGGA GTGAGGTAAG GGGTCGTCTC CCGCCCAGCC  -951
TTGGGATGGC ATATCCCAAG GCTTTTCCTA ACCACAGCTG GAACAGCAGG  -901
GGAAGTAGCG GAAAAAATGA GGCTATTCTC AGTGAGCAGG ACGCGCGTTA  -851
TCTAAGAGAT ATGCTTATTA CCTCCATCAA CAAACATTAG TGGAACTCAC  -801
AAGACCTGTG ATACTCCAGA CCGCAGGAAG ATTCGCGGCC ACATTCCTGG  -751
GCAAGACTAA CACCAGGAGA TTATCCAGTG GCTGCGATTT TCAAAAGTTT  -701
               STAT 6
CTTCTGTTCT GCTGAAGTGA AACGTCGGAA CTCCGAGAGC AGTGAGATTC  -651
                    STAT 1
CTTCAGGACG GCTTCCTTAG TTCCAGGACC CAAGCCCTTG TTGTCCAGTT  -601
                               USF
CCGGTGTACA GCTCCGGAAC CTCCGGGTAC ACGTGGGCGC TTTGCCGCTT  -551
       Sp1            Sp1              Sp1            Sp1
GAGGGGCGGG GCAGGGGCGG AACCAGGCAG GAGGGCGAGG CCGGCCCCGC  -501
                   Sp1
CTTCAGGGCT CCCATGGCTC CCCGCGCCCC ACGGCCAGAT GCTGACGTGT  -451
CCTTTCTTTC CCACTACACC TCCCGACACC ACCGGCTCAG ACTTCCCCGG  -401
   Sp1                        NF-Y
AAGTCCCGCC TCAAAATTTG CCCGTCACCG CAAAT CAGCT CTTTCAGCGC  -351
  STAT 1                            Sp1
TTCCTGGAGC TCCGCACCCG TTTTGGGACC AAGGGGCGGA GCTTCATAAA  -301
         NF-Y                             NF-Y
CAGAGTGCCT ATTGG CTTGA GTTCTAGCAG GTTCTGGACG CTA ATTGG CC -251
           →
AGAGGGAGGT GGCGCTCTGG CACGCTTGCG CGGCGAGTAG AACGTGTGGC  -201
GGCGGCCGAG ATCGCGTCTC CTCTTTTGCT TCCAGTCCCG CTGCTGTTAT  -151
AGTGTGCGCC TGGCGCTTCT ATCCAGTCCA CCGACGGTAG CCTGTCGTCC  -101
GCCTTCTCCT TGTCGCTGCA ACCTCTCACC AGCACCGGCA CCCTCTCTGA   -51
CACTTCCAGT CCGGCCGGCC GCCCACTCT CTGTCTTTCC AGCTCCACAC     -1
ATGGCCTCCG CGCAAAGCAG CGGCTCCTCC GGGGGACCCG CGGTCCCCAC   +51
```

FIGURE 1. Isolated promoter region of the rat SND p102 gene (GenBank AY957585). Translation start site (**A**) is considered as position +1 of the sequence. Initiation of transcription was localized at position −216 by RLM-RACE and is marked in the sequence with an *arrow*. Binding sites for some of the transcription factors predicted by bioinformatic analysis are underlined. Binding sequences for NF-Y confirmed by EMSA are shown in a box. HNF-4 hepatocyte nuclear factor 4; C/EBP CCAAT/enhancer binding protein; STAT signal transducer and activator of transcription; USF upstream stimulatory factor.

different transcription start sites (182 and 216 bases upstream translation start site) could be due to formation of a secondary structure in the 5' end of RNA and synthesis of two different fragments to cDNA. Taking all this into account, we may conclude that the transcription initiation site was situated 216 bases upstream of the translation start site.

Bioinformatic Analysis of the Rat SND p102 Gene Promoter Sequence

Bioinformatic analysis of the isolated promoter sequence was performed using TESS, Matinspector software from Genomatix, and Jaspar. As shown in FIGURE 1, the analysis revealed the presence of different putative binding sites for transcription factors involved in both basal and regulated transcription processes. The isolated sequence lacked TATA box, but presented CCAAT boxes, recognized by the transcription factor NF-Y, in positions (-257, -253), (-290, -286), and (-370, -366) as well as a region rich in GC boxes, recognized by the protein Sp1,[15] in positions (-547, -531), (-519, -500), (-482, -472), (-398, -390), and (-316, -312) of the promoter sequence. Putative binding sites for upstream stimulatory factor (USF),[16] CCAAT/enhancer binding protein (C/EBP),[17] and signal transducer and activator of transcription

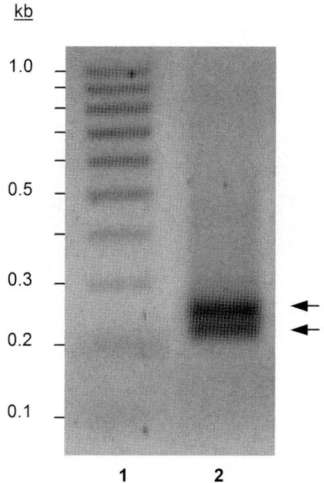

FIGURE 2. Identification of the transcription initiation site of SND p102 gene. RLM-RACE was performed and followed by two consecutive PCR with antisense oligonucleotides complementary to regions (+55, +77) and (-5, +17) of SND p102 cDNA. Two products, of 275 and 245 bases, were amplified in the second PCR and cloned into pUC19 vector for sequencing. Translation initiation site was assigned to base G located 216 bases upstream ATG start codon. *Lane* 1: DNA molecular weight marker. *Lane* 2: RLM-RACE product after nested PCR.

(STATs)[18] were also found. The analysis also revealed several DNA elements that matched the consensus binding sequences for liver-enriched transcription factors, such as hepatocyte nuclear factor 4 (HNF-4).[19]

Binding of NF-Y to the SND p102 Gene Promoter Region

Binding of transcription factor NF-Y to the isolated promoter sequence was analyzed by electrophoretic mobility shift and supershift assays. Oligonucleotides complementary to regions (-271, -239), (-303, -271), and (-384, -352), which contain CCAAT boxes recognized by NF-Y, were designed and labeled with digoxigenin. Incubation of the probe for region (-384, -352) with nuclear extracts of HepG2 (FIG. 3) or McA-RH7777 (FIG. 4) cells resulted in a shifted band. Competitive assays with increasing amounts of specific, nonspecific and mutated unlabeled probes were performed, confirming the shift (FIG. 3A, lanes 2–5; FIG. 4, lanes 4–9). The same binding pattern of NF-Y to region (-384, -352) of the promoter was observed for regions (-271, -239) and (-303, -271) (data not shown). To confirm the specificity of NF-Y binding to the three analyzed CCAAT boxes, supershift analysis was performed.

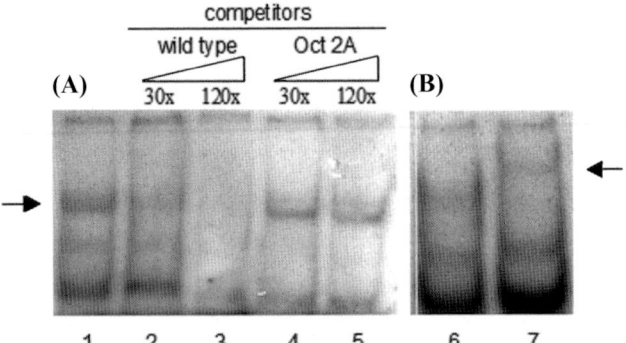

FIGURE 3. Electrophoretic mobility shift (**A**) and supershift (**B**) assays to analyze NF-Y binding to the CCAAT box located at (-370, -366) of the rat SND p102 gene promoter using nuclear extracts from HepG2 cells. Incubation of the double-stranded oligonucleotide labeled with digoxigenin, corresponding to the region (-384, -352) of the promoter, with nuclear extracts resulted in a shift (*lanes* 1 and 6). Incubation with a specific antibody for NF-Y resulted in a supershift (*lane* 7). Competition assays using 30- and 120-fold molar excess of specific (wild-type) and nonspecific (Oct2A) competitors were performed. Incubation with excess of specific unlabeled probe resulted in the absence of the DNA–protein complex formation (*lanes* 2 and 3), whereas incubation with excess of nonspecific probe for transcription factor Oct2A (*lanes* 4 and 5) did not inhibit the formation of the DNA–protein complex.

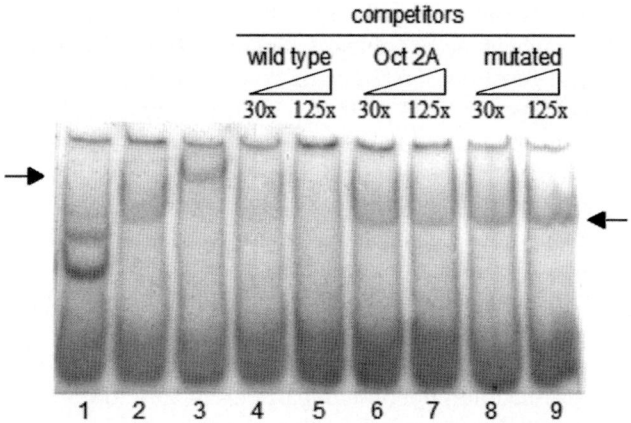

FIGURE 4. Electrophoretic mobility shift and supershift assays for NF-Y binding to the CCAAT box located at (−370, −366) of the rat SND p102 gene promoter. Nuclear extracts from HepG2 (*lane* 1) and McA-RH7777 (*lanes* 2–9) were incubated with double-stranded oligonucleotide corresponding to the region (−384, −352) of the promoter and labeled with digoxigenin. A DNA–protein complex resulting in a shifted band was observed in the presence of HepG2 (*lane* 1) or McA-RH7777 (*lane* 2) nuclear extracts. Incubation with a specific antibody for NF-Y resulted in a supershift (*lane* 3). Competition assays using 30- and 125-fold molar excess of different competitors were performed. Incubation with excess of specific unlabeled probe NF-Y (−384, −352).wt resulted in the absence of the complex formation (*lanes* 4 and 5), whereas incubation with excess of nonspecific probe for transcription factor Oct2A (*lanes* 6 and 7) or mutated specific probe NF-Y (−381, −353).mut (*lanes* 8 and 9) did not inhibit the formation of the complex.

The addition of a monoclonal antibody anti-NF-YA produced a single supershifted complex in HepG2 (FIG. 3B, lane 7) and McA-RH7777 (FIG. 4, lane 3) nuclear extracts.

DISCUSSION

In this report, we isolated and characterized a region of 1,688 bases corresponding to the promoter region of the rat SND p102 gene which was registered in GenBank with the accession number AY957585. The isolated promoter region was extensively analyzed with bioinformatic tools and some elements were identified as putative binding sites for transcription factors implicated in basal or regulated transcription. Homology studies showed the chromosomal location of the promoter region at the rat chromosome four genomic contig NW-047689.4. It matched 98% with the 5′ flanking region of the gene identified as Snd1. This is the name given by NCBI to the *Staphylococcal nuclease domain–containing 1* gene, including SND p102 and p105 coactivator genes (GenBank accession numbers AY697864 and U83883). The proteins encoded by these two genes displayed similar molecular weights and structural domains (four staphylococcal nuclease domains and one Tudor domain).

The biological function of SND p102 is yet poorly understood. As far as we know, homologues of SND p102 can act as transcriptional coactivators in the nuclei of cow, mouse and rat cells,[4,6] but additional functions have to be considered since several homologue proteins have been identified in subcellular fractions other than nuclei. In the cytosol of adipocytes and mammary gland and in the ER of mammary gland and liver, the p100 protein has been described as a participant in the formation, growth or secretion of lipid bodies.[4] This is consistent with recent evidence that pointed to a role for the ER SND p102 in VLDL generation and secretion.[2] It was found that adenovirus-mediated differential expression of SND p102 in rat hepatocytes modulated the amount of phospholipid secreted into lipoprotein particles of very low and high density. Assuming a potential involvement of this protein in lipid homeostasis in the liver, it is of great importance to characterize the requirements and mechanism for SND p102 gene transcription. Thus, bioinformatic analysis of the isolated sequence allowed us to identify several putative elements involved in both basal and regulated transcription. The promoter has no consensus TATA box or initiator element, but in contrast, there are GC-rich regions with putative binding sites for Sp1 transcription factor and several CCAAT boxes recognized by NF-Y. Both Sp1 and NF-Y are ubiquitous nuclear factors which usually function synergistically and are effective for basal transcription of several genes.[20–23]

TATA-less promoters are typical of housekeeping genes, together with the presence of GC-boxes and multiple transcriptional start sites. The SND p102 promoter matches some of the characteristics of these housekeeping gene, such as the lack of TATA box and the presence of GC-boxes. However, the multiplicity of transcriptional start sites is a point of discussion because we found a main initiation site at −216 bp and a secondary start point 30 bp downstream. RLM-RACE was used to localize the transcription initiation site of SND p102 gene. The amplification of two products in nested PCR could be probably due to the formation of an internal structure in the 5′ end of RNA, which presents a high percentage of GC and autocomplementarity, more than the presence of two transcription initiation sites. For this reason, we have assigned the origin of transcription to base G located 216 bases upstream translation initiation site.

In addition to GC-boxes recognized by Sp1, the promoter also contains several potential binding sites for NF-Y, which recognizes CCAAT boxes.[7] These are common elements in eukaryotic promoters that can act differently in promoters with or without TATA box. In the absence of TATA box, NF-Y can function as a recruiting factor for the general transcriptional machinery that allows polymerase II to find the start site.[24] In most of the TATA-less promoters, CCAAT boxes appear in reverse orientation, ATTGG, and are generally distributed in (−40, −80) region. This is consistent with the location of CCAAT boxes in the SND p102 promoter at positions −42 and −75. These two ATTGG elements appear in close proximity and have been demonstrated to be specific binding sites for NF-Y by EMSA experiments. Another CCAAT box is found in forward direction in a more distant region −165 which is also bound specifically by NF-Y. Because there are a wide range of proteins that recognize and bind CCAAT boxes, it is recommended that NF-Y be identified as the protein involved in the DNA–protein complex generated in the promoter sequence. When we tested nuclear extracts from different species, human HepG2 and rat McA-RH7777 hepatoma cells, different DNA–protein complexes resulted in EMSA competition experiments, depending on the source of nuclear proteins. However, in both cases a common band is directly supershifted by the anti-NF-Y antibody, demonstrating that NF-Y is really the protein bound to the CCAAT elements included in the probes. These results reinforce the idea that CCAAT/ATTGG boxes may represent functional promoter elements and that the binding of NF-Y could be essential for transcription of SND p102 gene. Nevertheless, further studies have to be performed to confirm the role of NF-Y and to clarify which other transcription factors might be crucial for the basal and regulated transcriptional activity of this promoter.

ACKNOWLEDGMENTS

The authors thank Montse Busto for her expert technical help with cell cultures. This work was supported by DGICYT (BMC2001-0067), FIS (Red G03-015) and the Basque Government (IE05-147). Lorena Rodríguez, and Nerea Bartolomé, are recipients of predoctoral fellowships from the Spanish Ministry of Education and Science and the Basque Government, respectively.

REFERENCES

1. CRISTÓBAL, S., B. OCHOA & O. FRESNEDO. 1999. Purification and characterization of a cholesteryl ester hydrolase from rat liver microsomes. J. Lipid Res. **40:** 715–725.
2. PALACIOS, L., B. OCHOA, M.J. GÓMEZ-LECHÓN, et al. 2006. Overexpression of SND p102, a rat homologue of p100 coactivator, promotes the secretion of lipoprotein phospholipids in primary hepatocytes. Biochim. Biophys. Acta **1761:** 698–708.
3. FRESNEDO, O., M. LÓPEZ DE HEREDIA, M.J. MARTÍNEZ, et al. 2001. Inmunolocalization of a novel cholesteryl ester hydrolase in the endoplasmic reticulum of murine and human hepatocyte. Hepatology **3:** 662–667.
4. KEENAN, T.W., S. WINTER, H.R. RACKWITZ & H.W. HEID. 2000. Nuclear coactivator protein p100 is present in endoplasmic reticulum and lipid droplets of milk secreting cells. Biochim. Biophys. Acta **1523:** 84–90.

5. CAUDY, A.A., R.F. KETTING, S.M. HAMMOND, *et al.* 2003. A micrococcal nuclease homologue in RNAi effector complexes. Nature **425**: 411–414.
6. BROADHURST, M.K., R.S.F. LEE, S. HAWKINS & T.T. WHEELER. 2005. The p100 EBNA-2 coactivator: a highly conserved protein found in a range of exocrine and endocrine cells and tissues in cattle. Biochim. Biophys. Acta **1681**: 126–133.
7. MANTOVANI, R. 1998. A survey of 178 NF-Y binding CCAAT boxes. Nucleic Acids Res. **26**: 1135–1143.
8. ROZEN, S. & H.J. SKALETSKY. 2000. Primer3 on the WWW for general users and for biologist programmers. *In* Bioinformatics Methods and Protocols: Methods in Molecular Biology. S. Krawetz & S. Misener, Eds.: 365–386. Humana Press, Totowa, NJ.
9. SCHUG, J. & G.C. OVERTON. 1997. TESS: Transcription Element Search Software on the WWW. Technical Report CBIL-TR-1997-1001-v0.0. School of Medicine. University of Pennsylvania.
10. CARTHARIUS, K., K. FRECH, K. GROTE, *et al.* 2005. MatInspector and beyond: promoter analysis based on transcription factor binding sites. Bioinformatics **21**: 2933–2942.
11. SANDELIN, A., W. ALKEMA, P. ENGSTRÖM, *et al.* 2004. JASPAR: an open access database for eukaryotic transcription factor binding profiles. Nucleic Acids Res. **32**: D91–D94 [Database issue].
12. LENHARD, B. & W. WASSERMAN. 2002. TFBS: computational framework for transcription factor binding site analysis. Bioinformatics **18**: 1135–1136.
13. ISUSI, E., P. ASPICHUETA, M. LIZA, *et al.* 2000. Short- and long-term effects of atorvastatin, lovastatin and simvastatin on the cellular metabolism of cholesteryl esters and VLDL secretion in hepatocytes. Atherosclerosis **153**: 283–294.
14. BRADFORD, M.M. 1976. A rapid and sensitive method for the quantitation of microgram quantities of protein utilizing the principle of protein-dye binding. Anal. Biochem. **72**: 248–254.
15. BOUWMAN, P. & S. PHILIPSEN. 2002. Regulation of the activity of Sp1-related transcription factors. Mol. Cell. Endocrinol. **195**: 27–38.
16. CORRE, S. & M.D. GALIBERT. 2005. Upstream stimulating factors: highly versatile stress-responsive transcription factors. Pigment Cell Res. **18**: 337–348.
17. ROESLER, W.J. 2001. The role of C/EBP in nutrient and hormonal regulation of gene expression. Annu. Rev. Nutr. **21**: 141–165.
18. BROMBERG, J. & J.E. DARNELL JR. 2000. The role of STATs in transcriptional control and their impact on cellular function. Oncogene **19**: 2468–2473.
19. SCHREM, H., J. KLEMPNAUER & J. BORLAK. 2002. Liver-enriched transcription factors in liver function and development. Part I: the hepatocyte nuclear factor network and liver-specific gene expression. Pharmacol. Rev. **54**: 129–158.
20. RODER, K., S.S. WOLF, K.J. LARKIN & M. SCHWEIZER. 1999. Interaction between the two ubiquitously expressed transcription factors NF-Y and Sp1. Genes **234**: 61–69.
21. SUNYAKUMTHORN, P., T. BOONSAEN, V. BOONSAENG, *et al.* 2005. Involvement of specific proteins (Sp1/Sp3) and nuclear factor Y in the basal transcription of the distal promoter of the rat pyruvate carboxylase gene in beta-cells. Biochem. Biophys. Res. Commun. **329**: 188–196.
22. TAPIAS, A., P. MONASTERIO, C.J. CIUDAD & V. NOE. 2005. Characterization of the 5′-flanking region of the human transcription factor Sp3 gene. Biochim. Biophys. Acta **1730**: 126–136.

23. XIONG, S., S.S. CHIRALA & S.J. WAKIL. 2000. Sterol regulation of human fatty acid synthase promoter I requires nuclear factor-Y and Sp1 binding sites. Proc. Natl. Acad. Sci. USA **97:** 3948–3953.
24. MANTOVANI, R. 1999. The molecular biology of the CCAAT-binding factor NF-Y. Gene **239:** 15–27.

The cAMP-Responsive Unit of the Human Insulin-Like Growth Factor–Binding Protein-1 Coinstitutes a Functional Insulin-Response Element

GHISLAINE SCHWEIZER-GROYER,[a] GUILLAUME FALLOT,[b] FRANÇOISE CADEPOND,[a] CHRISTELLE GIRARD,[a] AND ANDRÉ GROYER[b]

[a]*INSERM U.788, 94276 Le Kremlin-Bicêtre Cédex, France*

[b]*INSERM U.773, Faculté de Médecine Xavier Bichat, 75870 Paris Cédex 18, France*

> ABSTRACT: Insulin-like growth factor–binding protein-1 (IGFBP-1) is one of the genes involved in glucose homeostasis. *In vivo*, its level is increased by counter-regulatory hormones (glucocorticoids and glucagon via its second messenger cAMP) and decreased by insulin, these variations being primarily correlated with IGFBP-1 gene transcription. Previous reports described a functional insulin response element (IRE), immediately 5′- to the glucocorticoid response element (GRE). This IRE has been shown to mediate partial inhibition (1) of basal IGFBP-1 promoter activity and (2) of glucocorticoid-induced stimulation of gene transcription by insulin. In this work, using human HepG2 hepatoma cells as a model system, we showed: (1) that insulin inhibited both basal and cAMP-induced hIGFBP-1 promoter (nt-1 to -341) activity; (2) that in the absence of insulin, forkhead box class O (FOXO) transcription factors enhance constitutive hIGFBP-1 promoter activity without interfering with the stimulatory effect of cAMP; (3) that PI-3′ kinase signaling is involved in the inhibition of constitutive and cAMP-induced promoter activities by insulin; (4) that wild-type FOXO-1 mediates the inhibitory effect of insulin on the promoter, although FOXO-1$_{Ala3}$, a nonphosphorylatable mutant of FOXO-1, does not; (5) that the cAMP-responsive unit (CRU), that includes a putative IRE (nt-265 to -282) and a cAMP responsive element (CRE; nt-258 to -263), is sufficient per se to mediate both cAMP stimulation of a heterologous promoter, and inhibition of both basal and cAMP-induced promoter activities by insulin; and (6) that the inhibitory effects of insulin on the isolated CRU are mediated by the FOXOs. This

Address for correspondence: Dr. Ghislaine Schweizer-Groyer, Inserm U.788, 80, rue du Général Leclerc, 94276 Le Kremlin-Bicêtre Cédex, France. Voice: +33-1-49-59-18-93; fax: +33-1-45-21-19-40.
 e-mail: groyer@kb.inserm.fr

Ann. N.Y. Acad. Sci. 1091: 296–309 (2006). © 2006 New York Academy of Sciences.
doi: 10.1196/annals.1378.075

study is the first evidence for the occurrence of a second IRE within hIGFBP-1 promoter sequences, IRE_{CRU}, located 5′- to the CRE.

KEYWORDS: IGFBP-1; promoter; insulin-response element; FOXOs; Akt; human hepatoma cell

INTRODUCTION

The insulin-like growth factor–binding proteins (IGFBPs) are a family of proteins that bind IGF-I and IGF-II with high affinity and play a key role in their bioavailability, both in the serum and in the extracellular milieu at the target tissue level (reviewed in Ref. 1). IGFBP-1 is primarily synthesized and secreted by hepatocytes, and its circulating level is rapidly regulated by nutrient availability in the serum. In normal hepatocytes as well as in hepatoma cells, *IGFBP-1* gene transcription is enhanced by glucocorticoids and glucagon (via its second messenger cAMP), and inhibited by insulin in a dominant manner.[2,3] The sequences specifying both the liver specificity and the multihormonal regulation of *IGFBP-1* gene transcription are integrated within a promoter/proximal enhancer of ∼340 nt 5′- to the transcription initiation site (promoter thereafter).[4–7] A functional insulin response element (IRE) has been identified within the promoter, immediately 5′- to the glucocorticoid response element (GRE).[5–8] This IRE was shown to mediate partial inhibition (1) of basal IGFBP-1 promoter activity and (2) of glucocorticoid-induced stimulation of gene transcription by insulin. Studies on the mechanism mediating insulin inhibition of IGFBP-1 gene expression through this IRE (IRE_{GRU}) showed that it relies on the disruption of its interaction with a subfamily of winged-helix transcription factors, the forkhead box class O (FOXOs). In the absence of insulin the FOXOs interact with the IRE_{GRU} and enhance promoter activity, whereas in its presence FOXOs' phosphorylation by PKB/Akt, a downstream kinase in phosphatidylinositol-3′-kinase (PI-3′K) signaling, promotes their nuclear exclusion, and thus the suppression of their transactivation properties on *IGFBP-1* gene transcription.[9–13]

A cAMP-responsive element (CRE; nt-263 to -258) has also been identified in the human IGFBP-1 promoter,[14] and insulin exerts a dominant inhibitory effect on cAMP-induced *IGFBP-1* gene transcription.[3,6,14,15] We have previously noticed that the A/T-rich regions located immediately 5′-to the CRE (nt-265 to -285) and to the GRE (nt-102 to -117) interact with the same liver-enriched transcription factors (HNF-3α, DBP, and the C/EBP family of bZIP proteins).[16]

In the present study, we addressed the question of whether the A/T-rich-sequence located 5′- to the CRE could mediate the inhibitory effect of insulin on basal and cAMP-induced *hIGFBP-1* gene transcription. We report for the first time that this promoter sequence functions as an authentic IRE, and that FOXO transcription factors mediate the inhibitory effect of insulin.

MATERIALS AND METHODS

Plasmids

pIGFBP-1-Luc : Genomic hIGFBP-1 sequences (nt-341 to +92 relative to the cap site) were excised from pG1.15[17] and inserted at the Hind III restriction site of pFLITS (generous gift of Paul Steenbergh) after having engineered Hind III cohesive ends.

pCRU-Luc : hIGFBP-1 promoter sequences spanning the CRU (nt-330 to -217) were amplified by polymerase chain reaction (PCR). The purified amplicon was inserted between the Hind III and BamH I sites of pTk-Luc, a reporter vector containing thymidine kinase (Tk) promoter sequences (nt-83 to +57) upstream of the luciferase reporter gene.

pFOXO expression vectors : pFOXO-1 (wild-type), pFOXO-1 $_{Ala3}$ (mutated at positions Thr^{24}/Ala, Ser^{256}/Ala, Ser^{319}/Ala, which cannot be phosphorylated by Akt), and pFOXO-1$_{AS}$ (which drives the expression of a 511bp, antisense FOXO-1 transcript) were described elsewhere.[10]

Cell Culture and Transfection

HepG$_2$ human hepatoblastoma cells were grown in DMEM containing 4.5 g/L glucose, 10% fetal calf serum, 2 mM L-glutamine, 100 U/mL penicillin, and 100 μg/mL streptomycin. The cells were maintained in 5% CO$_2$ in air.

A total of 10^6 HepG$_2$ cells was transfected with pIGFBP-1-Luc (4 μg), or pCRU-Luc (4 or 6 μg), and with 3 μg of pSVF-CAT (an internal monitor of transfection efficiency) using polyethyleneimine.[18] When appropriate, pFOXO-1, pFOXO-1$_{Ala3}$ (0.25-1 μg), or pFOXO-1$_{AS}$ (10 μg) were added to the transfection mixture.

CAT and luciferase assays were performed according to standard procedures.[19,20] Luciferase activity was normalized relative to that of CAT in the same cell extract. Luc/CAT ratios were used to compare the variations in promoter activity.

RESULTS

Inhibition of Basal and cAMP-Induced hIGFBP-1 Promoter Activities by Insulin: Presence of a Putative IRE in the CRU

In/ex vivo, insulin exerts a dominant negative effect on basal and on glucocorticoid- or cAMP-induced *IGFBP-1* gene transcription. A 19-bp-long palindromic sequence (IRE$_{GRU}$), partially overlapping with the 5'-end of the GRE, has been shown to mediate most of the insulin effects.[5,8] In this connection, we have noticed that another 19 bp palindromic sequence, located

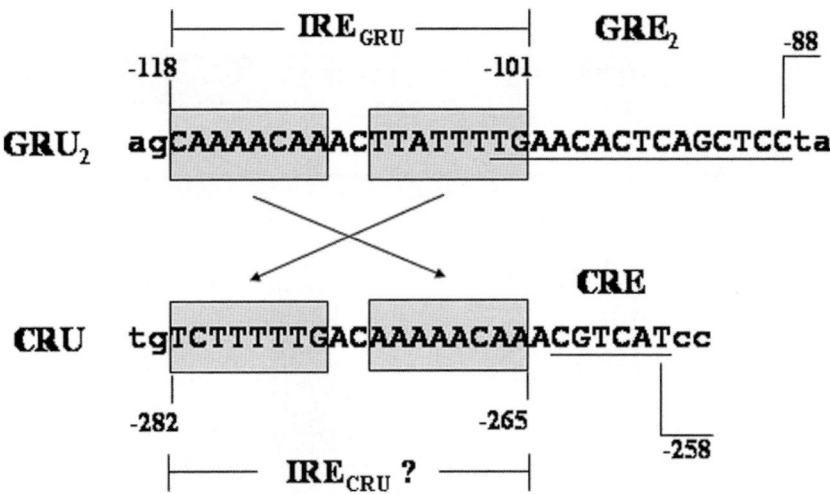

FIGURE 1. hIGFBP-1 glucocorticoid-response unit and putative cAMP-response unit. The glucocorticoid- and cAMP-response elements of the hIGFBP-1 gene are underlined. The two hemi-palindromes of the insulin-response element (IRE) located 5′ to the GRE and of the putative IRE located 5′ to the CRE are boxed and sequence similarities between them are highlighted.

immediately 5′- to the CRE, shares a high degree of similarity ($\geq 75\%$ for each half of the palindrome) with the IRE_{GRU} (FIG. 1). We decided to study the effects of insulin on cAMP-induced hIGFBP-1 promoter activity, and to investigate whether the 19-bp palindrome localized 5′- to the CRE was relevant as a functional IRE on both basal and hormone-stimulated promoter activity.

In a first set of experiments, the study was performed on pIGFBP-1-Luc-transfected HepG$_2$ cells (FIG. 2A). A significant luciferase activity was observed, demonstrating that promoter sequences (up to nt-341 relative to the transcription start site) were sufficient to yield a constitutive promoter activity in this cell line. This constitutive promoter activity was inhibited by insulin ($47.6 \pm 4.7\%$), consistent with previous observations.[2,7,8,21] When pIGFBP-1-Luc-transfected cells were treated with forskolin, an adenylate cyclase activator that increases intracellular cAMP levels, hIGFBP-1 promoter activity was enhanced 2.8 ± 0.3-fold. As expected, combined forskolin and insulin treatment inhibited the cAMP-dependent increase in promoter activity by $87.9 \pm 8.5\%$ (FIG. 2B; see also Refs. 2 and 6).

We have previously observed that HNF3 interacts with the palindromic sequence localized 5′- to the CRE,[16] as it binds to the IRE_{GRU} of the *IGFBP-1* and *PEPCK* genes.[16,22–24] However, in neither case was HNF3 able to mediate the negative effects of insulin on gene expression. Other forkhead transcription factors, the FOXOs, enhance basal *IGFBP-1* gene expression in an IRE_{GRU}-dependent manner, are phosphorylated in response to insulin, and mediate

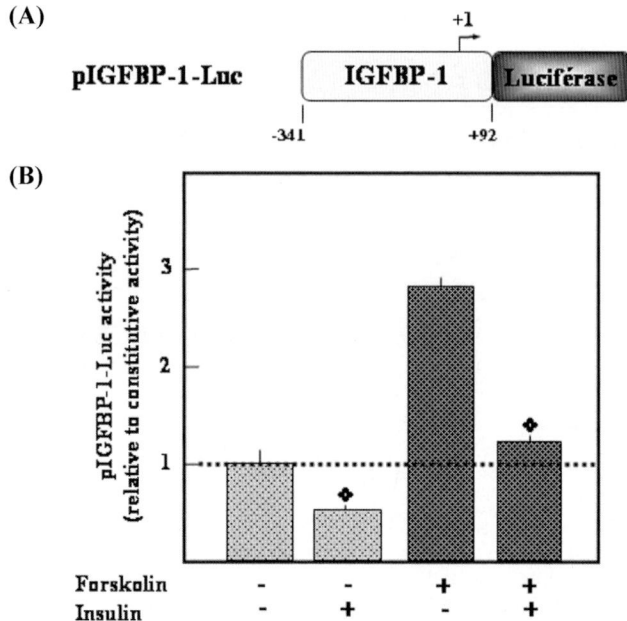

FIGURE 2. Inhibition of constitutive and cAMP-induced hIGFBP-1 promoter activity by insulin. HepG2 cells were transfected with pIGFBP-1-Luc alone. In pIGFBP-1-Luc, luciferase expression was controlled by nt-341 to +92 of the hIGFBP-1 gene promoter (**A**) Four hours post-transfection, the cells were either left untreated (addition of vehicle alone) or treated with insulin (100 nM), forskolin (10 μM), or a combination of insulin and forskolin for 20 h (**B**). ♦, $P < 0.05$ (Student's t-test) when compared to the appropriate control.

its inhibitory effect on basal IGFBP-1 transcription.[10,12,25] We thus decided to check the effect of the FOXOs on basal IGFBP-1 promoter activity, on its stimulation by cAMP, and on the inhibition of basal and cAMP-induced promoter activity by insulin.

FOXO Transcription Factors Enhance Constitutive hIGFBP-1 Activity and Are Not Involved in Its Stimulation by cAMP

When either pFOXO-1 (wild-type) or pFOXO-1$_{Ala3}$ (constitutive mutant) were co-transfected with pIGFBP-1-Luc in HepG2 cells, the basal promoter activity was enhanced >30-fold. This was not the case when pFOXO-1$_{AS}$ (encoding an antisense FOXO-1 transcript) was co-transfected (FIG. 3A).

It has been reported that IRE$_{GRU}$ sequences, although dispensable, contribute to the maximal stimulation of IGFBP-1 promoter activity by glucocorticoids

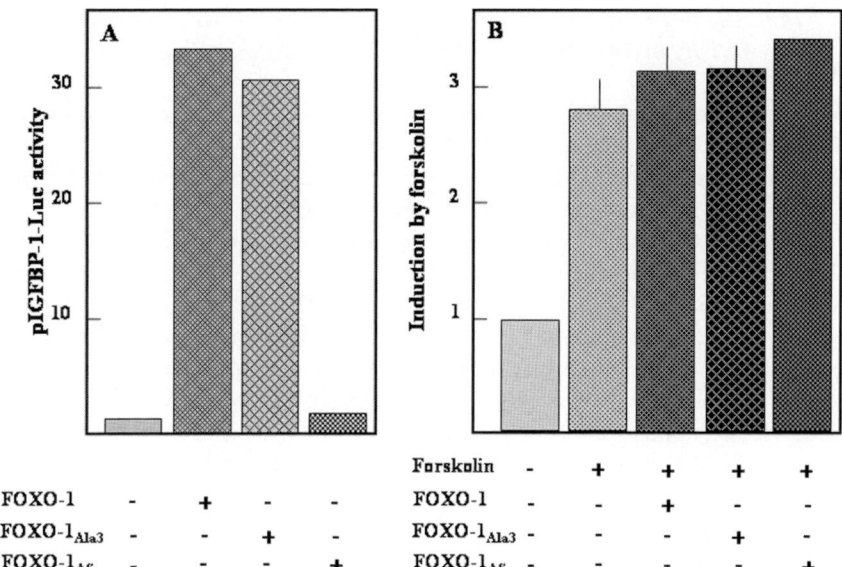

FIGURE 3. FOXO-1 enhances hIGFBP-1 promoter activity but does not interfere with cAMP stimulation. HepG2 cells were transfected with pIGFBP-1-Luc alone or in the presence of either pFOXO-1 or pFOXO-1$_{Ala3}$ or of pFOXO-1$_{AS}$. Four hours post-transfection, the cells were either left untreated (addition of vehicle alone) or treated with forskolin (10 μM) for 20 h. (**A**) Constitutive promoter activity measured in the presence of FOXO-1$_{wt}$ or of mutated FOXO-1 (FOXO-1$_{Ala3}$), or after co-transfection of a plasmid generating antisense FOXO-1 transcripts (FOXO-1$_{AS}$) was compared with that determined in their absence. (**B**) Induction of promoter activity by forskolin was computed relative to the cognate control (i.e., cells transfected with the same set of vector[s], and incubated with vehicle alone).

and that this requirement remained even in the presence of overexpressed FOXO-1$_{wt}$ or FOXO-1$_{Ala3}$.[6,26,27] Nevertheless, overexpression of FOXO-1$_{wt}$ or FOXO-1$_{Ala3}$ decreases the magnitude of glucocorticoid stimulation of the wild-type IGFBP-1 promoter (via the displacement of endogenous factors?), but not of its IRE$_{GRU}$ mutated variant.[27]

We wondered whether such cross-talk occurred for the stimulation of hIGFBP-1 promoter activity by cAMP. Forskolin induced hIGFBP-promoter activity by 2.8 ± 0.3- fold (mean ± SEM) and this induction factor was not significantly changed when either FOXO-1$_{wt}$ or FOXO-1$_{Ala3}$ was overexpressed in the transfected cells (3.3 ± 0.4, mean ± SEM, $P > 0.05$). The extent of induction was similar even when pFOXO$_{AS}$, which impairs endogenous FOXO-1 expression and activity, was co-transfected (FIG. 3B). Thus, neither overexpression of FOXO-1 nor inhibition of its expression seemed to interfere with the stimulation of wild-type hIGFBP-1 promoter activity by cAMP.

Inhibition of Constitutive and cAMP-Induced hIGFBP-1 Promoter Activity by Insulin is Dependent on PI-3' Kinase Signaling

Overexpression of FOXO-1$_{wt}$ did not alter the dominant negative effect of insulin on constitutive or cAMP-induced hIGFBP-1 promoter activity (81.8% ± 9.4 [mean ± SEM]) (FIG. 4). Thus subsequent experiments were performed in the presence of overexpressed FOXO-1$_{wt}$.

In HepG$_2$ cells, inhibition of constitutive hIGFBP-1 promoter activity by insulin was unchanged when Raf/MEK/Erk signaling was blocked by treatment with the specific MEK1/2 inhibitor PD98059 (PD), and this held true whether FOXO-1$_{wt}$ was overexpressed (FIG. 4) or not (not shown). Similarly, stimulation of promoter activity by cAMP was still blunted by insulin in the presence of PD (76.5% in the presence vs. 77.5% in the absence of the MEK inhibitor) (FIG. 4). In contrast, the specific PI3'-kinase inhibitor Ly294002 (Ly) suppressed the dominant negative effect of insulin on both constitutive and

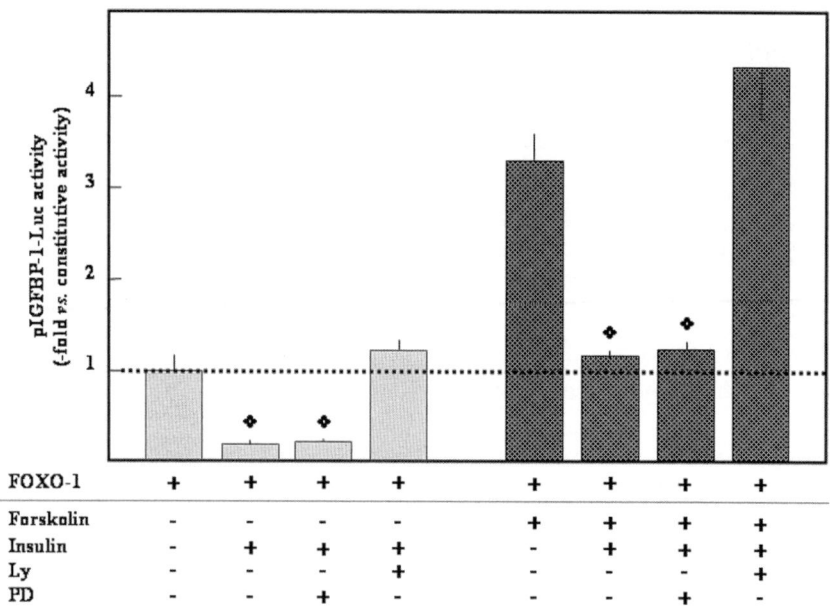

FIGURE 4. Inhibition of constitutive and cAMP-induced promoter activity by insulin is dependent on PI-3'K signaling. HepG2 cells were transfected with pIGFBP-1-Luc in the presence of wild-type FOXO-1. Four hours post-transfection, the cells were either left untreated (addition of vehicle alone) or treated with forskolin (10 μM) for 20 h. When appropriate, insulin (100 nM) either alone or in combination with PD98059 (50 μM) or with Ly294002 (50 μM) were also added to the culture medium. For combined treatments, PD98059 or Ly294002 were always added 10 min before insulin in the culture medium.◆, $P < 0.05$ (Student's t-test) when compared with the appropriate control.

cAMP-induced hIGFBP-1 promoter activity (FIG. 4). This remained true when only endogenous FOXOs were expressed in HepG$_2$ cells (not shown). These results suggest that PI-3′ kinase, but not Raf/MEK/Erk signaling, mediates inhibition of both constitutive (*see* also Refs. 9 and 10) and cAMP-induced hIGFBP-1 promoter activity by insulin (this study). Furthermore, a slight stimulation of promoter activity was observed (1.31 ± 0.07 fold vs. untreated cells; FIG. 4) in Ly-treated cells, irrespective of the presence or absence of forskolin. This suggested a modest activation of the PI3′-kinase signaling in cultured HepG$_2$ cells.

FOXO Transcription Factors Mediate the Inhibitory Effect of Insulin on cAMP-Induced hIGFBP-1 Promoter Activity via a Novel IRE : IRE$_{CRU}$

As mentioned above, stimulation of hIGFBP-1 promoter activity by cAMP was not altered when pFOXO-1$_{wt}$, pFOXO-1$_{Ala3}$, or pFOXO-1$_{AS}$ was

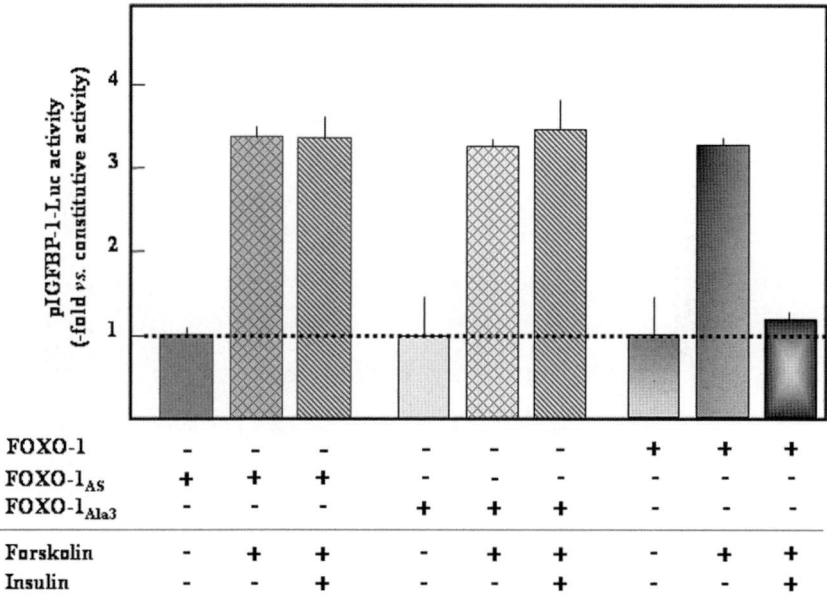

FIGURE 5. Inhibition of cAMP-induced promoter activity by insulin is dependent on FOXO-1. HepG2 cells were co-transfected with pIGFBP-1-Luc and either pFOXO-1, pFOXO-1$_{Ala3}$, or pFOXO-1$_{AS}$. Four hours post-transfection, the cells were either left untreated (addition of vehicle alone) or treated for 20 h either with forskolin (10 μM) or with both forskolin and insulin (100 nM). Luciferase activities were computed relative to the cognate control (i.e., cells transfected with the same set of vector(s) and treated with vehicle). The results represent the mean of at least three independent experiments performed in triplicate.

cotransfected with pIGFBP-1-Luc. In contrast, overexpression of FOXO-1$_{Ala3}$ or impairment of endogenous FOXO-1 expression and activity upon cotransfection of pFOXO-1$_{AS}$ blunted the inhibitory effect of insulin on cAMP-induced stimulation of hIGFBP-1 promoter activity (no inhibition vs. the 82% inhibition in cells co-transfected with wild-type FOXO-1) (FIG. 5). These experiments demonstrate that FOXO transcription factors mediate the inhibitory effect of insulin on cAMP-induced *hIGFBP-1* gene transcription, and that FOXOs' phosphorylation by Akt is mandatory in mediating insulin inhibition.

To check whether or not the 19-bp sequence located immediately 5'- to the CRE, which displays a high degree of sequence similarity with the previously described IRE$_{GRU}$ (see above), mediates the inhibitory effect of insulin on cAMP-induced hIGFBP-1 promoter activity (i.e., stands as an authentic IRE:

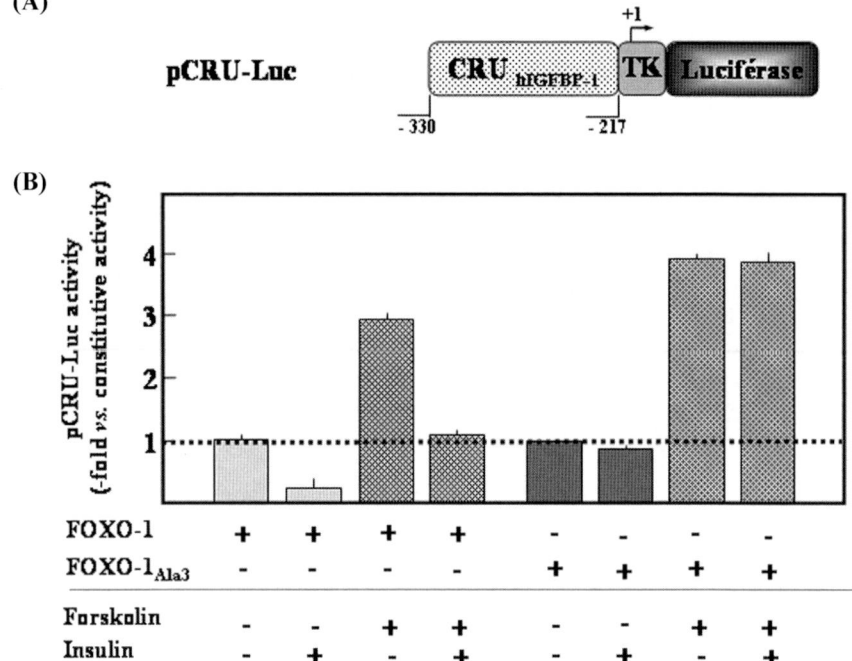

FIGURE 6. FOXO transcription factors mediate the inhibitory effect of insulin via a novel IRE: IRE$_{CRU}$. HepG2 cells were co-transfected with pCRU-Luc and either pFOXO-1 or pFOXO-1$_{Ala3}$. (**A**) In pCRU-Luc, Tk promoter activity was driven by the whole CRU (nt-330 to -217). (**B**) Four hours post-transfection, the cells were either left untreated (addition of vehicle alone) or treated for 20 h either with forskolin (10 μM), with insulin (100 nM), or with both forskolin and insulin. Promoter activities were computed relative to the cognate control (i.e., cells transfected with the same set of vector(s) and treated with vehicle). The results represent the mean of at least three independent experiments performed in triplicate.

IRE$_{CRU}$), HepG$_2$ cells were tranfected with pCRU-Luc (nt-330 to -217 of the hIGFBP-1 promoter (i.e., encompassing the CRE and the 19 bp sequence located immediately 5'- to the CRE) (FIG. 6A). IRE$_{GRU}$ was not included in this construct. We addressed the question of whether the dominant negative effect of insulin on cAMP-stimulated hIGFBP-1 promoter activity was mediated via the interaction of the putative IRE$_{CRU}$ with the FOXOs in relation to their phosphorylation status.

When HepG$_2$ cells were transfected with pCRU-Luc, and either FOXO-1$_{wt}$ or FOXO$_{Ala3}$, basal Tk promoter activity was very efficiently enhanced (by $\sim$50- and $\sim$180-fold, respectively, with 1 μg expression plasmid) (FIG. 6A). Forskolin treatment induced 3.0- and 4.0-fold additional increases in luciferase activities, respectively (FIG. 6B). Stimulations by forskolin were similar to those obtained with the complete proximal promoter (see above). In contrast, the dominant-negative effect of insulin on constitutive and on cAMP-induced promoter activity was only observed in cells overexpressing FOXO-1$_{wt}$, which could be phosphorylated by Akt after activation of PI-3'K signaling by insulin (75% and 90% inhibition, respectively) (FIG. 6B). The effect of insulin was abolished when the FOXO-1$_{Ala3}$ mutant (which could not be phosphorylated by Akt) was overexpressed in pCRU-Luc-transfected HepG$_2$ cells. These results show that the 19-bp sequence located immediately 5'- to the CRE behaves as a *bona fide,* functional IRE, and that the ability of FOXOs to be phosphorylated is mandatory for the ability of insulin to inhibit both basal and cAMP-induced hIGFBP-1 promoter activity.

DISCUSSION

Insulin exerts multiple effects on cell growth and metabolism in different cell types. In the liver, it regulates the transcription of genes involved in glucose production and utilization, among them being PEPCK, which catalyses the first committed step of gluconeogenesis, and IGFBP-1, which controls the bioavailability of the IGFs. The potent increase of IGFBP-1 during short-term fasting and the subsequent binding of IGFs is thought to limit their anabolic effect on amino acids, thus sparing them for gluconeogenesis. In the liver, *IGFBP-1* and *PEPCK* genes are regulated similarly: their expression is enhanced by glucocorticoids and cAMP, and dominantly inhibited by insulin.

A functional IRE (IRE$_{GRU}$) has already been described in the GRU of IGFBP-1 promoter. Nevertheless, Yeagley and co-workers[27] reported that mutation of IRE$_{GRU}$ did not completely impede inhibition of basal *IGFBP-1* gene transcription by insulin, and did not affect the inhibition of glucocorticoid-induced gene transcription by insulin. These data strongly suggested the existence of at least one additional IRE.

Accordingly, we report herein that a sequence closely related to IRE_{GRU} lies 5'- to the CRE. We termed it IRE_{CRU} and demonstrated that it is functionally relevant: it can mediate the dominant-negative effect of insulin on basal and cAMP-induced *hIGFBP-1* gene transcription/promoter activity. The inhibition by insulin of constitutive hIGFBP-1 gene transcription may thus be a consequence of the activity of two IREs: IRE_{GRU} and IRE_{CRU}. That inhibition of hormonal responses by insulin could be mediated by different IREs has also been reported for the PEPCK gene.[27–29]

We have shown that IRE_{CRU} acts as a potent enhancer sequence in the absence of insulin. Whatever was the promoter context (whole promoter or isolated CRU) the enhancer effect was dependent on the FOXO transcription factors, with FOXO-1$_{Ala3}$ (which cannot be phosphorylated by Akt) being more efficient than wild-type FOXO-1. Similar results were obtained for constitutive IGFBP-1 whole promoter activity, when the IRE_{GRU} was not mutated.[10,27]

Yeagley and co-workers[27] have reported that the magnitude of glucocorticoid stimulation was diminished when wild-type IGFBP-1 promoter activity was studied in the presence of overexpressed FOXO. In contrast, when the IRE_{GRU} (but not the IRE_{CRU}) was mutated, stimulation of IGFBP-1 promoter activity by glucocorticoids was still observed (albeit less effective than that observed with the wild-type IGFBP-1 promoter) and equivalent whether or not wild-type FOXO-1 or FOXO-1$_{Ala3}$ was overexpressed. The latter observation is closely related to ours with the CRU. Indeed, the extent of cAMP stimulation was altered neither by overexpression of wild-type FOXO-1 or FOXO-1$_{Ala3}$ nor by the suppression of endogenous FOXO-1 expression using an antisense RNA strategy. This suggests that in the CRU environment, no competition occurs between FOXO-1 and CREB for sequence-specific interaction with DNA or for the recruitment of CBP/p300, a co-activator that interacts with both FOXO-1 and CREB. Indeed, such competition mechanisms were suggested for the GRU.[29,30] In support of our results, O'Brien and co-workers[31] reported that in the PEPCK gene promoter, IRE_{CRU} does not contribute to the stimulation of promoter activity by cAMP.

We have furthermore demonstrated that the inhibitory effect of insulin on both constitutive and cAMP-induced *IGFBP-1* gene expression is dependent on the phosphorylation status of the FOXO transcription factors by the PI-3'K/Akt signaling. Inhibition of both constitutive and cAMP-induced reporter gene expression was only observed when wild-type FOXO-1, but not its non-phosphorylatable mutant FOXO-1$_{Ala3}$, was co-transfected (overexpressed) with pCRU-Luc in HepG$_2$ cells. Our results pinpoint differences between the GRU and the CRU in the molecular mechanisms of insulin action. Indeed, in contrast with our observations, FOXO-1$_{Ala3}$ yielded a partial inhibition of IRE_{GRU} activity in rat H4II hepatoma cells,[29] and inhibition of glucocorticoid-stimulated IGFBP-1 gene transcription by insulin was still observed when the

IRE$_{GRU}$ had previously been mutated, suggesting an "integrated" response mediated both by IRE$_{GRU}$-dependent and IRE$_{GRU}$-independent mechanisms.[27,32]

Finally, our results show for the first time that the 19-nt sequence located immediately 5′- to the CRE behaves as an enhancer in the presence of FOXO transcription factors, that it corresponds to an authentic IRE (IRE$_{CRU}$), and that the FOXOs mediate the inhibition of both basal and cAMP-induced IGFBP-1 promoter activity by insulin via IRE$_{CRU}$, provided that they could be phosphorylated by Akt subsequent to insulin receptor/PI-3′K signaling activation.

REFERENCES

1. CLEMMONS, D.R. 1997. Insulin-like growth factor binding proteins and their role in controlling IGF actions. Cytokine Growth Factor Rev. **8:** 45–62.
2. KACHRA, Z., C.R. YANG, L.J. MURPHY & B.I. POSNER. 1994. The regulation of insulin-like growth factor-binding protein 1 messenger ribonucleic acid in cultured rat hepatocytes: the roles of glucagon and growth hormone. Endocrinology **135:** 1722–1728.
3. UNTERMAN, T.G., D.T. OEHLER, L.J. MURPHY & R.G. LACSON. 1991. Multihormonal regulation of insulin-like growth factor-binding protein-1 in rat H4IIE hepatoma cells: the dominant role of insulin. Endocrinology **128:** 2693–2701.
4. BABAJKO, S., F. TRONCHE & A. GROYER. 1993. Liver-specific expression of human insulin-like growth factor binding protein 1: functional role of transcription factor HNF1 *in vivo*. Proc. Natl. Acad. Sci. USA **90:** 272–276.
5. GOSWAMI, R., R. LACSON, E. YANG, *et al.* 1994. Functional analysis of glucocorticoid and insulin response sequences in the rat insulin-like growth factor-binding protein-1 promoter. Endocrinology **134:** 736–743.
6. SUH, D.S., G.T. OOI & M.M. RECHLER. 1994. Identification of cis-elements mediating the stimulation of rat insulin-like growth factor-binding protein-1 promoter activity by dexamethasone, cyclic adenosine 3′,5′-monophosphate, and phorbol esters, and inhibition by insulin. Mol. Endocrinol. **8:** 794–805.
7. SUWANICHKUL, A., S.V. ALLANDER, S.L. MORRIS & D.R. POWELL. 1994. Glucocorticoids and insulin regulate expression of the human gene for insulin-like growth factor-binding protein-1 through proximal promoter elements. J. Biol. Chem. **269:** 30835–30841.
8. SUWANICKUL, A., S.L. MORRIS & D.R. POWELL. 1993. Identification of an insulin-responsive element in the promoter of the human gene for insulin-like growth factor binding protein-1. J. Biol. Chem. **268:** 17063–17068.
9. CICHY, S.B., S. UDDIN, A. DANILKOVICH, *et al.* 1998. Protein kinase B/Akt mediates effects of insulin on hepatic insulin-like growth factor-binding protein-1 gene expression through a conserved insulin response sequence. J. Biol. Chem. **273:** 6482–6487.
10. GUO, S., G. RENA, S. CICHY, *et al.* 1999. Phosphorylation of serine 256 by protein kinase B disrupts transactivation by FKHR and mediates effects of insulin on insulin-like growth factor-binding protein-1 promoter activity through a conserved insulin response sequence. J. Biol. Chem. **274:** 17184–17192.

11. KOPS, G.J., N.D. DE RUITER, A.M. DE VRIES-SMITS, *et al.* 1999. Direct control of the Forkhead transcription factor AFX by protein kinase B. Nature **398:** 630–634.
12. DURHAM, S.K., A. SUWANICHKUL, A.O. SCHEIMANN, *et al.* 1999. FKHR binds the insulin response element in the insulin-like growth factor binding protein-1 promoter. Endocrinology **140:** 3140–3146.
13. RENA, G., A.R. PRESCOTT, S. GUO, *et al.* 2001. Roles of the forkhead in rhabdomyosarcoma (FKHR) phosphorylation sites in regulating 14-3-3 binding, transactivation and nuclear targeting. J. Biochem. **354:** 605–612.
14. SUWANICHKUL, A., L.A. DEPAOLIS, P.D. LEE & D.R. POWELL 1993. Identification of a promoter element which participates in cAMP-stimulated expression of human insulin-like growth factor-binding protein-1. J. Biol. Chem. **268:** 9730–9736.
15. BABAJKO, S. 1995. Transcriptional regulation of insulin-like growth factor binding protein-1 expression by insulin and cyclic AMP. Growth Regul. **5:** 83–91.
16. NEAU, E., D. CHAMBERY, G. SCHWEIZER-GROYER, *et al.* 1995. Multiple liver-enriched trans-acting factors interact with the glucocorticoid- (GRU) and cAMP- (CRU) responsive units within the h-IGFBP-1 promoter. Prog. Growth Factor Res. **6:** 103–117.
17. BRINKMAN, A., C.A. GROFFEN, D.J. KORTLEVE & S.L. DROP. 1988. Organization of the gene encoding the insulin-like growth factor binding protein IBP-1. Biochem. Biophys. Res. Commun. **157:** 898–907.
18. BOUSSIF, O., F. LEZOUALC'H, M.A. ZANTA, *et al.* 1995. A versatile vector for gene and oligonucleotide transfer into cells in culture and *in vivo*: polyethylenimine. Proc. Natl. Acad. Sci. USA **92:** 7297–7301.
19. SLEIGH, M.J. 1986. A nonchromatographic assay for expression of the chloramphenicol acetyltransferase gene in eucaryotic cells. Anal. Biochem. **156:** 251–256.
20. DE WET, J.R., K.V. WOOD, M. DELUCA, *et al.* 1987. Firefly luciferase gene: structure and expression in mammalian cells. Mol. Cell Biol. **7:** 725–737.
21. ROBERTSON, D.G., E.M. MARINO, P.M. THULE, *et al.* 1994. Insulin and glucocorticoids regulate IGFBP-1 expression via a common promoter region. Biochem. Biophys. Res. Commun. **200:** 226–232.
22. UNTERMAN, T.G., A. FAREEDUDDIN, M.A. HARRIS, *et al.* 1994. Hepatocyte nuclear factor-3 (HNF-3) binds to the insulin response sequence in the IGF binding protein-1 (IGFBP-1) promoter and enhances promoter function. Biochem. Biophys. Res. Commun. **203:** 1835–1841.
23. O'BRIEN, R.M., E.L. NOISIN, A. SUWANICHKUL, *et al.* 1995. Hepatic nuclear factor 3- and hormone-regulated expression of the phosphoenolpyruvate carboxykinase and insulin-like growth factor-binding protein 1 genes. Mol. Cell Biol. 1995;**15:** 1747–1758.
24. WANG, J-C, P.E. STROMSTEDT, R.M. O'BRIEN & D.K. GRANNER. 1996. Hepatic nuclear factor 3 is an accessory factor required for the stimulation of phosphoenolpyruvate carboxykinase gene transcription by glucocorticoids. Mol. Endocrinol. **10:** 794–800.
25. ZHANG, X., L. GAN, H. PAN, *et al.* 2002. Phosphorylation of serine 256 suppresses transactivation by FKHR (FOXO1) by multiple mechanisms. Direct and indirect effects on nuclear/cytoplasmic shuttling and DNA binding. J. Biol. Chem. **277:** 45276–45284.
26. UNTERMAN, T., D. OEHLER, H. NGYUEN, *et al.* 1995. A novel DNA/protein complex interacts with the insulin-like growth factor binding protein-1 (IGFBP-1) insulin

response sequence and is required for maximal effects of insulin and glucocorticoids on promoter function [review]. Prog. Growth Factor Res. **6:** 119–129.
27. YEAGLEY, D., S. GUO, T. UNTERMAN & P.G. QUINN. 2001. Gene- and activation-specific mechanisms for insulin inhibition of basal and glucocorticoid-induced insulin-like growth factor binding protein-1 and phosphoenolpyruvate carboxykinase transcription. Roles of forkhead and insulin response sequences. J. Biol. Chem. **276:** 33705–33710.
28. YEAGLEY, D., J.M. AGATI & P.G. QUINN. 1998. A tripartite array of transcription factor binding sites mediates cAMP induction of phosphoenolpyruvate carboxykinase gene transcription and its inhibition by insulin. J. Biol. Chem. **273:** 18743–18750.
29. HALL, R.K., T. YAMASAKI, T. KUCERA, et al. 2000. Regulation of phosphoenolpyruvate carboxykinase and insulin-like growth factor-binding protein-1 gene expression by insulin. The role of winged helix/forkhead proteins. J. Biol. Chem. **275:** 30169–30175.
30. GHOSH, A.K., R. LACSON, P. LIU, et al. 2001. A nucleoprotein complex containing CCAAT/enhancer-binding protein beta interacts with an insulin response sequence in the insulin-like growth factor-binding protein-1 gene and contributes to insulin-regulated gene expression. J. Biol. Chem. **276:** 8507–8515.
31. O'BRIEN, R.M., P.C. LUCAS, T. YAMASAKI, et al. 1994. Potential convergence of insulin and cAMP signal transduction systems at the phosphoenolpyruvate carboxykinase (PEPCK) gene promoter through CCAAT/enhancer binding protein (C/EBP). J. Biol. Chem. **269:** 30419–30428.
32. GAN, L., H. PAN & T.G. UNTERMAN. 2005. Insulin response sequence-dependent and -independent mechanisms mediate effects of insulin on glucocorticoid-stimulated insulin-like growth factor binding protein-1 promoter activity. Endocrinology **146:** 4274–4280.

c-Jun and JunB Are Essential for Hypoglycemia-Mediated *VEGF* Induction

BJÖRN TEXTOR, MELANIE SATOR-SCHMITT,
KARL HARTMUT RICHTER, PETER ANGEL,
AND MARINA SCHORPP-KISTNER

Division of Signal Transduction and Growth Control A100, Deutsches Krebsforschungszentrum (DKFZ) Heidelberg, Im Neuenheimer Feld 280, D-69120 Heidelberg, Germany

> ABSTRACT: Physiological conditions like hypoxia or hypoglycemia trigger expression of *VEGF*, a key regulator of angiogenesis. To elucidate the molecular mechanism underlying the *VEGF* regulation of hypoglycemia, we investigated the role of AP-1 transcription factor subunits c-Jun and JunB. Using c-$jun^{-/-}$ and $junB^{-/-}$ mouse embryonic fibroblasts, we demonstrate that both c-Jun and JunB are required for the hypoglycemia-mediated induction of *VEGF* expression. This process is independent of the master regulator of hypoxic stress HIF-1, as HIF expression and stabilization are not affected by the loss of AP-1 subunits. Analysis of signaling cascades regulating c-Jun and/or JunB activity and/or transcription upon hypoglycemia by application of specific inhibitors of protein kinase C (PKC) or extracellular signal-regulated kinase (ERK) signaling revealed that hypoglycemia-mediated induction of c-Jun is regulated via a PKCα-dependent signaling pathway. In contrast, JunB is activated by the MAP kinase ERK for the AP-1 subunits c–Jun and JunB to mediate *VEGF* regulaltion of hypoglycemia.
>
> KEYWORDS: AP-1; c-Jun; JunB; *VEGF*; hypoglycemia

INTRODUCTION

Mammalian cells possess adaptive programs which enable them to respond to and to circumvent cellular stress, such as that caused by pathogenic infection, chemical insult, nutrient deprivation (hypoglycemia),[1] or reduced oxygen tension (hypoxia).[2] Increasing tissue mass due to both organogenesis and tumor growth results in hypoxic or hypoglycemic conditions in concert with

---

Address for correspondence: Dr. Marina Schorpp-Kistner, Division of Signal Transduction and Growth Control, Deutsches Krebsforschungszentrum (DKFZ), Im Neuenheimer Feld 280, D-69120 Heidelberg, Germany. Voice: 49-6221-42-4575; fax: 49-6221-42-4554.
  e-mail: marina.schorpp@dkfz.de

the induction of vascular endothelial growth factor (VEGF) expression, a key regulator of angiogenesis.[3] It is well established that *VEGF* induction is tightly controlled by the transcription factor HIF, consisting of the constitutively expressed HIF-1β and the alpha subunits that are stabilized upon hypoxic or hypoglycemic conditions.[4] Previously, it has been reported that hypoglycemia-mediated *VEGF* induction depends on PKC activity,[1] yet the exact molecular mechanism with regard to the involvement of individual PKC isoforms remains elusive. The PKC family consists of 12 phospholipid-dependent serine/threonine kinases that mediate signals from the cell surface to the nucleus and play key roles in cellular signaling pathways.[5] There is growing evidence that besides HIF, AP-1 is also required for maximal induction of *VEGF* in response to hypoxia.[6,7] The AP-1 transcription factor family is composed of dimeric protein complexes formed by products of the *jun, fos,* and *ATF* gene families and is a major nuclear target of mitogen- and stress-induced signal transduction cascades.[8,9] Intensive investigation of AP-1 function in genetically modified cell culture and mouse models led to the well-accepted perception that the individual AP-1 subunits may have independent functions as tissue-specific and signal-specific activators of AP-1-dependent genetic programs.[8,9] Despite the widespread expression pattern of Jun members, only the loss of JunB affects vascular development[10] and recently we could show that JunB is essential for hypoxia-mediated *VEGF* induction independent of HIF signaling (M. Schorpp–Kistner, unpublished data).[11] In the present work, we investigated the activation of JunB and c-Jun in response to hypoglycemia and their involvement in hypoglycemia-mediated expression of *VEGF*.

MATERIALS AND METHODS

Cell Lines and Culture Conditions

Wild-type $junB^{-/-}$ and $c\text{-}jun^{-/-}$ mouse embryonic fibroblasts (MEFs) were cultured as described.[10,12] Hypoglycemia was induced by incubating cells in glucose-free DMEM supplemented with 10% FCS (Sigma, Munich, Germany) for the indicated time points. Inhibitor studies were performed by using PKC inhibitors Gödecke 6976 (200 nM, Calbiochem), Ly333,531 (20 nM, Alexis Corp., Lausen, Switzerland), and ERK inhibitor PD98059 (50 μM, Calbiochem, Darmstadt, Germany). Cells were pretreated with the respective inhibitor for 15 min, washed, and subsequently incubated in glucose-free medium to induce hypoglycemia.

Protein and RNA Analyses

For the immunodetection of active AP-1, nuclear extracts were prepared as described.[11] A total of 50 μg of protein was separated by SDS-PAGE and blotted onto nitrocellulose; immunodetection was performed with an enhanced chemiluminescence system (Perkin Elmer Life Sciences, Boston, MA, USA). The following primary antibodies were used: anti-JunB (1:500; N17, Santa Cruz, Heidelberg, Germany), anti-c-Jun (1:250; sc-45, Santa Cruz), anti-phospho-c-Jun (1:1000, Cell Signaling, Boston, MA, USA), and anti-RCC1 (1:500, BD Biosciences, Heidelberg, Germany). For detection of HIF-1, whole cell extracts were prepared as described.[11,12] A total of 50 μg of protein was used for immunoblotting as described above. The following antibodies were used: anti-HIF-1α (1:200, Cayman Europe, Tallinn, Estonia), anti-HIF-1β (1:1000, Novus Biologicals, Littleton, CO, USA), and anti-HSC70 (1:10000, Stressgene, Victoria, BC, Canada).

For mRNA analyses, wild-type and genetically altered MEFs were incubated under normal or hypoglycemic conditions for the indicated time points and total RNA was subsequently isolated as previously described.[10] A total of 30 μg of total RNA was separated by gel electrophoresis and blotted onto a nitrocellulose membrane. The expression of target genes was investigated by using specific, radiolabeled DNA-probes against *junB*, *c-jun*, *VEGF*, and *18S* RNA as described previously.[10]

RESULTS

Hypoglycemia Leads to the Induction of c-Jun, JunB, and the Stabilization of HIF-1α

To investigate the role of Jun proteins in the cellular response to hypoglycemia, wild-type fibroblasts were incubated for the indicated time points with glucose-free media, and subsequently the expression of c-Jun and JunB on RNA and protein level, respectively, was analyzed (FIG. 1). JunD mRNA was not induced (data not shown). Increase in *c-jun* transcription is detected within 1 h under hypoglycemia, while *junB* induction occurs after 1–2 h (FIG. 1A). c-Jun and JunB protein levels correlate with their respective mRNA kinetics while posttranslational activation of c-Jun via phosphorylation (p-c-Jun) is already seen in the absence of elevated *c-jun* transcripts (FIG. 1A) at 30 min under hypoglycemia (FIG. 1B, p-c-Jun)

c-Jun and JunB Are Required for the Induction of VEGF *by Hypoglycemia*

Because VEGF is induced upon hypoglycemia[1] and JunB is involved in the hypoxia-mediated regulation of *VEGF*,[11] we next asked whether JunB and/or c-Jun are required for the transcriptional upregulation of *VEGF* upon hypoglycemia. Northern blot analyses of wild-type MEFs revealed that *VEGF*

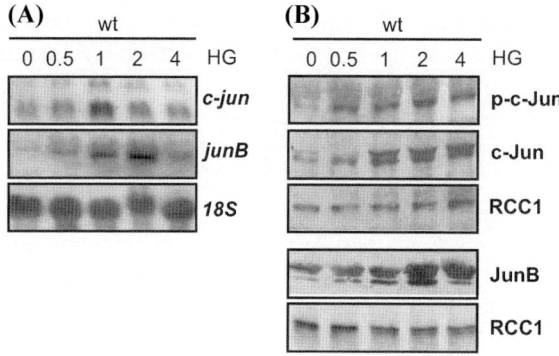

FIGURE 1. AP-1 subunits *c-jun* and *junB* are induced in hypoglycemia. (**A**) Wild-type MEFs were cultured under hypoglycemic conditions for the indicated time points. Northern blot analyses of 30 μg total RNA were performed using specific probes for *junB* and *c-jun* show respectively. *18S* hybridization shows equal loading. (**B**) Immunoblot analyses of c-Jun active phosphorylated c-Jun and JunB in wild-type MEFs maintained for the indicated time points under hypoglycemia. A total of 50 μg nuclear proteins was separated by SDS-gel electrophoresis, transferred onto a nitrocellulose membrane, and incubated with specific antibodies. RCC1 served as loading control.

mRNA was induced after 2 h of hypoglycemia (FIG. 2A). In contrast, the induction of *VEGF* in $c\text{-}jun^{-/-}$ as well as $junB^{-/-}$ MEFs was severely impaired (FIG. 2A), suggesting that both c-Jun and JunB are essential for *VEGF* induction of hypoglycemia.

To exclude the possibility that the loss of c-Jun or JunB in MEFs does not impair the expression and/or stabilization of the HIF-1 subunits HIF-1α and HIF-1β, immunoblot analyses were performed. Under hypoglycemic conditions, neither stabilization of HIF-1α nor expression of HIF-1β was affected in either c-Jun- or JunB-deficient MEFs (FIG. 2B).

Application of PKC or ERK Inhibitor Impairs Either c-jun *or* c-Jun *and* junB *Induction, Respectively, and Concomitant* VEGF *Expression*

In the past, it has been shown that hypoglycemia-induced *VEGF* expression is mediated by the PKC signaling pathway.[1] Because PKC can regulate the activity of MAP kinases[13,14] and MAP kinases themselves can act on the AP-1 members c-Jun and JunB,[15,16] we subsequently asked whether the inhibition of PKC or MAP kinases has an impact on *c-jun* or *junB* and concomitant *VEGF* expression. Thus, we incubated wild-type MEFs under hypoglycemic conditions in the presence of two different PKC inhibitors, Gödecke 6976 and Ly333,531, addressing either α and β[17] or specifically $β_I$ and $β_{II}$ isoforms,[18] respectively. Northern blot analyses revealed that pretreatment with Ly333,531 prior to the induction of hypoglycemia had a slight effect on the induction of *c-jun* and *VEGF* transcripts but none on *junB* compared to untreated cells

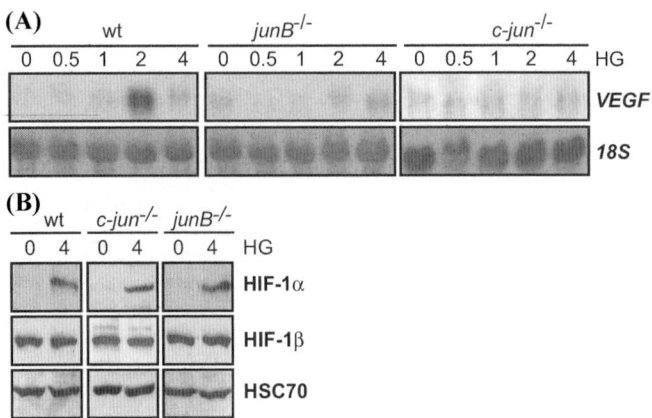

FIGURE 2. Induction of *VEGF* by hypoglycemia requires c-Jun and JunB, whereas stabilization of HIF-1 is not affected by the loss of c-Jun or JunB. (**A**) Wild-type, c-$jun^{-/-}$ and $junB^{-/-}$ MEFs were cultured under hypoglycemic conditions and harvested at the indicated time points. Northern blot analysis of 30 μg total RNA was performed by using a specific probe for *VEGF*. Hybridization against *18S* RNA was used as loading control. (**B**) Western Blot analyses of 50 μg whole-cell extracts from wild-type, c-$jun^{-/-}$ and $junB^{-/-}$ MEFs were performed with specific antibodies against HIF-1α and HIF-1β. HSC70 expression is shown as loading control.

(FIG. 3A). By contrast, inhibition of PKC by Gödecke 6976 led to a complete inhibition of *VEGF* and loss of *c-jun* mRNA expression, whereas the induction of *junB* was only mildly reduced (FIG. 3A), indicating that *c-jun* transcription is predominately controlled by a PKCα-dependent mechanism that most likely acts via JNK.[19] Because we could recently demonstrate that transcription factor NF-κB transactivates *junB* in response to hypoxia,[11] we analyzed hypoglycemia-dependent induction of *junB* mRNA in MEFs overexpressing a transdominant negative IκBα mutant protein, a so-called super-repressor (ΔN+) that impairs NF-κB activation.[11] In contrast to hypoxia signaling, *junB* was still inducible in the NF-κB super-repressor cells, indicating that NF-κB is not required for the activation of *junB* transcription in response to hypoglycemia (FIG. 3B). Because *junB* transcription is a preferred target of the MAP kinase ERK,[20] we analyzed *junB*, *c-jun*, and *VEGF* expression under hypoglycemic conditions in the presence of the specific ERK inhibitor PD98059. Preincubation of wild-type MEFs with PD98059 and subsequent incubation under hypoglycemia revealed that *junB* and *VEGF* induction are severely impaired (FIG. 3C), whereas *c-jun* activation was moderately affected (FIG. 3C), suggesting that upon hypoglycemia *junB* and concomitantly *VEGF* are controlled via the ERK signaling pathway. Hypoglycemia-dependent HIF-1α stabilization was not impaired in wild-type cells pretreated with PCK or ERK inhibitors, excluding a role for HIF in hypoglycemia-dependent *c-jun* and *junB* regulation (FIG. 3D).

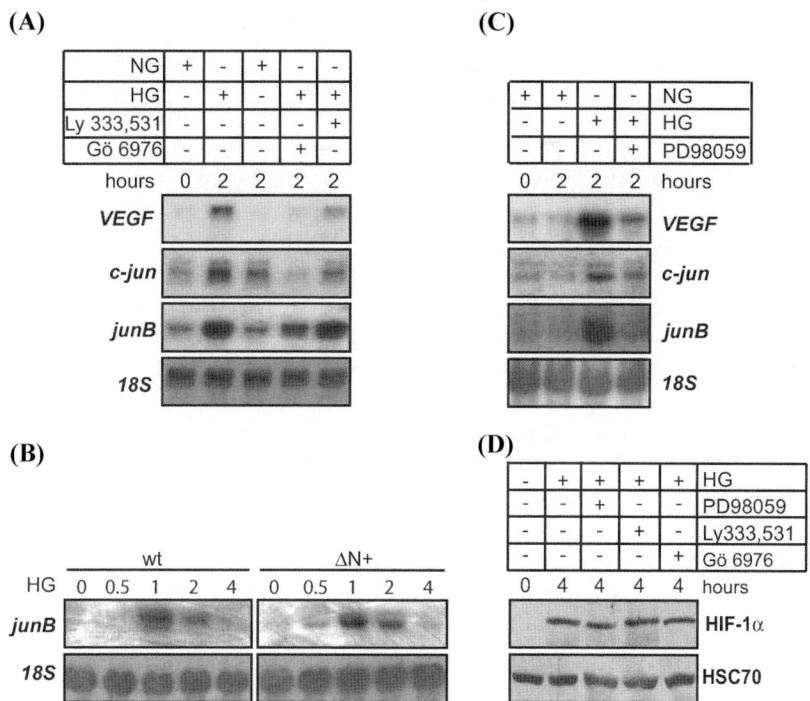

FIGURE 3. Inhibition of either protein kinase C (PKC) or the MAP kinase ERK leads to an inhibition of either *c-jun* or *junB*, respectively, resulting in loss of *VEGF* transcription despite stabilized HIF-1α protein. (**A**) Wild-type fibroblasts were cultured under hypoglycemic conditions (HG) in presence or absence of specific PKC inhibitors, as indicated on the top for the described time points. Northern blot analyses were performed using specific probes for *c-jun, junB,* and *VEGF*. *18S* hybridization signal is shown as loading control. (**B**) Wild-type MEFs and cells lacking NF-κB activity (ΔN+) were incubated by hypoglycemia and RNA was isolated after the indicated time points. Northern blot analyses of 30 μg total RNA was performed using a specific probe for *junB*. *18S* hybridization shows equal loading. (**C**) Wild-type fibroblasts were cultured under hypoglycemic conditions (HG) in presence or absence of the specific ERK inhibitor PD98059 as indicated on the top for the described time points. Northern blot analyses were performed as described in (A). (**D**) Western blot analyses of 50 μg total protein from wild-type MEFs preincubated with PKC and ERK inhibitors as depicted on the top. Cells were cultivated for the indicated time points under hypoglycemic conditions and were analyzed by using a specific antibody against HIF-1α. HSC70 expression is shown as a loading control.

DISCUSSION

Cells of multicellular organisms exhibit programs to adapt to hypoxic or to hypoglycemic conditions by induction of cellular processes like angiogenesis and glycolysis.[2] It is well accepted that under hypoxia and/or hypoglycemia, the transcription factor HIF induces expression of genes, such as *VEGF*, implicated in angiogenesis, erythropoiesis, and glycolysis. During the last years, there is growing evidence that HIF functionally cooperates with AP-1 in response to hypoxia.[7] While the AP-1 member c-Jun is bi-phasically induced, upon hypoxia, with an early HIF-independent and a late HIF-dependent response,[21] recent work in our laboratory shows that hypoxia-mediated induction of *junB* is achieved via NF-κB activation independently of HIF-1 signaling (M. Schorpper–Kistner, unpublished data).[11] Failure in JunB-mediated maximal VEGF induction in response to hypoxia is most likely causative for the early embryonic lethal phenotype of *junB*$^{-/-}$ embryos exhibiting vascular and angiogenic defects.[10] This concept is further supported by the absence of a vascular or angiogenic phenotype in mice with the genetic ablation of either c-Jun, JunD, or c-Fos.[8,9]

As hypoglycemia-mediated *VEGF* induction also requires HIF-1α, at least in MEFs,[1] the cellular response to hypoxia and hypoglycemia with regard to *VEGF* expression is considered to involve the same regulatory molecules. Yet under hypoglycemic conditions, in addition to HIF, activation of PKC and its impact on *VEGF* expression has been reported: Treatment of HepG2 human hepatoblastoma cells with the PKC inhibitor H-7 prior to the induction of hypoglycemia inhibited the expression of *VEGF*.[1] Nevertheless, the underlying regulatory network still remains elusive.

Our data obtained from different genetically altered mouse embryonic fibroblasts suggest that *VEGF* induction upon hypoglycemia requires both AP-1 subunits, c-Jun and JunB. While inhibition of PKCβ subunits by Ly333,531 had only a marginal effect, *VEGF* expression was lost in the presence of the inhibitor Gödecke 6976 addressing both α and β isoforms. This suggests on the one hand that *VEGF* induction is mediated by a classical PKCα-c-Jun-dependent signaling pathway, whereas on the other hand the ERK inhibitor PD98059 obstructed expression of both, *junB* and *VEGF*, emphasizing a requirement for ERK-dependent *junB* expression. Moreover, our data obtained with genetically modified MEFs including NF-κB super-repressor cells provide clear evidence that hypoglycemia-mediated *junB* regulation is achieved by a signaling pathway distinct from that implicated in the hypoxia response.

Taken together, our results demonstrate that *VEGF* is a target of a complex program of gene regulation addressed by different stimuli through distinct signaling pathways involving HIF and the AP-1 members c-Jun and JunB, which are orchestrating in conjunction the induction of *VEGF* in response to hypoglycemia.

ACKNOWLEDGMENTS

We are grateful to Drs. Claus Scheidereit and R. Schmidt-Ullrich for providing NF-κB super-repressor fibroblasts. We thank Drs. A.H. Licht,

B. Hartenstein, and J. Hess for critical reading of the manuscript. This work was supported by Training and Mobility of Researchers programs of the European Economic Community and by the Deutsche Forschungsgemeinschaft (Scho 365/3-2 and SFB-TR23).

REFERENCES

1. PARK, S.H. et al. 2001. Hypoglycemia-induced VEGF expression is mediated by intracellular Ca^{2+} and protein kinase C signaling pathway in HepG2 human hepatoblastoma cells. Int. J. Mol. Med. **7:** 91–96.
2. PUGH, C.W. & P.J. RATCLIFFE. 2003. Regulation of angiogenesis by hypoxia: role of the HIF system. Nat. Med. **9:** 677–684.
3. FERRARA, N. 2004. Vascular endothelial growth factor: basic science and clinical progress. Endocr. Rev. **25:** 581–611.
4. SEMENZA, G.L. 2004. Hydroxylation of HIF-1: oxygen sensing at the molecular level. Physiology (Bethesda) **19:** 176–182.
5. CASABONA, G. 1997. Intracellular signal modulation: a pivotal role for protein kinase C. Prog. Neuropsychopharmacol. Biol. Psychiatry **21:** 407–425.
6. DAMERT, A., E. IKEDA & W. RISAU. 1997. Activator-protein-1 binding potentiates the hypoxia-induciblefactor-1-mediated hypoxia-induced transcriptional activation of vascular-endothelial growth factor expression in C6 glioma cells. Biochem. J. **327:** 419–423.
7. LADEROUTE, K.R. 2005. The interaction between HIF-1 and AP-1 transcription factors in response to low oxygen. Semin. Cell Dev. Biol. **16:** 502–513.
8. EFERL, R. & E.F. WAGNER. 2003. AP-1: a double-edged sword in tumorigenesis. Nat. Rev. Cancer **3:** 859–868.
9. HESS, J., P. ANGEL & M. SCHORPP-KISTNER. 2004. AP-1 subunits: quarrel and harmony among siblings. J. Cell Sci. **117:** 5965–5973.
10. SCHORPP-KISTNER, M. et al. 1999. JunB is essential for mammalian placentation. EMBO J. **18:** 934–948.
11. ANDRECHT, S. et al. 2002. Cell cycle promoting activity of JunB through cyclin A activation. J. Biol. Chem. **277:** 35961–35968.
12. DIGNAM, J.D. et al. 1983. Eukaryotic gene transcription with purified components. Methods Enzymol. **101:** 582–598.
13. LANG, W. et al. 2004. Cooperation between PKC-alpha and PKC-epsilon in the regulation of JNK activation in human lung cancer cells. Cell Signal. **16:** 457–467.
14. UEDA, Y. et al. 1996. Protein kinase C activates the MEK-ERK pathway in a manner independent of Ras and dependent on Raf. J. Biol. Chem. **271:** 23512–23519.
15. DE GROOT, R.P. et al. 1991. Activation of junB by PKC and PKA signal transduction through a novel cis-acting element. Nucleic Acids Res. **19:** 775–781.
16. ANGEL, P. & M. KARIN. 1991. The role of Jun, Fos and the AP-1 complex in cell-proliferation and transformation. Biochim. Biophys. Acta **1072:** 129–157.
17. MARTINY-BARON, G. et al. 1993. Selective inhibition of protein kinase C isozymes by the indolocarbazole Go 6976. J. Biol. Chem. **268:** 9194–9197.
18. JIROUSEK, M.R. et al. 1996. (S)-13-[(dimethylamino)methyl]-10,11,14,15-tetrahydro-4,9:16, 21-dimetheno-1H, 13H-dibenzo[e,k]pyrrolo[3,4-h][1,4,13]

oxadiazacyclohexadecene-1,3(2H)-d ione (LY333531) and related analogues: isozyme selective inhibitors of protein kinase C beta. J. Med. Chem. **39:** 2664–2671.
19. LIU, X. *et al.* 1997. Hypoglycemia-induced c-Jun phosphorylation is mediated by c-Jun N-terminal kinase 1 and Lyn kinase in drug-resistant human breast carcinoma MCF-7/ADR cells. J. Biol. Chem. **272:** 11690–11693.
20. COFFER, P. *et al.* 1994. junB promoter regulation: Ras mediated transactivation by c-Ets-1 and c-Ets-2. Oncogene **9:** 911–921.
21. LADEROUTE, K.R. *et al.* 2002. The response of c-jun/AP-1 to chronic hypoxia is hypoxia-inducible factor 1 alpha dependent. Mol. Cell Biol. **22:** 2515–2523.

Altered Gene Expression Pattern in Peripheral Blood Leukocytes from Patients with Arterial Hypertension

A.V. TIMOFEEVA, L.E. GORYUNOVA, G.L. KHASPEKOV, D.A. KOVALEVSKII, A.V. SCAMROV, O.S. BULKINA, YU.A. KARPOV, K.A. TALITSKII, V.V. BUZA, V.V. BRITAREVA, AND R.SH. BEABEALASHVILLI

Russian Cardiology Reseach and Production Center, 121552 Moscow, Russia

ABSTRACT: The role of various inflammatory mechanisms and oxidative stress in the development of atherosclerosis and arterial hypertension (AH) has been increasingly acknowledged during recent years. Hypertension *per se* or factors that cause hypertension along with other complications lead to infiltration of activated leukocytes in the vascular wall, where these cells contribute to the development of vascular injury by releasing cytokines, oxygen radicals, and other toxic mediators. However, molecular mechanisms underlying leukocyte activation at transcriptional level in AH are still far from being clear. To solve this problem we employed cDNA microarray technology to reveal the differences in gene expression in peripheral blood leukocytes from patients with AH compared with healthy individuals. The microarray data were verified by a semi-quantitative RT-PCR method. We found 25 genes with differential expression in leukocytes from AH patients among which 21 genes were upregulated and 4 genes were downregulated. These genes are implicated in apoptosis (CASP2, CASP4, and CASP8, p53, UBID4, NAT1, and Fte-1), inflammatory response (CAGC, CXCR4, and CX3CR1), control of MAP kinase function (PYST1, PAC1, RAF1, and RAFB1), vesicular trafficking of molecules among cellular organelles (GDI-1 and GDI-2), cell redox homeostasis (GLRX), cellular stress (HSPA8 and HSP40), and other processes. Gene expression pattern of the majority of genes was similar in AH patients independent of the disease stage and used hypotensive therapy, but was clearly different from that of normotensive subjects.

KEYWORDS: peripheral blood leukocytes; arterial hypertension; cDNA microarray; gene expression; RT-PCR

Address for correspondence: A.V. Timofeeva, Russian Cardiology Reseach and Production Center, Laboratory of Genetic Engineering, 3rd Cherepkovskaya str. 15a, 121552 Moscow, Russia. Voice: +7-95-414-60-84; fax: +7-95-414-67-27.
e-mail: Angelica_T@cardio.ru

INTRODUCTION

The pathogenesis of hypertension is a multifactorial process that involves the interaction of genetic and environmental factors and results in elevated arterial blood pressure.[1] A wide variety of interdependent physiological systems has been found to regulate blood pressure level. Among these systems are baroreceptors, natriuretic peptides, renin–angiotensin–aldosterone system, kinin–kallikrein system, adrenergic receptor system, and factors produced by blood vessels that cause vasodilation or contraction. In varying degrees, abnormalities in the functioning of these systems contribute to enhanced cardiac ejection fraction, increased plasma volume, and vasoconstriction, resulting in the development of hypertension. In addition to the effects of vasoactive molecules, such as catecholamins, angiotensin II, and endothelin-1, demonstrating high levels in the circulation in hypertension and finally causing the remodeling of the arterial wall,[2-5] blood pressure elevation *per se* can be considered as a biomechanical stress for the vascular wall. In this process mechanical force initiates signal pathways usually used by growth factors, leading to vascular cell death and inflammatory response resulting in VSMC growth and proliferation.[6] It has been demonstrated that the same factors (chemical and mechanical) influence not only the vessel wall, but also blood cells.[7-9] This leads to activation of leukocytes that release ROS and mediators of proteolytic tissue degradation, finally contributing to oxidative stress, inflammation, and endothelial damage.[10] For example, B. Kristal and colleagues showed that leukocytes from hypertensive patients released superoxide anione faster than those of normotensives, and the redox state of plasma glutathione was twofold higher in AH, indicating a systemic oxidative stress.[11] Accordingly, we hypothesized that circulating blood cells may carry disease-specific information due to alterations in their local environment. In this regard, we employed cDNA microarray technology to identify the differences in gene expression in peripheral blood leukocytes from patients with arterial hypertension (AH) compared with normal individuals.

Because the pathophysiological mechanisms underlying the origin of AH are individual for the majority of hypertensive patients and generally it is impossible to reveal them in each particular case, it was interesting to determine common changes in gene expression pattern in leukocytes from AH patients independently of the disease stage and medical treatment used. In this connection we did not use strict criteria in AH patient selection. We suppose that such an approach will help us to better understand the molecular processes occurring in the leukocytes in hypertension and to find novel therapeutic strategies toward the AH patients in view of the fact that antihypertensive treatment at present reduces cardiovascular risk, but does not completely eliminate it.

MATERIALS AND METHODS

Subjects

Twenty arterial hypertension patients (7 women and 13 men) with different stages of (I–III), diverse cholesterol, glucose, and creatinine levels, and various end-organ damages and other AH complications were enrolled in the study (TABLE 1). Among them 7 patients did not receive any kind of treatment for 14 days prior to blood withdrawal, whereas 13 patients were treated either with valsartan ($n = 7$), an angiotensin II receptors antagonist, or with amlodipine ($n = 6$), a calcium channel blocker. Ten nonsmoking normotensive subjects (8 men and 2 women) devoid of end-organ damage (coronary heart disease, stroke, peripheral artery disease, microalbuminuria) with mean age of 37 ± 13 years served as control.

Collection of Blood Samples and RNA Isolation

Nine milliliters of peripheral blood was collected in Vacuette® tubes containing 3.8% sodium citrate. Total leukocyte fraction was isolated through buffy coat formation and erythrocytes were removed using red blood cell lysis buffer (155 mM ammonium chloride, 10 mM KHCO3, 1 mM EDTA, pH 7.4).

Leukocytes were stored at $-70°C$ until further processing. Total RNA was isolated from the samples by acid guanidine thiocyanate/phenol/chloroform method according to prescriptions of Clontech kit (Cat. No. K1038-1) and quantitated using a spectrophotometer. DNA contamination level was tested by semi-quantitative PCR using human DNA dilutions as standards. DNA contamination level of accepted RNA samples did not exceed 0.1%. The quality of total RNA samples was validated by denatured agarose gel-electrophoresis according to the integrity of 28S and 18S rRNAs.

cDNA-probe Preparation for Microarray Analysis

One and a half micrograms of total RNA was reverse-transcribed at 37°C for 60 min in a mixture (10 μL) containing 50 mM Tris-HCl (pH 8.3), 75 mM KCl, 3 mM MgCl$_2$, 5 mM DTT, 0.5 mM each of dCTP, dGTP, dTTP, and 5 μM dATP (USB, Cleveland, OH, USA), 35 μCi of [α-^{32}P]dATP (10 mCi/mL, 3,000 Ci/mmol, Obninsk, Russia), 100 units of MMLV reverse transcriptase (Promega, Madison, WI, USA) and 2.5 μM random primer (N$_9$). The resulting ^{32}P-cDNA probe was purified from unincorporated nucleotides by gel filtration on Sephadex G-50 columns, treated at 68°C for 20 min with denaturing solution (0.1 M NaOH, 10 mM EDTA) to remove RNA, and neutralized with 1 M NaH$_2$PO$_4$ (pH 7.0). The prepared cDNA-probe was incubated at 68°C for 10 min and immediately added to the hybridization mixture.

TABLE 1. Characteristics of hypertensive patients

Patient and Sample Name[a]	Age (yr)	Sex	SBP (mmHg)	DBP (mmHg)	Chol (mmol/L)	Glu (mmol/L)	Creat (μmol/L)	WBC × 10⁹/L	AH Stage	End-organ Damages and Complications	Treatment
H1/II-v	61	F	160	100	5	5.9	79	5.1	II	PAD, LVH	v
H2/III-a	69	M	190	115	4.2	4.3	107	5.3	III	CA, AR	a
H3/II-v	59	M	200	110	6.2	6.3↑	108	6.0	II	LVH	v
H4/III-v	68	F	170	100	6.6↑	6.3↑	89	6.1	III	TIA, PAD	v
H5/III-a	66	M	160	100	5.0	4.1	94	4.1	III	CHD, CE, PAD, LVH	a
H6/III-a	68	M	170	90	6.1	4.5	103	4.6	III	TIA, LVH	a
H7/III-a	64	M	150	90	4.4	4.5	124↑	7.0	III	CHD, AR, CEPAD	a
H8/III-a	66	M	180	100	7.6↑	5.3	118↑	5.4	III	CHD, PC, CE, TIA, PAD, LVH	a
H9/III-v	70	F	190	90	5.5	6.7↑	104	6.1	III	CHD, AR, PAD	v
H10/III-v	69	M	180	110	5.1	5.5	101	5.4	III	CHD, CE, PAD	v
H11/III-a	62	F	160	96	6.8↑	6.1	79	6.1	III	CHD, AR, CE TIA, LVH	a
H12/III-v	67	M	185	100	8.8↑	5.6	103	6.2	III	CHD, PAD	v
H13/II-v	65	M	140	80	7.4↑	4.7	107	4.9	II	UD	v
H14/I	63	M	180	100	4.1	5.8	80	4.9	I	UD	—
H15/II	53	M	240	120	6.1	5.9	82	4.1	II	PAD, LVH	—
H16/II	60	M	180	100	7.9↑	5.7	124↑	5.1	II	PAD	—
H17/I	47	F	160	100	6.1	5.9	95	5.5	I	UD	—
H18/III	57	F	170	110	6.8↑	6.5↑	92	5	III	CHD	—
H19/II	39	F	170	100	5.9	5.4	74	5	II	UD	—
H20/II	61	M	170	90	7.8↑	6.5↑	104	4.9	II	PAD	—

[a]Patients numbered from 1 to 20 with notification of AH stage (I–III) untreated or treated with valsartan (v) or amlodipine (a);

ABBREVIATIONS: SBP: systolic blood pressure; DBP: diastolic blood pressure; Chol: cholesterol; Glu: glucose; Creat: creatinine; WBC: white blood cells; TIA: transient ischemic attack; CHD: coronary heart disease; CA: cerebrovascular accident; AR: arrhythmia; CE: circulatory encephalopathy; PC: postinfarction cardiosclerosis; PAD: peripheral artery disease; LVH: left ventricular hypertrophy; UD: undetermined.

cDNA Microarray Hybridization

Gene expression was analyzed using Atlas™ Human cDNA Expression Arrays from Clontech (Human 1.2 and Human 1.2 II), each containing 1,176 human gene fragments. The membranes were prehybridized for 4 h at 42°C in hybridization solution consisting of 50% formamide (Sigma, St. Louis, MO, USA), 6×SSPE, 5×Denchardt solution, 0.2% SDS (Sigma), 2mg/mL heparin, and 0.5 mg/mL denatured herring sperm DNA (Fluka Chemie, Buchs, Switzerland). Purified and denatured labeled cDNA-probes were added to the hybridization solution and incubated at 42°C overnight. After hybridization the membranes were washed at increasing stringency: 2×20 min in $2\times$SSPE and 1% SDS at 42°C; 2×20 min in $2\times$SSPE and 1% SDS at 65°C; 2×20 min in $0.2\times$SSPE and 0.5% SDS at 65°C. The membranes were then exposed to PhosphorImager screens for 3–5 days. The images were scanned in a PhosphorImager SI system (Molecular Dynamics, USA) and analyzed using AtlasImage 2.0 software (Clontech Laboratory, Inc., Palo Alto, CA, USA). Areas of arrays with obvious artifacts were manually excluded. Normalization of signal intensities obtained from different hybridization experiments was based on the sum of background-subtracted signal data of all expressed genes except for of genes with expression signal exceeding fivefold mean value of the experiment.

Semi-Quantitative RT-PCR Analysis

One and a half micrograms of total RNA was reverse-transcribed in 20 μL of the reaction mixture containing 50 mM Tris-HCl (pH 8.3), 75 mM KCl, 3 mM MgCl2, 10 mM DTT, 1 mM each of the deoxynucleoside triphosphates (USB), 2.5 μM random primers (N9) and 200 U of MMLV reverse transcriptase (Promega). The reactions were performed at 37°C for 60 min and stopped by incubation at 95°C for 5 min with subsequent addition of 4 volumes (80 μl) of the stop solution (1 mM EDTA, 10 mM Tris-HCl, pH 8.0). The cDNA probes were stored frozen at −20°C. PCR amplification was carried out with 1 μL cDNA in 15 μL reaction mixture containing 10 mM Tris-HCl (pH 8.5), 50 mM KCl, 2 mM MgCl2, 200 μM each dNTP, 1 μM each gene-specific forward and reverse primers, 2.5 U of Taq polymerase (Biomaster, Moscow, Russia), "TaqStart" antibodies (Clontech) and 60 mM tetramethylammonium chloride (Fluka Chemie) using DNA Thermal Cycler (Hybaid Ltd., Ashford, Middlesex, UK). Gene-specific primers for PCR analysis were designed using Oli99 (Technogene, Russia) software for all except for GAPDH mRNA. For the latter case, sequences for PCR primers were obtained from Clontech. The sequences of PCR-primers and the number of PCR cycles used to obtain gene-specific PCR-product in exponential reaction phase are presented in TABLE 2. Thermal profile for each gene was 94°C for 5 min followed by desired number of cycles at 94°C for 30 sec, 60°C for 30 sec, and 72°C for 40 sec. A total of 5 μL each PCR mixture was run on 1.5% agarose gel in

TABLE 2. Nucleotide sequence of gene-specific primer pairs used in semi-quantitative RT-PCR analysis, number of PCR cycles, and the size of PCR products

Gene Symbol	Accession Number	Nucleotide Sequence of PCR Primer Pair		Number of PCR Cycles	PCR Product Size (bp)
		Forward (5'–3')	Reverse (5'–3')		
CXCR4	NM_003467	CAGCACATCATGGTTGGCCTTATCC	CTCGGTGATGGAAATCCACTTGTGC	22	260
CX3CR1	NM_001337	ACAGGGTGGCTGACTGGCAGATC	GCCAATGCAAAGATGACGGAGTAG	26	194
CAGC	NM_005621	TCTTCCACCAATACTCAGTTCGGAAG	ACCCTCATTGAGGACATTGCTGGG	22	284
OSM	NM_020530	ACAGAGGACGCTGCTCAGTCTGG	GGCGCTCCCTGCAGTGCTCTC	28	214
CASP4	NM_001225	GACAGAGGCTGTTCCCTATGG	GAAGCATTTGTCCTGCCATACG	25	233
CASP2	NM_032982	AGCTGTTGTTGAGCGAATTGTTAGAAC	GGAGACTCAAGTCGTAGTCACAGC	29	272
CASP8	NM_033358	TGGATTTGCTGATTACCTACCTAAACAC	ATGTCATCATCCAGTTTGCATTTGGAG	27	195
UBID4	NM_006268	AACAGTCGAGCGCGAAAGCGGATC	AGTGGGCATAGTGGTAACTGAGGC	26	243
RAFB1	NM_004333	TCTTCTAGCCTTTCAGTGCTACCTTC	CCAGTAAGCCAGGAAATATCAGTGTC	30	283
RAF1	NM_002880	AATAGTTCAGCAGTTTGGCTATCAGCG	CAATCAAAGACGCAGCATCAGTATTCC	26	274
EGR1	NM_001964	GGCGAGCAGCCCTACGAGCAC	GGCCGGTGGGTTGGTCATGCTC	33	230
FLI1	NM_002017	AGCCAGTGAGGGTCAACGTCAAGC	GGCCCACTCCAGCCATTGCCTC	26	256
P53	NM_000546	GAGAGCTGAATGAGGGCCTTGGAAC	GCTGATTGTAAACTAACCCTTAACTGC	29	456
PYST1	NM_001946	AGCAGCGACTGGAACGAGAATACG	CAGCCAAGCAATGTACCAAGACACC	25	582
PAC1	NM_004418	CCCGACAAGCCCGGCTCATGTG	CGGCCTCCGCTGTTCTTCACCC	29	448
GDI1	NM_001493	CACTGAGACTGAGGCCTTGGCTTC	GTACAACTTGATGCGGTTGACGGTC	27	273
GDI2	NM_001494	CTTTACAGTGAAATCTTTGGCAAGATATGG	GACTTGGTTCTGTGGAATAATGATCTG	27	347
GLRX	NM_002064	GGGCTTCTGGAATTTGTCGATATCAC	GAAGTGCTGTCATCATTTGCAATGCC	31	271
NAT1	NM_001418	AGCACTAGACGAGATGACAAACTCCG	GCATACAGTGAGCTATACTTTGGCTC	25	235
G6S	NM_002076	CTATCGGTTAATGATGTTACAGTCCTG	CTAGCTACAGACACCTACTACAATCAC	29	226
IGF2R	NM_000876	GAAGGCAATAGCTGGAATCTGGGTG	CAGCGCTCAGATGGTGTCGTTG	24	321
Fte-1	NM_001006	GCAGACAATGATTGAAGCTCACGTTG	CTACTGCCTTCACCATGAAGCTCC	25	359
MNDA	NM_002432	AGCTTTCATTTCTCAGCCCTTTACAAG	AAGTGATGGCATATCTTTGCAAGTTC	24	386
HSPA8	NM_006597	GTCCAGGCAGCCATCTTGTCTGG	CATCTACACTTGGTTGGCTAATCAAC	28	836
HSP40	NM_006145	CCGCGCAAGCGCGAGATCTTCG	CCAAAGTTCACGTTGGTGAAGCCAC	26	268
ERK2	NM_002745	CATCGCCGAAGCACCATTCAAGTTC	AAGCCAAGACGGGCTGGAGACAG	27	215
18S	X03205	CGTCTGCCCTATCAACTTTCGATGG	AGACTTGCCCTCCAATGGATCCTC	13	259
GAPDH	J04038	ACCACAGTCCATGCCATCAC	TCCACCACCCTGTTGCTGTA	21	452

Tris-borate-EDTA electrophoresis buffer with ethidium bromide (1 μg/mL). The fluorescence-stained bands corresponding to PCR products were quantified using an AlphaImagerTM System (Alpha Innotech Corporation, San Leandro, CA, USA). The PCR product amount of each studied gene obtained in exponential phase of the reaction was referred to that of 18S rRNA. Each PCR experiment was performed in double series and statistically analyzed applying the Mann–Whitney U-test to evaluate the significance of differences between gene expression mean values for control and AH groups. P value lower than 0.05 was considered to be statistically significant.

RESULTS

In the present study we investigated the gene expression profile in the total fraction of blood leukocytes from patients with AH. Hypertensive patients were differed in the stage of the disease (TABLE 1). Two subjects had the initial form of AH with transient elevation of systolic blood pressure reaching 160–180 values; seven subjects were diagnosed as having AH of the second stage, characterized by permanent rise of systolic blood pressure in the absence of medical treatment (140–240 mm Hg) and by end-organ damage such as peripheral artery disease and/or hypertrophy of the left heart ventricule. Eleven subjects had the third stage of the disease, complicated by coronary heart disease, cerebrovascular accident, cardiac infarction, circulatory encephalopathy, and transient ischemic attacks. At the moment of blood withdrawal seven patients did not receive any hypotensive therapy, whereas 13 patients were treated either with valsartan or amlodipine. Plasma levels of cholesterol, glucose, and creatinine were varied from patient to patient, being elevated in some of them, whereas the number of leukocytes was within the normal values for all patients.

For microarray experiment, three pooled total RNA samples representing two groups of AH patients (treated and untreated) and control group (donors) were obtained by mixing the same amounts of individual total RNAs. After hybridization of reverse-transcribed pooled total RNAs with cDNA microarrays, the expression pattern in each studied group was analyzed and the list of genes was selected in such a way that it was similar in two groups of AH patients but different from those of the control group. This approach revealed 40 genes, among which 31 genes were upregulated and 9 genes were downregulated in AH groups. To validate differential levels of mRNA abundance detected on the microarray we used a semi-quantitative RT-PCR method. In that case, the gene expression level of the selected group of genes was analyzed not in pooled RNA samples but in each individual RNA sample. After normalization RT-PCR data were averaged for each studied group of AH patients and donors to calculate fold changes of gene expression in AH; these were compared with the array data. The microarray results were found to be in concordance with the results of RT-PCR for 25 transcripts examined. The list of these genes with the corresponding expression level fold changes in AH is presented in TABLE 3.

TABLE 3. List of 25 genes differentially expressed in two groups of patients ("H+T" – all treated hypertensive patients, "H" – all untreated hypertensive patients).

Gene Symbol	Gene Name	H+T		H	
		Total Fold Change	P Value	Total Fold Change	P Value
I. Inflammatory response					
CXCR4	Chemokine, CXC motif, receptor 4	2.12↓	0.00025	2.64↓	0.00063
CX3CR1	Chemokine, CX3C motif, receptor 1	3.82↑	0.00025	5.40↑	0.00091
CAGC	S100 calcium-binding protein A12	1.86↑	0.00919	1.67↑	0.05372
OSM	Oncostatin M	1.19↓	0.10101	2.6↓	0.00041
II. Apoptosis					
CASP4	Caspase 4, apoptosis-related cysteine protease	3.61↑	0.00356	4.06↑	0.00629
CASP2	Caspase 2, apoptosis-related cysteine protease	1.74↑	0.00126	1.91↑	0.00091
CASP8	Caspase 8, apoptosis-related cysteine protease	3.88↑	0.00017	6.06↑	0.00021
UBID4	Requiem, apoptosis response zinc finger gene	1.84↑	0.00052	1.75↑	0.00249
III. Transcription regulation					
EGR1	Early growth response 1	4.9↓	0.00006	1.4↓	0.06371
FLI1	Friend leukemia virus integration 1	1.32↑	0.04215	1.59↑	0.00123
P53	Transformation-related protein 53	1.43↑	0.00292	1.76↑	0.00842
IV. Control of MAP kinase function					
PYST1	Dual-specificity phosphatase 6	2.88↑	0.00193	4.18↑	0.00064
PAC1	Dual-specificity phosphatase 2	4.12↓	1.7E-06	4.42↓	0.00011
RAFB1	V-RAF murine sarcoma viral oncogene homologue B1	1.62↑	0.01209	1.53↑	0.00966
RAF1	V-RAF-1 murine leukemia viral oncogene homologue 1	1.57↑	0.00647	1.59↑	0.00463

NOTE: Total fold change in gene expression level was calculated for each AH group relating to donors as a ratio of average values.

TABLE 3. (continued)

Gene Symbol	Gene Name	H+T		H	
		Total Fold Change	P Value	Total Fold Change	P Value
V. Vesicular trafficking					
GDI1	GDP dissociation inhibitor 1	1.34↑	0.01845	1.36↑	0.02480
GDI2	GDP dissociation inhibitor 2	2.82↑	0.00356	3.54↑	0.00842
VI. Cell redox homeostasis					
GLRX	Glutaredoxin	4.02↑	0.00081	4.59↑	0.00091
VII. Regulation of translation					
NAT1	Death-associated protein 5	1.41↑	0.02997	1.88↑	0.00249
VIII. Glycosaminoglycan catabolism					
G6S	N-acetylglucosamine-6-sulfatase	1.85↑	0.00291	1.69↑	0.01471
IX. Lysosomal enzyme trafficking					
IGF2R	Insulin-like growth factor 2 receptor	1.57↑	0.03032	1.35↑	0.04308
X. Protein biosynthesis					
Fte-1	Ribosomal protein S3a	2.69↑	0.00802	3.70↑	0.08782
XI. Cell differentiation					
MNDA	Myeloid cell nuclear differentiation antigen	2.35↑	0.01209	3.01↑	0.00123
XII. Cellular stress					
HSPA8	Heat-shock 70-kD protein 8	2.58↑	0.02136	4.54↑	0.00123
HSP40	Heat-shock 40-kD protein 1	1.72↓	0.00005	2.02↓	0.00010

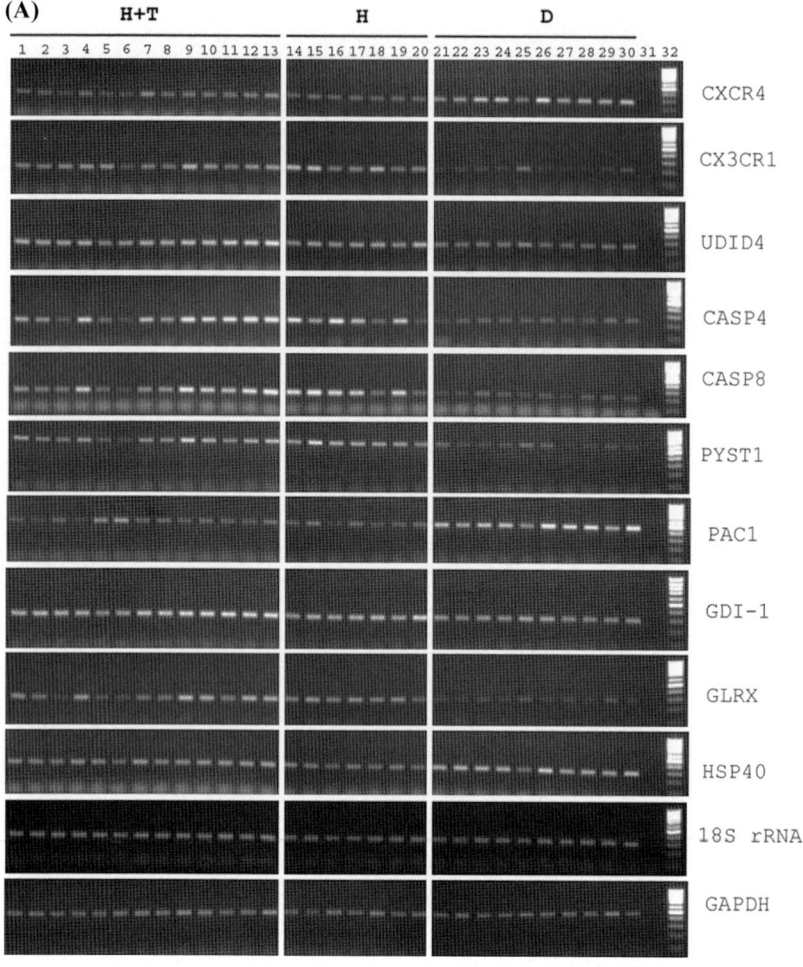

FIGURE 1. Semi-quantitative RT-PCR data analysis. **(A)** Electrophoregrams of PCR-products of the 10 differentially expressed genes in AH relative to donors and ubiquitously expressed genes, GAPDH and 18S rRNA, used for data normalization. *Lanes 1–20:* RT-PCR products synthesized on mRNA from AH patients (H1–H20, respectively; for description of AH patient see TABLE 1) forming H+T group (treated AH patients) and H group (untreated AH patients). *Lanes 21–30:* RT-PCR products synthesized on mRNA from control subjects (D1–D10, respectively). *Lane* 31 is negative PCR-control; *lane* 32 is 100-bp DNA ladder (Fermentas). Names of the studied genes are indicated. **(B)** Histogram of ratio of averaged band intensities of RT-PCR experiments for all members of AH groups and control group. The data are plotted after densitometrical analysis of the electrophoregrams from panel A. PCR-product amount of each studied gene was referred to that of 18S rRNA and values obtained were used to calculate the ratio between averaged gene expression level in AH group and donors. Ninety-five percent confidence interval of the mean was calculated and represented as whiskers on the plot. The logarithm of the ratio is plotted (*right scale*). *Left scale* shows fold scale of increase or decrease of correspondent gene expression level for AH group with respect to control group.

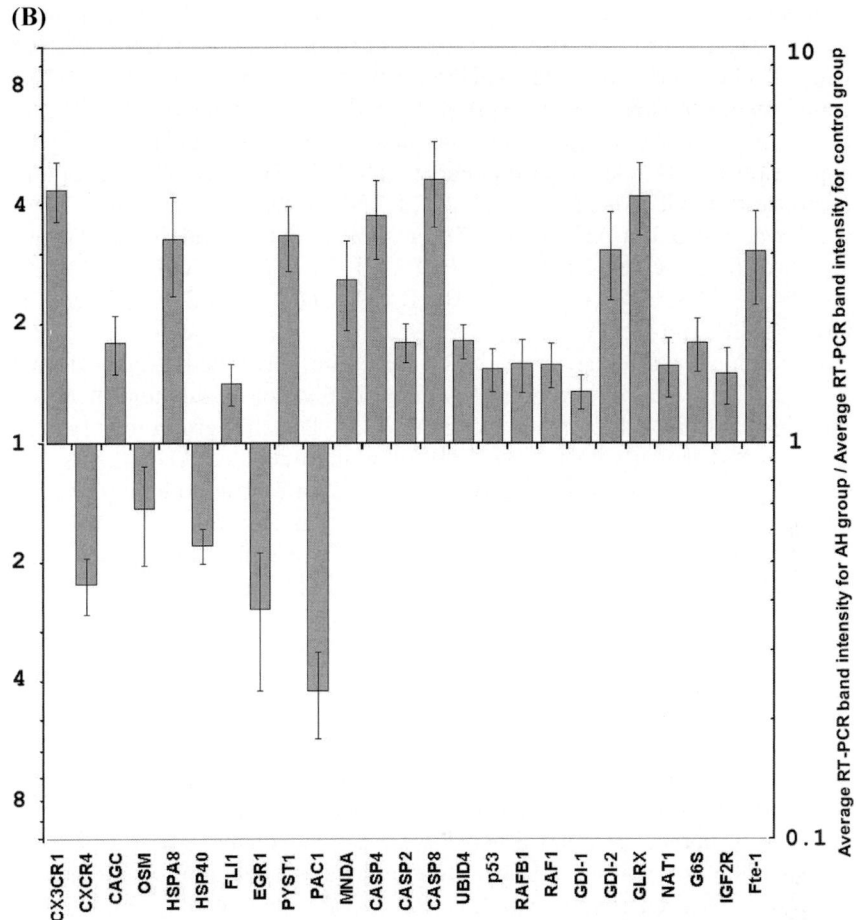

FIGURE 1. (continued)

The discrepancy between the two methods employed was observed for 15 transcripts, which led to exclusion of them from consideration as differentially expressed genes. As shown in TABLE 3, genes that are differentially expressed in AH are divided into 12 groups according to their functional role in a cell. According to the literature data, these genes are implicated in the programmed cell death signaling pathways, cellular stress, inflammatory response, control of MAP kinase functions, cell differentiation, transcription, and translation regulation, protein biosynthesis, vesicular and lysosomal enzyme trafficking, and glycosaminoglycan catabolism. CXCR4, CX3CR1, CASP4, CASP8, PYST1, PAC1, GDI-2, GLRX, MNDA, HSPA8, and HSP40 genes demonstrate 2–6-fold reliable changes ($P \leq 0.05$) in gene expression level in two AH groups as compared to donors. We have chosen 10 of 25 differentially expressed genes and two control genes (GAPDH and 18S rRNA) to demonstrate the results of RT-PCR experiment. As shown in FIGURE 1A, the selected genes display expression patterns discriminating donors and AH patients, while control genes are ubiquitously expressed. For the visualization of RT-PCR data for all 25 differentially expressed genes, the histogram of fold changes of gene expression in AH relative to donors was plotted (FIG. 1B). This histogram demonstrates that 5 of 25 genes (PAC1, HSP40, CXCR4, EGR1, and OSM) are downregulated in AH patients with respect to donors and 20 other genes are upregulated (CX3CR1, CAGC, HSPA8, FLI1, PYST1, MNDA, CASP2, CASP4, CASP8, UBID4, p53, RAFB1, RAF1, GDI-1, GDI-2, GLRX, NAT1, G6S, IGF2R, and Fte-1).

Hierarchical cluster analysis of RT-PCR data performed using Michael Eisen's, "Gene Cluster" program[12] resulted in division of all subjects tested into this study into three main groups (FIG. 2). The first group mainly consisted of AH patients treated with valsartan and untreated AH patients; the second group comprised three AH patients treated with amlodipine, one patient treated with valsartan, two untreated patients and one donor; and the third group included donors. It would be interesting to establish the true cause of such a division of AH patients by increasing the number of AH patients and sorting them according to strict criteria.

DISCUSSION

We have employed cDNA microarray technology and found 25 differentially expressed genes in leukocytes of patients with AH, among which 20 genes were upregulated and 5 genes were downregulated. These genes are grouped into different classes according to their possible implication in various processes, such as apoptosis (CASP2, CASP4, and CASP8, p53, UBID4, NAT1, and Fte-1), inflammatory response (CAGC, CXCR4, and CX3CR1), control of MAP kinase function (PYST1, PAC1, RAF1, and RAFB1), vesicular trafficking of molecules among cellular organelles (GDI-1 and GDI-2), cell redox homeostasis (GLRX), cellular stress (HSPA8 and HSP40), and others (TABLE 3). The group of genes implicated in the inflammation process is of primary concern from the viewpoint of their possible role in pathophysiological mechanisms causing inflammation in the vessel wall in AH. For example, we have demonstrated elevated expression of CAGC known to directly activate endothelial cells, mononuclear phagocytes, and lymphocytes[13] with subsequent induction of expression of many pro-inflammatory molecules such as various cytokines (IL-1β and TNF-α) or adhesion receptors.[14] Moreover, we have found an opposite effect in the regulation of two chemokine receptors: CXCR4 was downregulated and CX3CR1 was upregulated. According to the literature data, they are known to be expressed in different leukocyte subsets to provide leukocyte recruitment in the vessel. CXCR4 is

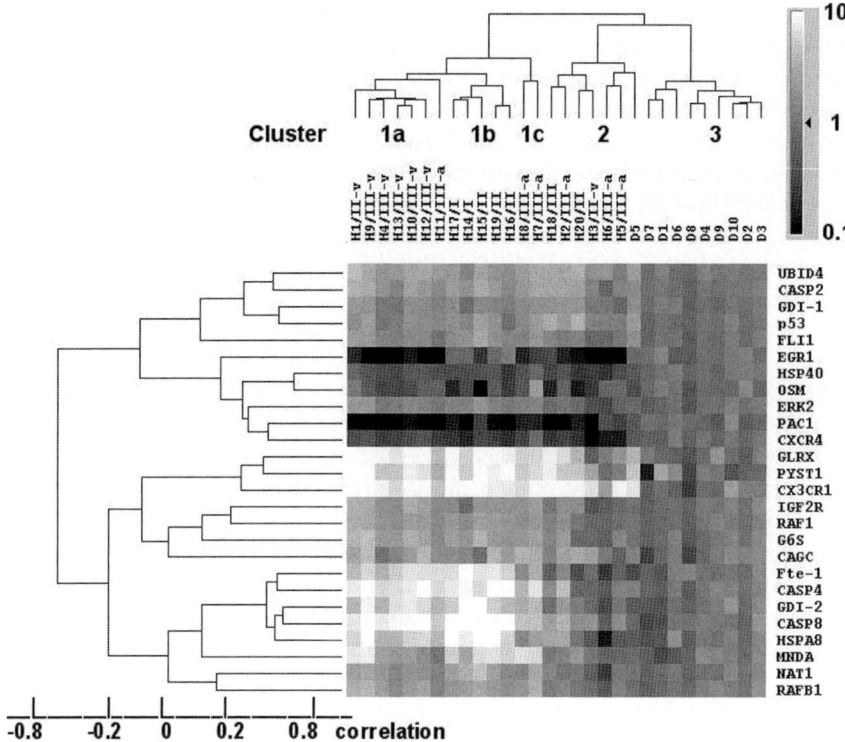

FIGURE 2. RT-PCR data clustering using Euclidean distance and weighted complete average linkage algorithm for 25 transcripts demonstrating differential expression. All gene expression data for each gene before clustering were normalized to the mean value of the same gene for Cluster 3, which is composed of all members of a control group except Donor 5. The logarithms of the relative expression value were used for clustering. Rows represent individual transcripts, columns represent different samples. Transcripts with elevated or repressed expression are represented in shades of gray scale.

expressed on the cell surface of most leukocyte populations, including the majority of T-lymphocyte subsets, and all B cells and monocytes, but only weakly on natural killer cells,[15] while CX3CR1 is mainly expressed on NK cells, cytotoxic T lymphocytes, and macrophages. In addition, it is important to note that unlike CXCR4 and other chemokine receptors, CX3CR1 is connected with its ligand in an integrin-independent manner, providing rapid leukocyte capture under high blood flow.[16] Thus, we suppose that elevated expression of CX3CR1 and decreased expression of CXCR4 in blood leukocytes from hypertensive patients provide adhesion and migration of certain types of leukocytes to the vascular wall, specifically, NK cells and cytotoxic T cells, whose excessive activity may damage tissues, including the endothelium.

Along with definition of the possible participants in the inflammation process in AH, the abnormalities in the regulation of different mitogen-activated protein kinases (MAPK) signaling pathways are of particular interest. MAPKs play an important role in a variety of cellular processes, including proliferation, differentiation, and apoptosis[17] and their function is dependent on the activity of dual-specificity phosphatases (DUSPs), being the inhibitors of MAPKs. In the present study we investigated the expression level of two members of DUSPs, PAC1, and PYST1, and the effect of their regulation was opposite. According to the literature data, PAC-1 and PYST1 display specificity for inactivation of different MAP kinases. Thus, studies using purified proteins and transfected mammalian cells have revealed that low concentrations of PYST1 completely inactivate ERK1 and ERK2, but not JNK/SAPK or p38 MAP kinases,[18] whereas PAC-1 shows limited specificity for ERK, but inactivates p38 and JNK/SAPK.[19] According to our data, PYST1 is significantly elevated in leukocytes of AH patients. Therefore, we assume that ERK1/ERK2 signaling pathway is blocked in these cells and growth-stimulating and survival signals of ERK are shut down. On the other hand, downregulation of PAC1 demonstrated in AH in our study suggests retention of JNK/SAPK and p38 MAP kinase activity, which controls production of inflammatory cytokines and apoptotic cell death.[20,21] On the whole, we suppose that regulation of PYST1 and PAC1 gene transcription combined with their differential binding of MAPK may provide a mechanism for rapid and targeted inactivation of ERK1/2 and activation of JNK and p38 MAPK pathways in leukocytes of AH patients. This imbalance between growth factor-activated ERK and stress-activated JNK-p38 pathways may be important in determining whether a cell survives or undergoes apoptosis and/or takes part in inflammation process.

Reduction in leukocyte survival in AH is confirmed by significant elevation in gene expression of apoptosis-related cystein proteases such as caspase-2, caspase-4, and caspase-8, transformation-related protein p53, apoptosis response zinc finger gene (UBID4), death-associated protein 5 (NAT1), and ribosomal protein S3a (Fte-1) found in this study. The latter is of interest because as a ribosomal protein involved in translation process, it was found to be implicated in the regulation of cell death. However, the views concerning its negative or positive implication in cell death are opposing. On the one hand, it has been demonstrated that RPS3a together with Bcl-2 prevents apoptosis by inhibiting poly (ADP-ribose) polymerase activity,[22] whereas on the other, enhanced RPS3a expression is regarded as priming a cell for apoptosis.[23]

Changes in leukocyte survival in AH can be also assumed by imbalance in the regulation of gene expression of the two heat-shock proteins, Hsc70 (HSPA8) and Hsp40 found in this investigation. As shown in our study, HSPA8 was upregulated and Hsp40 was downregulated. Hsc70/Hsp70 and co-chaperone Hsp40 represent one of the major chaperone systems for the folding of a large variety of cytosolic proteins in eukaryotes[24] in contrast to other chaperone systems assisting the folding of a more limited set of proteins. It has become increasingly clear that disruption of chaperoning mechanisms contributes to the possibility of intramolecular misfolding that may lead to aggregation. Misfolded, aggregated proteins are often cytotoxic and lead to cell death.[25] So we can speculate that the imbalance in gene expression of these chaperones may result in cell death in leukocytes in patients with AH. On the other hand, Hsc70 was found to be a major component of the transplasmamembrane oxidoreductase (PMO) complex involved in the cellular response against oxidative stress.[26] Moreover, it is well established that overexpression of Hsc70 confers resistance to oxidative stress generated by ROS, showing an antioxidant role for this protein and protecting cells from damages.[27]

Several lines of evidence suggest that oxidative stress is involved in the pathogenesis of cardiovascular diseases through various mechanisms such as stimulation of SMC proliferation, inactivation of endothelium-derived nitric oxide, and induction of redox-sensitive genes.[28]

To maintain redox status under reducing conditions, living cells possess two major systems: the thioredoxin (TRX)/thioredoxin reductase and the glutathione (GSH)/glutaredoxin (GLRX),[29] which catalyze reduction of disulfide bonds in a variety of proteins. The redox status of sulfhydryl groups is important for such cellular functions as synthesis and folding of proteins and regulation of the structure and activity of enzymes, receptors, and transcription factors. Thus, induction of one of these systems, in particular glutathione (GSH)/glutaredoxin (GLRX), as observed in the present study, may be a counterregulatory mechanism against increased oxidative stress in AH, as found by B. Kristal et al.[11] Moreover, GLRX plays an important role in cytoprotection against apoptosis.[30]

Taken together, we have found the differentially expressed genes in leukocytes of hypertensive patients that are known to be implicated in inflammation, apoptotic, and survival processes. We cannot attribute these changes to a certain type of leukocyte because we investigated the total leukocyte fraction. We hope that this study is helpful for better understanding of the molecular processes that occur in leukocytes from patients with AH.

REFERENCES

1. LIFTON, R.P., A.G. GHARAVI & D.S. GELLER. 2001. Molecular mechanisms of human hypertension. Cell **104:** 545–556.
2. OLSEN, M.H., K. WACHTELL, K.L. HERMANN, et al. 2002. Is cardiovascular remodeling in patients with essential hypertension related to more than high blood pressure? Am. Heart J. **144:** 530–537.
3. PLANTE, G.E. 2002. Vascular response to stress in health and disease. Metabolism **51:** 25–30.
4. DAO, H.H., F.M. MARTENS, R. LARIVIERE, et al. 2001. Transient involvement of endothelin in hypertrophic remodeling of small arteries. J. Hypertens. **19:** 1801–1812.

5. VAN ZWIETEN, P.A. 2000. The role of angiotensin II receptors and their antagonists in hypertension. Ann. Ital. Med. Int. **15:** 85–91.
6. SHAW, A. & Q. XU. 2003. Biomechanical stress-induced signaling in smooth muscle cells: an update. Curr. Vasc. Pharmacol. **1:** 41–58.
7. STEVENSON, J.R., J. WESTERMANN, P.M. LIEBMANN, et al. 2001. Prolonged alpha-adrenergic stimulation causes changes in leukocyte distribution and lymphocyte apoptosis in the rat. J. Neuroimmunol. **120:** 50–57.
8. HAHN, A.W., U. JONAS, F.R. BUHLER & T.J. RESINK. 1994. Activation of human peripheral monocytes by angiotensin II. FEBS Lett. **347:** 178–180.
9. OHKI, R., K. YAMAMOTO, H. MANO, et al. 2002. Identification of mechanically induced genes in human monocytic cells by DNA microarrays. J. Hypertens. **20:** 685–691.
10. SMEDLY, L.A., M.G. TONNESEN, R.A. SANDHAUS, et al. 1986. Neutrophil-mediated injury to endothelial cells. J. Clin. Invest. **77:** 1233–1243.
11. KRISTAL, B., R. SHURTZ-SWIRSKI, J. CHEZAR, et al. 1998. Participation of peripheral polymorphonuclear leukocytes in the oxidative stress and inflammation in patients with essential hypertension. Am. J. Hypertens. **11:** 921–928.
12. EISEN, M.B., P.T. SPELLMAN, P.O. BROWN & D. BOTSTEIN. 1998. Cluster analysis and display of genome-wide expression patterns. Proc. Natl. Acad. Sci. USA **95:** 14863–14868.
13. ROTH, J., T. VOGL, C. SORG & C. SUNDERKOTTER. 2003. Phagocyte-specific S100 proteins: a novel group of proinflammatory molecules. Trends Immunol. **24:** 155–158.
14. HOFMANN, M.A., S. DRURY, C. FU, et al. 1999. RAGE mediates a novel proinflammatoryaxis: a central cell surface receptor for S100/calgranulin polypeptides. Cell **97:** 889–901.
15. HORI, T., H. SAKAIDA, A. SATO, et al. 1998. Detection and delineation of CXCR-4 (fusin) as an entry and fusion cofactor for T-tropic [correction of T cell-tropic] HIV-1 by three different monoclonal antibodies. J. Immunol. **160:** 180–188.
16. UMEHARA, H., E.T. BLOOM, T. OKAZAKI, et al. 2004. Fractalkine in vascular biology. Arterioscler Thromb. Vasc. Biol. **24:** 34–40.
17. CHANG, L. & M. KARIN. 2001. Mammalian MAP kinase signalling cascades. Nature **410:** 37–40.
18. GROOM, L.A., A.A. SNEDDON, D.R. ALESSI, et al. 1996. Differential regulation of the MAP, SAP and RK/p38 kinases by Pyst1, a novel cytosolic dual-specificity phosphatase. EMBO J. **15:** 3621–3632.
19. CHU, Y., P.A. SOLSKI, R. KHOSRAVI-FAR, et al. 1996. The mitogen-activated protein kinase phosphatases PAC1, MKP-1, and MKP-2 have unique substrate specificities and reduced activity in vivo toward the ERK2 sevenmaker mutation. J. Biol. Chem. **271:** 6497–6501.
20. RAINGEAUD, J., S. GUPTA, J.S. ROGERS, et al. 1995. Pro-inflammatory cytokines and environmental stress cause p38 mitogen-activated protein kinase activation by dual phosphorylation on tyrosine and threonine. J. Biol. Chem. **270:** 7420–7426.
21. XIA, Z., M. DICKENS, J. RAINGEAUD, et al. 1995. Opposing effects of ERK and JNK-p38 MAP kinases on apoptosis. Science **270:** 1326–1331.
22. SONG, D., S. SAKAMOTO & T. TANIGUCHI. 2002. Inhibition of poly(ADP-ribose) polymerase activity by Bcl-2 in association with the ribosomal protein S3a. Biochemistry **41:** 929–934.

23. NAORA, H. & H. NAORA. 1999. Involvement of ribosomal proteins in regulating cell growth and apoptosis: translational modulation or recruitment for extraribosomal activity. Immunol. Cell Biol. **77:** 197–205.
24. ALBERTO, J.L., M.D. MACARIO & E. CONWAY DE MACARIO. 2005. Sick chaperones, cellular stress, and disease. N. Engl. J. Med. **353:** 1489–1501.
25. KAKIZUKA, A. 1998. Protein precipitation: a common etiology in neurodegenerative disorders? Trends Genet. **14:** 396–402.
26. BULLIARD, C., R. ZURBRIGGEN, J. TORNARE, et al. 1997. Purification of a dichlorophenol-indophenol oxidoreductase from rat and bovine synaptic membranes: tight complex association of a glyceraldehyde-3-phosphate dehydrogenase isoform, TOAD64, enolasegamma and aldolase C. Biochem. J. **324:** 555–563.
27. CHONG, K.Y., C.C. LAI, S. LILLE, et al. 1998. Stable overexpression of the constitutive form of heat shock protein 70 confers oxidative protection. J. Mol. Cell. Cardiol. **30:** 599–608.
28. KUNSCH, C. & M.M. MEDFORD. 1999. Oxidative stress as a regulator of gene expression in the vasculature. Circ. Res. **85:** 753–766.
29. HOLMGREN, A. 1989. Thioredoxin and glutaredoxin systems. J. Biol. Chem. **264:** 13963–13966.
30. DAILY, D., A. VLAMIS-GARDIKAS, D. OFFEN, et al. 2001. Glutaredoxin protects cerebellar granule neurons from dopamine-induced apoptosis by dual activation of the ras-phosphoinositide 3-kinase and jun n-terminal kinase pathways. J. Biol. Chem. **276:** 21618–21626.

Effects of AT_1 Receptor-Mediated Endocytosis of Extracellular Ang II on Activation of Nuclear Factor-κB in Proximal Tubule Cells

JIA L. ZHUO, OSCAR A. CARRETERO, AND XIAO C. LI

Laboratory of Receptors and Signal Transduction, Division of Hypertension and Vascular Research, Henry Ford Hospital, Detroit, Michigan 48202, USA

ABSTRACT: Angiotensin II (Ang II) exerts powerful proinflammatory and growth effects on the development of Ang II-induced hypertensive glomerulosclerosis and tubulo-interstitial fibrosis. The proinflammatory and growth actions of Ang II are primarily mediated by activation of cell surface type 1 receptors (AT_1) and the transcription factor nuclear factor-κB (NF-κB). However, binding of cell surface receptors by extracellular Ang II also induces receptor-mediated endocytosis of the agonist-receptor complex in renal cells. The purpose of the present study was to determine whether AT_1 receptor-mediated endocytosis of extracellular Ang II is required for Ang II-induced NF-κB activation and subsequent proliferation of rabbit renal proximal tubule cells. Expression of AT_1 (primarily AT_{1a} or human AT_1) receptors in these cells was confirmed by Western blot, showing that transfection of a human AT_1 receptor-specific 20-25 nucleotide siRNA knocked down more than 70% of AT_1 receptor protein ($P < 0.01$). Stimulation of proximal tubule cells by Ang II (1 nM) induced fourfold increases in NF-κB activity ($P < 0.01$). The Ang II-increased NF-κB activity was significantly attenuated by coadministration of losartan (10 μM), an AT_1 receptor-selective blocker, or colchicine (1 μM), a selective cytoskeleton microtubule inhibitor known to block receptor-mediated endocytosis ($P < 0.01$). Furthermore, Ang II significantly increased ^{3}H-thymidine incorporation (>55%, $P < 0.01$), an index of cell proliferation and DNA synthesis, and the effect was also attenuated by coadministration of losartan and colchicine ($P < 0.01$). Our results therefore suggest that AT_1 receptor-mediated endocytosis of extracellular Ang II may be required for Ang II-induced NF-κB activation and subsequent cell proliferation in renal proximal tubule cells.

KEYWORDS: angiotensin II; cell proliferation; chemokine; cytokines; kidney

Address for correspondence: Jia L. Zhuo M.D., Ph.D., Division of Hypertension and Vascular Research, Henry Ford Hospital, 2799 West Grand Boulevard, Detroit, MI 48202, USA. Voice: 1-313-916-0018; fax: 1-313-916-1479.
e-mail: jzhuo1@hfhs.org

INTRODUCTION

Angiotensin II (Ang II) has been recognized for several decades as a vasoactive peptide that plays an important role in physiological regulation of renal sodium and fluid reabsorption and blood pressure homeostasis. However, an increasing effort has recently been directed to studying the potential role of Ang II as a powerful proinflammatory cytokine and growth factor in the pathogenesis of many progressive renal diseases, including Ang II-induced hypertensive glomerulosclerosis and tubulo-interstitial fibrosis.[1-3] Indeed, there is accumulating evidence that Ang II can affect the transcription of genes related to cell growth and proliferation.[4] Ang II has been shown to induce expression of proto-oncogenes and genes for growth factors, extracellular matrix and hypertrophic markers.[4] Although not classified as gene transcription-modulating drugs, ACE inhibitors, which block Ang II formation, and AT_1 receptor antagonists, which block Ang II type 1 receptors, have been shown to prevent gene expression induced by Ang II, thereby supporting an important role of Ang II in promoting inflammation and cell growth. The growth-promoting and proliferative effects of Ang II may be partly mediated by activating cell surface AT_1 receptors or by intracellular Ang II acting on intracellular Ang II receptors.[5,6] It is interesting to note that intracellular Ang II may stimulate cytoplasmic receptors to activate a variety of intracellular kinases, leading to phosphorylation of many cytoplasmic and nuclear proteins, such as extracellular signal-regulated protein kinase(s) (ERKs),[7] JAK-STAT signaling,[8] and calcineurin phosphatase.[9]

Although Ang II is known to induce target organ damage in cardiovascular and renal tissues by activating a number of proinflammatory cytokines and growth factors, the precise mechanisms are far from fully understood.[1-3] Numerous studies suggest that nuclear factor-κB (NF-κB) may be the key factor in mediating Ang II-induced inflammatory responses and tissue injury. For example, infusion of Ang II in rats increased renal and vascular smooth muscle cell NF-κB binding activity, and activation of NF-κB was associated with increases in inflammatory cell infiltration and tubulo-interstitial inflammation.[10] In rats harboring both human renin and angiotensinogen genes and thereby producing high circulating and tissue Ang II,[3] NF-κB binding activity was substantially increased and inhibition of NF-κB was found to ameliorate Ang II-induced inflammatory target organ damage in the heart and kidney.[3,4] NF-κB is an important transcription factor in inflammatory diseases, and activation of NF-κB by Ang II in turn stimulates transcription of many cytokines and chemokines, including angiotensinogen, monocyte chemoattractant peptide-1 (MCP-1), transforming growth factor (TGF)β, and RANTES (regulated on activation normal T cell expressed and secreted).[11] When not activated, NF-κB exists in an inactive form in the cytoplasm, binding to inhibitory IκB proteins. Stimulation of cells results in phosphorylation and degradation of IκB proteins, which releases NFκB dimers. These dimers are translocated to the nucleus, where they activate appropriate target genes.[12,13] Although it is commonly presumed from *in vivo* studies that Ang II activates NF-κB through cell surface receptor–mediated signaling to induce tubulo-interstitial inflammation, it is not known whether intracellular Ang II plays any role in this response. In addition to activating cell surface receptors, extracellular Ang II is also internalized after binding to cell membrane AT_1 receptors.[14] In the present study, we used rabbit proximal tubule cells as a model to test the hypothesis that AT_1 receptor-mediated Ang II endocytosis is required for Ang II-stimulated NF-κB activation and promotion of its translocation into the nucleus.

MATERIALS AND METHODS

Proximal Tubule Cell Culture

Cultured proximal tubule cells were obtained from American Type Culture Collection (vEPT, ATCC) and subcultured as described previously.[15,16] These cells were derived from the S1 segment of rabbit kidney proximal tubules and have been shown to express major components of the renin–angiotensin system, including angiotensinogen, renin, angiotensin-converting enzyme, and Ang II receptors,[15] and response to Ang II.[16] In the present study, unless specified otherwise, proximal tubule cells were subcultured in 6-well plates in complete DMEM/F-12 growth medium supplemented with 50 nM hydrocortisone, 5% heat-inactivated fetal bovine serum, 100 U/mL penicillin and 100 μg/mL streptomycin. Cells were maintained at 37°C and 95% O_2/5% CO_2 and fed every 2–3 days. Serum was removed from the medium for 24 h before the experiments began.[15–17]

Western Blot of AT_{1A} Receptor Protein Expression

To confirm that proximal tubule cells express AT_1 receptors, subconfluent, serum-starved cells were either treated with medium alone or transfected with an AT_{1A} receptor-specific 20–25 nucleotide siRNA (AT_1R siRNA) (Santa Cruz Biochemicals, Santa Cruz, CA, USA) for 48 to 72 h as described previously.[17,18] After treatment, the cells were washed twice with ice-cold phosphate-buffered saline and lysed with a modified RIPA buffer (50 mM Tris-HCl, 50 mM, 1% NP-40, 0.25% Na-deoxycholate, 150 mM NaCl, 1 mM EDTA, 1 mM PMSF, 1 μg/mL each of aprotinin, leupeptin, and pepstatin, 1 mM Na_3VO4, and 1 mM NaF, pH 7.4). Proteins were extracted and samples (10 μg each) were electrophoretically separated on 8–16% Tris-glycine gels as described previously.[16,17] After SDS separation, proteins were transferred to Millipore Immobilon-P membranes, and the membranes were blotted overnight at 4°C with 5% nonfat dry milk and incubated for 3 h at room temperature with a primary rabbit anti-AT_1 receptor polyclonal antibody raised against the N-terminal extracellular domain of AT_1 receptors (1:200; SC-1173, Santa Cruz).[16,17] To ensure equal protein loading, the same membranes were treated with a stripping buffer (Pierce Biotechnology) for 20 min, blotted with 5% nonfat dry milk, and reprobed with a mouse anti-β-actin monoclonal antibody at 1:2000 (Sigma-Aldrich, St. Louis, MO, USA). Western blot signals were detected using enhanced chemiluminescence (Amersham Piscataway, NJ, USA) and analyzed using a microcomputer imaging device with a digital camera (MCID, Imaging Research, St. Catherines, Ontario, Canada).

Measurement of Activated Nuclear Transcription Factor NF-κB in Nuclear Extracts

To determine whether AT_1 receptor-mediated endocytosis of extracellular Ang II is required for activation of Ang II on NF-κB in proximal tubule cells, we studied the effects of the endocytotic machinery inhibitors or the AT_1 receptor blocker losartan on NF-κB activity in cell nuclear extracts. Subconfluent cells plated in 6-well plates were treated for 60 min at 37°C according to the following protocols: (i) medium alone; (ii) lipopolysaccharide (LPS) only (5 μg/mL), a well-known NF-κB activator that served as a positive control[13]; (iii) Ang II only (1 nM); (iv) Ang II + colchicine (1 μM), which inhibits endocytosis in a non-Ang II-selective fashion by disrupting cytoskeleton microtubules;[17,19] (v) Ang II + PAO (10 μM), which blocks cell surface AT_1 receptor internalization by inhibiting tyrosine phosphatase;[17,20] (vi) Ang II + losartan (10 μM), which blocks cell surface AT_1 receptor-mediated Ang II internalization;[9,14] and (vii) Ang II + PD 123319 (10 μM), which inhibits AT_2 receptors and has been reported to mediate Ang II-induced NF-κB activation in VSMCs and COS-7 cells.[21,22] After treatment, nuclear extracts were prepared using a commercial kit (Active Motif, Carlsbad, CA, USA). In brief: cells were washed with a PBS/phosphatase inhibitor solution, scraped into tubes, and lysed in a lysis buffer containing a DTT/protease inhibitor cocktail. Nuclear fractions were collected for measurements of activated NF-κB using commercial Trans NF-κB ELISA for p65 and p50 (Active Motif).

Measurement of Proximal Tubule Cell Proliferation

To determine whether Ang II stimulates proximal tubule cell proliferation, we measured [^{3}H]-thymidine incorporation, an index of DNA synthesis and cell proliferation.[23-25] Cells were split into 6-well plates and incubated in DMEM/F12 growth medium for 2–3 days (10^5/well). Subconfluent cells were treated with serum-free medium (control; $n = 6$ wells), Ang II (1 nM, $n = 6$ wells), Ang II plus the AT_1 receptor blocker losartan (10 μM, $n = 6$ wells),

Ang II plus the AT_2 receptor blocker PD123319 (10 μM, $n = 6$ wells), or Ang II plus the endocytotic inhibitor colchicine (1 μM), and pulsed with 1 μCi/mL [^{3}H]-thymidine for 24 h.[17,19] After treatment, the medium was removed and cells were washed with ice-cold phosphate-buffered saline (PBS) and lysed with 200 mM NaOH and 0.25% SDS. The contents were transferred to tubes, vacuum-filtered, and washed four times with 5 mL ethanol/TCA (70%/5%) using a Brandel cell harvester. Incorporated [^{3}H]thymidine was counted in vials containing 4 mL scintillation cocktail as described previously.[17,19]

STATISTICAL ANALYSIS

Results are expressed as mean ± SEM. Unless otherwise specified, six samples from two separate experiments were collected for each treatment and assayed in duplicate, including Western blot. Differences between two treatments were compared by Student's unpaired t-test, while differences between more than two treatments were analyzed by one-way analysis of variance, followed by a Newman-Keuls test for multiple comparisons. A $P < 0.05$ was considered significant.

RESULTS AND DISCUSSION

Chronic administration of Ang II in rats is characterized by development of Ang II-dependent hypertension, accompanied by accumulation of Ang II in renal cortical endosomes[5] and tubulo-interstitial inflammatory injury in the kidney.[1] NF-κB has been suggested to play an important role in recruiting inflammatory cells to the tubulo-interstitium in Ang II-induced hypertensive renal injury,[3,4,10] but it is not known whether Ang II induces activation of NF-κB and stimulates proliferation of cultured proximal tubule cells. In the present study, we used rabbit proximal tubule cells as an *in vitro* model to study the potential role of AT_1 receptor-mediated Ang II endocytosis in activation of NF-κB and the cellular mechanisms involved. FIGURE 1 shows Western blots of AT_1 receptor protein expression in rabbit proximal tubule cells. The specificity of the receptor protein in these cells was confirmed using a specific Ang II receptor subtype siRNA. Because transfection of an AT_1 receptor-specific siRNA effectively knocked down over 70% of AT_1 receptor protein expression (FIG. 1), our results suggest that rabbit proximal tubule cells express predominantly AT_1 receptors, equivalent to human AT_1 receptors. Expression of endogenous AT_1 receptors is essential for studying the effects of AT_1 receptor-mediated endocytosis of extracellular Ang II on activation of NF-κB in these cells. In a previous study in Ang II-infused rats, we showed that extracellular Ang II is accumulated in the intracellular endosomal compartments via AT_1 receptor-mediated endocytosis.[6] However, we do not know whether blockade of AT_1 receptor-mediated endocytosis of extracellular

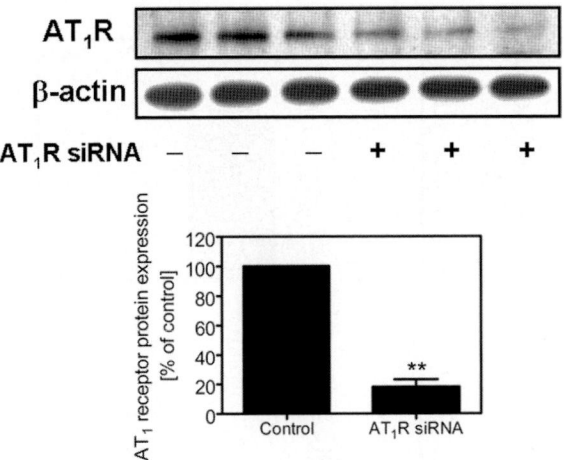

FIGURE 1. Expression of specific AT_{1a} receptor protein in cultured rabbit proximal tubule cells. *Top panel* shows Western blots of AT_1 receptor protein (~42 kDa) and β-actin, with the latter used to confirm equal loading. *Bottom panel* shows semi-quantitative data from control cells and cells transfected with an AT_1 receptor siRNA, which specifically knocks down AT_1 receptor protein expression. ** $P < 0.01$ versus control.

Ang II has any biological relevance with respect to induction of NF-κB activation.

FIGURE 2 shows the effects of Ang II on NF-κB activity and the potential role(s) of AT_1 and AT_2 receptors and the endocytotic machinery in Ang II-induced NF-κB activation in proximal tubule cells. Stimulation by Ang II (1 nM) resulted in fourfold increases in NF-κB activity. The Ang II-induced increase in NF-κB activity was significantly attenuated by coadministration of losartan (10 μM), an AT_1 receptor-selective blocker, or colchicine (1 μM), a selective cytoskeleton microtubule inhibitor.[17,19] Because losartan blocks not only cell surface AT_1 receptor-mediating signaling, but also AT_1 receptor-mediated endocytosis,[9,14,17] we believe the effect of losartan on Ang II-increased NF-κB activity was mediated at least in part by inhibiting Ang II endocytosis. Although AT_2 receptors have been shown to mediate NF-κB activation in vascular smooth muscle cells, endothelial cells or COS-7 cells expressing mutant AT_2 receptors,[10,22] we found that blocking AT_2 receptors with PD123319 (10 μM) had no significant effect on Ang II-induced NF-κB activity. The reasons underlying the differences between proximal tubule cells expressing endogenous receptors and those transfected with mutant receptors are not known, because at 10 μM this compound almost completely blocks AT_2 receptor binding in both nonrenal and renal cells expressing AT_2 receptors.[9] Furthermore, AT_2 receptors have been reported to oppose, rather than induce AT_1 receptor-mediated responses.[9]

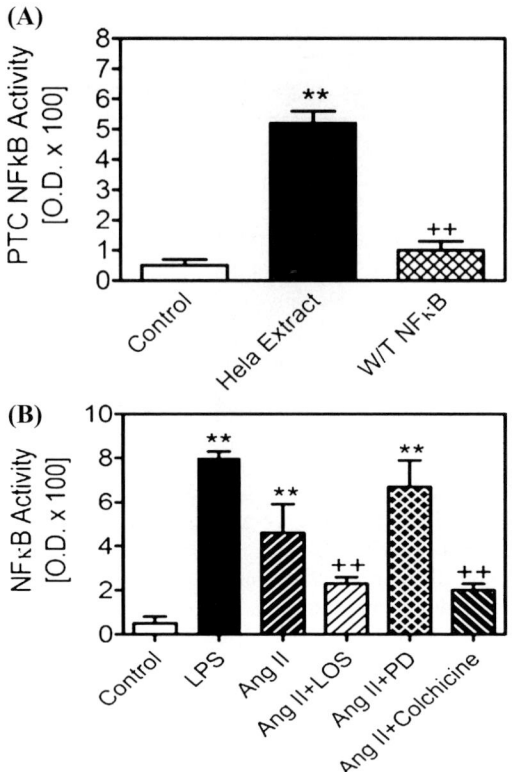

FIGURE 2. Effects of Ang II and endocytotic inhibitors on Ang II-induced activation of nuclear factor-κB (NF-κB) in cultured rabbit proximal tubule cells. The *top panel* shows the specificity of Trans NFκB ELISA for measurement of activated p65 and p50 (Active Motif). HeLa extract and lipopolysaccharide (LPS) were used as a positive control and wild-type NF-κB for a negative control. The *bottom panel* shows that both LPS and Ang II increased NF-κB activity, and the effect of Ang II was attenuated by losartan and colchicine, both of which block AT_1 receptor-mediated Ang II endocytosis. ** $P < 0.01$ versus control; + $P < 0.01$ versus Ang II alone.

The effect of colchicine on Ang II-induced NF-κB activation further suggests that AT_1 receptor-mediated endocytosis may play a role in inducing NF-κB activation by Ang II. While colchicine does not have a direct effect on AT_1 receptor binding, it selectively inhibits endocytosis and intracellular trafficking of various proteins by disrupting cell cytoskeleton microtubules.[17,19] Colchicine also inhibits protein kinase C and phospholipase C activation by Ang II after endocytosis[26] and blocks intracellular Ang II accumulation in proximal tubule cells.[17] Our results therefore suggest that cytoskeleton microtubules play an important role in AT_1 receptor-mediated endocytosis of extracellular Ang II and subsequent activation of NF-κB.

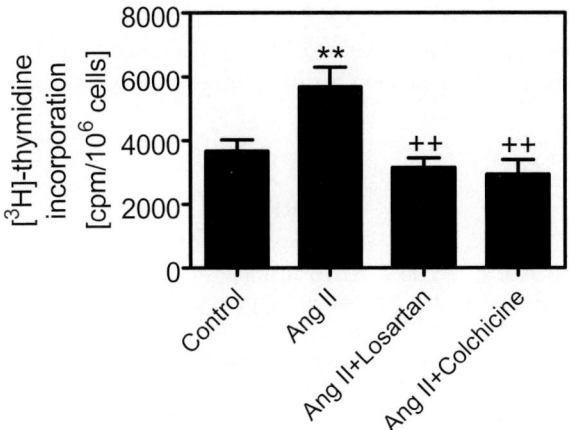

FIGURE 3. Effect of Ang II and endocytotic inhibitors losartan and colchicine on Ang II-induced increases in [^{3}H]-thymidine incorporation in cultured rabbit proximal tubule cells. Note that Ang II increased [^{3}H]-thymidine incorporation, which serves as an index of DNA synthesis and cell proliferation, and that losartan and colchicine significantly attenuated Ang II-induced cell proliferation. ** $P < 0.01$ versus control; $+ P < 0.01$ versus Ang II alone.

Activation of NF-κB and its subsequent translocation to the cell nucleus play an important role in promoting transcription of growth factors and proliferative cytokines, which stimulates cell growth and proliferation.[12,13] In the present study, we first determined whether Ang II-induced activation of NF-κB is associated with stimulation of proximal tubule cell proliferation and then studied whether AT$_1$ receptor-mediated endocytosis is involved. FIGURE 3 shows that Ang II increased ^{3}H-thymidine incorporation by 55%, and the effect of Ang II was significantly attenuated by coadministration of the AT$_1$ receptor blocker losartan and the cytoskeleton microtubule inhibitor colchicine. Again, these results highlight the importance of the Ang II endocytosis/NF-κB activation signaling cascade in promoting proximal tubule cell proliferation. Our results are consistent with previous studies in Ang II-infused rats or rats with overexpression of human renin and angiotensinogen showing that Ang II increased NF-κB activity and induced tubulo-interstitial inflammation and fibrosis in the kidney.[3,10] Blockade of NF-κB activation and translocation to the cell nucleus may therefore provide therapeutic benefits in preventing and treating experimental and/or clinical Ang II-dependent hypertensive renal injury.

ACKNOWLEDGMENTS

This work was supported by a National Institute of Diabetes, Digestive, and Kidney Diseases Grant (RO1DK067299, to J.L.Z.), an American Heart

Association Grant-in-Aid (0355551Z, to J.L.Z.), and a National Kidney Foundation of Michigan Grant-in-Aid (to J.L.Z.). Dr. Carretero is supported by a National Heart, Lung, and Blood Institute Program Project Grant (HL-28982).

REFERENCES

1. JOHNSON, R.J., C.E. ALPERS, A. YOSHIMURA, et al. 1992. Renal injury from angiotensin II-mediated hypertension. Hypertension **19:** 464–474.
2. SUZUKI, Y., M. RUIZ-ORTEGA & J. EGIDO. 2000. Angiotensin II: a double-edged sword in inflammation. J. Nephrol. **13** (Suppl. 3): S101–S110.
3. MULLER, D.N., R. DECHEND, E.M. MERVAALA, et al. 2000. NF-κB inhibition ameliorates angiotensin II-induced inflammatory damage in rats. Hypertension **35:** 193–201.
4. RUIZ-ORTEGA, M., O. LORENZO & J. EGIDO. 2000. Angiotensin III increases MCP-1 and activates NF-κB and AP-1 in cultured mesangial cells and mononuclear cells. Kidney Int. **57:** 2285–2298.
5. ZHUO, J.L., J.D. IMIG, T.G. HAMMOND, et al. 2002 Ang II accumulation in rat renal endosomes during Ang II-induced hypertension: role of AT(1) receptor. Hypertension **39:** 116–121.
6. DUFF, J.L., B.C. BERK & M.A. CORSON. 1992. Angiotensin II stimulates the p44 and p42 mitogen-activated protein kinases in cultured rat aortic smooth muscle cells. Biochem. Biophys. Res. Commun. **188:** 257–264.
7. KURTZ, T.W. & D.G. GARDNER. 1998. Transcription-modulating drugs: a new frontier in the treatment of essential hypertension. Hypertension **32:** 380–386.
8. SCHMITZ, U., T. ISHIDA, M. ISHIDA, et al. 1998. Angiotensin II stimulates p21-activated kinase in vascular smooth muscle cells: role in activation of JNK. Circ. Res. **82:** 1272–1278.
9. DE GASPARO, M., K.J. CATT, T. INAGAMI, et al. 2000. International Union of Pharmacology. XXIII. The angiotensin II receptors. Pharmacol. Rev. **52:** 415–472.
10. RUIZ-ORTEGA, M., O. LORENZO, M. RUPEREZ, et al. 2001. Systemic infusion of angiotensin II into normal rats activates nuclear factor-kappaB and AP-1 in the kidney: role of AT(1) and AT(2) receptors. Am. J. Pathol. **158:** 1743–1756.
11. MEZZANO, S.A., M. RUIZ-ORTEGA & J. EGIDO. 2001. Angiotensin II and renal fibrosis. Hypertension **38:** 635–638.
12. HOFFMANN, A., A. LEVCHENKO, M.L. SCOTT & D. BALTIMORE. 2002. The IkappaB-NF-kappaB signaling module: temporal control and selective gene activation. Science **298:** 1241–1245.
13. STANCOVSKI, I. & D. BALTIMORE. 1997. NF-kappaB activation: the I kappaB kinase revealed? Cell **91:** 299–302.
14. FERGUSON, S.S. 2001. Evolving concepts in G protein-coupled receptor endocytosis: the role in receptor desensitization and signaling. Pharmacol. Rev. **53:** 1–24.
15. ROMERO, M.F., J.G. DOUGLAS, R.L. ECKERT, et al. 1992. Development and characterization of rabbit proximal tubular epithelial cell lines. Kidney Int. **42:** 1130–1144.

16. ZHUO, J.L., X.C. LI, J.L. GARVIN, et al. 2006. Intracellular angiotensin II induces cytoplasmic Ca^{2+} mobilization by stimulating intracellular AT_1 receptors in proximal tubule cells. Am. J. Physiol. Renal Physiol. **290:** F1382–F1390.
17. LI, X.C., O.A. CARRETERO, L.G. NAVAR & J.L. ZHUO. 2006 AT_{1a} receptor-mediated accumulation of extracellular angiotensin II in proximal tubule cells: role of cytoskeleton microtubules and tyrosine phosphatase. Am. J. Physiol. Renal Physiol. **291:** F375–F383.
18. VAZQUEZ, J., M.F. ADJOUNIAN, C. SUMNERS, et al. 2005 Selective silencing of angiotensin receptor subtype 1a ($AT_{1a}R$) by RNA interference. Hypertension **45:** 1–5.
19. ELKJAER, M.L., H. BIRN, P. AGRE, et al. 1995 Effects of microtubule disruption on endocytosis, membrane recycling and polarized distribution of aquaporin-1 and gp330 in proximal tubule cells. Eur. J. Cell Biol. **67:** 57–72.
20. ANDERSON, K.M. & M.J. PEACH. 1994. Receptor binding and internalization of a unique biologically active angiotensin II-colloidal gold conjugate: morphological analysis of angiotensin II processing in isolated vascular strips. J. Vasc. Res. **31:** 10–17.
21. RUIZ-ORTEGA, M., O. LORENZO, M. RUPEREZ, et al. 2000. Angiotensin II activates nuclear transcription factor kappaB through AT(1) and AT(2) in vascular smooth muscle cells: molecular mechanisms. Circ. Res. **86:** 1266–1272.
22. WOLF, G., U. WENZEL, K.D. BURNS, et al. 2002. Angiotensin II activates nuclear transcription factor-kappaB through AT1 and AT2 receptors. Kidney Int. **61:** 1986–1995.
23. WOLF, G., F.N. ZIYADEH, G. ZAHNER & R.A. STAHL. 1995. Angiotensin II-stimulated expression of transforming growth factor beta in renal proximal tubular cells: attenuation after stable transfection with the c-mas oncogene. Kidney Int. **48:** 1818–1827.
24. LI, X.C., O.A. CARRETERO, Y. SHAO & J.L. ZHUO. 2006 Glucagon receptor-mediated extracellular signal-regulated kinase 1/2 phosphorylation in rat mesangial cells: role of protein kinase A and phospholipase C.. Hypertension **47:** 580–585.
25. RUIZ-ORTEGA, M. & J. EGIDO. 1997 Angiotensin II modulates cell growth-related events and synthesis of matrix proteins in renal interstitial fibroblasts. Kidney Int. **52:** 1497–1510.
26. SCHELLING, J.R., A.S. HANSON, R. MARZEC & S.L. LINAS. 1992. Cytoskeleton-dependent endocytosis is required for apical type 1 angiotensin II receptor-mediated phospholipase C activation in cultured rat proximal tubule cells. J. Clin. Invest. **90:** 2472–2480.

Retinoic Acid and Histone Deacetylase Inhibitor BML-210 Inhibit Proliferation of Human Cervical Cancer HeLa Cells

VERONIKA V. BORUTINSKAITE,[a,b] RUTA NAVAKAUSKIENE,[a] AND KARL-ERIC MAGNUSSON[b]

[a]*Department of Developmental Biology, Institute of Biochemistry, LT-08662 Vilnius, Lithuania*

[b]*Division of Medical Microbiology, Department of Molecular and Clinical Medicine, Linköping University, SE-581 85 Linköping, Sweden*

ABSTRACT: Human papillomavirus (HPV) infection is believed to be the central cause of cervical cancer. The viral proteins E6 and E7 from high-risk HPV types prevent cells from differentiating apoptosis and inducing hyperproliferative lesions. Human cervical carcinoma HeLa cells contain integrated human papillomavirus type 18 (HPV-18). Retinoic acid (RA) is a key regulator of epithelial cell differentiation and a growth inhibitor *in vitro* of HeLa cervical carcinoma cells. Cellular responses to RA are mediated by nuclear retinoic acid receptors (RARs) and retinoid X receptors. On the other hand, histone deacetylase inhibitors have been shown to be chemopreventive agents for the treatment of cancer cells. In this article, we have examined the antiproliferative effect of RA and histone deacetylase inhibitor BML-210 on HeLa cells, and particularly the effects on protein expression that may be involved in the cell cycle control and apoptosis. Our data suggest that a combination of RA and BML-210 leads to cell growth inhibition with subsequent apoptosis in a treatment time-dependent manner. We confirm that BML-210 alone or in combination with RA causes a marked increase in the level of p21. The changes in the p53 level are under the influence of p38 phosphorylation. We also discovered that the histone deacetylase inhibitor BML-210 causes increased levels of anti-apoptotic protein Bcl-2 and phosphorylated p38 MAP Kinase; the latter link in cell cycle arrest with response to extracellular stimuli. Our results suggest that RA and BML-210 are involved in different signaling pathways that regulate cell cycle arrest and lead to apoptosis of HeLa cells.

KEYWORDS: histone deacetylase inhibitor; proliferation; p53; p21

Address for correspondence: V.V. Borutinskaite, Department of Developmental Biology, Institute of Biochemistry, LT-08662 Vilnius, Lithuania. Voice: +370-52-7291; fax: +370-52-729196.
e-mail: verbu@imk.liu.se

INTRODUCTION

The human papillomavirus (HPV) group presently comprises 67 different virus types that infect epithelial cells and induce hyperproliferative lesions. Oncogenic viruses often exploit the machinery of the cells they infect in order to promote their own survival and replication. As a part of this process, they interfere with normal cellular control mechanisms, leading to abnormal cell growth, genetic alterations, and even malignant transformation. The high-risk HPV types encode three oncoproteins: E5, E6, and E7 where E6 and E7 are involved in cellular immortalization.[1-3] E6 and E7 genes are transcribed from the HPV early promotor, which is designated p97 in HPV-16 and HPV-31 and p105 in HPV-18.

Moreover E6 and E7 function to stimulate cell proliferation and do so by interfering with the functioning of the regulatory proteins in cells, including the p53 and pRb tumor suppressor gene products.[3,4] Normally, p53 and Rb control signaling pathways that regulate the cell cycle, but also help monitor and protect cell proliferation and the integrity of hereditary traits on the chromosomes. p53 is a transcription factor activating the expression and transcription of a variety of genes, such as p21, a known p53 responsive gene. p21 is a key regulator of the cell cycle and acts as a universal inhibitor of CDK.[5] The high-risk HPV E6 protein can combine with p53 and, as a result it promotes the ubiquitination and degradation of p53.[6]

Retinoids are a group of vitamin A-related compounds that exert potent influences on cell differentiation, proliferation, and development.[7] They can block tumor growth by inhibiting uncontrolled cell proliferation, inducing apoptosis of abnormal cells, and promoting normal cell differentiation.[8] Retinoic acid (RA) can inhibit HPV-16 and -18 in cervical cancer cell lines *in vitro* by suppressing the transcription of E6 and E7. Previous investigations have shown that retinoic acid receptors (RAR)-mediated repression of HPV-18 E6/E7 transcription occurs through interference with AP-1-mediated activation by RARs. AP-1 is an essential activator of the HPV-18 and -16 enhancers. It is known that AP-1-mediated activation can be repressed by RARs, and conversely RAR-mediated activation can be blocked by AP-1. Overall, these studies support a direct role of RA in the chemoprevention of HPV-induced cancers.[9,10]

The addition of HDAC inhibitors (HDACI) has been shown to reduce proliferation and to stimulate differentiation and apoptosis in transformed cells *in vitro* and *in vivo*.[11] The biological effects of HDACI are mediated by catalyzing the removal of acetyl groups on the NH_2-terminal lysine residues of core nucleosomal histone; this activity is generally associated with transcriptional repression.

This article was designed to examine the biological effects of HDACI on the cancer cell line HeLa. We focused particularly on BML-210, as a potentially new therapeutic agent in cancer treatment, and its combination with RA. We examined whether these compounds were able to mediate inhibition of cell growth and cell cycle arrest via expression of proteins involved in these processes in a variety of cancer cell lines like HeLa.

MATERIAL AND METHODS

Cell Culture

The HeLa cell line was cultured in Dubecco's modified Eagle's medium (DMEM) (Gibco, Invitrogen, Paisley, UK) containing 10% fetal bovine serum (FBS), 100 U/mL penicillin, 100 μg/mL streptomycin (Gibco), in 37°C in humidified air, and 5% CO_2. Cultures were seeded at a density of 4×10^5 cells/mL.

Chemicals

RAs were purchased from Sigma Chem. Co. (St. Louis, MO) and added to the medium from solution of 500 μM to a final concentration of 1 μM. RA was dissolved in ethanol. BML-210 (BIOMOL, Plymouth Meeting, PA) was dissolved in DMSO (stock solution 10 mM) and added to the medium to a final concentration 5 μM or 10 μM. The final DMSO concentration in the medium was 0.001%.

Primary Antibodies

Anti-Bcl-2, -p21, -p53, and -Sp1 were from Santa Cruz Biotechnology, Inc. (Santa Cruz, CA), and anti-phospho-p38 mitogen-activated protein (MAP) kinase (Thr 180/Tyr 182) and anti-p38 MAP kinase were from Cell Signalling Technology, Inc. (Beverly, CA).

Cell Treatment with All-Trans RA and BML-210

After 24 h from seeding, 1 μM of RA or 10 μM of BML-210 alone or the combination of 1 μM of RA or 5 μM BML-210 were added to the medium, and the cells were further grown to 6–7 days.

Immunoblot Analysis

HeLa cells were seeded and grown in the absence or presence of RA and/or BML-210 as indicated. Previously cells were washed twice in ice-cold PBS, were incubated with ice-cold Nuclei EZ lysis buffer (Sigma), scraped off and put on ice for 5 min. Nuclei were collected by centrifugation at 500 g at 4°C for 5 min and a clear cytoplasmic fraction was obtained by centrifugation of supernatant at 10,000 g at 4°C for 30 min. Nuclei were then incubated with extraction buffer (20 mM HEPES [pH 7.9], 1.5 mM $MgCl_2$, 0.42 mM NaCl, 0.2 mM EDTA, 0.5 mM DTT, 0.05% NP-40, 25% glycerol) on ice.

Approximately 20 μg of protein was fractionated on 8–16% polyacrylamide gradient gel in Tris–glycine electrophoresis buffer. After SDS-PAGE, proteins were transferred to Immobilon™ PVDF membranes (Millipore, Solna, Sweden) and blocked by incubating with 3% BSA dissolved in PBS supplemented with 0.1% Tween-20 at 4°C overnight. Then the membranes were incubated for 1 h at room temperature with primary antibodies in PBS containing 0.1% Tween-20, 0.35 M NaCl, and 1% BSA. They were subsequently washed for 3 × 10 min with PBS-Tween-20 and incubated with a horseradish peroxidase-conjugated secondary antibody (DAKO, Copenhagen, Denmark) diluted 1:2000 in PBS-Tween-20 for 1 h at room temperature. Thereafter, the filters were washed as described above, and immunoreactive bands were detected by enhanced chemiluminescence (ECL™ Western blotting detection reagents; GE Health Care Amersham Biosciences, Uppsala, Sweden), according to the manufacturer's instructions. And for chemiluminescence detection of p53 and p21 we used SuperSignal West Pico Chemiluminescent Substrate (PIERCE, Rockford, IL), according to the manufacturer's instructions. Reprobing of the membranes, if needed, was done according to the standard ECL Western blotting protocols (GE Health Care Amersham Biosciences).

RESULTS

Distinct Effects of RA and BML-210 on the Proliferation of HeLa Cells

As shown in FIGURE 1, RA, BML-210, and its combination inhibited proliferation of HeLa cells in a time-dependent manner. A 4-day treatment with 1 μM RA resulted in about 71% growth inhibition and this inhibition process starts after the second day of treatment. BML-210 alone did not suppress cell growth in comparison with RA-treated cells; cell growth inhibition after 3 days of treatment was observed, compared to untreated control cells. In our experiments, we have examined whether a combination of RA and BML-210 exert a growth-inhibitory effect on Hela cells. When RA and BML-210 were co-administered at concentrations of 1 μM and 5 μM, respectively, RA boosted the growth-inhibitory effect of BML-210 in HeLa cells so that growth inhibition had already started after 2 days of combined treatment.

Effects of RA and BML-210 on Bcl-2 Expression

Based on previous results showing that RA and BML-210 or in combination inhibit the proliferation of HeLa cells, we investigated the protein expression

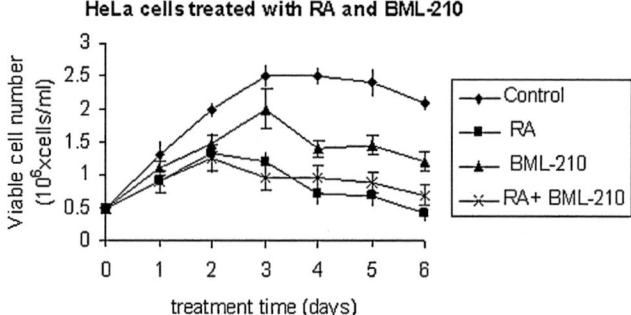

FIGURE 1. Growth inhibition in HeLa cells treated with RA and BML-210. HeLa cells were treated with 1 μM RA, 10 μM BML-210, or in combination (1 μM RA and 5 μM BML-210) for different time periods. Cells were washed twice in ice-cold PBS and the number of viable cells was count after trypsinization with trypsin-EDTA. Results are mean ± SEM ($n = 3$).

level that could be involved in growth inhibition of HeLa cells. We first analyzed the expression of Bcl-2, which is supposed to have a role in the regulation of cell proliferation and apoptosis. HeLa cells were treated with 1 μM RA, 10 μM BML-210, or a combination of 1 μM RA and 5 μM BML-210, collected at different time points and analyzed by Western blotting. As shown in FIGURE 2 A the expression level of Bcl-2 in cytoplasmic fraction is rather constant after treatment with RA alone. However, a combination of RA and BML-210 or BML-210 alone the expression level of Bcl-2 treatment is time-dependent. Bcl-2 protein expression increased between 24 and 48 h and decreased at 96 h of the treatment with BML-210 alone, but already increased at 8–24 h and decreased at 48 h after combination of RA and BML-210.

RA and BML-210 Affects Sp1 Protein Expression Level

HeLa cells were treated with 1 μM RA, 10 μM BML-210, and a combination of 1 μM RA with 5 μM BML-210, collected at different time points and analyzed by Western blotting. A representative Western blot analysis of the Sp1 protein in the cytoplasmic fraction is shown in FIGURE 2 A Sp1 protein migrates as two bands with molecular masses of 95 and 105 kDa. The two species are the result of differential posttranslational modification of the Sp1 polypeptide, corresponding to an unphosphorylated (lower band) and phosphorylated (upper band) protein.[12]

Clear differences in Sp1 expression between control cells and cells treated with different agents (RA, BML-210) cells were observed. We observed increased level of Sp1 at 24 h after RA and BML-210 treatment, and at 8 h of combinatorial treatment. To see changes in Sp1 expression level in the nucleus,

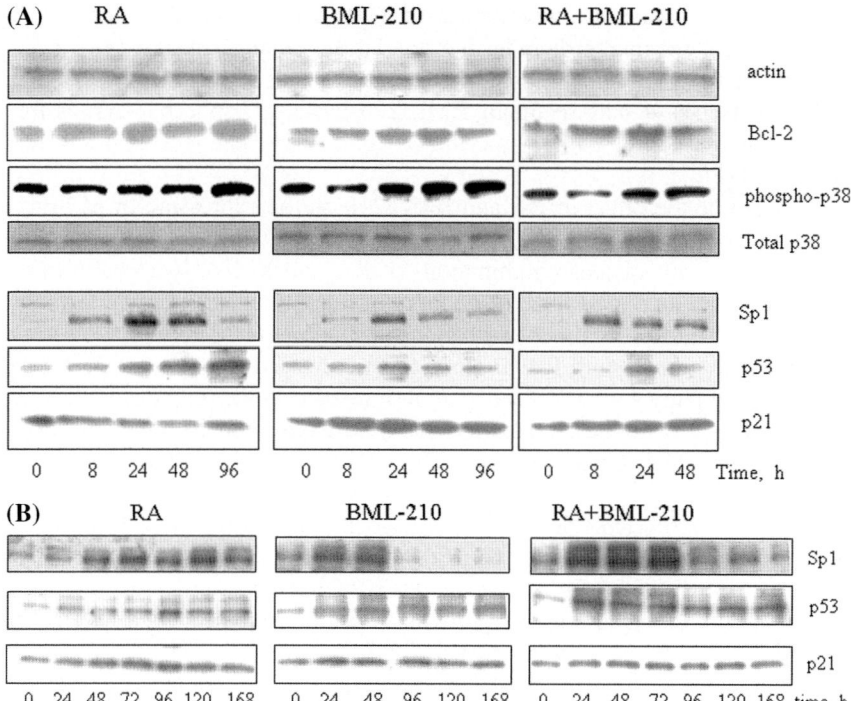

FIGURE 2. Expression of p21, p53, Bcl-2, Sp1, and p38 proteins, in HeLa cells after treatment with RA and BML-210 alone or in combination. Cytoplasmic (**A**) and nuclear (**B**) proteins were isolated from HeLa cells treated with 1 μM RA, 10 μM BML-210, or in combination (1 μM RA and 5 μM BML-210) for different time periods. The proteins were fractionated by SDS-PAGE on an 8–16% acrylamide gel gradient and transferred onto Immobilon PVDF membrane. The membranes were analyzed with primary antibodies against p21, p53, Bcl-2, Sp1, p38, phospho-p38, and developed with an enhanced chemiluminescence detection system.

we prepared the nuclear fraction of HeLa cells after treatment with RA and BML-210 at different time points. We observed an increased level of Sp1 between 48 and 72 h of the treatment with RA, but at 24–48 h during BML-210 treatment and at 24–72 h with RA and BML-210 combination.

Differential Effects of RA and BML-210 on p53, p38, and p21 Protein Expression

Since MAP kinases are involved in many different cellular processes, we examined the p38 protein expression after HeLa cells treatment with RA and BML-210. Using polyclonal antibody raised against total p38 and phospho-p38, we demonstrated that p38 in cytoplasmic fraction of HeLa cells become phosphorylated after 24-h treatment with BML-210, on a combination of RA and BML-210 the total protein level of total p-38 remains constant (FIG. 2 A). With RA the level of phospho-p38 was rather constant between 8 and 24 h and then increased at 48 h of treatment. BML-210 caused phospho-p38 expression level to increase after 24 h, as did the combinatorial treatment with RA and BML-210 (FIG. 2 A).

To examine the effect of RA and BML-210 on the expression of p53 protein, which is known as a phospho-p38 downregulated protein, HeLa cells were treated with various combinations of RA and BML-210. As shown in FIGURE 2 B an increased level of p53 expression in nuclear fraction was observed at 24 h after treatment with BML-210 alone or in combination with RA. After treatment with RA, the p53 level increased slowly after 24 h and the highest intensity we observed at 96 h of treatment, like in the cytoplasmic fraction (FIG. 2 A) In the cytoplasmic fraction we observed the highest level of p53 protein at 24 h of treatment with BML-210 alone or in combination with RA (FIG. 2 A).

Since the cyclin-dependent protein kinase inhibitory protein, p21, has been shown to be transcriptionally regulated by p53, we examined whether p53 accumulation also induced changes in p21 protein levels.

FIGURE 2 A, B displays the p21 protein in untreated control cells. We found increased p21 protein expression level after treatment with BML-210 alone or in combination with RA at 24–48 h in the cytoplasmic fraction (FIG. 2 A). We determine an increased p21 protein level at 96 h with RA alone (FIG. 2 A). The p21 expression level in nuclear fraction had increased at 24 h and then remained more constant (FIG. 2 B).

DISCUSSION

We first analyzed the effect of RA and BML-210 on growth of HeLa cells, and found that RA and BML-210 alone or in combination induced growth suppression of HeLa cells. With RA present HeLa cell growth was inhibited to a greater extent than with HDACI alone.

To further examine the molecular mechanism of the growth suppression by RA and BML-210, we analyzed the expression of some of the proteins that are involved in a processes such as the inhibition of proliferation and the induction of apoptosis.

Bcl-2 is perhaps the best-characterized member of the *bcl-2* family of apoptosis suppressor genes in various cell lines. While the precise physiological role of Bcl-2 has not yet been identified, its overexpression has been shown to protect cells from a variety of otherwise death-inducing stimuli. It has been proposed to play a central role in the regulation of cell growth and apoptotic signaling.[13–15] We found that a combination of RA and BML-210 led to an earlier increase in Bcl-2 protein expression in comparison with RA or

BML-210 alone. It was furthermore shown that in addition to promoting cell cycle arrest, HDACI dramatically induces apoptosis and downregulates bcl-2 expression in lymphomas. The repression of bcl-2 by HDACI occurred at the transcriptional level. Western blot and quantitative chromatin immunoprecipitation (ChIP) assays performed by Duan and co-workers showed that even though HDACI increased overall acetylation of histones, localized histone H3 deacetylation occurred at both bcl-2 promotors.[16] We found that BML-210 after 48 h with RA combination and BML-210 alone decreased Bcl-2 protein level in a cytoplasmic fraction, respectively. These results correlate with the increased p53 protein level in HeLa cells.

p53 is another protein that is very important in cell growth and apoptosis regulation. Since HPV-16 or -18 E6 mediate in the ubiquitin/proteasome-dependent degradation of p53 protein, it was pertinent to examine the effects of RA or BML-210 on the p53 expression.[4] Thus, we checked p53 protein expression in the cytoplasmic and nuclear fractions of HeLa cells. HeLa cells contain very little functional p53 activity, but earlier studies have suggested that RA can suppress the transcription of E6 and E7 and increase p53 protein level as well.[17] Our data shows that p53 level indeed increased after treatment with RA, BML-210, and in combination compared to control HeLa cells. Earlier studies have demonstrated that phospho-p38 protein migrates to the nucleus and activates p53 gene transcription.[18] We displayed that increased level of phospho-p38 protein correlates with increased level of p53 protein for the same treatments. Furthermore, we found a correlation between an increased level of p53 and decreased level of Bcl-2 proteins, which could suggest that p53 regulates the degradation of Bcl-2 protein.

MAP kinase cascades are important mediators of the cellular response to a wide variety of extracellular signals, including mitogens, growth factors, cytokines, and cellular stress. The p38 MAP kinases are activated primarily in response to proinflammatory cytokines and cellular stress, such as UV irradiation and osmotic and chemical shock. Activation of p38 occurs by dual phosphorylation on Thr and Tyr amino acids in the p38 kinase domain by an upstream MAP kinase.[19]

Since the Sp1 transcription factor can be involved in HPV genes transcription, we analyzed the effect of RA and BML-210 on the Sp1 expression level in HeLa cells. Indeed, the Sp1 protein level changed during treatment with RA, BML-210, or in combination. In both the cytoplasmic and nuclear fractions we detected two bands of Sp1, which represent unphosphorylated and phosphorylated protein forms. Furthermore, Sp1 expression increased after treatment with BML-210 or in combination with RA at 24 h and then decreased after 72 h, in contrast to the constant level of Sp1 after RA treatment. The correlation between cytoplasmic increased level of Bcl-2 at 8 h and nuclear Sp1 at 24 h suggests that the combination of RA and BML-210 activated these signaling proteins earlier than RA or BML-210 alone. By others it has been shown that DNA binding activity of Sp1 correlates with increase in Sp1 phosphorylation, and that Sp1 can be phosphorylated by a number of cellular kinases, including protein kinase A, different members of the protein kinase C family, and ERK2.[12]

The gene-encoding p21 is regulated by at least three classes of signals that result in arrest of cell growth: (*a*) the tumor suppressor protein p53, (*b*) extracellular growth factors acting in a p53-independent mechanism,[20] and (*c*) factors that induce cellular differentiation of many cell types such as myoblasts, keratinocytes, intestinal epithelial cells, and monocytes.[21] Other investigators have recently reported the capacity of Sp1 to downregulate p21(WAF1/Cip1) production, thereby reducing p21(WAF1/Cip1)-cyclin D1-Cdk4 complex formation and inhibiting vascular SMC proliferation.[22] The close correlation between the effect of Bcl-2 on both RA-induced growth arrest and RA-induced p21 gene expression suggests the possibility that Bcl-2 affects cell growth through the action of p21.[23] Our data revealed a correlation between Bcl-2 and p21 expression in cytoplasmic fractions of treated HeLa cells after different treatments. For RA the p21 activity correlated with Sp1 activity, but for BML-210 or its combination with RA p53 the increased level correlates with the p21 as well as with the Sp1 level.

In summary, we conclude that BML-210 and a combination of RA with BML-210 affect in a distinct manner the level of proteins that are involved in cell cycle control and growth of HeLa cells. Thus, both BML-210 and RA inhibit cell growth of human cervical cancer HeLa cells and could be important as chemopreventive agents for the treatment of cancer cells.

ACKNOWLEDGMENTS

The research was supported by Visby program (01681/2005), Swedish Research Council (Medicine), and also V.V. Borutinskaite was supported by a Marie Curie fellowship.

REFERENCES

1. HAWLEY-NELSON, P. *et al.* 1989. HPV16 E6 and E7 proteins cooperate to immortalize human foreskin keratinocytes. EMBO J. **8:** 3905–3910.
2. HUDSON, J.B. *et al.* 1990. Immortalization and altered differentiation of human keratinocytes *in vitro* by the E6 and E7 open reading frames of human papillomavirus type 18. J. Virol. **64:** 519–526.
3. MUNGER, K. *et al.* 1989. Complex formation of human papillomavirus E7 proteins with the retinoblastoma tumor suppressor gene product. EMBO J. **8:** 4099–4105.
4. MIETS, I.A. *et al.* 1992. The transcriptional transactivation function of wild-type p53 is inhibited by SV40 large T-antigen and by HPV-16 E6 oncoprotein. EMBO J. **11:** 5013–5020.
5. XIONG, Y. *et al.* 1993. p21 is a universal inhibitor of cyclin kinases. Nature **366:** 701–704.

6. SCEFFNER, M. et al. 1990. The E6 oncoprotein encoded by human papillomavirus types 16 and 18 promotes the degradation of p53. Cell **63:** 1129–1136.
7. GUDAS, L.J. 1994. Retinoids and vertebrate development. J. Biol. Chem. **269:** 15399–15402.
8. NILES, R.M. 2000. Recent advances in the use of vitamin A (retinoids) in the prevention and treatment of cancer. Nutrition **16:** 1084–1089.
9. NICHOLSON, R.C. et al. 1990. Negative regulation of the rat stromelysin gene promoter by retinoic acid is mediated by an AP1 binding site. EMBO J. **9:** 4443–4454.
10. SCHULE, R. et al. 1991. Retinoic acid is a negative regulator of AP-1- responsive genes. Proc. Natl. Acad. Sci. USA **88:** 6092–6096.
11. STRAHL, B.D. & C.D. ALLIS. 2000. The language of covalent histone modifications. Nature (Lond.) **403:** 41–45.
12. TRISCIUOGLIO, D. et al. 2004. bcl-2 Induction of urokinase plasminogen activator receptor expression in human cancer cells through Sp1 activation. J. Biol. Chem. **279:** 6737–6745.
13. DENG, G. & E.R. PODACK. 1993. Suppression of apoptosis in a cytotoxic T-Cell line by interleukin 2- mediated gene transcription and deregulated expression of the protooncogene bcl-2. Proc. Natl. Acad. Sci. USA **90:** 2189–2193.
14. KORSMEYER, S.J. 1992. Bcl-2 initiates a new category of oncogenes: regulators of cell death. Blood **80:** 879–886.
15. NUNEZ, G. et al. 1990. Bcl-2 maintains B cell memory. Nature **353:** 71–73.
16. DUAN, H. et al. 2005. Histone deacetylase inhibitors down-regulate bcl-2 expression and induce apoptosis in t(14;18) lymphomas. Mol. Cell. Biol. **25:** 1608–1619.
17. UM, S.J. et al. 2000. Antiproliferative effects of retinoic acid/interferon in cervical carcinoma cell lines: cooperative growth suppression of IRF-1 and p53. Int. J. Cancer **85:** 416–423.
18. SANCHEZ-PRIETO, R. et al. 2000. A role for the p38 mitogen-activated protein kinase pathway in the transcriptional activation of p53 on genotoxic stress by chemotherapeutic agents. Cancer Res. **60:** 2464–2472.
19. BEARDMORE, V.A. et al. 2005. Generation and characterization of p38β (MAPK11) gene-targeted mice. Mol. Cell. Biol. **25:** 10454–10464.
20. BIGGS, J.R. et al. 1996. The role of transcriptional factor Sp1 in regulating the expression of the WAF1/CIP1 gene in U937 leukemic cells. J. Biol. Chem. **271:** 901–906.
21. LIU, M. et al. 1996. Transcriptional activation of the human $p21^{WAF1/CIP1}$ gene by retinoic acid receptor. J. Biol. Chem. **271:** 31723–31728.
22. KAVURMA, M.M. & L.M. KHACHIGIAN. 2003. ERK, JNK, and p38 MAP kinases differentially regulate proliferation and migration of phenotypically distinct smooth muscle cell subtypes. J. Biol. Chem. **278:** 32537–32543.
23. CHOU, H.K. 2000. Bcl-2 accelerates retinoic acid-induced growth arrest and recovery in human gastric cancer cells. Biochem. J. **348:** 473–479.

Effects of Histone Deacetylase Inhibitors, Sodium Phenyl Butyrate and Vitamin B3, in Combination with Retinoic Acid on Granulocytic Differentiation of Human Promyelocytic Leukemia HL-60 Cells

RASA MERZVINSKYTE,[a] GRAZINA TREIGYTE,[a]
JURATE SAVICKIENE,[a] KARL-ERIC MAGNUSSON,[b]
AND RUTA NAVAKAUSKIENE[a]

[a]*Department of Developmental Biology, Institute of Biochemistry, LT-08662 Vilnius, Lithuania*

[b]*Division of Medical Microbiology, Department of Molecular and Clinical Medicine, Linköping University, SE-58185 Linköping, Sweden*

ABSTRACT: Water-soluble vitamin B3, niacin, and its related compounds were suggested to be applicable for medical use. In this article, we examined the anti-leukemic effects of two distinct histone deacetylase (HDAC1 and Sir2) inhibitors, sodium phenyl butyrate (PB) and vitamin B3, respectively, on human promyelocytic leukemia cells HL-60, using HDACIs alone and in combination with all *trans* retinoic acid (RA). We demonstrated that the HDACI combinations exert different effects on cell cycle arrest and differentiation as determined by nitro blue reduction and the expression of the early myeloid differentiation marker CD11b. The most beneficial effects were found by use of 6-h pretreatment with PB and vitamin B3 before the exposition to RA alone or in combination with vitamin B3, showing significant acceleration and a high level of granulocytic differentiation. The effects were associated with a rapid histone H4 acetylation and later histone H3 modifications. Our results suggest that the use of two HDACI altogether before the induction of differentiation and acting via chromatin remodeling may be promising for the treatment of acute promyelocytic leukemia.

KEYWORDS: sodium phenyl butyrate; vitamin B3; leukemia; granulocytic differentiation

INTRODUCTION

Histone deacetylase inhibitors (HDACI) comprise a new class of potential anticancer agents for the treatment of solid and hematological malignancies. They have generated significant interest due to their ability to cause growth arrest, terminal differentiation, and/or apoptosis in cancer cells. HDAC inhibition causes acetylated nuclear histones to accumulate in both tumor and normal tissues. The effects of HDACI on gene expression are highly selective, leading to transcriptional activation/deactivation of certain genes.[1-3] HDAC inhibition also results in the activation of the cyclin-dependent kinase inhibitor p21 (WAF1/CIP1) and inhibition of transcription factors, such as p53, GATA-1, and estrogen receptor-α.[4,5] The functional significance of protein acetylation and the precise mechanisms, whereby HDACI induce tumor cell growth arrest, differentiation, and/or apoptosis, are currently the focus of intensive research. Several HDACI, e.g., sodium phenyl butyrate (PB), valproic acid, and novel compounds, such as depsipeptide, have shown impressive antitumor activity *in vivo* with remarkably little toxicity in preclinical studies and have currently entered the clinical arena.[6]

Acute promyelocytic leukemia (APL) is characterized by the expansion of malignant myeloid cells blocked at the promyelocytic stage of hematopoietic development and is invariably associated with reciprocal chromosomal translocations involving the retinoic acid receptor-α (RARα) gene.[7] New understanding in APL pathogenesis is obtained through novel therapeutic strategies with, for instance, HDACI and inorganic arsenicals, such as As_2O_3. These are currently being tested in murine leukemia models as well as in humans with APL.[8] Agents, such as all-*trans* retinoic acid (RA), sodium PB, and As_2O_3, which induce differentiation or apoptosis in APL, have been shown to successfully achieve remission. Recently, it was shown that several niacin-related compounds, such as vitamin B3 (vitB3), isonicotinamide and others, belonging to Sir2 family of HDAC, had a differentiation activity in leukemia cells.[9]

In our study, we used the histone deacetylase HDAC1 and Sir2 inhibitors, sodium PB and vitB3, respectively, alone and in combination with the differentiation agent, all-*trans* RA, in human promyelocytic leukemia cell line HL-60. We found the most promising combination for differentiation therapy defined by 6-h pretreatment with PB and vitB3 before exposition to RA alone or with vitB3. Such treatment significantly accelerated and increased cell differentiation (up to 95%) during the 48-h treatment.

MATERIALS AND METHODS

Cell Culture

The human promyelocytic leukemia cell line HL-60 was maintained in RPMI 1640 medium supplemented with 10% fetal bovine serum, 100 U/mL penicillin, and 100 μg/mL streptomycin (Gibco, Grand Island, NY). Cells were grown at 37°C in a humidified 5% CO_2 atmosphere and used for assays during the exponential phase of growth. Sodium PB (Calbiochem, San Diego, CA) was dissolved in water at a concentration of 1 mM and stored at –20°C. VitB3 nicotinamide (Sigma, St. Louis, MO) was dissolved in RPMI medium at a concentration 1 mM and stored at +4°C. All-*trans*-RA (Sigma) was dissolved in ethanol at a concentration 500 μM and stored at –20°C. Cells were harvested for further analysis either before or after various time points of treatment with 3 mM PB, 5 mM vitB3, and 1 μM RA or their combinations.

Assessment of Cell Viability and Differentiation

Viable cells were detected by exclusion of 0.2% trypan blue and counted in a hemocytometer. Granulocytic differentiation of HL-60 cells was determined by NBT reduction.[10]

Cell Cycle Analysis

Cell cycle progression was monitored by quantitating cellular DNA content after staining with propidium iodide (PI). Untreated, HDACI- and RA-treated cells were collected by centrifugation, suspended in PBS, and fixed in ice-cold 70% ethanol (ratio 1:10) for 24 h at –20°C. After centrifugation at 500 g for 5 min, cells were suspended in PBS containing PI (50 μg/mL) and RNAse (0.2 mg/mL) and incubated at room temperature for 30 min. The tubes were then maintained at 4°C in the dark until analysis using a flow cytometer (Becton-Dickinson FACSCalibur, San Jose, CA). The percentage of cells in G0/G1, S, and G2/M phases was evaluated with CellQuest software (Becton-Dickinson FACSCalibur).

Cell Surface Antigen Analysis

Control and drug-treated HL-60 cells (5×10^5 cells/sample) were collected, washed twice with PBS, and suspended in 50 μL PBS, pH 7.4. Then 5 μL of monoclonal mouse anti-human CD11b, C3bi receptor/RPE (DakoCytomation, Glostrup, Denmark) antibody was added to the sample and gently suspended and incubated in the dark at 4°C for 30 min. Cells were washed with PBS containing 2% bovine serum albumin, fixed in 4% paraformaldehyde for 15–30 min on ice, and the pellet was resuspended in PBS. Eight thousands events were analyzed for each sample by immunofluorescence using flow cytometry. Proliferating cells with and without anti-CD11b antibodies were used as a control.

Isolation of Cell Nuclei

Cells (5×10^6 to 10^7) were harvested by centrifugation at 500 g for 6 min, washed twice in ice cold PBS, and suspended in Nuclei EZ lysis buffer (Sigma). Nuclei were isolated according to the manufacturer's protocol (Sigma). Nuclei were washed in the same cold Nuclei EZ buffer, vortexed briefly and set on ice for 5 min, then pelleted at 500 g for 5 min, completely resuspended in nuclei EZ storage buffer (Sigma) and frozen at $-70°C$.

Isolation and Fractionation of Histones

Histones were extracted, as described previously by Treigyte and Gineitis.[11] Isolated nuclei were suspended in 5 vol of 0.4 N H_2SO_4 by stirring and incubated overnight at $0°C$. The supernatant was collected by centrifugation at 15,000 g for 10 min at $+2°C$ and the sediment was extracted once more. After centrifugation, both extracts were combined and histones precipitated by adding 5 vol of ethanol at $-20°C$ overnight. Precipitated histones were collected by centrifugation, washed several times with ethanol, and kept in $-20°C$ until analysis. Histone electrophoresis was carried out essentially as described by Hurley.[12] Histones (5 μg) were dissolved in a buffer containing 0.9 M acetic acid, 10% glycerol, 6.25 M urea, and 5% β-mercaptoethanol, and run on a 15% polyacrylamide gel containing 6 M urea and 0.9 M acetic acid for 3 h at 12 mA and 500 V by using 0.9 M acetic acid as buffer. Histones were transferred to Immobilon™ PVDF transfer membrane (Millipore, Bedford, MA) and analyzed with antibodies against hyperacetylated histone H4 (Penta), phosphorylated histone H3 (Ser10) (Upstate, Lake Placid, NY) and Phospho (Ser10) Acetyl (Lys9) histone H3 (Abcam plc, Cambridge, UK). Thereafter, the membranes were subsequently washed with PBS–Tween-20 and incubated with horseradish peroxidase-conjugated secondary antibody (DAKO, A/S, Glostrup, Denmark) for 1 h at room temperature. Then, the filters were washed as described earlier and immunoreactive bands detected by enhanced chemiluminescence using ECL™ Western blotting detection reagents (Amersham Pharmacia, Uppsala, Sweden), according to the instructions of the manufacturer.

RESULTS

Effects of HDACIs and RA on HL-60 Cell Growth and Granulocytic Differentiation

In order to examine the effects of HDACIs and the differentiation inducer RA on HL-60 cell growth and differentiation, the cells were treated with PB, vitB3, or RA alone or in combination for 1–5 days (FIG. 1 A, D). As seen in FIGURE 1 A, all used agents and their combinations, inhibited cell growth. FIGURE 1 B presents the cell differentiation level caused by PB, vitB3, or RA alone, and their combinations. RA caused differentiation up to 90% after 5 days of treatment. PB induced it up to 30%, while vitB3 did not induce granulocytic differentiation.

To examine the differentiating effects of HDACIs and RA combinations, cells were first pretreated with PB alone or together with vitB3 for 6 h, and then treated with RA alone or together with vitB3 (FIG. 1 C, D). The combined HDACI pretreatment before the exposition to RA alone or with vitB3 resulted in the elevation of differentiation up to 65% (at 24 h) and 95% (at 48 h). These results demonstrate that the combined pretreatment with two different classes of HDACI, PB and vitB3, before the induction of differentiation by RA causes the acceleration and augmentation of HL-60 cell differentiation to granulocytes.

Influence of Combinatorial Treatment of HDACIs and RA on Cell Cycle Progression and the Expression of the Early Myeloid Differentiation Marker CD11b in HL-60 Cells

FIGURE 2 presents the distribution of HL-60 cells treated with various combinations of HDACI and RA in cell cycle phases. Cell cycle analysis revealed

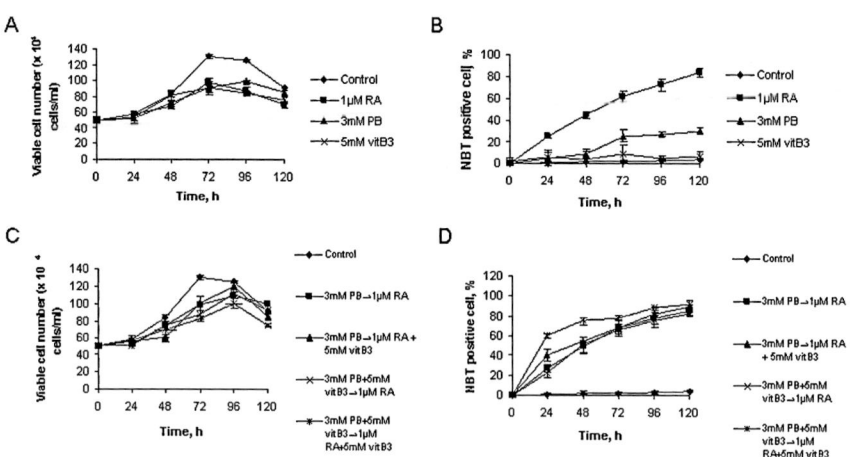

FIGURE 1. Effects of PB and vitB3 alone and in combinations on HL-60 cell growth and granulocytic differentiation. Cells were exposed to 5 mM vitB3, 3 mM PB, and 1 μM RA alone for 24, 48, 72, 96, or 120 h or pretreated with 5 mM vitB3 and 3 mM PB for 6 h before treatment with 1 μM RA alone (vitB3 + PB→ RA) or together with 5 mM vitB3 (vitB3 + PB→ RA+vitB3) for 120 h. At each time point indicated, aliquots of the cultures were subjected to counting following staining with 0.2% trypan blue for the determination of the total number of viable cells (**A, C**) or NBT-positive cells in culture (**B, D**). Results are given as mean ($\pm$ SEM; $n = 3$).

that treatment with PB, vitB3, and RA alone or in combination caused a decrease in the S and increase in the G0/G1 phases. When cells were pretreated with a combination of PB and vitB3 followed by treatment with RA alone or together with vitB3, a considerable increase up to 65% at 24 h (data not shown) with further increase up to 79% at 48 h was detected in the G0/G1 phase (FIG. 2).

We further examined the influence of HDACIs and RA on the early myeloid differentiation marker CD11b expression in HL-60 cells (FIG. 3). A 6-h pretreatment with PB and vitB3 before a 24-h treatment with RA alone or together with vitB3 caused about 85% cells expressing CD11b in both cases. After 48 h of such treatments, the CD11b expression diminished up to 74% and 56%, respectively. In contrast to other treatments presented in FIGURE 3, HDACIs and RA combinations increased to a considerable extent the cell number entering to differentiation program during the first 24 h of treatment.

Thus, these results demonstrate that combinatorial treatment of two HDACI and RA caused cell cycle arrest in the G0/G1 phase and marked increase in the expression of early myeloid differentiation marker CD11b in HL-60 cells. This combination could be most promising for the differentiation therapy of leukemia.

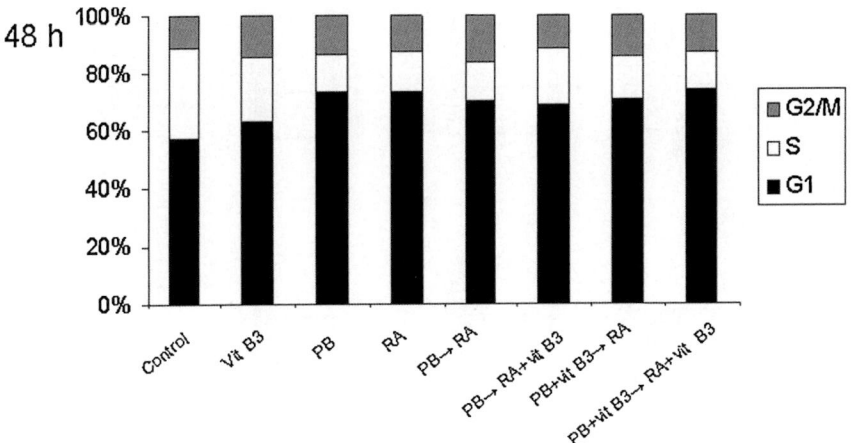

FIGURE 2. Effects of PB and vitB3 alone and in combinations on HL-60 cell cycle phase distribution. HL-60 cells were treated with 5 mM vitB3, 3 mM PB, or 1 μM RA for 48 h or pretreated with 5 mM vitB3 and 3 mM PB for 6 h before 48 h treatment with 1 μM RA alone (vitB3 + PB → RA) or together with 5 mM vitB3 (vitB3 + PB → RA + vitB3). The cell cycle phase distribution (%) was determined from the DNA frequency distribution histograms of PI stained cells. Results are from one representative experiment of three where standard deviations (not shown) were less than ±10%.

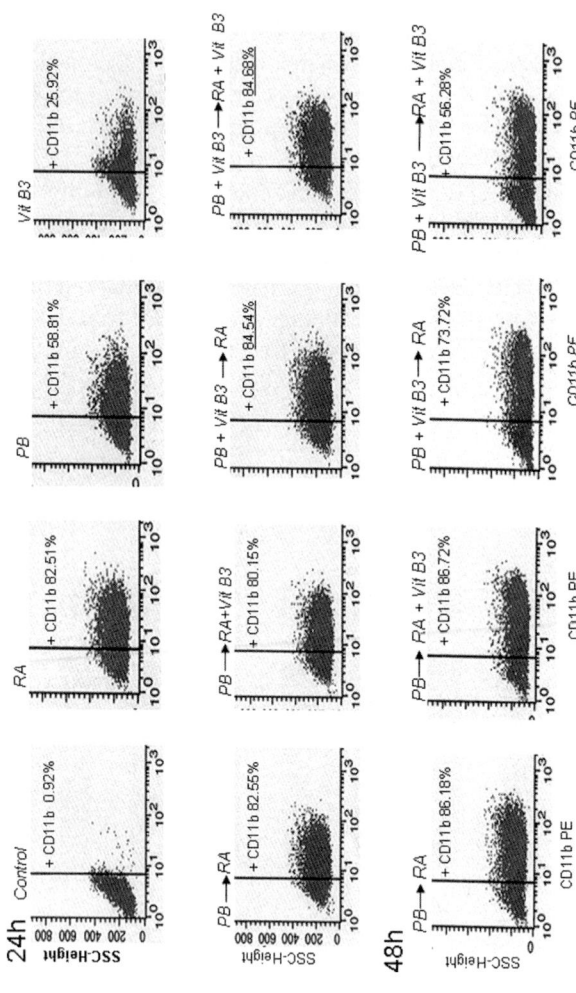

FIGURE 3. Influence of HDACIs and RA or their combinations on the expression of the early myeloid cell surface marker CD11b in HL-60 cells. Control cells and treated with 3 mM PB, 5 mM vitB3, and 1 μM RA alone or in indicated combinations for 24–48 h were analyzed for the expression of CD11b marker by flow cytometry.

Histone H3 and H4 Modifications Caused by HDACIs and RA Treatments

The differentiation agent RA and two HDACIs, PB and vitB3, were used to study their influence on histone H4 hyperacetylation in HL-60 cells. As can be seen in FIGURE 4, in control-proliferating HL-60 cells histone H4 is present in Ac0 and Ac1 forms. After treatment with vitB3 and RA, histone H4 becomes acetylated and is present at Ac0, Ac1, Ac2, and Ac3 forms. Only PB caused histone H4 hyperacetylation, which occurred during first 3–6 h of treatment. When PB and vitB3 were used in combination (FIG. 5), histone H4 mainly displayed the Ac4 form. During the first hours of such treatment, the phosphorylation (Ser10) and acetylation (Lys9) /phosphorylation (Ser10) of histone H3 was not detected or was detected only at a very low level (FIG. 5).

Thus, the combined HDACIs and RA treatment caused acetylation of histone H4 (Ac2-4 forms) and low level of histone H3 modifications.

DISCUSSION

Although the standard approach to treating myeloid leukemia remains chemotherapy, recent advances in the understanding of the biology of these disorders have lead to the development of targeted treatment strategies. The organization of chromatin is crucial for the regulation of gene expression.

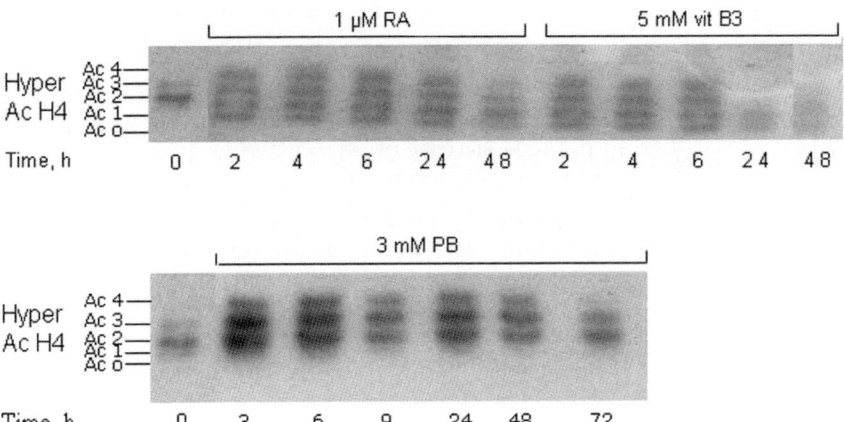

FIGURE 4. Time course analysis of histone H4 acetylation in response to RA and HDACIs. HL-60 cells were treated with 1 μM RA, 5 mM vitB3, or 3 mM PB for 72 h. Histones from control and treated HL-60 cells for indicated time points were resolved on 15% polyacrylamide–acetic acid–urea (AU) gels and examined by Western blot analysis by using antibodies against hyperacetylated histone H4 forms. The acetylation state of histones is indicated representing un-acetylated (Ac 0), mono- (Ac 1), di- (Ac 2), tri- (Ac 3), and tetra- (Ac 4) acetylated forms.

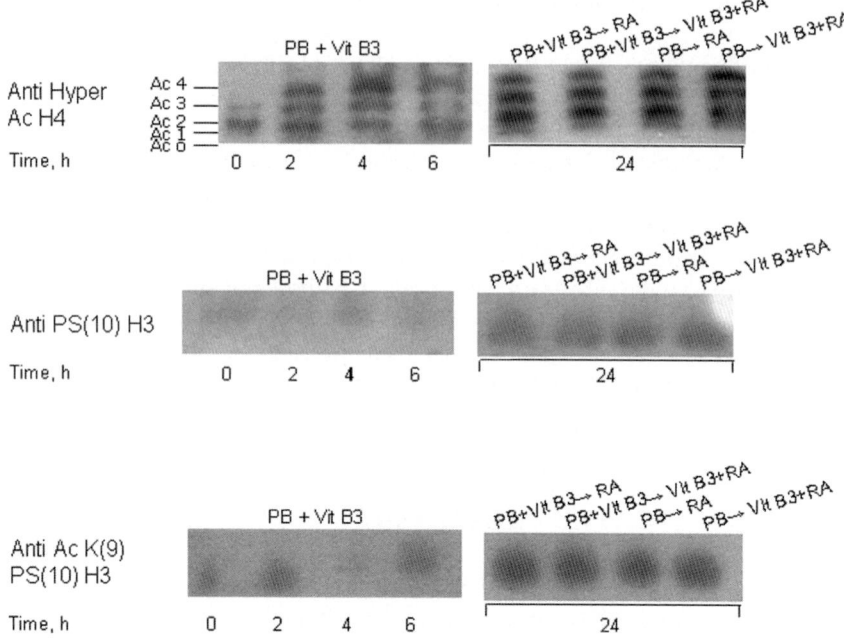

FIGURE 5. Modifications of histone H4 and H3 in response to HDACIs and RA. HL-60 cells were treated with 5 mM vitB3 together with 3 mM PB (vitB3 + PB) for indicated time points or pretreated with 5 mM vitB3 and 3 mM PB for 6 h before 48 h treatment with 1 μM RA alone (vitB3 + PB→ RA) or together with 5 mM vitB3 (vitB3 + PB→ RA + vitB3). Histones from control or treated HL-60 cells were resolved on 15% polyacrylamide–acetic acid–urea (AU) gels and examined by Western blot analysis using specific antibodies against Acetyl (Lys9) Phospho (Ser10) H3 and against Phospho (Ser10) H3. The acetylation state of histone H4 is indicated representing un-acetylated (Ac 0), mono- (Ac 1), di- (Ac 2), tri- (Ac 3), and tetra- (Ac 4) acetylated forms.

In particular, both the positioning and properties of nucleosomes influence promotor-specific transcription in response to extracellular or intracellular signals.[13] Several leukemogenic transcription factors repress expression of specific genes because of aberrant recruitment of HDACs. This repression of gene expression appears to be an important step in the action of these transcription factors, and aberrant recruitment of HDAC activity has been reported in cell lines derived from patients with APL.[14–16] Furthermore, resistance to the differentiating actions of all *trans* RA in patients with APL has been overcome by co-treatment with inhibitors of HDAC.[17]

In our study, we investigated the *in vitro* antileukemic effects of inhibitors of two different classes (HDAC1 and Sir2) of HDAC, PB and vitB3, respectively, or in combination with RA on human promyelocytic leukemia HL-60 cell line. Both HDACIs, PB and vitB3, interfered with HL-60 cell proliferation and differentiation. Thus, PB and vitB3, as well as RA, inhibited the proliferation of promyelocytic leukemia HL-60 cells. The growth inhibition induced by HDACIs, PB and vitB3, and RA alone or in various their combinations was associated in particular with cell cycle arrest in the G0/G1 phase. Cell cycle regulation in the G1 phase has attracted a great deal of attention as a promising target for research on cancer treatment. Many of the important genes associated with G1 regulation have been shown to play a key role in proliferation, differentiation, and oncogenic transformation and programmed cell death. Currently, a variety of cytostatic agents that affect G1 progression or G1/S transition, are being evaluated in clinical trials.[5,18,19]

Furthermore, we demonstrated that 6-h pretreatment with PB and vitB3 before exposition to RA alone or together with vitB3 significantly accelerated and increased cell differentiation (up to 95%) during 48 h of treatment; this is considered to be associated with the G1 block in cell cycle progression. Our results show that HL-60 cells became committed to differentiation already after the first 24 h of treatment, when 85% of cells expressed the early differentiation marker CD11b. After 48 h, the expression of CD11b reduced up to 56%, indicating that HL-60 cells become differentiated into granulocyte-like cells.

Posttranslational modifications of histones, such as acetylation, phosphorylation, and methylation, are common modes for chromatin remodeling. We observed rapid histone H4 acetylation and subsequent histone H3 phosphoacetylation in HL-60 cells after treatment with HDACI alone or in combination with RA. The increased level of histone H4 acetylation was associated with the initiation and maturation stages of HL-60 cell differentiation. In HL-60 and CCRF-CEM cells treated with HDACI trichostatin A, Chambers and co-workers[20] found by microarray analyses the functional link between histone acetylation and the expression profiles of approximately 12,000 genes. It comprised at least 9% of the genome.

The data about nucleosomes, containing phosphorylated and acetylated H3, point to the fact that both modifications may cooperate to facilitate transcription of certain genes.[21] Phosphorylated and/or acetylated H3 may promote gene activation—these modifications may serve as recognition sites for the recruitment of transcription factors or regulatory complexes.[22] Li and others[23] demonstrated that the treatment of NB4 cells with arsenic trioxide markedly increased histone H3 phosphorylation at serine 10, an event that is associated with acetylation of the lysine 14 residues. Furthermore, arsenic trioxide is highly effective for the treatment of APL, even in patients who are unresponsive to RA therapy. Our results show that histone H3 phosphoacetylation slightly increased after the combined treatment of HDACIs and RA, which could be associated with altered gene transcription during granulocytic differentiation.

Taken together, the combination defined by us (i.e., 6-h pretreatment with two classes of HDACI, PB and vitB3, before the exposition to RA alone or in combination with vitB3) may be promising for differentiation therapy acting via chromatin remodeling.

ACKNOWLEDGMENTS

This research was supported by the Swedish Institute (Visby Program No. 01361/2006), the Swedish Research Council, and the Lithuanian State Science and Studies Foundation (No.V-73/06 and No. T-37/06).

REFERENCES

1. KOUZARIDES, T. 2003. Wellcome Trust Award Lecture. Chromatin-modifying enzymes in transcription and cancer. Biochem. Soc. Trans. **31**: 741–743.
2. YANG, X.Y. & E. SETO. 2003. Collaborative spirit of histone deacetylases in regulating chromatin structure and gene expression. Curr. Opin. Gen. Dev. **13**: 143–153.
3. YOSHIDA, M. *et al.* 1990. Potent and specific inhibition of mammalian histone deacetylase both *in vivo* and *in vitro* by trichostatin A. J. Biol. Chem. **265**: 17174–17179.
4. MUNRO, J. *et al.* 2004. Histone deacetylase inhibitors induce a senescence-like state in human cells by a p16-dependent mechanism that is independent of a mitotic clock. Exp. Cell Res. **295**: 525–538.
5. SAKAJIRI, S. *et al.* 2005. Histone deacetylase inhibitors profoundly decrease proliferation of human lymphoid cancer cell lines. Exp. Hematol. **33**: 53–61.
6. DOKMANOVIC, M. & A.P. MARKS. 2005. Prospects: histone deacetylase inhibitors. J. Cell Biochem. **96**: 293–304.
7. GRIGNANI, F. *et al.* 2000. PML/RAR alpha fusion protein expression in normal human hematopoietic progenitors dictates myeloid commitment and the promyelocytic phenotype. Blood **96**: 1531–1537.
8. SHINJO, K. *et al.* 2005. Delayed recovery of normal hematopoiesis in arsenic trioxide treatment of acute promyelocytic leukemia: a comparison to all-*trans* retinoic acid treatment. Intern. Med. **44**: 818–824.
9. IWATA, K., S. OGATA, K. OKUMURA & H. TAGUCHII. 2003. Induction of differentiation in human promyelocytic leukemia HL-60 cell line by niacin – related compounds. Biosci. Biotechnol. Biochem. **67**: 1132–1135.
10. COLLINS, S. 1987. The HL-60 promyelocytic leukemia cell line: proliferation, differentiation and cellular oncogene expression. Blood **70**: 1233–1244.
11. TREIGYTE, G. & A. GINEITIS. 1979. Specific changes in the biosynthesis and acetylation of nucleosomal histones in the early stages of embryogenesis of sea urchin. Exp. Cell Res. **121**: 127–134.
12. HURLEY, C.K. 1977. Electrophoresis of histones: a modified Panyim and Chalkley system for slab gels. Anal. Biochem. **80**: 624–626.
13. GRUNSTEIN, M. 1997. Histone acetylation in chromatin structure and transcription. Nature **389**: 349–352.
14. GRIGNANI, F. *et al.* 1998. Fusion proteins of the retinoic acid receptor-alpha recruit histone deacetylase in promyelocytic leukaemia. Nature **391**: 815–818.
15. HE, L.-Z. *et al.* 2001. Histone deacetylase inhibitors induce remission in transgenic models of therapy-resistant acute promyelocytic leukemia. J. Clin. Invest. **108**: 1321–1330.
16. LIN, R.J. *et al.* 1998. Role of the histone deacetylase complex in acute promyelocytic leukemia. Nature **391**: 811–814.

17. WARRELL, R.P., JR., et al. 1998. Therapeutic targeting of transcription in acute promyelocytic leukemia by use of an inhibitor of histone deacetylase. J. Natl. Cancer Inst. **90:** 1621–1625.
18. SHARABANI, H. et al. 2006. Cooperative antitumor effects of vitamin D (3) derivatives and rosemary preparations in a mouse model of myeloid leukemia. Int. J. Cancer **118:** 3012–3021.
19. TADDEI, A. et al. 2005. The effects of histone deacetylase inhibitors on heterochromatin: implications for anticancer therapy. EMBO Rep. **6:** 520–524.
20. CHAMBERS, A.E. et al. 2003. Histone acetylation-mediated regulation of genes in leukaemic cells. Eur. J. Cancer **39:** 1165–1175.
21. CHEUNG, P. et al. 2000. Signaling to chromatin through histone modifications. Cell **103:** 263–271.
22. MIZZEN, C.A. & C. D. ALLIS. 2000. New insights into an old modification. Science **289:** 2290–2291.
23. LI, J. et al. 2002. Arsenic trioxide promotes histone H3 phosphoacetylation at the chromatin of CASPASE-10 in acute promyelocytic leukemia cells. J. Biol. Chem. **277:** 49504–49510.

The Histone Deacetylase Inhibitor FK228 Distinctly Sensitizes the Human Leukemia Cells to Retinoic Acid-Induced Differentiation

JURATE SAVICKIENE,[a] GRAZINA TREIGYTE,[a]
VERONIKA BORUTINSKAITE,[a] RUTA NAVAKAUSKIENE,[a]
AND KARL-ERIC MAGNUSSON[b]

[a]*Department of Developmental Biology, Institute of Biochemistry, LT-08662 Vilnius, Lithuania*

[b]*Division of Medical Microbiology, Department of Molecular and Clinical Medicine, Linköping University, SE-58185 Linköping, Sweden*

ABSTRACT: FK228 (depsipeptide) is a novel histone deacetylase inhibitor (HDACI) that has shown therapeutical efficacy in clinical trials for malignant lymphoma. In this article, we examined *in vitro* effects of FK228 on human leukemia cell lines, NB4 and HL-60. FK228 alone (0.2–1 ng/mL) inhibited leukemia cell growth in a dose-dependent manner and induced death by apoptosis. FK228 had selective differentiating effects on two cell lines when used for 6 h before induction of granulocytic differentiation by retinoic acid (RA) or in combination with RA. These effects were accompanied by a time- and dose-dependent histone H4 hyper-acetylation or histone H3 dephosphorylation and alterations in DNA binding of NF-κB in association with cell death and differentiation. Pifithrin-α (PFT), an inhibitor of p53 transcriptional activity, protected only NB4 cells with functional p53 from FK228-induced apoptosis and did not interfere with antiproliferative activity in p53-negative HL-60 cells. In NB4 cells, PFT inhibited p53 binding to the p21 (Waf1/Cip1) promotor and induced DNA binding of NF-κB leading to enhanced cell survival. Thus, beneficial effects of FK228 on human promyelocytic leukemia may be exerted through the induction of differentiation or apoptosis via histone modification and selective involvement of transcription factors, such as NF-κB and p53.

KEYWORDS: differentiation; histone deacetylase inhibitor; leukemia; NF-κB; p53

Address for correspondence: Jurate Savickiene, Department of Developmental Biology, Institute of Biochemistry, LT-08662 Vilnius, Lithuania, Voice: 370-5-272-91-87; fax: 370-5-272-91-96.
e-mail: jurate_savickiene@yahoo.com

INTRODUCTION

Histone deacetylase inhibitors (HDACI) are a novel class of chemotherapeutic agents. They induce histone hyper-acetylation and transcription regulation through chromatin remodeling, leading to selective activation of genes associated with cell growth, differentiation, and survival.[1,2] In recent years, HDACI, like sodium butyrate, trichostatin, apicidin, trapoxin, oxamflatin, suberoylanilide hydroxamic acid (SAHA), valproic acid, and several others, have been shown to inhibit cell proliferation and induce differentiation and apoptosis in a variety of human cell lines.[2,3] FK228 (known as FR901228 or depsipeptide) is a member of the cycle peptide class of inhibitors and shows potent antitumor activity against human and murine tumors.[4–6] FK228 is used in clinical trials for acute lymphocytic leukemia, small lymphocytic lymphoma, T cell lymphoma, cutaneous T cell lymphoma, or progressive small-cell or non-small cell lung cancer.[7,8] Preclinical studies with FK228 in chronic lymphocytic leukemia (CLL) and acute myeloid leukemia (AML) demonstrated effective HDAC inhibition and antitumor activity, but its use was limited by progressive constitutional symptoms and needs alternative administration schedules.[9]

Acute promyelocytic leukemia (APL) is characterized by a chromosomal translocation $t(15; 17)$, which fuses the PML gene to the retinoic acid receptor α (RARα) gene, resulting in the expression of chimeric gene product PML–RARα. This oncoprotein represses the transcription of retinoic acid (RA)-responsive genes by association with a co-repressor complex containing HDAC activity.[10] The aberrant recruitment of HDAC interferes with normal cell growth and differentiation. Thus, the addition of therapeutical doses of RA causes the conformational change of co-repressor complex and recruitment of a transcriptional activator complex leading to transcriptional activation and re-initiation of differentiation. Currently, all-*trans*-RA is used successfully in differentiation therapy of APL.[10,11] A rare variant of APL, which is associated with $t(11; 17)$, has failed to respond to chemotherapy and treatment with RA.[12] The combined treatment of RA and HDACI, such as sodium phenylbutyrate and FK228, can induce *in vitro* differentiation of RA-resistant APL cells.[13–15] However, HDACI efficacy and selectivity toward cancer cells remain still poorly known.

In this article, we investigated the *in vitro* activities of FK228 alone or in combination with RA using APL cell line NB4, positive by chromosomal translocation (15; 17) and functional p53 status, and negative one, HL-60. Our findings demonstrate selective efficacy of FK228 in APL cell lines, including cell survival and sensitivity to RA-induced granulocytic differentiation through mechanisms involving H3 and H4 histone modifications, and activity modulation of transcription factors, such as NF-κB and p53.

MATERIALS AND METHODS

Materials

All chemicals used were of analytical grade and were purchased from Sigma Chemical Co. (St. Louis, MO). Oligonucleotides were synthesized by MWG-Biotech AG (Ebersberg, Germany). Polyclonal anti-acetylated histone H4 antibody and polyclonal anti-phosphorylated histone H3 at serine 10 antibody were from Upstate (Lake Placid, NY). FK228 was obtained from Fujisawa Pharmaceutical Co., Ltd. (Osaka, Japan), and pifithrin α (PFT) from Calbiochem (Mannheim, Germany).

Cell Cultures

The human promyelocytic leukemia HL-60 and APL NB4 cells were cultured in RPMI 1640 medium supplemented with 10% fetal bovine serum, 100 U/mL penicillin, and 100 μg/mL streptomycin (Gibco, Grand Island, NY) at 37°C in a humidified 5% CO_2 atmosphere and used for assays during exponential phase of growth.

Cell Viability and Growth

Cell viability was assayed by exclusion of 0.2% trypan blue. Cell number was determined by counting cells in suspension in a hemocytometer. For a dose–response determination, FK228 alone or in combination were added to a final volume of 5 mL.

Cell Differentiation and Apoptosis

The degree of differentiation was assayed by the ability of cells to reduce nitro blue tetrazolium (NBT) to insoluble blue-black formazan after stimulation by PMA.[16] Cell suspension (100 μL) was mixed with an equal volume of 0.2% NBT in phosphate-buffered saline (PBS) containing PMA (40 ng/mL) and incubated at 37°C for 30 min. NBT-positive cells were counted using a hemocytometer. At least 200 cells were scored for each determination. Apoptotic cell morphology was evaluated using fluorescence microscopy. At the end of each incubation, cells were pelleted at 500 g for 5 min, resuspended in 100 μL PBS (5×10^6 cells/mL) and stained with 0.01% acridine orange–0.01% ethidium bromide (AO/EtBr) mixture (1:1, v/v), 6 μL for 100 μL cell suspension.[17]

Assessment of the Early Myeloid Differentiation Marker CD11b

NB4 or HL-60 cells (5×10^5 cells/sample) were collected, washed twice with PBS, and suspended in 50 μL PBS, pH 7.4. Then 5 μL of monoclonal mouse anti-human CD11b, C3bi receptor/RPE (DakoCytomation, Glostrup, Denmark) antibody was added to the sample, gently suspended, and incubated in the dark at 4°C for 30 min. Cells were washed with PBS containing 2% bovine serum albumin, fixed in 4% para formaldehyde for 15–30 min on ice and the pellet was resuspended in PBS. Ten thousands events were analyzed for each sample by immunofluorescence using flow cytometry. Proliferating cells with and without CD11b antibodies were used as a control.

Preparation of Nuclear Extracts

Cells were harvested and pelleted at 500 g for 6 min, and washed twice in ice-cold PBS. Nuclei were prepared using Nuclei Isolation Kit (Sigma) according the manufacturer's recommendation. Nuclei were completely suspended in Nuclei EZ storage buffer and frozen at −70°C. Nuclear protein extracts for EMSA were prepared by lysis of nuclei in buffer, containing 20 mM Tris–HCl, pH 8,0, 200 mM EDTA, 2 mM EGTA, 20% glycerol, 400 mM NaCl, and inhibitors: 1 mM PMSF, 3 mM DTT, and protease inhibitor cocktail (Roche, Basel Switzerland). After incubation for 1 h on ice, the extracts were centrifuged at 18,000 g for 20 min, and were used immediately. Protein concentrations were measured using commercial RCDC Protein Assay (BioRad, Munich, Germany).

Electrophoretic Mobility Shift Assay (EMSA)

The probes used were synthetic oligonucleotides representing binding sites: (5′-AAGCCTGGGCAACATAGAAAGTCCCCATCTGTACAAAA-3′) NF-κB from the FasL promotor; (5′-AGTTGAGGGGACTTTCCCAGGC-3′) NF-κB consensus motif; (5′-ATCAGGAACATGTCCCAACATGTTGAG CTCT-3′) p53 from the p21 promoter. Complementary oligonucleotides were annealed and labeled at their 5′ ends using [γ-^{32}P-ATP] (Amersham Biosciences, Buckinghamshire, England) and T4 polynucleotide kinase (MBI Fermentas Inc., Vilnius, Lithuania). Standard DNA reactions were performed with 15 μg nuclear extracts in a 20 μL of reaction buffer (10 mM HEPES pH 7.9, 3 mM MgCl$_2$, 0,1 mM EDTA, 40 mM NaCl, 10% glycerol) containing 2 μg BSA, 1 μg poly(dI-dC), and 1 pM labeled oligonucleotide for 30 min at room temperature.[18] When desired, an unlabeled competitor oligonucleotide was added to the protein extracts at a 100-fold molar excess for a 15-min pre-incubation. DNA–protein complexes were resolved on 6% polyacrylamide gel containing $1 \times$ Tris–borate buffer. After electrophoresis, the gels were dried and then exposed to X ray films.

Isolation of Histones and Western Blot Analysis

Histones were extracted as described previously.[19] Isolated nuclei were suspended in 5 vol of 0.4 N H_2SO_4 by stirring and incubated overnight at 0°C. The supernatant was collected by centrifugation at 15,000 g for 10 min at +2°C and the sediment was extracted once more. After centrifugation, both extracts were combined and histones were precipitated by adding 5 vol of ethanol at −20°C overnight. Precipitated histones were collected by centrifugation, washed several times with ethanol, and stored at −20°C until analysis. Histone electrophoresis was carried out essentially as described by Hurley.[20] Shortly, histones (5 μg) were dissolved in a buffer containing 0.9 M acetic acid, 10% glycerol, 6.25 M urea, and 5% β-mercaptoethanol, and run on a 15% polyacrylamide gel containing 6 M urea and 0.9 M acetic acid by using 0.9 M acetic acid as a buffer. After electrophoresis, gels were stained with Brilliant Blue G-colloidal (Sigma) or transferred to Immobilon™ PVDF transfer membrane (Millipore, Bedford, MA) for the evaluation of acetylated histone H4 (H4 Ac4) and phosphorylated histone H3 forms (H3 PS10) using specific antibodies. Immunoreactive bands were visualized by ECL chemiluminescence detection (Amersham Pharmacia, Uppsala, Sweden) according to the instructions of the manufacturer.

RESULTS

FK228 Induces Growth Inhibition and Death in Leukemia Cells

Different concentrations of FK228 (0.25, 0.5, 1 ng/mL) were used for the treatment of NB4 and HL-60 leukemia cells during 5 days. In both cell lines, FK228 as a single agent inhibited cell growth and induced cell death in a dose- and time-dependent manner (FIG. 1 A, B). In HL-60 cells, HDACI produced a greater growth inhibition and cell death at a shorter drug exposure time. Apoptosis was the main form of depsipeptide-induced cell death, as determined by staining with AO/Et Br. FK228 (1 ng/mL) induced massive leukemia cell apoptosis. Exposure of NB4 cells for 4–8 h to FK228 at a concentration 10 ng/mL was sufficient to induce irreversible cell death (data not shown). In both cell lines, the combined treatment with FK228 (0.25, 0.5 ng/mL) and RA (1 μM) caused the additive antiproliferative and apoptotic effects, which were more prominent in HL-60 cells (FIG. 1 C).

Distinct Effects of FK228 on RA-Induced Differentiation in Leukemia Cells

To study the effects of depsipeptide on leukemia cell differentiation, we used low doses of FK228 for co-treatment with RA or for 6-h pretreatment before the

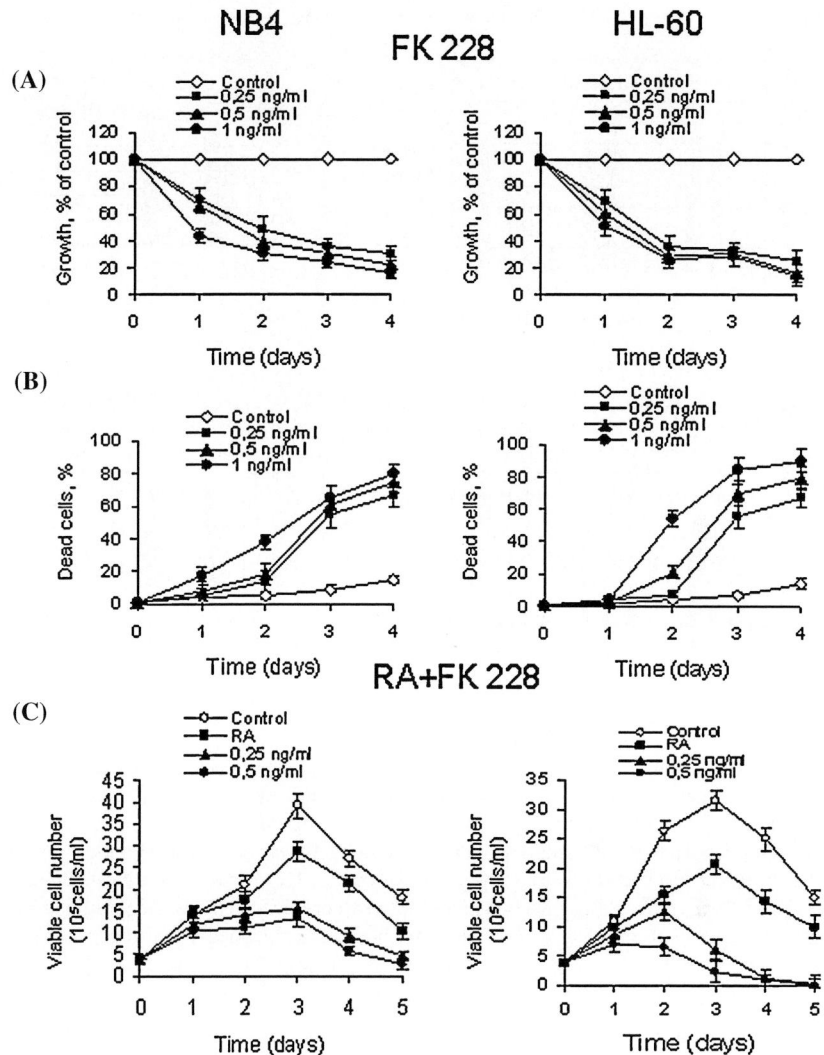

FIGURE 1. Dose-dependent effects of FK228 alone and in combination with RA on leukemia cell growth and viability. NB4 and HL-60 cells were exposed to the indicated concentrations of FK228 alone (**A, B**) and in combination with 1 µM RA (**C**) for 5 days. Aliquots of the cultures were subjected to counting following staining with 0.2% trypan blue for the determination of viable and dead cells.

induction of granulocytic differentiation by RA. In the HL-60 cell population, at 0.1 ng/mL FK228, a dose that does not induce apoptosis in combination with RA, the NBT-positive cell number increased more than twice on days 1 and 2 compared to RA alone (FIG. 2 A). In NB4 cells, the potentiating effect

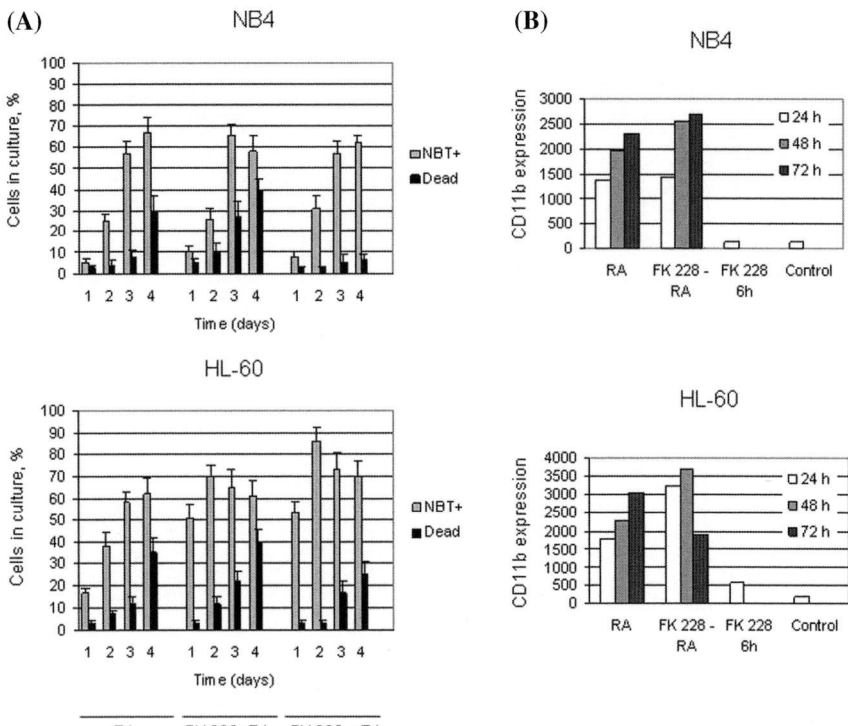

FIGURE 2. Time-dependent effects of FK228 on leukemia cell differentiation. HL-60 and NB4 cells were co-treated with RA (1 μM) and FK228 (0.1 and 0.25 ng/mL, respectively) for 4–5 days or pretreated with FK228 (0.25 ng/mL) for 6 h before the induction of differentiation by RA. Granulocytic differentiation (**A**) was determined by the ability of cells to reduce NBT. Results are mean ± SEM; $n = 3$. Expression of the differentiation marker CD11b, indicated as mean channel FL2-H fluorescence (**B**), was determined by a flow cytometry analysis. Data are representative of three independent experiments that gave comparable results.

was observed using RA with 0.25 ng/mL FK228 (an optimal dose for cell survival [FIG. 1] in such a combination) on day 3 only. FK228 alone failed to induce differentiation of NB4 cells and partially (to about 20%) induced HL-60 cell differentiation (data not shown). However, after 24-h exposure to RA following 6-h pretreatment with 0.25 ng/mL FK228 (an optimal dose and time for histone H4 acetylation in FIG. 3), HL-60 cell differentiation increased threefold compared to control (RA) and reached a maximum (about 90%) on day 2 with minimal cytotoxicity (FIG. 2 A). The same pretreatment caused substantial increase in the early differentiation marker CD11b expression in HL-60 cells and no marked changes in NB4 cells during a commitment stage of granulocytic differentiation (FIG. 2 B).

FK228 Induces Histone H4 and H3 Modifications

We analyzed where a short-time (2–8 h) exposure to FK228 at different concentrations (0.25–10 ng/mL) could promote histone H4 acetylation. Analysis of changes in histone acetylation by staining with brilliant blue G-colloidal (FIG. 3 A) or by immunoblotting (FIG. 3 B) demonstrated that FK228 induced a dose- and time-dependent increase in histone H4 acetylation in NB4 and HL-60 cells. In untreated cells, histones were predominantly un-acetylated and mono-acetylated (FIG. 3 A). Maximum histone H4 hyper-acetylation at FK228 concentrations of 5–10 ng/mL was achieved after 4 h; however, those doses caused high cytotoxic effect and marked cell death. Data in FIGURE 3 B clearly showed the moving of histone H4 into the highly acetylated isoforms at HDACI doses higher than 1 ng/mL. In HL-60 cells, the level of acetylated histone H4 to tri- and tetra-isoforms occurred at lower HDACI concentration (0.25 ng/mL) than in NB4 cell line during 4 h of treatment.

Phosphorylation is another important histone modification that is often associated with chromatin condensation during mitosis and induction of apoptosis or DNA damage.[21] Next, we examined histone H3 phosphorylation at serine 10 in leukemia cell response to FK228. In HL-60 cells, increasing doses of FK228 resulted in a decrease of histone H3 phosphorylation in parallel with

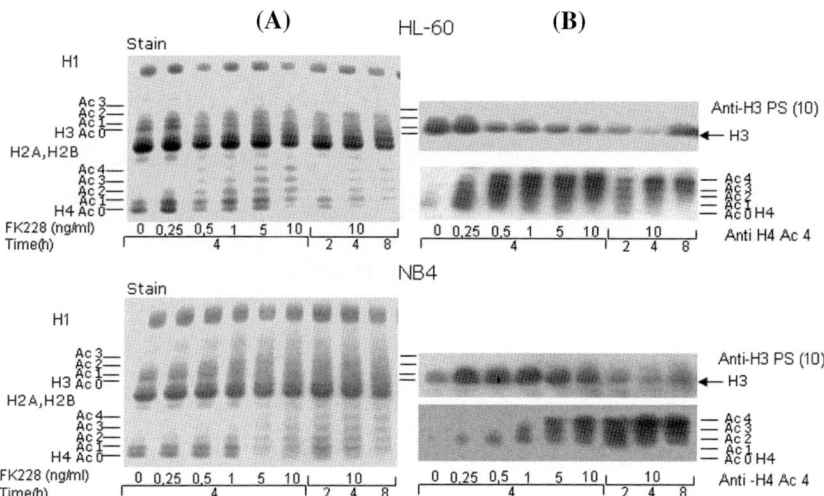

FIGURE 3. FK228 induces dose- and time-dependent histone H4 and H3 modifications. Histones from nuclear proteins of untreated or HDACI-treated NB4 and HL-60 cells were subjected to AU (15% polyacrylamide, acetic acid, urea) electrophoresis, gel staining with brilliant blue G-colloidal (**A**) and immunoblotting (**B**) with anti-acetylated histone H4 and anti-phosphorylated histone H3 at serine 10 antibodies. The five acetylation states of histone are indicated, representing un-acetylated (Ac 0), mono- (Ac 1), di- (Ac 2), tri- (Ac 3), and tetra- (Ac 4) acetylated forms.

an increase of histone H4 hyper-acetylation (FIG. 3 B), indicating a tight link between the aceylation/phosphorylation status of histone H3 and H4. The accumulation of histone H3 in phosphorylated form was noticed after 8 h treatment with 10 ng/mL FK228 (an apoptogenic dose). In less apoptosis-sensitive NB4 cells, the changes in histone H3 phosphorylation were similar, but occurred more slowly.

Modulation of Transcription Factor NF-κB Binding Activity by FK228

The nuclear transcription factor NF-κB acts as a survival factor and is required for the proliferation and differentiation of different cancer cells.[22] Therefore, we performed EMSA to study the NF-κB binding activity in response to FK228. As shown in FIGURE 4 A, 4-h exposure to different doses of FK228

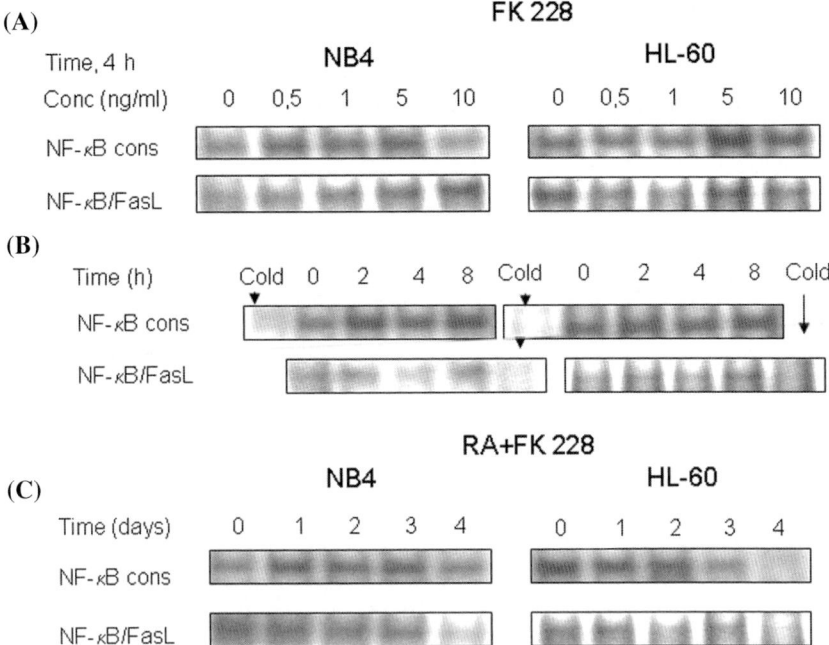

FIGURE 4. Alterations in NF-κB binding activity in response to FK228 alone or in combination with RA. Nuclear extracts were prepared from control cells and treated with different doses of FK228 during 4 h (**A**) or treated with 5 ng/mL FK228 for a different time (**B**), and in combination with RA (1 μM) and FK228 (0.25 ng/mL) during 4 days (**C**). EMSA was performed using a total of 15 μg protein from each nuclear extract and oligonucleotides, containing NF-κB consensus motif or NF-κB binding site from the FasL promotor. Specific DNA complexes with NF-κB were eliminated competitively by the addition of 100-fold molar excess of unlabeled oligonucleotide (cold).

caused cell line-specific increase in the intensity of the NF-κB–DNA complex formation at concentrations of 0.5–5 ng/mL and decrease at the apoptotic dose 10 ng/mL in both cell lines. HDACI caused a dose-dependent increase in NF-κB binding to the FasL promotor. In both cell lines, DNA binding to NF-κB was altered in response to 5 ng/mL FK228 (FIG. 4 A) and more prominent at 8 h exposure (FIG. 4 B), which is consistent with histone H4 hyper-acetylation occurring at 4–8 h (FIG. 3). The combined treatment with FK228 (0.25 ng/mL) caused the maintenance of elevated NF-κB binding activity during cell differentiation and suppression during differentiation-leading apoptosis on day 3 or 4 in HL-60 and NB4 cells, respectively (FIGS. 2 and 4 C). However, in this case, no upregulation of NF-κB binding to the FasL promotor was found in cells undergoing apoptosis.

PFT Protects p53-Positive NB4 Cells from Apoptosis

To determine whether HDACI-mediated apoptosis depends on p53 function, we used PFT, a transcriptional inhibitor of p53, and FK228 at doses that cause a similar level of NB4 and HL-60 cell death (FIG. 1). PFT (30 μM) itself inhibited p53-negative HL-60 cell growth at the same level as in combination with FK228 (0.25 ng/mL). PFT failed to inhibit p53-positive NB4 cell growth and increased cell survival in combination with 0.5 ng/mL FK228 (FIG. 5 A). Next, we exposed HL-60 and NB4 cells to PFT for 4 h, and after drug washing the cells were treated without or with low (0.2 ng/mL) and high (1 ng/mL) concentrations of FK228 for 4 days. As shown in FIGURE 5 B, PFT increased the survival of NB4 cells much better than of HL-60, and did not attenuate HL-60 cell death induced by a high dose of depsipeptide.

Induction of DNA damage by apoptogenic agents is known to activate p53, which in turn acts as a transcriptional regulator of several target genes, including CDK inhibitor p21 (Waf1/Cip1).[23] EMSA revealed that in NB4 cells with functional p53, a high dose (10 ng/mL) of FK228 caused maximal induction of p53 binding to the p21 promotor at 8 h of incubation (FIG. 6 A). Long-term treatment for 4 days with a moderate dose of FK228 (0.5 ng/mL) showed constitutive p53 binding activity, which was upregulated in NB4 cells on days 2–4. As shown in FIGURE 6 B, p53 binding to the p21 promotor was maintained during 4 h of FK228 (1 ng/mL) treatment and suppressed following 4 h treatment with PFT itself and in combination with HDACI. The same treatment decreased NF-κB binding activity in HL-60 cells, but enhanced it in NB4 cells.

DISCUSSION

Here, we have examined the *in vitro* activities of FK228 for APL cell lines, NB4 and HL-60, which exhibit promyelocytic phenotype and respond to RA

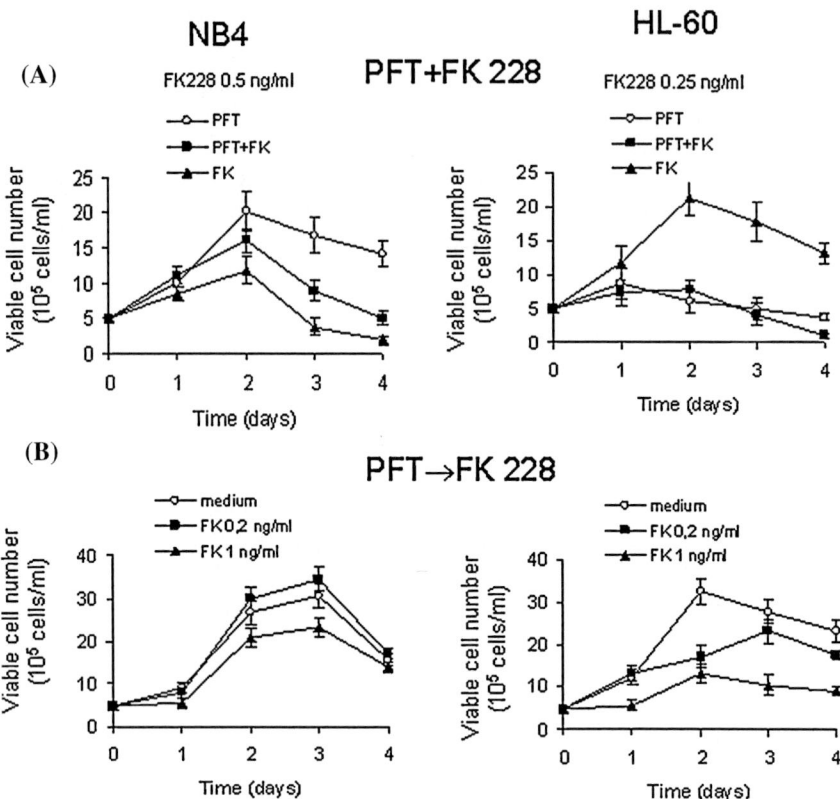

FIGURE 5. PFT protects p53-positive NB4 cells from apoptosis. **(A)** NB4 and HL-60 cells were incubated with 30 μM PFT itself or with 0.5 ng/mL and 0.25 ng/mL FK228, respectively, without or with PFT for 4 days or **(B)** exposed to PFT for 4 h and then transferred onto a fresh medium without or with different concentrations of FK228 for 4 days. Aliquots of the cultures were subjected to counting following staining with 0.2% trypan blue for the determination of the total number of viable cells. Results are mean ± SEM ($n = 3$).

that induces a terminal differentiation, followed by natural apoptosis of malignant cells.[24] We demonstrated that FK228 exhibits distinct antiproliferative and cytotoxic effects on both cell lines. This may be due to a difference in a set of genes that are independently controlled in each cell line.[25] For instance, the apoptotic process in differentiating NB4 cells is temporarily inhibited by the upregulation of the apoptosis inhibitor, survivin.[24] In human cancer cells, the protective role of p53 against apoptosis was shown in the upregulation of its downstream gene, p21,[25,26] which controls cell cycle progression.[27,28] HDACI is known to upregulate the p21$^{Waf1/Cip1}$ gene inducing the accumulation of acetylated histones in the p21 gene promotor in many cancer cells.[29,30] Furthermore, upregulation of p21, Fas, FasL,

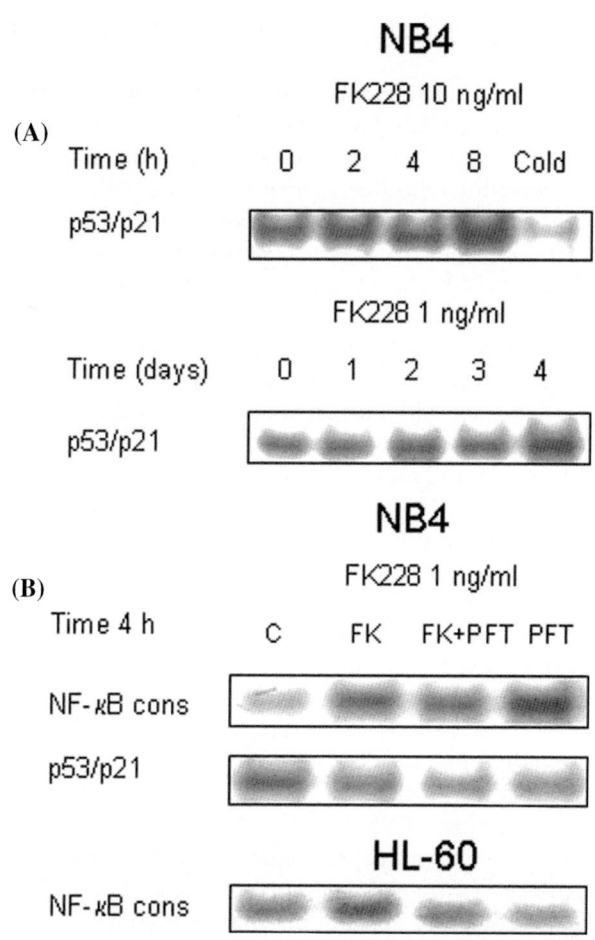

FIGURE 6. PFT affects NF-κB and p53 binding activity in leukemia cells with different functional statuses of p53. (**A**) Nuclear extracts were prepared from control NB4 cells and treated with FK228 (10 ng/mL) for 2–8 h, or with FK228 (1 ng/mL) for 4 days; and (**B**) NB4 and HL-60 cells were treated with PFT (30 μM) alone and co-treated with PFT and FK228 (1 ng/mL) for 4 h. EMSA was performed using a total 15-μg protein from each nuclear extract and oligonucleotides, containing NF-κB consensus motif and p53 binding site from the p21 promotor.

downregulation of antiapoptotic Bcl-X_L, and activation of caspase-3 were reported in adult T cell lymphoma, uveal melanoma, and neuroblastoma cells treated with HDACI.[7,31–33] Our EMSA results demonstrate the activation of NF-κB binding to the FasL promotor by FK228, suggesting the importance of the Fas/FasL system in depsipeptide-induced cell death. However, this activity decreases in the presence of RA, supporting the idea that the Fas/FasL pathway is likely not involved in the apoptosis of terminally differentiated leukemia cells.[18,34,35]

FK228 was effective in inducing apoptosis in cell lines that are different by p53 status. Recently, a synthetic compound PFT has been reported to be a specific inhibitor of p53 transactivation and was proposed in cancer treatment.[36] However, PFT shows different effects; for instance, the inhibition of cell growth and induction or enhancement of p53-dependent apoptosis in mouse epidermal cells JB6[37] or, in contrast, the protection of neurons and cardiomyocytes against apoptosis induced by DNA-damaging agents and doxorubicin, respectively.[38,39] In our study, we demonstrate that p53-defective HL-60 cells displayed enhanced apoptotic potential compared to NB4 upon treatment with 30 μM PFT alone. PFT-pretreated NB4 cells with functional p53 exhibited increased cell survival after subsequent treatment with FK228. One of the postulated mechanisms of PFT activity is an inhibition of p53 translocation from the cytoplasm to the nuclei and prevention of its binding to specific DNA sites.[38] The latter suggestion is illustrated in our study by the reduced p53 binding activity to the p21 promotor in NB4 cells treated with PFT alone or together with FK228. PFT was shown also to reduce the activation of p53-regulated genes, including bax, cyclin G, and mdm2 that control the degradation rate of p53.[36,39] Furthermore, PFT-induced apoptosis may be mediated through a p53-independent pathway by the involvement of p38 and extracellular kinase activities of ERKs, or by the upregulation of proapoptotic bax and activation of caspases.[37,40] In our study, PFT induces apoptosis in p53-negative HL-60 cells and does not interfere with the antiproliferative action of FK228.

Another important fact is that FK228 enhanced and accelerated differentiation in HL-60 cells more effectively than in NB4 cells even at lower doses. Moreover, 6-h treatment with depsipeptide was sufficient to potentiate RA-mediated granulocytic differentiation in HL-60 cells. These effects were accompanied by different intensities of histone H4 acetylation and histone H3 dephosphorylation. Recent data point to a molecular link between both histone modifications. In general, increased histone acetylation (hyper-acetylation) triggers chromatin remodeling and transcriptional activation, while histone phosphorylation is often associated with chromatin condensation that includes mitosis and apoptosis.[21,41] Our results also reveal a coordinated correlation between the dynamic of histone H4 acetylation/histone H3 dephosphorylation (at serine 10) or H4 deacetylation/H3 phosphorylation upon treatment with FK228. The responses to HDACI, as a consequence of histone acetylation, may influence the pattern of gene expression. A finite subset of genes (about 9% of the genome) was found to be regulated in HL-60 cells upon treatment with the HDACI, trichostatin, over 50% of which were transcription factors or transcription augmenters. A number of genes were altered in expression, whereas others had opposite regulatory transcription profiles.[42] HDACI and RA alone induce distinct subsets of genes (about 3%), and as was shown in NB4 cells, the majority of genes were induced by the combination of RA and

HDACI, SAHA.[43] The cellular responses to HDACI, such as differentiation or apoptosis, often depend on the intrinsic characteristics of cancer cells.[44] The selective differentiation effects in HL-60 cells, obtained by 6-h treatment with FK228 followed by RA, may be explained by intensive histone H4 acetylation and, possibly, a rapid downregulation of transcription of primary response genes, such as myb and c-myc, that are amplified in HL-60 cells.[4,42] In certain circumstances the transcriptional activity of NF-κB activity could explain the effectiveness of HDACI to trigger differentiation or apoptosis. In certain cell types, HDACI can activate NF-κB–dependent gene expression resulting in the upregulation of anti-apoptotic genes, such as Bcl-X$_L$,[44,45] and in others it can suppress NF-κB activation.[7,46] Our EMSA results demonstrate time- and dose-dependent differences in the transcription factor NF-κB binding activity between HL-60 and NB4 cell lines in association with cell survival and maturation to granulocytes. FK228 reduced NF-κB binding activity during apoptosis in both the leukemia cell lines that have been demonstrated in ATL cells too.[47] Thus, this transcription factor could be considered as one of the general targets of FK228.

In summary, our data have provided evidence for myeloid cell line-specific, differential activity of FK228 in the enhancement of RA-mediated differentiation that is associated with the regulation of gene expression mediated through chromatin remodeling.

ACKNOWLEDGMENTS

The study was supported by the Swedish Research Council, Lithuanian State Sciences and Studies Foundation (No. T-37/06), and the Swedish Institute (Visby program No. 01361/2006).

REFERENCES

1. YOSHIDA, M., T. SHIMAZU, M. NISHIYAMA, et al. 2003. Protein deacetylases: enzymes with functional diversity as novel therapeutic targets. Progr. Cell Cycle Res. **5:** 269–278.
2. RICHON, V. & J.P. O'BRIAN. 2002. Histone deacetylase inhibitors: a new class of potential therapeutic agents for cancer treatment. Clin. Cancer Res. **8:** 662–664.
3. MARKS, P., R.A. RIFKIND, A. RICHON, et al. 2001. Histone deacetylases and cancer: causes and therapies. Nature Rev. Cancer **1:** 194–202.
4. UEDA, H., H. NAKAJIMA, Y. HORI, et al. 1994. Action of FR901228, a novel antitumor bicyclic depsipeptide produced by *Chromobacterium violaceum* no. 968, on Ha-ras transformed NIH3T3 cells. Biosci. Biotechnol. Biochem. **58:** 1579–1583.
5. UEDA, H., T. MANDA, S. MATSUMOTO, et al. 1994. FR901228, a novel antitumor bicyclic depsipeptide produced by *Chromobacterium violaceum* no. 968. III. Antitumor activities on experimental tumors in mice. J. Antibiot. (Tokyo) **47:** 315–323.

6. NAKAJIMA, H., Y.B. KIM, H. TERANO, et al. 1998. FR901228, a potent antitumor antibiotic, is a novel histone deacetylase inhibitor. Exp. Cell Res. **241:** 126–133.
7. MORI, N., T. MATSUDA, M. TADANO, et al. 2004. Apoptosis induced by the histone deacetylase inhibitor FR901228 in human T-cell leukemia virus type 1-infected T-cell lines and primary adult T-cell leukemia cells. J. Virol. **78:** 4582–4590.
8. PIEKARZ, R.L., R. ROBEY, V. SANDOR, et al. 2001. Inhibitor of histone deacetylation, depsipeptide (FR901228), in the treatment of peripheral and cutaneous T-cell lymphoma: a case report. Blood **98:** 2865–2868.
9. BYRD, C., G. MARCUCCI, M.R. PARTHUN, et al. 2005. A phase I and pharmacodynamic study of depsipeptide (FK228) in chronic lymphocytic leukemia and acute myeloid leukemia. Blood **105:** 959–967.
10. ALTUCCI, L. & H. GRONEMEYER. 2001. The promise of retinoids to fight against cancer. Nat. Rev. Cancer **1:** 181–193.
11. DEGOS, L., H. DOMBRET, C. CHOMIENNE, et al. 1995. All-*trans*-retinoic acid as a differentiating agent in the treatment of acute promyelocytic leukemia. Blood **85:** 2643–2363.
12. LICHT, J.D., C. CHOMIENNE, A. GOY, et al. 1995. Clinical and molecular characterization of a rare syndrome of acute promyelocytic leukemia associated with translocation (11; 17). Blood **85:** 1083–1094.
13. KOSUGI, H., M. TOWARI, S. HATANO, et al. 1999. Histone deacetylase inhibitors are potent inducer/enhancer of differentiation in acute myeloid leukemia: a new approach to anti-leukemia therapy. Leukemia **13:** 1346–1324.
14. KITAMURA, K., S. HOSHI, M. KOIKE, et al. 2000. Histone deacetylase inhibitor but not arsenic trioxide differentiates acute promyelocytic leukaemia cells with t(11; 17) in combination with all-*trans*-retinoic acid. Br. J. Haematol. **108:** 696–702.
15. WARREL, R.P., J.R. HE, Z. RICHON, et al. 1998. Therapeutic targeting of transcription in acute promyelocytic leukemia by use of an inhibitor of histone deacetylase. J. Natl. Cancer Inst. **90:** 1621–1625.
16. COLLINS, S. 1987. The HL-60 promyelocytic leukemia cell line: proliferation, differentiation and cellular oncogene expression. Blood **70:** 1233–1244.
17. MERCILLE, S. & B. MASSIE. 1994. Induction of apoptosis in nutrient-deprived cultures of hybridoma and myeloma cells. Biotechnol. Bioeng. **44:** 1140–1154.
18. SAVICKIENE, J., G. TREIGYTE, K.E. MAGNUSSON, et al. 2005. p21 (Waf1/Cip1) and FasL gene activation via Sp1 and NFκB is required for leukemia cell survival but not for cell death induced by diverse stimuli. Intern. J. Biochem. Cell Biol. **37:** 784–796.
19. TREIGYTE, G. & A. GINEITIS. 1979. Specific changes in the biosynthesis and acetylation of nucleosomal histones in the early stages of embryogenesis of sea urchin. Exp. Cell Res. **121:** 127–134.
20. HURLEY, C.K. 1977. Electrophoresis of histones: a modified Panyim and Chalkley system for slab gels. Anal. Biochem. **80:** 624–626.
21. CHEUNG, P., K.G. TANNER, W.L. CHEUNG, et al. 2000. Synergistic coupling of histone H3 phosphorylation and acetylation in response to epidermal growth factor stimulation. Mol. Cell **5:** 905–915.
22. GLOSH, S. 1999. Regulation of inducible gene expression by the transcription factor NF-κB. Immunol. Res. **19:** 183–189.
23. GARTEL, A.L. & A.L. TYNER. 1999. Transcriptional regulation of the p21 (WAF1/CIP1) gene. Exp. Cell Res. **246:** 280–289.

24. DREXLER, H.G., H. QUENTMEIR, R.A.F. MACLEOD, et al. 1995. Leukemia cell lines: in vitro models for the study of acute promyelocytic leukemia. Leuk. Res. **19**: 681–691.
25. LEE, K.-H., M.-Y. CHANG, J.-I. AHN, et al. 2002. Differential gene expression in retinoic acid-induced differentiation of acute promyelocytic leukemia cells, NB4 and HL-60 cells. Biochem. Biophys. Res. Commun. **296**: 1125–1133.
26. GOPOSPE, M., C. CIRIELLI, X. WANG, et al. 1997. p21 (Waf1/Cip1) protects against p53-mediated apoptosis of human melanoma cells. Oncogene **14**: 929–935.
27. KUO, P.-C., H.-F. LIU & J.-I. CHAO. 2004. Survivin and p53 modulate quercetin-induced cell growth inhibition and apoptosis in human lung carcinoma cells. J. Biol. Chem. **279**: 55875–55885.
28. EL-DEIRY, W.S., J.W. HARPER, P.M. O'CONNOR, et al. 1994. WAF1/CIP1 is induced in p53-mediated G arrest and apoptosis. Cancer Res. **54**: 1169–1174.
29. HARPER, J.W., G.R. ADAMI, N. WEI, et al. 1993. The p21 Cdk-interacting protein Cip1 is a potent inhibitor of G1 cyclin-dependent kinases. Cell **75**: 806–816.
30. RICHON, V.M., T.W. SANDHOFF, R.A. RIFKIND, et al. 2000. Histone deacetylase inhibitor selectively induces p21 WAF1 expression and gene-associated histone acetylation. Proc. Natl. Acad. Sci. USA **97**: 10014–10019.
31. KIM, J.S., S. LEE, T.W. LEE & J.B. TREPEL. 2001. Transcriptional activation of p21 WAF1/CIP1 by apicidin, a novel histone deacetylase inhibitor. Biochem. Biophys. Res. Commun. **281**: 866–871.
32. GLICK, R.D., S.I. SWENDEMAN, D.C. COFFEY, et al. 1999. Hybrid polar histone deacetylase inhibitor induces apoptosis and CD95/CD95 ligand expression in human neuroblastoma. Cancer Res. **59**: 4392–4399.
33. KLISOVIC, D.D., S.E. KATZ, D. EFFRON, et al. 2003. Depsipeptide (FR901228) inhibits proliferation and induces apoptosis in primary and metastatic human uveal melanoma cell lines. Invest. Res. Vis. Sci. **44**: 2390–2398.
34. KWON, S.H., S.H. AHN, Y.K. KIM, et al. 2002. Apicidin, a histone deacetylase inhibitor, induces apoptosis and Fas/Fas ligand expression in human acute promyelocytic cells. J. Biol. Chem. **277**: 2073–2080.
35. KIKUCHI, H., R. ILIZUKA, R. SUGIYAMA, et al. 1996. Monocytic differentiation modulated apoptotic response to cytotoxic anti-Fas antibody and tumor necrosis factor α in human monoblast hematopoietic cells. J. Leukoc. Biol. **60**: 778–783.
36. SALIH, H.R., G.C. STARLING, S. BRANDL, et al. 2002. Differentiation of promyelocytic leukemia: alterations in Fas (CD95/Apo-1) and Fas ligand (CD178) expression. Br. J. Haematol. **117**: 76–85.
37. KOMAROV, P.G., E.A. KOMAROVA, R.V. KONDRATOV, et al. 1999. A chemical inhibitor of p53 that protects mice from the side effects of cancer therapy. Science **285**: 1733–1737.
38. KAJI, A., Y. ZHANG, M. NOMURA, et al. 2003. Pifithrin-α promotes p53-mediated apoptosis in JB6 cells. Mol. Carcinog. **37**: 138–148.
39. CULMSEE, C., X. ZHU, Q.S. YU, et al. 2001. A synthetic inhibitor of p53 protects neurons against death induced by ischemic and excitoxic insults, and amyloid beta-peptide. J. Neurochem. **77**: 220–228.
40. LIU, X., C.C. CHUA, J. GAO, et al. 2004. Pifithrin-α protects against doxorubicin-induced apoptosis and acute cardiotoxicity in mice. Ann. J. Physiol. Heart Circ. Physiol. **286**: H933–H939.
41. LORENZO, E., C. RUIZ-RUIZ, A.J. QUESADA, et al. 2002. Doxorubicin induces apoptosis and CD95 gene expression in human primary endothelial cells through a p53-independent mechanism. J. Biol. Chem. **277**: 10883–10892.

42. HAKE, S.B., A. XIAO & C.D. ALLIS. 2004. Linking the epigenetic "language" of covalent histone modifications to cancer. Br. J. Cancer **90:** 761–769.
43. CHAMBERS, A.E., S. BANERJEE, T. CHAPLIN, et al. 2003. Histone acetylation-mediated regulation of genes in leukaemic cells. Eur. J. Cancer **39:** 1165–1175.
44. HE, L., T. TOLENTINO, P. GRAYSON, et al. 2001. Histone deacetylase inhibitors induce remission in transgenic models of therapy-resistant acute promyelocytic leukemia. J. Clin. Invest. **108:** 1321–1330.
45. MAYO, M.W., C.E. DENLINGER, R.M. BROAD, et al. 2003. Ineffectiveness of histone deacetylase inhibitors to induce apoptosis involves the transcriptional activation of NF-κB through the Akt pathway. J. Biol. Chem. **278:** 18980–18989.
46. QUIVY, V., E. ADAM, Y. CALLOTE, et al. 2002. Synergistic activation of human immunodeficiency virus type I promoter activity by NF-κB and inhibition of deacetylases: potential perspectives for the development of therapeutic strategies. J. Virol. **76:** 11091–11103.
47. YIN, L., G. LAEVSKY & C. GIARDINA. 2002. Butyrate suppression of colonycyte NF-kappa b activation and cellular proteosome activity. J. Biol. Chem. **276:** 1714–1719.

Effect of Valproic Acid, a Histone Deacetylase Inhibitor, on Cell Death and Molecular Changes Caused by Low-Dose Irradiation

DARINA ZÁŠKODOVÁ,[a] MARTINA ŘEZÁČOVÁ,[a] JIŘINA VÁVROVÁ,[b] DORIS VOKURKOVÁ,[c] AND ALEŠ TICHÝ[b]

[a]*Department of Medical Biochemistry, Charles University in Prague, Faculty of Medicine in Hradec Králové, Šimkova 870, 500 01 Hradec Králové, Czech Republic*

[b]*Department of Radiobiology, Faculty of Military Health Sciences, University of Defense, Třebešská 1575, 500 01 Hradec Králové, Czech Republic*

[c]*Institute of Clinical Immunology and Allergology, Charles University in Prague, Faculty of Medicine in Hradec Králové and University Hospital, 500 05 Hradec Králové, Czech Republic*

ABSTRACT: Valproic acid (VA), a histone deacetylase inhibitor (HDACI), *in vitro* induces differentiation of promyelocyte leukemia cell (HL-60) and proliferation arrest and apoptosis of various leukemia cell lines. In MOLT-4 cells (human T lymphocyte leukemia) the cell cycle arrest is caused by 2 mM VA, while 4 mM VA induces mainly apoptosis. In our work we studied effect of VA on molecular mechanisms responsible for cell cycle arrest (2 mM VA) or apoptosis induction (4 mM VA). The aim of our article was to evaluate a cotreatment by low (cytostatic) concentrations of VA with ionizing radiation and an effect of this combination on apoptosis induction in tumor cells MOLT-4. We prove that 24-h long incubation with VA causes acetylation of histones H3 and H4 in concentration-dependent manners. During first hours after the beginning of cultivation with VA in both studied concentrations (2 and 4 mM) an increase of p53 and its phosphorylation on serine 392 is detected, as well as a phosphorylation of Mdm2 on serine 166. After 8 and 24 h after the beginning of cultivation with 2 mM VA we detect p21, which is not observed after exposure to 4 mM VA. Cleavage of lamin B to 46 kDa fragment as an indicator of apoptosis was apparent after 24-h long incubation with 4 mM VA. In this article we prove radiosensitizing effect of VA. After 3-days long cultivation of cells with 2 mM VA the D_0 value decreased from 0.7 to 0.2 Gy. Also the EC70 value fell from 0.97 to 0.38 mM

Address for correspondence: Darina Záškodová, Department of Medical Biochemistry, Charles University in Prague, Faculty of Medicine in Hradec Králové, Šimkova 870, 500 01 Hradec Králové, Czech Republic. Voice: +420-495-816-166; fax: +420-495-512-715.
e-mail: zaskodovad@lfhk.cuni.cz

when the cells were irradiated with a dose of 1 Gy before the continual cultivation with VA. Continual cultivation of MOLT-4 cells irradiated by the dose of 1 Gy with VA caused during 14 days after irradiation significant increase of apoptotic cells in comparison to the cells exposed to only one factor. As a conclusion it can be postulated that continual exposure of MOLT-4 cells to VA increases apoptosis and decreases colony-forming capacity of the cells irradiated with small dose of radiation.

KEYWORDS: valproic acid; ionizing radiation; apoptosis; p53; Mdm2; p21

INTRODUCTION

Valproic acid (VA) (2-propylpentanoic acid) is a short chain fatty acid with anticonvulsant properties. Moreover, during the past years it has become evident that VA possesses antitumor properties of an inhibitor of histone deacetylases (HDAC).[1]

Histone acetyltransferases and HDAC are two antagonistic enzymatic groups responsible for chromatin remodeling. In transcriptionally silent cells, the nuclear histones are hypoacetylated on lysine residues of their NH_2 terminal tails, so that the chromatin forms a compact mass inaccessible for transcription factors. Whereas the acetylation on the lysine residues neutralizes the positive charge, which makes the chromatin unfolded and the transcription can proceed. Histone acetylation constitutes a key mechanism controlling chromatin remodeling and gene regulation. Thus the histone deacetylases inhibitors (HADCI) influence the expression of approximately 2% of genes, especially genes important for cell cycle regulation and induction of differentiation and/or apoptosis.

Three days long incubation of leukemic cells HL-60 and MOLT-4 with VA inhibits proliferation, decreases number of cells in S phase of the cell cycle and induces apoptosis.[2] EC50 value measured by colony-forming efficiency is 1.8 mM VA for both leukemic cell lines.

Also ionizing radiation is an important apoptosis inductor at leukemic cell lines MOLT-4 and HL-60. In contrary to the HL-60 cells (which are p53 negative), the MOLT-4 cells express functional p53 and exhibit greater radiosensitivity.[3,4] It is generally accepted that ionizing radiation causes double-strand break (DSB) of DNA, which trigger Ataxia-telangiectasia mutated kinase (ATM kinase) activation.[5] Activated ATM kinase triggers phosphorylation of other targets, mainly p53, Mdm2, Chk1, and Chk2. These phosphorylations are followed by further processes related to cell cycle arrest and reparation of radiation-induced damage or apoptosis induction. During the first hours (2–6 h) after the irradiation by a high lethal dose of 7.5 Gy we proved accumulation of p53 and its phosphorylation on serine 15 and serine 392. The p21 upregulation followed (4–6 h) the p53 phosphorylation process.[3]

It is known from literature[6] that accumulation of p21 appears after treatment of tumor cells by HDACI and that this accumulation is p53 independent, as it was detected also at cells lacking p53.[7,8] In unstressed cells, p53 is maintained at low levels through target degradation by its key negative regulator Mdm2.[9] Mdm2 is an E3 ligase that ubiquitylates a defined set of lysine residues at the C terminus of p53 and thus triggers its rapid degradation by proteasomes or promotes its nuclear export.[10]

It is presumed that apoptosis induced in tumor cells after exposure to HDACI is mediated by mitochondrial pathway. Kawagoe et al.[11] described effect of VA on apoptosis induction at leukemic cell line MV411 and they found that VA induces apoptosis by initiation of mitochondrial pathway through cytochrome c release from mitochondria and further activation of caspase 8, 9, and 3. A caspase inhibitor, zVAD-FMK, inhibited the DNA fragmentation by VA but not cell death.

In our study we evaluated effect of cytostatic (2 mM) and cytotoxic (4 mM) concentration of VA on posttranslational modifications of histones H3 and H4 and proteins p53 and Mdm2. While acetylation of H3 and H4 and phosphorylation of p53 and Mdm2 was detected after both concentrations, p21 was upregulated only after cytostatic (2 mM) concentration.

Because of many tumor cells exhibit aberrant regulation of gene expression and primary activity of HDACI is to regulate gene expression, one of possible approaches in antitumor therapy is their combination with other antitumor agents.[12] Rosato and co-worker[12] proved synergic interactions between HDACI SAHA (suberoylanilide hydroxamic acid) or sodium butyrate (NaB) and TRAIL (tumor necrosis apoptosis inducing ligand) at leukemic cell lines HL-60, Jurkat, and U937. Simultaneous administration of TRAIL and HDACI potently induces mitochondrial damage, caspase activation and apoptosis in human leukemia cells. Kim et al.[13] proved dose-dependent reduction in survival and radiosensitization with another HDACI—TSA (trichostatin A) treatment in human glioblastoma cells, where TSA decreased clonogenical survival of the cells when applied before as well as after the irradiation. Munshi et al.[14] described radiosensitizing effect of NaB at two melanoma cell lines; this effect was not observed at normal fibroblasts. Radiosensitizing effect of NaB is caused mainly by decrease of reparation capacity of tumor cells.

The aim of our article was to determine whether low cytostatic concentrations of VA can potentiate radiation-caused apoptosis induction at T lymphocyte cell line MOLT-4.

MATERIALS AND METHODS

Cell Cultures and Culture Conditions

The human T lymphocyte leukemia cells MOLT-4 from American Type Culture Collection (ATCC, Manassas, VA) have been cultured in Iscove's modified

Dulbecco's medium (Sigma-Aldrich s.r.o., Prague, Czech Republic) supplemented with a 20% fetal calf serum in a humidified incubator at 37°C and a controlled 5% CO_2 atmosphere. The cultures have been divided every second day by a dilution to a concentration of 2×10^5 cells/mL. The cell counts have been performed with a hemocytometer, the cell membrane integrity has been determined by using the Trypan blue exclusion technique. The cell lines in the maximal range of up to 20 passages have been used for this study.

Gamma Irradiation

The exponentially growing MOLT-4 cells have been suspended at a concentration of 2×10^5 cells/mL in a complete medium. Aliquots of 10 mL have been plated into 25 cm^2 flasks (Nunc GmbH & Co., Wiesbaden, Germany) and irradiated using a ^{60}Co γ-ray source with a dose-rate of 0.4 Gy/min. After the irradiation the flasks have been placed in a 37°C incubator with 5% CO_2 and the aliquots of the cells have been removed at various times after irradiation for analysis. The cells have been counted and cell viability was determined with the Trypan blue exclusion assay.

VA

VA (Sigma-Aldrich) has been added into the cultivation flask for various time (1–120 h) in a final concentration of 0.5–10 mM. The VA (sodium salt, Sigma-Aldrich) was dissolved in phosphate-buffered saline solution (PBS) to a stock concentration of 100 mM and stored at $-20°C$.

In Vitro *Clonogenic Survival Assay*

The survival curves have been generated using an *in vitro* clonogenic assay. The untreated control (10^2/mL) and the irradiated and treated MOLT-4 cells (10^2–10^5/mL) have been mixed in the Iscove's modified Dulbecco's medium supplemented with a 0.9% methylcellulose, a 30% fetal calf serum. A total of 1 mL of the plating mixture has been dispersed into 35 mm tissue culture Petri dishes. The colonies (containing 40 or more cells) have been counted after 14 days of the incubation in 5% CO_2 and 5% O_2 at 37°C and the curves have been generated. All semisolid cultures have been performed in duplicates. Two independent experiments (four measurements) have been performed.

CD7 Antibody, Apoptosis Detection

For apoptosis detection we used APOPTEST-FITC (Dako, Brno, Czech Republic). During apoptosis, cells expose phosphatidylserine at the cell surface. Annexin V is a phospholipid binding protein, which in the presence of calcium ions binds selectively and with high affinity to phosphatidylserine. For detection of cell surface markers in MOLT-4 cells we used PE-conjugated anti-human CD7-PE (8H8.1, IgG2b-IM1429) 49) obtained from Immunotech (Marseille, France).

Flow Cytometric Analysis

The ability of the cells to scatter light in a forward direction (FS) correlates with cell volume, while their ability to scatter light in side direction (SS) correlates with cell granularity. We analyzed VA-treated and irradiated cells for changes in the intensity of FS and SS as compared to untreated control cells. Flow cytometric analysis was performed on a Coulter Epics XL flow cytometer equipped with a 15mW argon-ion laser with excitation capabilities at 488 nm (Coulter Electronic, Hialeah, FL). A minimum of 10,000 cells was collected for each 2-color sample in a list mode file format. List mode data was analyzed using Epics XL System II software (Coulter Electronic).

Western Blotting

At various time after VA treatment and radiation the MOLT-4 cells have been washed with a PBS. The lysates containing an equal amount of protein (30 μg) have been loaded into an each lane of a polyacrylamide gel. After electrophoresis, the protein has been transferred to a polyvinylidene fluoride (PVDF) membrane. The membranes were blocked in Tris-buffered saline containing 0.05% Tween 20 and 5% nonfat dry milk and then incubated with primary antibody (p21–Sigma-Aldrich; acetylated histone H3 and H4, p53— Cell Signaling Technology, Inc., Boston, MA; p53 phosphorylated at serine 392 and lamin B—Calbiochem-Merck Biosciences, Darmstadt, Germany) at 4°C overnight. After washing, the blots were incubated with secondary antibody (Dako) and the signal was developed with a chemiluminiscence (ECL) detection kit (Boehringer, Boehring Ingelneim, Germany).

Statistical Analysis

The results have been statistically evaluated with a Student's t-test. The values represent the mean ± SD (a standard deviation of the mean). The statistical significance of the difference of means in comparable sets is indicated.

RESULTS

Clonogenity

The cells have been cultivated with 2 mM VA for 3 days before the irradiation and 14 days later we determined the colony-forming efficiency. The D_0 value (the dose reducing cell survival to 37%) decreased from 0.7 Gy (control) to 0.2 Gy (precedent treatment with 2 mM VA) (FIG. 1 A).

The cells were irradiated with the dose of 1 Gy and the colony-forming efficiency in the continual presence of VA during 14 days was established. The EC70 value (the concentration after which 70% of cells retain their clonogenic capacity and form colonies) decreased from 0.97 mM in the case of cultivation with alone VA to 0.38 in the case of treatment with precedent irradiation (FIG. 1 B).

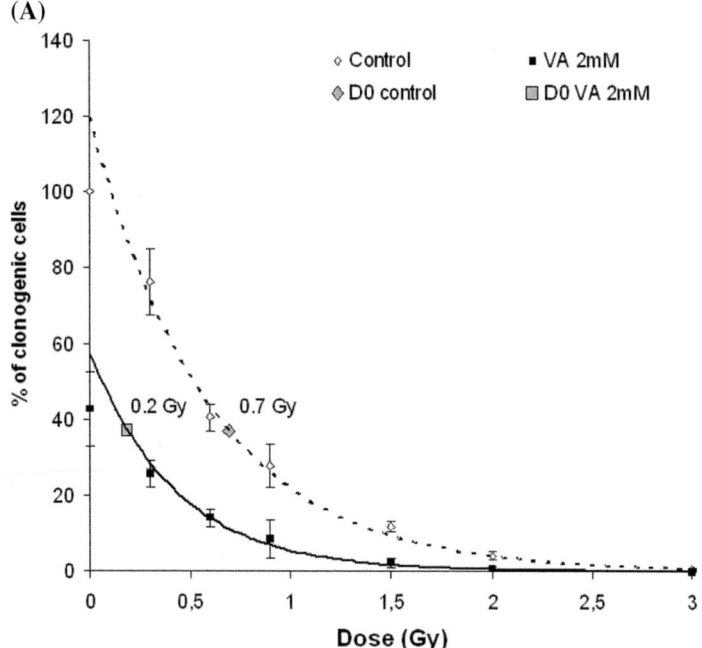

FIGURE 1. (**A**) The effect of 72-h-long incubation with 2 mmol/L VA and consecutive irradiation on clonogenic survival of MOLT-4 cells. For the clonogenic survival data, each point is a mean of four measurements from two experiments ± SD. The clonogenicity test has been started immediately after the end of the irradiation. Equations of generated curves: Control: $y = 119.54e^{-1.6941x}$ ($R^2 = 0.9936$); VA: $y = 57.348e^{-2.3473x}$ ($R^2 = 0.9868$). (**B**) The effect of irradiation by the dose of 1 Gy on clonogenic survival of MOLT-4 cells in continuous presence of VA. For the clonogenic survival data, each point is a mean of four measurements from two experiments ± SD. The clonogenicity test has been started immediately after the end of the irradiation and the cells were cultivated in medium containing VA. Equations of generated curves: Control: $y = -0.8304x^3 + 9.8211x^2 - 37.134x + 42.75$ ($R^2 = 1$); 1 Gy: $y = 4.5384x^3 + 12.842x^2 - 87.525x + 98.707$ ($R^2 = 0.9921$).

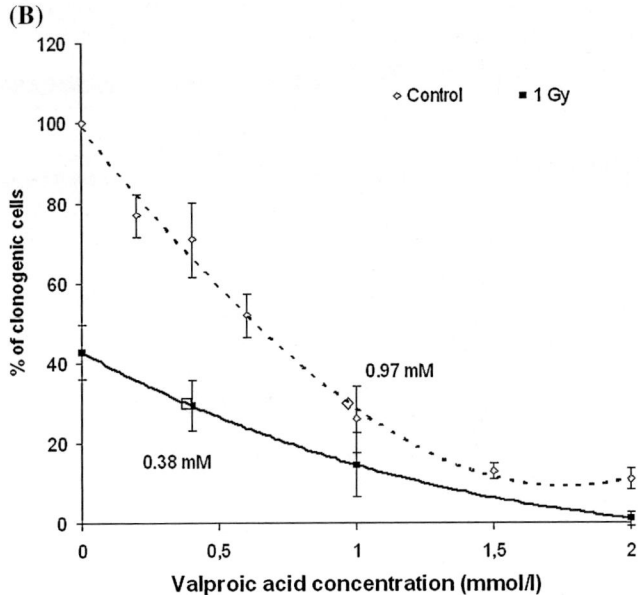

FIGURE 1. Continued.

Electrophoresis and Western Blotting

FIGURE 2 shows changes in protein expression after treatment with VA. The addition of 2 mM and 4 mM VA to MOLT-4 culture provoked in both the cases elevated expression of protein p53 in wild form as well as its phosphorylation at serine 392 in a short period of treatment. Elevated level of protein p53 induced consequently the upregulation of p21, but only after treatment with 2 mM, when probably the injury is not lethal and the cell undergoes repair processes during arrest in G1/S phase. After treatment with 4 mM VA we did not record increased p21 expression, the damage is more extensive and the cell seems to start apoptosis directly without any effort of reparation.

In a later period the acetylation of nuclear histones H3 and H4 proceeded. Histone H4 was more sensitive to the action of VA, its acetylation occurred already 24 h after 0.5 mM VA addition. Histone H3 required at least 2 mM concentration for 24 h to be acetylated (FIG. 3). The cleavage of lamin B as a proof of apoptosis occurred after 4 mM VA treatment, a fragment with molecular weight 46 kDa appeared 24 h after the beginning of incubation (FIG. 2).

Flow Cytometric Analysis

Both irradiation of MOLT-4 cells by the dose of 1 Gy and incubation with 0.5 mM VA caused apoptosis induction and increase of number of cells of

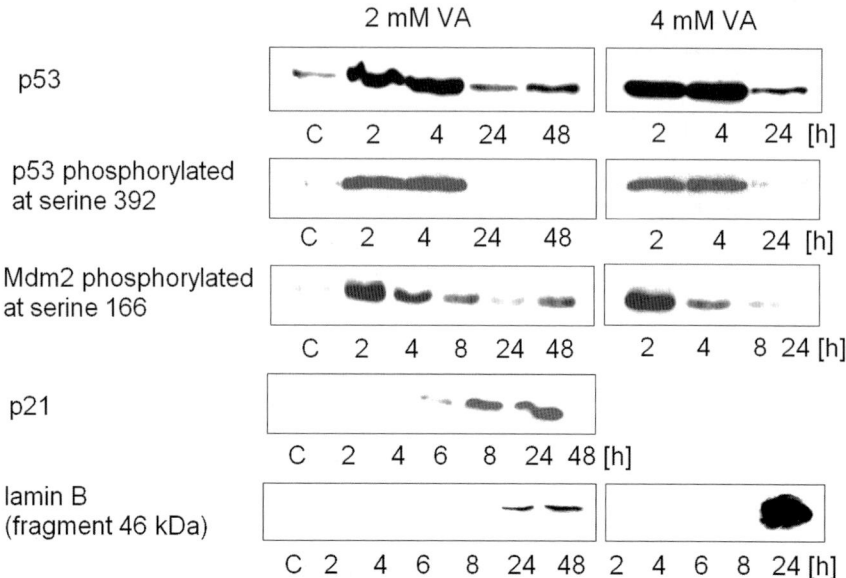

FIGURE 2. The effect of VA on the expression of protein p53 (wild form as well as phosphorylation at serine 392) and induction of protein p21. MOLT-4 cells were incubated with 2 mM and 4 mM VA for indicated periods, washed with PBS and lysed. SDS-PAGE and Western blotting were performed.

smaller size and increased granularity. After the irradiation maximal increase of this subpopulation could be seen on days 1 and 3, during later intervals the number of cells with low SS and high FS significantly decreased back to control level. Incubation with VA caused slow increase of number of apoptotic cells with maximum on day 14. In the case of combined effect of irradiation and VA synergic effect of both factors could be seen (FIG. 4).

It is apparent from the figure that the combination of irradiation by the dose of 1 Gy and 0.5 mM VA had synergic effect from the point of view of apoptosis induction. While during first 3 days after irradiation the main inductor of apoptosis was ionizing radiation, on day 6 both factors contributed equally to apoptosis induction and on day 14 most of the cells in group treated by combination of 1 Gy + 0.5 mM VA were apoptotic. (FIG. 5 A and B).

DISCUSSION

In our study we proved that combination of VA and ionizing radiation decreases clonogenical survival of the leukemic cells when VA is applied before as well as after the irradiation. During continuous cultivation (14 days) with VA after the irradiation even relatively low concentrations of VA were effective. During 14-days long cultivation of MOLT-4 cells with 0.5 mM VA after the

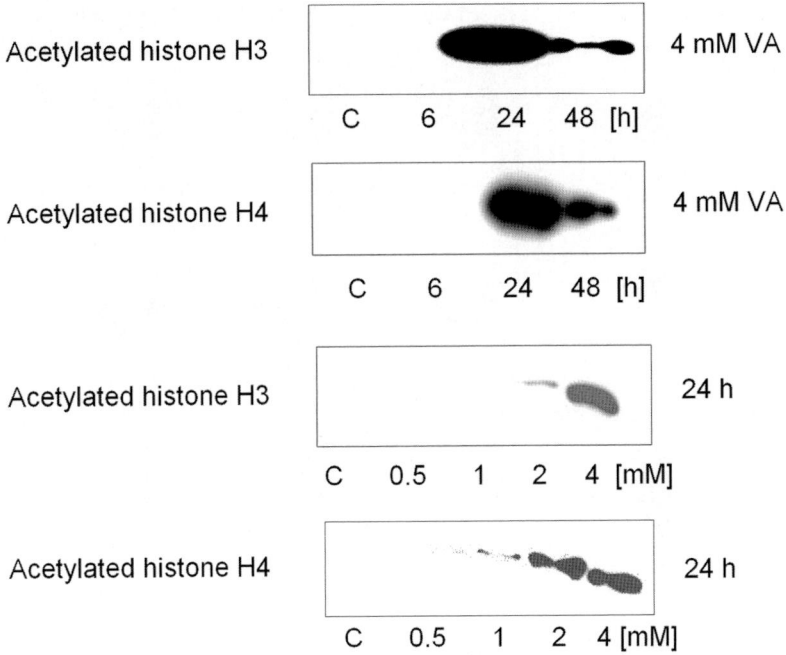

FIGURE 3. The time and dose response of acetylation of nuclear histones H3 and H4. MOLT-4 cells were treated with 4 mM VA for indicated period (time response) and with different concentrations of VA for 24 h (dose response). Immunoblots were performed.

irradiation by the dose of 1 Gy apoptosis induction is significantly increased in comparison to only irradiated cells, where 14 days after the irradiation by the dose of 1 Gy only live cells are detected. Similar effect was described by Kim et al.[13] during their studies of combined effect of TSA and ionizing radiation at glioma cells.

Mechanism of joint action of VA and ionizing radiation still remains an open question—it can include interactions with repair mechanisms or transduction cascade signals. VA chemically belongs to the group of short chain carboxylic acids, such as butyrate or phenylbutyrate. Similarly as most of HDACI[15] VA causes acetylation of histones H3 and H4.[16] Also in our work with leukemic cells we proved this acetylation, but not before 24-h long incubation, while increased levels of p53 and phosphorylation of p53 on serine 392 we detected already 2 and 4 h after the beginning of the incubation. A total of 24 h after the beginning of incubation the changes of p53 disappeared. Changes of p53 therefore precede changes observed on histones. These results were agreeing for both concentrations of VA—2 and 4 mM.

Under normal circumstances p53 is quickly degraded by ubiquitination catalyzed by Mdm2 protein. It seems that phosphorylations of p53 caused by

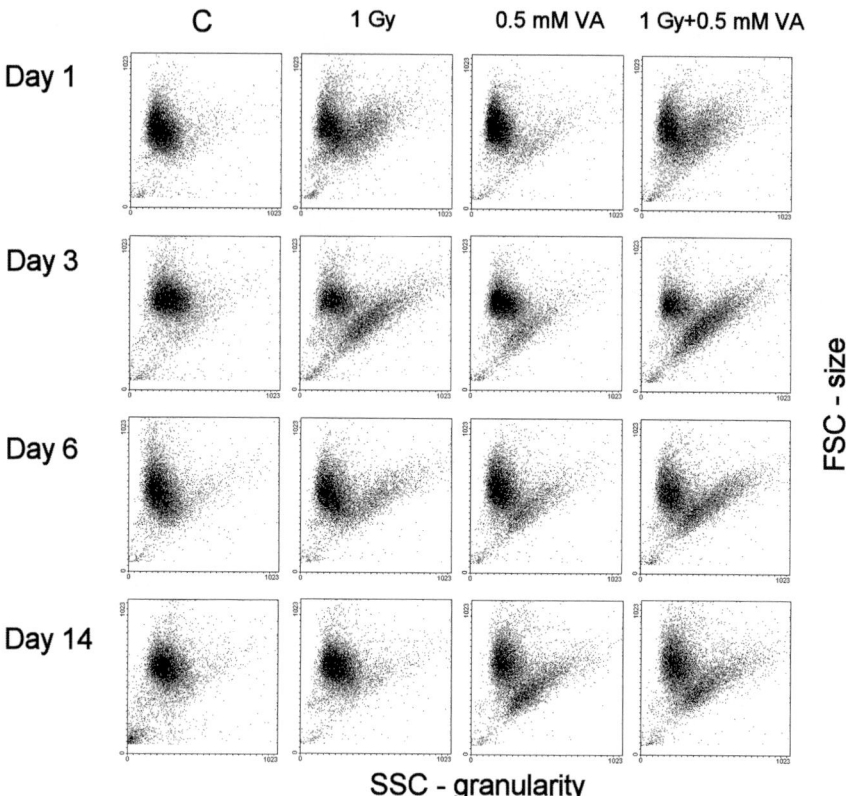

FIGURE 4. Flow cytometric analysis of light scattering properties of MOLT-4 leukemia cells. Intensity of light scattered in forward direction correlates with cell size, side scatter correlates with granularity. Aboundance of apoptic cells reaches the maximum from 3–6 days after irradiation with the dose of 1 Gy, in cells treated with 0.5 mM VA the peak appears later, in the case of cells treated with combination, the effects of both the two insults are potentiated.

ionizing radiation prevent binding of Mdm2 to p53 and therefore decrease p53 degradation and simultaneously they enable p53 acetylation. Acetylation protects p53 from export into the cytoplasm and from degradation, as it blocks ubiquitin binding sites.[17] Pretreatment of cells with TSA presumably enhances the p53 mechanism and radiosensitizes tumor cells.[13] Contrarywise, presence of Mdm2 induces HDAC1 activity, which cleaves acetyl residues from the ubiquitin binding sites, which enables p53 degradation.[10] In our experiments we evaluated activating phosphorylation of Mdm2 on serine 166 and we obtained similar results as in the case of phosphorylation of p53.

At different tumor cells, even those lacking functional p53, an increase of p21 was found 6–24 h after beginning of HDACI treatment.[7,8] Also at MOLT-4

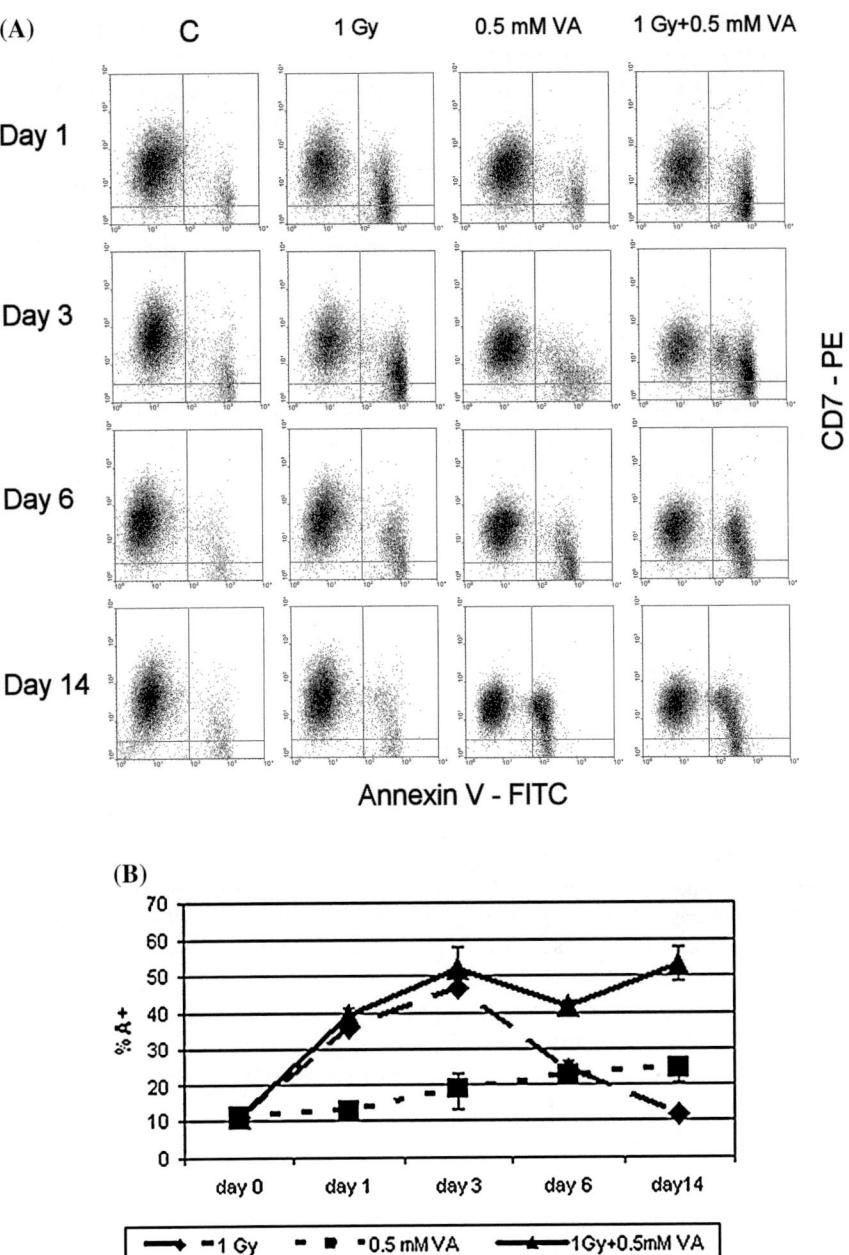

FIGURE 5. (A, B): Percentage of annexin V positive cells. MOLT-4 cells were irradiated with 1 Gy or treated with 0.5 mM VA or with combination of both factors. In indicated time periods cells of three groups were analyzed by flow cytometric analysis.

cells after incubation with 2 mM VA expression of p21 was observed already 6 h after the beginning of the cultivation. The increase of p21 reached its maximum after 24 h of treatment, after 48 h it was normalized again, and no p21 was detected during incubation with 4 mM VA. Importance of p21 is in cell cycle arrest in G1 phase and is therefore related to cytostatic effect VA. Longer time of cultivation or higher doses of VA induce apoptosis, which is p21 independent. The question remains, whether the p21 expression after treatment of leukemic cells by VA is related to increased expression and phosphorylation of p53, as it was also proved after HDACI treatment of cells lacking p53.[18,19]

Munshi et al.[14] proved acetylation of histone H4 after exposure to NaB in both, melanoma cell lines and normal fibroblasts. However, a radiosensitizing effect of NaB they proved only at melanoma cell lines. Therefore it seems that radiosensitizing effect is not dependent on histone acetylation status. They also do not link radiosensitizing effect with changes of cell cycle distribution and in p21 expression, which are apparent in both, melanoma cells (p53 mutant) and normal fibroblasts (p53 wild type). However, in melanoma cell lines they proved increase of proapoptic protein Bax, which indicates amplification of mitochondrial pathway of apoptosis induction after combined effect of HDACI and ionizing radiation.

Reparation processes proceeding in response to ionizing radiation are of great importance in clonogenic survival of tumor cells after combined influence of HDACI and ionizing radiation. Some results suggest[14] that HDACI decrease reparative capacity of tumor cells. Phosphorylation of histone H2AX appears very early after irradiation at many tumor and nontumor cells. NaB alone does not cause phosphorylation of H2AX. When these cells are irradiated by the dose of 2 Gy after 24-h long preincubation with NaB, the phosphorylation of histone H2AX 30 min after irradiation is increased and also the duration of the phosphorylation is extended,[14] which proves decreased ability of the tumor cells to repair DSB. Some papers prove[6] increase of nuclear clusterin after incubation with HDACI. Nuclear clusterin is an ionizing radiation-inducible protein that binds Ku-70 (protein important in nonhomologous end joining reparation) and triggers apoptosis when overexpressed in tumor cells.[20] Thus it seems that radiosensitizing effect of HDACI is closely related to molecular mechanisms which decrease ability of the cells to repair DSB caused by ionizing radiation.

ACKNOWLEDGMENTS

The authors would like to thank Grant Agency of Czech Republic (Grant Project No. 202/04/0598) and Ministry of Education of Czech Republic (Project No. MSM 0021620820) for financial support.

REFERENCES

1. BLAHETA, R.A. & J. CINATL, JR. 2002. Anti-tumor mechanisms of valproate: a novel role for an old drug. Med. Res. Rev. **22:** 492–511.
2. ŘEZÁČOVÁ, M., J. VÁVROVÁ, D. VOKURKOVÁ & D. ZÁŠKODOVÁ. 2006. Effect of valproic acid and antiapoptic cytokines on differentiation and apoptosis induction of human leukemia cells. Gen. Physiol. Biophys.**25(1):** 65–79.
3. SZKANDEROVÁ, S., J. VÁVROVÁ, M. ŘEZÁČOVÁ, et al. 2003. Gamma irradiation results in phosphorylation of p53 at serine-392 in human T-lymphocyte leukaemia cell line MOLT-4. Folia. Biol. (Praha) **49:** 191–196.
4. VÁVROVÁ, J., M. MAREKOVÁ & D. VOKURKOVÁ. 2001. Radiation-induced apoptosis and cell cycle progression in TP53-deficient human leukemia cell line HL-60. Neoplasma **48:** 26–33.
5. BAKKENIST C.J. & M.B. KASTAN. 2003. DNA damage activates ATM through intermolecular autophosphorylation and dimer dissociation. Nature **421:** 499–506.
6. GLASER, K.B., M.J. STAVER, J.F. WARING, et al. 2003. Gene expression profiling of multiple histone deacetylase (HDAC) inhibitors: defining a common gene set produced by HDAC inhibition in T24 and MDA carcinoma cell lines. Mol. Cancer Ther. **2:** 151–163.
7. GUI, C.-Y., L. NGO, W.S. XU, et al. 2004. Histone deacetylase (HDAC) inhibitor activation of p21^{WAF} involves changes in promoter-associated proteins, including HDAC1. Proc. Natl. Acad. Sci. USA **101:** 1241–1246.
8. RICHON, V.M., T.W. SANDHOFF, R.A. RIFKIND & P.A. MARKS. 2000. Histone deacetylase inhibitor selectively induces p21^{WAF1} expression and gene-associated histone acetylation. Proc. Natl. Acad. Sci. USA **97:** 10014–10019.
9. APPELLA, E. & C.W. ANDERSON. 2001. Post-translation modifications and activation of p53 by genotoxic stresses. Eur. J. Biochem. **268:** 2764–2772.
10. ITO, A., Y. KAWAGUCHI, C.-H. LAI, et al. 2002. MDM2-HDAC1-mediated deacetylation of p53 is required for its degradation. EMBO J. **21:** 6236–6245.
11. KAWAGOE, R., H. KAWAGOE & K. SANO. 2002. Valproic acid induces apoptosis in human leukemia cells by stimulating both caspase-dependent and independent apoptic signaling pathways. Leuk. Res. **26:** 495–502.
12. ROSATO, R. & S. GRANT. 2004. Histone deacetylase inhibitors in clinical development. Expert Opin. Investig. Drugs **13:** 21–38.
13. KIM, J.H., J.H. SHIN & I.H. KIM. 2004. Susceptibility and radiosensitization of human glioblastoma cells to trichostatin A, a histone deacetylase inhibitor. Int. J. Radiat. Oncol. Biol. Phys. **59:** 1174–1180.
14. MUNSHI, A., J.F. KURLAND, T. NISHIKAWA, et al. 2005. Histone deacetylase inhibitors radiosensitize human melanoma cells by suppressing DNA repair aktivity. Clin. Cancer Res. **11:** 4912–4922.
15. YOSHIDA, M., M. KIJIMA, M. AKITA & T. BEPPU. 1990. Potent and specific inhibition of mammalian histone deacetylase both *in vivo* and *in vitro* by trichostatin A. J. Biol. Chem. **265:** 17174–17179.
16. GOTTLICHER, M., S. MINUCCI, P. ZHU, et al. 2001. Valproic acid defines a novel class of HDAC inhibitors inducing differentiation of transformed cells. EMBO J. **20:** 6969–6978.
17. CHEHAB, N.H., A. MALIKZAY, E.S. STAVRIDI & T.D. HALAZONETIS. 1999. Phosphorylation of Ser-20 mediates stabilization of human p53 in response to DNA damage. Proc. Natl. Acad. Sci. USA **96:** 13777–13782.

18. VRANA, J.V., R.H. DECKER, C.R. JOHNSON, et al. 1999. Induction of apoptosis in U937 human leukemia cells by suberoylanilide hydroxamic acid (SAHA) proceeds through pathways that are regulated by Bcl-2/Bcl-X_L, c-Jun, and p21^{CIP1}, but independent of p53. Oncogene **18:** 7016–7025.
19. BLAGOSKLONNY, M.V., R. ROBEY, D.L. SACKETT, et al. 2002. Histone deacetylase inhibitors all induce p21 but differetially cause tubulin acetylation, mitotic arrest and cytotoxicity. Mol. Cancer Ther. **1:** 937–941.
20. LESCOV, K.S., D.Y. KLOKOV, J. LI, et al. 2003. Synthesis and functional analyses of nuclear clusterin, a cell death protein. J. Biol. Chem. **278:** 11590–11600.

Protein Folding Information in Nucleic Acids which Is Not Present in the Genetic Code

JAN C. BIRO

Homulus Foundation, San Francisco, California 94195, USA

ABSTRACT: Nucleic acid subsequences comprising the 1st and/or 3rd codon residues in mRNAs express significantly higher free folding energy (FFE) than the subsequence containing only the 2nd residues ($P < 0.0001$, $n = 81$). This periodic FFE difference is not present in introns. The FFE in the 1st and 3rd residues is additive, which suggests that these residues contain a significant number of complementary bases and contribute to selection for local mRNA secondary structures. This periodic, codon-related structure forming of mRNAs indicates a connection between the structure of exons and the corresponding (translated) proteins. The folding energy dot plots of RNAs and the residue contact maps of the coded proteins are indeed similar. Residue contact statistics using 81 different protein structures confirmed that amino acids that are coded by partially reverse and complementary codons (Watson–Crick base pairs at the 1st and 3rd codon positions and translated in reverse orientation) are preferentially colocated in protein structures.

KEYWORDS: codon; translation; protein folding; RNA folding; specific protein interaction; complementarity; protein design; Anfinsen; protein structure; folding energy

INTRODUCTION

The protein folding problem has been one of the grand challenges in computational molecular biology. The problem is to predict the native three-dimensional (3D) structure of a protein from its amino acid sequence. It is widely believed that the amino acid sequence contains all the necessary information to make up the correct 3D structure, because protein folding is apparently thermodynamically determined; that is, given a proper environment, a protein will fold up spontaneously. This is called Anfinsen's thermodynamic principle.[1]

Address for correspondence: Jan C. Biro, 88 Howard, #1205, San Francisco, CA 94195. Voice: +1-414-777-1443; fax: +1-415-777-1443.
e-mail: jan.biro@sbcglobal.net

The thermodynamic principle has been confirmed many times on many different kinds of proteins *in vitro*. Critics says that the *in vivo* chemical conditions are different from those *in vitro*, the correct folding is determined by interactions with other molecules (chaperons, hormones, substrate, etc.) and protein folding is much more complex than renaturation of denatured poly amino acids. The fact that many naturally occurring proteins fold reliably and quickly to their native state, despite the astronomical number of possible configurations, has come to be known as Levinthal's Paradox.[2]

Anfinsen's principle was formulated in the 1960s using purely chemical experiments and a lot of intuition. Today, we have a lot of sequences and structures available to establish a logical and understandable link between sequence, structure, and function. But it is still not possible to correctly predict the structure (or a range of possible structures) purely from the sequence, *ab initio* and *in silico*.[3]

There are two potential, external sources of additional and specific protein folding information: (*a*) the chaperons (other proteins that assist in the folding of proteins and nucleic acids[4]; and (*b*) the protein coding nucleic acid sequences themselves (which are templates of the protein syntheses, but are not defined as chaperons).

The idea that the nucleotide sequence itself could modulate translation and hence affect cotranslational folding and assembly of proteins has been investigated in a number of studies.[5–7] Studies on the relationships between synonymous codon usage and protein secondary structural units are especially popular.[8–10] The genetic code is redundant (61 codons code 20 amino acids) and as many as 6 synonymous codons can code the same amino acid (Arg, Leu, Ser). The "wobble" base has no effect on the meaning of most codons but still the codon usage (wobble usage) is not randomly defined[11,12] and there are well-known, stable species-specific differences in the codon usage. It seems to be logical to search for some meaning (biological purpose) of the wobble bases and try to associate them with protein folding.

Another observation concerning the code redundancy dilemma is that there is a widespread selection (preference) for local RNA secondary structure in protein coding regions.[13] A given protein can be encoded by a large number of distinct mRNA species, potentially allowing mRNAs to simultaneously optimize desirable RNA structural features in addition to their protein coding function. The immediate question is whether there is some logical connection between the possible, optimal RNA structures and the possible, optimal biologically active protein structures.

MATERIALS AND METHODS

Single-stranded RNA molecules can form local secondary structures through the interactions of complementary segments. Watson–Crick (WC) base pair formation lowers the average free energy, dG, of the RNA and the magnitude of change is proportional to the number of base pair formations. Therefore the free folding energy (FFE) is used to characterize the local complementarity of nucleic acids.[13] The free folding energy is defined as FFE = (dG_{shuffled} − dG_{native})/L × 100, where L is the length of the nucleic acid, that is, free energy difference between native and shuffle (randomized) nucleic acids per 100 nucleotides. Higher positive values indicate stronger bias toward secondary structure in the native mRNA, and negative values indicate bias against secondary structure in the native mRNA.

We used a nucleic acid secondary structure predicting tool, the *mfold*[14] to obtain dG values and the lowest dG was used to calculate the FFE. The mfold also provided the folding energy dot plots, which are very useful to visualize the energetically most favored structures in a two-dimensional (2D) matrix.

A series of JAVA tools were used: SeqX to visualize the protein structures in 2D as amino acid residue contact maps[15]; SeqForm for selection of sequence residues in predefined phases (every third in our case).[16] Structural data were downloaded from PDB,[17] NDB,[18] and from a wobble base oriented database called Integrated Sequence–Structure Database (ISSD).[19]

Structures were generally randomly selected regarding species and biological function (a few exceptions are mentioned in the Results). Care was taken to avoid very similar structures in the selections. A propensity for alpha helices was monitored during selection and structures with very high and very low alpha helix content were also selected to make sure of a wide range of structural representation.

Linear regression analyses and Student's t-tests were used for statistical analyses of the results.

RESULTS

A selection of 81 different protein structures, the corresponding protein, and coding sequences were used to these observations. These 81 proteins were represented different (randomly selected) species and different (also randomly selected) protein functions and therefore the results might be regarded as more generally valid. The propensity of different secondary structure elements was recorded (as annotated in different databases). The proportion of alpha helices varied from 0 to 90% in the 81 proteins and showed a significant negative correlation to the proportion of beta sheets (not shown). The coding sequences were phase separated by SeqForm into three subsequences, each containing only the 1st, 2nd, or 3rd letters of the codons. Similar phase separation was made for intronic sequences immediately before and after the exon. There are, of course, no known codons in the intronic sequences, therefore we continued the same phase that we applied for the exon, assuming that this kind of selection is correct and maintained the name of the phase denotation even for noncoding

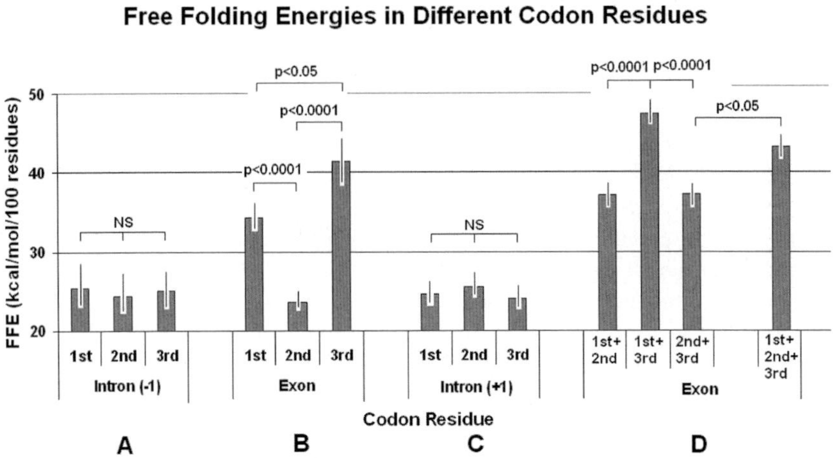

Free Folding Energies in Different Codon Residues

FIGURE 1. FFEs in different codon residues. FFEs were determined in phase-selected subsequences of 81 different genes. The original nucleic acids contained the intact three-letter codons (1st + 2nd + 3rd). Subsequences were constructed by periodical removal of one letter from the codon and maintaining the other two (1st + 2nd, 1st + 3rd, 2nd + 3rd) or removing two letters and maintaining only one (1st, 2nd, 3rd). Distinction was made between exons (B and D) and the preceding (−1, A) and following (+1, C) sequences (introns). The dG values were determined by *mfold* and the FFE was calculated. Each bar represents the mean ± SEM, $n = 81$.

regions. Subsequences corresponding to the 1st and 3rd codon letters in the coding regions had significantly higher FFEs than subsequences corresponding to the 2nd codon letters. No such difference was seen in noncoding regions (FIG. 1 A–C).

Higher FFE in subsequences of 1st and 3rd codon residues than in the 2nd indicates the presence of a larger number of complementary bases at the right positions of these subsequences. However, this might be the case only because the first and last codons form simpler subsequences and contain longer repeats of the same nucleotide than the 2nd codons. This would not be surprising for the 3rd (wobble) base but would not be expected for the 1st residue, even though it is known that the central codon letters are the most important to distinguish between amino acids (as shown in the in the *Common Periodic Table of Codons and Amino Acids*[20]). It is more significant to see that the FFEs in 1st and 3rd residues are additive and together they represent the entire FFE of the intact mRNA (FIG. 1 D).

There is a correlation between the protein structure and the FFE associated with codon residues. The correlation is especially prominent when the FFE ratios are compared to the helix/sheet ratios (FIG. 2).

The unique, codon-related FFE pattern and its correlation to alpha helix content suggested some similarity between protein structures and the possible structures of the coding sequences.

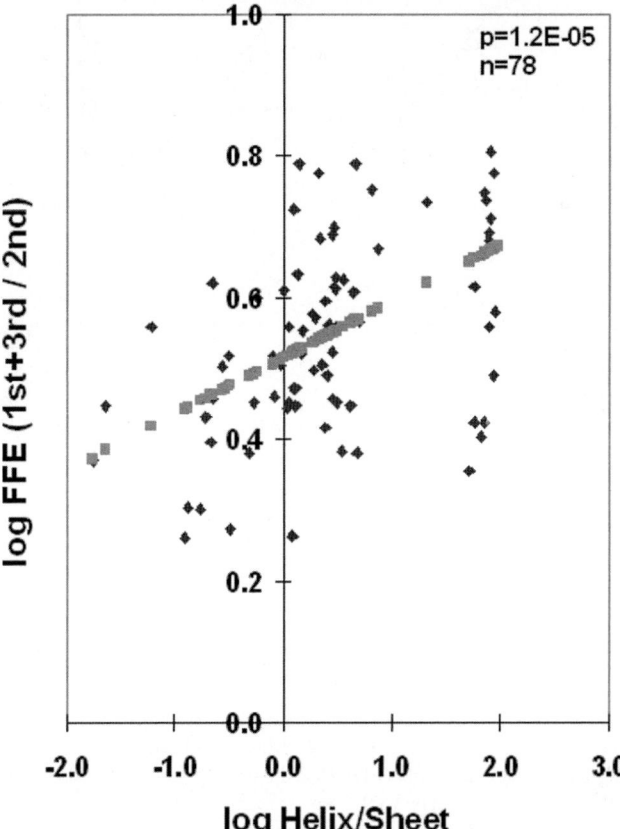

FIGURE 2. FFE associated with codon positions versus protein structure. FFEs associated with 1st, 2nd and 3rd codon residues in 78 different mRNA sequences were calculated and compared to the helix/sheet ratios of the corresponding protein structures. Linear regression analyses where squares represent the linear regression line.

This possibility was examined by visual comparison of 16 randomly selected protein residue contact maps and the energy dot plots of the corresponding RNAs. We could see similarities between the two different kinds of maps (FIG. 3). However, this type of comparison is not quantitative and statistical evaluation is not directly possible.

Another similar, but still not quantitative, comparison of protein and coding structures was performed on four proteins that are known to have very similar 3D structures but their primary structure (the sequence) is less than 30%

Comparison of Protein and Corresponding mRNA Structures

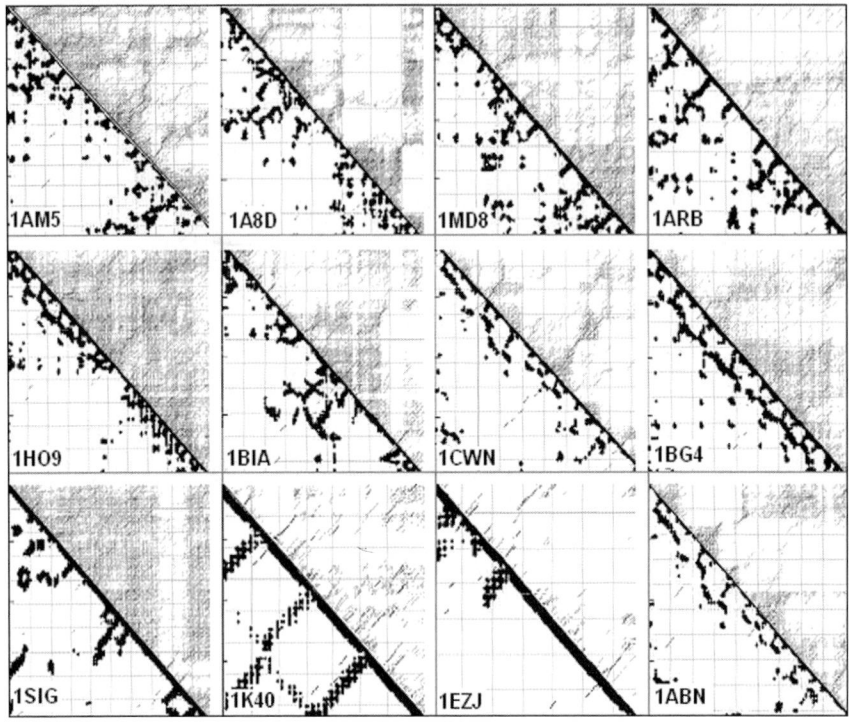

FIGURE 3. Comparison of protein and corresponding mRNA structures. Residue contact maps (RCM) were obtained from the PBD files of protein structures using the SeqX tool (left triangles). Energy dot plots (EDP) for the coding sequences were obtained using the mfold tool (right triangles). The two kinds of maps were aligned along a common left diagonal axis to make an easy visual comparison of the different kind of representations possible. The black dots in the RCMs indicate amino acids that are within 6 Å of each other in the protein structure. The colored (grass-like) areas in the EDPs indicate the energetically mostly likely RNA interactions.

similar, as well as the sequence of their mRNA. These four proteins are examples of the fact that the tertiary structure of proteins is much more conserved than the amino acid sequence. We asked the question whether this is true for the RNA structures and sequence. We found that there are signs of conservation even of the RNA secondary structure (as indicated by the energy dot plots) and there are similarities between the protein and nucleic acid structures (FIG. 4).

Comparisons of the protein residue contact map with the nucleic acid folding maps suggest similarities between the 3D structures of these different kinds of molecules. However, this is a semiquantitative method.

Comparison of the Protein and mRNA Secondary Structures

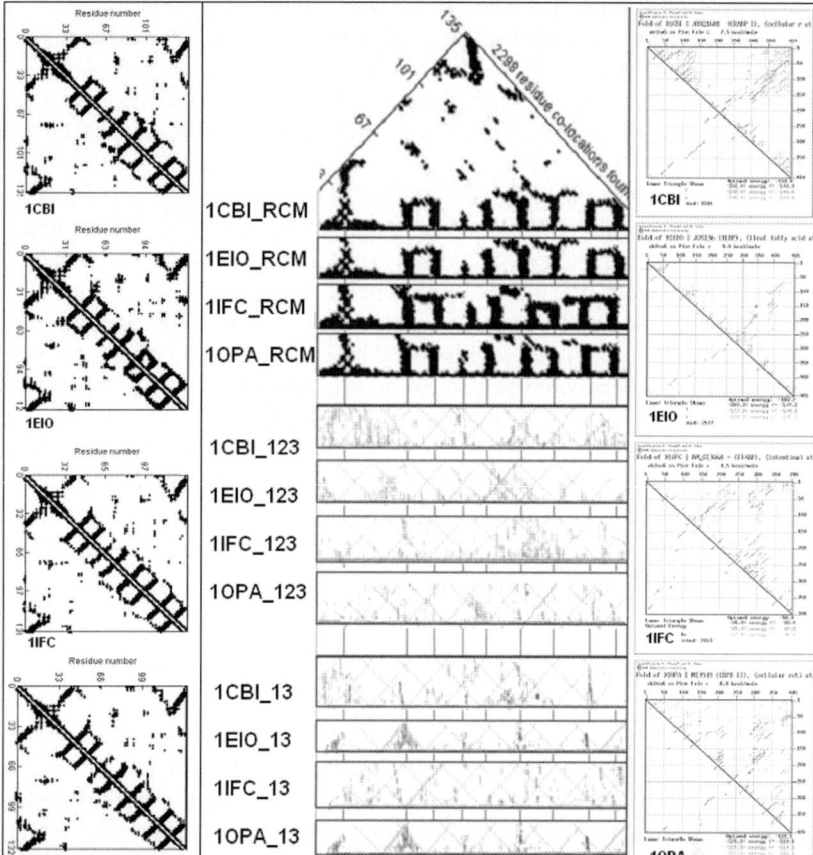

FIGURE 4. Comparison of the protein and mRNA secondary structures. Residue contact maps (RCM) ware obtained from the PBD files of four protein structures (1CBI, 1EIO, 1IFC, 1OPA) using the SeqX tool (left column). Energy dot plots (EDP) for the coding sequences were obtained using the mfold tool (right column). The left diagonal portion of these two kinds of maps was compared in the central part of the figure. Blue horizontal lines in the background correspond to the main amino acid colocation sites in the RCM. Intact RNA (123) as well as subsequences containing only the 1st and 3rd codon letters (13) are compared. The black dots in the RCMs indicate amino acids that are within 6 Å of each other in the protein structure. The colored (grass-like) areas in the EDPs indicate the energetically most likely RNA interactions.

A more direct statistical support might be obtained by analyzing and comparing residue colocations in these structures. Assume that the structural unit of mRNA is a tri-nucleotide (codon) and the structural unit of the protein is the amino acid. The codon may form a secondary structure by interacting with other codons accordingly to the WC base complementary rules, and contribute to the formation of a local double helix. The 5′-A1U2G3-3′ sequence (Met, M codon) forms a perfect double string with the 3′-U3A2C1-5′ sequence (His, H codon, reverse and complementary reading). Suboptimal complexes are 5′-A1×2G3-3′ partially complemented by 3′-U3×2C1-5′ (AAG, Lys; AUG, Met; AGG, Arg; ACG, Pro; and CAU, His; CUU, Leu; CGU, Arg; CCU, Pro, respectively).

Our experiments with FFE indicate that local nucleic acid structures are formed under this suboptimal condition, that is, when the 1st and 3rd codon residues are complementary but the 2nd is not. If this is the case, and there is a connection between nucleic acid and protein 3D structure, one might expect that the four amino acids coded by 5′-A1×2G3-3′ codons will preferentially colocate with other four amino acids coded by 3′-U3×2C1-5′ codons. We have constructed 8 different complementary codon combinations and found that the codons of colocating amino acids are often complementary at the 1st and 3rd positions and follow the D-1 × 3/RC-3 × 1 formula but not the seven other formulas (FIG. 5 A and B). It means that amino acids that are coded by partially reverse and complementary codons (WC base pairs at the 1st and 3rd codon positions and translated in reverse orientation) are preferentially colocated in protein structures.

DISCUSSION

It is well known that coding and noncoding DNA sequences (exon/intron) are different and this difference is somehow related to the asymmetry of the codons, that is, that the 3rd codon letter (wobble) is poorly defined. Many Markov models have been formulated to find this asymmetry and *de novo* predict coding sequences (genes). These *in silico* methods work rather well but not perfectly and some scientists remain unconvinced that the codon asymmetry explains the exon–intron differences satisfactorily.

Another codon-related problem is that the well-known, nonoverlapping, triplet codon translation is extremely phase dependent and there is theoretically no tolerance for any phase shift. There are famous examples of how single nucleotide deletion might destroy the meaningful translation of a sequence and which are incompatible with life. However, considering the magnitude and complexity of the eukaryotic proteome, the precision of translation is astonishingly good. Such physical precision is not possible without massive and consistent physico-chemical fundamentals. Therefore, discovery of the existence of secondary structure bias (folding energy differences) in coding regions of many organisms[13] was a very welcome observation because it differentiates exons from introns on a physico-chemical basis.

Our experiments with FFE confirmed that this bias exists. In addition, there is a very consistent and very significant pattern of FFE distribution along

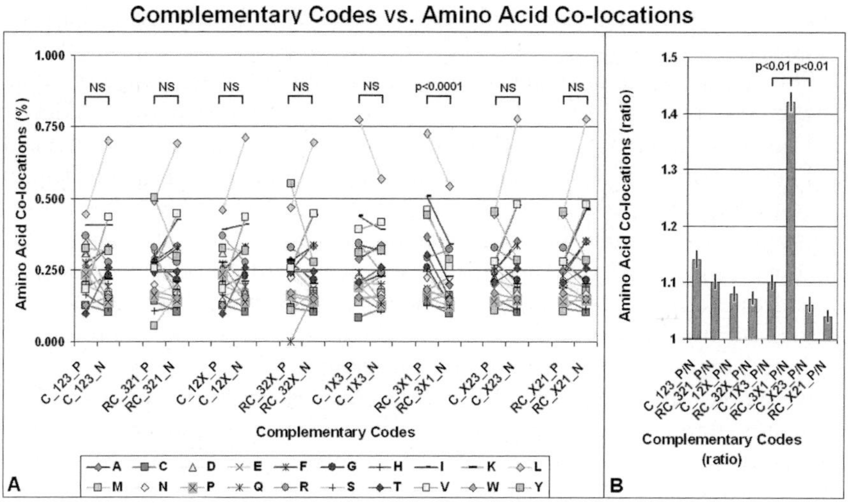

FIGURE 5. Complementary codes versus amino acid colocations. (A part) The propensity of the 400 possible amino acid pairs was monitored in 81 different protein structures with the SeqX tool. The tool detected colocations when two amino acids were closer than 6 Å to each other (neighbors on the same strand were excluded). The total number of colocations was 34,630. Eight different complementary codes were constructed for the codons (two optimal and six suboptimal). In the two optimal codes all three codon residues (123) were complementary (C) or reverse-complementary (RC) to each other. In the suboptimal codes only two of three codon residues were C or RC to each other (12, 13, 23), while the third was not necessarily complementary (X) (e.g., complementary code RC_3×1 means that the 1st and 3rd codon letters are always complementary (to D_1×3), but not the 2nd, and the possible codons are read in reverse orientation). The 400 colocations were divided into 20 subgroups corresponding to 20 amino acids (one of the colocating pairs), each group containing the 20 amino acids (corresponding to the other amino acid in the colocating pairs). If the codons of the amino acid pairs followed the predefined complementary code, the colocation was regarded as positive (P); if not, the colocation was regarded as negative (N). Each symbol represents the mean frequency of P or N colocations corresponding to the indicated amino acid. Paired Student's t-test, $n = 20$. The ratio (B part) of positive (P) and negative (N) colocations was calculated on data from part A. Each bar represents the mean ± SEM, $n = 20$.

the nucleotide sequence. Comparing the FFE of phase-selected subsequences, subsequences comprised of only the 1st or only the 3rd codon letters showed significantly higher FFE than those consisting only of the 2nd letters. This FFE difference was not present in intronic sequences preceding and following the exons, but it was present in exons from different species including viruses. This is an interesting observation because these phenomena might not only distinguish between exons and introns on a physico-chemical basis, but it might even clearly define the tri-nucleotide codons and thus the phase of the translation. This codon-related phase-specific variation in FFE may explain why mRNAs have greater negative FFEs than shuffled or codon choice randomized sequences.[21]

FFE in nucleic acids is always associated with WC base pair formation. Higher FFE indicates more WC pairs (presence of complementarity) and lower FFE indicates fewer WC pairs (less complementarity). The FFE in the 1st and 3rd codon positions was additive, while the 2nd letter did not contribute to the total FFE; the total FFE of the entire (intact) nucleic acid was the same as subsequences containing only the 1st and 3rd codon letters (2nd deleted). This is an indication for that the local RNA secondary structure bias is caused by complementarity of the 1st and 3rd codon residues in local sequences. This partial, local complementarity is more optimal in reverse orientation of the local sequences as expected with loop formations.

It is known that single-stranded RNA molecules can form local secondary structures through the interactions of complementary segments. The novel observation here is that these interactions preferentially involve the 1st and 3rd codon residues. This connection between the RNA secondary structure and codons immediately directed attention toward the question of protein folding and its long suspected connection to RNA folding.[22,23]

Only about one-third (20/64) of the genetic code is used for protein coding, that is, there is a great excess of information in the mRNA. At the same time, the information carried by amino acids seems to be insufficient (as stated by some scientists) to complete unambiguous protein folding. Therefore, it is believed that the 3rd codon residue (wobble base) carries some additional information to that already present in the genetic code. A specialized wobble base oriented database, the ISSD,[19] was established in an effort to connect different features of protein structure to wobble bases[24] with more or less success.

We found a significant correlation between FFE ratios and the helix/sheet content of protein structures. It was possible to make direct visual comparison of mRNA structure (as statistically predicted by mfold energy dot-plot) and protein structures (as 2D residue contact maps). This method suggests similarity between nucleic acid and protein structures.

It is known that some complex protein structures are very similar even if there is less than 30% sequence similarity. It was interesting to see that the same principle might apply for nucleic acids, and structural similarity might exist even when the sequence similarity is low. Furthermore, significant similarity between nucleic acid and protein structures might exist even without translational connection. Structure seems to be more preserved, even in nucleic acids, than sequence. However, even if the matrix comparisons are suggestive, they remain semiquantitative methods. Better support was necessary.

A working hypotheses grew out of these observations, namely that (*a*) partial, local reverse-complementarity exists in nucleic acids that form the nucleic acid structure; (*b*) there is some degree of similarity between the folding of nucleic acids and proteins; (*c*) protein structure determines the amino acid colocations; and (*d*) as a consequence, amino acids coded by the interacting (partially reverse complementary) codons might show preferential colocations in the protein structures.

And it seems to be the case: codons which contain complementary bases at the 1st and 3rd positions and are translated in reverse orientation result in amino acids which are preferentially colocated (interacting) in the 3D protein structure. Other complementary residue combinations or translation in the same (not reverse) direction (as much as seven combinations in total) did not result in any preferentially colocating subset of amino acid pairs.

Construction of residue contact maps for protein structures and statistical evaluation of residue colocations is a frequently used method for visualization and analyses of spatial connections between amino acids.[25–27] The amino acid colocations in real protein structures is clearly not random[28,29] and therefore residue colocation matrices are often used to assist in the prediction of novel protein structures.[30,31] We have carefully examined the physico-chemical properties of specifically interacting amino acids in and between protein structures, and we concluded that these interactions follows the well-known physicochemical rules of size, charge and hydrophobe compatibility (unpublished data) well in line with Anfinsen's prediction. The recent study supports the fact that there is a previously unknown connection between the codons of specifically interacting amino acids; those codons are complementary at the 1st and 3rd (but not the 2nd) codon positions.

The idea that sequence complementarity might explain the nature of specific protein–protein interactions is not new and was suggested already in 1981.[32] I was never able to experimentally confirm my own original theory, which suggested a perfect complementarity between codons of interacting amino acids,[32,33] while others were more successful.[34] The explanation is that the codon complementarity is suboptimal and does not involve the 2nd codon residue. Experimental *in vitro* confirmation is required to validate this recent theoretical and *in silico* prediction.

REFERENCES

1. ANFINSEN, C.B., R.R. REDFIELD, W.I. CHOATE, *et al.* 1954. Studies on the gross structure, cross-linkages, and terminal sequences in ribonuclease. J. Biol. Chem. **207:** 201–210.
2. LEVINTHAL, C. 1969. How to fold graciously in Mossbauer spectroscopy in biological systems. *In* Proceedings of a Meeting held at Allerton House, Monticello, IL. P. Debrunner, J.C.M. Tsibris, & E. Munck, Eds.: 22–24. University of Illinois Press. Urbana, IL.
3. KLEPEIS, J.L. & A.C. FLOUDAS. 2003. ASTRA-FOLD: a combinatorial and global optimization framework for ab initio prediction of three-dimensional structures of proteins from the amino acid sequence. Biochem. J. **85:** 2119–2146.
4. WALTER, S. & J. BUCHNER. 2002. Molecular chaperones—cellular machines for protein folding. Angew Chem. Int. Ed. Engl. **41:** 1098–1113.

5. KOMAR, A.A., A. KOMMER, I.A. KRASHENINNIKOV & A.S. SPIRIN. 1997. Cotranslational folding of globin. J. Biol. Chem. **272:** 10646–10651.
6. THANARAJ, T.A. & P. ARGOS. 1996. Protein secondary structural types are differentially coded on messenger RNA. Protein Sci. **5:** 1973–1983.
7. BRUNAK, S. & J. ENGELBRECHT. 1996. Protein structure and the sequential structure of mRNA: alpha-helix and beta-sheet signals at the nucleotide level. Proteins **25:** 237–252.
8. GUPTA, S.K., S. MAJUMDAR, T.K. BHATTACHARYA & T.C. GHOSH. 2000. Studies on the relationships between the synonymous codon usage and protein secondary structural units. Biochem. Biophys. Res. Commun. **269:** 692–696.
9. CHIUSANO, M.L., F. ALVAREZ-VALIN, M. DI GIULIO, et al. 2000. Second codon positions of genes and the secondary structures of proteins. Relationships and implications for the origin of the genetic code. Gene **261:** 63–69.
10. GU, W., T. ZHOU, J. MA, X. SUN, Z. LU, et al. 2004. The relationship between synonymous codon usage and protein structure in Escherichia coli and Homo sapiens. Biosystems **73:** 89–97.
11. ERMOLAEVA, O. 2001. Synonymous codon usage in bacteria. Curr. Issues Mol. Biol. **3:** 91–97.
12. BIRO, J.C., J.M. BIRO & A.M. BIRO. 2005. Hidden massages in hidden subsequences: a study on collagens. In 30th FEBS Congress—9th IUBMB Conference, Budapest, Hungary, 2–7 July 2005 [Abstract].
13. KATZ, L. & C.B. BURGE. 2003. Widespread selection for local RNA secondary structure in coding regions of bacterial genes. Genome Res. **13:** 2042–2051.
14. ZUKER, M. 2003. Mfold web server for nucleic acid folding and hybridization prediction. Nucleic Acids Res. **31:** 3406–3415.
15. BIRO, J.C. & G. FORDOS. 2005. SeqX: a tool to detect, analyze and visualize residue co-locations in protein and nucleic acid structures. BMC Bioinformatics **6:** 170. Available at http://www.janbiro.com/downloads.
16. BIRO, J.C. 2005. SeqForm. Available at http://www.janbiro.com/downloads.
17. BERMAN, H.M., J. WESTBROOK, Z. FENG, et al. 2000. The Protein Data Bank. Nucleic Acids Res. **28:** 235–242. Available at http://www.pdb.org/.
18. BERMAN, H.M., W.K. OLSON, D.L. BEVERIDGE, et al. 1992. The Nucleic Acid Database: a comprehensive relational database of three-dimensional structures of nucleic acids. Biophys J. **63:** 751–759. Available at http://ndbserver.rutgers.edu/index.html.
19. ADZHUBEI, I.A. & A.A. ADZHUBEI. 1999. ISSD Version 2.0: taxonomic range extended. Nucleic Acids Res. **27:** 268–271. Available at http://www.protein.bio.msu.su/issd/.
20. BIRO, J.C., B. BENYO, C. SANSOM, et al. 2003. A common periodic table of codons and amino acids. Biochem. Biophys. Res. Commun. **306:** 408–415.
21. SEFFENS, W. & D. DIGBY. 1999. mRNA has greater negative folding free energies than shuffled or codon choice randomized sequences. Nucleic Acids Res **27:** 1578–1584.
22. ORESIC, M., M. DEHN, D. KORENBLUM & D. SHALLOWAY. 2003. Tracing specific synonymous codon-secondary structure correlations through evolution. J. Mol. Evol. **56:** 473–484.
23. D'ONOFRIO, G., T.C. GHOSH & G. BERNARDI. 2002. The base composition of the genes is correlated with the secondary structures of the encoded proteins. Gene **300:** 179–187.

24. XIE, T. & D. DING. 1998. The relationship between synonymous codon usage and protein structure. FEBS Lett. **434:** 93–96.
25. KUMAREVEL, T.S., M.M. GROMIHA & M.N. PONNUSWAMY. 2001. Distribution of amino acid residues and residue-residue contacts in molecular chaperons. Prep. Biochem. Biotechnol. **31:** 163–183.
26. EILERS, M., A.B. PATEL, W. LIU & S.O. SMITH. 2002. Comparison of helix interactions in membrane and soluble alpha-bundle proteins. Biochem. J. **82:** 2720–2736.
27. GLASER, F., D.M. STEINBERG, I.A. VAKSER & N. BEN-TAL. 2001. Residue frequencies at protein-protein interfaces. Proteins Struct. Funct. Genet. **43:** 89–102.
28. NAOR, D., D. FISHER, R.L. JERNIGAN, et al. 1996. Amino acid pair interchanges at spatially conserved locations. J. Mol. Biol. **256:** 924–938.
29. AZARYA-SPRINZAK, E., D. NAOR, H.J. WOLFSON & R. NUSSINOV. 1997. Interchanges of spatially neighboring residues in structurally conserved environment. Protein Eng. **10:** 1109–1122.
30. SINGER, M.S., G. VRIEND & R.P. BYWATER. 2002. Prediction of protein residue contacts with a PDB-derived likelihood matrix. Protein Eng. **15:** 721–725.
31. SHAO, Y. & C. BYSTROFF. 2003. Predicting inter-residue contacts using templates and pathways. Proteins Struct. Funct. Genet. **53:** 497–502.
32. BIRO, J. 1981. Comparative analysis of specificity in protein-protein interactions. Part II: the complementary coding of some proteins as the possible source of specificity in protein-protein interactions. Med. Hypotheses **7:** 981–993.
33. SEGERSTEEN, U., H. NORDGREN & J.C. BIRO. 1986. Frequent occurrence of short complementary sequences in nucleic acids. Biochem. Biophys. Res. Commun. **139:** 94–101.
34. HELA, J.R., G.W. ROBERTS, J.G. RAYNES, et al. 2002. Specific interactions between sense and complementary peptides: the basics for the proteomic code. Chembiochem **3:** 136–151.

Antigens and Cytokine Genes in Antitumor Vaccines

The Importance of the Temporal Delivery Sequence in Antitumor Signals

MARÍA JOSÉ HERRERO,[a] RAFAEL BOTELLA,[b] FRANCISCO DASÍ,[a] ROSA ALGÁS,[c] MARÍA SÁNCHEZ,[a] AND SALVADOR F. ALIÑO[a]

[a]Gene Therapy Group, Department of Pharmacology, Faculty of Medicine, University of Valencia, 46010 Valencia, Spain

[b]Dermatology Unit, Instituto Valenciano de Oncología, 46009 Valencia, Spain

[c]Radiotherapy Unit, Hospital Clínico Universitario, 46010 Valencia, Spain

ABSTRACT: Studies against cancer, including clinical trials, have shown that a correct activation of the immune system can lead to tumor rejection whereas incorrect signaling results in no positive effects or even anergy. We have worked assuming that two signals, GM-CSF (granulocyte and macrophage colony-stimulating factor) and tumor antigens are necessary to mediate an antitumor effective response. To study which is the ideal temporal sequence for their administration, we have used a murine model of antimelanoma vaccine employing whole B16 tumor cells or their membrane protein antigens (TMPs) in combination with *gm-csf* transfer before or after the antigen delivery. Our results show that: (i) When *gm-csf* tisular transfection is performed before TMP delivery, a tumor growth inhibition is observed, but with a limit effect when administering high antigen doses; in contrast, when signals are inverted, the limited effect is lost and greater antitumor efficacy is obtained. (ii) A similar behavior, but with stronger positive results, is observed employing *gm-csf* transfection and whole tumor cells as antigens. While negative results are obtained with *gm-csf* before cells, the best results (total survival of treated mice) are obtained when GM-CSF is administered in transfected cells. We conclude that optimal antitumoral response can be obtained when the antigen signal is given before (or simultaneous with) GM-CSF production, while the inversion of the signals could result in the undesired inhibition or anergy of the immune response.

KEYWORDS: cell vaccine; tumor antigen; gene transfer; GM-CSF; melanoma

Address for correspondence: María José Herrero, Gene Therapy Group, Department of Pharmacology, Faculty of Medicine, University of Valencia, Av. Blasco Ibáñez, 15, 46010 Valencia, Spain. Voice: +34-9-6-3864621; fax: +34-9-6-3864972.
e-mail: mariajoseherrero@ono.com

Ann. N.Y. Acad. Sci. 1091: 412–424 (2006). © 2006 New York Academy of Sciences.
doi: 10.1196/annals.1378.084

INTRODUCTION

Genetically modified tumor cells have been demonstrated to be very efficient in various models of antitumoral vaccines.[1–3] Still the main problem in these models is the great difficulty of obtaining tumor cells that we also would have to keep in culture, transfect and then return to the same patient, which is not easy, even employing other proposed methods, not only with tumor cells, but also with syngenic or allogenic cells.[4,5] All these efforts made for many years in the field of cancer vaccine research have converged, in the idea that the best results are obtained by vaccinating not only with the tumor antigens, but also with the genes that the immune system would itself use to achieve a good response: the cytokines. However, the functioning of the immune system is very complex and combining these two components has not always been successful,[6,7] leading to immune suppression or even anergy. It has not been established yet which is the correct temporal order to supply the two signals. The most effective way of administering these two components is by introducing cells, tumoral or not, that produce the cytokine themselves,[8] and since then, many studies have achieved promising results and clinical trials have been set up. However, the right sequence for injection if we want to separate the two components remains to be elucidated.

To understand what kind of antigens work best and how they invoke a really effective response, we wanted to work with the tumor cell antigens instead of whole cells. We tested the different combinations regarding cytokine temporal administration, studying which is the best dose of each component and which is the best temporal sequence for its injection. We have worked in a murine model of melanoma, using the two mentioned components to make an antitumoral vaccine. As antigens, we have used irradiated whole tumor cells (B16 cells) or an extract of its hydrophilic antigens (tumor membrane proteins, TMP). As cytokine, we have used the m-GMCSF encoding gene, which has largely been shown to be a very useful antitumoral tool.[1,4,6,9] In this work, these three possibilities have been combined in order to understand better in which sequence these signals must be given to the immune system to achieve good antitumoral results, thereby avoiding the signals' order that lead to the inhibition or anergy of the immune response. Our experiments show that the best results are obtained when *gm-csf* is administered after or simultaneously with the antigen signal.

MATERIALS AND METHODS

Cells and Transfection

B16 murine melanoma cells, syngenic in C57BL/6 mice, were used in all our experiments as a source of antigen to vaccinate, either as whole cells or used to get an extract of its membrane proteins (TMP). B16 cells were also used in every case to inoculate the tumors in vaccinated and control groups of mice, injecting 10^5 freshly detached cells in its usual culture medium.

Cells were grown in DMEM (Dulbecco's modified Eagle's medium) (Sigma, Madrid, Spain), supplemented with 10% heat-inactivated fetal bovine serum (FBS, Biomedia, Barcelona, Spain), penicillin (100 U/mL) and streptomycin (100 μg/mL). They were kept in a humidified incubator with 5% CO_2 at 37°C. B16 are adherent cells that are detached from flasks with trypsin-EDTA. When whole cells are used for vaccinating, they are irradiated with a 150-Gy dose after being detached from their flasks and then frozen in DMSO 5% in FBS to be kept at –80 or –150°C until their use.

The fresh (no selection) transfection procedure for the experiments with GM-CSF (granulocyte and macrophage colony-stimulating factor)-producing B16 cells was chemical, preparing PEI 25 KDa (polyethyleneimine; Sigma, Spain) polyplexes or DOTAP (Roche, Valencia, Spain) lipoplexes with 7.5 μg/mL plasmid dose.[10] Cells were transfected when more than 80% confluence was reached in their flasks and they were irradiated with 150-Gy dose, frozen in DMSO 5% in FBS, and kept at –80 or –150°C until their use.

Plasmids

pcDNA3 plasmid from Invitrogen (Barcelona, Spain) was used in some of the experiments, with cytomegalovirus promoter and resistance to neomycin and ampicillin.

pMok m-gmcsf was generously given by Dr. A. Koenig (Mologen, Berlin, Germany), containing the murine *gm-csf* gene, controlled by cytomegalovirus promoter and with resistance to kanamycin. The plasmids were amplified in *Escherichia coli* DH5α in selective LB broth (Pronadisa, Valencia, Spain) and extracted with the Qiagen Giga Endofree kit (Quiagen, Barcelona, Spain), quantified by spectrophotometry and tested by electrophoresis to confirm integrity.

Obtaining TMPs

The mixture of hydrophilic membrane proteins from B16 cells was performed following a protocol based on C. Bordier's work.[11] In brief: after the irradiated cells are detached and counted, the pellet was resuspended in 1 mL of extraction buffer, for every 10 million cells, containing Tris 10 mM, $MgCl_2$ was 2 mM,0 Triton X-114 0.5% (Sigma, Spain) and PMSF 0.1 mM (Sigma, Spain), pH 7.2. The mixture was incubated in ice for 45 min, shaking gently every 10 min. Then it was centrifuged at 5,300 rpm, for 15 min at 4°C. The supernatant was recovered and put

750 μL above 250 μL of centrifuge buffer (sucrose 10% in extraction buffer) in microcentrifuge tubes. After incubating for 3 min at 37°C, it was centrifuged again for 5 min at 2,100 rpm. Supernatant (750 μL) was recovered and Triton X-114 added to a final concentration of 0.5% before putting the tube in ice again over 250 μL of centrifuge buffer. After incubation 3 min more at 37°C and centrifugation 5 min more at 2,100 rpm, two different phases were obtained in the microcentrifuge tube, the upper one containing the hydrophilic proteins, which are kept in aliquots at −20°C until their use.

The proteins were quantified with the CBQCA protein quantitation kit (Molecular Probes-Invitrogen, Barcelona, Spain), following the manufacturer's instructions.

Determination of m-GMCSF Production

For the GM-CSF production of the transfected B16 cells, an enzyme-linked immunosorbent assay (ELISA) was carried out on a supernatant sample of the culture medium, 72 h post transfection and prior to the cells Madrid detachment and irradiation, using BD OptEIA ELISA kit for m-GMCSF (Pharmingen, BD Biosciences, Madrid, Spain). The time point of 72 h was chosen on the basis of a prior experiment made studying cytokine production over time, using the referred transfection conditions,[9,10] in order to achieve the adequate production level according to the literature.[4] Then B16 cells were transfected in their flasks after reaching more than 80% confluence with PEI/DNA polyplexes or DOTAP/DNA lipoplexes, with a final concentration of 7.5 μg/mL pMok m-gmcsf dose. Supernatant samples were taken every 24 h over 6 days, cleaned by centrifugation (3,000 rpm, 5 min), and used for ELISA with BD OptEIA ELISA kit for m-GMCSF.

For the tissue transfection experiments, a previous assay was carried out assay injecting only the plasmid pMok m-gmcsf at two different doses, 10 or 50 μg per animal, in 200 μL saline in the same way it would be done in the vaccination experiments. The animals were sacrificed and tissue samples of the injected area were analyzed by quantitative reverse transcription polymerase chain reaction (RT-PCR) to confirm mGM-CSF production. The pieces of tissue were homogenized in Trizol (Invitrogen, Spain) according to the manufacturer's protocol in order to obtain the RNA from the samples, which was later quantified by spectrophotometry, confirming a good purity level. A total of 5 ng of each RNA was taken to obtain the reverse transcriptase reaction and later to perform the quantitative PCR, using Applied Biosystems' conditions and reactives, TaqMan RTreagents, Taq Man Assay on Demand for *m-gmcsf,* Abi Prism 7700 Software (Applied Biosystems, Barcelona, Spain). The results of the qPCR are expressed in $2^{-\Delta Ct}$, as already described in the literature.[12]

Vaccination Procedures

C57BL/6 mice (Harlan, Ibérica, Barcelona, Spain) were used in all experiments. The animals were kept under standard laboratory conditions and housed five mice per cage. All the experiments were approved by the Biological Research Committee of the Faculty of Medicine of the University of Valencia (Spain).

All the vaccinations followed the same pattern: animals were vaccinated at weeks -3, -1, and $+1$ regarding the tumor implantation (day 0), when 10^5 wild-type freshly detached B16 cells were injected in the left leg of the animals in 100 μL DMEM. The vaccinations were always performed in the right leg.

For experiments of tissue transfection, 50 μg of the naked pMok m-gmcsf plasmid or pcDNA3 plasmid were injected in 200 μL saline (only saline in controls), before or after the antigens, depending on the experiment, in the right groin area. Two antigen doses (in 100 μL PBS 1×) were administered 24 and 72 h after plasmid injection. As antigens we have used whole B16 irradiated cells (2×10^6 cells/dose) or the corresponding TMP (from 20 to 200 μg/dose, divided in the two injections).

For the inverse tissue transfection, the inverse process is performed, injecting first one dose of antigen, the second 48 h later, and plasmid 72 h later. Two different doses of plasmid were tested in this experiment, 10 and 50 μg pMok m-gmcsf.

For experiments with GM-CSF-producing B16 cells, 2×10^5 freshly transfected B16 cells (with PEI/pMok m-gmcsf polyplexes, 7.5 μg/mL) were injected per dose, per mouse, in 100 μL DMEM, in the right leg in weeks -3, -1, and $+1$ with respect to tumor implantation, day 0 (10^5 wild-type B16 cells in left leg). The transfection percentage of the cells is between 20 and 40% of total cells (data not shown). Control and B16* group followed the same pattern, but only with 100 μL DMEM or 2×10^5 B16 wild-type irradiated cells in 100 μL DMEM, respectively.

Tumor Growth Measurement and Statistical Analysis

Tumor growth in mice was monitored visually and measured with a caliper in two dimensions, A, the long diameter and B, the short one. The tumor volume was calculated with the formula $V = (A \times B^2)/2$ and expressed in mm^3. To statistically compare the results of the different treatment groups, two-way ANOVA statistical analysis were performed (FIGS. 3, 4, and 5), with Bonferroni post tests (95% confidence interval), expressing the statistically significant differences with $^*P < 0.05$, $^{**}P < 0.01$, $^{***}P < 0.001$ when comparing with control group, and the same but with +, ++, +++ when comparing with saline (no-plasmid) group.

RESULTS

qRT-PCR of GM-CSF

After injecting 10 μg ($n = 4$) or 50 μg ($n = 4$) pMok m-gmcsf ($n = 4$, per group) in 200 μL saline in the right groin area of C57BL/6 mice, the animals were sacrificed, two at once, and the others on day +2 and day +4. Tissue samples of the injected area and the equivalent in the contralateral leg (as control) were homogeneized in Trizol to obtain RNA and perform the qRT-PCR assay to detect mGM-CSF expression, using Applied Biosystems Assay on Demand for *m-gmcsf* and their protocols and other reagents. The Ct results in FIGURE 1 are expressed in $2^{-\Delta Ct}$ (2 exp (–[Ct from mGMCSF –Ct from 18S ribosomal]), as previously used in the literature.[12] The differences between treated and control tissue are bigger at day +2 (FIG. 1A), reaching about one order of magnitude, while at day +4 the two groups still show a difference, but a smaller one (FIG. 1B). In both graphs, the different GM-CSF production is higher when injecting 50 μg of pMok m-gmcsf, which means that in those groups a higher net production of *m-gmcsf* has been achieved.

ELISA of m-GMCSF

B16 cells were transfected with PEI/DNA polyplexes or DOTAP/DNA lipoplexes, using 7.5 μg/mL final concentration of pMok m-gmcsf and 80% culture confluence conditions. Samples were taken from culture medium every 24 h over 6 days, centrifuged at 3,000 rpm for 5 min to remove any cells or debris, and then used for ELISA assays to measure mGM-CSF production employing BD OptEIA Kit. The results in FIGURE 2 are expressed in nanograms of mGM-CSF protein secreted per 10^6 cells per 24 h. The comparison of the two transfection complexes shows that PEI/pMok m-gmcsf polyplexes are much more effective than DOTAP/pMok m-gmcsf lipoplexes, which is the reason why PEI/pMok m-gmcsf polyplexes were chosen for the following experiments, discarding DOTAP. In the vaccination experiments, at 72 h post-transfection cells are detached and irradiated, because at 72 h the recommended level in the literature[4] for GM-CSF production was reached and it was a reasonable time to keep cells in culture safely (FIG. 2).

TMP-Based Vaccines

The results of tumor growth in vaccinated mice, employing two different protocols, based on a different temporal sequence of TMP and *gm-csf* delivery, are presented in FIGURES 3 and 4. C57BL/6 mice ($n = 5$) were treated with 50 μg plasmid in 200 μL saline solution in the right groin area (only saline in control groups) and then treated 24 and 72 h later with 10 μg TMP (FIG. 3A)

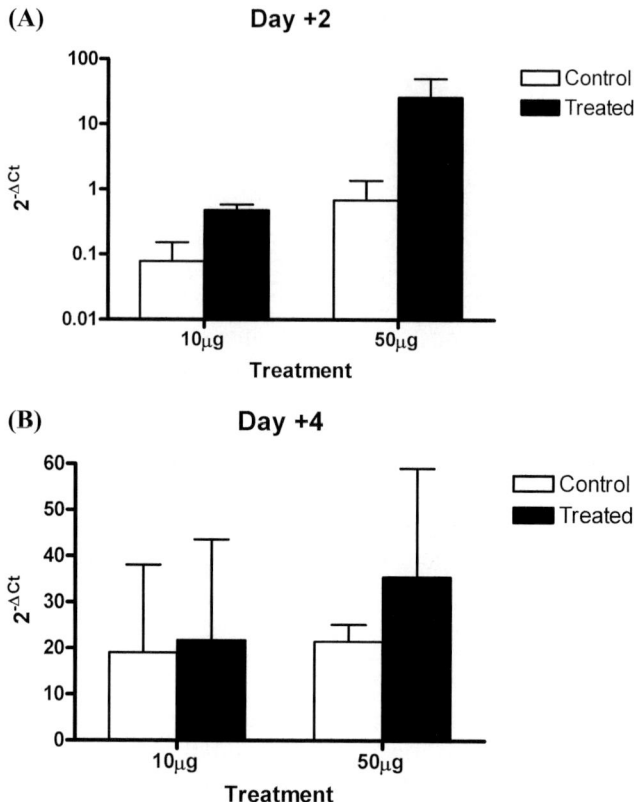

FIGURE 1. qRT-PCR of mGM-CSF. C57BL/6 mice ($n = 4$) were injected with 10 or 50 μg pMok m-gmcsf in 200 μL saline in the right groin area and their tissue samples were homogeneized in Trizol to obtain RNA and perform the qRT-PCR to detect m GM-CSF expression, values in $2^{-\Delta Ct}$. (**A**) results of the mice sacrificed 48 h after plasmid injection and (**B**) results of those sacrificed 96 h after transfection.

or 100 μg TMP (FIG. 3B). This treatment pattern was performed in weeks −3, −1, and +1, with respect to tumor implantation on day 0 (10^5 wild-type B16 cells), in the left leg. The tumor size was calculated by measuring two tumor dimensions, A, the long diameter and B, the short one, with a caliper, using the formula $V = (A \times B^2)/2$. FIGURE 3 shows the results when *gm-csf* was administered before TMP. At low doses of TMP (FIG. 3A), significant inhibition of tumor growth was obtained with at least $P < 0.01$, while no significant inhibition was observed when TMP was injected alone or associated with the control plasmid (pcDNA3). However, the antitumor efficacy of *gm-csf* was lost when higher doses of TMP were employed (FIG. 3B); thus we observe that the best inhibitory effect was obtained employing TMP alone or associated with the empty plasmid (pcDNA3). Interestingly, when the inverse treatment was purposed, *gm-csf* after TMP (FIG. 4), again the best tumor growth inhibition

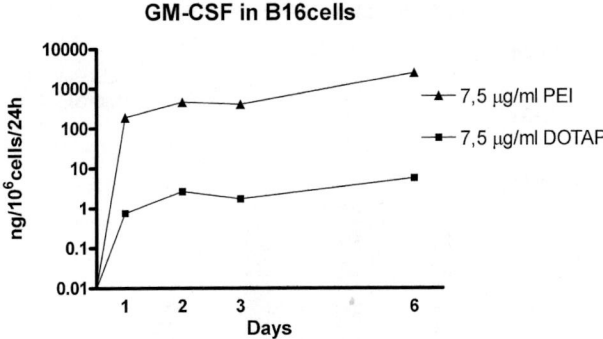

FIGURE 2. ELISA of mGM-CSF. B16 cells were transfected with PEI/DNA polyplexes or DOTAP/DNA lipoplexes, using 7.5 μg/mL final concentration of pMok m-gmcsf. Medium samples were taken every 24 h over 6 days, centrifuged at 3,000 rpm for 5 min to remove any cells or debris, and used to perform ELISA assays to quantify mGM-CSF production with Pharmingen's OpTEIA Kit.

was obtained administering pMok m-gmcsf, this time after the TMP doses and specially when reducing the plasmid dose from 50 to 10 μg (FIG. 4, $P < 0.001$ at day +20 in TMP+pGM-CSF 10 group).

Cell-Based Vaccines

The aim of these experiments was to compare the tumor growth inhibition efficacy when administering GM-CSF before (FIG. 5A) or after (FIG. 5B) cell administration. The results obtained are comparable to those obtained with TMP. In FIGURE 5A, C57BL/6 mice ($n = 5$) were treated following the same pattern as for FIGURE 3, but administering 2×10^6 B16 wild-type irradiated cells instead of TMP. Here, when pMok m-gmcsf is administered before the cells, which represents a high antigen dose, the tumor growth inhibition results are worse than those obtained using B16 irradiated cells alone (FIG. 5A, at least $P < 0.01$), as already seen in FIGURE 3B, where high TMP dose was injected. In FIGURE 5B, on the contrary, when GM-CSF were administered via transfected cells, thus, after or simultaneously to the cells, the best results of all the experiments were obtained: no tumor was observed at least in the following month.

DISCUSSION

Our experimental model combining GM-CSF and antigen signals in different doses and different order of administration shows that the best results in an antitumor vaccine are obtained when GM-CSF is administered after the antigen signal.

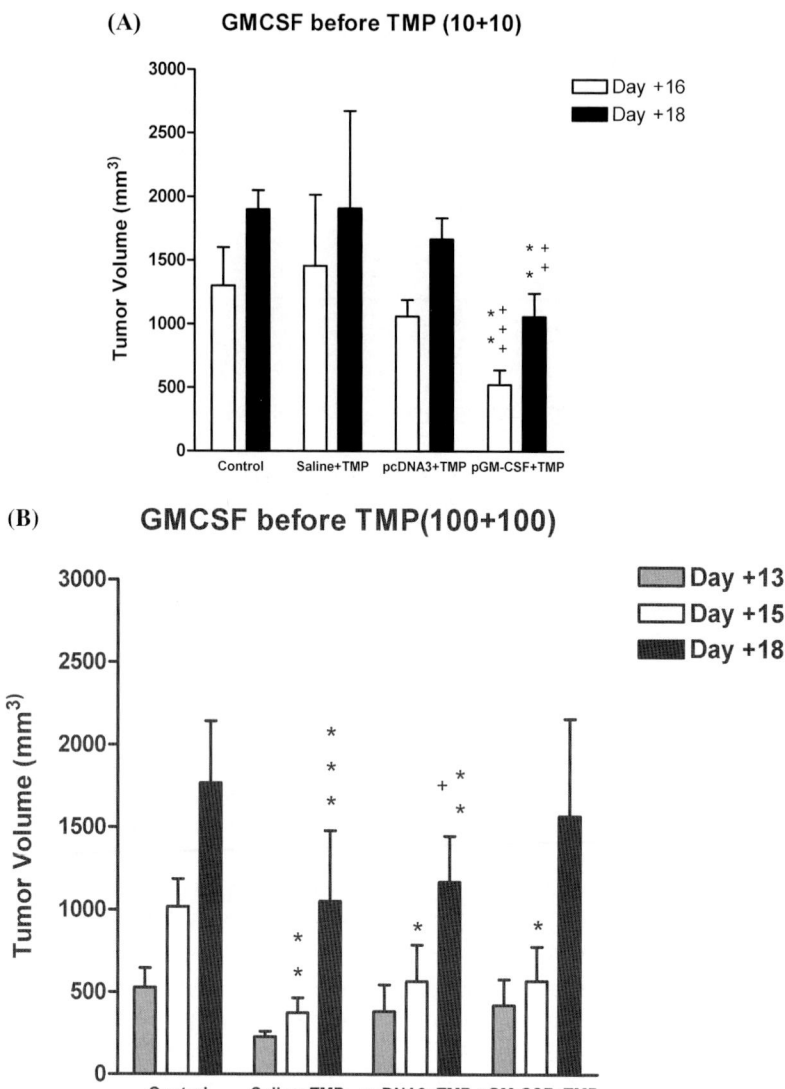

FIGURE 3. Tissue transfection plus TMP-based vaccines. C57BL/6 mice were treated with 50 μg plasmid in 200-μL saline solution in the right groin area (only saline in control groups) and then treated 24 and 72 h after plasmid injection with 10 μg TMP (**A**) or 100 μg TMP (**B**). This treatment pattern was performed in weeks −3, −1, and +1, with respect to tumor implantation on day 0 (10^5 wild-type B16 cells, implanted in left leg). Tumor volume results from measuring the tumor with a caliper in two dimensions—A, the long diameter and B, the short one—and applying the formula V = (A × B^2)/2. Two-way ANOVA statistical analysis was performed, with a Bonferroni post test (95% CI), $*P < 0.05$, $**P < 0.01$, $***P < 0.001$ when comparing with control group, and the same but with +, ++, +++ when comparing with saline (without plasmid) group.

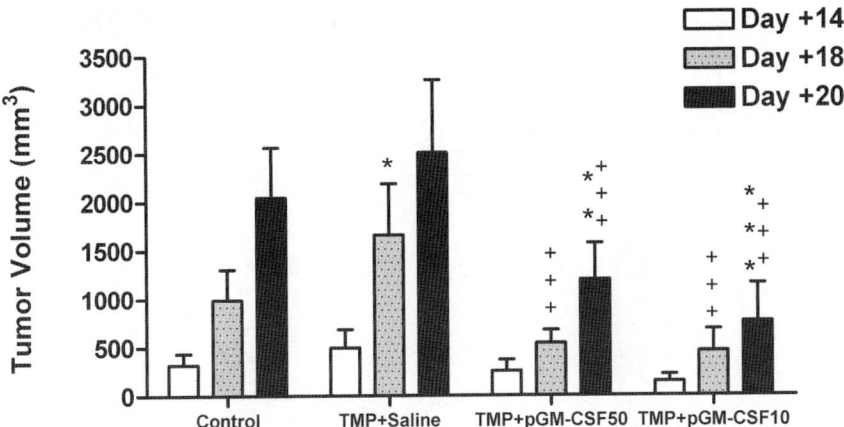

FIGURE 4. TMP-based vaccines plus tisular transfection. C57BL/6 mice received the same treatment as in FIGURE 3 but with two differences, the sequence of the pattern was inverted, thus, it began with antigen treatment, 100 μg TMP, on first day and day 48 h later, and then plasmid injection (or saline in control groups) was performed after 72 h. Another difference was the plasmid dose: two doses were tested, 10 and 50 μg. $*P < 0.05$, $**P < 0.01$, $***P < 0.001$ when comparing with control group, and the same but with +, ++, +++ when comparing with saline (no plasmid) group.

Antitumor vaccines are very powerful tools, especially genetic vaccines, which have reached the best results, even in clinical trials.[1-5] However, in some cases the vaccines have not mediated immune activation, but rather anergy.[6,7] The immune system can limit its response if a threshold level of stimulation is exceeded and that is why it is very important to know how cytokine genes and antigens can cooperate in a synergistic way in an activation or, on the contrary, in the anergy of a immune response. We have evaluated the efficiency of the temporal sequence for the signal delivery in the system, following two different models: antigens plus gene or gene plus antigen. In both cases, the vaccinations were performed using isolated TMP as antigens or the tumor cells bearing these antigens themselves. In the experiments in which GM-CSF was administered before TMP for vaccination purpose, we observed that a good tumor growth inhibition was obtained. The effect was dose-dependent, achieving good results with a single 10 μg TMP/dose (data not shown) and improving with 10 + 10 μg TMP. When we keep on increasing the TMP dose, we arrive at a limit effect, obtaining the opposite result—worse tumor inhibition when administering the cytokine. Interestingly, we observed that this limited antitumor effect at higher antigen dose (100 + 100 μg TMP) can be broken by inverting the signal sequence of the vaccination components. Thus, the highest tumor growth inhibition was obtained when injecting TMP

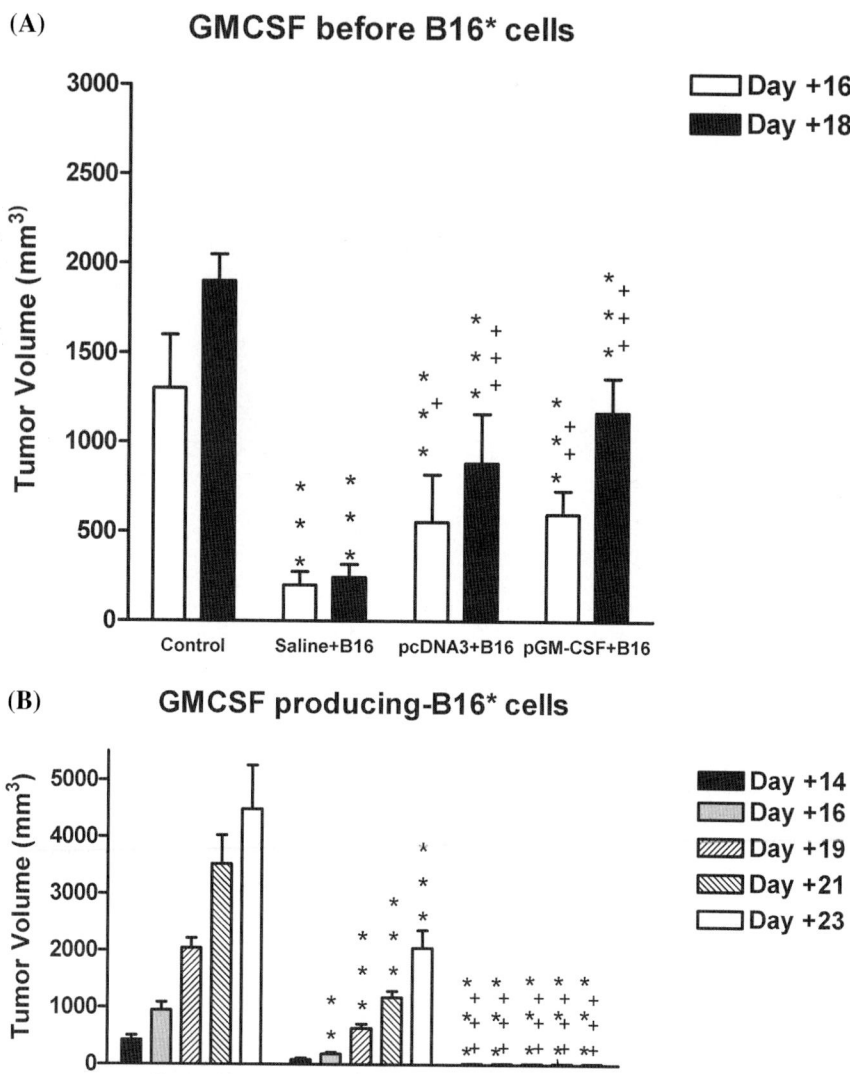

FIGURE 5. Cell-based vaccines, tissue transfection, and GM-CSF-producing B16 cells. Comparison of the tumor growth inhibition efficacy when administering GM-CSF before (**A**) or after (**B**) the cell administration. In (**A**) C57BL/6 mice were treated in the same way as in FIGURE 3, but giving 2×10^6 B16 wild-type irradiated cells as antigens in the vaccine. In (**B**) 2×10^5 wild-type freshly transfected B16 cells (with PEI/pMok m-gmcsf polyplexes, 7.5 μg/mL) were injected per dose (in 100 μL DMEM), per mouse, in the right leg in weeks −3, −1, and +1, with respect to tumor implantation, day 0 (10^5 wild-type B16 cells in left leg). Control and B16* group followed the same pattern, but only with 100 μL DMEM or 2×10^5 B16 wild-type irradiated cells in 100 μL DMEM, respectively. *$P < 0.05$, **$P < 0.01$, ***$P < 0.001$ when comparing with control group, and the same but with +, ++, +++ when comparing with B16* group.

first, 100 + 100 μg, and then *m-gmcsf*, especially at 10 μg plasmid dose. The limited effect is lost when we invert the order of signal sequence, recovering and even improving the tumor inhibitory effect. So, the synergy is probably established when the first signal to arrive is the antigen and the second, GM-CSF. In order to confirm these results, additional experiments were performed in which tumor cells were employed as antigen signals. Our model, translated to cells, also confirmed the previous results employing TMP as antigens. Genetically modified cells have been demonstrated to be the most effective[8] and that is the reason why we wanted to corroborate our hypothesis by also working with them. In the experiments in which GM-CSF was administered before B16 nontransfected irradiated cells, the highest tumor growth inhibition was obtained using the cells alone, indicating that the previous tissue transfection of *m-gmcsf* decreases the cell vaccine efficacy, as already observed in TMP experiments. In addition, the small inhibitory effects seen in both pcDNA3 and pMok m-gmcsf groups were not significantly different, suggesting that the low tumor inhibition was not cytokine-dependent, and was most probably due to the lack of antigen signal in the first moments of interaction with the immune system. In contrast, when we worked with genetically modified cells that produce GM-CSF themselves, we achieved the best results. In this case, no tumor development was observed in any mouse, at least for 1 month, and 60% of the treated mice survived (data not shown). These data are consistent with the rest of TMP experiments, as the first signal is the antigen, and immediately after, GM-CSF is produced.

To summarize, in the context of a GM-CSF–based antitumor vaccine, the results obtained with TMP and also with cells suggest that the optimal sequence for achieving the best efficacy in tumor growth inhibition is presenting the antigen prior to the GM-CSF signal, whereas the inversion of these signals could result in inhibition or even anergy of the immune system.

ACKNOWLEDGMENTS

The present work has been partially supported by Projects FIS PI 021740 from Instituto de Salud Carlos III and SAF 2004–08161 from MEC.

REFERENCES

1. DUNUSSI-JOANNOPOULOS, K., G. DRANOFF, H.J. WEINSTEIN, *et al.* 1998. Gene immunotherapy in murine acute myeloid leukemia: granulocyte-macrophage colony-stimulating factor tumor cell vaccines elicit more potent antitumor immunity compared with B7 family and other cytokine vaccines. Blood **91:** 222–230.
2. MASTRANGELO, M.J. & E.C. LATTIME. 2002. Virotherapy clinical trials for regional disease: *in situ* immune modulation using recombinant poxvirus vectors. Cancer Gene Ther. **9:** 1013–1021.

3. SE-HEON, K., J.F. CAREW, D.A. KOOBY, et al. 2000. Cancer Gene Ther. **7:** 1279–1285.
4. BORRELLO, I. & D. PARDOLL. 2002. Cytokine and Growth Factor Rev. **13:** 185–193.
5. JAFFEE, E.M., R. ABRAMS, J. CAMERON, et al. 1998. A phase I clinical trial of lethally irradiated allogeneic pancreatic tumor cells transfected with the GM-CSF gene for the treatment of pancreatic adenocarcinoma. Human Gene Ther. **9:** 1951–1971.
6. SERAFINI, P., R. CARBLEY, K.A. NOONAN, et al. 2004. High-dose granulocyte-macrophage colony-stimulating factor-producing vaccines impair the immune response through the recruitment of myeloid suppressor cells. Cancer Res. **64:** 6337–6343.
7. RODRÍGUEZ-LECOMPTE, J.C., S. KRUTH, et al. 2004. Cell-based cancer gene therapy: breaking tolerance or inducing autoimmunity? Anim. Health Res. Rev. **5:** 227–234.
8. SHI, F.S., S. WEBER, J. GAN, et al. 1999. Granulocyte-macrophage colony-stimulating factor (GM-CSF) secreted by cDNA-transfected tumor cells induces a more potent antitumor response than exogenous GM-CSF. Cancer Gene Ther. **6:** 81–88.
9. MORET-TATAY, I., J. DIAZ, F.M. MARCO, et al. 2003. Complete tumor prevention by engineered tumor cell vaccines employing nonviral vactors. Cancer Gene Ther. **10:** 887–897.
10. GUILLEM, V.M. & S.F. ALIÑO. 2004. Transfection pathways of nonspecific and targeted PEI-polyplexes. Gene Ther. Mol. Biol. **8:** 369–384.
11. BORDIER, C. 1981. Phase separation of integral membrane proteins in Triton X-114 solution. J. Biol. Chem. **25:** 1604–1607.
12. BOTELLA-ESTRADA, R., F. DASÍ, D. RAMOS, et al. 2005. Cytokine expression and dendritic cell density in melanoma sentinel nodes. Melanoma Res. April: 99–106.

Arrest of Cancer Cell Proliferation by dsRNAs

TATYANA O. KABILOVA, AL`BINA V. VLADIMIROVA, ELENA L. CHERNOLOVSKAYA, AND VALENTIN V. VLASSOV

Institute of Chemical Biology and Fundamental Medicine SB RAS, Novosibirsk 630090, Russia

ABSTRACT: Inhibition of *c-myc* and *N-myc* genes by dsRNAs in carcinoma and neuroblastoma cells was investigated. siRNA-Ex3 targeted to the third exon of *c-myc* gene was found to decrease the level of *c-myc* but not *N-myc* mRNA and decrease the rate or even arrest proliferation of *c-myc* overexpressing cell lines KB-3-1 and SK-N-MC. This siRNA did not affect proliferation of IMR-32 (which overexpress *N-myc*). siRNA-Ex2 corresponding (with 1–2 mismatches) to the conservative region of the second exon of both *c*- and *N-myc* was able to downregulate both genes and to reduce proliferation of KB-3-1, SK-N-MC, and IMR-32 cells. Long dsRNA corresponding to the 3 exon of *c-myc* gene (dsMyc), poly(I:C), and GU-rich siRNA-I, corresponding to the intron sequence of human *MDR1* gene demonstrated high antiproliferative activity in experiments with KB-3-1 cells. Short-term elevation of *PKR* or/and *OAS1* mRNA levels was detected in the cells affected by interferon inducer poly(I:C). dsMyc, poly(I:C), and even siRNA-I, which could not affect *c-myc* mRNA by RNA interference mechanism, were found to inhibit proliferation of the KB-3-1 cells and to decrease the mRNA level of interferon-sensitive genes *c-myc* and β-*actin*.

KEYWORDS: dsRNA; siRNA; *c-myc*; *N-myc*; neuroblastoma; interferon response

INTRODUCTION

The *myc* family transcription factors (c-Myc, N-Myc, and L-Myc) participate in control of cell proliferation, differentiation, and tumorogenesis. These factors appear to be especially attractive targets for genetic and pharmacological approaches because they are functionally downstream of most of the known oncogenes regulation pathways and are overexpressed/deregulated in most tumor cell types.[1,2] A lack of downregulation of total *myc* (*N-myc* and/or *c-myc*) expression plays a key role in the development of neuroblastoma, the

---

Address for correspondence: Elena L. Chernolovskaya, ICBFM SB RAS, Lavrentiev avenue, 8, Novosibirsk 630090, Russia. Voice: +7-383-333-3761; fax: +7-383-333-3677.
e-mail: elena_ch@niboch.nsc.ru

most frequent tumor of the childhood. Conventional therapeutic schemes for neuroblastoma include treatment by α-interferon and retinoic acid that inhibit *myc* expression and induce differentiation of cancer cells, however, in the case of mutations affecting *c-myc* regulatory regions, tumors become resistant to this type of treatment.[3,4] Identification of selective inhibitors of *myc* genes could be important for the development of antineuroblastoma therapeutics.

Double-stranded RNA (dsRNA)-mediated interference (RNAi) provides a powerful approach for sequence-specific silencing of gene expression.[5,6] Physiologically, RNAi is a conserved cellular pathway, activated by dsRNA precursors that are rapidly processed into short RNA duplexes (siRNAs) of 21–25 nucleotides in length by the cellular enzyme Dicer. These siRNAs are incorporated into a protein complex that recognizes and cleaves its target RNAs.[5] In mammalian cells, specific silencing of gene expression could be achieved by introducing 21-nt siRNAs duplexes.[7]

Silencing of *c-myc* gene expression by vectors expressing siRNAs or shRNAs, siRNA prepared chemically or enzymaticaly in human lung carcinoma A549, hepatoma HepG2, GT38, HeLa, and breast cancer adenocarcinoma MCF-7 cell lines was reported.[8–12] Our studies[13,14] demonstrated that siRNA-Ex3 (targeted to the third exon of *c-myc* gene) prepared enzymatically can effectively silence *c-myc* gene expression in KB-3-1 and SK-N-MC cells and subsequently reduce their proliferation rate. In the current study the ability of different siRNAs and long dsRNA to regulate *myc* genes expression and proliferation of cancer cells via RNA interference mechanism or interferon response induction was investigated. The reliability of commonly used markers of interferon response is discussed.

MATERIALS AND METHODS

siRNA and dsRNA Duplex Preparation

siRNA duplexes with the following sense strands were used: siRNA-Ex3 corresponded to nt 1450–1470 relative to the start codon of *c-myc* mRNA (5'-GAGGUGCCACGUCUCCACAUU-3'); siRNA-Ex2 corresponded to nt 697–715 of *c-myc* mRNA and 302–320 nt of N-myc mRNA with two mismatches indicated by bold underlined letters (5'-GAGGA**U**AUCUGGAAGAA**A**UUU-3'); siRNA-I corresponded to the sequence in the first intron of human *MDR1* gene (5'-AAAUCUGAAAGCCUGACACUU-3'); siRNA-Scr—the scrambled negative control, which has no significant homology to any known gene sequences from mouse, rat, or human (5'-AAAUCUGAAAGCCUGACACUU-3'). The siRNA duplexes were prepared by *in vitro* transcription with T7 RNA polymerase on linear dsDNA-templates according to Ambion technology (available at http://www.ambion.com/techlib/tn/103/2.html) with some modifications. Briefly, in separate reactions, two DNA oligonucleotides, which contain 21 nt antisense sequence of each transcribed RNA and 8 nt leader sequence, optimized for maximal RNA yield, were hybridized to the universal T7 promotor primer (5′-TAATACGACTCACTATAGGAGACAGG-3′) that contains T7 promotor sequence and 8 nt complementary to the leader sequence. To create double-stranded siRNA transcription templates, 3′ ends of hybridized oligonucleotides were extended by the Klenow fragment of *Escherichia coli* DNA polymerase I. Polymerization assays were carried out in the presence of 10 μM oligonucleotides in 50 mM Tris-HCl buffer (pH 7.6), containing 10 mM $MgCl_2$, 5 mM DTT, 0.25 mM dNTP, and 0.125 u/μL Exo-Klenow fragment (SibEnzyme, Novosibirsk, Russia) at 25°C for 1 h. Subsequent transcription assays were done under standard reaction conditions (at 37°C for 4 h) in reaction mixture containing 0.5 μg of DNA template, 40 mM Tris-HCl (pH 8.1), 16 mM $MgCl_2$, 2 mM spermidine, 50 mM NaCl, 0.1 mg/mL BSA, 10 mM DTT, 4 mM NTP and 5 u/μL T7 RNA polymerase. In transcription assay we used natural uridine 5′-triphosphate (UTP) instead of modified UTP proposed by Ambion (Ambion, CA). Additional 5′-terminal universal leader sequences were removed from transcripts by T1 RNase after hybridization of sense and antisense strands. The RNA was incubated in the presence of 0.8 u/μL of RNAse T1 (Fermentas Vilnius, Lithuania) in 1x buffer (50 mM Tris-HCl, pH 7.0; 2 mM EDTA; 200 mM NaCl) at 37°C for 2 h. Obtained siRNA was purified by reverse HPLC on Symmetry C-18 column (4.6 × 150 mm), which removes excess nucleotides, short oligomers, proteins, and salts from the reaction mixture. Long double-stranded dsMyc corresponding to 1159–1631 nt of *c-myc* mRNA was prepared by *in vitro* transcription with T7 RNA polymerase on PCR fragments amplified with the use of T7 promotor sequence containing primers. The obtained dsMyc was purified by gel-filtration.

Cell Culture and siRNA Transfection

Human oral carcinoma cell line KB-3-1 and human neuroblastoma cell lines IMR-32 and SK-N-MC were obtained from Institute of Cytology RAS (St. Petersburg, Russia). The cells were grown in Dulbecco's modified Eagle's medium supplemented with 10% fetal bovine serum, 100 u/mL penicillin, 100 μg/mL streptomycin, and 0.25 μg/mL amphotericin at 37°C in a humidified atmosphere containing 5% CO_2/95% air. Cells in exponential phase of growth were plated in 96-well plates (KB-3-1 at density 10^3 cells/well, SK-N-MC and IMR-32 at density 10^4 cells/well) or 24-well plates (KB-3-1 at density 7.5 × 10^3 or 3 × 10^4 cells/well, SK-N-MC at density 6 × 10^4 or 10^5 cells/well, and IMR-32 at density 6 × 10^4 cells/well) 1 day before experiment and exposed to siRNA using Oligofectamine (Invitrogen, Carlsbad, CA) according to the manufacturer's protocol. *c-myc* mRNA level was examined 1–12 days after the siRNA treatment. Untreated or treated with Oligofectamine only cells were used as controls.

Reverse Transcription-PCR

Quantification of specific mRNAs by reverse transciption polymerase chain reaction (RT-PCR) was done essentially as described.[15] Briefly, total RNA was isolated from KB-3-1, SK-N-MC or IMR-32 cells monolayers using SDS/phenol-based standard method.[16] The reverse transcription reaction was performed in a final volume of 20 μL containing 1 μg of total RNA, 5 μM oligo(dT)$_{15}$ primers, 50 mM Tris-HCl (pH 8.3), 75 mM KCl, 3 mM MgCl$_2$, 0.5 mM dNTP, 5 mM DTT, and 10 u of murine leukemia virus reverse transcriptase (MuLV RT). The reaction mixture was incubated at 42°C for 1 h.

The newly synthesized cDNA was amplified by PCR in a total volume 20 μL containing 2 μL of cDNA template, 10 mM Tris-HCl (pH 8.3), 50 mM KCl, 1.5 mM MgCl$_2$, 0.01% Tween-20, 0.25 mM dNTP, 0.25 μM each primer, and 2 u of Taq polymerase. Amplification cycles were the following: one cycle at 95°C for 3 min, 58°C for 1 min, 72°C for 1 min, and subsequently 25 cycles at 95°C for 1 min, 58°C for 1 min, 72°C for 1 min. Values for the expression of each gene were normalized to the expression of β$_2$-*microglobuline* (*beta$_2$-m*). Sequences of the primers used for RT-PCR were the following: c-myc forward: 5'-CTCCTCGGTGTCCGAGGACC-3' and c-myc reverse: 5'-GTTCGCCTCTTGACATTCTCC-3'; N-myc forward: 5'-TCTGTCGGT TGCAGTGTTG-3' and N-myc reverse: 5'-TTCTCAAGCAGGATCTCCG-3'; OAS1 forward: 5'-GATGTGCTGCCTGCCTTTGATG-3' and OAS1 reverse: 5'-CCCTGGGCTGTGTTGAAATGTG-3'; PKR forward: 5'-GCGT GAAGTAAAAGCATTGG-3' and PKR reverse: 5'-GCTCATGTATCGCAA AGTTCC-3'; *beta$_2$-m* forward: 5'-ATCTTCAAACCTCCATGATG-3' and *beta$_2$-m* reverse: 5'-ACCCCCACTGAAAAAGATGA-3'. RT-PCR products (455 bp from *c-myc*, 385 bp from *N-myc*, 320 bp from 2',5'-oligoadenylate synthetase (*OAS1*), 531 bp from dsRNA-induced protein kinase R (*PKR*), and 120 bp from *beta$_2$-m* genes) were subjected to electrophoresis on 1.5% agarose gel and visualized by ethidium bromide staining. Agarose gels were photographed and band intensities were determined using Gel-Pro Analyzer 4.0 program.

MTT Assay

Cell number was determined using a colorimetric assay based on reduction of the dye 3-[4,5-dimethylthiazol-2-yl]-2,5-diphenyltetrazolium bromide (MTT) by living cells.[17] Exponentially growing cells were plated at density 10^3 cells/well for KB-3-1 and 10^4 cells/well for SK-N-MC and IMR-32 cells in 96-well microtiter plates (100 μL/well) and transfected with different concentrations of siRNA (25–300 nM) as described above. After 24–120 h incubation at 37°C, MTT was added to final concentration of 0.5 mg/mL. Three hours later culture medium was removed, MTT formosan crystals in each well were solubilized in 100 μL DMSO, and absorbance was measured spectrophotometrically at 570 nm. Index of proliferation reflecting relative number of living cells normalized to the number of untreated cells 24 h after transfection was used for data presentation.

RESULTS

Silencing of c- *and* N-myc *Expression by dsRNAs*

In our experiments we used 21 nt siRNA-Ex3 with 3' overhanging uridine dinucleotides targeted to the sequence (1452–1470) of *c-myc* mRNA for inhibition of *c-myc* gene expression. siRNA-Ex2 was designed to target both *c-* and *N-myc* genes simultaneously, sequence of this siRNA corresponds to homologous region nt 697–715 of *c-myc* mRNA and 302–320 nt of *N-myc* mRNA with two mismatches (one mismatch compared to each gene). The scrambled siRNA-Scr, which has no significant homology to any known gene sequences from mouse, rat, or human, was used as a negative control. siRNA-I corresponded to the sequence of the 1 intron of *MDR1* gene was selected as control, because it has no mRNA target in the studied cell lines. Recent experimental data[18,19] indicate that sequences of siRNAs are much more important for the efficacy of silencing than accessibility of the target sequence in the secondary structure of mRNA, which is crucial in the antisense technology. Our siRNAs were designed according to the rules established in Reference 18: (i) A/U at the 5' end of the antisense strand; (ii) G/C at the 3' end of the sense strand; (iii) A/U-rich 5' terminal one-third of the antisense strand; and (iv) the absence of long GC stretches. The target sequence was aligned to the human genome database in a BLAST search to eliminate significant homology to other genes. Long double-stranded dsMyc corresponded to 1159–1631 nt of *c-myc* mRNA was designed to act at two levels of gene regulation: as interfering RNA and as interferon inducer. The siRNAs and long dsMyc duplexes were prepared by *in vitro* transcription with T7 RNA polymerase on dsDNA-templates as described in the Materials and Methods section. Commercially available interferon inducer poly(I:C) was used as a control.

Activity of siRNAs, dsMyc, and poly(I:C) was assayed in experiments with human epidermal oral carcinoma KB-3-1 and neuroblastoma SK-N-MC cell lines, which are characterized by elevated *c-myc* mRNA expression level[20,21] and *N-myc* overexpressing neuroblastoma IMR-32.[22] Expression levels of *c-* and *N-myc* genes in tested cell lines were evaluated by RT-PCR, using *beta$_2$-m* as internal standard. To investigate the effect of short and long dsRNAs on *c-myc* mRNA level, siRNAs in concentration 150 nM (~210 ng/mL) and dsRNA in concentration 50 ng/mL complexed with Oligofectamine were incubated with the cells and *c-myc* mRNA level was assayed by RT-PCR 24, 48, and 72 h after the exposure. The data were quantified and normalized to *beta$_2$-*

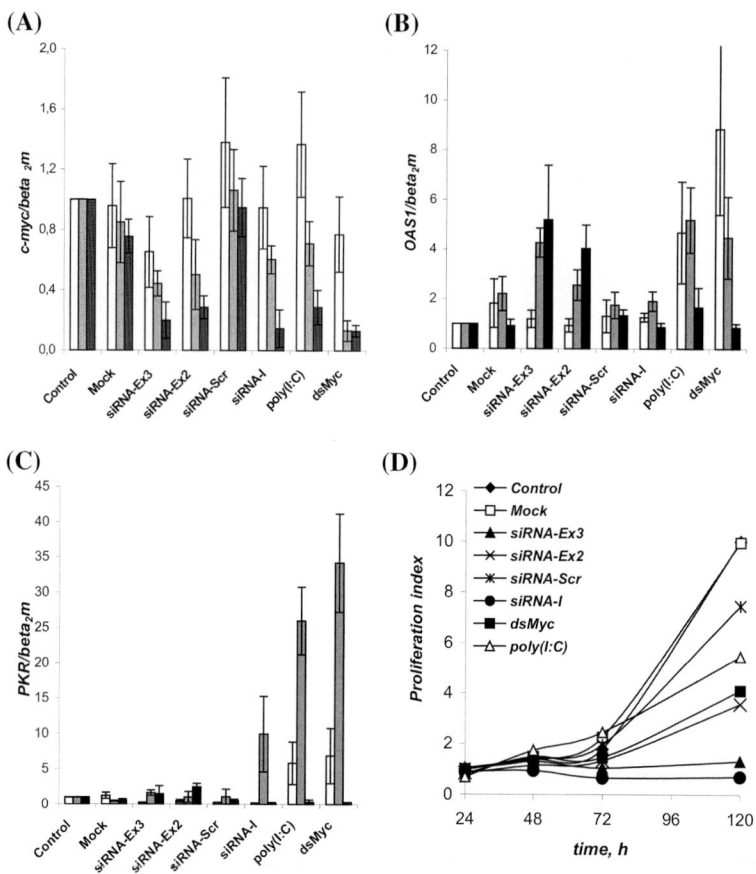

FIGURE 1. Changes in c-myc (**A**), OAS1 (**B**), PKR (**C**) mRNA levels, and proliferation index (**D**) after transfection of KB-3-1 cells with siRNAs (150 nM) or dsRNA (50 ng/mL). The level of specific mRNAs was measured by RT-PCR 24 h (white bars), 48 h (gray bars), and 72 h (black bars) after transfection. Relative gene expression levels were normalized to $beta_2$-m mRNA. Error bars indicate the standard deviation of three individual sets of experiments. Cells proliferation rate was analyzed using MTT-assay; index of proliferation is the relative numbers of living cells. Amounts of living cells in controls (untreated cells) 24 h after transfection were set at 1. Presented data are averages of three independent experiments. Standard deviations were less then 15%.

m mRNA levels. It was found (FIG. 1 A) that siRNA-Ex3 and Ex2 induce efficient reduction of c-myc mRNA level in carcinoma KB-3-1 (FIG. 1 A) and neuroblastoma SK-N-MC (data not presented) cell lines. Incubation of the cells during 72 h with 150 nM siRNA-Ex3 decreases c-myc mRNA level to 20% of the initial value in KB-3-1 and to 40% in SK-N-MC cells, the N-myc mRNA level in IMR-32 cells was not affected (FIG. 2 A). The efficacy of

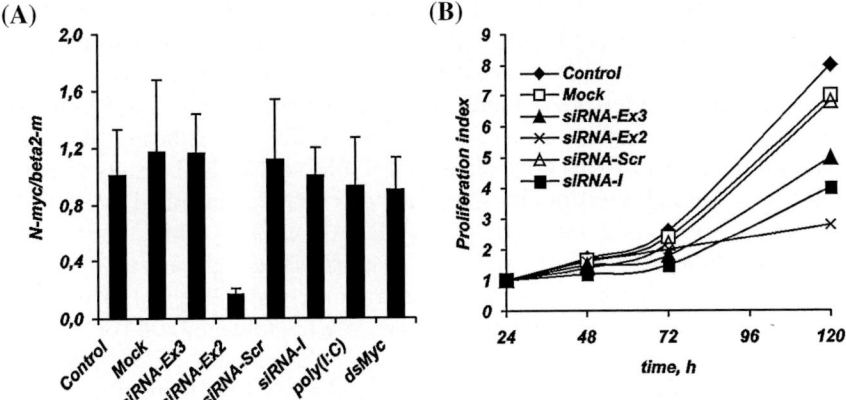

FIGURE 2. Changes in *N-myc* mRNA level (**A**) and proliferation index (**B**) after transfection of IMR-32 cells with siRNAs (150 nM) or dsRNA (50 ng/mL) 72 h after transfection. Relative *N-myc* expression level was normalized to $beta_2$-*m* mRNA level determined by RT-PCR. Error bars indicate the standard deviation of three individual sets of experiments. Cells proliferation rate was analyzed using MTT-assay; index of proliferation is the relative number of living cells. Amounts of living cells in controls (untreated cells) 24 h after transfection were set at 1. Presented data are averages of three independent experiments. Standard deviations were less then 15%.

siRNA-Ex2-induced silencing of *c-myc* expression in KB-3-1 and SK-N-MC was similar to the activity of siRNA-Ex3 (FIG. 1 A for KB-3-1). Transfection of IMR-32 cells with 150 nM siRNA-Ex2 resulted in decrease of *N-myc* mRNA level to 15% from the level in control cells. No decrease of *c-myc* mRNA level in KB-3-1 and SK-N-MC cells or *N-myc* mRNA level in IMR-32 cells treated with siRNA-Scr (FIG. 1 A) or with transfection reagent only was observed (FIGS. 1 A and 2 A). It was found that dsRNAs which has no significant homology to target genes: siRNA-I in concentration 150 nM and poly(I:C) in concentration 50 ng/mL were able to reduce the level of *c-myc* mRNA with comparable efficacy. The most efficient inhibition of *c-myc* expression was observed in KB-3-1 cells transfected with 50 ng/mL dsMyc (FIG. 1A), 48 h after transfection *c-myc* mRNA level was already fivefold lower than in control cells. The response of IMR-32 cells to dsRNA preparations was dramatically different: non of unspecific siRNA or dsRNA was able to change significantly the level of *N-myc* gene.

Antiproliferative Activity of dsRNA

c-Myc oncogene plays an important role in cell growth regulation, its overexpression leads to tumorigenesis and uncontroled proliferation. Disrupting the expression of the *c-myc* gene should lead to inhibition of cell growth. It has been shown, that silencing of *c-myc* gene expression leads to inhibition of cell growth in *c-myc* overexpressing cell lines.[8–10,14] To compare antiproliferative activity of investigated siRNA and dsRNAs, we monitored the number of living cells after transfection at 24 h intervals using the MTT assay. The number of untreated cells 24 h after transfection was set at 1. The data indicates (FIG. 1 D) that the treatment of KB-3-1 cells with 150 nM siRNA-Ex3 results in complete arrest of cell division, the same activity was demonstrated by 150 nM siRNA-I. Antiproliferative activity of siRNA-Ex2 and dsMyc was lower: threefold decrease of proliferation rate was observed 4 days after the treatment with these preparations in concentrations, respectively, 150 nM and 50 ng/mL. Poly(I:C) was even less effective, only 1.8-fold inhibition of proliferation was caused by 50 ng/mL poly(I:C). Neuroblastoma SK-N-MC displayed similar, but less pronounced response to si- and dsRNAs (data not presented). Extent of proliferation inhibition in KB-3-1 and SK-N-MC correlated with the level of *c-myc* silencing in these cells. The most efficient inhibitor of IMR-32 cells proliferation was siRNA-Ex2. Transfection of this siRNA in concentration 150 nM reduced the proliferation rate by factor 3.5–4 (FIG. 2 B). Less efficient inhibition was caused by siRNA-I and siRNA-Ex3 which did not exert any influence on the *N-myc* mRNA level.

Dynamics of PKR *and* OAS1 *mRNA Levels Variation in the Cells after dsRNA Transfection*

The induction of interferon response under action of dsRNA was investigated using *OAS1* and *PKR* as markers of interferon response. The levels of *OAS1* and *PKR* mRNA in KB-3-1 cells were measured by RT-PCR, $beta_2$-*m* mRNA was used as internal standard. The data indicate (FIG. 1 B) that 24 h after treatment of KB-3-1 cells with 50 ng/mL dsMyc or poly(I:C) or the expression of *OAS1* gene was induced five- to eightfold, whereas the treatment with 150 nM siRNAs had no significant effect on the expression of this gene. The elevated level of *OAS1* mRNA was detected up to 48 h after transfection with dsRNAs; after 72 h of incubation the level of *OAS1* mRNA did not differ significantly from that in untreated cells or cells, treated with Oligofectamine only. In the cells transfected with 150 nM siRNA-I or siRNA-Scr the level of *OAS1* mRNA was not affected by the treatment at all. In the case of siRNA-Ex3 or siRNA-Ex2 the increase in *OAS1* mRNA level was detected later (48 h after transfection) than in the case of long dsRNAs. At the same time no significant variations in *PKR* mRNA level (FIG. 1 C) was observed after transfection of siRNA-Ex3, -Ex2, and -Scr in KB-3-1 cells. In the case of potential interferon inducers, the level of this gene increase 10-, 25- and 35-fold 48 h after transfection of siRNA-I, poly(I:C) and dsMyc, respectively. Twenty-four hours after transfection a moderate increase in *PKR* mRNA level (five- to sixfold) was observed in the case of long dsRNAs, 72 h after transfection with long dsRNA the *PKR* mRNA level in the cells did not exceed the level in control cells.

DISCUSSION

Deregulation of genes encoding components of signaling pathways are considered as a key factor in the development of different types of tumors in humans. We investigated inhibition of *MYC* genes by short and long double-stranded RNAs in KB-3-1, SK-N-MC, and IMR-32 cells. For comparison of specific and nonsequence specific activity of dsRNAs the following concentrations were used: 150 nM concentration for siRNAs and concentration 50 ng/mL for long dsRNA. Concentration of siRNA was selected according to our previous study,[13,14] where concentration dependence and kinetics of inhibition of *c-myc* expression in KB-3-1 and SK-N-MC by siRNA-Ex3 have been investigated. Relatively high concentration of siRNAs required for efficient silencing could be explained by short half-life of *c-myc* mRNA (0.5–1 h in different cells).[23] Concentration of long dsRNA was determined in control experiments. Moderate concentration dependence was observed under action of dsMyc and poly(I:C) at concentrations 50–250 ng/mL, at higher concentrations toxic effects of long dsRNA caused high extent of cell death.

Sequence-specific siRNA-Ex3 targeted to the third exon of *c-myc* gene was found to decrease the level of *c-myc* in KB-3-1 (FIG. 1 A) and SK-N-MC, but not *N-myc* mRNA in IMR-32 cells (FIG. 2 A). siRNA-Ex3 decreased the rate or even arrested the proliferation of *c-myc* overexpressing cell lines KB-3-1 (FIG. 1 D) and SK-N-MC, but did not affect significantly the proliferation of IMR-32 (FIG. 2B), which expresses high level of *N-myc* mRNA and relatively low level of *c-myc* mRNA.[14] Because overexpression of *c-* or *N-myc* is responsible for the formation of different types of neuroblastomas, we designed single siRNA to target both *c-* and *N-myc* genes simultaneously. The sequence of this siRNA-Ex2 corresponds to homologous region nt 697–715 of *c-myc* mRNA and 302–320 nt of *N-myc* mRNA with two mismatches (one mismatch compared to each gene). The obtaned data indicate, that this siRNA is able to downregulate both genes and to inhibit proliferation of KB-3-1, SK-N-MC, and IMR-32 cells. Our data support multiple observations reported in the literature,[24] that mismatches, outside the center of siRNA are well tolerated and do not abolish silencing activity of siRNA.

Long dsRNAs and some short dsRNA can stimulate innate cytokine responses in mammals, which can induce proliferation blockage, differentiation or apoptosis in cancer cells.[25,26] Long dsRNAs: dsMyc homologous to the 3 exon of *c-myc* gene were found to inhibit proliferation and to decrease the mRNA level of interferon-sensitive genes *c-myc* (FIG. 1 D) and *beta-actin* (our unpublished data) when delivered into KB-3-1 carcinoma and SK-N-MC neuroblastoma cells. The level of downregulation was higher than that of synthetic interferon inducer poly(I:C).

Two control siRNAs: siRNA-I corresponding to the sequence in the first intron of human *MDR1* gene and having no mRNA target in the studed cell lines and siRNA-Scr—the scrambled negative control, which has no significant homology to any known gene sequences in mouse, rat, or human were tested on the same cells. siRNA-Scr did not affect c- and *N-myc* expression and proliferation of the studied cells. siRNA-I inhibited *c-myc* expression in KB-3-1 cells (FIG. 1 A) and proliferation (FIG. 1 D) of the cells. The silencing and antiproliferative activity of this siRNA in KB-3-1 cells was even higher, than that of sequence-specific inhibitor siRNA-Ex3. No obvious homology to immune-stimulating motives in siRNA reported previously[25,26] was found in its sequence. The fact, that the level of *N-myc* mRNA in IMR-32 was not affected by siRNA-I can be explained by the properties of the cell line. IMR-32 neuroblastoma is known to be resistant to all-*trans*-retinoic acid and α–interferon due to the corruption of interferon-response element in *N-myc* promotor.[22] The obtained data suggest, that siRNA-I acts as an interferon inducer and its ability to inhibit proliferation of IMR-32 is realized via action on other than *N-myc* genes responsible for cell growth, including *c-myc*, because siRNA-Ex3 targeted *c-myc* gene caused twofold reduction of IMR-32 proliferation rate (FIG. 2 B).

To separate sequence-specific and sequence-nonspecific action of dsRNAs we investigated the expression of two key interferon responsive genes, *OAS1* and *PKR* in KB-3-1 cells that were treated with 150 nM siRNAs or 50 ng/mL dsRNAs. Dynamics of *PKR* and *OAS1* mRNA levels variation revealed that long dsRNAs induce efficient, but short-term activation of expression of these genes. The highest level of specific mRNA was detected 24–48 h for *OAS1* and 48 h after transfection for *PKR*. Seventy-two hours after transfection the expression level reverted to the level in untreated cells. In the cells transfected with siRNA-I, no significant increase in *OAS1* mRNA was found, the short-term activation of *PKR* expression was 2.5–3 times less efficient than in the case of long dsRNA (dsMyc and poly[I:C]). The dynamics of *OAS1* mRNA level variations caused by sequence-specific inhibitors acting via RNA interference mechanism differs from the dynamics caused by of interferon inducers. The observed three- to fivefold activation 48–72 h after transfection with siRNA-Ex3 or Ex2 can be attributed to some off-target effects of the siRNAs or to effect of *c-myc* mRNA level reduction on the expression of *OAS1*. We concluded that expression of the genes negatively regulated by interferon normalized to the level of "interferon-resistant" gene can be used as reliable markers of interferon response. All interferon inducers examined in the present research caused concentration dependent inhibition of *c-myc* expression in KB-3-1- and SK-N-MC cells and reduction of their proliferation rate. In this case *c-myc* was not only the marker of interferon response, but also a regulator of cell growth. Our preliminary results (data not shown) indicate that the level of another interferon-sensitive gene—*beta-actin*, normalized to the level of $beta_2$-m—correlates well with antiproliferative activity of interferon inducers. Taking into account, that rearrangements in the promotor region can result in the loss of interferon sensitivity of the gene as it happened in the case of *N-myc* in IMR-32 cells, more than one marker should be used for evaluation of interferon response as side effect in RNA interference experiments.

Results of the present study evidence that dsRNAs can serve as antiproliferative agents, causing both, sequence-specific silencing of specific target mRNAs and activating innate immunity response.

ACKNOWLEDGMENT

This work was supported by RAS programs "Physico-Chemical Biology" and "Fundamental Science for Medicine," Interdisciplinary grant from SB RAS 5.10.

REFERENCES

1. SPENCER, C.A. & M. GROUDINE. 1991. Control of c-myc regulation in normal and neoplastic cells. Adv. Cancer Res. **56:** 1–48.
2. NESBIT, C.E., J.M. TERSAK & E.V. PROCHOWNIK. 1999. MYC oncogenes and neoplastic disease. Oncogene **18:** 3004–3016.
3. REYNOLDS, C.P., Y. WANG, L.J. MELTON, *et al.* 2000. Retinoic-acid-resistant neuroblastoma cell lines show altered MYC regulation and high sensitivity to fenretinide. Med. Pediatr. Oncol. **35:** 597–602.
4. LIPPMAN, S.M., B.S. GLISSON, J.J. KAVANAGH, *et al.* 1993. Retinoic acid and interferon combination studies in human cancer. Eur. J. Cancer **5:** 9–13.
5. MEISTER, G. & T. TUSCHL. 2004. Mechanisms of gene silencing by double-stranded RNA. Nature **431:** 343–349.
6. MELLO, C.C. & D. CONTE. 2004. Revealing the world of RNA interference. Nature **431:** 338–342.
7. ELBASHIR, S.M., J. HARBORTH, W. LENDECKEL, *et al.* 2001. Duplexes of 21-nucleotide RNAs mediate RNA interference in cultured mammalian cells. Nature **411:** 494–498.
8. HOSONO, T., H. MIZUGUCHI, K. KATAYAMA, *et al.* 2004. Adenovirus vector-mediated doxycycline-inducible RNA interference. Human Gene Ther. **15:** 813–819.
9. WANG, Y., S. LIU, G. ZHANG, ZHOU, *et al.* 2005. Knockdown of c-Myc expression by RNAi inhibits MCF-7 breast tumor cells growth *in vitro* and *in vivo*. Breast Cancer Res. **7:** R220–R228.
10. GAO, X., H. WANG, & T. SAIRENJI. 2004. Inhibition of Epstein-Barr virus (EBV) reactivation by short interfering RNAs targeting p38 mitogen-activated protein kinase or c-myc in EBV-positive epithelial cells. J. Virol. **78:** 11798–11806.
11. DEMETERCO, C., P. ITKIN-ANSARI, B. TYRBERG, *et al.* 2002. c-Myc controls proliferation versus differentiation in human pancreatic endocrine cells. J. Clin. Endocrinol. Metab. **87:** 3475–3485.

12. SHEN, L., C. ZHANG, J.L. AMBRUS, *et al*. 2005. Silencing of human c-myc oncogene expression by poly-DNP-RNA. Oligonucleotides **15:** 23–35.
13. KABILOVA, T.O., E.L. CHERNOLOVSKAYA, A.V. VLADIMIROVA, *et al*. 2004. Silencing of c-myc expression in tumor cells by siRNA. Nucl. Nucl. Nucleic Acids **23:** 867–872.
14. KABILOVA, T.O., E.L. CHERNOLOVSKAYA, A.V. VLADIMIROVA, *et al*. 2006. Inhibition of human cancer cell proliferation by anti c-myc siRNA. Oligonucleotides **3:** 153–157.
15. GOSWAMI, P.C., L.D. ALBEE, D.R. SPITZ, *et al*. 1997. A polymerase chain reaction assay for simultaneous detection and quantitation of proto-oncogene and GAPD mRNAs in different cell growth rates. Cell Prolif. **30:** 271–282.
16. CHATTOPADHYAY, N., R. KHER & M. GODBOLE. 1993. Inexpensive SDS/phenol method for RNA extraction from tissues. Biotechniques **15:** 24–26.
17. CARMICHAEL, J., W.G. DEGRAFF, A.F. GAZDAR, J.D. MINNA, *et al*. 1987. Evaluation of a tetrazolium-based semiautomated colorimetric assay: assessment of chemosensitivity testing. Cancer Res. **47:** 936–942.
18. UI-TEI, K., Y. NAITO, F. TAKAHASHI, *et al*. 2004. Guidelines for the selection of highly effective siRNA sequences for mammalian and chick RNA interference. Nucleic Acids Res. **32:** 936–948.
19. HOHJOH, H. 2004. Enhancement of RNAi activity by improved siRNA duplexes. FEBS Lett. **557:** 193–198.
20. FAGOT, D., C. BUQUET-FAGOT, F. LALLEMAND, *et al*. 1994. Antiproliferative effects of sodium butyrate in adriamycin-sensitive and -resistant human cancer cell lines. Anticancer Drugs **5:** 548–556.
21. FEO, S., C. DI LIEGRO, T. JONES, *et al*. 1994. The DNA region around the c-myc gene and its amplification in human tumor cell lines. Oncogene **9:** 955–961.
22. HEMMI, H., K. YAMADA, U. YOON, *et al*. 1995. Coexpression of the myc gene family members in human neuroblastoma cell lines. Biochem. Mol. Biol. Int. **36:** 1135–1141.
23. HANN, S.R. & R.N. EISENMAN. 1984. Proteins encoded by the human c-myc oncogene: differential expression in neoplastic cells. Mol. Cell Biol. **4:** 2486–2497.
24. DU, Q., H. THONBERG, J. WANG, *et al*. 2005. A systematic analysis of the silencing effects of an active siRNA single-nucleotide mismatched target sites. Nucleic Acids Res. **33:** 1671–1677.
25. JUDGE, A.D., V. SOOD, J.R. SHOW, *et al*. 2005. Sequence-dependent stimulation of the mammalian innate immune response by synthetic siRNA. Nat. Biotechnol. **23:** 457–462.
26. SIOUD, M. 2005. Induction of inflammatory cytokines and interferon responses by double-stranded and single-stranded siRNAs is sequence-dependent and requires endosomal localization. J. Mol. Biol. **348:** 1079–1090.

Design and Functional Activity of Phosphopeptides with Potential Immunomodulating Capacity, Based on the Sequence of Grb2-Associated Binder 1

AKOS KERTESZ,[a] BALAZS TAKACS,[a] GYORGYI VARADI,[b] GABOR K. TOTH,[b] AND GABRIELLA SARMAY[a]

[a]*Department of Immunology, Lorand Eotvos University, Budapest 1117, Hungary*
[b]*Department of Medical Chemistry, University of Szeged, Szeged 6720, Hungary*

> ABSTRACT: A cell membrane permeable phosphopeptide corresponding to the SHP-2 binding motif of Grb2-associated binder 1 (Gab1) interferes with the Gab1 adaptor-dependent functions and modulates B cell receptor-triggered intracellular signaling in B cell tumors.
>
> KEYWORDS: Burkitt lymphoma; Gab1; immunomodulation; phosphopeptides

INTRODUCTON

Grb2-associated binder 1 (Gab1) is a member of a scaffolding/adaptor protein family involved in the signal transduction pathways of growth factors, cytokines, and antigen receptors. Gab adaptor proteins have several tyrosine residues which are phosphorylated upon ligand-mediated tyrosine kinase activation and bind signaling molecules with SH2 domains, such as SHP-2 protein tyrosine phosphatase, phosphatidyl inositol 3-kinase (PI3-K), and phospholipase Cγ (PLCγ).[1,2] Gab adaptors also have pleckstrin homologous (PH) domain, which interacts with phosphatidyl inositol trisphosphate (PIP3) in the cell membrane, and proline-rich sequences providing binding sites for Grb2. The adaptor may bind to the signaling receptor directly via its Met binding domain, such as in case of epidermal growth factor receptor (EGFR), indirectly via Grb2 or Grb2/SHC complexes, and its membrane location is probably stabilized by the PH domain/PIP3 interaction.[3] Thus binding to Gab adaptors

---

Address for correspondence: Gabriella Sarmay, Eotvos Lorand University, Pazmany Peter setany 1/c, 1117 Budapest, Hungary. Voice: 36-1-2090-555; ext.: 8662; fax: 36-1-3812-176.
e-mail: sarmayg@elte.hu

alters the location of signaling enzymes, recruitment to the cell membrane brings them to the vicinity of their substrates and may increase their activity.

Catalytic activity of SHP-2 tyrosine phosphatase is stimulated by the engagement of its SH2 domains with specific phosphotyrosine containing motifs, in the case of EGF signaling, these are provided by Gab1 627Y and 659Y. It is now well accepted that SHP-2 has a positive role in signals generated by activated membrane receptor complexes, and the Ras/Erk pathway represents the major signaling module positively regulated by this phosphatase.[4,5] It was recently found that the negative regulatory protein of Ras, RasGAP has a binding site on Gab1 and this YXXP motif is a potential target for SHP-2. Dephosporylation of this motifs by SHP-2 inhibit RasGAP recruitment to cell membrane thus inducing a sustained Ras activation and downstream activation of protein kinase Erk1/2, major regulators of cell proliferation, differentiation, and survival.[6]

The p85 subunit of PI3-K binds to the tyrosine phosphorylated YVPM motifs of Gab1 (447Y, 472Y) upon a number of stimuli including B cell receptor (BCR), T cell receptor (TCR), and EGFR. Thus PI3-K is transported to the plasma membrane, to the vicinity of its substrate and becomes activated, resulting in the production of PIP3, that provides additional binding sites for PH domain containing proteins, such as Gab1 itself and PI3-K-dependent kinase (PDK). In turn PDK activates the serine/threonine kinase Akt/PKB. It was recently shown that Gab1 can be the primary mediator of EGF stimulated activation of the PI3-K/Akt cell survival pathway.[7–9]

The positive regulatory role of SHP-2 and PI3-K in signaling cascades leading to cell growth suggests their involvement in tumorigenesis, raising the possibility that SHP-2 and PI3-K may be excellent targets of the treatment of some forms of cancer.

We hypothesized that phosphopeptides representing the tyrosine-phosphorylated motifs of Gab1 may interfere with the appropriate signaling molecules in B cells, thus modulating the cell activation. Our main goal is to design cell membrane permeable phosphopeptides based on the sequence of Gab1 that would regulate the B cell response, ultimately inhibiting proliferation of B cell tumors. To this end first we tested an octaarginine peptide vector, and an octanoylated variant, which are potentially able to translocate into living cells and then coupled the phosphopeptide cargo to the optimal membrane permeable carrier. We have shown here that a phosphopeptide corresponding to the SHP-2 binding motif of Gab1 coupled to an octanoyl-R8 (OR8) carrier is able to get across the plasma membrane and modulate intracellular signaling.

MATERIAL AND METHODS

Burkitt lymphoma cell line BL41 was used in all of experiment. The cells were cultured in RPMI 1640 supplemented by 10% fetal calf serum (FCS).

Phosphopeptides were synthesized by solid phase method on TenteGel resin using Fmoc strategy. The products were purified by reverse phase-high performance-liquid chromatography (RP-HPLC), their structure and homogeneity were assessed by mass spectrometry (MS) and RP-HPLC methods. The link between the carrier and the phosphopeptide was made via Michael addition in liquid phase between the SH group of a cysteine in the C terminal end of the carrier peptide and a maleinimido group coupled to the N terminal end of the corresponding phosphopeptide. The same kind of reaction was used for the labeling of phosphopeptide with the fluorescent dye, Bodipy-FL (Molecular Probes/Invitrogen, Eugene, OR). The eight arginine containing short peptide (R8) was N terminally modified by an octanoyl group (OR8).

The transfection efficiency of Bodipy-FL-labeled carriers and OR8-phosphopeptide conjugate was analyzed by flow cytometry (FACSCalibur; Beckton-Dickinson Biosciences, San Jose, CA) and laser scanning confocal microscopy (Olympos Hamburg Europe GmbH, Germany).

To identify binding partners of phosphorylated G^{621}DKQVEYp LDLDLD^{633} (GDLD) Gab1 motif pull down assay was applied, by using biotinylated-phosphopeptide coupled to immobilized avidin-beads. The peptide-coated beads were incubated in 1% Triton X-100 containing detergent extracts of BL41 cells, the bound proteins were separated by SDS-PAGE, and then tested by Western blot analysis.

The SHP-2 activating capacities of the phosphopeptides were compared in an ELISA-based phosphatase assay.[10] Briefly, biotinylated phosphopeptides were added to avidine-coated plates, incubated with recombinant SHP-2, washed and then the residual level of peptide-phosphorylation was monitored by adding anti-phosphotyrosine.

Intracellular free [Ca^{2+}] was measured in cells loaded with Fluo-3AM by flow cytometry.

The degree of phosphorylation of intracellular proteins was monitored in BL41 cells stimulated by anti-IgM for 2 min. The cells were pretreated with various doses of OR8-GDLDpe for different time intervals, and then anti-IgM was added for 2 min. The cells were immediately frozen in liquid nitrogen then detergent extracts were prepared, separated by SDS-PAGE and analyzed by Western blotting using phosphotyrosine-specific antibodies. Erk and Akt activation was followed by probing the blots with phospho-Erk- and phospho-Akt-specific antibodies.

RESULTS AND DISCUSSION

First we analyzed the proteins interacting with the Gab1 phosphopeptides in the detergent extract of BL41 cells. GDLD phosphopeptide (GDLDpe) bound SHP-2, PLCγ, and Lyn, while the EL-PN phosphopeptide selectively bound the p85 regulatory subunit of PI3-K. In order to test whether the Gab1 phosphopeptides can activate SHP-2, phosphatase assays were applied. The results showed that GDLDpe did not only bound to, but efficiently activated SHP-2 phosphatase, while the other peptides were less effective. We have chosen this phosphopeptide, GDLD for further experiments.

To check the *in vivo* effect of the phosphopeptide we had to transport it through the cell membrane. To this end first we compared the transfection efficiency of a group of membrane-permeable peptide-based small molecular entities. One of the most typical peptide vectors is a short, arginine-rich segment derived from human immunodeficiency virus (HIV)-1 Tat.[11,12] A short peptide corresponding to residues 35–62 containing several Arg was first found to be able to transport an antibody fragment into tumor cells.[13] It was also shown that N-terminal stearylation of the octaarginine peptide increased its transfection efficiency.[14] On the basis of these previous studies first we tested the transfection efficiency of the eight arginine containing peptide, R8 and its derivative, OR8. Confocal microscopic studies showed that the Bodipy-FL-labeled peptides penetrated through the plasma membrane into the cells, and stained unevenly distributed small particles in the cytoplasm. We further analyzed the intracellular localization of the peptides by double staining, using $DiIC_{18}(3)$ (Molecular Probes) to detect membrane lipids, and LYSOTRACKER (Molecular Probes) to distinguish lysosomes, respectively. The results demonstrated that R8 and OR8 did not associate with membrane lipids, but located inside the cell, and showed a considerably high colocalization with the LYSOTRACKER, indicating that a significant fraction of the penetrated peptide can be found in lysosomes. OR8 seemed to be somewhat better as compared its transfection efficiency to R8, thus it was used in further experiments. Next the Gab1 phosphopeptide GDLD was coupled to the OR8 carrier and the peptide delivery into the cells was examined. OR8-GDLD, similarly to the OR8 carrier, penetrated into the cells and about 50% of the staining was found to be located in lysosomes (data not shown).

The kinetics and dose dependence of the peptide transport through the plasma membrane was studied by flow cytometry. The peptide uptake was dose dependent and relatively fast, we could not see any difference in staining intensity of samples treated with the OR8-GDLD for 5 or 20 min. The penetration of the peptide into the cells was independent of temperature, samples treated at 4°C and 37°C showed identical staining (FIG. 1). These data collectively indicate that the mechanism of uptake cannot be endocytosis, but it is most probably mediated by electrostatic interactions between the positively charged arginine residues and the polar heads of lipids in the plasma membrane, followed by inverted micelle formation.[15] Micelles may fuse with lysosomes, alternatively, may reopen inside the cells and the content is release into the cytoplasm.

Next we tested whether the GDLDpe delivered to the BL41 cells induce any functional change in the cells. As an early sign of cell activation the modulation of intracellular free $[Ca^{2+}]$ concentration was compared in untreated and OR8-GDLD treated BL41 cell samples. The carrier construct, OR8 had no effect

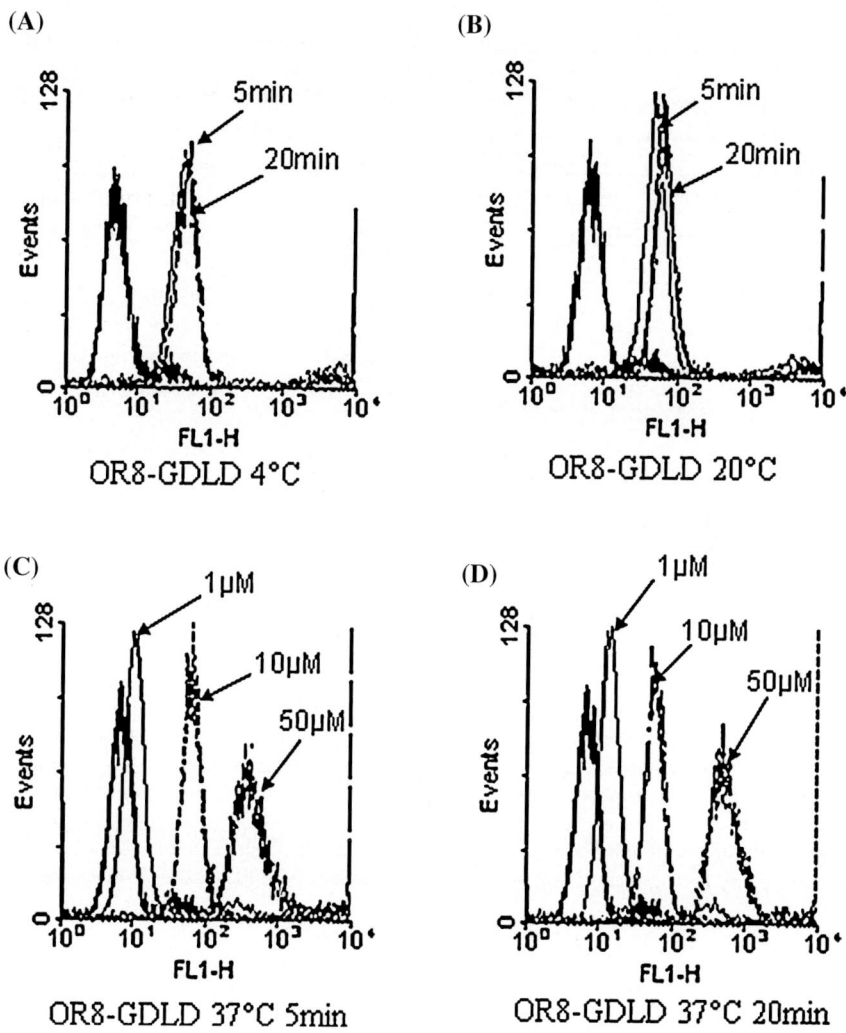

FIGURE 1. Flow cytometric analysis of the intracellular transport of the Bodipy-FL-labeled OR8-GDLD phosphopeptide conjugate. BL41 cells were treated with OR8-GDLD conjugate for 5 or 20 min at 4°C (**A**) and 20°C (**B**), respectively. BL41 cells were incubated with different doses of OR8-GDLD conjugate at 37°C for 5 min (**C**) and for 20 min (**D**), respectively.

on the cells, while the OR8-GDLD induced a significant [Ca^{2+}] mobilization, and the response was only partially inhibited by EGTA, suggesting that the GDLD phosphopeptide induces the release of [Ca^{2+}] from intracellular pools, followed by [Ca^{2+}] influx (FIG. 2 A). When the cells were treated with anti-IgM and the OR8-GDLD together, the phosphopeptide enhanced the release of

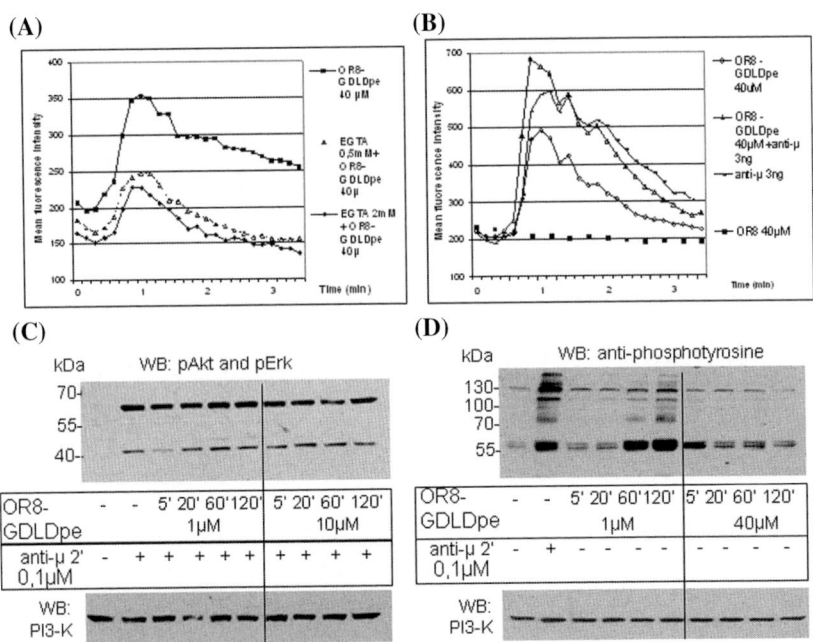

FIGURE 2. The effect of OR8-GDLD phosphopeptide on the intracellular $[Ca^{2+}]$ mobilization (**A, B**) and on protein phosphorylation (**C, D**). BL41 cells were pretreated with 1 μM or 10 μM OR8-GDLDpe for various time intervals, and then stimulated by anti-IgM for 2 min. The membranes were probed with antibodies specific for the phosphorylated form of Erk and Akt, respectively (**C**). The cells were incubated in the presence of 1μM and 40 μM OR8-GDLDpe, respectively, for different time intervals. Tyrosine phosphorylation was monitored by phophotyrosine-specific antibodies (**D**). Positive control samples were stimulated by anti-IgM for 2 min probing the membranes with PI3-K shows equal loading of samples.

$[Ca^{2+}]$ from intracellular pool, the amplitude of the anti-IgM-induced response increased approximately by 20% (FIG. 2 B). The same phosphopeptide bound to PLCγ as shown by the pull-down experiments, thus we suppose that it may activate the enzyme, inducing the cleavage of PIP2 and the release of inositol trisphosphate, which in turn binds to its receptor on endoplasmic reticulum and triggers $[Ca^{2+}]$ release.

GDLD phosphopeptide also bound to SHP-2 in the pull-down experiments, furthermore, it activated the phosphatase *in vitro*, so we supposed that it may influence Erk activation, which is regulated by SHP-2. It was shown earlier that SHP-2 may dephosphorylate Gab1 at the site that binds the negative regulatory protein, RasGAP, inducing thereby a sustained Erk activation.[6] Thus the effect of OR8-GDLDpe on Erk and Akt activation was tested. Adding 1 μM OR8-GDLD to the BL41 cells 5 min before anti-IgM stimuli reduced anti-IgM-triggered Erk phosphorylation, but this decreased phosphorylation was not detected after longer incubation time, or when higher dose of the phosphopeptide was added (FIG. 2 C). The same treatment had no effect on BCR-induced Akt phosphorylation. These data suggest that the GDLD phosphopeptide delivered into the cells may interfere with the phospho-Gab1/SHP-2 interaction, probably releasing SHP-2 from the Gab1 bound multi-molecular complex, and hampering thereby Erk activation. Surprisingly, adding the phosphopeptide to the cells 20 or more min before BCR clustering or at a higher dose had no detectable effect on Erk activation. We suppose that during the longer preincubation time SHP-2 might became activated and dephosphorylated the peptide, while interacting with the higher dose (10 μM) of the phosphopeptide SHP-2 might became activated earlier, leading to the same result.

To see whether the GDLD phosphopeptide has any effect on the tyrosine phosphorylation of intracellular proteins, the BL41 cells were treated with the OR8-GDLDpe and the tyrosine phosphorylation was tested by Western blots. Incubating the cells with the phosphopeptide for 60 and 120 min induced phosphorylation of several proteins, showing a similar pattern to the sample triggered by anti-IgM, while high dose (40 μM) of the peptide had no significant effect (FIG. 2 D). We suppose that again the low dose of phosphopeptide may inhibit SHP-2, allowing basic tyrosine phosphorylation, while the high dose may activate the phosphatase. Thus the GDLD phosphopeptide has a dose-dependent biphasic effect of SHP-2 activity in living cells.

Taking together these data indicate that Gab1 phosphopeptide GDLD delivered into the cells by membrane-permeable peptide-based vector, interferes with the signaling enzymes PLCγ and SHP-2, dose dependently enhancing intracellular Ca^{2+} mobilization, modulating the phosphorylation of intracellular proteins and resulting in a decreased Erk activation. We suppose that based on these data phosphopeptidomimetics may be designed, which may target Erk pathway in B cells.

ACKNOWLEDGMENTS

This work was supported by the Hungarian Scientific Research Fund (OTKA TS 044711) and by the National Research and Development Programs (NKFP 1A040-04).

REFERENCES

1. HOLGADO-MADRUGA, M., D.R. EMLET, D.K. MOSCATELLO, et al. 1996. A Grb2-associated docking protein in EGF- and insulin-receptor signalling. Nature **379:** 560–564.
2. INGHAM, R.J., M. HOLGADO-MADRUGA, C. SIU, et al. 1998. The Gab1 protein is a docking site for multiple proteins involved in signaling by the B cell antigen receptor. J. Biol. Chem. **273:** 30630–30637.

3. GU, H & B.G. NEEL. 2003. The "Gab" in signal transduction. Trends Cell Biol. **13:** 122–130.
4. AGAZIE, Y.M. & M.J. HAYMAN. 2003. Molecular mechanism for a role of SHP2 in epidermal growth factor receptor signaling. Mol. Cell Biol. **23:** 7875–7886.
5. ZHANG, S.Q., W. YANG, M.I. KONTARIDIS, *et al.* 2004. Shp2 regulates SRC family kinase activity and Ras/Erk activation by controlling Csk recruitment. Mol. Cell **13:** 341–355.
6. MONTAGNER, A., A. YART, M. DANCE, *et al.* 2005. A novel role for Gab1 and SHP2 in epidermal growth factor-induced Ras activation. J. Biol. Chem. **280:** 5350–5360.
7. HOLGADO-MADRUGA, M., D.K. MOSCATELLO, D.R. EMLET, *et al.* 1997. Grb2-associated binder-1 mediates phosphatidylinositol 3-kinase activation and the promotion of cell survival by nerve growth factor. Proc. Natl. Acad. Sci. USA **94:** 12419–12424.
8. YART, A., M. LAFFARGUE, P. MAYEUX, *et al.* 2001. A critical role for phosphoinositide 3-kinase upstream of Gab1 and SHP2 in the activation of Ras and mitogen-activated protein kinases by epidermal growth factor. J. Biol. Chem. **276:** 8856–8864.
9. MATTOON, D.R., B. LAMOTHE, I. LAX & J. SCHLESSINGER. 2004. The docking protein Gab1 is the primary mediator of EGF-stimulated activation of the PI3-K/Akt cell survival pathway. BMC Biol. **2:** 24.
10. KONCZ, G., G.K. TOTH, G. BOKONYI, *et al.* 2001. Co-clustering of Fcγ and B cell receptors induces dephosphorylation of the Grb2 associated binder 1 docking protein. Eur. J. Biochem. **268:** 3898–3906.
11. FUTAKI, S., T. SUZUKI, W. OHASHI, *et al.* 2001. Possible existence of common internalization mechanisms among arginine-rich peptides. J. Biol. Chem. **276:** 5836–5840.
12. WADIA, J.S. & S.F. DOWDY 2003. Modulation of cellular function by TAT mediated transduction of full length proteins. Curr. Prot. Pept. Sci. **4:** 97–104.
13. FUTAKI, S. 2005. Membrane-permeable arginine-rich peptides and the translocation mechanisms. Adv. Drug Deliv. Rev. **57:** 547–558.
14. FUTAKI, S., W. OHASHI, T. SUZUKI, *et al.* 2001. Stearylated arginine-rich peptides: a new class of transfection systems. Bioconjug. Chem. **12:** 1005–1011.
15. JOLIOT, A. & A. PROCHIANTZ. 2004. Transduction peptides: from technology to physiology. Nature Cell Biol. **6:** 189–196.

Distinct Activity of Peptide Mimetic Intracellular Ligands (Pepducins) for Proteinase-Activated Receptor-1 in Multiple Cells/Tissues

SATOKO KUBO,[a] TSUYOSHI ISHIKI,[a] ICHIKO DOE,[a] FUMIKO SEKIGUCHI,[a] HIROYUKI NISHIKAWA,[b] KENZO KAWAI,[b] HIROFUMI MATSUI,[c] AND ATSUFUMI KAWABATA[a]

[a]*Division of Physiology and Pathophysiology, School of Pharmaceutical Sciences, Kinki University, 3-4-1 Kowakae, Higashi-Osaka 577-8502, Japan*

[b]*Research and Development Center, Fuso Pharmaceutical Industries Limited, Osaka 536-8523, Japan*

[c]*Institute of Clinical Medicine, University of Tsukuba, Ten-nohdai, Tsukuba, Ibaraki 305-8575, Japan*

ABSTRACT: Proteinase-activated receptor-1 (PAR1), a G protein–coupled receptor (GPCR) for thrombin, can be activated not only by PAR1-activating peptides (PAR1APs) based on the N-terminal cryptic tethered ligand sequence but also by an N-palmitoylated (Pal) peptide, Pal-RCLSSSAVANRSKKSRALF-amide (P1pal-19), based on the intracellular loop 3 of PAR1, designated pepducin, in human platelets or PAR1-transfected cells. The present article evaluated the actions of P1pal-19 and also the shorter peptide, Pal-RCLSSSAVANRS-amide (P1pal-12), known as a possible PAR1 antagonist, in multiple cells/tissues that naturally express PAR1. P1pal-19 as well as a PAR1AP, TFLLR-amide, evoked cytosolic Ca^{2+} mobilization in cultured human lung epithelial cells (A549) and rat gastric mucosal epithelial cells (RGM1). P1pal-19 and TFLLR-amide, but not a PAR2-activating peptide, SLIGRL-amide, caused delayed prostaglandin E_2 formation in RGM1 cells. P1pal-19, like TFLLR-amide, produced endothelial NO-dependent relaxation in rat aorta and epithelial prostanoid-dependent relaxation in mouse bronchus. The P1pal-19-induced relaxation remained constant even after desensitization of PAR1 with TFLLR-amide in either tissue. P1pal-19 failed to mimic the contractile effects of TFLLR-amide in the endothelium-denuded preparations of rat aorta or superior mesenteric artery and the rat gastric longitudinal smooth muscle strips. P1pal-12 partially inhibited the vasorelaxation caused by TFLLR-amide and

Address for correspondence: Atsufumi Kawabata, Division of Physiology and Pathophysiology, School of Pharmaceutical Sciences, Kinki University, 3-4-1 Kowakae, Higashi-Osaka 577-8502, Japan. Voice: +81-6-6721-2332; ext.: 3863; fax: +81-6-6730-1394.
e-mail: kawabata@phar.kindai.ac.jp

P1pal-19, but not SLIGRL-amide, in the rat aorta. Our data thus indicate that P1pal-19 is capable of mimicking the effects of PAR1APs in the endothelial and epithelial, but not smooth muscle, cells/tissues, and suggest that P1pal-12 may act as a PAR1 antagonist in the vascular endothelium.

KEYWORDS: proteinase-activated receptor-1 (PAR1); thrombin; pepducin; intracellular ligands; palmitoylated peptide

INTRODUCTION

Proteinase-activated receptors (PARs), a family of G protein–coupled receptors (GPCRs), mediate a variety of cellular functions of certain serine proteinases. Among four members of the PAR family, PAR1, a thrombin receptor, and PAR2, a receptor for trypsin, mast cell tryptase and coagulation factors VIIa and Xa, in particular, appear to play critical and extensive roles in a variety of organs/tissues/cells including the gastrointestinal, respiratory, neuronal, and also circulatory systems.[1–3] Activation of PARs is triggered by proteolytic unmasking of the cryptic tethered ligand present in the N-terminal extracellular domain of the receptors. Exogenously applied synthetic peptides as short as 5–6 amino acids based on the tethered ligand sequence of PAR1, PAR2, or PAR4, but not PAR3, are capable of causing specific, nonenzymatic activation of each parent receptor by directly binding to the extracellular loop 2.[1–3]

Intriguingly, Covic et al.[4] have shown th N-palmitoylated peptides based on the intracellular loop 3 of certain GPCRs cause activation and/or inhibition of G protein signaling only in the presence of the parent GPCR. These peptides, termed *pepducins*, are considered to act as receptor-modulating agents by targeting the intracellular surface of GPCRs. An N-palmitoylated peptide, Pal-RCLSSSAVANRSKKSRALF-amide (P1pal-19), based on the intracellular loop 3 of human PAR1, has been demonstrated to cause G_q protein signaling, such as inositol phosphate production and cytosolic Ca^{2+} mobilization, in a manner dependent on the C-terminal of PAR1 in platelets or PAR1-transfected cells.[4,5] P1pal-19 is thus considered an intracellular agonist for PAR1. In contrast, a shorter N-palmitoylated peptide, Pal-RCLSSSAVANRS-amide (P1pal-12), based on the intracellular loop 3 of PAR1, appears to act as an intracellular antagonist for PAR1.[5,6] Similarly, PAR4-based pepducins have also been developed, and, most interestingly, the anti-PAR4 pepducin appears to act as an PAR4 antagonist even *in vivo*.[5–8] Given poor evidence for biological activity of PAR1 pepducins in cells/tissues other than human platelets and recombinant systems, the present article investigated and characterized the action of the PAR1-activating pepducin P1pal-19 in multiple cells/tissues that naturally express PAR1, as compared with the PAR1-activating peptide (PAR1AP)

TFLLR-amide, that is, an extracellular agonist. We also tested if P1pal-12 could act as a PAR1 antagonist in the vascular tissue.

MATERIALS AND METHODS

Animals

Male Wistar rats (7–9 weeks old) and male ddY mice (22–28 g) were purchased from Japan SLC, Inc. (Shizuoka, Japan). All animals were used with approval by the Kinki University School of Pharmaceutical Sciences' Committee for the Care and Use of Laboratory Animals, on the basis of Guiding Principles for the Care and Use of Laboratory Animals Approved by The Japanese Pharmacological Society.

Assay of Cytosolic Ca^{2+} Mobilization in Pulmonary and Gastric Epithelial Cell Lines

We used two distinct epithelial cell lines from the human pulmonary type II-like epithelium (A549) and from the rat normal gastric mucosal epithelium (RGM1), known to express PARs.[9,10] A549 cells were cultured in Dulbecco's modified Eagle's medium (DMEM) (Sigma-Aldrich, St. Louis, MO) supplemented with 10% heat-inactivated fetal bovine serum (FBS) (Thermo, Melbourne, Australia) and antibiotics, while RGM1 cells were in DMEM Nutrient Mixture F-12 HAM medium (Sigma-Aldrich) supplemented with 20% FBS in the presence of antibiotics. A549 and RGM1 cells (1.5×10^5 cells) were seeded and grown for 24 h on four round glass coverslips (13.2 mm in diameter) coated with collagen (Cellmatrix Type I-A; Nitta Gelatin, Inc., Osaka, Japan) in a tissue culture dish (35 mm in diameter) containing the above-mentioned standard medium. After additional 24-h incubation in the serum-free medium, the cells were loaded with Fura 2-AM (Dojindo, Kumamoto, Japan) at 10 μM for 1 h at room temperature in a buffer of the following composition (mM): NaCl, 150; KCl, 3; $CaCl_2$, 1.5; $MgCl_2$, 1.0; HEPES, 20; D-glucose, 10. The glass coverslips with the cells were then washed with the HEPES buffer and set on a holder. The holder was mounted in a quartz cuvette containing 2 mL of the HEPES buffer gassed with 95% O_2/5% CO_2, which was then placed in an Intracellular Ion Analyzer (CAF-110; Japan Spectroscopic Co., Tokyo, Japan). The cuvette was maintained at 25°C with constant stirring throughout the experiment. Cytosolic Ca^{2+} levels were measured as a ratio of fluorescence intensity at an emission wavelength of 500 nm with excitation at wavelengths of 340 and 380 nm applied alternately at a frequency of 128 Hz. The data were recorded on a computer at an acquisition rate of 100 Hz using a Power Lab

Recording System (AD Instruments, New South Wales, Australia). The cells were stimulated with the intracellular PAR1 agonist P1pal-19 or extracellular agonists, such as the PAR1AP, TFLLR-amide, and the PAR2-activating peptide (PAR2AP), SLIGRL-amide. The maximal increase in cytosolic Ca^{2+} levels was determined at the end of each experiment by adding 20 μM ionomycin, and changes in cytosolic Ca^{2+} levels caused by each agonist are expressed as the percentage of the ionomycin-evoked response.

Determination of Prostaglandin E_2 Formation in Rat Normal Gastric Mucosal Epithelial Cells

RGM1 cells were grown to 90% confluence in a CO_2 incubator maintained at 5% CO_2 and 37°C. The cells (1.5×10^5 cells/well) were seeded and grown in the above-mentioned standard medium for 24 h in a 6-well culture plate (well size: 33 mm in diameter), and then cultured in the FBS-free medium overnight. The cells were stimulated with the extracellular and/or intracellular agonists for PAR1 and PAR2, 1 h after exchange of the solution with the fresh FBS-free medium (1 mL). After stimulation for 18 h, the culture medium was collected, and the amount of prostaglandin E_2 in the diluted samples was determined spectrophotometrically using an EIA kit (Cayman Chemical Company, Ann Arbor, MI).

Tissue Preparation and Isometric Tension Recording

Animals were sacrificed by decapitation under urethane (1.5 g/kg, i.p.) anesthesia. The thoracic aorta, superior mesenteric artery, and stomach were excised from the rats, and the main bronchus was from the mice. Ring segments of the rat aorta (4 mm in length), rat superior mesenteric artery (3 mm in length), and mouse bronchus (2 mm in length), and strips of rat gastric longitudinal smooth muscle (15 mm × 5 mm) were prepared in an ice-cold Krebs-Henseleit solution (composition in mM: NaCl, 4.7; $CaCl_2$, 2.5; $MgCl_2$, 1.2; $NaHCO_3$, 25; KH_2PO_4, 1.2; glucose, 10). In some experiments, the aortic and arterial rings were stripped of the endothelium by gently rubbing the inner surface with a cotton string. The ring preparations were mounted between two triangle wire hooks, while the gastric strips were set with two clips. The preparations were suspended in organ baths containing 2 mL (ring preparations) or 3 mL (gastric strips) of Krebs-Henseleit solution maintained at 37°C and bubbled with 95% O_2/5% CO_2 to keep the pH at 7.4. The vascular and gastric segments were allowed to equilibrate for 60 min under a resting tension of 10 mN, and the bronchial rings were equilibrated for 45 min under a load of 2.5 mN. Isometric tension was recorded through a force-displacement transducer (UL-10GR, Minebea, Japan). The integrity of the vascular and gastric preparations was monitored a few times by measuring the contractile response to a high K^+

(50 mM)-containing Krebs-Henseleit solution in which the corresponding molar equivalent of NaCl was removed, and the contractility of the bronchial rings were checked by application of carbachol at 0.2–10 μM. The relaxant effects were determined in the rat aortic rings and mouse bronchial rings precontracted by phenylephrine and carbachol at 1 μM, respectively. The contractile responses are expressed as a percentage (% KCl) of the contraction induced by 50 mM KCl, and the relaxation responses are expressed as a percentage (% papaverine) of the relaxation to 100 μM papaverine. In inhibition experiments, N^G-nitro-L-arginine methyl ester hydrochloride (L-NAME), an inhibitor of NO synthase, at 100 μM, was added to the bath 5 min before addition of phenylephrine in the rat aortic rings, and indomethacin, a cyclooxygenase inhibitor, at 10 μM was applied 5 min before carbachol in the mouse bronchial rings. P1pal-12 was applied to the aortic rings 30 min before phenylephrine in the rat aorta. Desensitization experiments were conducted in the rat aorta and mouse bronchus. For PAR2 and/or PAR1 desensitization, SLIGRL-amide at 100 μM and/or TFLLR-amide at 100 μM were applied twice, respectively, to the rat aorta and mouse bronchus tissues precontracted with phenylephrine and carbachol, respectively, followed by application of P1pal-19. Complete desensitization of each PAR was confirmed by observing no response to the second application of the corresponding extracellular PAR agonist.

Drugs

P1pal-19 and P1pal-12 were prepared by Kurabo Industries, Ltd. (Osaka, Japan), and we ourselves verified their composition and purity by mass spectrometry. SLIGRL-amide and TFLLR-amide were synthesized by a solid-phase method and purified by high-performance liquid chromatography (HPLC), followed by determination of the composition and purity by mass spectrometry. Phenylephrine hydrochloride and acetylcholine chloride were obtained from Acros (Morris Plains, NJ) and Tokyo Kasei Kogyo (Tokyo, Japan), respectively, and indomethacin and papaverine were from Wako (Osaka, Japan). L-NAME and ionomycin were purchased from Sigma (St. Louis, MO) and Calbiochem (Darmstadt, Germany), respectively. Indomethacin was dissolved in 5 mM Na_2CO_3 just before the use, and P1pal-19, P1pal-12, ionomycin, and Fura 2-AM were dissolved in DMSO. All other chemicals were dissolved in distilled water.

Statistical Analysis

Data are shown as mean with SEM. Statistical analysis was performed by Student's t-test for two-group comparisons and by ANOVA followed by Tukey's test for multiple comparisons. Significance was set at a $P < 0.05$ level.

RESULTS

Cytosolic Ca^{2+} Mobilization Caused by Intracellular and Extracellular Agonists for PAR1 and/or PAR2 in Human Lung Epithelial Cells and Rat Normal Gastric Epithelial Cells

In the human lung epithelial cell line A549, the PAR1AP, TFLLR-amide, and the PAR2AP, SLIGRL-amide, that are extracellular agonists, at maximally effective concentrations (100 and 30 μM, respectively) elevated cytosolic Ca^{2+} levels (FIG. 1 A), as reported previously.[9] The intracellular PAR1 agonist P1pal-19 even at lower concentrations, 1–10 μM, caused more profound elevation in cytosolic Ca^{2+} levels in a concentration-dependent manner (FIG. 1 B). In the rat gastric mucosal epithelial cell line RGM1, TFLLR-amide and SLIGRL-amide at 10 μM, that was a maximally effective concentration in this cell line, produced Ca^{2+} signals (FIG. 1 C). P1pal-19 at 1–10 μM caused much greater increase in cytosolic Ca^{2+} levels in a concentration-dependent manner (FIG. 1 D).

Effects of Intracellular and Extracellular Agonists for PAR1 and/or PAR2 on Prostaglandin E_2 Formation in RGM1 Cells

TFLLR-amide at 100 μM caused prostaglandin E_2 release for 18 h, while SLIGRL-amide at 100 μM had no such effect (FIG. 1 E), in agreement with our recent evidence.[10] P1pal-19 even at 1 and 3 μM produced much greater prostaglandin E_2 for 18 h (FIG. 1 F).

Relaxant Effects of the Intracellular PAR1 Agonist in Rat Aortic Rings

In the aortic preparation precontracted with phenylephrine at 1 μM, P1pal-19 caused persistent relaxation (FIG. 2 A), which was concentration-dependent in a range of 1–30 μM (FIG. 2 B). It is of note that TFLLR-amide produced transient relaxation followed by contraction in the precontracted aortic rings (see FIG. 2 D tracing in the bottom). The P1pal-19-caused relaxation was abolished by pretreatment with L-NAME, a NO synthase inhibitor, at 100 μM or by removal of the endothelium (FIG. 2 C), a property corresponding to the mechanism for TFLLR-amide-triggered aortic relaxation.[11,12] Surprisingly, desensitization of PAR1 by applying TFLLR-amide twice failed to attenuate P1pal-19-induced relaxation responses in the precontracted aortic preparations (FIG. 2 D and E).

Relaxant Effects of the Intracellular PAR1 Agonist in Mouse Main Bronchus

P1pal-19 at 3–30 μM caused relaxation responses in a concentration-dependent manner in mouse main bronchus, as did the PAR1AP,

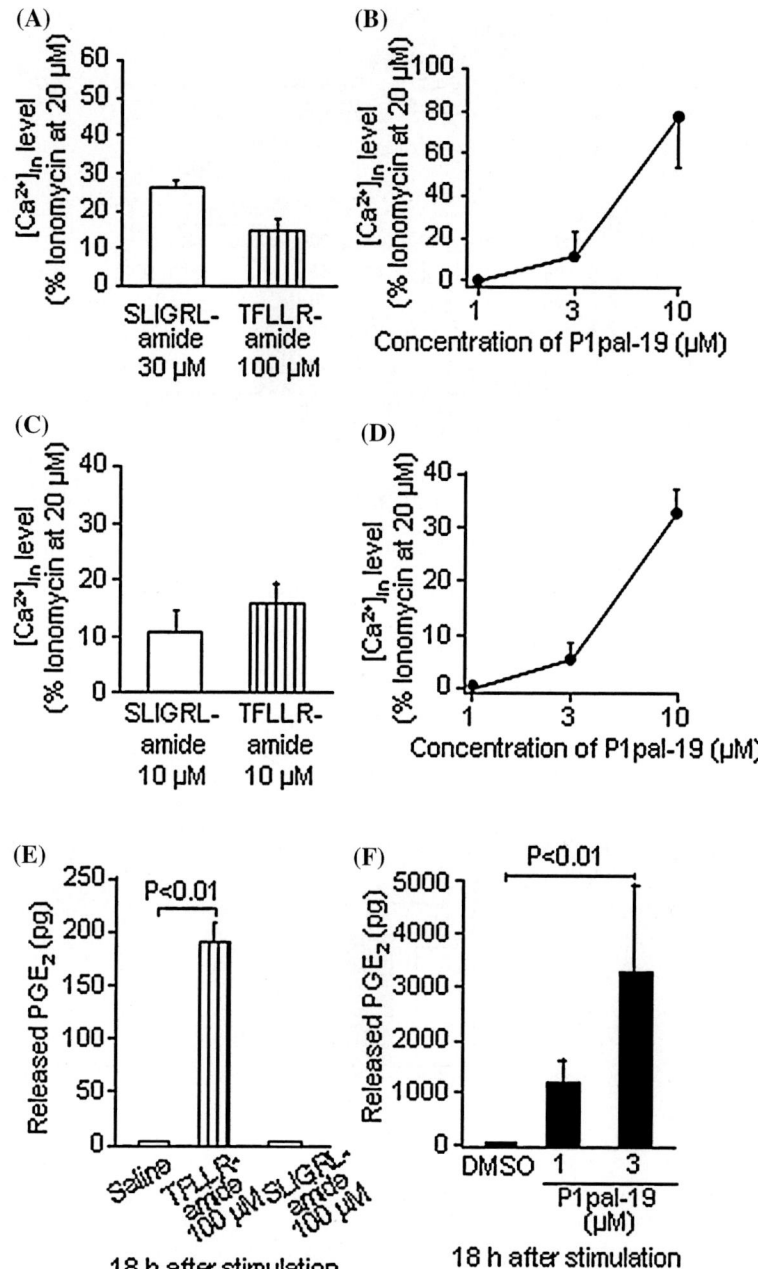

FIGURE 1. Effects of the extracellular PAR2 agonist SLIGRL-amide, the extracellular PAR1 agonist TFLLR-amide and the intracellular PAR1 agonist P1pal-19 on intracellular Ca^{2+} ($[Ca^{2+}]_{in}$) levels and release of prostaglandin E_2 (PGE_2) in A549 cells (**A, B**) or RGM1 cells (**C–F**). Data show the mean with SEM from 4–6 (**A**), 3 (**B**), 11–16 (**C**), 3–5 (**D**), and 4–9 (**E, F**) experiments.

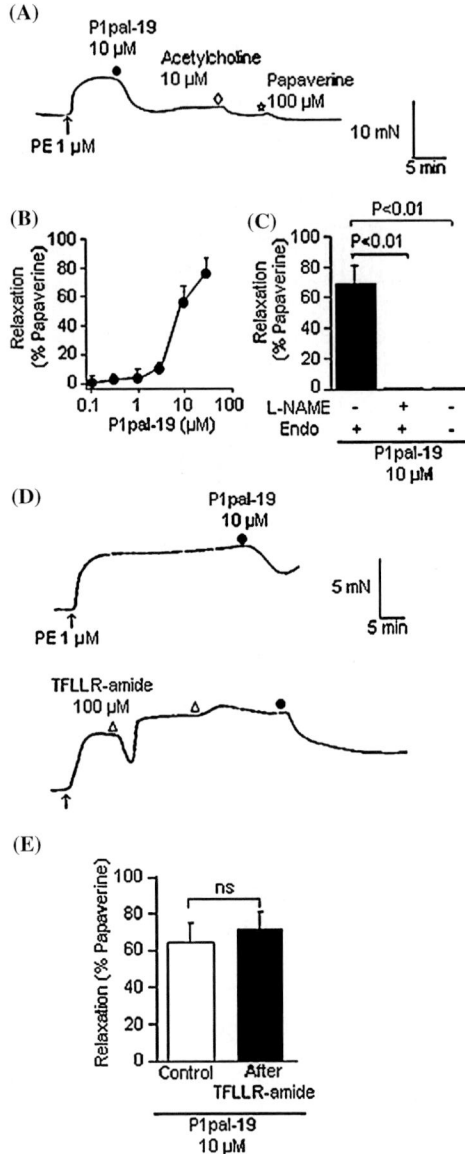

FIGURE 2. Endothelial NO-dependent relaxant effects of the intracellular PAR1 agonist P1pal-19 in rat aorta. Relaxation responses were observed in the rat aorta precontracted with 1 μM phenylephrine (PE). (**A**) Typical recording of P1pal-19-induced relaxation. (**B**) Concentration-related relaxant effect of P1pal-19. (**C**) Effect of N^G-nitro-L-arginine methyl ester (L-NAME) or removal of the endothelium on P1pal-19-evoked relaxation. L-NAME at 100 μM was added to the bath 5 min before 1 μM PE. (**D**) and (**E**) Desensitization experiments. P1pal-19 at 10 μM was applied after adding 100 μM TFLLR-amide twice in the aorta precontracted with 1 μM PE. Data show the mean with SEM from 3–5 (**B, C**) or 6–7 (**E**) experiments. ns = not significant.

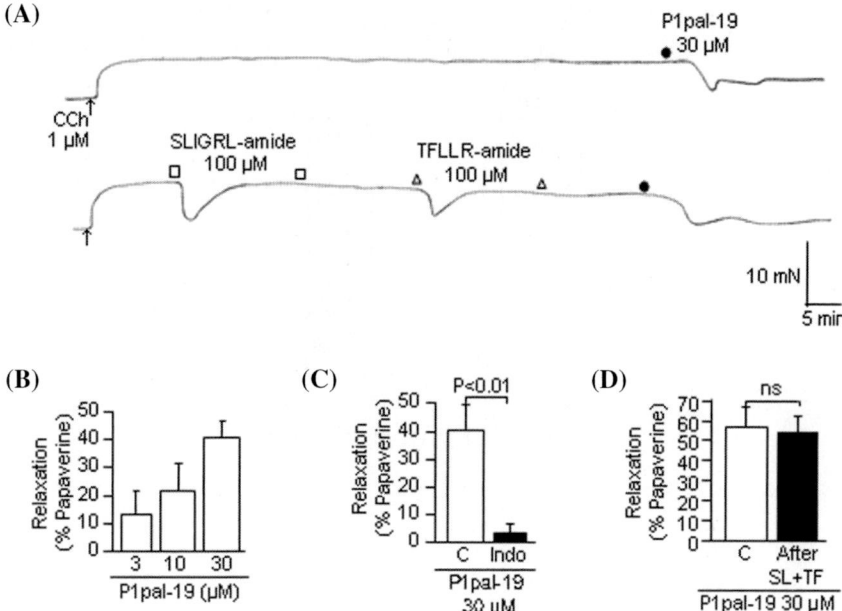

FIGURE 3. Indomethacin-sensitive relaxant effect of the intracellular PAR1 agonist P1pal-19 in mouse main bronchus. Relaxation responses were observed in the main bronchus precontracted with 1 μM carbachol (CCh). (**A**) Typical tracing. (**B**) Concentration-related effect of P1pal-19. (**C**) Indomethacin (Indo) at 10 μM was added 5 min before carbachol. (**A**) and (**D**) Desensitization experiments. P1pal-19 at 30 μM was applied after adding SLIGRL-amide and TFLLR-amide at 100 μM twice. Data show the mean with SEM from 5–6 experiments. ns = not significant. C = control.

TFLLR-amide, and the PAR2AP, SLIGRL-amide, at 100 μM (FIG. 3 A and B). The relaxation response to P1pal-19 at 30 μM was abolished by pretreatment with indomethacin, a cyclo-oxygenase inhibitor at 10 μM (FIG. 3 A and C) property corresponding to the mechanism for TFLLR-amide-triggered airway relaxation.[13,14] Nonetheless, desensitization of both PAR1 and PAR2 by applying their extracellular agonists twice failed to abolish the relaxant effect of P1pal-19 (FIG. 3 A and D).

Contractile Activities of the Extracellular and Intracellular Agonists for PAR1 in Distinct Smooth Muscle Tissues

As described previously,[11,12,15,16] TFLLR-amide caused contractile responses in the endothelium-denuded rat aorta or superior mesenteric artery (FIG. 4 A and B) and in the rat gastric longitudinal smooth muscle (FIG. 4 C). In contrast, P1pal-19 had no or only slight contractile effects in those smooth muscle preparations (FIG. 4).

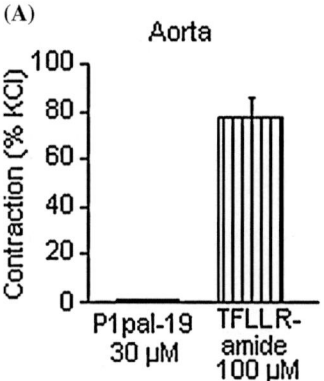
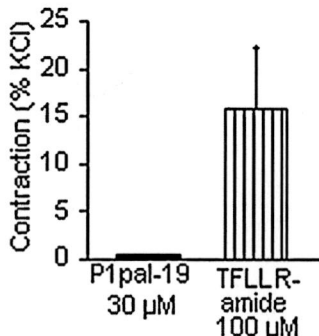
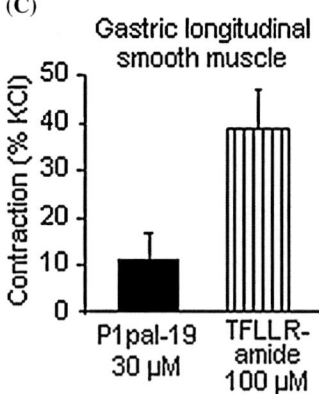

FIGURE 4. Lack of contractile effects of the intracellular PAR1 agonist P1pal-19 in the endothelium-denuded aorta (**A**) and superior mesenteric artery (**B**), and gastric longitudinal smooth muscle (**C**) from rats. Data show the mean with SEM from 3–8 (**A**) or 5 (**B**, **C**) experiments.

Inhibitory Effect of the Intracellular PAR1 Antagonist P1pal-12 on the Vasorelaxation Caused by Intracellular and Extracellular Agonists for PAR1 and/or PAR2 in Rat Aortic Rings

In the precontracted rat aortic rings, P1pal-12, a possible intracellular PAR1 antagonist, at 10 µM partially inhibited the relaxation caused by either TFLLR-amide at 3–30 µM or P1pal-19 at 10 µM (FIG. 5 A). In contrast, P1pal-12 at the same concentration did not inhibit the relaxation caused by SLIGRL-amide at 0.1–100 µM (FIG. 5 B). Inhibitory effect of P1pal-12 at higher concentrations was not examined, because P1pal-12 itself at concentrations over 10 µM caused small relaxation (data not shown).

DISCUSSION

In the present study, the potent agonistic activity of P1pal-19 for PAR1 was confirmed in A549 and RGM1 cells that naturally express functional PAR1, as assessed by determining cytosolic Ca^{2+} mobilization and/or prostaglandin E_2 production. As did the PAR1AP, TFLLR-amide, P1pal-19 caused endothelial NO-dependent relaxation in rat aorta and epithelial prostanoid-dependent relaxation in mouse main bronchus, suggesting its possible agonistic activity for PAR1 in these tissues. Nonetheless, the specificity of P1pal-19 for PAR1 in the endothelial and epithelial tissues is still open to question, because desensitization of PAR1 by repeated application of the PAR1AP unaffected the P1pal-19-induced aortic and bronchial relaxation. On the other hand, P1pal-19 appears incapable of showing agonistic activity for PAR1 in smooth muscle tissues, because it failed to mimic the contractile activity of the PAR1AP in the endothelium-denuded arterial rings and gastric longitudinal strips. Furthermore, our data provide evidence for the antagonistic efficacy of P1pal-12 for PAR1 in the vascular endothelium, as suggested previously in human platelets and PAR1-transfected cells.[4]

Covic et al.[4] have shown that $G_{q/11}$ protein signaling by P1pal-19 is dependent on the presence of the C-terminal of PAR1 molecules, because P1pal-19 did not activate C-tail-deleted PAR1 mutant. Therefore, it was of our surprise that P1pal-19-evoked vasorelaxation and bronchodilation were resistant to PAR1 desensitization by the extracellular PAR1 agonist, that is, PAR1AP. It has been demonstrated that P1pal-19 causes inositol triphosphate production in COS7 cells transiently transfected with PAR1, whereas it produced no such responses in the cells transfected with PAR4, cholecystokinin A and B receptors, somatostatin receptors or NK_1 receptors, in the absence of PAR1.[4] However, the possibility can not be ruled out that unknown GPCRs other than the above-described receptors, present in the endothelial and/or epithelial cells, might compensate PAR1 that was desensitized by the extracellular agonist,

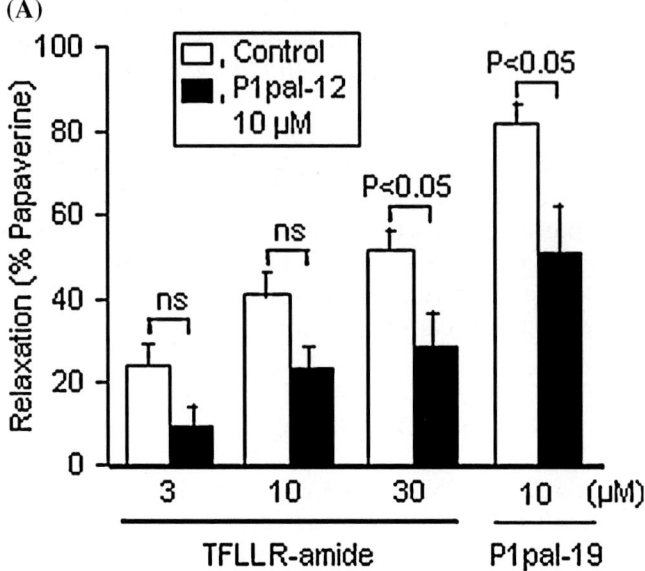
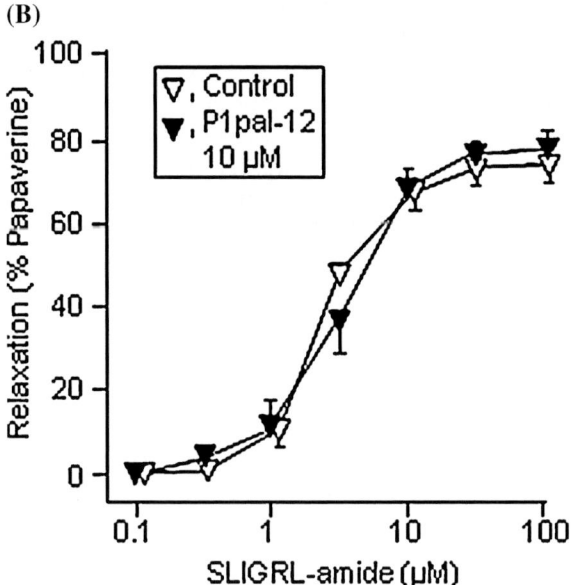

FIGURE 5. Effects of the intracellular PAR1 antagonist P1pal-12 on the relaxation caused by the extracellular and intracellular PAR1 agonists, TFLLR-amide and P1pal-19, respectively (**A**) or the extracellular PAR2 agonist SLIGRL-amide (**B**) in rat aorta. TFLLR-amide, P1pal-19 and SLIGRL-amide were applied to the preparations precontracted with 1 μM phenylephrine. P1pal-12 at 10 μM was added to the bath 30 min before phenylephrine. Data show the mean with SEM from 5–14 experiments. ns = not significant.

mediating the relaxant activity of P1pal-19 in rat aorta and mouse bronchus in the present study. Our data may thus predict that P1pal-19 in isolated tissues might be less specific for PAR1 than in cultured cells transfected with certain specific receptors.

Stimulation of PAR1 present in the vascular endothelium and airway epithelium causes release of NO and prostaglandin E_2, respectively, leading to smooth muscle relaxation.[11,13,14,16] In contrast, direct stimulation of PAR1 expressed in vascular and gastric smooth muscle cells produces contractile responses.[11,12,15,17] Our present data imply that P1pal-19 is capable of stimulating PAR1 in the endothelial and epithelial cells, but not in the smooth muscle cells. Although this discrepancy remains to be interpreted, it is speculated that P1pal-19 could easily reach and activate PAR1 on the monolayers, such as the endothelium and epithelium, but may not be easily accessible to PAR1 present in the deep layer of the smooth muscle. Apart from the specificity, these characteristics of P1pal-19, if any, may be beneficial in certain situations. For example, P1pal-19, administered intravenously, might specifically stimulate PAR1 in the vascular endothelium without affecting PAR1 in the vascular smooth muscle, leading monophasic hypotension. Alternatively, an inhalation of P1pal-19 might specifically activate PAR1 in the airway epithelium, leading to improvement of respiratory functions.

The long-lasting relaxant effect of P1pal-19 in the aortic preparations, as compared with TFLLR-amide (see FIG. 2 A and D), may be, in part, explained by lack of the contractile effect of P1pal-19 on the aortic smooth muscle; that is, the contractile activity of TFLLR-amide would overcome and shorten the preceding relaxant effect. However, the long-lasting bronchorelaxation caused by P1pal-19 (see FIG. 3A), as compared with TFLLR-amide, would reflect actual persistent activation of PAR1, because TFLLR-amide had no contractile activity in the resting bronchial preparations (data not shown). It is also of note that P1pal-19 caused more persistent Ca^{2+} signals than TFLLR-amide in A549 and RGM1 cells (data not shown). The persistent activation of PAR1 by P1pal-19 may interpret much greater production of prostaglandin E_2 following stimulation with P1pal-19 for 18 h, than caused by TFLLR-amide, in RGM1 cells. It is likely that desensitization mechanisms of PAR1 might hardly work following stimulation with P1pal-19, leading to long-lasting cellular signaling.

It is noteworthy that P1pal-12, a presumed intracellular PAR1 antagonist, actually inhibited the relaxant effect of TFLLR-amide, a PAR1AP, and also P1pal-19, an intracellular PAR1 agonist, in the rat aorta (see FIG. 5 A). The specificity of P1pal-12 for PAR1 can be supported by the finding that it unaffected the vasorelaxation caused by SLIGRL-mide, a PAR2AP (FIG. 5 B). P1pal-12 could be considered a partial agonist rather than a full antagonist, because P1pal-12 at high concentrations over 10 μM exerted small relaxant effect in the aortic preparations and also caused slight Ca^{2+} signaling in both A549 and RGM1 cells (for this reason, we could not examine the antagonistic effect of P1pal-12 in these cells). Further molecular modification of P1pal-12 would be necessary to obtain an ideal pepducin antagonist for PAR1, although our data provided significant information for future development of such compounds.

In summary, our present study identified interesting characteristics of PAR1 pepducins, P1pal-19 and P1pal-12, in isolated tissue assay systems, although further molecular modification would be necessary for their use as experimental tools and/or medicines.

REFERENCES

1. KAWABATA, A. 2003. Gastrointestinal functions of proteinase-activated receptors. Life Sci. **74:** 247–254.
2. OSSOVSKAYA, V.S. & N.W. BUNNETT. 2004. Protease-activated receptors: contribution to physiology and disease. Physiol. Rev. **84:** 579–621.
3. HOLLENBERG, M.D. 2005. Physiology and pathophysiology of proteinase-activated receptors (PARs): proteinases as hormone-like signal messengers: PARs and more. J. Pharmacol. Sci. **97:** 8–13.
4. COVIC, L. et al. 2002. Activation and inhibition of G protein-coupled receptors by cell-penetrating membrane-tethered peptides. Proc. Natl. Acad. Sci. USA **99:** 643–648.
5. KULIOPULOS, A. & L. COVIC. 2003. Blocking receptors on the inside: pepducin-based intervention of PAR signaling and thrombosis. Life Sci. **74:** 255–262.
6. COVIC, L. et al. 2002. Pepducin-based intervention of thrombin-receptor signaling and systemic platelet activation. Nat. Med. **8:** 1161–1165.
7. HOLLENBERG, M.D. et al. 2004. Proteinase-activated receptor-4: evaluation of tethered ligand-derived peptides as probes for receptor function and as inflammatory agonists in vivo. Br. J. Pharmacol. **143:** 443–454.
8. HOULE, S. et al. 2005. Neutrophils and the kallikrein-kinin system in proteinase-activated receptor 4-mediated inflammation in rodents. Br. J. Pharmacol. **146:** 670–678.
9. KAWAO, N. et al. 2005. Signal transduction for proteinase-activated receptor-2-triggered prostaglandin E2 formation in human lung epithelial cells. J. Pharmacol. Exp. Ther. **315:** 576–589.
10. SEKIGUCHI, F. et al. 2005. Intracellular signaling for prostaglandin E2 production caused by activation of the thrombin receptor, PAR-1, in the rat gastric mucosal epithelial cells. J. Pharmacol. Sci. **97:** 108P.
11. KAWABATA, A. et al. 2004. Distinct roles for protease-activated receptors 1 and 2 in vasomotor modulation in rat superior mesenteric artery. Cardiovasc. Res. **61:** 683–692.
12. LANIYONU, A.A. & M.D. HOLLENBERG. 1995. Vascular actions of thrombin receptor-derived polypeptides: structure-activity profiles for contractile and relaxant effects in rat aorta. Br. J. Pharmacol. **114:** 1680–1686.
13. COCKS, T.M. et al. 1999. A protective role for protease-activated receptors in the airways. Nature **398:** 156–160.
14. KAWABATA, A. et al. 2004. Proteinase-activated receptor-2-mediated relaxation in mouse tracheal and bronchial smooth muscle: signal transduction mechanisms and distinct agonist sensitivity. J. Pharmacol. Exp. Ther. **311:** 402–410.

15. KAWABATA, A. *et al.* 2004. A protective role of protease-activated receptor 1 in rat gastric mucosa. Gastroenterology **126:** 208–219.
16. HOLLENBERG, M.D. *et al.* 1997. Proteinase-activated receptors: structural requirements for activity, receptor cross-reactivity, and receptor selectivity of receptor-activating peptides. Can. J. Physiol. Pharmacol. **75:** 832–841.
17. SEKIGUCHI, F. *et al.* 2005. Mechanisms for modulation of mouse gastrointestinal motility by proteinase-activated receptor (PAR)-1 and -2 *in vitro*. Life Sci. **78:** 950–957.

A New Method to Assess Drug Sensitivity on Breast Tumor Acute Slices Preparation

PEDRO MESTRES,[a] ANDREA MORGUET,[a] WERNER SCHMIDT,[b] AXEL KOB,[c] AND ELKE THEDINGA[c]

[a]*Department of Anatomy and Cell Biology, University of Saarland, University Hospital, Homburg/Saar, Germany*

[b]*Clinic for Gynecology, University of Saarland, University Hospital, Homburg/Saar, Germany*

[c]*Bionas, Inc., D-18119 Rostock, Germany*

> ABSTRACT: A method for assessing tumor drug sensitivity is described that is based on preparation of tissue slices and use of silicon chips equipped with electrochemical sensors (multisensor array). The tumor slices (200–300 μM thick) are prepared after surgery and incubated in a medium for recovery after slicing. The advantage, compared to other preparations, is that the original three-dimensional structure is retained. Multisensor arrays measure: (*a*) pericellular acidification (anaerobic metabolism) and (*b*) oxygen consumption (respiration). The innovative aspect is that such measurements can be made online, as opposed to using a large battery of endpoint tests on cell vitality and proliferation. Electron microscopy of slices serves to determine cell density and structure and induction of apoptosis/necrosis. Slices of more than 200 breast tumors were used. Metabolic activity was inhibited by sodium fluoride, which reduces glycolysis, and potassium cyanide, which inhibits respiration. These changes are thus reflected in the curves of acidification and oxygen consumption. In other experiments the cytostatic Taxol, an anticytoskeletal agent, was used showing dose and time-dependent effects on acidification and oxygen consumption. In conclusion, the method presented here, is able to provide information on drug sensitivity of a tumor, which aids in designing individualized therapy and is used for drug screening.
>
> KEYWORDS: breast cancer; tumor slices; silicon-sensor chips; extracellular acidification; cell respiration; metabolic activity; drug screening; chemosensitivity; toxicity

Address for correspondence: Prof. Dr. Pedro Mestres, Department of Anatomy, and Cell Biology, University of Saarland, University Hospital, Building 61, D-66421, Homburg/Saar (Germany), Voice: +4968411626141; fax: +4968411626293.
  e-mail: anpmes@uniklinikum-saarland.de

INTRODUCTION

Tumors react in different ways against anticancer drugs. In the clinic it is therefore important to determine tumor drug sensitivity or resistance as early as possible in order to establish a tumor- and patient-specific therapy (individualized treatment).

With this aim, a large number of tests have been developed, the majority of them based on the detection of metabolic activity of tumor cells.[1,2] In general, such tests require the preparation of tumor specimens obtained by surgery or biopsy for cell culture.[3] However, this preparation causes many alterations in tissue structure and cell activity of the original tissue and moreover, with this method, certain cell types may be missed. Certainly one of the more severe changes concerns the three-dimensional arrangement of cells and extracellular matrix, which is destroyed by enzymatic and/or mechanical dissociation, a step otherwise necessary for creating cell suspensions and cultures.

This handicap can be avoided using tissue slices instead of cell suspensions.[4] With slicing, the original tissue architecture of the tumor is retained, and thus slices can be considered as a small representation of the initial tumor; with a single slice one can speak of a "microtumor" *in vitro*.

In the present study tumor slices and a silicon-sensor chip device have been used to measure the metabolic activity of acute prepared breast tumor slices under control conditions and submitted to the effects of metabolic inhibitors and cytostatics. The silicon-based sensor-chip technology presents an important advantage when compared with the majority of viability tests (endpoint tests)—the online monitoring of the cell and tissue activity. The potential of this approach to assess chemosensitivity and chemoresistence of human tumors *ex vivo* will be discussed.

MATERIAL AND METHODS

Reagents

Low buffered medium (DMEM/F12), L-glutamine, penicillin–streptomycin, and NEA were purchased from Invitrogen/Gibco (Karlsruhe, Germany). Fetal calf sera (FCS) were purchased from PAA Laboratories (Coelbe, Germany). Hydrocortisone, epidermal growth factor (EGF), and Triton-X 100 were purchased from Sigma-Aldrich (Schnelldorf, Germany). Sodium fluoride (NaF) was purchased from Merck (Darmstadt, Germany). Cyanide (KCN) and paclitaxel-Taxol® were purchased from Riedel-de-Haen.

Breast Tumor Slices

More than 200 breast tumors were examined. The specimens were obtained surgically and immediately transferred in a medium solution (DMEM/F12 and penicillin-streptomycin). After dissection under a stereomicroscope, the tissue sample was attached on a Teflon plate with a nontoxic acryl adhesive, covered with medium (DMEM/F12 + 10% FCS + insulin + EGF + hydrocortisone), and transferred to a tissue chopper (Ivan Sorvall, New York, USA). Slices of 250–300-µM thick were obtained. The slices were collected first in fresh medium (see above) and then dissected again under stereomicroscopy. Some slices were placed immediately into the sensor-chip device (BIONAS 2500® analyzing system, Bionas, Rostock, Germany); others were incubated overnight or maintained 7 or more days in culture. For histologic evaluation (tissue and cell structure and apoptotic and necrotic cellular changes) some slices of each tumor were selected for light microscopy and transmission electron microscopy.

Sensor-Chip Assays

The chemicals used were sodium fluoride (NaF) and potassium cyanide (KCN) and Taxol, an anticancer agent that acts against the cytoskeleton. Different concentrations of Taxol were used, ranging from 0.01 mg/mL to 0.1 mg/mL.

The effects of these substances on the tumor slice metabolism were measured in a Si-sensor-chip device named BIONAS 2500® analyzing system. Detection of acidification and oxygen consumption were realized in a special sensor chip by the readout of five ionic-sensitive field effect transistors (ISFETs) and an oxygen electrode on the multisensor array. By ISFET sensor the concentration of hydrogen ions were measured and the oxygen sensor registered the diminution of solved oxygen in the extracellular environment in the perfusion chamber (BIONAS) caused by living cells. The BIONAS system delivers on-line data on global metabolic activity of the specimen and is supported by proper software.[4]

RESULTS AND DISCUSSION

Several metabolic pathways, depending on their activity, yield protons in varying amounts that are released from the cell, causing an extracellular acidification.[2,5] Together with the measurement of oxygen consumption, the monitoring of extracellular pH provides reliable information on the metabolic activity of cells.[6] The sensor-chip device is composed of a perfusion chamber in which the multisensor-chip is placed on the bottom, a fluidic system that supplies the chamber with culture medium, and a control unit for the fluidic system and for data processing.[4] The slices are maintained in the perfusion chamber (approximately six µL volume) at controlled temperature (FIG. 1). The six modules (sensor-chip and perfusion chamber) available in the BIONAS system were used for our measurements.

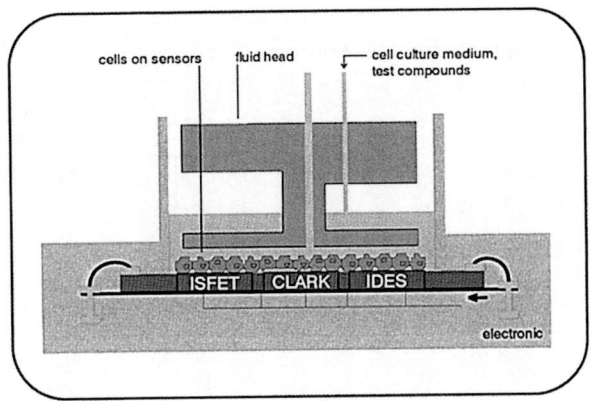

- ISFET → Acidification (glycolysis, respiration)
- CLARK → Oxygen consumption (respiration)
- IDES → Adhesion (confluence)

FIGURE 1. Diagram of the perfusion chamber of the sensor-chip device BIONAS 2500®.

Modified, slightly buffered medium (RM [running medium]) is fed over the cells at a pump rate of 56 μL/min. The media flow is stopped periodically. The modus of stop and go phases is freely programmable. During the go phase RM is pumped into the chamber. During the stop phase the slice or sample interacts with the medium and the metabolic effects are registered.[7–10]

Measurements are recorded online as changes in sensor potential in volts (acidification) and changes in sensor signal in amperes (oxygen consumption). In the example presented, slices of a breast tumor were incubated through several cycles with medium alone before using a medium containing sodium fluoride (NaF). NaF is an inhibitor of glycolysis, an important source of protons, and shows less influence on respiration. The decrease in the voltage amplitude during stop phases shows that the concentration of protons has been diminished, which indicates that the slice is responsive (FIG. 2). The software allows a further processing of the raw data: The output parameters (rates) are the calculated slopes of changes in acidification and oxygen measurements during the stop phase of fluidic modus. The rates can be normalized as follows: The rates occurring during measurement phase with medium containing NaF (or other substances) were divided through a rate (middle of rates from two stop phases) resulting from the initial measurement period without any substance. The resulting values express the relative change in acidification or oxygen consumption. For these computations we use BIONAS software and for final presentation we used Origin 7.0 (Origen Lab, Additive GmbH, Friedrichsdorf, Germany).

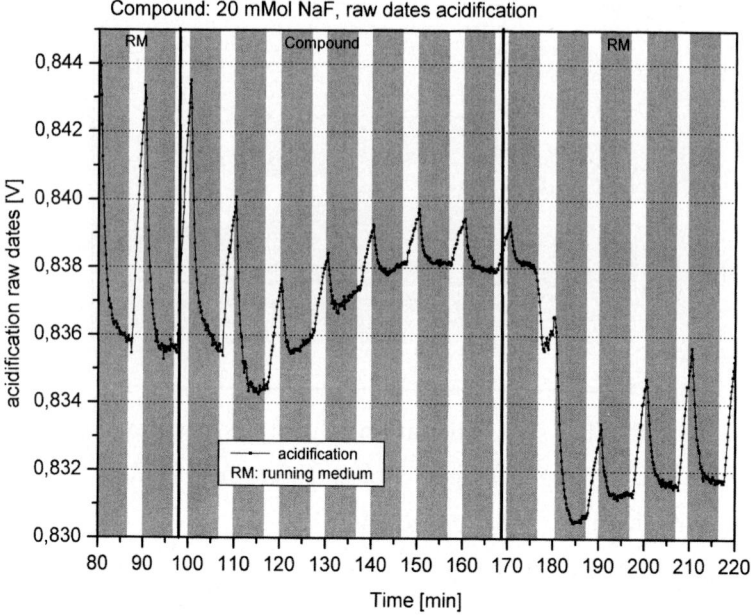

FIGURE 2. Acidification curve (raw data) of an experiment with slices of breast cancer. The pump "go" (*gray fields*) and "stop" (*white fields*) phases are indicated. Under the influence of NaF the acidification activity becomes reduceds; see the amplitude of the curve before (RM), during (+compound) and after RM application of NaF.

Tissue slices of breast tumors were placed in the Si-sensor device and studied primarily under control conditions, that is, without any chemical influence. This initial period is important because, during this time, the slice becomes adapted to the conditions in the chamber. This initial period takes 4–6 h and is characterized by an unusual noise and a low acidification rate. After that, the slices develop a regular rhythm of acidification, which is a sign of normal metabolic activity and, at the same time, provides information on slice vitality. If no drugs were applied, the curve remains quite constant for a long time (many hours up to days), indicating a stable metabolic activity of the specimen.

A notable difference in the response was observed between acute prepared slices and those that were cultivated overnight or for several days or weeks. The best results were obtained with slices maintained in medium overnight as well as with those cultured for 7 or more days. Acute prepared slices very often failed, probably due to the lack of a recovery period after slicing. Microscopy studies have confirmed maintenance of the three-dimensional tissue structure, even in slices cultured for more than 1 week (FIG. 3).

Experiments performed with inhibitors of metabolism (cyanide, sodium fluoride) have shown that such substances can diffuse into the slice tissue and consistently influence the metabolic activity of the cells. The study with the

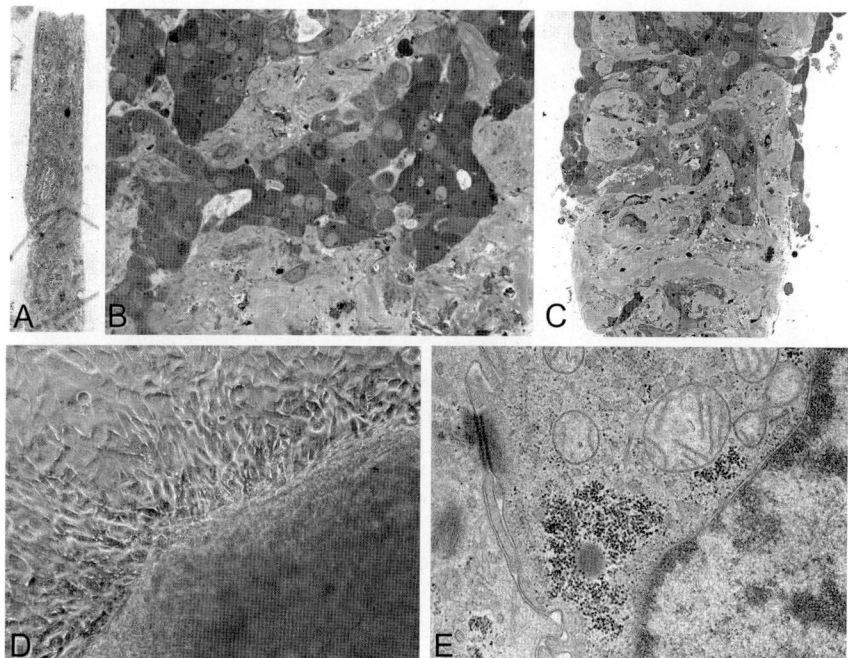

FIGURE 3. Breast tumor slices in semithin-cross section. (**A**) Slice 317-μM thick in overview (obj. 4×). (**B**) Slice 1 day in culture. Note the arrangement of tumor cells in middle part of the section (obj. 40×). (**C**) Slice after 7 days in culture. Note the distribution of tumor cells at the surface, but also in deep parts of slice (obj. 20×). (**D**) Slice of a breast tumor (CST p0st0) after 11 days in culture, showing a strong outgrowth (obj. 20×). (**E**) Transmission electron microscopy of a tumour slice treated with NaF. The cells accumulate abundant glycogen (dark granules).

transmission electron microscope reveals that in cells of slices treated with NaF the utilization of glucose decreases, accumulating in the cytoplasm in the form of glycogen (FIG. 3).

The application of cyanide causes a remarkable decrease in respiratory activity, while the application of sodium fluoride only produces a minor diminution of the oxygen consumption (FIG. 4). Both effects are in this case reversible after washing out of the chemicals by RM. Higher doses or longer exposure to the drug cause irreversible effects, and this technique is able to reveal them, when the experiments run for a long enough time.

The metabolic patterns of breast cancer slices treated with Taxol are characterized by a progressive decrease of metabolic activity until a very low level, depending on the dose or concentration. For instance, an application of 0.1 mg/mL Taxol for 18 h provokes a diminution of the acidification of approximately 70%, but there is an initial increase of the activity for a duration of approximately

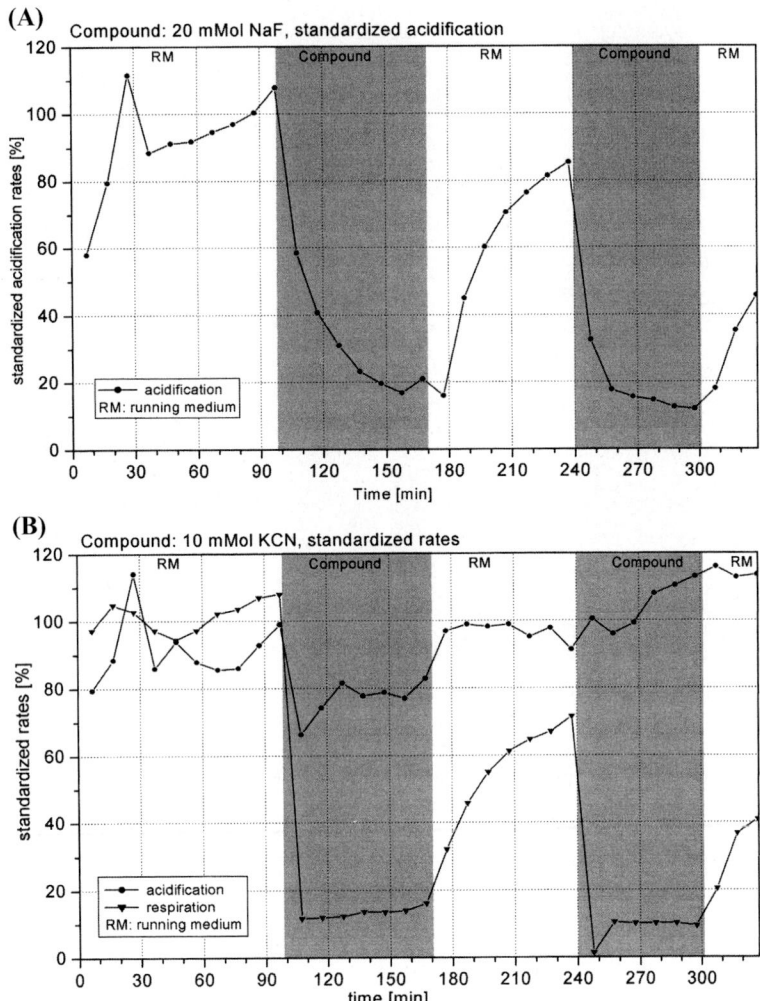

FIGURE 4. Breast tumour slices treated with NaF and KCN. (**A**) The diagram shows remarkable effects of NaF on acidification rates (normalized data of FIG. 2). (**B**) Acidification and respiration rates in a slice treated with KCN. Respiration becomes depressed in presence of KCN and recovers if the compound is removed by perfusion of the chamber only with RM. In contrast the acidification remains practically unchanged.

2 h (FIG. 5). In contrast, the 0.01 mg/mL dose causes an evident increase of metabolic activity at the beginning for a duration of 6 or 7 h before a slow reduction of metabolic activity takes place (FIG. 5). Although the level of acidification at the end of the experiment was low in both cases, treatment with Triton X-100 was necessary to demonstrate whether the slices were still alive. The

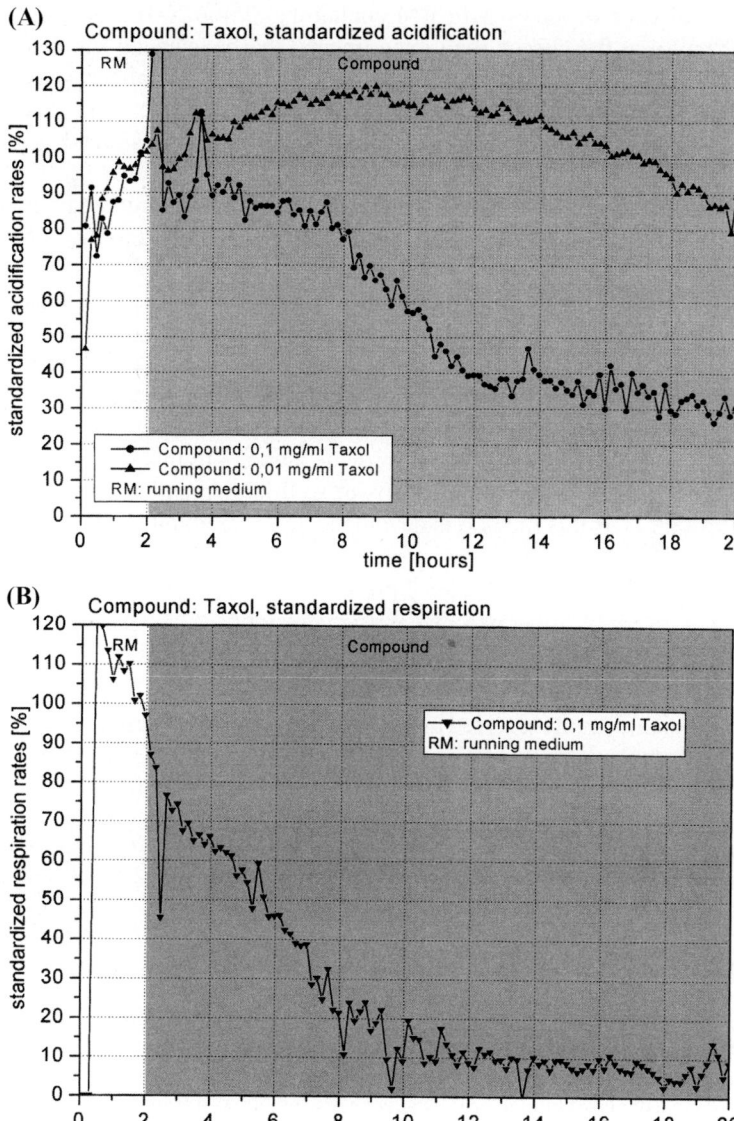

FIGURE 5. Breast tumor slice treated with Taxol. Both figures first show an initial measurement phase with RM only (*white fields*). The additions of compound (gray fields) were done after 2-h initial phase. (**A**) Acidification rates: with a low dose of Taxol (0.01 mg/mL) the slice reacts, at first increasing slightly the metabolic activity and decreasing it progressively after approximately 6 h continuous application of compound. With a higher dose (0.1 mg/mL), an increase can be observed, but only during the first 2 h followed by a fast decrease. (**B**) Respiration rates corresponding to the dose 0.1 mg/mL Taxol. Note that in this case no initial increase in activity can be seen.

treatment of cells or slices with RM containing Triton X-100 is a standard to test vitality at the end of such experiments, that is, causing the detergent to destroy cell membranes with consequent decrease of metabolic activity to zero. This behavior can only be observed when vital cells are in the chamber. These patterns of metabolic activity against Taxol can be considered as a sign of sensitivity to the drug.

According with our protocol, the tumor fragments can be prepared (slicing), quickly transferred into the sensor-chip device, and submitted to the influence of anticancer drugs. The examples presented show that such specimens remain vital, developing a quite normal metabolic activity (vitality), making it possible to analyze the responsiveness to the drugs and to attach correlative investigations with other methods. Although this approach is very promising, whether the data obtained with the sensor-chip technique are clinically relevant will depend on the correlation with results seen in the clinical follow-up of the patients. Studies in progress will provide answers to this central question as well as information about the effects of well-known anticancer drugs and other substances as this method is developed and accredited.

ACKNOWLEDGMENTS

The authors thank Gabi Kiefer, Birgit Leis, Norbert Pütz, and Holger Summa for the technical support received in this investigation. This work was supported by grants from the Federal Ministry for Education and Research (BMBF) and the Saving Bank Union of Saarland (SV Saar) to Pedro Mestres.

REFERENCES

1. MESTRES, P. 2001. Anticancer Drugs Chemosensitivity testing of human tumors using Si-sensor chips. **12** (Suppl. 4): A3.
2. MESTRES, P. 2003. Chemosensitivity testing of human tumors using Si-sensor chips. Recent Results Cancer Res., **161:** 26–38.
3. KRASNA, L., I. NETIKOVA, A. CHALOUPKOVA, *et al.* 2003. Assessment of *in vitro* drug resistance of human breast cancer cells subcultured from biopsy specimens. Anticancer Res. **23:** 2593–2600.
4. MESTRES, P., A. MORGUET, A. SCHOFER, *et al.* 2005. Applications of silicon sensor technologies to tumor tissue *in vitro*. *In* Methods in Molecular Medicine. Vol. 111: Chemosensitivity–Vol. 2: *In vivo* Models, Imaging and Molecular Regulators. R. D. Blumenthal, Ed.109–126. Humana Press, Inc. Totowa, NJ.
5. HAFNER, F. 2000. Cytosensor microphysiometer: technology and recent applications. Biosens. Bioelectron. **15:** 149–158.
6. EKLUND, S.E., D. TAYLOR, E. KOZLOV, *et al.* 2004. A microphysiometer for simultaneous measurement of changes in extracellular glucose, lactate, oxygen and acidification rate. Anal. Chem. **76:** 519–527.

7. BAUMANN, W., M. LEHMANN, M. BITZENHOFER, et al. 1999. Microelectronic sensor system for microphysiological application on living cells. Sensor Actuators B**55**: 77–89.
8. LEHMANN, M., W. BAUMANN, M. BRISCHWEIN, et al. 2000. Noninvasive measurement of cell membrane associated proton gradients by ionsensitive field effect transistor arrays for microphysiological and bioelectronical applications. Biosens. Bioelectron. **15**: 117–124.
9. LEHMANN, M., W. BAUMANN, M. BRISCHWEIN, et al. 2001. Simultaneous measurement of cellular respiration and acidification with a single CMOS ISFET. Biosens. Bioelectron. **16**: 195–203.
10. EHRET, R., W. BAUMANN, M. BRISCHWEIN, et al. 2001. Multiparametric microsensor chips for screening applications. Fresenius J. Anal. Chem. **369**: 30–35.

Process Simulation in a Mechatronic Bioreactor Device with Speed-Regulated Motors for Growing of Three-Dimensional Cell Cultures

MINA MIHAILOVA,[a] VASSIL TRENEV,[b] PENKA GENOVA,[b] AND SPIRO KONSTANTINOV[c]

[a]*IPF of TU-Sofia, 8800 Sliven, 59 Bourgasko Chaussee Str., Bulgaria*

[b]*CLMI, Bulgarian Academy of Sciences, 1 Acad. G. Bonchev Str., 1113 Sofia, Bulgaria*

[c]*Lab for Experimental Chemotherapy, Department of Pharmacology, Faculty of Pharmacy, Medical University of Sofia, 2 Dunav Str., 1000 Sofia, Bulgaria*

> ABSTRACT: Tissue engineering is a new scientific research field that allows the establishment of tissue equivalents rising from isolated cells in combination with biocompatible materials and cultivation in more or less sophisticated bioreactor systems. Such systems gave the unique opportunity to perform *in vitro* investigations of transcription and translation, cell growth, biochemistry and mechanics of healthy normal organs as well as those affected by malignant tumors, infections, and immune deficiency under controlled conditions. In rotating vessel bioreactors under microgravity and defined medium content, cells proliferate, stay abundant to each other, and form three-dimensional structures, assigned as spheroids. Such spheroids might be grown on microcarriers. A wide spectrum of different cell culture experiments involving normal and transformed human cells indicates that: in the rotating bioreactor system miniPERM no complete lack of gravity could be reached; a great part of the seeded cell material does not proliferate at the beginning; and the appearance of bigger spheroids is rather random. We describe the acquisition of spheroids from HD-MY-Z and Neuro-2A tumor cells. Spheroids of 100 and more cells were obtained from HD-MY-Z and Neuro-2A cells. Interestingly, chronic myeloid leukemia LAMA-84 cells did not form any cell clumps and they kept a completely undifferentiated phenotype despite their semiadherent manner of growth under conventional conditions. A detailed theoretical and virtual simulation study of the influence of every component of gravitation, inertia, and hydrodynamic force fields was performed. Therefore, a new concept for mechatronic bioreactor device with active electronic control was developed and virtually tested.

Address for correspondence: Vassil Trenev, CLMI, Bulgarian Academy of Sciences, 1 Acad. G. Bonchev Str., 1113 Sofia, Bulgaria. Voice: +35-92-979 2416; fax: +3592-9723571.
e-mail: vtrenev@clmi.bas.bg

KEYWORDS: rotating wall vessel bioreactors; miniPERM; dynamical model; virtual simulation; three dimensional cell culture

INTRODUCTION

In the human body there are more than 200 types of cells that form different organs such as skin, bones, and muscles, among others. These organs represent unique mixtures of different types of cells. More than 70 years ago Edmund B. Wilson wrote in his book entitled *The Cell in Development and Heredity*: the key to every biological problem has to be found in the cell itself.[1] The growing of cells outside the body enables the study of many basic biological and physiological phenomena such as controlling the cell cycle and intercellular interactions. Conventional methods for cell cultivation are based on methods for attaching the cells to the bottom of culture dishes using gravity. These cells are growing as monolayers and do not fulfill all the original functions of the tissue from which they are originating. In contrast, the cellular constructs within the mammalian body are three-dimensional. Therefore, methods for establishing three-dimensional upscaled cultures represent a new, very promising focus of scientific research efforts.[1] In this article, we will review the main types of rotating wall vessel bioreactors. In the Rotary Cell Culture System (RCCS™; FIG. 1) cells are suspended by randomizing the gravity vector so that they are in continuous free-fall through the culture medium.[2] Cells could grow in all directions. Destructive stress forces (e.g., shear stress) are insignificant because the system has no propellers, airlifts, bubbles, or agitators. Because the cells are maintained in a gentle fluid orbit, the cells are allowed to co-localize, communicate, and form three-dimensional structures. More than 50 cell types have been successfully cultured in the RCCS. Cells not only grew, but also developed into functional tissue organoids (www.synthecon.com). Very similar to the RCCS is the bioreactor system developed in connection with the NASA program, the so-called high-aspect ratio vessel (HARV), which was coupled with synchronously rotating CCD camera (FIG. 1). The rotary culture max (RCM) is a new version of RCCS, where the central tubular perfusion filter is replaced by large flat filter that covers the inside surface, thus creating an unobstructed culture chamber. The volumes of this chamber can be scaled up to grow larger three-dimensional organoids (FIG. 1). The rotating wall vessel (RWV) and the Synthecon's rotating wall-perfused vessel (RWPV) systems were designed to allow the cultivation of shear stress-sensitive mammalian cells in a microgravity environment.[2,3] The last system provides capabilities for replenishing fresh medium, as well as monitoring and controlling oxygen, pH, and temperature. In this system fluid flow is near solid body or laminar at most operating conditions. If the inner cylinder and the outer cylinder rotate at the same angular velocity, then the laminar flow fluid velocity gradient [rad] is minimized.[3]

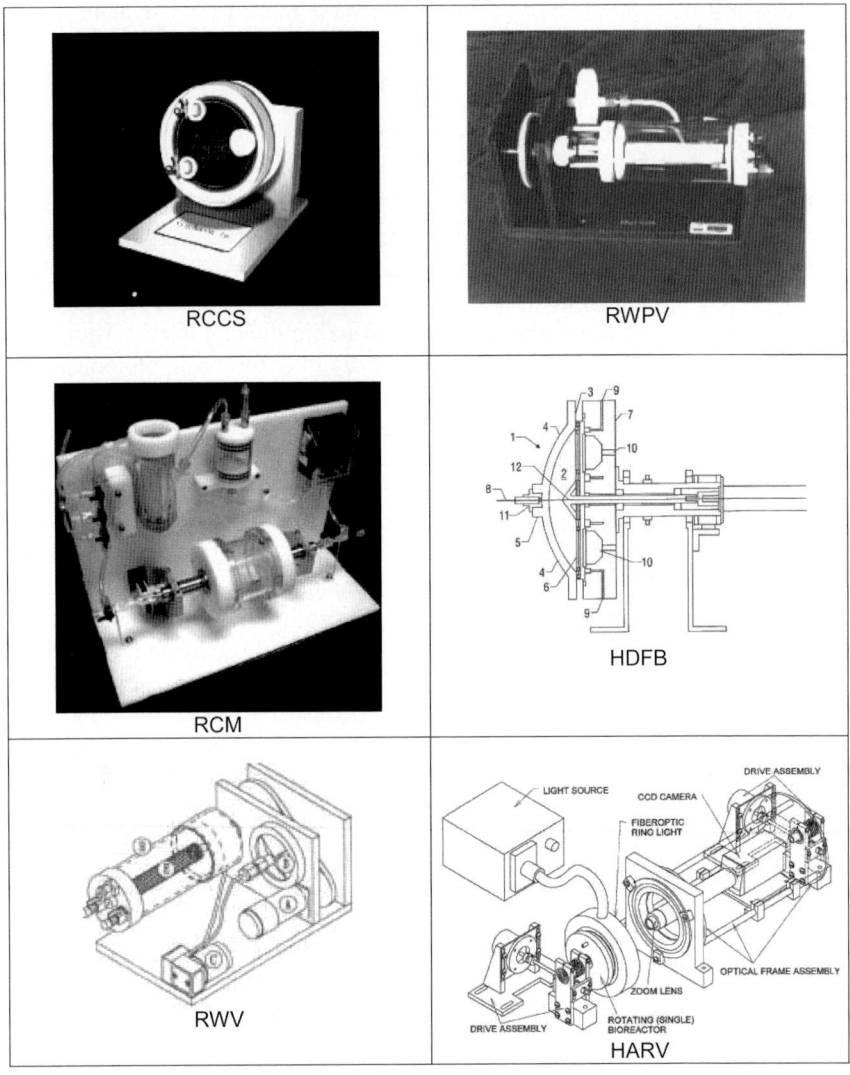

FIGURE 1. Different types of rotating wall vessel bioreactor systems. RCCS = rotary cell culture system; RWPV = rotating wall-perfused vessel; RCM = rotary culture max; HDFB = hydrodynamic focusing bioreactor; RWV = rotating wall vessel; and HARV = high aspect ratio vessel.

The hydrodynamic focusing bioreactor (HDFB; FIG. 1) lacks the limitations of RWV, for example, formation of air bubbles, and offers possibilities for controlling the location of the cell formations.[2] What the rotating wall vessel achieves in two-dimensional space, the random-positioning machine (RPM) achieves in three-dimensional space: the weight vector experienced by the cells changes its direction in three-dimensional space with a frequency of more than 1 Hz using a cardanic framework, the rotation of which is performed by independent motors.

This work deals with the creation and numerical testing of dynamical models for the movement behavior of cellular spheroids in the rotating wall vessel bioreactor system miniPERM. In addition, the acquisition of cell spheroids from neuronal and hematopoietic tumor cells is described in the same bioreactor system.

MATERIALS AND METHODS

Dynamical Modeling Background

Kinematics and force characteristics of the cellular spheroid movement in a rotating wall vessel bioreactor miniPERM (IVSS Ltd., Germany) are summarized in FIGURE 2. The bioreactor is driven by an electrical step motor and its angular speed can be selected within the interval from 0.1 up to 4/s using the controlling device. The rotation of the central shaft is brought to the bioreactor unit(s) by frictional contact. The production module (PM) has to be completely filled with culture medium, thus ensuring its rotation like solid body. The growing cells eventually form spheroids and complete complex, nearly plane movement. They rotate with the vessel together (frame system OXY) and perform a second relative movement (coordinate system Oxy) (FIG. 3A, B). Fluid and gas exchange occurs through special membranes to the bioreactor tank and to the atmosphere in the incubator. The components of the absolute acceleration are: launch components (normal a^{en} and tangential a^{et}) and relative components (a^r and Coriolis' a^c), as indicated in FIGURE 3B. The index p is used for the spheroid, and f for the fluid. These accelerations are forming the field of inertia forces where the movement of the spheroids takes place. By n_{tr} and τ_{tr} the normal and the tangential of the trajectory of the particle are indicated within a coordinate frame OXY. The mass of the particle is denoted as m_p and the fluid removed by it denoted m_f. Both masses are proportional to their respective densities ρ_p and ρ_f.

Generally, the dynamic model of this nonholonomic system is composed according to the equations of Newton–Euler (FIG. 3) and corresponds to the following system of differential equations with general coordinates x and y (independent parameters):

$$\ddot{x} - 2\dot{y}\dot{\varphi} + b\dot{x} - \mu\left(x\dot{\varphi}^2 + y\ddot{\varphi}\right) = -g\mu\sin\varphi$$
$$\ddot{y} + 2\dot{x}\dot{\varphi} + b\dot{y} + \mu\left(-y\dot{\varphi}^2 + x\ddot{\varphi}\right) = -g\mu\cos\varphi \quad (1)$$
$$\varphi = \varphi(t) \quad \mu = 1 - \frac{\rho_f}{\rho_p} \quad \dot{\varphi} = \omega(t)$$

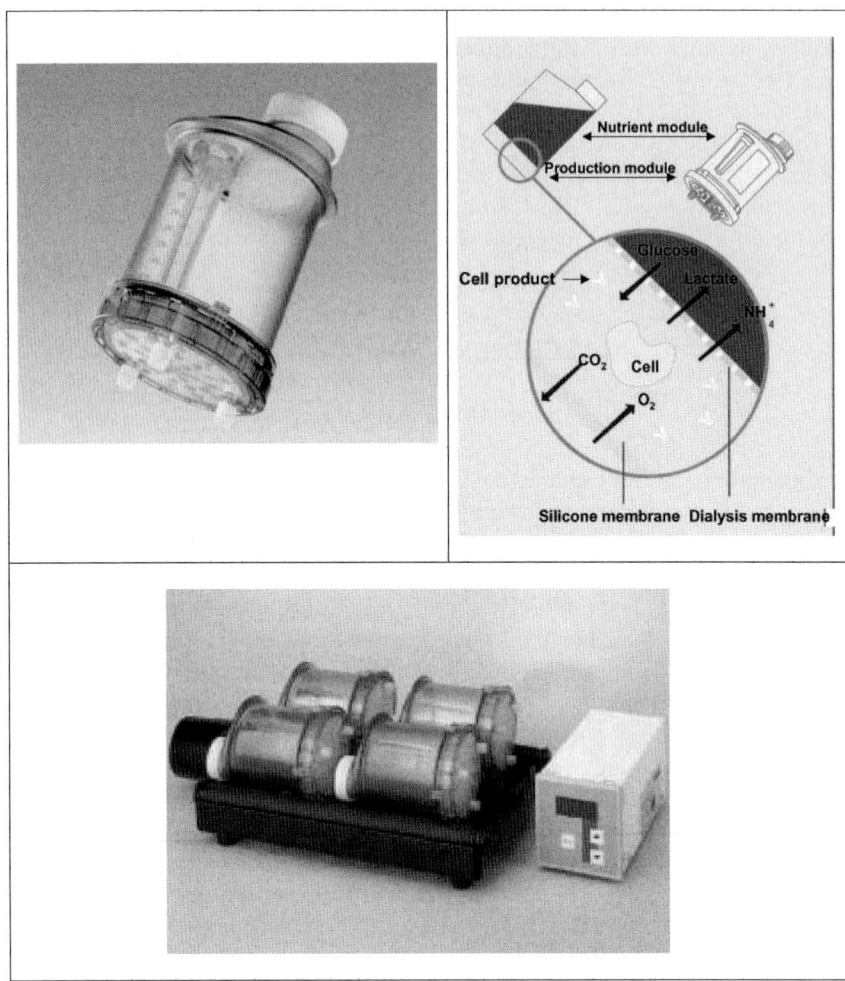

FIGURE 2. The miniPERM bioreactor system (bioreactor unit, consisting of a tank with production module; principles of functioning; turning device).

The physical sense of the members in the left part of the equations (1) is to describe the projections of the linear relative, Coriolis' and centrifugal accelerations, which are forming the fields of the centrifugal and centripetal forces. The resistance forces were assumed as linear functions of the relative velocity. The value of the coefficient b was experimentally determined by the Stokes' method.[4] The components of the gravitational and Archimedes' (buoyancy) accelerations are in the right side of the equations. The differential equations (1) are linear, nonhomogenous, and with variable coefficients. They can be numerically solved using the MATLAB 6.1™ software. When it is assumed that

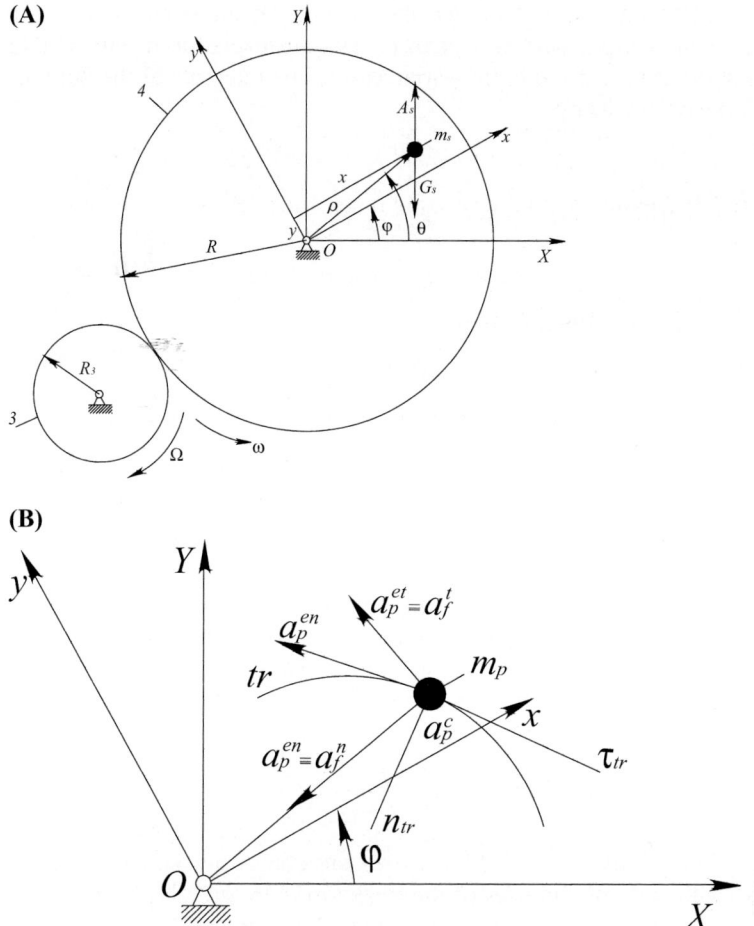

FIGURE 3. Descarte's frame and relative coordinate system for the movement of the particle.

$\omega = const$, representing the case of existing bioreactors, the differential equations became:[5]

$$\ddot{x} - 2\dot{y}\omega + b\dot{x} - x\mu\omega^2 = -g\mu \sin \omega t$$
$$\ddot{y} + 2\dot{x}\omega + b\dot{y} + y\mu\omega^2 = -g\mu \cos \omega t \quad (2)$$

These equations have theoretical and analytical solutions, but the characteristic equation is of fourth degree and has to be numerically solved. That is why the system was modeled and solved by the use of MATLB SIMULING 6.1. The influence of different inertia and gravitational fields on the trajectories of the spheroids and on the microgravity can be evaluated by the consequent investigation of the dynamic models proposed.

Stability conditions according to the theory of Routh–Hurwitz are as follows:

The homogenous part of system (2) is transformed to a normal equation of first order in relation to the variables y_k. The integrals of the homogenous system have the form:

$$y_k = A_k e^{r_i t} \quad k = 1, 2 \ldots 4 \tag{3}$$

where k is identical with the variable index and r_i are solutions of the characteristic equation:

$$a_4 + a_3 r + a_2 r^2 + a_1 r^3 + a_0 r^4 = 0 \tag{4}$$

The constants in Equation 4 are:

$$\begin{aligned}
a_4 &= \mu^2(\dot{\varphi}^4 + \ddot{\varphi}^2); \\
a_3 &= 2\mu\dot{\varphi}(2\ddot{\varphi} - b\dot{\varphi}); \\
a_2 &= b^2 + (4 - \mu^2)\dot{\varphi}^2; \\
a_1 &= 2b; \\
a_0 &= 1
\end{aligned} \tag{5}$$

When all Hurwitz's determinants in the polynomial are positive, according to the Ruth–Hurwitz theory, all real parts of the solutions will be negative; thus y_k tends to zero for $t \to +\infty$, or, in other words, the system is stable. All coefficients in a_k have to be positive. For $\dot{\varphi} = \omega = const$ from Equation 5 follows:

$$\begin{aligned}
\ddot{\varphi} \neq 0 \quad & a_3 = 2\mu\dot{\varphi}(2\ddot{\varphi} - b\dot{\varphi}) > 0 \\
\ddot{\varphi} = 0 \quad & a_3 = -2\mu b\dot{\varphi}^2 < 0
\end{aligned} \tag{6}$$

and a_3 has negative values. The system is unstable on account of the centrifugal forces although the radiuses of the trajectories are growing very slowly. It is assumed that the cellular spheroids are big enough and, when colliding with the vessel wall, are rejected and continue their movement from a new starting position.

Rules for active electronic control: The above described dynamical and mathematical model can be used for the development of computer-aided rotational speed controlling for the miniPERM bioreactor turning device. In order to optimize the controlling algorithm, the following functions were analyzed:

$$\omega = \pm A^{-Bt^2} \cdot \sinh(C + t) + D \tag{7}$$

$$\dot{\varphi} = -Ae^{-B \sin t} sh(\sin t + C) + D \tag{8}$$

Motor controlling rules for achieving a defined trajectory of spheroid movement (first dynamics' task) can be determined. If the adequate trajectory is $F(x, y) = 0$, then the velocity and acceleration of the spheroid are known. According to these conditions, the coordinates, velocities, and accelerations of the relative movement are defined. When exchanging all these parameters in the differential equations (1), the rules for motor controlling can be found. Their solution is polyvalent and requires the use of optimization methods.

Cultivation of Cells in miniPERM

Mouse neuroblastoma Neuro-2A cells: The culture medium was DMEM supplemented with nonessential MEM amino acids, 2% L-glutamine, and 10–15% fetal bovine serum (FBS). Before starting the bioreactor cultivation, cells were passaged 3–4 times as conventional culture by trypsinization. Initially, Neuro-2A cells tend to grow better in tissue culture plates than in tissue culture flasks.[6] The following time schedule was used for stepwise increase of the rotation velocity of the bioreactor unit:

Day	0	5	7	9	10	13	15–21
Velocity (rpm)	2	3	6	7	9	12	12

Starting on day 6 every 2 days probes from the production module were taken and microscopically analyzed for the presence of three-dimensional particles, the so-called spheroids. After reaching the maximal angular velocity each second day half of the production module volume (about 2.5 mL) was exchanged with fresh serum-containing medium.

Chronic myeloid leukemia-derived LAMA-84 cells: These cells are grown in RPMI-1640 medium supplemented with 10% FBS and 2% L-glutamine. They form a culture of suspension cells and fibroblast-like semiadherent cells. The following time schedule was used for the increase of the rotation frequency:

Day	0	3	5	6	10
Velocity (rpm)	2	5	8	12	12

RESULTS

Numerical Experiments

The analysis was performed using the following limit ranges of the physical parameters:

$$0.091 < \mu < 0.375; 0 < \varphi < k(2p); 0.1 < \omega < 4/s; b = 13,000/s$$
$$\text{for } \mu = 0.091; \text{ and } b = 800/s \; \mu = 0.375; x_0 = y_0 = 0.03 [m]. \quad (9)$$

A detailed theoretical and virtual simulation consists of the following components:

(a) movement caused by the gravity field only
The polar radius of the trajectory is obtained from

$$\rho^2 = \left[x_0 - \frac{\mu g (1 - \cos\varphi)}{\omega^2}\right]^2 + \left[y_0 - \frac{\mu g (\varphi - \sin\varphi)}{\omega^2}\right]^2 \quad (10)$$

For the limit values of the angular speed and for $\varphi = 0.1$ [rad] the polar radius has inadmissible values ($0.4 \leq \rho \leq 5.0$); in other words it is not possible to speak about microgravity at all. In addition, for these small values of φ the rotation of the bioreactor does not have any influence. In FIGURE 4 the trajectory of the spheroid is shown for $\omega = 1.0/s$ and $\varphi = 690°$. The trajectory has a spiral form:

(b) movement under the influence of gravitational and centrifugal fields.
The equation for the polar radius is

$$\rho^2 = \frac{1}{4}[(x_0 - a)e^{ct} + (x_0 + a)e^{-ct} + 2a\sin\varphi]^2$$
$$+ \frac{1}{4}[(y_0 - a)(e^{ct} + e^{-ct}) + 2a\cos\varphi]^2 \quad (11)$$

where

$$a = \frac{g\mu}{(1+\mu)\omega^2}; \quad c = \omega\sqrt{\mu} \quad (12)$$

For all four combinations of limit values of μ and ω, the polar radius for $\varphi = 1$ [rad] was calculated as well as the respective time for each variant. Results are summarized in TABLE 1.

For the first and third variant it is not possible to speak about microgravity at all. In 10 sec the polar radius rises over 5 m. It is noteworthy that the third variant is not used in practice, because with the growth of the cellular spheroids

TABLE 1. Polar radii for extreme values of the physical parameters

Parameters		Variants			
ω	0.1	4.0	0.1	4.0	1.0
μ	0.091	0.091	0.375	0.375	0.2
b	800	800	13,000	13,000	5,500
T	10	0.25	10	0.25	1.0
A	83.41	0.052	272.73	0.1705	1.667
C	0.0302	1.207	0.0612	2.449	0.447
ρ	5.853	0.0637	11.18	0.0583	0.5887
ρ_c[m] eq.(10)	0.0423	0.0422	0.0404	0.0377	0.042
$\|a_c\|$ [m] eq.(10)	0.261×10^{-2}	0.411×10^{-2}	1.25×10^{-2}	0.179×10^{-2}	0.464×10^{-2}
$\|a_s\|$ [m] eq.(10)	0.261×10^{-2}	0.413×10^{-2}	1.31×10^{-2}	0.208×10^{-2}	0.467×10^{-2}
ρ_c[m] eq.(2)	0.0416	0.0424	0.0422	0.0424	0.0424

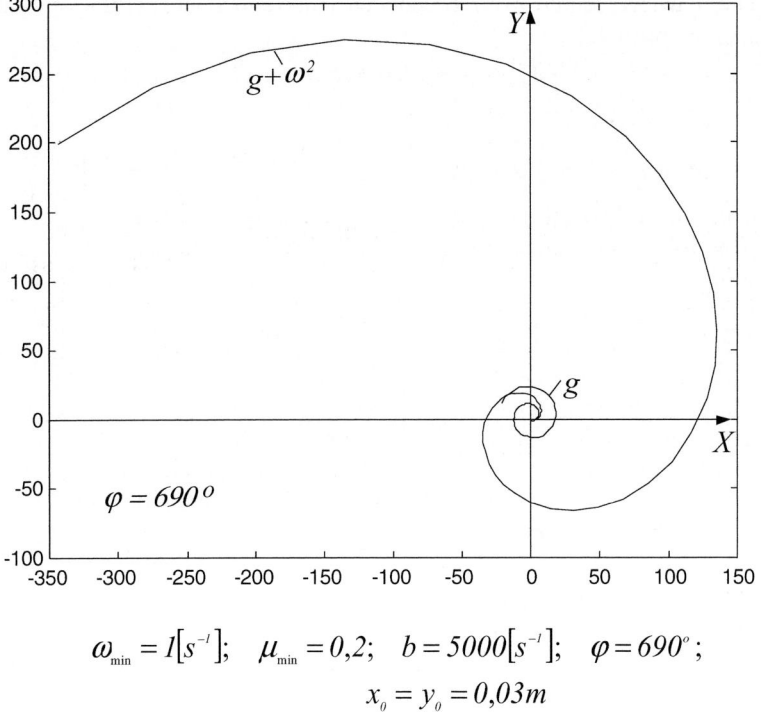

$$\omega_{min} = 1[s^{-1}]; \quad \mu_{min} = 0{,}2; \quad b = 5000[s^{-1}]; \quad \varphi = 690°;$$
$$x_0 = y_0 = 0{,}03m$$

FIGURE 4. Trajectories of the particle for variants 1 and 2 (TABLE 1).

the angular speed has to increase. The second variant is also inadmissible (high rotation frequency for small cellular spheroids). The fourth variant is possible to be used. For 0.25 sec ρ increases from the initial value of 0.0424 to 0.0583. In addition, the frequently used fifth variant was studied. In this case the resulting trajectory is of spiral form and is approximately the same as in the above-mentioned case (*a*) (FIG. 4).

(c) movement under the influence of gravity, Coriolis' and hydrodynamic force fields:

The equation for the polar radius is

$$\rho^2 = [A_1 + e^{-bt}(A_3 \cos 2\omega t + A_4 \sin 2\omega t) + C_1 \cos \omega t + C_2 \sin \omega t]^2 + \\ + [B_1 + e^{-bt}(B_3 \cos 2\omega t + B_4 \sin 2\omega t) + D_1 \cos \omega t + D_2 \sin \omega t]^2 \quad (13)$$

The constants in Equation 13 depend on three parameters: μ, ω, and on the coefficient of the hydrodynamic resistances *b*. Taking into account that for less than 0.1 sec the members multiplied by e^{-bt} get insignificantly minor, for ρ the equation is:

$$\rho^2 = [A_1 + C_1 \cos \varphi + C_2 \sin \varphi]^2 + [B_1 + C_2 \cos \varphi - C_1 \sin \varphi]^2 \quad (14)$$

which can be presented by one constant (time-independent) and harmonic member as follows:

$$\rho^2 = \rho_c^2 + 2\left[(A_1C_1 + B_1C_2)\cos\varphi + (A_1C_2 - B_1C_1)\sin\varphi\right]$$
$$\rho_c^2 = A_1^2 + B_1^2 + C_1^2 + C_2^2; \qquad (15)$$
$$\rho_h^2 = a_c \cos\varphi + a_s \sin\varphi$$

From the data in TABLE 1 it is evident that in all variants the constant value of the polar radius remains unchanged and is very similar to the starting value. The amplitudes of the harmonic members vary from 2.6 up to 4.6 mm. The third variant represents an exception, because the combination of bigger spheroid sizes with lower rotation frequency is inadmissible from the biological point of view. The second variant is a combination of small-size spheroids that are rotated at higher speeds and therefore pronounced shear stress damage is expected. The solutions in all three described cases are analytically found.

The influence of the coefficient of hydrodynamic resistance b (1–15,000/s), as well as the starting position of the spheroids (polar angle 0–270° and polar radius 0–0.03 m) on the trajectory form, has been investigated in detail. The hydrodynamic resistance coefficient has a strong impact on the stability of the trajectory and on the microgravity generation (FIG. 5).

In a previous publication the following velocity functions were studied:

$$\omega = \bar{\omega} + \frac{E}{\bar{\omega}}\sin(\bar{\omega}t) \qquad (16)$$

The analysis of Equation 16 showed that the polar radius decreases at the beginning, but increases after reaching a threshold value. In FIGURE 6 the dense spot results from closely passing to each other twice-repeated trajectories.

The function (7) is nonperiodical. The velocity tends to a constant value equal to D. According to Equation 8 the changes of the trajectory radius are very slow and small (FIGS. 7, 8, and 9). Despite those unimportant radius drifts, the coefficient a_3 from Equation 5 does not reach positive values (FIG. 10), even at angular acceleration values not equal to zero according to the rule (8). So, the trajectory keeps its spiral form approximately. The adequacy of the proposed dynamic model was experimentally evaluated by growing cellular spheroids as indicated below.

Cells Cultivated in miniPERM

Neuro-2A cells are slowly growing semiadherent cells that spontaneously differentiate in up to 30% in neuronal-like cells (FIG. 11A). First cell clumps were seen on the sixth day and the probe contained many damaged cells that failed to adapt to the shear stress conditions in the bioreactor system (FIG. 11B). After the tenth day the microscopic picture changed dramatically: round non-differentiated and viable suspension cells together with bigger spheroids could

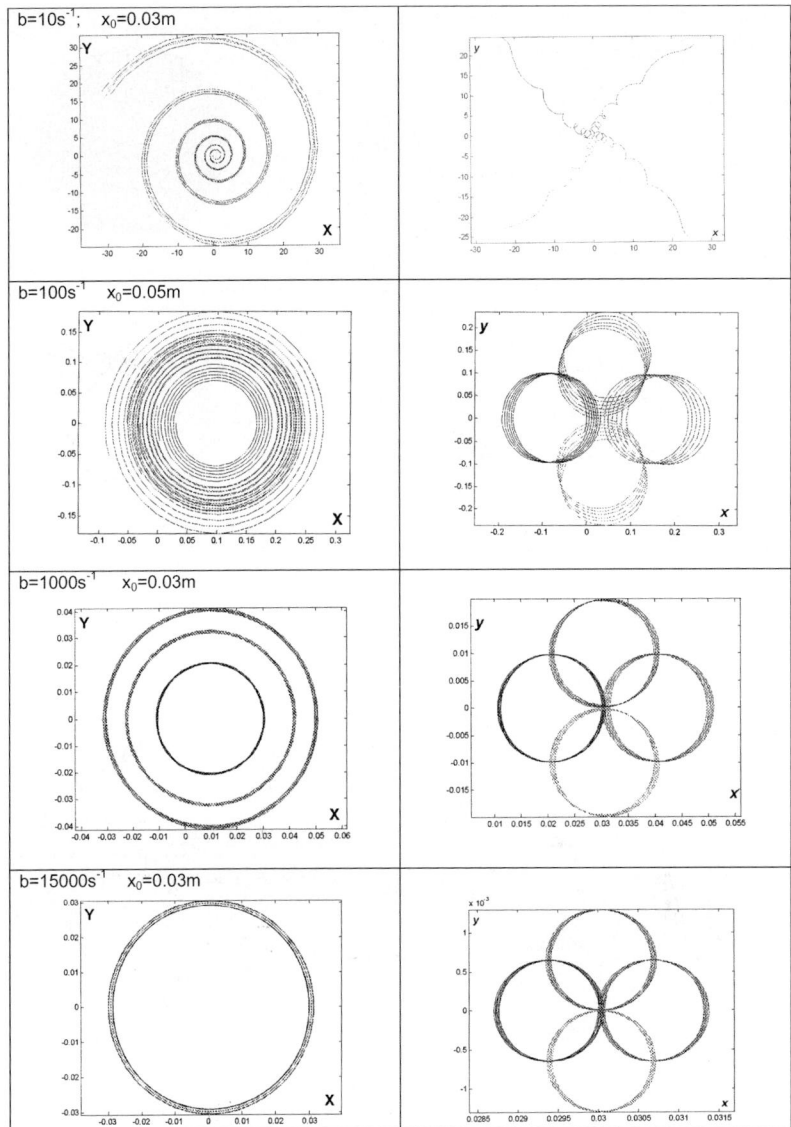

FIGURE 5. Influence of the hydrodynamic resistance on the trajectory of the particle.

be seen (FIG. 11C). From day 15 to the end of bioreactor operation spheroids of more than 100 cells were recorded (FIG. 11D). The total biomass obtained was over 1.1 g in weight.

Chronic myeloid leukemia (CML)–derived loosely adherent LAMA-84 cells failed to form any multicellular formations and grew as single cell suspensions.

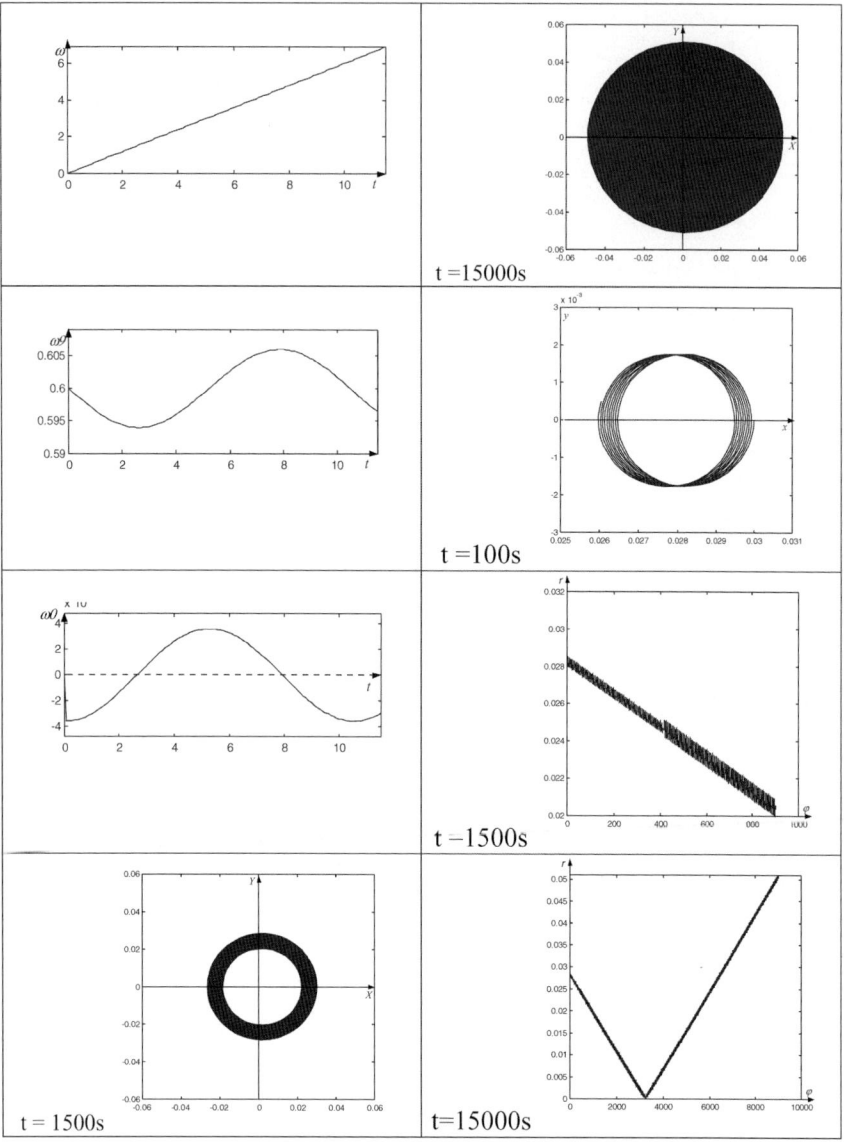

FIGURE 6. Trajectories of the particle according to the conditions in Equation 16: $E = 0.0036/s^2$ $\omega = 0.6/s$.

DISCUSSION

Already performed space experiments indicated that under weightless conditions three-dimensional cell cultivation is possible. On earth the rotation of

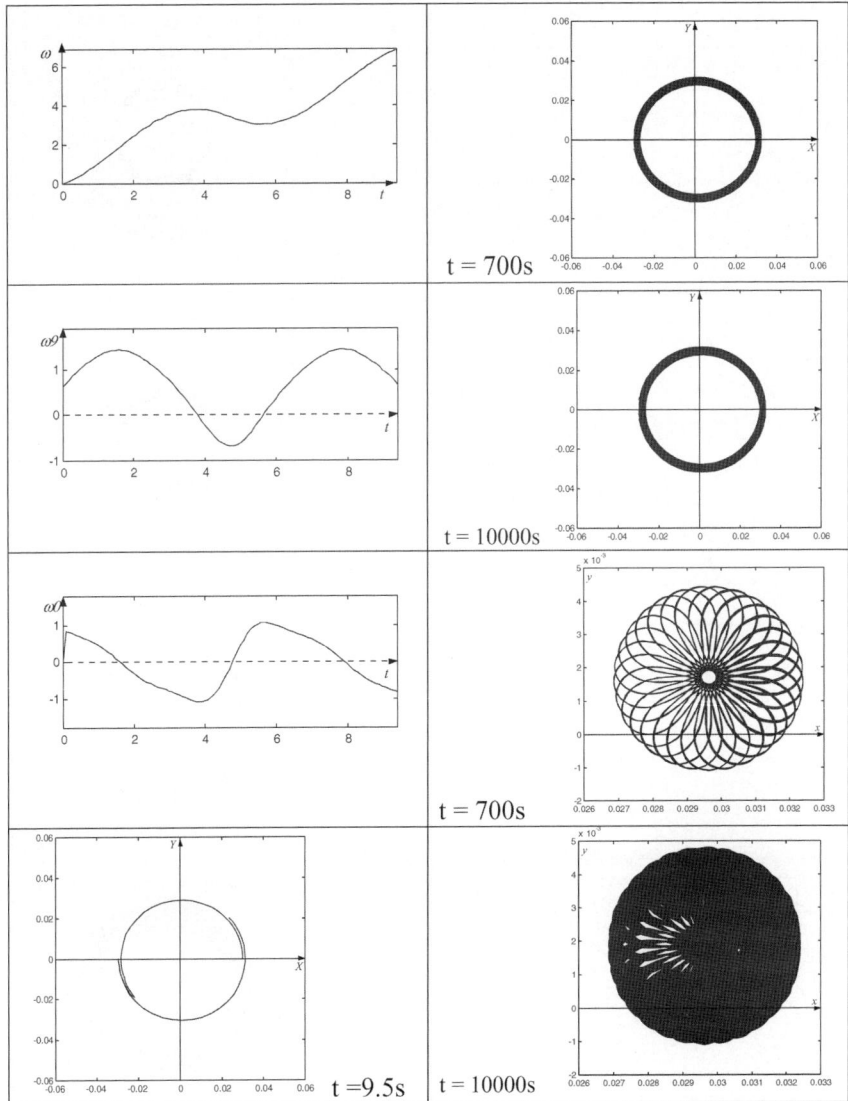

FIGURE 7. Trajectories of the particle according to the conditions in Equation 7: $A = 1$; $B = 0.5$; $C = 6$; $D = 1$.

the bioreactor partially compensates for the Earth's acceleration g, thus leading to simulated artificial microgravity. The microgravity concept is rather old: clinostats are the first described devices for this purpose.[7] The use of a clinostat for studying how gravity affects plant growth dates back well before the advent of space flight, possibly to the late 1800s.[7]. Only during the 2-year

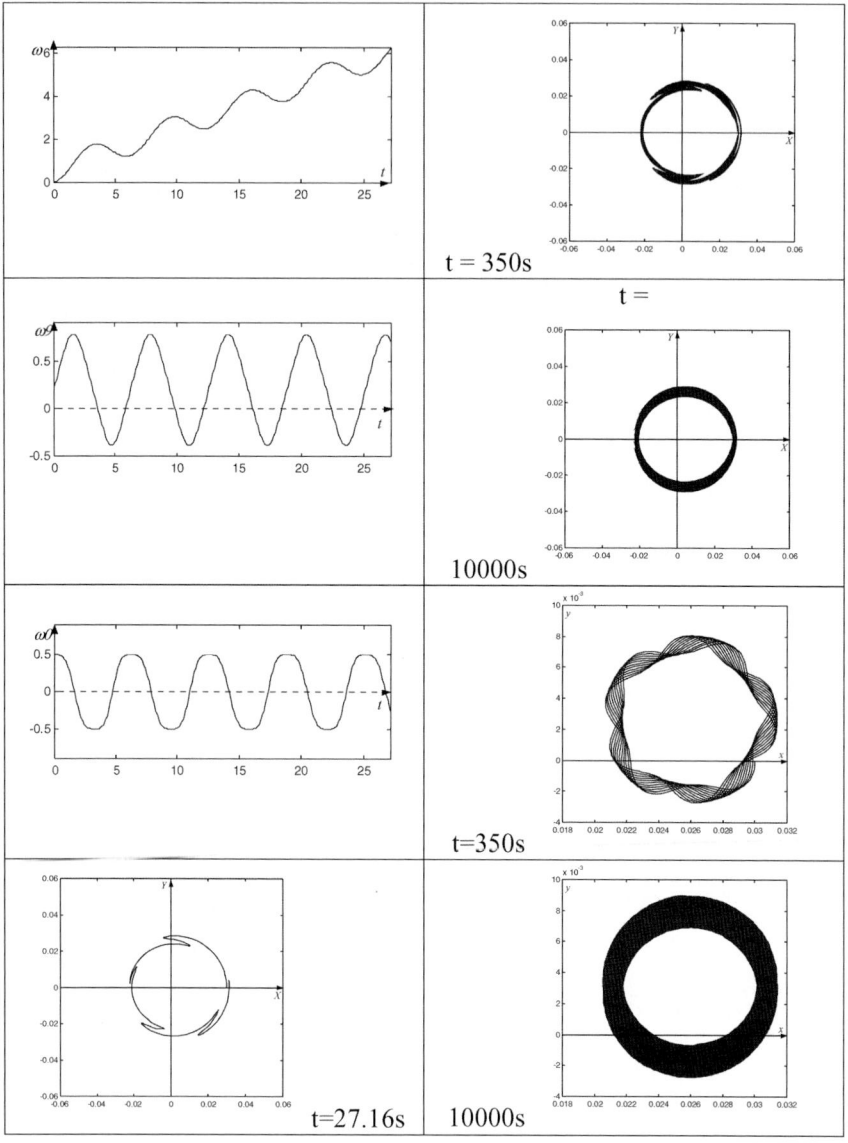

FIGURE 8. Trajectories of the particle according to the conditions in Equation 7: $A = -0.5; B = 0.2; C = 0.4; D = 1$.

period between 2001 and 2002 did NASA sponsor 47 research proposals in the field of cellular biotechnology on earth and in space. More than 50 papers were published.[8] For the RWV and RWPV bioreactors two dynamic models have been used: a hydrodynamic model[9–12] and a model concerning the movement

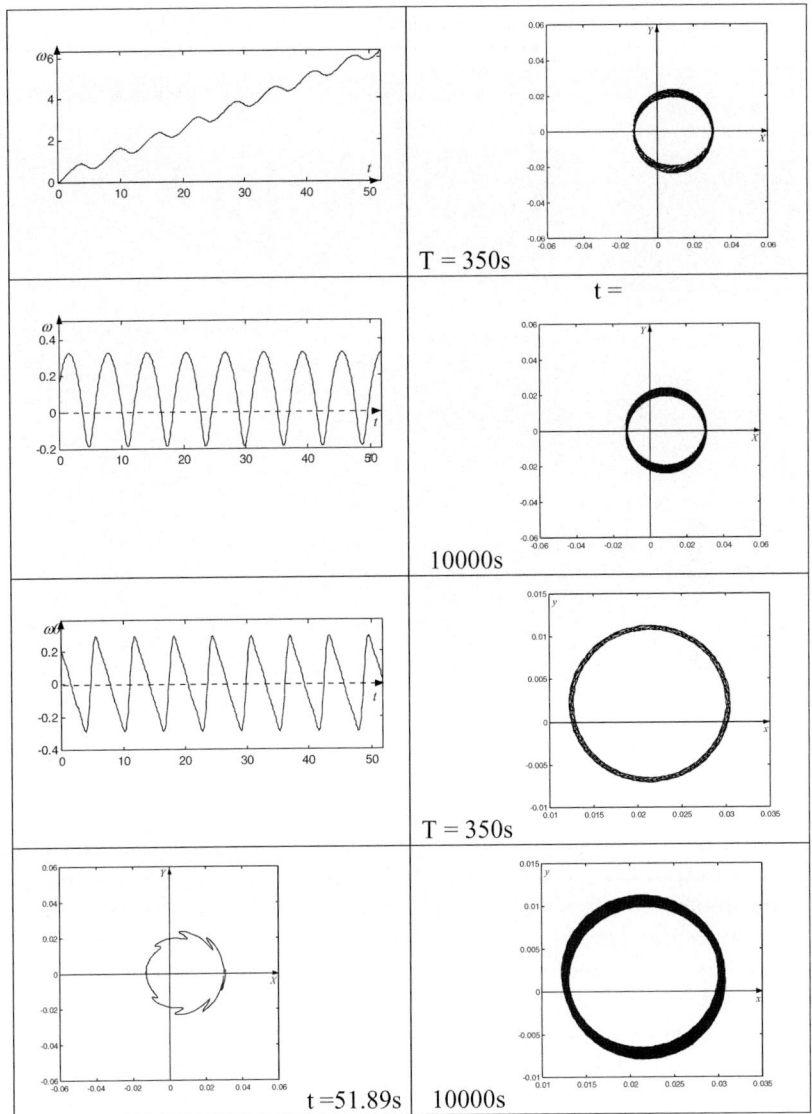

FIGURE 9. Trajectories of the particle according to the conditions in Equation 7: $A = 0.5; B = 0.2; C = 0.4; D = 1$.

of a body in resistive environment.[13–16] Turbulences in the laminar flow were experimentally confirmed, too.[14,17–20] Fluid flow turbulences are not taken into account which, in our opinion, leads to insufficiently correct determination of all parameters of the particle's movement. Moreover, these parameters have special impact on the formation of shear stress. Maximal values of shear stress

$\ddot{\varphi} \neq 0 \quad a_3 = 2\mu\dot{\varphi}(2\ddot{\varphi} - b\dot{\varphi}) \quad \ddot{\varphi} = -Ae^{-B\sin t}sh(\sin t + C) + D, \quad 2(\dot{\varphi}\ddot{\varphi} - b\dot{\varphi}^2) = f_{t\to\infty}(t)$

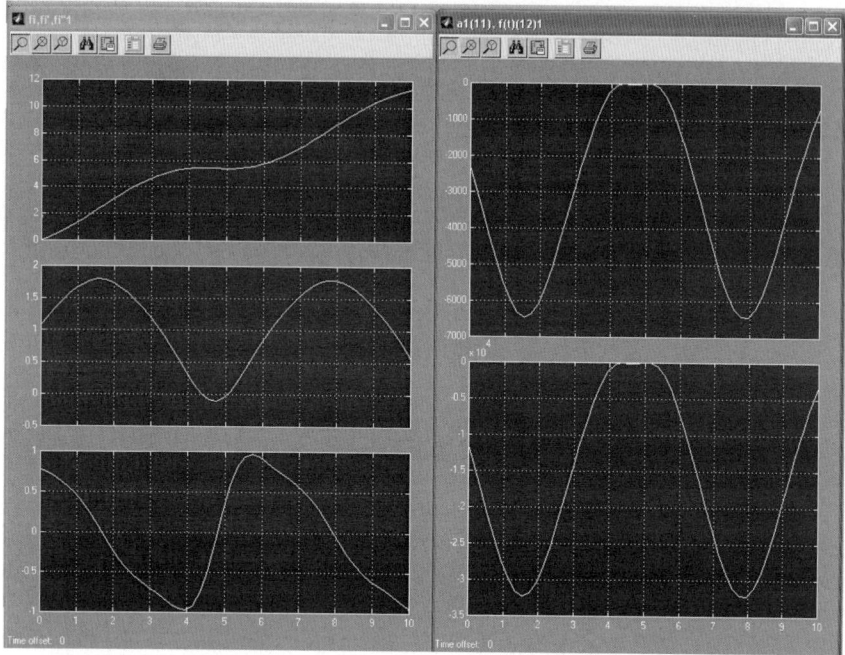

FIGURE 10. Graphs of the functions given in equations (6).

were determined by terminal velocity[3] or by the components of the shear stress force in a cylindrical coordinate system.[9] Unfortunately, the results are difficult to compare. It seems that the problem is not sufficiently solved in theory or experimentally. The dynamic model for bioreactors of the miniPERM type described by us is closer to the real situation because there is no laminar flow and therefore there is no turbulence.

Recent developments allow that development of the tissue can be monitored through the incorporation of sophisticated technical tools for online micro- and macroscopic observation of the structural properties of the bioreactor culture (e.g., video microscopy, magnetic resonance imaging, and computerized microtomography). All collected inputs could be analyzed by a microprocessor unit and fed back to the bioreactor system in order to optimize the control of the culture parameters at predefined levels.[21]

The analysis of the above-mentioned dynamic models indicates that the spheroid movement is unstable and that the polar radius is continuously increasing. Under defined functions for angular velocity change ($\omega(t)$), stable trajectories are obtained for a longer period of time. These trajectories can be repeated many times, if adequate electronic motor controlling is used.

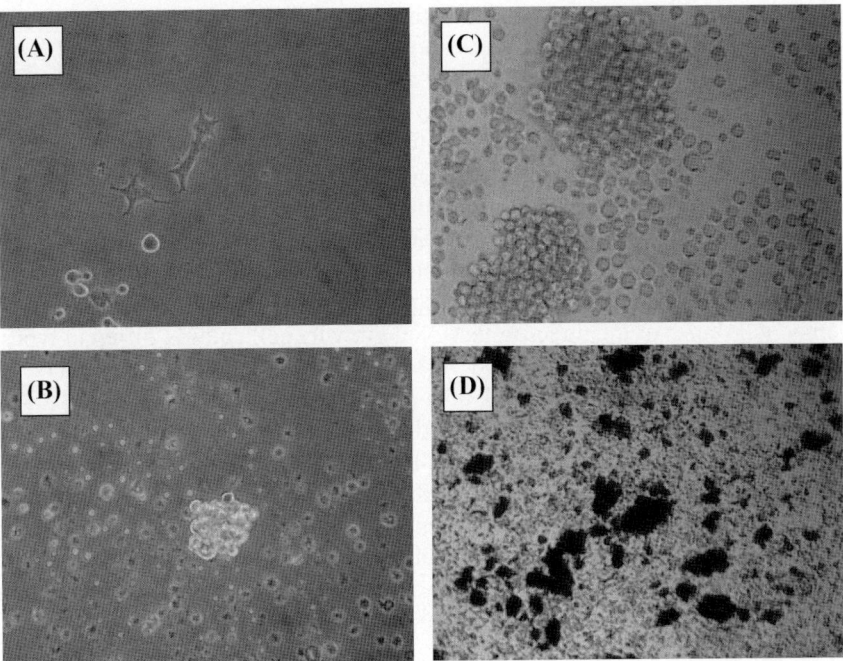

FIGURE 11. Microscopic pictures of cultivated mouse neuroblastoma Neuro-2A cells: (**A**) conventional culture of Neuro-2A cells (magnification 200 ×; about 30% of the cells show neuronal phenotype, spontaneously); (**B**) cellular spheroid of 30–40 cells on day 6 of cultivation in HDC5 production module of miniPERM; single cells and damaged cell rests are visible (magnification 200 ×); (**C**) cellular spheroids and single cell suspension on day 10 of cultivation in HDC5 production module of miniPERM (magnification 200 ×); (**D**) cellular spheroids and single cell suspension on day 15 of cultivation in HDC5 production module of miniPERM (prestained with MTT magnification 50 ×).

Adequate trajectories ensuring optimal cell growth in terms of intensive nutrient and gas exchange at minimal shear stress can be approximately achieved by techniques based on the first task of the dynamics and by using optimization procedures for motor controlling.

As previously described, human Hodgkin's lymphoma HD-MY-Z cells formed big three-dimensional spheroids of more than 100 cells in the miniPERM system, too.[22] CML LAMA-84 cells failed to form multicellular clumps possibly because of keeping the nondifferentiated stage of stem cells. Similarly, neuroblastoma cells grew as suspension culture in miniPERM, but simultaneously a significant part of them formed spheroids of different sizes. The number of these three-dimensional formations started to increase first after adaptation of the Neuro-2A cells to the bioreactor conditions during the period between days 6 and 8.

Taken together our experiments with hematopoietic and neuronal cells indicate that the miniPERM system allows three-dimensional growth of both types of cells and could serve for the acquisition of large cell amounts. The formation of spheroids is not the only means for cell growth in this bioreactor system and depends more or less on the cell type. Differentiation stage and formation of extracellular matrix components have to be studied in detail and the preservation of the undifferentiated status of stem cells (if confirmed) represents a new perspective on direction of research in terms of finding new cellular and tissue sources for transplantation purposes.

Recently performed and described experiments indicate that at certain developmental stages of the three-dimensional growing cell culture, different types of bioreactors could be used. These bioreactors differ in their constructive and mechanical characteristics. Under starting conditions stationary bioreactors may be advantageous because of the avoidance of shear stress and signal transduction disturbance of cultured cells. The next stages of three-dimensional growth may need bioreactors of the miniPERM type that use semipermeable membrane technologies. The last stages of upscaled cultures would grow better in bioreactors of the rotating wall vessel type (RWPV and RCM) of NASA that combine microgravity conditions with continuous perfusion with fresh culture medium.

REFERENCES

1. RATCLIFFE, A. & L. NIKLASON. 2002. Bioreactors and bioprocessing for tissue engineering. Ann. N. Y. Acad. Sci. **961:** 210–215.
2. GUIDI, A. 2001. Development of new bioreactor for mechanobiologic research in a microgravity environment. <www.angelfire.com/space/andreaguidi/carlogavazzispace/reportbioreactor.pdf>.
3. HAMMOND, T.G. & J.M. HAMMOND. 2001. Optimized suspension culture: the rotating wall vessel. Am. J. Physiol. Renal Physiol. **281:** F12–F25.
4. GENOVA, P., S. KONSTANTINOV, P. GOSPODINOV, et al. 2003. Theoretical and experimental determination of coefficients of hydrodynamic resistance in liquid medium for cell culture. Mechanics of Machines **46:** 103–106.
5. GENOVA, P., S. KONSTANTINOV, P. GOSPODINOV, et al. 2003. Cellular spheroid movement in a rotating bioreactor. Mechanics of Machines **46:** 107–110.
6. VUNJAK-NOVAKOVIC, G.N. SEARBY, J. DE LUIS, et al. 2002. Microgravity studies of cells and tissues. Ann. N. Y. Acad. Sci. **974:** 504–517.
7. KLAUS, D. 2001. Clinostats and bioreactors. Gravit. Space Biol. Bull. **14:** 55–64
8. NASA Physical Sciences Research Division. Annual Report 2001–2002: 18–27. <http://exploration.nasa.gov/documents/reports/UG_Annual_Report_02.pdf>.
9. BEGLEY, C. & S.J. KLEIS. 2000. The fluid dynamic and shear environment in the NASA/JSC rotating-wall perfused-vessel bioreactor. Biotechnol. Bioeng.. **70:** 32–40.
10. BEGLEY, C. & S.J. KLEIS. 2002. RWPV bioreactor mass transport: earth-based and in microgravity. Biotechnol. Bioeng. **80:** 465–476.

11. GREGORIADES, N., J. CLAY, N. MA, et al. 1999. Cell damage of microcarrier cultures as a function of local energy dissipation created by a rapid extensional flow. Biotechnol. Bioeng. **69:** 171–181.
12. LAPPA, M. 2003. Organic tissues in rotating bioreactors: fluid-mechanical aspects, dynamic growth models, and morphological evolution. Biotechnol. Bioeng. **84:** 518–532.
13. POLLACK, S., D. MEANEY, E. LEVIN, et al. 2000. Numerical model and experimental validation of microcarrier motion in a rotating bioreactor. Tissue Eng. **6:** 519–530.
14. LIO, T., X. LI, X. SUN, et al. 2004. Analysis on forces and movement of cultivated particles in a rotating wall vessel bioreactor. Biochem. Eng. J. **18:** 97–104.
15. ROBERTS, G. D. KORNFELD & W. FOWLIS. 1991. Particle orbits in a rotating liquid. J. Fluid Mechanics **229:** 555–567.
16. WOLF, D. & R. SCHWARZ. 1991. Analysis of gravity-induced particle motion and fluid perfusion flow in the NASA-designed rotating zero-head-space tissue culture vessel. NASA Technical Paper **3143:** 1–12.
17. RADIN, S., P. D. DUCHEYNE, P. AYYASWAMY, et al. 2001. Biotechnol. Bioeng. **75:** 369–378.
18. BOTCHWEY, E., S. POLLACK, E. LEVINE, et al. 2004. Quantitative Analysis of Three-Dimensional Fluid Flow in Rotating Bioreactors for Tissue Engineering. Published online at Wiley-Interscience <www.interscience.wiley.com> New York.
19. ANDERECK, D., S LIU & H. SWINNEY. 1986. Flow regimes in a circular Couette system with independently rotating cylinders. Fluid Mech. **164:** 155–183.
20. WOLF, D. & R. SCHWARZ. 1992. Experimental measurement of the orbital paths of particles sedimenting within a rotating viscous fluid as influenced by gravity. NASA Technical Paper **3200:** 1–15.
21. MARTIN, I.V., D. WENDT & M. HEBERER. 2004. The role of bioreactors in tissue engineering [review]. Trends Biotechnol. **22:** 80–86.
22. KONSTANTINOV, S., M.M. MINDOVA, P.T. GOSPODINOV, et al. 2004. Three-dimensional bioreactor cultures: a useful dynamic model for the study of cellular interactions. Ann. N.Y. Acad. Sci. **1030:** 103–115.

Animal Model of Drug-Resistant Tumor Progression

NADEZDA MIRONOVA,[a] OLGA SHKLYAEVA,[a] EKATERINA ANDREEVA,[b] NELLY POPOVA,[b] VASILYI KALEDIN,[b] VALERYI NIKOLIN,[b] VALENTIN VLASSOV,[a] AND MARINA ZENKOVA[a]

[a]*Institute of Chemical Biology and Fundamental Medicine Siberian Branch of Russian Academy of Sciences, 630090, Novosibirsk, Russian Federation*

[b]*Institute of Cytology and Genetics Siberian Branch of Russian Academy of Sciences, 630090, Novosibirsk, Russian Federation*

ABSTRACT: Experimental animal model of tumor progression based on mice lymphosarcoma (LS) and resistant lymphosarcoma (RLS) has been developed. LS tumor displays high sensitivity to cyclophosphamide, which is widely used in anticancer therapy. RLS tumor was derived from LS by passaging in mice receiving low concentration of cyclophosphamide (20 mg/kg) and display resistance to cyclophosphamide (up to dose 150 mg/kg). The primary cultures of LS and RLS tumors display different expression levels of the genes related to apoptosis and multiple drug-resistant phenotype: in RLS tumor high levels of *mdr1b* and *bcl-2* genes and low level of *p53* gene expression were found. A total of 10% of cells in RLS primary culture display multiple drug-resistant phenotype and survive even at high dose of cytostatics. Cultivation of RLS primary culture in the presence of increasing vinblastine concentrations gives RLS_{40} cell culture, which exhibits high levels of *mdr1a/1b* genes expression as compared to RLS and 20-fold increase of resistance to cytostatics. Drug-resistant RLS_{40} cells were transplanted into CBA mice and sensitivity of the tumors to anticancer drugs was tested. RLS_{40} tumors were resistant to a number of cytostatics used in anticancer therapy (cyclophosphamide, cysplatin, vinblastine, rubomycinum). Thus, RLS_{40} tumor can be used as model, which corresponds to tumor status observed in patients after one or several courses of chemotherapy and can be useful for testing conventional therapy alone or together with newly developed gene-targeted therapeutics.

KEYWORDS: multiple drug resistance phenotype; animal model; apoptosis

Address for correspondence: Dr. Marina Zenkova, Institute of Chemical Biology and Fundamental Medicine SB RAS, 8, Lavrentiev avenue, 630090, Novosibirsk, Russian Federation. Voice: +7-383-3333761; fax: +7-383-3333677.
e-mail: marzen@niboch.nsc.ru

INTRODUCTION

Development of oncology is impossible without experiments, which are not allowed to be carried out on humans. Different animal models are available for preclinical testing, and the choice of the adequate model is important for investigation and drug development.[1,2] The models used in experimental oncology include spontaneous, transplantable, and inducible tumors in animals, xenografts of human cancer cells, and molecular-genetic models (transgenic mice).

Each of experimental models mentioned above has some advantages and disadvantages. Inducible tumors in animals adequately reflect processes of cancerogenesis. It is known that aniline induces bladder cancer,[3] radon—lung cancer,[4] nitrosomethyl urea—lymphosarcoma (LS).[5] However, a number of compounds—cancerogenes capable to induce a definite type of tumor is limited. Similarly, some human carcinogens are not carcinogenic to rodent and vice versa[6] that restricts the area of application of inducible animal neoplasias. Spontaneous neoplasias in animals arise with a low frequency[6-8] and in the various age periods[9] that is an essential disadvantage of this model. Carcinogenesis in transgenic animal is provided by overexpression of oncogenes and/or silencing of tumor suppressor genes.[10-12] However, high levels of expression of genes observed in some transgenic animal models are never found in normal animal or human tumors.[13]

The choice of animal tumor allograft or human tumor xenograft as models is still debated. There are some models of immunodeficient mice—(*a*) obtained artificially by a combination of surgery (thymectomy), radiation, and/or drug treatment; and (*b*) mutant mice—nude and severe combined immunodeficiency (SCID)[1]—in which human xenograft tumors can be transplanted. Xenografted human tumors in immunodeficient mice have two main advantages: human origin of tumor and desired pathological type. The disadvantages of these models are that tumor is under the pressure of xenoorganism, it is mostly ectopic,[1,14] and it cannot be used for testing of immune-based therapeutics.[2] Some of the mentioned disadvantages can be overcome in SCID/Hu mice by implanting the SCID mouse fetal human liver fragments under the renal capsule in addition to subcutaneous implants of the fetal human bone.[15,16] However, this protocol is laborious and the mice are very expensive for routine analysis.

Investigations of animal allografted tumors revealed that there is no full similarity between the results obtained in animal models and to humans.[17] However, experiments on these models are useful because the models display important features of the process of tumor progression under pressure of organism. The allografted tumors can be considered as models for testing of gene-targeted therapeutics, especially for those targeted to the animal homologs of human genes.

In the present article we developed a new experimental animal model of tumor progression based on mice LS and resistant lymphosarcoma (RLS),[5] which display high drug resistance. This model corresponds to tumor status observed in patients after several courses of chemotherapy and can be useful for testing conventional therapy and for developing gene-targeted therapeutics (siRNAs, antisense oligonucleotides, and ribozymes) affecting expression of *mdr1a/1b* and *bcl-2* genes.

MATERIALS AND METHODS

Primary Cell Cultures of Tumors LS and RLS

Ascites (0.2 mL) containing LS tumor cells at concentration 5×10^6 cells/mL and RLS at concentration 5×10^5 cells/mL were transplanted intraperitoneally to CBA mice. At the eighth day of tumor development, saline solution (0.5 mL) was injected into the tumor to separate cells from peritoneal section. Mice were sacrificed by translocation of cervical vertebrates, and ascite (0.5–0.8 mL) was collected by syringe. Ascite liquids were diluted up to 5 mL with phosphate-buffered saline solution (PBS) buffer and separated on 3 mL of Lymphocyte Separation Medium (LSM) by centrifugation at 1000 rpm for 15 min. Cell layer containing lymphocytes was collected, suspended in 3 mL of PBS buffer and pelleted by centrifugation at 1000 rpm for 5 min. Collected cells were suspended in 5 mL Iscove's Modified Dulbecco Medium (IMDM) supplemented with 10% fetal bovine serum (FBS), 1% antibiotic-antimycotic solution (10,000 μg/mL streptomycin, 10,000 I.U./mL penicillin, amphotericin 25 μg/mL; MP Biomedicals, Irvine, CA). The cells were cultured at 37°C in a humidified incubator at 5% CO_2 for 48 h up to total adhesion of macrophages to flask bottom. Then cell suspension containing lymphocytes was transferred to new flask and cultured under the same conditions for 2–3 months.

RNA Isolation

Total RNA was isolated from cells of primary cultures according to published protocol.[18] Concentration of the RNA samples was evaluated by absorbance at 260 and 280 nm.

Reverse Transcription Polymerase Chain Reaction (RT-PCR) Analysis

cDNA synthesis was performed in 20 μL of 50 mM Tris-HCl, pH 8.3, containing 75 mM KCl, 3 mM $MgCl_2$, 0.01 M DTT, 2 μg of total cellular RNA, 0.5 mM of each dNTP, 12 pmol of random primer, and 10 u of MoMLV reverse transcriptase (Institute of Chemical Biology and Fundamental Medicine Siberian Branch of Russian Academy of Sciences, Novosibirsk, Russia). The reaction was incubated at 37°C for 1 h. Samples were either used immediately for PCR or stored at –20°C. cDNA derived from 0.1 μg of total cellular RNA was used for PCR with "primer dropping."[19] To perform PCR the following primers were used: *mdr1a*-specific primers GCAGGTTGGC-TAGACAGGTTGT (forward, accession number M33581, residues 259–281) and GAGCGCCACTCCATGGATAA (reverse, residues 309–328), yielding a 69 bp product; *mdr1b*-specific primers CTGCTGTTGGCGTATTTGGG (forward, accession number M14757, residues 201–220) and TGGCAGAATAC TGGCTTCTGCT (reverse, residues 349–370), yielding a 121 bp product; *bcl-2*-specific primers TCGCAGAGATGTCCAGTCAGC (forward, accession number NM_0097415, residues 1523–1543) and CATCCCAGCCTCC GTTATCC (reverse, residues 1758–1777), yielding a 255 bp product; *p53*-specific primers GAACCGCCGACCTATCCTTAC (forward, accession number AY212017, residues 840–860) and GTTTGGGCTTTCCTCCTTGAT (reverse, residues 1231–1251), yielding a 412 bp product; two pairs of β-*actin*-specific primers (i) AGCCATGTACGTAGCCATCCA (forward, accession number NM_007393, residues 469–490) and TCTCCGGAGTCCATCACA ATG (reverse, residues 529–550), yielding a 81 bp product, and (ii) CTCTC-CCTCACGCCATCCT (forward, residues 589–607) and TCCGCCTAGAAG-CACTTGC (reverse, residues 1195–1213), yielding a 625 bp product. *mdr1a* and *mdr1b*-specific products were amplified for 28 cycles, while β-*actin*-specific primers (*a*) were added at the fifth cycle. *bcl-2* and *p53*-specific products were amplified for 31 cycles, while β-*actin*-specific primers (*b*) were added to the reaction at the sixth cycle. Amplification was performed as follows: initial step 95°C for 5 min, then 28 or 31 cycles of 94°C for 1 min, 57°C for 1 min, and 72°C for 1 min. PCR products were analyzed by electrophoresis in 8% polyacrylamide gel for *mdr1a* and *mdr1b* products and in 2% agarose gel for *bcl-2* and *p53* products, visualized by ethidium bromide staining, photographed in UV light and densitometrically quantified using Gel-Pro Analyzer 4.0 (Media Cybernetics, Silver Spring, MD). The data were presented as ratio of specific gene expression level to β-*actin* expression level. The absence of contaminants was routinely checked by RT-PCR assay of negative control samples where the RNA sample was replaced by sterile water or MoMLV reverse transcriptase was absent in the reaction mixture.

Cells Viability in the Presence of Cytostatics

The viability of the obtained primary cultures LS, RLS, RLS_5, RLS_{10}, RLS_{20}, and RLS_{40} to cytostatics (vinblastine, cytarabine, and doxorubicine) was tested *in vitro* using the MTT assay.[20] Optical density was evaluated at the wavelength 570 nm (background was measured at 620 nm) by multichannel spectrophotometer.

The Study of Growth Inhibition of Tumors LS, RLS, and RLS_{40} In Vivo

Three-month-old male CBA/Lac (CBA) mice were housed in groups of 8–10 individuals in plastic cages. The mice had free access to food and water and the daylight conditions were normal. Tumor cells (5×10^6 cells/mL for LS and RLS$_{40}$, 5×10^5 cells/mL for RLS) in 0.2 mL of saline buffer were transplanted into the thigh muscles, where they formed a solid tumors. Single injection of following cytostatics was performed at the 11th day of tumor progression: cyclophosphamide, 100 mg/kg, intraperitoneally; cisplatin, 10 mg/kg, intraperitoneally; vinblastine, 1.5 mg/kg, intravenously; rubomycinum, 15 mg/kg, intravenously. The cytostatics were dissolved in saline buffer just before injection. Control animals were injected with saline buffer. Mice were sacrificed by dislocation of the cervical vertebrates at the seventh day after drugs injection. Tumor weight was determined as difference between the weight of the amputated tumor-bearing stump and the tumor-free contralateral stump. The inhibition of tumor growth was estimated as follows: [mean tumor weight$_{control}$ − mean tumor weight$_{experiment}$] /mean tumor weight$_{control}$.

Statistics

The comparison between groups was conducted with Student's t-test (two-tailed, unpaired). $P < 0.01$ was considered to be statistically significant.

RESULTS

From In Vivo to In Vitro

Tumor Types

Molecular mechanisms of drug resistance of malignant cells were shown to be diverse.[21] They can result from alteration of anticancer drug transport through cell membrane, which are associated with changes in the expression levels of genes encoding family of transmembrane transporters (*mdr* family),[22] and lead to decreased of intracellular drug accumulation, or/and canceling of the apoptosis induced by drugs (changes in the expression levels of *p53*, *bcl-2* genes involved in execution of apoptosis).[23,24]

Two types of LS-based tumors obtained by passaging in CBA mice display different sensitivity toward cyclophosphamide, which metabolite, processed into liver, is known to induce apoptosis. LS is sensitive to treatment with cyclophosphamide at concentrations 100–150 mg/kg of animal weight. Tumors RLS, which were derived from LS by passaging in mice administrated sequentially with 10–20 mg/kg of cyclophosphamide, are resistant even to high doses of cyclophosphamide and do not regress via apoptosis. These two types of tumors were chosen to develop an experimental model of ascitic or solid tumor exhibiting (*a*) resistance to different anticancer therapeutics and (*b*) resistance toward cyclophosphamide.

Expression levels of genes *mdr1a* and *mdr1b* encoding *p*-glycoprotein (multiple drug resistance phenotype), pro- and antiapoptotic genes (*p53* and *bcl-2*)

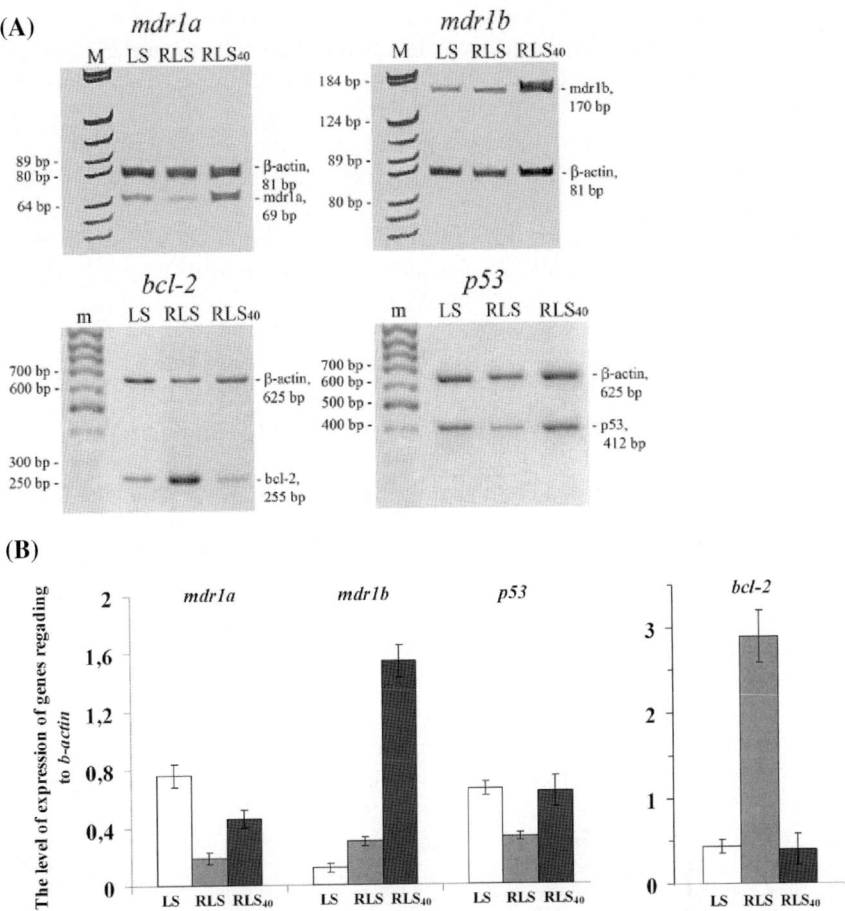

FIGURE 1. Expression of genes—markers of apoptosis and multiple drug resistance in LS, RLS, and RLS$_{40}$ cell lines. (**A**) Analysis of RT-PCR products by electrophoresis in 8% PAAG (*mdr1a* and *mdr1b*) or in 2% agarose gel (*bcl-2* and *p53*). β-*actin* gene was used as internal control. M—pBR322 DNA/Bsu RI (HaeIII) molecular weight marker, m—GeneRuler^{TM} (Fermentas, Lithuania) 50 bp DNA Ladder. Bonds, which correspond to specific PCR products, are indicated on the left. (**B**) Expression levels of *mdr1a*, *mdr1b*, *p53* and *bcl-2* genes in LS, RLS and RLS$_{40}$ cells obtained by quantification of RT-PCR results. The quantitative analysis was carried out using GelPro Software (Gel-Pro Analyzer 4.0; Media Cybernitics); the expression levels are measured as ratio of IOD (integral optical density) of gene-specific PCR-products to IOD of β-*actin*-specific product.

in LS and RLS primary cell cultures prepared as described in the Materials and Method section were evaluated using RT-PCR (FIG. 1). In RLS primary culture, expression of *mdr1b* gene was 2.5-times higher and *bcl-2* was 6.7-times higher, while expressions of *mdr1a* and *p53* genes were 4- and 2-times lower, respectively, as compared with LS one.

To define which primary culture—LS or RLS—can be used for selection of highly resistant cell line, sensitivity of LS and RLS cells toward cytostatics conventionally used in anticancer therapy was studied. IC_{50} (concentration of a cytostatic that cause 50% mortality of cells) values of vinblastine for LS and RLS were found to be 4 and 6 nM, and in the case of doxorubicine—82 and 234 nM for LS and RLS cells, respectively. The cell population of RLS primary culture was found to be heterogenic: about 10% of cells survive even at cytostatic concentration much higher then IC_{50}. Apparently, these higher resistant cells define multiple drug-resistant phenotype of RLS tumor. Thus, primary culture RLS can be used for selection of cell line exhibiting resistance to different anticancer therapeutics.

Selection of RLS_{40} Cells on Vinblastine-Supplemented Medium

RLS primary culture was cultivated in IMDM medium supplemented with 10% FBS, 1% antibiotic-antimycotic solution (standard conditions), and vinblastine at concentrations ranged from 5 nM to 40 nM. The first round of selection comprised two stages: (i) cultivation of RLS cells under standard conditions but in the presence of 5 nM vinblastine for 48 h; and (ii) separation of viable lymphocytes on LSM (see Materials and Methods). Then, collected lymphocytes were cultured under the same conditions and vinblastine concentration to produce amount of cells sufficient for the second round of selection. Isolation of viable lymphocytes was repeated each 72 h. After each round of selection, concentration of vinblastine in the medium was stepwise increased up to 40 nM. Intermediate cell cultures RLS_5, RLS_{10}, RLS_{20} and RLS_{40} obtained at the end of each round of selection were frozen and kept until use. From part of the cells RNA was isolated and used for RT-PCR analysis.

Sensitivity of Cell Lines RLS_5, RLS_{10}, RLS_{20}, and RLS_{40} toward Cytostatics and Levels of Genes Expression

Sensitivity of selected cell lines toward cytostatics (vinblastine, cytarabine, and doxorubicine) was evaluated using MTT assay. Cells were cultured under standard conditions in the presence of different concentrations of cytostatics for 48 h. IC_{50} values of vinblastine were found to be 11, 20, 50, and 320 nM for cell cultures RLS_5, RLS_{10}, RLS_{20}, and RLS_{40}, respectively. Thus, IC_{50} of vinblastine for selected cell culture RLS_{40} was 50-times higher then for parental cell culture RLS. IC_{50} values of cytarabine and doxorubicine for RLS_{40} cell culture were 3-times higher then for the parental RLS cells.

Expression levels of *mdr1a*, *mdr1b* genes, antiapoptotic marker *bcl-2*, and proapoptotic gene *p53* in RLS_{40} cell culture were analyzed by RT-PCR using β-*actin* mRNA as internal standard (FIG. 1). LS and RLS tumor cells were used as controls. It is seen that in RLS_{40} cells expression levels of *mdr1a* and *p53* genes are increased 2-times and *mdr1b* gene increased 4-times, while expression level of *bcl-2* gene is decreased 10-times as compared to the parental tumor RLS.

From In Vitro to In Vivo

Transplantation of RLS_{40} Tumor Cells to CBA Mice

RLS_{40} cells were transplanted into CBA mice intraperitoneally (5×10^5 cells per 0.2 mL of saline buffer) to form ascitic tumors or into thigh muscle (10^6 cells per 0.2 mL of saline buffer) to form solid tumors. The selected RLS_{40} cell line form both ascitic and solid tumors *in vivo* with the growth rate similar to the parental RLS tumors (primary data not shown). At the eighth day of each of the passages of RLS_{40} tumors *in vivo*, ascitic liquid was collected for preparation of primary cell cultures (see Materials and Methods) and for analysis of expression level of *mdr1a*, *mdr1b*, *p53*, and *bcl-2* genes by RT-PCR. Obtained data show similar profile of gene expression in the tumor cells obtained at the first three passages *in vivo*, which is evidence of genetic stability of RLS_{40} cell culture.

Sensitivity of Tumors LS, RLS, and RLS_{40} to Cytostatics In Vivo

Sensitivity of RLS_{40} tumor to cytostatics conventionally used in anticancer therapy was studied *in vivo* and compared with sensitivity of parental tumors LS and RLS. Tumors LS, RLS, and RLS_{40} were simultaneously transplanted in thigh muscle and the mice were administrated with a single injected of cytostatics (cyclophosphamide, cisplatin, rubomycinum and vinblastine) at the 11th day of tumor progression. Tumor weight was determined at the seventh day after injection (see Materials and Methods). The data on inhibition of tumor growth by cytostatics are shown in TABLE 1.

The obtained data clearly show that LS tumor is entirely regressed upon treatment with cyclophosphamide (100 mg/kg) and regressed with remission observed on the 12th day after injection of 25 mg/kg cyclophosphamide. Cisplatin

TABLE 1. Inhibition of LS, RLS, and RLS_{40} tumor growth by cytostatics

Drug	Growth inhibition (%)		
	LS	RLS	RLS_{40}
Cyclophosphamide	100	16.5	5
Cisplatin	93	44	20
Vinblastine	20	No effect	No effect
Rubomycinum	89	51	13

NOTE: Growth inhibition was calculated as described in Materials and Methods.

and rubomycinum induced essential inhibition of LS tumor growth (TABLE 1) but the tumor did not disappear entirely. RLS tumor regressed insignificantly (16.5%) upon treatment with cyclophosphamide at a dose 100 mg/kg whereas decreased doses (25–50 mg/kg) had no effect on the tumor growth. Treatment of RLS tumor by cisplatin and rubomycinum resulted in twofold reduction of RLS tumor growth. RLS_{40} tumor was highly resistant to therapeutics (TABLE 1): vinblastine and cyclophosphamide have only minor effect on the tumor growth, and inhibition of tumor growth observed in the case of rubomycinum and cisplatin was 13% and 20%, respectively.

DISCUSSION

In our study we characterized three types of LS: (i) LS induced by injection of nitrosomethyl urea and then consecutively transplanted into CBA mice, which displays high sensitivity to cyclophosphamide (50–150 mg/kg) and regress by apoptosis; (ii) RLS obtained from LS by passaging *in vivo* on low doses of cyclophosphamide, which acquires resistance to high doses of cyclophosphamide; and (iii) RLS_{40} obtained by selection of RLS tumor cells *in vitro* on medium supplemented with cytostatics, which displays 20-fold increase of resistance to cytostatics *in vitro* and *in vivo*. Analysis of expression of genes providing multiple drug resistance phenotype in LS, RLS, and RLS_{40} tumor cells revealed different profile of gene expression. LS tumor cells are characterized by high level of *p53* gene expression and noticeable expression of *mdr1a* gene. RLS tumors obtained from LS by passaging *in vivo* at low doses of cyclophosphamide, overexpressed *mdr1b* and *bcl-2* genes, while expression of *p53* is inhibited. Selection of RLS primary culture in the presence of vinblastine *in vitro* gives a cell culture (RLS_{40}), which displays significantly increased expression levels of *mdr1a* and *mdr1b* genes, while expression of *bcl-2* gene is decreased as compared to RLS tumors. RLS_{40} cell culture was shown to be stable upon passaging in CBA mice and give rise to tumors, which exhibit high resistance to cytostatics *in vivo*. Thus, RLS_{40} tumor can be considered as an animal model of tumor progression, characterized by high drug resistance and poor prognosis of treatment. This model corresponds to tumor status observed in patients after several courses of chemotherapy and can be used for testing conventional therapy alone or combined with gene-targeted therapeutics (siRNAs, antisense oligonucleotides, and ribozymes) affecting expression of *mdr1a/1b* and *bcl-2* genes.

ACKNOWLEDGMENTS

This work was supported by RAS programs Molecular and cellular biology and Science to medicine, FCSTP grant RI-012/001/254.

REFERENCES

1. KIM, J., M. O'HARE & R. STEIN. 2004. Models of breast cancer: is merging human and animal models the future? Breast Cancer Res. **6:** 22–30.
2. BROWN, P. 2003. The mouse in preclinical trials: transgenic, carcinogen-induced, or xenograph models—which to use? Breast Cancer Res. **5:** 27.
3. RADOMSKI, J. 1979. The primary aromatic amines: their biological properties and structure-activity relationships. Ann. Rev. Pharmacol. Toxicol. **19:** 129–157.
4. DARBY, S., D. HILL, A. AUVINEN, et al. 2005. Radon in homes and risk of lung cancer: collaborative analysis of individual data from 13 European case-control studies. Br. Med. J. **330:** 223–228.
5. KALEDIN, V., V. NIKOLIN, T. AGEEVA, et al. 2000. Cyclophosphamide-induced apoptosis of murine lymphosarcoma cells in vivo. Vopr. Onkol. **46:** 588–593.
6. ANISIMOV, V., S. UKRAINTSEVA & A. YASHIN. 2005. Cancer in rodent: does it tell us about cancer in humans? Nat. Rev. Cancer **5:** 807–819.
7. WALSH, K. & J. POTERACKI. 1994. Spontaneous neoplasms in control Wistar rats. Fundam. Appl. Toxicol. **22:** 65–72.
8. TILLMANN, T., K. KAMINO & U. MOHR. 2000. Incidence and spectrum of spontaneous neoplasms in male and female CBA/J mice. Exp. Toxicol. Pathol. **52:** 221–225.
9. FREDRICKSON, T. 1987. Ovarian tumors of the hen. Environ. Health Perspect. **73:** 35–51.
10. TAKAYAMA, H., W. LAROCHELLE & R. SHARP. 1997. Diverse tumorigenesis associated with aberrant development in mice overexpressing hepatocyte growth factor/scatter factor. Proc. Natl. Acad. Sci. USA **94:** 701–706.
11. STEWART, T., P. PATTENGALE & P. LEDE. 1984. Spontaneous mammary adenocarcinomas in transgenic mice that carry and express MTV/myc fusion genes. Cell **38:** 627–637.
12. GIOVANNI, M., E. ROBANUS-MAANDAG & Van DER VALK. 2000. Conditional biallelic Nf2 mutation in the mouse promotes manifestations of human neurofibromatosis type 2. Genes Dev. **14:** 1617–1630.
13. GREEN, J., K. DESAI & Y. YE. 2004. Genomic approaches to understanding mammary tumor progression in transgenic mice and responses to therapy. Clin. Cancer Res. **10:** 385s–390s.
14. ROFSTAD, E. 1992. Comparative sensitivity of cells from human tumors and derivative tumor xenografts to radiation and heat treatments. J. Natl. Cancer Inst. **84:** 1517–1524.
15. YACCOBY, S., C. JOHNSON, S. MAHAFFEY S, et al. 2002. Antimyeloma efficacy of thalidomide in the SCID-hu model. Blood **100:** 4162–4168.
16. EPSTEIN, J. & S. YACCOBY. 2005. The SCID-hu myeloma model. Methods Mol. Med. **113:** 183–190.
17. VOSKOGLOU-NOMIKOS, T., J. PATER J & L. SEYMOUR. 2003. Clinical predictive value of the in vitro cell line, human xenograft, and mouse allograft preclinical cancer models. Clin. Cancer Res. **9:** 4227–4239.
18. CHATTOPADHYAY, N., R. KHER & M. GODBOLE. 1993. Inexpensive SDS/phenol method for RNA extraction from tissues. Biotechniques **15:** 24–26.
19. WONG, H., W. ANDERSON, T. CHENG & K. RIABOWOL. 1994. Monitoring mRNA expression by polymerase chain reaction: the "primer-dropping" method. Anal. Biochem. **223:** 223–258.

20. PARK, J., B. KRAMER, S. STEINBERG, *et al.* 1987. Chemosensitivity testing of human colorectal carcinoma cell lines using a tetrazolium-based colorimetric assay. Cancer Res. **47:** 5875–5879.
21. STAVROVSKAYA, A. 2000. Cellular mechanisms of multidrug resistance of tumor cells. Biochemistry (Mosc.) **65:** 95–106.
22. CORDON-CARDO, C., J. O'BRIEN, J. BOCCIA, *et al.* 1990. Expression of the multidrug resistance gene product (P-glycoprotein) in human normal and tumor tissues. J. Histochem. Cytochem. **38:** 1277–1287.
23. O'CONNOR, P., J. JACKMAN, I. BAE, *et al.* 1997. Characterization of the p53 tumor suppressor pathway in cell lines of the National Cancer Institute anticancer drug screen and correlations with the growth-inhibitory potency of 123 anticancer agents. Cancer Res. **57:** 4285–4300.
24. DIVE, C. 1997. Avoidance of apoptosis as a mechanism of drug resistance. J. Intern. Med. **740:** 139–145.

Preparation and Characterization of Recombinant Chicken Growth Hormone (chGH) and Its Putative Antagonist chGH G119R Mutein

HELENA E. PACZOSKA-ELIASIEWICZ,[a] GILI SALOMON,[b]
SHAY REICHER,[b] EUGENE E. GUSSAKOWSKY,[c] ANNA HRABIA,[a]
AND ARIEH GERTLER[b]

[a]*Department of Animal Physiology, University of Agriculture, Kraków, Poland*

[b]*Faculty of Agricultural, Food and Environmental Quality Sciences, The Hebrew University of Jerusalem, Jerusalem, Israel*

[c]*Department of Botany, University of Manitoba, Winnipeg, MB R3T 2N2, Canada*

ABSTRACT: Synthetic cDNA of chicken GH (chGH) and its G119R mutein was synthesized after being optimized for expression in *E. coli*. The respective cDNAs were inserted into expression vector, expressed and found almost entirely in the insoluble inclusion bodies (IBs). The IBs were isolated, the proteins solubilized in 4.5 M urea, at pH 11.3 in presence of cysteine, refolded, and purified to homogeneity by anion-exchange chromatography on Q-Sepharose. The overall yields were 400 to 500 mg from 5 L of fermentation. Both proteins were > 98% pure, as evidenced by SDS-PAGE, and contained at least 95% monomers, as documented by gel-filtration chromatography under non-denaturing conditions. Circular dichroism analysis revealed that both proteins have identical secondary structure characteristic of cytokines, namely > 50% of alpha helix content. Chicken GH was capable of forming a 1:2 complex with recombinant oGH receptor extracellular domain, but its affinity, as determined by RRA, was 11-fold lower than that of ovine GH (oGH). Correspondingly, its bioactivity, assessed using FDC-P1 3B9 cells stably transfected with rabbit GHR, was 30–40-fold lower, whereas chGH G119R mutant did not bind to oGHR-ECD and was devoid of any biological activity in FDC-P1 3B9 cells. However, in binding experiments that were carried out using chicken liver membranes, both oGH and chGH showed similar IC_{50} values in competition with ^{125}I-oGH, while the IC_{50} of G119R mutein was 10-fold higher. These results emphasize the importance of species specificity and indicate the possibility of antagonistic activity of chGH G119R.

KEYWORDS: recombinant; growth hormone; chicken; mutein

Address for correspondence: Dr. Anna Hrabia, Department of Animal Physiology, University of Agriculture, al. Mickiewicza 24/28, 30-059 Kraków, Poland. Voice: 48-12 633 38 24; fax: 48-12-662-41-07.
e-mail: rzhrabia@cyf-kr.edu.pl

INTRODUCTION

Rodent[1] and chicken growth hormone (chGH)[2] play a major role in growth promotion and metabolism. A recently published paper has indicated that the chGH gene has high diversity and affects growth and carcass traits.[3] Forty-six single nucleotide polymorphisms (SNPs) were found, but only five were located in exons. Two of those were nonsynonymous, one in the signal peptide (A13T) and another in mature protein (R59H), but neither was associated with any growth or metabolic parameters.[3] In order to study the function of chGH by possible pharmacological intervention and eventual biotechnological use, a simple nonexpensive method for its preparation is needed. Single point mutation in many mammalian GHs, by replacement of Gly located in exon 3 with bulky positively charged Arg or Lys, yielded an analogue exhibiting antagonistic properties.[4,5] In the case of human GH such a molecule was further mutated, pegylated, and approved as a drug (Pegvisomant) for treatment of acromegaly and pituitary tumors.[6] To prepare the corresponding mutein of chGH we have identified the analogous Gly (G119) and mutated it into Arg.

MATERIALS AND METHODS

Materials

Recombinant ovine (o)GH, rat (r)GH, and oGH receptor extracellular domains (oGHR-ECD) were obtained from Protein Laboratories Rehovot (PLR) Ltd. (Israel), and DNA constructs from Entelechon GmbH (Regensburg, Germany). Molecular weight marker for SDS-PAGE was purchased from Bio-Rad (Hercules, CA, USA). SDS-PAGE reagents, nalidix acid, lysozyme, and cysteine were purchased from Sigma-Aldrich (St. Louis, MO, USA) and Bio-Lab Ltd. (Jerusalem, Israel). RPMI-1640 medium, fetal calf serum (FCS), and horse sera were purchased from Biological Industries (Kibbutz Beit HaEmek, Israel). Antibiotic antimycotic solution (5 × 10^4 U/mL penicillin, 50 mg/mL streptomycin, 0.125 mg/mL fungisone), NaCl and Tris were purchased from Bio-Lab Ltd. (Jerusalem, Israel) and Bacto-tryptone, Bacto-yeast extract, glycerol, EDTA, Triton X-100, and urea were obtained from ENCO Diagnostics Ltd. (Petah-Tikva, Israel).

Preparation of Expression Plasmids Encoding Chicken GH and its G119R Mutein

Synthetic cDNA encoding the sequences of chicken GH and its G119R mutant (NCI Gene Database Accession Number–NP˙989690) were modified to ensure better codon usage and expression in *Escherichia coli (E. coli)*. The cDNA in pTOPO was digested with *Nco*I and *Hind*III, extracted, and ligated into linearized pMon3401 expression vector. *E. coli* MON105 competent cells were transformed with the new expression plasmid and plated on LB-agar plates containing 75 μg/mL spectinomycin for plasmid selection. Four *E. coli* colonies were isolated and confirmed to contain chGH or chGH G119R cDNA by digestion with *Nco*I/*Hind*III restriction enzymes. All colonies were positive and one of them was sequenced.

Expression, Refolding, and Purification of chGH and chGH G119 Mutein

E. coli (strain MON105) cells transformed with the respective plasmids were grown in 2.5-L flasks in 500 mL TB medium (1.2% Bacto-tryptone, 2.4% Bacto-yeast extract, 0.4% [v/v] glycerol in ddH$_2$O) (200 rpm, 37°C) to an absorbance of 0.9 at 600 nm. Freshly prepared nalidixic acid (2.5 mL, 10 mg/mL) was then added. The cells were grown for an additional 4 h, pelleted for 10 min at 7,000 g, and stored at –20°C. Frozen precipitates from 5 L fermentation were thawed and resuspended in 800 mL cold 10 mM EDTA, pH 8.0, 10 mM Tris-HCl, pH 8.0, in the presence of 0.5 mg/mL lysozyme. Following 30-min incubation on stirrer at 4°C, the cells were sonicated for 5 min at 50%-cycle program and pelleted for 20 min at 10,000 g (4°C). The pellet was then suspended in cold ultra pure water and sonicated and centrifuged as mentioned above. The procedure was repeated twice. The pellet was then suspended in 1% Triton X-100, sonicated, and centrifuged. The last 5–6 washes were carried out with cold ultra pure water; sonication and centrifugation were applied as mentioned above. Finally, the pellet containing either chGH or chGH G119 protein was solubilized in a total volume of 400 mL containing 4.5 M urea buffered with 40 mM Tris Base. Cysteine was added to a final concentration of 0.1 mM. The pH was increased to 11.0 with NaOH and the clear solution was gently stirred at 4°C for 1 h. Then it was diluted with 2 volumes of water and dialyzed against 5 × 10 L of 10 mM Tris-HCl, pH 8.0 for 48 h. The resultant solution was then applied onto Q-Sepharose Fast Flow anion exchange resin (30-mL bead volume) preequilibrated with 10 mM Tris-HCl, pH 8.0 at 4°C. Absorbance at 280 and 260 nm was read until it decreased to low values. Elution was carried out using increasing NaCl solutions in the same buffer (pH 8.0), and 50-mL fractions were collected. Protein concentration was determined by absorbance at 280 nm and monomer content by gel filtration chromatography on Superdex 75HR 10/30 column (Amersham Biosciences AB, Uppsala, Sweden). The pooled fractions containing the monomeric protein were pooled, dialyzed against NaHCO$_3$, to ensure a 4:1 protein-to-salt ratio, and lyophilized. All procedures were carried out at 4°C.

Determination of Purity and Monomer Content

SDS-PAGE was carried out according to the method of Laemmli[7] in a 15% polyacrylamide gel under reducing and nonreducing conditions. The gel was stained with Coomassie Brilliant Blue R. Gel-filtration chromatography was performed on a Superdex^{TM} 75 HR 10/30 column with 0.2-mL aliquots of the Q-Sepharose column-eluted fraction using TN buffer (25 mM Tris-HCl, 150 mM NaCl, pH 8).

Determination of CD Spectra and Secondary Structure

The CD spectra (in millidegrees) were measured with an AVIV model 62A DS CD spectrometer (Lakewood, NJ, USA) using a 0.020-cm rectangular QS Hellma cuvette. The absorption spectra were measured with an AVIV model 17DS UV-visible IR spectrophotometer using a 1.000-cm QS cuvette and correction for light scattering. Collection of the data and calculation of the results were carried out as described previously.[8] Recombinant rat GH served as reference.

Various Binding Assays

Determination of chGH:oGHR-ECD complex stoichiometry by gel-filtration chromatography and competitive binding assays using radioactive oGH as a ligand and oGHR-ECD or chicken liver membrane fraction were carried out according to the procedures described previously,[9] except that addition of mAb F296 was omitted. Iodination of oGH was described earlier.[10]

Biological Activity in Vitro *in the FDC-P1-3B9 Bioassay*

FDC-P1-3B9 cells were grown as suspension cultures in 75-cm^2 tissue-culture flasks. For maximal growth and routine maintenance, cells were cultured in RPMI-1640 medium containing 5% FCS supplemented with antibiotic antimycotic solution and 100 ng/mL recombinant hGH was added to each flask to promote growth. The cells were incubated under a humidified atmosphere of 95% air and 5% CO_2 at 37°C. Before seeding, the cells were washed three times with phosphate-buffered saline (PBS) and centrifuged at 1000 g for 4 min. Cells then were resuspended in RPMI-1640 medium containing 5% horse serum. The experiment was performed using 96-well plates in which 0.1-mL suspensions containing 1.5×10^4 cells/well were seeded. In all assays increasing concentrations of chGH or oGH were added and 48 h after hormone addition the proliferation was determined by MTT assay. The growth curves were drawn using the Prizma (4.0) nonlinear regression sigmoidal dose-response curve[11] and the EC_{50} values were calculated.

RESULTS

Purification and Physicochemical Characterization of chGH and chGH G119R Mutein

Both proteins were purified by anion-exchange chromatography (see MATERIALS AND METHODS). Fractions eluted with 50 mM NaCl (chGH) or 100 mM NaCl (mutein) containing monomers were pooled, dialyzed against $NaHCO_3$ to ensure 3:1 protein:salt ratio, and lyophilized. The yields varied from 400 to 500 mg from 5l bacterial cultures. The purity and homogeneity of the purified mutants were documented by SDS-PAGE under reducing conditions, yielding only one band of $\sim$ 22 kDa (FIG. 1A). As expected, in the absence of the reducing agent, the mobility of all four proteins was slightly higher, indicating a globular structure. Gel filtration at pH 8 under native conditions yielded a single monomeric peak consisting of at least 95% monomer and corresponding to a molecular mass of 22 kDa (not shown). The secondary structures of the two purified proteins as compared to recombinant rat GH were calculated from the CD spectra performed at pH 7.5 FIG.1 B). The respective values for chGH, chGH G119R, and rat GH (mean $\pm$ SD) were: α-helix (54 $\pm$ 0.3, 54 $\pm$ 0.3, 49 $\pm$ 0.4,%), β-sheets (10 $\pm$ 0.7, 10 $\pm$ 0.7, 13 $\pm$ 0.8%), β-turns (16 $\pm$ 0.4, 16 $\pm$ 0.3, 18 $\pm$ 0.3%), and remainder (19 $\pm$ 0.9, 20 $\pm$ 0.7, 20 $\pm$ 0.5%), characteristic to all known GHs, indicating proper refolding. Specific extinction coefficients at 280 nm for a 0.1% solution, assuming an extra Ala at the N terminus, were calculated according to Pace et al.,[12] yielding the values of 0.806 and

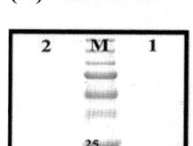

FIGURE 1. (**A**) SDS-PAGE (15%) of chGH (*lane 2*) and chGH G119R (*lane 1*) mutein. Middle lane shows molecular mass markers (in kDa). Aliquots of 5 μg of each protein were applied per lane. (**B**) CD spectra of purified recombinant chGH, chGH G119R mutein, and rGH. The proteins were dissolved in 50 mM KH_2PO_4 buffer, pH 7.6. For other details see text.

0.802 for chGH and chGH G119R mutein, respectively. The stability of both proteins in solution was tested at 4°C at pH 8. Both could be stored as sterile 0.1% solutions for 30 days, at 4°C without undergoing any changes in their monomeric content and retaining their activity in the PDF-P1-3B9 bioassay.

Detection of chGH-oGHR-ECD Complexes by Gel Filtration

To characterize the binding stoichiometry between the various GHs and oGHR-ECD, the respective ligands and oGHR-ECD were mixed in 1:1 or 1:2 molar ratios and separated by gel filtration using an analytical Superdex 75 column to determine the molecular mass of the binding complex under nondenaturing conditions. The experiments were performed using a constant 5 µM of the respective ligand. Both ovine and chicken GHs formed a 1:2 molar ratio complex with oGHR-ECD. This stoichiometry was evidenced by the appearance of a single main peak for the complex with shorter retention time (RT), (11.1 to 11.3 min), as compared to the higher retention times of oGHR-ECD (13.6 to 14.1 min) or of oGH/chGH 22K (14.1 to 14.4 min, respectively). This main peak appeared also when the components were mixed at a 1:1 molar ratio, indicating that the 1:1 complex is unstable and undergoes fast conversion to 1:2 complex even in a relative excess of the hormone. The calculated molecular mass of the complexes, based on standard marker peaks RT, were 69–72 kDa, lower than the predicted value of 76 to 78 kDa 1:2 complexes of chGH or oGH with oGHR-ECD. The differences between the calculated and predicted values are likely related to the differences between the molecular mass and Stock radius of the complex. Such discrepancies were reported earlier by us[13] and others[14] and were suggested to originate from more compact structure of the complex. In contrast to chGH, the chGH G119R did not form any complex with oGHR-ECD.

Binding Experiments and Bioassay

Iodinated oGH served as the ligand in all competitive experiments and the respective nonlabeled oGH, chGH, and chGH G119R as competitors. Recombinant oGHR-ECD (4 ng/tube) or chicken liver membrane preparation (400 µg protein/tube) was used as receptor source. The inhibition curves and the IC_{50} values were calculated by using nonlinear regression, best fit for one-site competition Prism 4.0 program.[11] Using the oGHR-ECD the respective IC_{50} values for oGH and chGH obtained from experiments were respectively (mean ± SD, calculated from two experiments) 2.02 ± 0.34 and 23.5 ± 3.22 nM, while the chGH G119R mutein did not bind at all. In contrast, when homologous assay using chicken liver membranes was performed, the respective IC_{50} values for oGH, chGH, and chGH G119R (mean ± SEM, calculated from three experiments) were 0.50 ± 0.05, 0.46 ± 0.17, and 4.61 ± 1.24, respectively. The differences between the two GHs were obviously not significant, but differed from chGH G119R mutein. Bioassay measuring the somatogenic activity of oGH, chGH, and chGH G119R using a cell line stably transfected with rabbit GH receptors (FDC-P1 3B9) was employed. Both chGH and oGH reached the same maximal activity and the respective EC_{50} (mean ± SD, calculated from two experiments) were 0.18 ± 0.05 and 6.36 ± 0.05 nM, respectively. The G119R was not active at all and could not inhibit chGH-dependent cell proliferation in this system.

CONCLUSION

The present results demonstrate our ability to produce gram quantities of recombinant chGH and its G119R mutein. Both proteins are properly refolded, as indicated by CD spectra, elecrophoretically pure, and under nondenaturing conditions consist of > 95% monomers. Binding experiments and bioassay using cell line possessing heterologous GH receptors emphasize the importance of species specificity. The chGH G119R mutein in contrast to wild-type chGH was devoid of any bioactivity, but bound to homologous GH receptors; a possibility that it can exhibit an antagonistic activity in homologous system is likely. However, additional experiments using homologous *in vitro* or *in vivo* bioassays are needed to prove this assumption.

ACKNOWLEDGMENT

This study was supported by Grant 2 P06D 34 28 from Polish State Committee for Scientific Research (KBN) to A.H.

REFERENCES

1. BYATT, J.C., N.R. STATEN, W.J. SALSGIVER, *et al.* 1993. Stimulation of food intake and weight gain in mature female rats by bovine prolactin and bovine growth hormone. Am. J. Physiol. **264:** E986–E992.
2. VASILATOS-YOUNKEN, R., Y. ZHOU, X. WANG, *et al.* 2000. Altered chicken thyroid hormone metabolism with chronic GH enhancement *in vivo*: consequences for skeletal muscle growth. J. Endocrinol. **166:** 609–620.
3. NIE, Q., B. SUN, D. ZHANG, *et al.* 2005. High diversity of the chicken growth hormone gene and effects on growth and carcass traits. J. Hered. **96:** 698–703.
4. CHEN, W.Y., N.Y. CHEN, J. YUN, *et al.* 1994. In vitro and in vivo studies of antagonistic effects of human growth hormone analogs. J. Biol. Chem. **269:** 15892–15897.
5. GOFFIN, V., S. BERNICHTEIN, O. CARRIERE, *et al.* 1999. The human growth hormone antagonist B2036 does not interact with the prolactin receptor. Endocrinology **140:** 3853–3856.

6. ROELFSEMA, F., N.R. BIERMASZ, J.A. ROMIJN & A.M. PEREIRA. 2005. Current pharmacotherapy for acromegaly: a review. Expert Opin. Pharmacother. **6:** 2393–2405.
7. LAEMMLI, U.K. 1970. Cleavage of structural proteins during the assembly of the head of bacteriophage T4. Nature **227:** 680–685.
8. PRIVENCHER, S.W. & J. GLÖCKNER. 1981. Estimation of globular protein secondary structure from circular dichroism. Biochemistry **20:** 33–37.
9. EMANE, M.N., C. DELOUIS, P.A. KELLY & J. DJIANE. 1986. Evolution of prolactin and placental lactogen receptors in ewes during pregnancy and lactation. Endocrinology **118:** 695–700.
10. GERTLER, A., A. ASHKENAZI & Z. MADAR. 1984. Binding sites of human growth hormone and ovine prolactin in the mammary gland and the liver of lactating dairy cow. Mol. Cell. Endocrinol. **34:** 51–57.
11. PRISMA. 2003. GraphPad Prism TM Version 4.0. GraphPad Software Inc., San Diego, CA.
12. PACE, C.N., F. VAJDOS, L. FEE, *et al*. 1995. How to measure and predict the molar absorption coefficient of a protein. Protein Sci. **4:** 2411–2423.
13. BIGNON, C., E. SAKAL, L. BELAIR, *et al*.1994. Preparation of recombinant extracellular domain of rabbit prolactin receptor expressed in *Escherichia coli* and its interaction with lactogenic hormones. J. Biol. Chem. **269:** 3318–3324.
14. FRITZ, H., I. TRAUTWHOLD & E. WERLE. 1965. Determination of the molecular weight of new trypsin inhibitors with the aid of sephadex gel filtration. Hoppe Seylers Z. Physiol. Chem. **342:** 253–263.

Induction of Apoptosis of Osteoclasts by Targeting Transcription Factors with Decoy Molecules

ROBERTA PIVA,[a] LETIZIA PENOLAZZI,[a] MARGHERITA ZENNARO,[b] ERCOLINA BIANCHINI,[a] EROS MAGRI,[c] MONICA BORGATTI,[a] ILARIA LAMPRONTI,[a] ELISABETTA LAMBERTINI,[a] ELISA TAVANTI,[a] AND ROBERTO GAMBARI[a,b]

[a]*Department of Biochemistry and Molecular Biology, Section of Molecular Biology, Università degli Studi di Ferrara, Ferrara, Italy*

[b]*ER-GenTech, Università degli Studi di Ferrara, Ferrara, Italy*

[c]*Department of Experimental and Diagnostic Medicine, Section of Pathological Anatomy, Università degli Studi di Ferrara, Ferrara, Italy*

ABSTRACT: We review the effects of two transcription factor decoy oligonucleotides on apoptosis of human osteoclasts (OCs). The first decoy molecule was designed to inhibit nuclear factor kappa-B (NF-κB) binding to target sequence, the second to increase estrogen receptor (ER) α expression. We found that both decoy molecules are potent inducers of apoptosis of human OCs, associated with increase of caspase 3 activity and decrease of interleukin 6 expression. In addition, we provide evidence indicating that these oligonucleotides are active *in vivo* in inducing OCs apoptosis. Because OCs are essential for skeletal development and remodeling throughout the life of animal and man, the approach described is of potential clinical importance.

KEYWORDS: NF-κB; estrogen receptor; PNA; gene therapy; transcription; osteoclast

TRANSCRIPTION FACTORS AND DIFFERENTIATION OF OSTEOCLASTS

The delivery of double-stranded oligodeoxynucleotides (ODNs) containing the consensus-binding sequence of a transcription factor is a promising approach to treat several diseases by modulating transcriptional regulation.[1] This transcription factor decoy (TFD) strategy results in the attenuation of the

Address for correspondence: Prof. Roberto Gambari, Department of Biochemistry and Molecular Biology, Section of Molecular Biology, Via Fossato di Mortara 74, 44100 Ferrara, Italy. Voice: +39-0532-424443; fax: +39-0532-424450.
e-mail: gam@unife.it

authentic *cis–trans* interaction, leading to the removal of transcription factors from the endogenous *cis*-element. The technique has been proven effective both *in vitro* and *in vivo*, suggesting its use in therapy.[2] As far as applications of the TFD approach, osteoclastogenesis appears to be of great interest. Osteoclasts (OCs) are giant multinucleated cells formed by the fusion of hematopoietic mononuclear precursor cells of the macrophage/monocyte lineage through a multistep process. OCs are the sole bone-resorbing cells, basing their action on mechanisms which are only in part characterized. Following attachment to the bone and consequent activation, OCs are active in demineralization, bone matrix degradation, and resorption. The functions of OCs are regulated by several signaling events, based on different kinds of hormones, cytokines, inflammatory, and transcription factors.[3] An understanding of the molecules involved both in stimulation and inhibition of OC activity is crucial to gain insight into the regulation of OC differentiation and function at the molecular level.

A growing body of studies reports that nuclear factor kappa-B (NF-κB) is involved.[4,5] In differentiating OCs, receptor activator of NF-κB ligand (RANKL) binds to its receptor RANK, leading to activation of several signaling cascades that stimulate differentiation and activation of OCs and inhibit OC apoptosis. RANK is also able to bind to a soluble decoy receptor osteoprotegerin (OPG), which, reducing RANKL availability, opposes RANK signaling. In addition to NF-κB, the activated signaling pathways include extracellular signal-regulated kinase (ERK), c-Jun terminal kinase (JNK), and p38 mitogen-activated protein kinase through recruitment of the tumor necrosis factor (TNF) receptor-associated factors (TRAFs). OC environment shows also specific sensitivity to most hormones and cytokines that influence bone resorption through modulation of the balance of RANKL and OPG.[3,6]

These pathways described briefly here turn on transcription factors that regulate OC-specific gene expression program, and consequently OC differentiation, activation, survival, and apoptosis. Several reports demonstrate that, in addition to NF-κB, other transcription factors including PU.1, the major components of AP-1 transcription factor complex (c-Jun, c-Fos, and Fra-1), Mitf, and estrogen receptor (ER) are involved in regulation of various aspects of OC differentiation.[6]

In this scenario, a critical role is played by estrogen, which acts as an inhibitor of OC activity.[7,8] In addition to estrogen, the inhibitory molecules of OC activity include TGF-α, interferon (IFN)-α, IFN-β, and IFN-γ, calcitonin, prostaglandin E2 (PGE2), IL-4, IL-18, bisphosphonates, nitric oxide, and osteoprotegerin. Estrogen suppresses OC function mainly by inhibiting OC formation and differentiation, suppressing the production of IL-6 and IL-1 bone-resorbing cytokines, inhibiting RANK-mediated JNK activation and c-Jun activation and expression, and by promoting OC apoptosis through a receptor-mediated genomic action. In addition, it has been demonstrated that

ERs interact with NF-κB heterodimers, preventing them from binding to their response elements. Downregulation of this interaction leads to acceleration in bone loss, as seen in postmenopausal women.

Of particular interest is the role of the ER alpha (ERα) isoform in protection of bone mass, as indicated, for example, by association between osteoporosis and ERα loss, as in menopause.

Although the processes involved in the initiation of bone resorption are relatively well understood, the events that may arrest them potentiating OC apoptosis are not. In this context, we are interested in the identification of potential drug targets toward the treatment of osteopenic bone diseases such as osteoporosis. Therefore, in order to contribute to the development of novel drugs that could be used to control osteoporotic bone destruction by the OCs, we investigated the possibility of inhibiting osteoclastogenesis through specific TFD strategies.[6]

THE TFD APPROACH FOR ALTERATION OF GENE EXPRESSION

The TFD approach is based on the competition for transacting factors between endogenous *cis*-elements present within the regulatory regions of the target gene and exogenously added DNA decoys (or functionally active DNA analogues) mimicking the specific *cis*-elements.[1,2,9–11] The objective of this molecular intervention is to cause an attenuation of the authentic interactions of transfactors with their *cis*-elements, leading to a removal of the transfactors from the endogenous *cis*-element inside the cell.[1,9–11] The most important feature of potential decoy molecules is the ability to tightly bind to target transcription factors. In this case, the expression of genes directly regulated by the targeted transcription factors will be deeply altered.

OLIGONUCLEOTIDE DECOYS TARGETING NF-κB INDUCE APOPTOSIS OF HUMAN OCs

Under the experimental conditions established in our laboratory and previously reported, NF-κB activity was markedly reduced by using a specific decoy strategy against NF-κB in human primary OCs from peripheral blood.[12] A functional consequence was a very efficient induction of OC apoptosis (a typical apoptotic OC is shown in FIGURE 1A) associated with a significant decrease of expression of IL-6,[12] a typical target of NF-κB. The NF-κB decoy-mediated effects (FIG. 1B, left) were, as expected, associated with a sharp increase of caspase 3 activity.[12]

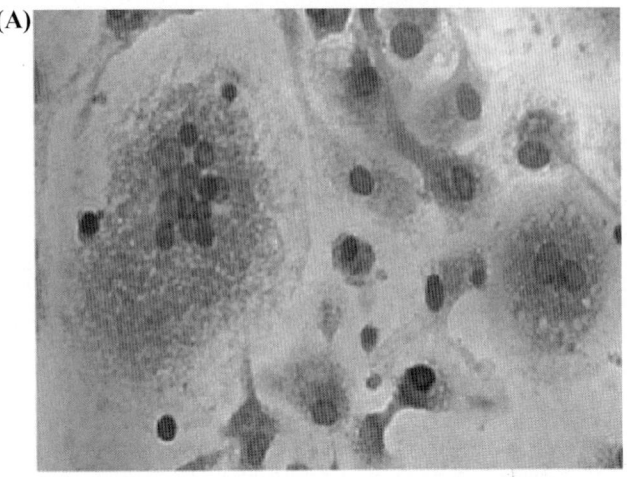

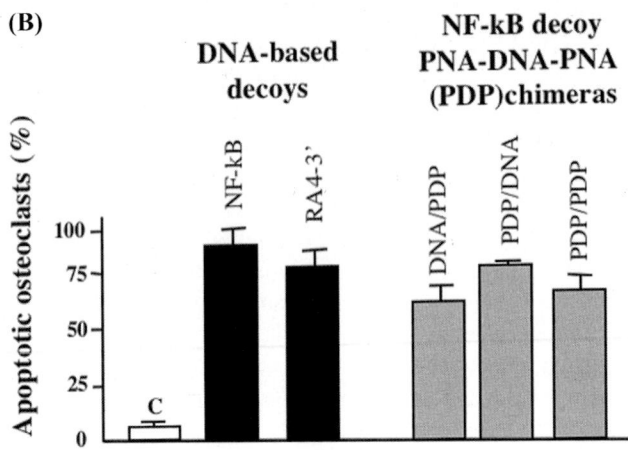

FIGURE 1. (**A**) Apoptotic osteoclasts. (**B**) Levels of induction of apoptosis in OCs treated with ODN- (*black bars*) and PDP- (*gray bars*) based decoy molecules are indicated. The *white bar* represents the percentage of untreated apoptotic OCs. These data are derived from Penolazzi et al.,[12] Penolazzi et al.,[25] and Piva et al.[6]

INDUCTION OF OC APOPTOSIS BY USING A DECOY OLIGODEOXYNUCLEOTIDE MIMICKING A REGION OF DISTAL PROMOTER C OF ERα GENE

In order to study therapeutic strategies to control bone formation, we investigated the possibility of inhibiting osteoclastogenesis through a specific decoy strategy designed to alter the expression of the human ERα gene.[6] We obtained an increase of ERα gene expression by interfering with the activity of a negative transcription factor and by removing it with a decoy oligonucleotide (RA4-3′) mimicking a region of distal promoter C of this gene and probably binding NFAT transcription factor.[6]

We demonstrated that this decoy was able to induce apoptosis in human primary OCs (FIG. 1B, left), but not in osteoblasts, in an estrogen-dependent manner, increasing also caspase 3 and Fas receptor levels.[6] These findings may be of relevance for a possible therapeutic approach for osteopenic diseases, tumors, and bone metastasis.

INDUCTION OF OC APOPTOSIS *IN VIVO*

Novel contributions toward understanding the field of OC biology, and consequently the possibility of developing new drugs for the treatment of several bone diseases, suffer from the lack of suitable experimental models.

Clinical orthodontics offers a good opportunity to study *in vivo* the efficacy of the TFD decoy approach for induction of apoptosis of OCs. In fact, the increased number and activity of OCs is involved in the regulation of alveolar bone resorption during orthodontic tooth movement, and also in the origin of dental problems in diseases of OC activation affecting the maxillo-mandibular bone. We designed *in vivo* experiments aimed at regulating alveolar bone resorption. Six Wistar male rats were subjected to orthodontic forces, in combination or not with NF-κB or RA4-3′ decoy treatment, by using a split-mouth design. Examination of paraffin sections of the excised molars showed that orthodontic forces caused a percentage of apoptotic OCs that appear to be highly potentiated by NF-κB[13] and RA4-3′,[14] but not by scramble ODN. These data confirm the results obtained *in vitro* with human OCs from peripheral blood, suggesting an important correlation between NF-κB ERα and the possibility of modulating OC functions.

DESIGN OF NOVEL TFD MOLECULES MIMICKING DNA

As for other ODN-based therapeutic strategies, the successful use of TFD will almost always depend on (i) stability of the employed molecules and (ii) efficient delivery to target tissues, cells, and cellular compartments.[15] In addition, decoy ODNs might require nuclear localization if they are to prevent the transactivation of their target genes. Unfortunately, the endocytotic pathways translocate most of the employed ODNs into lysosomal compartments, where they are efficiently degraded.[11] Therefore, the search for decoy biomolecules exhibiting on the one hand efficient decoy activity, and on the other hand resistance in serum and cellular extracts appears to be a major issue in pharmacological research. Examples are modified oligonucleotides, LNA- (locked nucleic acids) and (peptide nucleic acids) PNA-based transcription factor decoys.

For instance, PNAs are recently described DNA mimics in which the sugar-phosphate backbone is replaced by N-(2-aminoethyl)glycine units.[16] These molecules efficiently hybridize with complementary DNA, forming Watson–Crick double helices. While the possible use of PNAs for targeting mRNA molecules and genomic sequences are well documented,[17–19] the published reports on the possible use of PNA-based double-stranded molecules to target transcription factors indicate that these molecules are not efficient.[20] In sharp contrast, PNA-DNA chimeras[15] are of great interest for TFD pharmacotherapy. These molecules are composed of a part of PNA and a part of DNA. The PNA-DNA chimeras obey the Watson–Crick rules on binding to complementary DNA and RNA. We have designed and tested PNA-DNA-PNA (PDP) chimeras mimicking NF-κB sites,[21,22] and found an efficient decoy activity of these molecules. Recently published observations by our research group show that PDP chimeras are more resistant than the DNA counterpart to endonucleases and when incubated in the presence of serum or cellular extracts.[23,24] When PDP chimeras targeting NF-κB were used on human osteoclasts *in vitro*, we were able to demonstrate that PDP/DNA, DNA/PDP, and PDP/PDP are all active in inducing apoptosis of OCs at a level similar to that obtained with DNA (FIG. 1B, right).[25]

CONCLUSIONS

OCs are essential for skeletal development and remodeling throughout the life of animal and man, so molecular characterization of OC differentiation is of potential clinical importance.

Overactivity of OCs are described in several human bone diseases such as osteoporosis, Paget's disease of bone, expansile skeletal hyperphosphatasia, and familiar expansile osteolysis. The clinical phenotype of these diseases is one of uncontrolled bone remodelling, with presence of many, often enlarged, OCs. Even if the majority of the genes involved in these diseases are yet to be identified, evidence suggests that a critical role is played by genes linked to the NF-κB and ER signaling pathways.

In addition, the effects of the two decoy molecules designed to inhibit NF-κB and to increase ERα expression, respectively, are in agreement with recent studies demonstrating molecular cross-talk between NF-κB and ER transcription factors in which the ER mediates inhibition of NF-κB activity at several levels. Such cross-talk between these important regulators of the endocrine and immune systems is particularly relevant to the study of bone physiology, inflammation, cancer, and autoimmune diseases. Even if the molecular mechanisms of inhibition of NF-κB by ER are complex and probably specific to cell type and context, the induction of apoptosis of OCs with the decoy strategies here described might have relevance for the treatment of some human bone diseases.

ACKNOWLEDGMENTS

This work was supported by MIUR COFIN-2005, by UE ITHANET, by AIRC, by Associazione Veneta per la Lotta alla Talassemia, by the Italian Cystic Fibrosis Research Foundation, and by Fondazione CARIPARO. E.L. is a recipient of a fellowship from Fondazione CARIEF.

REFERENCES

1. MORISHITA, R. et al. 1997. In vivo transfection of cis element "decoy" against nuclear factor-kB binding site prevents myocardial infarction. Nat. Med. **3:** 894–899.
2. MANN, M.J. & V.J. DZAU. 2000. Therapeutic application of transcriptional factor decoy oligonucleotides. J. Clin. Invest. **106:** 1071–1075.
3. BOYLE, W.J., W.S. SIMONET & D.L. LACEY. 2003. Osteoclast differentiation and activation. Nature **423:** 337–342.
4. CHAISSON, M.L. et al. 2004. Osteoclast differentiation is impaired in the absence of inhibitor of kappa B kinase alpha. J. Biol. Chem. **279:** 54841–54848.
5. BOYCE, B.F. et al. 2005. Roles for NF-kappaB and c-Fos in osteoclasts. J. Bone Miner. Metab. **23:** 11–15.
6. PIVA, R. et al. 2005. Induction of apoptosis of human primary osteoclasts treated with a transcription factor decoy mimicking a promoter region of estrogen receptor alpha. Apoptosis **10:** 1079–1094.
7. KRASSAS, G.E. & P. PAPADOPOULOU. 2001. Oestrogen action on bone cells. J. Musculoskelet. Neuronal Interact. **2:** 143–151.
8. SYED, F. & K. SUNDEEP. 2005. Mechanisms of sex steroid effects on bone. Biochem. Biophys. Res. Commun. **328:** 688–696.
9. TOMITA, T. et al. 2000. Transcription factor decoy for NF-kB inhibits cytokine and adhesion molecule expressions in synovial cells derived from rheumatoid arthritis. Rheumatology **39:** 749–757.
10. BORGATTI, M. et al. 2004. Peptide nucleic acids (PNA)-DNA chimeras targeting transcription factors as a tool to modify gene expression. Curr. Drug Targets **5:** 735–744.
11. GAMBARI, R. 2004. New trends in the development of transcription factor decoy (TFD) pharmacotherapy. Curr. Drug Targets **5:** 419–430.
12. PENOLAZZI, L. et al. 2002. Decoy oligodeoxynucleotides targeting NF-kappaB transcription factors: induction of apoptosis in human primary osteoclasts. Biochem. Pharmacol. **66:** 1189–1198.
13. PENOLAZZI, L. et al. 2006. Local in vivo administration of a decoy oligonucleotide targeting NF-kappa B induces apoptosis of osteoclasts after application of orthodontic forces to rat teeth. Int. J. Mol. Med. **18:** 807–811.
14. PENOLAZZI, L. et al. 2006. In vivo local transfection of a cis element decoy mimicking an estrogen receptor alpha gene promoter region induces apoptosis of osteoclasts following application of orthodontic forces to rat teeth. Apoptosis **11:** 1653–1656.
15. GAMBARI, R. 2004. Biological activity and delivery of peptide nucleic acids (PNA)-DNA chimeras for transcription factor decoy (TFD) pharmacotherapy. Curr. Med. Chem. **11:** 1253–1263.

16. NIELSEN, P. *et al*. 1991. Sequence-selective recognition of DNA by strand displacement with a thymine-substituted polyamide. Science **254:** 1497–1500.
17. NIELSEN, P.E. & M. EGHOLM. 1999. An introduction to peptide nucleic acid. Curr. Issues Mol. Biol. **1:** 89–104.
18. EGHOLM, M. *et al*. 1993. PNA hybridizes to complementary oligonucleotides obeying the Watson-Crick hydrogen-bonding rules. Nature **365:** 566–568.
19. NIELSEN, P.E. 2004. PNA technology. Mol. Biotechnol. **26:** 233–248.
20. MISCHIATI, C. *et al*. 1999. Interaction of the human NF-kappaB p52 transcription factor with DNA-PNA hybrids mimicking the NF-kappaB binding sites of the human immunodeficiency virus type 1 promoter. J. Biol. Chem. **274:** 33114–33122.
21. ROMANELLI, A. *et al*. 2001. Molecular interactions with nuclear factor kappaB (NF-kappaB) transcription factors of a PNA-DNA chimera mimicking NF-kappaB binding sites. Eur. J. Biochem. **268:** 6066–6075.
22. BORGATTI, M. *et al*. 2003. Transcription factor decoy molecules based on a peptide nucleic acid (PNA)-DNA chimera mimicking Sp1 binding sites. J. Biol. Chem. **278:** 7500–7509.
23. BORGATTI, M. *et al*. 2002. Cationic liposomes as delivery systems for double-stranded PNA-DNA chimeras exhibiting decoy activity against NF-kappaB transcription factors. Biochem. Pharmacol. **64:** 609–616.
24. BORGATTI, M. *et al*. 2003. Resistance of decoy PNA-DNA chimeras to enzymatic degradation in cellular extracts and serum. Oncol. Res. **13:** 279–287.
25. PENOLAZZI, L. *et al*. 2004. Peptide nucleic acid-DNA decoy chimeras targeting NF-kappaB transcription factors: induction of apoptosis in human primary osteoclasts. Int. J. Mol. Med. **14:** 145–152.

Competition Effects Shape the Response Sensitivity and Kinetics of Phosphorylation Cycles in Cell Signaling

CARLOS SALAZAR AND THOMAS HÖFER

Theoretical Biophysics, Institute for Biology, Humboldt University, 10115 Berlin, Germany

ABSTRACT: Phosphorylation cycles are a core component of cell signaling networks. The response sensitivity and kinetics of these cycles are controlled by thermodynamic, kinetic, and structural factors, including binding affinities, catalytic activities, and the phosphorylation order of multiple sites. Based on mathematical models, we interpret the role of these factors in terms of competition effects. For the regulation of a single phosphorylation site, two kinds of competition effects turn out to shape behavior: the competition between kinase and phosphatase to bind the substrate, and the competition between the distinct phosphorylation forms of the substrate for binding to either enzyme. Depending on the concentrations and mutual affinities of the enzymes and the target, the response function can be graded, ultrasensitive, or biphasic. In multiply phosphorylatable proteins, additional factors generating competition effects are present and more complex responses can be obtained. For example, the combination of a cooperative kinetics with the conditions for zero-order ultrasensitivity may yield a bistable response. We show that a repeated competition between kinase and phosphatase for binding the substrate and/or between the phosphorylation and dephosphorylation reactions at each phosphorylation site generally result in a threshold response. The phosphorylation time is also strongly affected by the kinetic design of the cycle. In particular, threshold responses are generally associated with very long phosphorylation times. We also argue here that a description in terms of elementary binding and reaction steps is required for an appropriate analysis of these cycles in cell signaling.

KEYWORDS: phosphorylation cycles; stimulus response curves; ultrasensitive responses; graded responses; bistability; cooperativity

Address for correspondence: Carlos Salazar, Theoretical Biophysics, Institute for Biology, Humboldt University, Invalidenstr. 42, 10115 Berlin, Germany. Voice: +49-30-2093-8694; fax: +49-30-2093-8813.

e-mail: carlos.salazar@biologie.hu-berlin.de

INTRODUCTION

An appropriate response of the cell to its environment depends not only on the presence or absence of a signal, but also on its magnitude and duration. Cellular responses are frequently controlled by signaling proteins and transcription factors that are regulated by phosphorylation cycles and cascades. In resting cells, cycles of reversible protein phosphorylation exist in a steady state. External stimuli that become transduced into activity changes of kinases and/or phosphatases will shift the phosphorylation state of the target protein. The kinetics of this process as well as the relationship between the phosphorylation state of the substrate and the concentration (or activity) of the modifying enzymes have become a major topic of mathematical modeling during the last decades.[1–8]

The elucidation of the molecular and cellular mechanisms underlying the response sensitivity and kinetics of phosphorylation cycles and cascades is critical for understanding their function and for the development of therapeutic strategies. Here we argue that many of these mechanisms have in common the fact that they can be interpreted in terms of competition effects. Kinase and phosphatase can, for example, compete for binding the target protein. Competition may also exist between unphosphorylated and dephosphorylated substrate for binding the enzymes. Competition effects are the result of the interplay of structural, kinetic, and thermodynamic factors. Some of these factors are the existence of regulatory feedback loops, the multi-step nature of covalent modifications, and the relative abundance, binding affinity, and catalytic activity of the modifying enzymes.[8] In this article we discuss how some of these factors contribute to generate diverse competition effects and in which manner they regulate the response sensitivity and kinetics of phosphorylation cycles and cascades.

ULTRASENSITIVE AND GRADED RESPONSES

Processes driving cell proliferation and cell fate decisions involve discrete all-or-none responses to external stimuli.[5,9] Such ultrasensitive switch-like behavior is often characterized by a sigmoid stimulus–response curve. Bistability is an extreme case of ultrasensitive response, where signaling molecules switch between two discontinuous stable states and intermediate responses are not observed.[10,11] Bistable response in mitogen-activated protein kinase (MAPK) signaling has been demonstrated experimentally by the progesterone-induced maturation of *Xenopus* oocytes.[5,9] Ultrasensitive signaling allows the filtering out of weak signals, while the cell responds with maximal activation once the stimulus has exceeded a threshold level. Other biological events may need, however, a more graded regulation. Incremental controls appear to be critical during development; for example, when cells adopt different phenotypes depending on their position relative to certain morphogenetic gradients.[12]

Graded responses are characterized by a hyperbolic (Michaelis–Menten-type) stimulus–response curve. At the molecular level, these may arise when protein function is regulated by a single phosphorylatable amino acid residue. In contrast, activation or repression of target proteins by multiple phosphorylation events can occur in a switch-like form. For example, multisite phosphorylation of the CDK inhibitor, Sic1, creates an ultrasensitive biological switch for the onset of DNA replication.[13] Sic1 must be phosphorylated on at least six of nine phosphorylatable sites to stably associate with the ubiquitin system and to be thereafter eliminated. The Nuclear Factor of Activated T Cells (NFAT) is also regulated in an allosteric manner at several phosphorylation sites.[14–16] Cooperative dephosphorylation of NFAT on 13 serine residues by the calcium-dependent phosphatase calcineurin induces global conformational changes in the protein structure, triggering a threshold for the import into the nucleus, the binding to DNA, and to obtain a maximal transcriptional activity.

A recent article shows, however, that multisite phosphorylation can control protein activity in a gradual way, serving as a rheostat for cell signaling. Multiple Ca^{2+}-dependent phosphorylation sites within the transcriptional activator Ets-1 are thought to act additively to produce incremental DNA binding affinity.[12] Serine to alanine mutations in one, two, or three critical residues, which mimics a reduction in the phosphorylation level, gradually increases the DNA affinity and lowers Ets-1 autoinhibition. The occurrence of graded responses in multiply phosphorylatable proteins can be of great importance in cell signaling for accumulation under physiological conditions of partially phosphorylated forms that perform distinct functions.[17]

CONCENTRATIONS AND AFFINITIES CONTROL THE RESPONSE SENSITIVITY

Phosphorylation cycles may exhibit different kinds of stimulus–response curve, even when only a single phosphorylatable residue is involved.[1,3,7,8] Two key parameters that control the response sensitivity are (*a*) the concentrations of the modifying enzymes relative to the target protein and (*b*) the affinities of the modifying enzymes for the different phosphorylation states of the target. By changing these parameters, a great variety of cellular responses can be obtained, from gradual to steeply sigmoid to biphasic.

We recently showed that the effects of enzyme concentrations and target affinities can be understood in terms of competition.[8] When the enzyme concentrations are large enough so that they are not saturated by the target, the target's response is shaped by how kinase and phosphatase compete for binding to the target (FIG. 1 A, upper panel). In other words, phosphorylation is competitively inhibited by the phosphatase and dephosphorylation

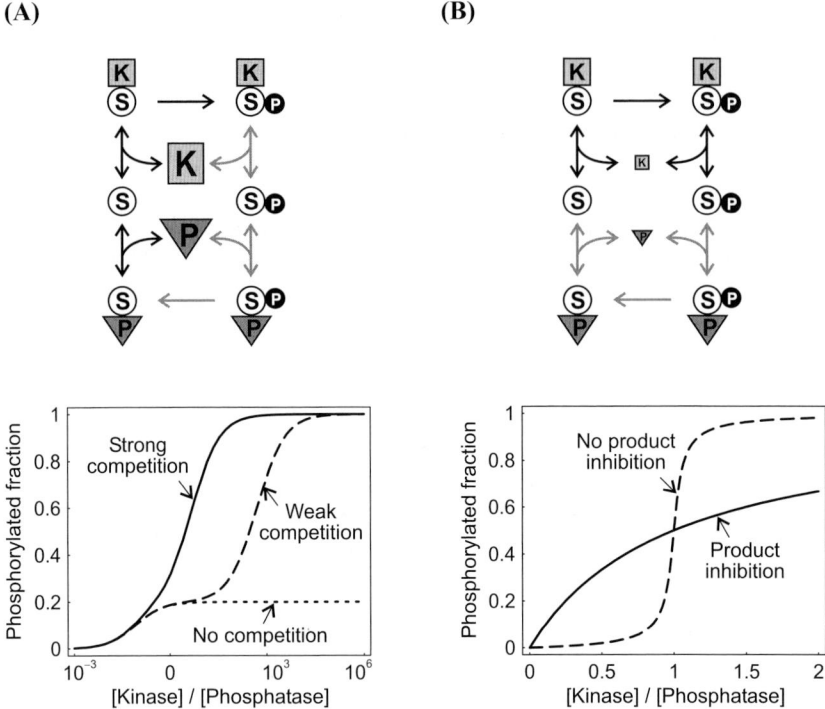

FIGURE 1. Concentrations and affinities shape the stimulus–response curve. (**A**) At a large enzyme concentration, the phosphorylation and dephosphorylation rates are determined by the competition between the enzymes for binding the unphosphorylated and phosphorylated forms of the substrate, respectively (black and gray arrows in the upper panel). The phosphorylated fraction is plotted versus the ratio of kinase to phosphatase concentration (stimulus strength) for the cases of no competition (dotted curve), weak competition (dashed curve), and strong competition (solid curve). (**B**) For low enzyme concentration, the phosphorylation and dephosphorylation rates are determined by the competition between the phosphorylated and dephosphorylated forms of the substrate for binding the kinase and phosphatase, respectively (black and gray arrows in the upper panel). The stimulus–response curve is shown for the cases of no competition, known as zero-order ultrasensitivity (dashed curve), and strong competition (solid curve). In the upper panels representing phosphorylation cycles, substrate, kinase, and phosphatase are denoted through white circles, gray squares, and gray triangles, respectively. Black circles denote phosphorylated residues in the substrate.

by the kinase. This effect strongly influences the stimulus–response relation (FIG. 1 A, lower panel). When the unphosphorylated target interacts only with the kinase and the phosphorylated target only with the phosphatase, such that there is no competition, the target protein cycles continuously between phosphorylated and dephosphorylated states. The stimulus–response curve is then a hyperbolic function, and the maximal phosphorylation level only depends on the catalytic rates of kinase and phosphatase (dotted curve). Notably, the phosphorylated fraction of the target always remains below 100%, because the phosphatase has unhindered access (in the sample calculation shown, the maximal rate of the phosphatase was taken to be five times the maximal rate of the kinase, so that at most 20% of the target can be phosphorylated at steady state).

When the kinase acquires a weak affinity to the phosphorylated target, the response curve becomes biphasic, because at high kinase concentrations the kinase competes with the phosphatase for binding to the phosphorylated target. The target proteins sequestered by the kinase are protected from dephosphorylation (dashed curve). High affinity of the kinase for the phosphorylated substrate leads to the kinase sequestering substrate at lower kinase concentrations, resulting again in a monophasic response curve (solid curve). The formation of kinase–target complexes that are protected from dephosphorylation now make 100% target phosphorylation possible. Moreover, the response curve is somewhat steeper than the hyperbolic curve obtained without enzyme competition.[8]

When the concentration of the target protein is large so that the enzymes become saturated, the behavior depends on how the different phosphorylation states of the substrate compete for binding the enzymes (FIG. 1 B, upper panel). In this case, the phosphorylated target can exert an inhibitory effect on phosphorylation when the kinase remains bound to it. Similarly, dephosphorylation can be inhibited by an unphosphorylated target sequestering the phosphatase. This kind of product inhibition occurs in many enzymatic reactions. Interestingly, in combination with saturation of the enzymes, it leads to a simple hyperbolic–response curve (FIG. 1 B, lower panel, solid curve). However, in the absence of product inhibition, the saturation of both kinase and phosphatase with their target protein yields sigmoid stimulus–responses curves (FIG. 1 B, lower panel, dashed curve). Because saturation implies that the enzyme kinetics are approximately zero order for a large range of target concentrations, the associated phenomenon of a steep threshold has been termed as *zero-order ultrasensitivity*.[3]

The relation between enzyme concentration and binding affinity defines the fraction of target protein bound to the modifying enzyme. A third key parameter is given by the kinetics of the catalytic step. Assuming rapid binding and dissociation of the enzymes, the phosphorylation rate of the substrate can be mechanistically interpreted as the product of the catalytic rate constant of phosphorylation and the kinase fraction bound to the target protein. This indicates two distinct strategies for controlling the phosphorylation state of the target: a change in the substrate affinity of the kinase or in its catalytic rate.[18] As a consequence, a high catalytic rate constant may compensate for weak binding affinity, so that effective (de)phosphorylation may be achieved even when there are no stable complexes between kinase (phosphatase) and target protein. In T cells, for example, stable complexes between NFAT1 and the kinase CK1 have been found, while a complex between the phosphatase calcineurin and NFAT1 could not be isolated from the cells under the same conditions.[19] Nevertheless, NFAT1 dephosphorylation by calcineurin proceeds rapidly and appears to be faster than the rephosphorylation of the protein by multiple kinases after stimulus-induced calcineurin activity recedes.[14,20]

MECHANISMS OF MULTISITE PHOSPHORYLATION AND RESPONSE SENSITIVITY

The above-discussed features are valid for proteins with single and multiple phosphorylation sites. In particular, the response curves for processive multisite phosphorylation, where the kinase has to bind only once to a protein to phosphorylate all its target residues, are expected to be very similar to the response curves for a single-site target considered thus far. However, multisite phosphorylation can increase considerably the possibilities for protein–protein interactions and phosphorylation sequence.[21,22] One common effect observed in multisite phosphorylation is cooperativity among the phosphorylation sites (FIG. 2 A, upper panel), that is, phosphorylation of the first residues accelerates the remaining phosphorylation steps. Such cooperative kinetics can cause a considerable ultrasensitivity of the response (FIG. 2 A, lower panel, dashed curve). As a rule of thumb, we found that the Hill coefficient that characterizes the degree of cooperativity can be nearly as large as the number of phosphorylation sites.[15] In the absence of a cooperative kinetics multisite protein phosphorylation can give rise to an activation threshold but, at the same time, produces a poor switch.[23] Above the threshold, the response does not switch abruptly but increases in a hyperbolic manner (FIG. 2 A, lower panel, solid curve).

Cooperative kinetics can even result in bistable response (FIG. 2 B). This phenomenon has been previously associated with regulatory couplings, such as positive feedbacks.[5,9] However, phosphorylation cycles can also exhibit bistability in the absence of feedback interactions.[10] This is the case for cooperative kinetics of multisite (de)phosphorylation when, in addition, the conditions for zero-order ultrasensitivity (saturation of the enzymes, absence of product inhibition; see foregoing Section) are present.[8,10] Bistability of phosphorylation cycles may represent an alternative for the generation of molecular memory, normally obtained from feedback loops in gene expression, or serve sustained signal propagation.

Structural factors, such as the order in which phosphorylation and dephosphorylation of multiple residues proceed, are additional "degrees of freedom" that affect the stimulus–response relation (FIG. 3 A). One or several kinases and phosphatases may process their target sites in a strictly ordered sequence. This is often the case when the aminoacid residues occur in repetitive motifs (e.g., S/T-X-X-S/T for CKI).[24,25] A special case is obtained when kinase and

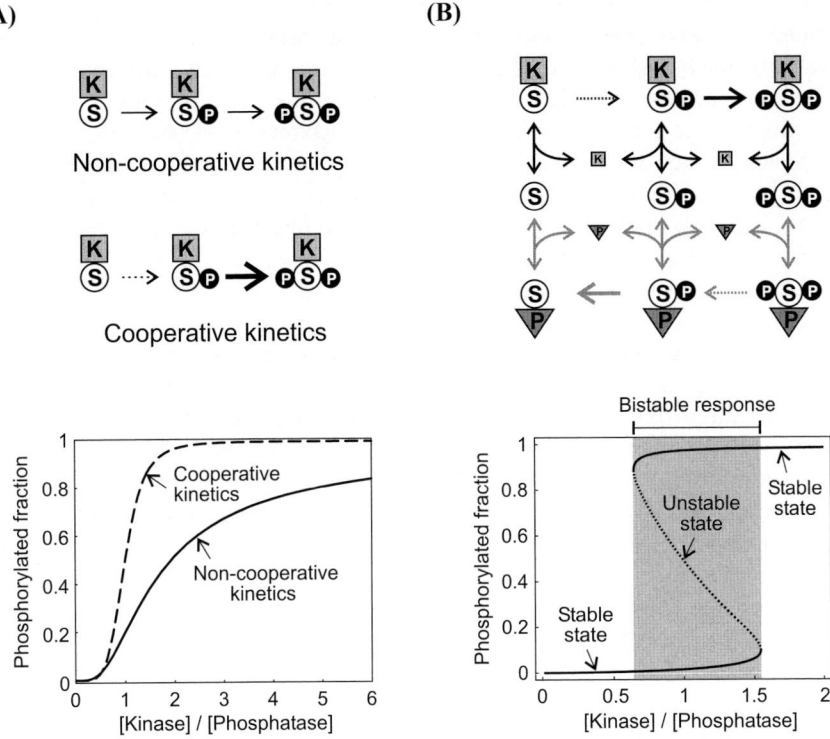

FIGURE 2. Cooperative kinetics and bistable responses. (**A**) In a cooperative kinetics, early phosphorylation steps (dashed arrow) enhance the phosphorylation rate of the late steps (thick arrow). Cooperativity increases the steepness of the threshold in the stimulus–response curve (compare dashed and solid curves). (**B**) The combination of the conditions for zero-order ultrasensitivity (enzyme saturation and no product inhibition) with a cooperative kinetics may result in a bistable response. In the lower panel, within the shadowed region, the response is bistable. Depending on the initial conditions, the substrate can attain one of the two stable steady states (solid curves).

phosphatase modify the substrate strictly sequentially in opposite directions (in particular, the last residue to be phosphorylated is the first to be dephosphorylated).[15,23] Alternatively, the first residue to be phosphorylated can also be the first to become dephosphorylated. Such a cyclical mechanism has been observed for the multiple phosphorylation of rhodopsin.[26] A third possibility is random phosphorylation and dephosphorylation, proceeding independently of the phosphorylation state of the other sites.[27,28] Both cyclical and random mechanisms exhibit shallower responses of complete phosphorylation with respect to the relative kinase concentration (FIG. 3 A, lower panel). At the same time, however, the cyclical mechanism is more responsive to a small concentration of kinase than either sequential or random mechanisms. For both cyclical and random mechanisms, the intermediate phosphorylation states generally attain higher concentrations than for the sequential mechanism. Thus, in these cases a much higher ratio of kinase activity to phosphatase concentration is needed to achieve complete phosphorylation of the target protein.

Stimulus–response curves of multiply phosphorylatable proteins are also affected by the formation of multiprotein complexes that can include kinases, phosphatases, their substrates, scaffold proteins, and other regulatory factors (FIG. 3 B, upper panel). For a pair of kinase and phosphatase, there can be two binary complexes, substrate–kinase and substrate–phosphatase, and the ternary substrate–kinase—phosphatase complex. This case may result when the binding mechanisms of the enzymes are independent. Kinases and phosphatases have indeed been observed to occur in the same complexes, which usually involve additional scaffold proteins.[29] However, the two enzymes would also not be able to bind together and thus compete for their substrate, in which case a ternary complex would not be possible.[30] An ultrasensitive response to changes in the ratio of kinase to phosphatase concentration is obtained when the enzymes compete for the substrate (FIG. 3 B, lower panel, solid curve). By contrast, scaffold-mediated binding of kinase and phosphatase together reduces the sensitivity of the response (FIG. 3 B, lower panel, dashed curve).

To produce a threshold and/or a switch in the response, competition must be iterated. This occurs in a distributive mechanism of enzyme binding (FIG. 3 C, upper panel), where the enzyme binds, processes one residue, dissociates, and then the same or another enzyme molecule rebinds to process the next residue. A distributive mode of phosphorylation may occur as a consequence of structural requirements (e.g., replacement of adenosine 5′bisphosphate (ADP) by adenosine 5′-triphosphate (ATP) only after release of the substrate) or due to kinetic factors (e.g., when dissociation of the kinase is faster than the phosphorylation reaction).[13,31,32] An alternative mechanism is the processing (phosphorylation or dephosphorylation) of two or more sites during a single encounter between enzyme and substrate. This so-called processive mechanism occurs, for example, with kinases containing SH2 or SH3 domains.[33–35] In the extreme case of completely processive (de)phosphorylation, the competition between the kinase and phosphatase would be restricted to a single binding event. Processive phosphorylation is generally characterized by a hyperbolic response to changes in the kinase concentration. By contrast, with a distributive mechanism, the enzymes compete several times for binding the substrate, which results in a sigmoid stimulus–response curve (FIG. 3 C, lower panel).

COMPETITION EFFECTS AND THE RESPONSE KINETICS

The response of a phosphorylation cycle will not only depend on the final state approached after activation (or inactivation) of kinase or phosphatase, but also on the time taken for this transition. The time course for the distinct

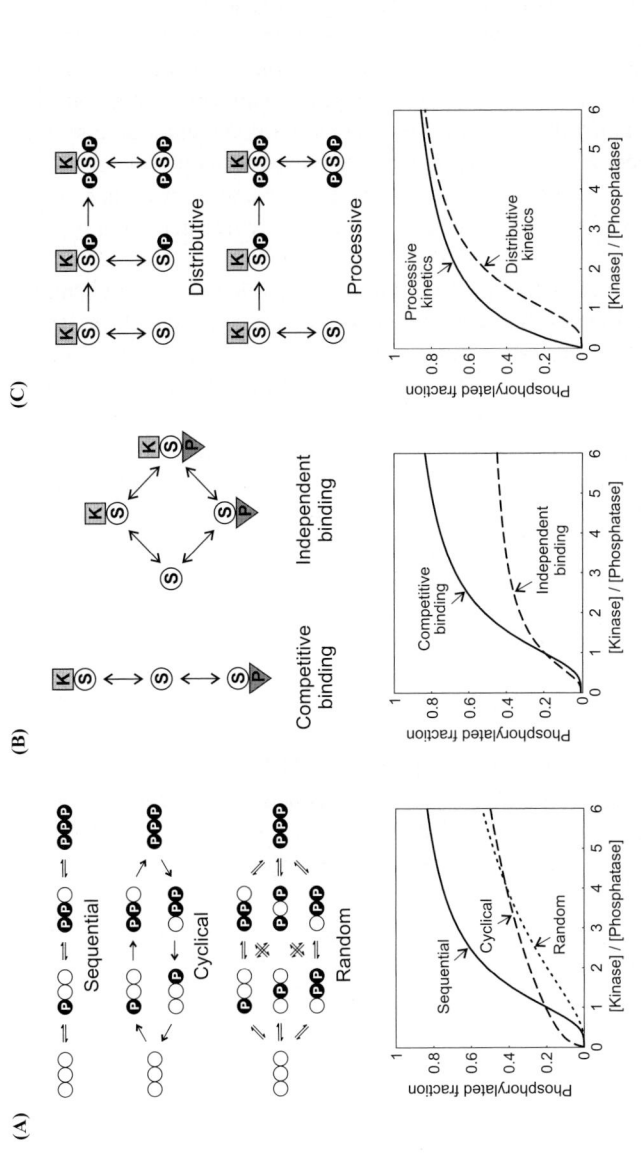

FIGURE 3. Structural and kinetic factors affect the response sensitivity. Schemes representing diverse structural and kinetic mechanisms of multisite phosphorylation and the corresponding response curves are shown. (**A**) Order of phosphate processing: kinase and phosphatase may process the residues sequentially in opposite directions (sequential: solid curve), in the same direction (cyclical: dashed curve), or each residue may be modified regardless of the state of the other residues (random: dotted curve). (**B**) Enzyme recruitment: the substrate can be regulated in the context of a multiprotein complex including the kinase and phosphatase (independent binding: dashed curve), or the enzymes can compete for their substrate (competitive binding: solid curve). (**C**) Catalytic mechanism: the enzyme may remain bound to its substrate until it modifies all phosphorylation sites (processive: solid curve), or several collision events between substrate and enzyme may be required for complete phosphorylation (distributive: dashed curve). A threshold response is obtained for the cases of sequential order, competitive binding, and distributive kinetics.

phosphorylation forms of a multiply phosphorylatable protein in the presence of a constant stimulation is displayed in FIGURE 4 A, where the transition time with respect to the fully phosphorylated target is highlighted.[8,36]

The kinetic design of the phosphorylation cycle, in particular, the occurrence of repeated competitions between kinase and phosphatase or between the substrate molecules when binding the enzymes, can strongly affect the transition time. As a general tendency, the responses are fast when kinase and phosphatase are abundant and do not too strongly bind to their respective reaction products (FIG. 4 B, dotted curve). In this case, the target cycles rapidly between phosphorylated and dephosphorylated states, which accounts for the fast response.

The saturation of the enzymes with the target protein leads to significantly slower kinetics (FIG. 4 B, solid and dashed curves; note the logarithmic time scale). In particular, the transition time corresponding to the case of zero-order ultrasensitivity (dashed curve) exhibits a maximum at the position of the threshold in the response curve where the kinase and phosphatase activities balance. Slower phosphorylation for near-threshold stimuli also occurs in the case of a cooperative kinetics or enzyme binding. This kinetic effect will be more pronounced with an increasing number of phosphorylation sites.

Concerning the order of phosphate processing, there is a profound qualitative difference in the response kinetics between the random, cyclical, and sequential mechanisms (FIG. 4 C). A random mechanism (dotted curve) gives rise to a faster phosphorylation than a sequential one (solid curve). This property

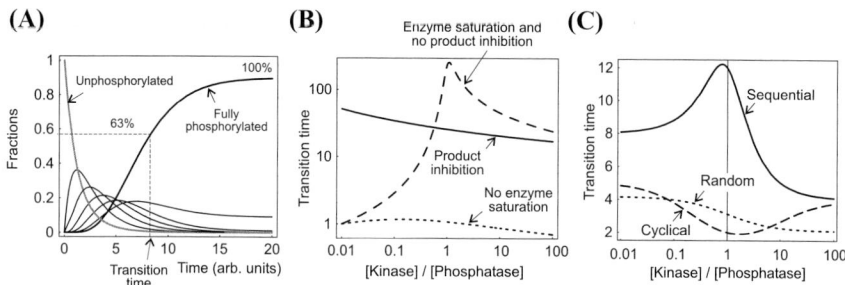

FIGURE 4. Competition effects affect the kinetics of phosphorylation kinetics. (**A**) Time course of the phosphorylation forms of a multiply phosphorylatable protein. The transition time as a measure of the time needed to reach the steady state has been indicated. (**B**) Transition time versus the stimulus strength for the cases of enzyme saturation with product inhibition, known as zero-order ultrasensitivity (solid curve), and without product inhibition (dashed curve) and for the case of no enzyme saturation and no product inhibition (dotted curve). (**C**) Transition time versus the stimulus strength for the sequential (solid curve), cyclical (dashed curve), and random (dotted curve) mechanisms. A slower phosphorylation occurs for zero-order ultrasensitivity and sequential multisite phosphorylation for near-threshold stimuli.

results from the fact that there are many more phosphorylation routes available for the random mechanism, where, in particular, phosphorylation can start from any site. By contrast, in the sequential case an intermediate site becomes a substrate for the kinase only after all the preceding sites have been phosphorylated. While the sequential mechanism exhibits a considerable slowing down of phosphorylation at intermediate kinase concentrations, the cyclical mechanism (dashed curve) shows a strong minimum of the transition time. The transition time of the random mechanism is a monotonically decreasing function of the kinase concentration.

There is an interesting relation between the transition-time curves (FIG. 4) and the steady-state response curves (FIG. 3). When the latter exhibits a pronounced threshold (ultrasensitivity), there is typically a corresponding maximum in the transition time: threshold effects are associated with slow kinetics. This maximum in the transition time for near-threshold stimuli is found irrespective of the molecular mechanism that causes the threshold and/or switch response, such as zero-order ultrasensitivity, multiple phosphorylation steps, or cooperativity. This slowing of the response may place an additional constraint on the processing of temporally varying input stimuli.

MATHEMATICAL MODELING

The mathematical description of phosphorylation cycles has been remarkably heterogeneous, from greatly simplified kinetics for the enzymes, e.g., linear or Michaelis–Menten kinetics, to models that partially include elementary binding and reaction steps. Some challenging problems during the modeling of these cycles are the various possibilities of protein–protein interactions and mechanisms of multisite phosphorylation, which can create a very large number of complexes and phosphorylation states.[21] An additional problem is the limited applicability for signal transduction of traditional enzyme kinetics. While in many metabolic reactions, enzymes occur at much lower concentrations than their substrates, justifying the use of the Michaelis–Menten rate law, such strict concentration hierarchies in signal transduction are unlikely. In signaling pathways, the target proteins of phosphorylation cycles are frequently enzymes, such as in kinase cascades, or transcription factors, both of which can be present in low concentrations.

We developed in a recent work a general framework for the mathematical description of phosphorylation cycles, which addresses these challenges.[8] Our approach starts from the description of elementary protein–protein interactions and catalytic steps. To treat protein–protein interactions systematically, we made use of the fact that often the protein associations occur on faster time scales than the phosphorylation–dephosphorylation reactions. This allows the application of a rapid-equilibrium approximation, which implies that only the protein–protein affinities and not the individual association and dissociation rate constants need to be considered. This simplification facilitates both analytic treatment and numerical analysis of the kinetic equations. This approach should prove useful also for a systematic analysis of more complex systems such as kinase cascades and networks.

CONCLUDING REMARKS

The steady-state and kinetic behavior of phosphorylation cycles and cascades can be interpreted in terms of competition effects. Generally, repeated competition between the components and/or steps of the cycles result in a threshold or switch-like response and in a slower kinetics for near-threshold stimuli. These properties are found irrespective of the molecular mechanism that causes the competition effects, such as zero-order ultrasensitivity, multiple phosphorylation steps, or cooperativity. A combination of these effects can extremely increase the degree of ultrasensitivity and even result in the emergence of more complex properties such as a bistable response. Quantitative aspects of the protein–protein interactions such as the relative concentration and binding affinities of the enzymes are also determinants of the steady-state and kinetic properties of phosphorylation cycles. Although these parameters are currently not known for most signaling pathways, large-scale measurements of cellular protein concentrations are already appearing, and affinities and kinetic parameters are beginning to be determined (e.g., see Ref. 37,38). We envisage that mathematical modeling will help establish links between structural and kinetic information on the phosphorylation cycles that regulate signaling proteins, and the functional characterization of the pathways.

REFERENCES

1. STADTMAN, E.R. & P.B. CHOCK. 1977. Superiority of interconvertible enzyme cascades in metabolic regulation: analysis of monocyclic systems. Proc. Natl. Acad. Sci. USA **74:** 2761–2765.
2. STADTMAN, E.R. & P.B. CHOCK. 1977. Superiority of interconvertible enzyme cascades in metabolite regulation: analysis of multicyclic systems. Proc. Natl. Acad. Sci. USA **74:** 2766–2770.
3. GOLDBETER, A. & D.E. KOSHLAND, JR. 1981. An amplified sensitivity arising from covalent modification in biological systems. Proc. Natl. Acad. Sci. USA **78:** 6840–6844.
4. HUANG, C.Y. & J.E. FERRELL, JR. 1996. Ultrasensitivity in the mitogen-activated protein kinase cascade. Proc. Natl. Acad. Sci. USA **93:** 10078–10083.
5. FERRELL, J.E., JR. & E.M. MACHLEDER. 1998. The biochemical basis of an all-or-none cell fate switch in Xenopus oocytes. Science **280:** 895–898.
6. HEINRICH, R., B.G. NEEL & T.A. RAPOPORT. 2002. Mathematical models of protein kinase signal transduction. Mol. Cell **9:** 957–970.

7. LEGEWIE, S., N. BLUTHGEN, R. SCHAFER, et al. 2005. Ultrasensitization: switch-like regulation of cellular signaling by transcriptional induction. PLoS Comput. Biol. **1:** e54.
8. SALAZAR, C. & T. HÖFER. 2006. Kinetic models of phosphorylation cycles: a systematic approach using the rapid equilibrium approximation for protein-protein interactions. Biosystems **83:** 195–206.
9. POMERENING, J.R., E.D. SONTAG & J.E. FERRELL, JR. 2003. Building a cell cycle oscillator: hysteresis and bistability in the activation of Cdc2. Nat. Cell Biol. **5:** 346–351.
10. MARKEVICH, N.I., J.B. HOEK & B.N. KHOLODENKO. 2004. Signaling switches and bistability arising from multisite phosphorylation in protein kinase cascades. J. Cell. Biol. **164:** 353–359.
11. ANGELI, D., J.E. FERRELL, JR & E.D. SONTAG. 2004. Detection of multistability, bifurcations, and hysteresis in a large class of biological positive-feedback systems. Proc. Natl. Acad. Sci. USA **101:** 1822–1827.
12. PUFALL, M.A., G.M. LEE, M.L. NELSON, et al. 2005. Variable control of Ets-1 DNA binding by multiple phosphates in an unstructured region. Science **309:** 142–145.
13. NASH, P., X. TANG, S. ORLICKY, et al. 2001. Multisite phosphorylation of a CDK inhibitor sets a threshold for the onset of DNA replication. Nature **414:** 514–521.
14. OKAMURA, H., J. ARAMBURU, C. GARCÍA-RODRÍGUEZ, et al. 2000. Concerted dephosphorylation of the transcription factor NFAT1 induces a conformational switch that regulates transcriptional activity. Mol. Cell **6:** 539–550.
15. SALAZAR, C. & T. HÖFER. 2003. Allosteric regulation of the transcription factor NFAT1 by multiple phosphorylation sites: a mathematical analysis. J. Mol. Biol. **327:** 31–45.
16. SALAZAR, C. & T. HÖFER. 2005. Activation of the transcription factor NFAT1: concerted or modular regulation? FEBS Lett. **579:** 621–626.
17. SPRINGER, M., D.D. WYKOFF, N. MILLER, et al. 2004. Partially phosphorylated Pho4 activates transcription of a subset of phosphate-responsive genes. PLoS Biol. **1:** 261–270.
18. ADAMS, J.A. 2003. Activation loop phosphorylation and catalysis in protein kinases: is there functional evidence for the autoinhibitor model? Biochemistry **42:** 601–607.
19. OKAMURA, H., C. GARCÍA-RODRÍGUEZ, H. MARTINSON, et al. 2004. A conserved docking motif for CK1 binding controls the nuclear localization of NFAT1. Mol. Cell Biol. **24:** 4184–4195.
20. SHAW, K.T., A.M. HO, A. RAGHAVAN, et al. 1995. Immunosuppressive drugs prevent a rapid dephosphorylation of transcription factor NFAT1 in stimulated immune cells. Proc. Natl. Acad. Sci. USA **92:** 11205–11209.
21. HLAVACEK, W.S., J.R. FAEDER, M.L. BLINOV, et al. 2003. The complexity of complexes in signal transduction. Biotechnol. Bioeng. **84:** 783–794.
22. HOLMBERG, C.I., S.E. TRAN, J.E. ERIKSSON, et al. 2002. Multisite phosphorylation provides sophisticated regulation of transcription factors. Trends Biochem. Sci. **27:** 619–627.
23. GUNAWARDENA, J. 2005. Multisite phosphorylation makes a good threshold but can be a poor switch. Proc. Natl. Acad. Sci. USA **102:** 14617–14622.
24. ROACH, P.J. 1991. Multisite and hierarchal protein phosphorylation. J. Biol. Chem. **266:** 14139–14142.

25. NISHIMURA, I., Y. YANG & B. LU. 2004. PAR-1 kinase plays an initiator role in a temporally ordered phosphorylation process that confers tau toxicity in Drosophila. Cell **116:** 671–682.
26. KENNEDY M.J., K.A. LEE, G.A. NIEMI, *et al.* 2001. Multiple phosphorylation of rhodopsin and the *in vivo* chemistry underlying rod photoreceptor dark adaptation. Neuron **31:** 87–101.
27. WAGNER, P.D., N.D. VU & J.N. GEORGE. 1985. Random phosphorylation of the two heads of thymus myosin and the independent simulation of their actin-activated ATPases. J. Biol. Chem. **260:** 8084–8089.
28. FERRELL, J.E., JR. & R.R. BHATT. 1997. Mechanistic studies of the dual phosphorylation of mitogen-activated protein kinase. J. Biol. Chem. **272:** 19008–19016.
29. IKEDA, S., S. KISHIDA, H. YAMAMOTO, *et al.* 1998. Axin, a negative regulator of the Wnt signaling pathway, forms a complex with GSK-3beta and beta-catenin and promotes GSK-3beta-dependent phosphorylation of beta-catenin. EMBO J. **17:** 1371–1384.
30. TANOUE, T., M. ADACHI, T. MORIGUCHI, *et al.* 2000. A conserved docking motif in MAP kinases common to substrates, activators and regulators. Nat. Cell. Biol. **2:** 110–116.
31. WAAS, W.F., H.H. LO & K.N. DALBY. 2001. The kinetic mechanism of the dual phosphorylation of the ATF2 transcription factor by p38 mitogen-activated protein (MAP) kinase. J. Biol. Chem. **276:** 5676–5684.
32. BURACK, W.R. & T.W. STURGILL. 1997. The activating dual phosphorylation of MAPK by MEK is nonprocessive. Biochemistry **36:** 5929–5933.
33. MAYER, B.J., H. HIRAI & R. SAKAI. 1995. Evidence that SH2 domains promote processive phosphorylation by protein-tyrosine kinases. Curr. Biol. **5:** 296–305.
34. PELLICENA, P. & W.T. MILLER. 2001. Processive phosphorylation of p130Cas by Scr depends on SH3-polyproline interactions. J. Biol. Chem. **276:** 28190–28196.
35. AUBOL, B.E., S. CHAKRABARTI, J. NGO, *et al.* 2003. Processive phosphorylation of alternative splicing factor/splicing factor 2. Proc. Natl. Acad. Sci. USA **100:** 12601–12606.
36. LLORENS, M., J.C. NUÑO, Y. RODRÍGUEZ, *et al.* 1999. Generalization of the theory of transition times in metabolic pathways: a geometrical approach. Biophys. J. **77:** 23–36.
37. WU, J.Q. & T.D. POLLARD. 2005. Counting cytokinesis proteins globally and locally in fission yeast. Science **310:** 310–314.
38. LEE, E., A. SALIC, R. KRÜGER, *et al.* 2003. The roles of APC and Axin derived from experimental and theoretical analysis of the Wnt pathway. PLoS Biol. **1:** e10.

Preparation of Leptin Antagonists by Site-Directed Mutagenesis of Human, Ovine, Rat, and Mouse Leptin's Site III

Implications on Blocking Undesired Leptin Action *In Vivo*

GILI SOLOMON,[a] LEONORA NIV-SPECTOR,[a] DANA GONEN-BERGER,[a] ISABELLE CALLEBAUT,[b] JEAN DJIANE,[c] AND ARIEH GERTLER[a]

[a]*Faculty of Agricultural, Food, and Environmental Quality Sciences, The Hebrew University of Jerusalem, Rehovot, 76100, Israel*

[b]*LMCP, CNRS UMR7590, Universities Paris 6 & Paris 7, Paris F-75005, France*

[c]*Institut National de la Recherche Agronomique, NOPA, Jouy-en-Josas Cedex F-78352, France*

ABSTRACT: Six muteins of human, ovine, rat, and mouse leptins mutated to Ala in amino acids 39–41 or 39–42 were prepared by site-directed mutagenesis of the putative site III, which does not affect binding but is necessary for receptor activation, then expressed, solubilized in 4.5 M urea, at pH 11.3 in presence of cysteine, refolded and purified to homogeneity by anion-exchange chromatography on Q-Sepharose or combination of anion-exchange chromatography followed by gel filtration. The overall yields were 400–800 mg from 5 L of fermentation. All proteins were >98% pure as evidenced by SDS-PAGE and contained at least 95% monomers as documented by gel-filtration chromatography under nondenaturing conditions. Circular dichroism analysis revealed that all six muteins have identical secondary structure characteristic of nonmutated leptins, namely 52–63% of alpha helix content. All muteins formed a 1:1 complex with chicken leptin binding domain, (chLBD) and bound chLBD or membrane-embedded leptin receptor with affinity identical to WT leptins. Muteins were devoid of any biological activity in several bioassays but were potent competitive antagonists. Some muteins were pegylated using 40 kDa PEG. Although pegylation decreased the *in vitro* activity, increasing circulation half-life can recompensate this deficit, so pegylated antagonists are expected to be more potent *in vivo*.

Address for correspondence: Arieh Gertler, Faculty of Agricultural, Food, and Environmental Quality Sciences, The Hebrew University of Jerusalem, POB 12, Rehovot, 76100, Israel. Voice/fax: 972-89-489-006.
  e-mail: gertler@agri.huji.ac.il

KEYWORDS: human leptin; ovine leptin; mouse leptin; rat leptin; leptin antagonists; site-directed mutagenesis; leptin site III; pegylation

INTRODUCTION

Leptin is a pleiotropic hormone cloned from and secreted mainly by adipose tissue[1] as well as by gut, placenta and other tissues.[2] Blocking leptin action by classical replacement therapy is thus not feasible. Leptin plays also an important role in immune response. Leptin deficient *ob/ob* mice (unless given leptin) do not develop autoimmune response, suggesting that antagonizing leptin may become a possible therapy in autoimmune diseases.[3-5] Leptin upregulation in heart failure and the ability of cardiac tissue to produce leptin and to express leptin receptors (LEPR) was recently documented[6,7] and endogenously produced leptin contributes to post infarction hypertrophy and heart failure which can be blocked by leptin antagonists (M. Karmazyn, personal communication). In conclusion, leptin antagonists not only offer a novel potent tool for studying leptin-related metabolic processes and for elucidating the role of leptin in mammalian physiology and pathology but have a potential of becoming a drug.

MATERIALS AND METHODS

Materials

Recombinant human, ovine, rat, and mouse leptins and chicken leptin binding domain (chLBD)[8] were obtained from Protein Laboratories Rehovot (PLR) Ltd. (Rehovot, Israel). Molecular-weight marker for SDS-PAGE was purchased from Bio-Rad (Hercules, CA). SDS-PAGE reagents, nalidixic acid, lysozyme, and cysteine were purchased from Sigma-Aldrich (St. Louis, MO) and Bio-Lab Ltd. (Jerusalem, Israel). RPMI 1640 medium, fetal calf (FCS) and horse sera were purchased from Biological Industries (Kibbutz Beit HaEmek, Israel). Antibiotic antimycotic solution (5 × 10^4 U/mL penicillin, 50 mg/mL streptomycin, 0.125 mg/mL fungisone), NaCl, and Tris were purchased from Bio-Lab Ltd. and Bacto-tryptone, Bacto-yeast extract, glycerol, EDTA, Triton X-100, and urea from ENCO Diagnostics Ltd. (Petah-Tikva, Israel). Branched polyethylene glycol (PEG) N- hydroxysuccinimide (NHS), molecular weight 40 kDa was used for pegylation. Polyethylene glycol (PEG-40) was purchased from Nektar Therapeutics Co. (available at www.nectar.com) Nectar Co.

Preparation of Expression Plasmids Encoding Mutated Leptins

Synthetic cDNA encoding the sequences of mouse (GenBank accession no. AAA64213) and rat (GenBank accession no. NM_013076) leptins modified to ensure better codon usage and expression in *Escherichia coli* were ordered from Entelechon GmbH (Regensburg, Germany). The cDNA in pTOPO was digested with *Nco*I and *Hind*III, extracted, and ligated into linearized pMon3401 expression vector. *E. coli* MON105 competent cells were transformed with the new expression plasmid and plated on LB-agar plates containing 75 μg/mL spectinomycin for plasmid selection. Four *E. coli* colonies were isolated and confirmed to contain mouse and rat leptin cDNA by digestion with *Nco*I/*Hind*III restriction enzymes. All colonies were positive and one of them was sequenced. Plasmids expressing ovine[9] and human[10] leptins were described previously. To prepare the leptin mutants, the pMon3401 expression plasmids encoding the wild-type (WT) leptins were used as starting material. The leptin inserts were modified with the Stratagene Quick Change mutagenesis kit (La Jolla, CA) according to manufacturer's instructions, using two complementary primers (TABLE 1). The primers were designed to contain base changes (marked in bold), in order to obtain the respective mutations but still conserve the appropriate amino-acid sequence, and to modify a specific restriction site (underlined) for colony screening. The procedure included 18 PCR cycles using *Pfu* polymerase. The mutated construct was then digested with *Dpn*I restriction enzyme, which is specific to methylated and hemi-methylated DNA (target sequence: 5'-G^{m6}ATC-3') in order to digest the template and select for mutations containing synthesized DNA. XL-1 competent cells were transformed with the mutated plasmids and grown in 5–10 mL LB medium and the plasmids were isolated. Five colonies of each mutant were screened for mutation, using the specific designed restriction site, and revealed at least 80% efficiency. Two colonies of each mutant were sequenced and confirmed to contain the mutation but no unwanted misincorporation of nucleotides. Mon105 competent cells were then transformed with the plasmids and used for expression.

Expression, Refolding, and Purification of Human, Ovine, Rat, and Mouse Leptin Muteins

The recombinant muteins of human, ovine, rat, or mouse leptins with an extra methionine-alanine (methionine is cleaved by the bacteria) at the N terminus were expressed in a 2.5 l culture, inclusion bodies (IBs) were then prepared as described previously[9,10] and frozen. Subsequently, IBs obtained from 2.5 L of bacterial culture were solubilized in 300 mL of 4.5 M urea, 40 mM Tris base containing 10 mM cysteine. In the case of ovine leptin or its mutants, the IBs obtained from 1.0 L of bacterial culture were solubilized in a similar manner in 200 mL. The pH of the solution was adjusted to 11.3 with NaOH. After 2 h of stirring at 4°C, three volumes of 0.67 M arginine were added to a final concentration of 0.5 M and stirred for additional 1.5 h. Then, the solution was dialyzed against 1.0 L of 10 mM Tris-HCl, pH 9 (human, rat, and mouse leptin muteins) or 10 mM Tris-HCl, pH 8 (ovine leptin muteins) for 60 h, with five

TABLE 1. Primers used for the preparation of human, ovine, rat, and mouse leptin muteins

Species	Primer	Primer sequence[a]	Restriction site[b]
Human	L39A/D40A/F41A-5	S 5′-CCAAACAGAAAGTCACT**GGTGCGGCCGCC**ATTCCTGGGCTC-3′	(−)
	L39A/D40A/F41A-5	A 5′-GAGCCCAGGAAT**GGCGGCCGC**ACCAGTGACTTTCTGTTTGG-3′	(−)
Human	L39A/D40A/F41A/I42A-5	S 5′-CCAAACAGAAAGTCACT**GGTGCGGCCGC**CTCCTGGGCTCCACC-3′	(−)
	L39A/D40A/F41A/I42A-5	A 5′-GGTGGAGCCCAGGAGGCGGCCGC**A**CCAGTGACTTTCTGTTTGG-3′	(−)
Ovine	L39A/D40A/F41A-5	S 5′-TCCAAACAGAGGGTCACC**GGTGCTGCAGCT**ATCCCTGGGCTCCACCC-3′	(+)
	L39A/D40A/F41A-5	A 5′-GGGTGGAGCCCAGGGATA**GCTGCAGCACC**GGTGACCCTCTGTTTGGA-3′	(+)
Ovine	L39A/D40A/F41A/I42A-5	S 5′-CAGAGGGTCACC**GGTGCTGCTGCT**GCTCCCGGCTCCACCC-3′	(+)
	L39A/D40A/F41A/I42A-5	A 5′-GGGTGGAGCCCGGGAGCAGCAGC**AGCAGC**GGTGACCCTCTG-3′	(+)
Rat/	L39A/D40A/F41A-5	S 5′- CAGCGTGTTACC**GGCGCGGCTGC**CATCCCGGGCCTGC -3′	(−)
Mouse	L39A/D40A/F41A-3	A 5′- GCAGGCCCGGGATGGCAGCCGCGCCGGT**AACACGCTG** -3′	(−)

[a]S, sense primer; A, antisense primer; all mutations are in bold letters.
[b]Successful mutations were monitored by disappearance (−) or appearance (+) of the respective restriction site *BshTI* (underlined), followed by DNA sequencing.

or six external solution exchanges. In the case of human and ovine muteins the protein was then applied at maximal flow rate (400–500 mL/h) onto a Q-Sepharose column (30-mL bead volume), pre-equilibrated with the respective buffer. The breakthrough fraction, which contained no leptin, was discarded, and the absorbed protein was eluted in a stepwise manner (50, 100, 150, and 400 mM of NaCl in 10 mM Tris-HCl, pH 9 or pH 8). A total of 50-mL fractions were collected and protein concentration was determined by absorbance at 280 nm. In the case of rat and mouse muteins NaCl was added to the dialysed solution to a final concentration of 150 mM which were then applied at maximal flow rate (400–500 mL/h) onto a Q-Sepharose column (30-mL bead volume), pre-equilibrated with the 10 mM Tris-HCl, pH 9, containing 150 mM NaCl. The breakthrough fraction, which contained the respective leptin or leptin mutein, was collected and concentrated to 3–4 mg protein/mL. Then 12-mL portions were applied to a preparative Superdex™ 75 column (Pharmacia Ltd., Uppsala, Sweden) (2.6 × 60 cm) pre-equilibrated with 10 mM Tris-HCl, pH 9, containing 150 mM NaCl. Fractions containing the monomeric protein were pooled, dialyzed against $NaHCO_3$ to ensure a 4:1 protein-to-salt ratio and lyophilized. Pegylation was carried out according to the protocol used for human growth hormone.[11]

Determination of Purity, Monomer Content, and Secondary Structure

SDS-PAGE was carried out according to Laemmli[12] in a 15% polyacrylamide gel under reducing and nonreducing conditions. The gel was stained with Coomassie Brilliant Blue R. Gel-filtration chromatography was performed on a Superdex 75 HR 10/30 column with 0.2-mL aliquots of the Q-Sepharose-column-eluted fraction using TN buffer (25 mM Tris-HCl, 150 mM NaCl, pH 8). The CD spectra in the wavelength range of 200–240 nm were measured with an Jasco J-810 Spectropolarimeter (Tokyo, Japan) using a 0.020-cm rectangular QS Hellma cuvette with a spectral resolution of 1 nm and signal-to-noise ratio of about 1% at 210–220 nm. Solutions were prepared dissolving the lyophilized samples in 50 mM phosphate buffer, pH 7.6 followed by centrifugation. Secondary structure of proteins was determined applying procedure and computer program CONTIN[13] to calculate α-helices, β-sheets, and β-turns as a percentage of amino acids involved in these ordered forms.

Binding and Bioassays

Binding assays using radioactive leptin were carried out according to procedures described in our recent manuscript.[8] The proliferation rate of leptin-sensitive BAF/3 cells stably transfected with the long form of human leptin receptor (hLEPR) was used to estimate the antagonistic activity of leptin muteins as described previously.[14] To determine antagonistic activity, 6.25×10^{-11} M WT homologous leptin was added to each well, which also contained different concentrations of muteins. The average absorbance in wells without leptin (negative control) was used as a blank value and subtracted from other absorbance values to yield the corrected absorbance values. The average absorbance in wells with WT leptin after subtracting the negative control was used as a positive control to calculate percent inhibition. The inhibition curves were drawn using the Prism (4.0) nonlinear regression one-site inhibition program[15] and the IC_{50} values were calculated. It should be pointed out that all mammalian leptins are capable of activating hLEPR to almost identical degree.[8,10,16]

RESULTS

Purification and Physico-Chemical Characterization of Leptins and Leptin Muteins

Human and ovine L39A/D40A/F41A and L39A/D40A/F41A/I42A muteins were purified by anion-exchange chromatography and rat and mouse L39A/D40A/F41A muteins by consecutive anion-exchange and gel-filtration chromatography. The fractions containing pure monomer were pooled, dialyzed against $NaHCO_3$, pH 8, at a 4:1 or 3:1 (w/w) protein: salt ratio and lyophilized. The yields of various muteins varied from 400 to 800 mg from 5 L of bacterial culture. The purity and homogeneity of the purified muteins were documented by three independent methods. SDS-PAGE yielded only one band of ~16 kDa, under both reducing and nonreducing conditions (not shown), reverse-phase chromatography also yielded a single peak (not shown) and gel filtration at pH 8 under native conditions yielded a single monomeric peak consisting of over 95% monomer, corresponding to a molecular mass of 16 kDa. The secondary structures of mouse and rat leptins and their muteins calculated from the CD spectra are similar to nonmutated leptins,[14,17] as shown by high content of α-helix (52–63%), 8–11% of β-sheets, and 14–18% β-turns, indicating proper refolding. The stability of all muteins in solution was tested at 4°C and 37°C. All six proteins could be stored at both temperatures as sterile 0.2 mM solutions for at least 30 days at pH 6 or 8 without undergoing any changes in their monomeric content and retaining their activity in the BAF/3 bioassay.

Binding Experiments and Bioassay

Iodinated human leptin served as the ligand in all competitive experiments and the respective nonlabeled WT leptins and muteins as competitors. Freshly prepared homogenate of BAF/3 cells stably transfected with the long form of hLEPR was used as the receptor source. Homogenate from 1.6×10^6 cells

TABLE 2. Binding to chLBD and antagonistic activity of human, ovine, rat, and mouse leptin muteins in BAF/3 cells stably transfected with the long form of hLEPR

Mutein[a]		IC_{50} (nM), mean ± SEM[b]		Agonistic activity
		Binding	Bioactivity	
Human	L39A/D40A/F41A	1.43 ± 0.91 (3)	11.9 ± 1.94 (3)	None
	L39A/D40A/F41A/I42A	1.85 ± 0.80 (3)	9.3 ± 1.60 (5)	None
Ovine	L39A/D40A/F41A	2.21 ± 0.74 (3)	18.0 ± 6.2 (3)	None
	L39A/D40A/F41A/I42A	1.99 ± 0.48 (3)	9.5 ± 4.6 (3)	None
Rat	L39A/D40A/F41A	1.83 ± 0.78 (2)[c]	33.0 ± 8.2 (2)[c]	None
Mouse	L39A/D40A/F41A	2.22 ± 1.13 (2)[c]	29.5 ± 9.2 (2)[c]	None

[a]Cells were stimulated with 0.0625 nM leptin.
[b]Numbers in parentheses indicate the number of performed experiments.
[c]Mean ± SD.

per tube gave 6–7% specific binding. The IC_{50} values, calculated by Prism 4.0 program[15] (nonlinear regression, best fit for one-site competition), shown in TABLE 2 indicate that all six muteins do not differ in affinity, which is also similar to that of nonmutated leptins (not shown). All muteins formed a 1:1 complex with chLBD (not shown). Human, ovine, rat, and mouse leptins exhibited almost identical activity in the BAF/3 bioassay with their respective EC_{50} values of 20.5, 22.5, 29.7, and 24.5 pM, comparable to human leptin in previously published papers,[10,16] whereas all six muteins were devoid of any agonistic activity (TABLE 2). To determine the antagonistic activity, BAF/3 cells were stimulated with 62.5 pM of human leptin in the presence (8–250 nM) or absence of the respective muteins. All six leptin muteins exhibited antagonistic activity and no significant differences between them were observed (TABLE 2). To verify the specificity of inhibition, the proliferation of BAF/3 cells was stimulated by 55 pM IL3 in the presence or absence of human, ovine, mouse, and rat leptin muteins. No inhibition was observed, even at a 1.25 μM concentration for all tested muteins (not shown).

Pegylation

Preliminary results indicated that both human, and rat and mouse leptin were double pegylated with molecular mass of ~100 kDa. They were 3- to 15-fold less active than the WT. Although the pegylation sites have not yet been finally identified, preliminary results indicate their location at the N-terminal part of the molecule, namely at N terminus, or lysines 5, 11, or 15. Similar experiments performed with the antagonists indicated that pegylation of human, rat, and mouse L39A/D40A/F41A leptin muteins resulted in double-pegylated species of ~100 kDa. Their inhibitory activities tested in BAF/3 cells bioassay as compared to the nonpegylated antagonists were respectively 1.1, 15.9, and 8.5%.

DISCUSSION

Several recent reports indicate that leptin exhibit undesired effects in autoimmune diseases,[3–5] heart failure,[3–7] and possibly in several types of cancers.[18] Therefore, preparation of reagents capable of abolishing leptin action is timely. As no structural information of the 3D structure of LEPR is available, the model of interleukin-6 (IL-6) was applied. In that model, a hexameric complex is formed gradually, first by IL-6 molecule which interacts with the IL-6R-alpha, then with gp130 forming an inactive trimer which subsequently dimerizes forming an active hexamer, whose formation is achieved due to interaction of IL-6 bound in one trimer (through its site III) with the immunoglobulin domain (IGD) of gp130 of the other trimer.[19] We have identified the putative leptin's binding site III by modeling LEPR, on the basis of its alignment with gp130, and fitting leptin on IL-6 in the IL-6/gp130 complex as leptin's amino acids 39–42 (LDFI), which are preserved in all leptin species. To verify this hypothesis and to test its generality we have prepared and purified to homogeneity human, ovine, rat and mouse triple (L39A/D40A/F41A) and quadruple (L39A/D40A/F41A/I42A, human and ovine only) leptin muteins. All six muteins had typical cytokine secondary structure, acted as true antagonists, namely they interacted with LEPR with affinity similar to the WT hormone (as evidenced by RRA), were devoid of biological activity in several leptin-responsive bioassays and specifically inhibited leptin action *in vitro*. An alternative way of preparing leptin antagonist was described recently by Tavernier and co-workers who modified the putative leptin's site III by mutating the T120 and S121 to Ala.[20] Both our six and their one mutein (A. Gertler, unpublished results) can be prepared in gram amounts and thus serve as a novel tool of studying leptin function *in vitro* and *in vivo*. To prolong their life in circulation some muteins were pegylated using 40 kDa polyethylene glycol. Although pegylation decreased the affinity, increasing circulation half-life can recompensate this deficit, so pegylated antagonists are expected to be more potent *in vivo*. Furthermore, the pegylation experiments were preliminary and after optimization pegylated analogs with higher inhibitory potential are expected. In conclusion, as antagonizing leptin has been suggested as a possible therapy in autoimmune diseases and heart failure, we have shown that large scale preparation of leptin antagonists is feasible.

ACKNOWLEDGMENT

This work was supported by the Israeli Science Foundation, grant no. 594/02 to Arieh Gertler.

REFERENCES

1. ZHANG, Y., R. PROENCA, M. MAFFEI, *et al.* 1994. Positional cloning of the mouse *obese* gene and its human homologue. Nature **372**: 425–432.

2. BADO, A., S. LEVASSEUR, S. ATTOUB, et al. 1998. The stomach is a source of leptin. Nature **394:** 790–793.
3. LA CAVA, A. & G. MATARESE. 2004. The weight of leptin in immunity. Nat. Rev. Immunol. **4:** 371–379.
4. MATARESE, G., S. MOSCHOS & C.S. MANTZOROS. 2005. Leptin in immunology. J. Immunol. **173:** 3137–3142.
5. PEELMAN, F., W. WAELPUT, H. ISERENTANT, et al 2004. Leptin: linking adipocyte metabolism with cardiovascular and autoimmune diseases. Prog. Lipid Res. **43:** 283–301.
6. PURDHAM, D.M., M.X. ZOU, V. RAJAPUROHITAM & M. KARMAZYN. 2004. Rat heart is a site of leptin production and action. Am. J. Physiol. **287** H2877–H2884.
7. ZEIDAN, A., D.M. PURDHAM, V. RAJAPUROHITAM, et al. 2005. Leptin induces vascular smooth muscle cell hypertrophy through angiotensin II and endothelin-1-dependent mechanisms and mediates stretch-induced hypertrophy. J. Pharmacol. Exp. Therp. **215:** 1075–1084.
8. NIV-SPECTOR, L., N. RAVER, M. FRIEDMAN-EINAT, et al. 2005. Mapping leptin-interacting sites in recombinant leptin-binding domain (LBD) subcloned from chicken leptin receptor. Biochem. J. **290:** 475–484.
9. GERTLER, A, J. SIMMONS & D.H. KEISLER. 1998. Large-scale preparation of biologically active recombinant ovine obese protein (leptin). FEBS Lett. **422:** 137–140.
10. RAVER, N., E. VARDY, O. LIVNAH, et al. 2002. Comparison of R128Q mutations in human, ovine, and chicken leptins. Gen. Comp. Endocrinol. **126:** 52–58.
11. CLARK, R. K. OLSON, G. FUH, et al. 1996. Long-acting growth hormones produced by conjugation with polyethylene glycol. J. Biol. Chem. **271:** 21969–21977.
12. LAEMMLI, U.K. 1970. Cleavage of structural proteins during the assembly of the head of bacteriophage T4. Nature **227:** 680–685.
13. PRIVENCHER, S.W. & J. GLÖCKNER. 1981. Estimation of globular protein secondary structure from circular dichroism. Biochemistry **20:** 33–37.
14. NIV-SPECTOR, L., D. GONEN-BERGER, I. GORDOU, et al. 2005. Identification of the hydrophobic strand in the A-B loop of leptin as major binding site III: implications for large-scale preparation of potent recombinant human and ovine leptin antagonists. Biochem. J. **291:** 221–230.
15. PRISM. 2003. GraphPad Prism TM Version 4.0. GraphPad Software, Inc. San Diego, CA.
16. RAVER, N., E.E. GUSSAKOVSKY, D.H. KEISLER, et al. 2000. Preparation of recombinant bovine, porcine, and porcine W4R/R5K leptins and comparison of their activity and immunoreactivity with ovine, chicken, and human leptins. Protein Expr. Purif. **19:** 30–40.
17. SALOMON, G., L. NIV-SPECTOR, E.E. GUSSAKOVSKY & A. GERTLER. 2006. Large-scale preparation of biologically active mouse and rat leptins and their L39A/D40A/F41A muteins which act as potent antagonists. Protein Expr. Purif. **47:** 128–136.
18. SOMASUNDAR, P., D.W. MCFADDEN, S.M. HILEMAN & L. VONA-DAVIS. 2004. Leptin is a growth factor in cancer. J. Surg. Res. **116:** 337–349.
19. BOULANGER, M.J., D.C. CHOW, E.E. BREVNOVA & K.C. GARCIA. 2003. Hexameric structure and assembly of the interleukin-6/IL-6 alpha-receptor/gp130 complex. Science **300:** 2101–2104.
20. PEELMAN, F., K. VAN BENEDEN, L. ZABEAU, et al. 2004. Mapping of the leptin binding sites and design of a leptin antagonist. J. Biol. Chem. **279:** 41038–41046.

Dual Activity of Phosphorothioate CpG Oligodeoxynucleotides on HIV

Reactivation of Latent Provirus and Inhibition of Productive Infection in Human T Cells

CARSTEN SCHELLER,[a] ANETT ULLRICH,[a] STEFAN LAMLA,[a] ULF DITTMER,[b] AXEL RETHWILM,[a] AND ELENI KOUTSILIERI[a]

[a]*Institute of Virology and Immunobiology, University of Wuerzburg, 97078 Wuerzburg, Germany*

[b]*Institute of Virology, UK Essen, 45122 Essen, Germany*

ABSTRACT: CpG oligodeoxynucleotides (CpG ODNs) bind to toll-like receptor-9 (TLR-9) and activate immune cells with antigen-presenting activity, including B cells and dendritic cells. Here we show that treatment of the latently human immunodeficiency virus (HIV)-infected T cell line ACH-2 with the CpG ODNs 2006 or 2040 triggers activation of viral gene expression, demonstrating that CpG-signaling activity can also be found in T cells. The CpG ODNs g12AAC and g12GTC had no effect on virus reactivation. In contrast to the stimulating effects on viral gene expression in latently infected cells, CpG ODNs potently suppressed HIV replication in productively infected MT4 T cells or PBLs. Inhibition of virus replication was not related to the CpG motif but similarly occurred with non-CpG phosphorothioate (PTO)-ODNs. Thus, virus inhibition was likely caused by the PTO backbone of the CpG ODNs, probably by interfering with events prior to integration of the viral cDNA into the host genome. The ability of CpG PTO-ODNs to trigger reactivation of latent HIV in combination with their antiviral activity on productive infection makes this substance class an interesting candidate for further test to asses their potential as supplements in HIV therapy.

KEYWORDS: CpG; TLR9; HIV; phosphorothioate oligodeoxynucleotides

INTRODUCTION

The toll-like receptor-9 (TLR-9) belongs to the family of pattern recognition receptors (PRRs) that play an important role in innate immunity.[1] TLR-9 recognizes bacterial DNA.[2] Oligodeoxynucleotides (ODNs) containing unmethylated CG dinucleotides within a specific sequence context, the so-called CpG motif, mimic bacterial DNA by the absence of methylated cytosine residues and similarly activate TLR-9. Synthetic CpG ODNs (length of 20–30 nucleotides) usually contain phosphorothioate (PTO) modifications in their backbone to enhance stability and reactivity. TLR-9 activates an IL-1 receptor-like pathway via MyD88 and IRAK that eventually leads to activation of p38, JNK and NF-κB.

In contrast to antigen-presenting cells, little is known about TLR-9 signaling in T cells, although direct effects on T cells have been reported.[3,4] We have published previously that CpG ODNs trigger signaling events in the A3.01 T cell line that activate NF-κB and reactivation of human immunodeficiency virus (HIV) in latently infected ACH-2 cells,[5] a subclone of A3.01 T cells.

The HIV is the causative agent of AIDS. HIV infects $CD4^+$ cells of the immune system, such as helper T cells or monocytes, and causes a slow-progressing disease that is characterized by a progredient decline of helper T cells and the development of immune deficiency.[6]

After binding to CD4 and one of the HIV-coreceptors CXCR4 or CCR5, HIV enters the cell and releases its RNA genome into the cytoplasm.[7] The viral RNA is in turn reverse transcribed into cDNA. After an obligate integration step into the host genome, the integrated HIV cDNA (the so-called provirus) is transcribed and translated into viral proteins. Expression of HIV genes is mediated by various cellular transcription factors, one of which is NF-kB.[8]

After assembly and release of new virus particles, the new virions can infect other cells. Virus-producing cells are called productively infected cells. Productively infected cells die in general within several days due to the cytotoxicity of the viral infection or clearance by the immune system.[9] However, some of these cells survive the productive replication cycle and fall back in an inactive state in which no viral genes are expressed. These latently infected cells differ from uninfected cells only by the presence of viral cDNA and have a normal life span. Latently infected memory T cells represent the clinically most important cell type, as they can survive for probably 60 years.[10] Latently infected cells can resume virus production at any time once the cell becomes activated again by an external stimulus. In contrast to productively infected cells, latently infected cells cannot be targeted by antiviral drugs, so that HIV patients require lifelong treatment and cannot be cured from HIV infection by current antiretroviral therapy.

In this article we report about a dual activity of CpG PTO-ODNs on HIV. Whereas the CpG motif is a strong activator of viral transcription in latently infected T cells, the PTO modification potently suppresses HIV replication in productively infected cells, targeting a step prior to integration of the viral cDNA into the host genome.

MATERIALS AND METHODS

Cells, Viruses, Cell Culture, and Reagents

The latently HIV-infected T cell line ACH-2 and the uninfected T cell lines A3.01 and MT4 were cultured at 37° in 5% CO_2 atmosphere in RPMI 1640 (Invitrogen, Karlsruhe, Germany), containing 10% FCS, penicillin, and streptomycin. Peripheral blood mononuclear cells (PBMC) were prepared by Ficoll/Paque density-gradient centrifugation from heparinized human blood. T cells were stimulated for 8 days with PHA at 2 μg/mL and cultured in RPMI, 10% FCS, containing 100 U/mL IL-2 (Proleukin, Eurocetus, Frankfurt, Germany). PHA-stimulated PBL, A3.01 and MT4 cells were infected with HIV-1 strain IIIB/LAI at an m.o.i. of 1.0 $TCID_{50}$ per cell.

PTO-ODNs

PTO-modified ODNs (MWG-Biotech AG, Ebersberg, Germany) were of HPSF purified quality and were dissolved in H_2O at a concentration of 1 mg/mL. The following sequences were used (CpG-motifs are underlined):

CpGs:
ODN 2006: TCGTCGTTTTGTCGTTTTGTCGTT;
ODN 2040: CTGTCGTTTTGTCGTTTTGTCTGG;
ODN g12AAC: GGGGGGGGGGGGGAACGTTGGGGGGGGGGGG;
ODN g12GTC: GGGGGGGGGGGGGGTCGTTGGGGGGGGGGGG;
non-CpGs:
ODN 2006 4×TG: TTGTTGTTTTGTTGTTTTGTTGTT;
ODN 2041: CTGGTCTTTCTGGTTTTTTCTGG.

Flow Cytometry

To determine expression of HIV p24, cells were fixed with 4% formalin in PBS and permeabilized with PBS containing 5% BSA and 0.5% saponin. Expression of HIV-p24 was detected with the mouse anti-HIVp24 mAb 183-H12-5C (AIDS Research and Reference Reagent Program). As a second antibody we used a PE-labeled anti-mouse IgG antibody (DAKO, Hamburg, Germany). Cells were analyzed by flow cytometry using a FACScan flow cytometer (Becton Dickinson, Heidelberg, Germany). Markers were set according to staining with an isotype-matched control antibody (DAKO).

HIV-p24 ELISA

Cells were cultured in a 96-well flat bottom plate (10^5 cells/well) in a total volume of 200 µL. Concentration of released HIV-p24 was determined by ELISA using 100 µL of the supernatants. The ELISA was performed according to the instructions of the manufacturer (Roche Diagnostics GmbH, Mannheim, Germany).

RESULTS

CpG ODNs Reactivate Latent HIV in ACH-2 Cells

The latently HIV-infected T cell line ACH-2[11] was incubated with various CpG ODNs for 24 h and production of the HIV gag protein p24 was determined by flow cytometry. As depicted in FIGURE 1, incubation with the CpG ODNs 2006 and 2040 reactivated latent HIV as detected by intracellular p24-staining. In contrast, the CpG sequences g12AAC and g12GTC that have been reported to have no effects on T cell stimulation[4] did not reactivate HIV from latency. We have reported previously that CpG-induced virus reactivation is accompanied by activation of the cellular transcription factor NF-κB, a downstream target of the TLR9-signaling cascade.[5]

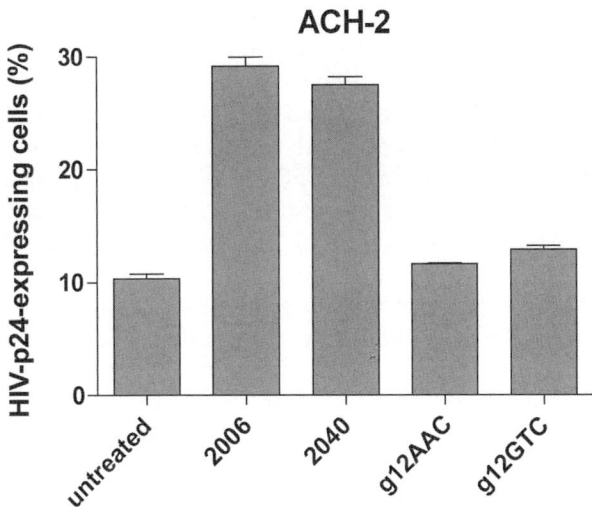

FIGURE 1. CpG ODNs reactivate HIV in latently infected ACH-2 T cells. ACH-2 cells were treated with medium alone (untreated) or with the CpG ODNs 2006, 2040, g12AAC, and g12GTC, respectively, for 24 h. Expression of HIV-p24 was detected by intracellular staining and flow cytometry. Values represent means ± SD from triplicate analyses.

Effects of CpG Sequences on Productive HIV Infection

CpG ODNs enhance productive HIV infection in A3.01 T lymphoblasts,[5] consistent with the stimulating effects observed in the latently infected subclone ACH-2. We have now compared the effects of CpG ODNs on productive HIV replication in different T cell lines. As depicted in FIGURE 2, the CpG ODN 2006 slightly enhanced HIV replication in A3.01 T cells, whereas the non-CpG sequence 2006 TpG had no effect on viral replication (FIG. 2 A). In contrast, ODN 2006 markedly inhibited HIV replication in MT4 T cells (FIG. 2 B). This inhibitory effect was even more pronounced with the non-CpG sequence 2006 TpG. Similarly, both CpG and non-CpG ODNs inhibited HIV replication in primary T cells isolated from healthy donors (FIG. 2 C). These results demonstrate that the inhibitory effect of the used PTO-ODNs is independent from the CpG motif but rather related to the chemical structure of the PTO-DNA backbone.

DISCUSSION

In the current article we show that CpG ODNs trigger TLR9 signaling in T cells that eventually leads to activation of NF-κB. In cells latently infected with HIV, TLR9 signaling induced reactivation of viral gene expression. Moreover, PTO-ODNs suppressed HIV replication in productively infected T cells. Virus inhibition was not dependent on the presence of the CpG motif as it was even more pronounced with ODNs without the CpG motif. We hypothesize that virus suppression was caused by the inhibitory effect of the PTO modification of the ODN backbone with the viral enzyme reverse transcriptase or with binding of HIV gp120 to the CD4 receptor, which has been reported before.[12,13] This is consistent with our observation that CpG PTO-ODNs activate latent infection but inhibit productive infection, suggesting that the inhibitory effects of PTO-ODNs are mediated by events prior to integration of the viral DNA into the host genome.

PTO-ODNs have been reported to suppress HIV replication in cultured tonsillar tissue.[14] Of note, inhibition was weaker when PTO-ODNs contained a CpG motif.[14] Also in our study the inhibitory effects were weaker with CpG ODNs compared to non-CpG sequences, probably as a result of TLR9-mediated activation of NF-κB and enhancement of HIV gene transcription.

We observed also a cell-specific difference in the effects of PTO-ODNs on HIV replication. Whereas PTO-ODNs showed significant inhibition of virus replication in MT4 T cells and primary blood lymphocytes, no inhibitory effects were seen with A3.01 T cells. In contrast, PTO-ODNs containing a CpG motif even enhanced virus replication in these cells. The reasons for these differences still remain to be elucidated but differences in cellular DNA-uptake could attribute for the different effects.

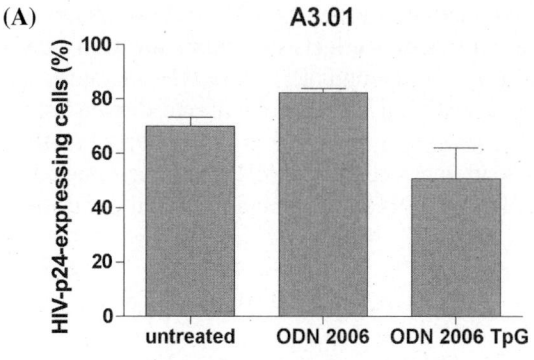
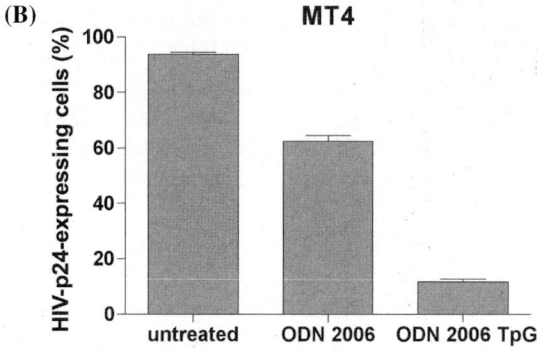
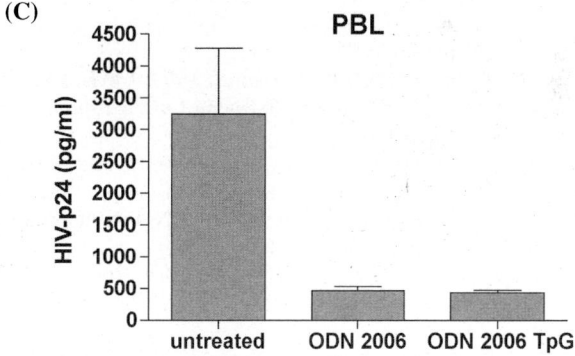

FIGURE 2. Effects of CpG PTO-ODNs on productive HIV infection. (**A**) A3.01 T cells were infected with HIV and treated with CpG (2006) and non-CpG (2006 TpG) ODNs for 7 days. Expression of HIV-p24 was detected by intracellular staining and flow cytometry. (**B**) MT4 T cells were infected with HIV and treated with CpG (2006) and non-CpG (2006 TpG) ODNs for 3 days. Expression of HIV-p24 was detected by intracellular staining and flow cytometry. (**C**) PBL were isolated from healthy donors, and stimulated with PHA/IL-2 for 8 days. Cells were infected with HIV and treated with CpG (2006) and non-CpG (2006 TpG) PTO-ODNs for 5 days. Supernatants were analyzed by HIV-p24-ELISA. (**A–C**) Values represent means ± SD from triplicate analyses.

Both the antiviral activity of the CpG PTO-ODNs observed on productive infection and the stimulating effects on latent infection may be interesting characteristics of these compounds for future HIV therapies. Latent infection is so far a therapeutically unsolved problem and directed stimulation of the latently infected reservoir in combination with effective antiviral treatment could eventually lead to eradication of HIV from the infected body. Thus, the potential of CpG PTO-ODNs in HIV therapy should be further explored.

ACKNOWLEDGMENTS

We thank Ingeborg Euler-Koenig and Simone Schimmer for excellent technical assistance. The following reagents were obtained through the AIDS Research and Reference Reagent Program, Division of AIDS, NIAID, NIH: A3.01 and ACH-2 from T. Folks; the HIV-1 p24 Hybridoma (183-H12-5C) from B. Chesebro and H. Chen. This work was supported by a grant from the Bundesministerium für Bildung, Wissenschaft, Forschung und Technologie, Germany (BMBF 01 KI 0211). C.S. and E.K. were supported by the "Competence Network HIV/AIDS, Germany"; U.D. was supported by a grant from the DFG (Di 714/6–1 and 6–2).

REFERENCES

1. IMLER, J.L. & J. A. HOFFMANN. 2000. Toll and toll-like proteins: an ancient family of receptors signaling infection. Rev. Immunogenet. **2:** 294–304.
2. HEMMI, H., O. TAKEUCHI, T. KAWAI, et al. 2000. A toll-like receptor recognizes bacterial DNA. Nature **408:** 740–745.
3. BENDIGS, S., U. SALZER, G.B. LIPFORD, et al. 1999. CpG-oligodeoxynucleotides co-stimulate primary T cells in the absence of antigen-presenting cells. Eur. J. Immunol. **29:** 1209–1218.
4. IHO, S., T. YAMAMOTO, T. TAKAHASHI & S. YAMAMOTO. 1999. Oligodeoxynucleotides containing palindrome sequences with internal 5′-CpG-3′ act directly on human NK and activated T cells to induce IFN-gamma production in vitro. J. Immunol. **163:** 3642–3652.
5. SCHELLER, C., A. ULLRICH, K. MCPHERSON, et al. 2004. CpG oligodeoxynucleotides activate HIV replication in latently infected human T cells. J. Biol. Chem. **279:** 21897–21902.
6. LEVY, J.A. 1994. HIV and the Pathogenesis of AIDS. ASM Press. Washington, DC.
7. GOMEZ, C. & T.J. HOPE. 2005. The ins and outs of HIV replication. Cell Microbiol. **7:** 621–626.
8. DUH, E.J., W.J. MAURY, T.M. FOLKS, et al. 1989. Tumor necrosis factor alpha activates human immunodeficiency virus type 1 through induction of nuclear factor binding to the NF-kappa B sites in the long terminal repeat. Proc. Natl. Acad. Sci. USA **86:** 5974–5978.
9. PERELSON, A.S., A.U. NEUMANN, M. MARKOWITZ, et al. 1996. HIV-1 dynamics in vivo: virion clearance rate, infected cell life-span, and viral generation time. Science **271:** 1582–1586.

10. CHUN, T.W. & A.S. FAUCI. 1999. Latent reservoirs of HIV: obstacles to the eradication of virus. Proc. Natl. Acad. Sci USA **96:** 10958–10961.
11. FOLKS, T.M., K.A. CLOUSE, J. JUSTEMENT, *et al.* 1989. Tumor necrosis factor alpha induces expression of human immunodeficiency virus in a chronically infected T-cell clone. Proc. Natl. Acad. Sci. USA **86:** 2365–2368.
12. STEIN, C.A., R. PAL, A.L. DEVICO, *et al.* 1991. Mode of action of 5′-linked cholesteryl phosphorothioate oligodeoxynucleotides in inhibiting syncytia formation and infection by HIV-1 and HIV-2 in vitro. Biochemistry **30:** 2439–2444.
13. MATSUKURA, M., K. SHINOZUKA, G. ZON, *et al.* 1987. Phosphorothioate analogs of oligodeoxynucleotides: inhibitors of replication and cytopathic effects of human immunodeficiency virus. Proc. Natl. Acad. Sci. USA **84:** 7706–7710.
14. SCHLAEPFER, E., A. AUDIGE, B. VON BEUST, *et al.* 2004. CpG oligodeoxynucleotides block human immunodeficiency virus type 1 replication in human lymphoid tissue infected ex vivo. J. Virol. **78:** 12344–12354.

Prostanoids with Cyclopentenone Structure as Tools for the Characterization of Electrophilic Lipid–Protein Interactomes

KONSTANTINOS STAMATAKIS AND DOLORES PÉREZ-SALA

Departamento de Estructura y Función de Proteínas, Centro de Investigaciones Biológicas, C.S.I.C., 28040 Madrid, Spain

ABSTRACT: Electrophilic eicosanoids arise from the free radical-induced peroxidation of arachidonic acid or its metabolites. These reactive species may play an important role in pathophysiological processes associated with inflammation and oxidative stress. Cyclopentenone prostaglandins (cyPG) and isoprostanes are reactive eicosanoids that can form covalent adducts with cysteine residues in proteins through Michael addition. In pharmacological studies, cyPG have shown potent protective effects in experimental models of inflammation and tissue injury, and they have been proposed to contribute to inflammatory resolution. An important mechanism for the anti-inflammatory effects of cyPG is the covalent modification of critical cysteine residues in proteins involved in the modulation of inflammation, such as transcription factors NF-κB and AP-1. In recent years, analogs of electrophilic prostanoids have been used in various approaches to identify biologically relevant protein targets for this modification. Prostanoids with cyclopentenone structure have been shown to target a defined subproteome that is beginning to be characterized. Structural studies suggest that diverse cyPG may modify distinct proteins selectively. Functional studies put forward a dual role for these compounds in the cellular response to inflammation or stress. Therefore, a detailed knowledge of targets of electrophilic eicosanoids and the functional consequences of their modification will contribute to the understanding of their mechanism of action and help assess whether these endogenous mediators can be exploited as the basis for the development of novel therapeutic strategies. In this article we discuss the recent advances in this rapidly growing field.

KEYWORDS: cyclopentenone prostaglandins; proteomics; posttranslational modification; inflammation; cysteine

Address for correspondence: Dr. Dolores Pérez-Sala, Centro de Investigaciones Biológicas (C.S.I.C.), Ramiro de Maeztu, 9, 28040 Madrid, Spain. Voice: 34-91-8373112; ext.: 4212; fax: 34-91-5360432.
 e-mail: dperezsala@cib.csic.es

INTRODUCTION

The formation of reactive species in living cells is a process cosubstantial with normal cell activity and it can be increased under pathophysiological conditions. These reactive intermediates are required for normal cell metabolism and gene expression and constitute important players in cell signaling. However, they can also damage cellular components, like DNA and proteins.[1] Reactive intermediates can be derived from the use of oxygen by living cells, like superoxide anion or hydrogen peroxide, nitric oxide (NO), and peroxynitrite. These species can act through a wide array of mechanisms including their reaction with other cellular components, such as amino acids, lipids, or sugars, which gives rise to various electrophiles with diverse biological activities. The electrophiles generated can interact with chemical groups of nucleophilic character, such as the thiol groups present in cysteine residues of proteins or in the tripeptide glutathione (GSH). The modification of cysteine residues by these diverse species completely alters their physicochemical properties and has important functional consequences. For these reasons much attention has been devoted in recent years to the identification and characterization of the structural and functional consequences of the modification of cysteine residues by diverse reactive species under physiological or pathophysiological conditions. The group of proteins involved in the regulation of cellular functions through the modification of their cysteine residues could be termed *sulfoproteome*. Methods for the detection of specific cysteine modifications by thiolation or nitrosylation have been developed.[2,3] These methods have allowed the identification of proteins which are targets for these modifications and in some cases, the identification of novel signaling pathways. In addition to its role in biological pathways, the modification of cysteine residues has been considered of great importance for the design of novel antiviral and anticancer therapies.[4] During the past 5 years several reports have highlighted the importance of the modification of cysteine residues by electrophilic eicosanoids. This type of modification has remarkable pharmacological consequences and its potential physiological relevance is beginning to be unveiled. The available evidence indicates that the direct modification of cysteine residues contributes to the effects of electrophilic eicosanoids on the onset and the resolution of inflammation, and it could provide connections between inflammatory situations and increased proliferation. An understanding of these processes requires the identification and characterization of the electrophilic eicosanoid interactome.

EICOSANOIDS AND THEIR IMPORTANCE IN INFLAMMATION

Eicosanoids are lipidic mediators derived from the essential fatty acid arachidonic acid that are involved in the regulation of multiple cellular processes, ranging from the regulation of vascular tone, to the modulation of gene expression and cell proliferation. Arachidonic acid is mainly found esterified to membrane phospholipids. A variety of stimuli, including inflammatory agents or tissue injury, activate phospholipases, mainly phospholipase A_2, that release arachidonic acid from membranes. Arachidonic acid can then be transformed through enzymatic and nonenzymatic reactions into a wide variety of biologically active compounds. Cyclooxygenase (COX) enzymes, catalyze the limiting step in the synthesis of prostaglandins (PG), that is, the formation of PGH_2. This compound is then transformed by various PG synthases into other PG with specific functions, and thromboxane A_2 (reviewed in Ref. 5). Arachidonic acid can be transformed into leukotrienes and lipoxins through the action of lipoxygenase (LOX) enzymes.[6] In addition, arachidonic acid and its derivatives can suffer various nonenzymatic transformations, like free radical-induced peroxidation, which gives rise to a series of compounds that are known as isoprostanes.[7] It is interesting to note that significant amounts of PG can also be generated through nonenzymatic mechanisms.[8] PG, in turn, can suffer nonenzymatic dehydration to yield PG with cyclopentenone structure (cyPG).[9]

Inflammation is a complex response characterized by multiple interactions between soluble and cellular factors that tend to defend the organism against traumatic, infectious, or autoimmune injury. At the site of tissue injury, a complex array of substances can be released, including neuropeptides, intracellular materials, or microbial antigens, which activate resident cells, such as mast cells and macrophages. These cells in turn release inflammatory mediators (histamine, cytokines, NO, PG, etc.) that induce vasodilatation and edema and contribute to attract neutrophils, which produce large amounts of proteases and reactive oxygen species, and lymphocytes that contribute to activate macrophages. During inflammation there are marked alterations in gene expression. Proinflammatory stimuli induce the expression and activation of proteins involved in the synthesis of inflammatory mediators. Inducible NO synthase and COX-2 play a key role in the increased synthesis of NO and PG during inflammation. Both types of mediators can amplify the response to proinflammatory stimuli at the early stages of the inflammatory response.[10,11] Thus, there are various self-amplifying mechanisms, which operate during the onset of the inflammatory response and promote a controlled destruction of tissue in order to remove the damaging agent. However, for this situation to lead to healing, the mechanisms for tissue repair must be in place. The mechanisms of resolution of inflammation are active and tightly regulated processes that favor the safe disposal of inflammatory cells and the restoration of the inflamed tissue to its physiological situation.[12] During the resolution phase of inflammation an important event takes place, referred to as "class switching," that affects the nature of the inflammatory mediators. This implies a reduction in proinflammatory and an increase in anti-inflammatory mediators that contribute to limit the amplitude of the response. In addition, several well-known inflammatory mediators can display dual effects, that is, they exert mainly proinflammatory actions during the early phases of the inflammatory response and inhibitory actions at late stages. An example of this is NO, for which we described recently a biphasic effect on the response of mesangial cells to cytokine stimulation.[13] A dual nature has also been reported for other mediators like tumor necrosis factor-α (TNF-α), interferon-γ (IFN-γ), transforming growth factor-β (TGF-β), and some eicosanoids, such as PGE_2.[14]

Eicosanoids play a key role both in the onset and in the resolution of inflammation. PG and leukotrienes display potent proinflammatory effects, including vasodilatation, edema, fever, activation of neutrophil chemotaxis, increased expression of adhesion molecules, etc.[5] Isoprostanes and some of their derivatives are considered an index of oxidative stress and lipid peroxidation.[15] Some isoprostane-derived products can contribute to oxidative damage through the modification of amino acid residues in proteins that may result in protein aggregation.[16] In contrast, cyPG and lipoxins appear to display mostly anti-inflammatory effects, including the inhibition of proinflammatory cytokine synthesis and neutrophil chemotaxis and transmigration, and the induction of neutrophil clearance through phagocytosis by monocytes-macrophages.[17] Eicosanoid production plays an important role in the process of "class switching." It has been reported that cyPG are generated in increased amounts at the late stages of inflammation.[18] In addition, lipoxins are generated through the transcellular metabolism of arachidonic acid, that is, arachidonic acid or its products generated by a class of cells are taken up by a different class, which can convert them in lipoxins through the action of 12- or 15-LOX.[12,17] This represents a mechanism by which cell–cell interactions favor a transition in the profile of the eicosanoids produced, from a proinflammatory to an anti-inflammatory nature, and this has important implications for the resolution of inflammation.

CYCLOPENTENONE PROSTAGLANDINS

cyPG are electrophilic eicosanoids that have elicited great interest recently. Their common feature is the presence of an unsaturated carbonyl group in the cyclopentane ring (cyclopentenone, see FIG. 1). CyPG are generated by spontaneous dehydration of other PG. Thus, PGA_2 arises by dehydration of PGE_2, while cyPG of the J series, including PGJ_2 and 15-deoxy-$\Delta^{12,14}$-PGJ_2 (15d-PGJ_2) arise from PGD_2. It has been known for over 20 years that cyPG exert antiviral and antiproliferative effects. However, a renewed interest in this area rose when 15d-PGJ_2 was identified as an endogenous ligand of the transcription factors of the PPAR (peroxisome proliferator activated receptor family),[19] which play an important role in the regulation of lipid metabolism and immune response. In fact, activation of PPAR has been reported to mediate some of the beneficial effects of

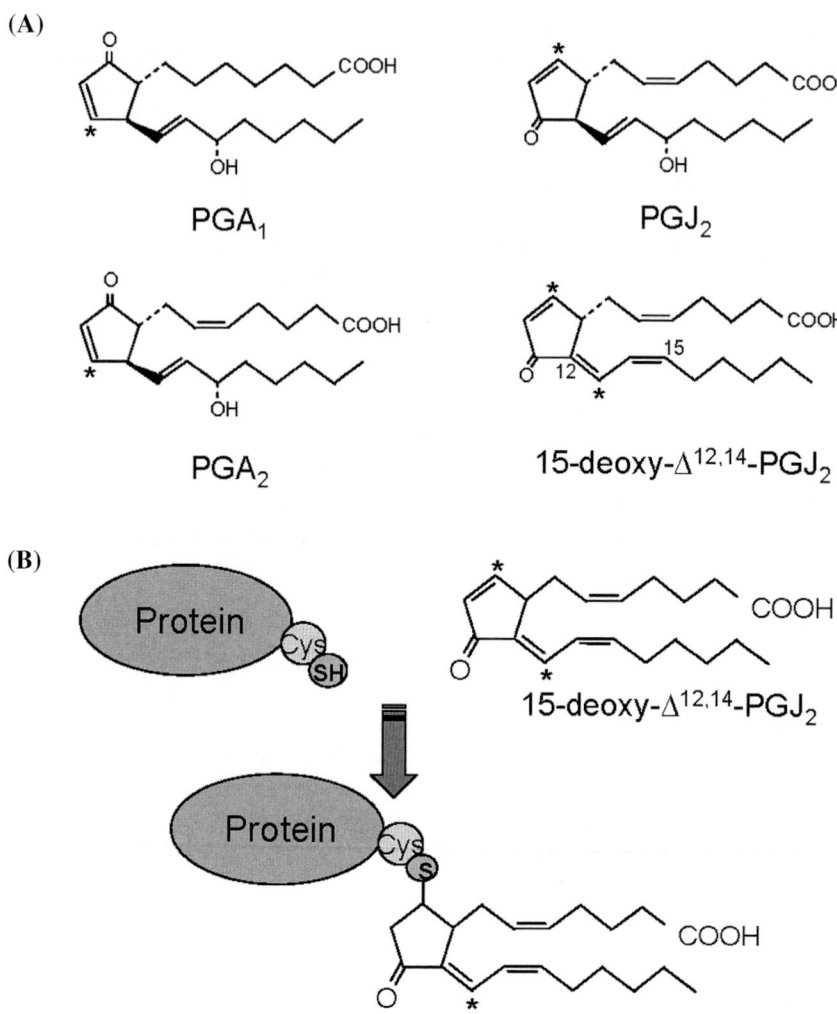

FIGURE 1. Structure of cyPG and formation of Michael adducts with proteins. (**A**) Structures of several cyPG of the A series (left) and of the J series (right). Electrophilic carbons are marked by asterisks. (**B**) Schematic representation of the formation of a Michael adduct between the electrophilic carbon in the cyclopentenone ring of 15d-PGJ$_2$ (carbon 9) and the sulphydryl group of a protein cysteine residue.

15d-PGJ$_2$ in experimental models of inflammation. However, it has been found that many of the anti-inflammatory effects of cyPG can occur by mechanisms independent from the activation of PPAR.[9] These effects include the inhibition of transcription factor NF-κB and of the induction of several proinflammatory genes, including iNOS, COX-2, or ICAM-1.[20,21] Some of the effects of cyPG could be observed in cell-free systems[22] and required the presence of critical cysteine residues in NF-κB or in IKK.[23,24] For these reasons it was postulated that these inhibitory effects could be mediated by the direct interaction of cyPG with cysteine residues by Michael addition. In this context, our group provided the first direct evidence of the interaction between cyPG and a protein target, the p50 subunit of transcription factor NF-κB, which we demonstrated by the use of a biotinylated analog of 15d-PGJ$_2$ combined with mass spectrometry analysis.[24] Several research groups have employed similar strategies to identify other targets of cyPG or of compounds with similar reactivity. Some of the identified targets play important roles in the modulation of inflammatory and proliferative processes. Anti-inflammatory cyPG may interact with the NF-κB activation pathway at multiple levels (FIG. 2), including the inhibition of the activity of the proteasome,[25] the modification of critical cysteines in IKKα and β,[22,23] and also modification of cysteines present in the DNA binding

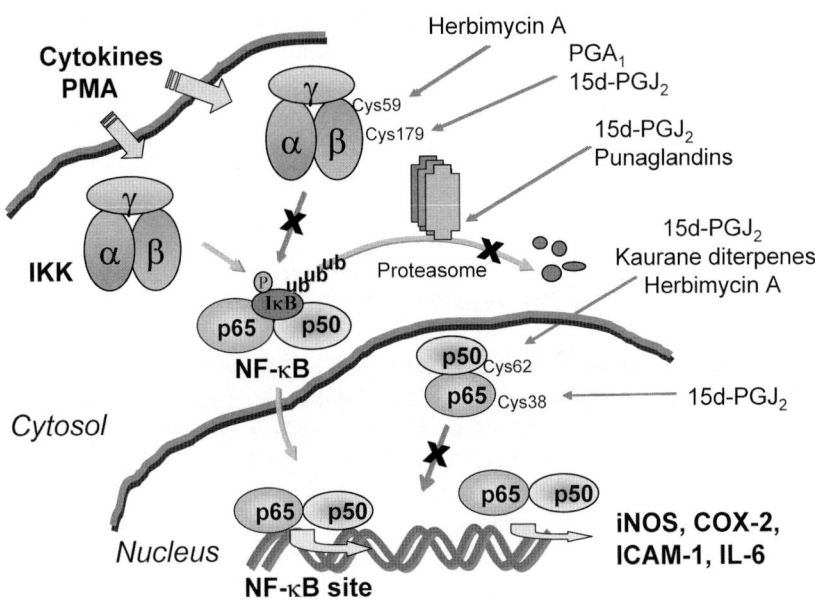

FIGURE 2. Schematic view of the NF-κB activation pathway showing the sites of interaction of several Michael receptors. The transcription factor NF-κB plays a key role in the inflammatory response. In a very schematic view, the pathway leading to NF-κB activation involves the stimulation of a complex with kinase activity (IKK), which phosphorylates the inhibitory subunit of NF-κB, IκB. This triggers IκB ubiquitination and degradation by the proteasome, which promotes the nuclear accumulation of the active dimer, represented by p65/p50, which activates the transcription of target genes. The NF-κB activation pathway has been the subject of numerous studies in search for anti-inflammatory molecules. Shown is the site of action of several molecules that have been reported to interfere with NF-κB activation through the formation of covalent adducts with critical proteins of this pathway by Michael addition.

domains of the subunits of this transcription factor (p65 and p50).[24,26] As a result of these interactions, cyPG can inhibit the phosphorylation and/or the degradation of the inhibitory subunit IκB, and block the translocation or the nuclear accumulation of the active factor. In addition, they may directly inhibit the binding of NF-κB proteins to DNA. The predominant effect would be dependent on the cell type. 15d-PGJ$_2$ can also target the cysteine residue present in the DNA binding domain of c-Jun, and thus, inhibit the DNA binding activity of AP-1.[27] In addition, some cyPG can bind to and inactivate thioredoxin and thioredoxin reductase,[28,29] which are involved in the control of cellular redox status and play a role in maintaining the cysteines of NF-κB and AP-1 in their reduced state, which is required for function. Therefore, the modification of these redox-controlling proteins could also have consequences for the modulation of the inflammatory response. Several electrophilic lipids, including 15d-PGJ$_2$ can also activate the transcription factor Nrf2, through the modification and inhibition of the function of Keap1, a cytoplasmic protein that acts as an inhibitor of Nrf2. Through this mechanism cyPG can induce the expression of hemooxygenase and of antioxidant genes.[30,31] Other identified targets of cyPG include H-Ras and LKB1/STK11 (serine/threonine kinase 11), proteins involved in the control of cellular proliferation.[32,33] For these reasons, cyPG have been proposed to participate in the interactions linking chronic inflammation and increased cell proliferation.

cyPG: Are they Physiologically or Pharmacologically Relevant?

The characterization of the biological effects of cyPG has evolved in parallel with the exploration of their potential therapeutic use in various experimental models of inflammation and tissue injury. 15d-PGJ$_2$ has been reported to display protective effects against renal ischemia-reperfusion injury,[34] adjuvant-induced arthritis,[35] experimental autoimmune encephalomyelitis,[36] restenosis after angioplasty,[37] etc. In some of these models, the beneficial effects have been found to correlate with the inhibition of the activity of transcription factors NF-κB and AP-1, and with the ability of cyPG to covalently modify proteins. Interestingly, animals deficient in PGD$_2$ synthase show exacerbated inflammation, thus pointing toward a protective role of PGD$_2$ metabolites in inflammation.[38] However, the physiological relevance of the effects of cyPG is not established. This is due to the fact that the concentrations of cyPG required to elicit anti-inflammatory effects are generally higher than those measured *in vivo*. This discrepancy could be due to the high reactivity of cyPG that readily form adducts with thiol groups, making the detection of their total levels in biological samples inaccurate. It also should be taken into account that the protective effects of cyPG are not general. In some tissues, cyPG can have toxic effects,[39] or induce unwanted reactions, such as allergy or cell proliferation. For these reasons, the potential therapeutic use of cyPG requires the identification of their targets and the assessment of their selectivity.

CYCLOPENTENONE ISOPROSTANES

In contrast to cyPG, the available information regarding the generation of isoprostanes in biological systems is more comprehensive. Roberts and Morrow first reported that free radical-induced peroxidation of arachidonic acid could lead to the nonenzymatic formation of bioactive PG F_2-like compounds, F_2-isoprostanes, *in vivo*. Later, several compounds were identified that could be formed by the isoprostane pathway (reviewed in Ref. 15). It is known that isoprostane levels are greatly increased in various tissues under conditions of oxidative stress or chronic inflammation, including Alzheimer, atherosclerosis, and rheumatoid arthritis. In a way analogous to that described for cyPG, nonenzymatic dehydration of isoprostanes generates compounds with cyclopentenone structure, which are highly reactive and have been found in tissues at near micromolar concentrations.[40] A_2 and J_2 isoprostanes are isomers of the corresponding cyPG. In a recent work,[41] A and J series isoprostanes have been found to mimic the effects of cyPG on the inflammatory response in macrophages. These compounds can inhibit NF-κB, induce apoptosis of macrophages, and in the case of the J series, activate PPARγ. These observations suggest that cyclopentenone isoprostanes could act as negative modulators of inflammation. In a way analogous to 15d-PGJ_2, the isoprostane 15-A_2-isoprostane has been reported to modify Keap1 and activate Nrf2.[30] The availability of isoprostanes through chemical synthesis will grant a spur in the research into their biological effects.

LIPIDIC MEDIATORS AND OTHER COMPOUNDS WITH UNSATURATED CARBONYL GROUPS

There are other lipidic mediators, that can arise under situations of inflammation or oxidative stress, which possess unsaturated carbonyl groups and may form covalent adducts with thiol groups. Among these compounds are 5- and 15-eicosatetraenoid acid, 9- and 13-oxo-octadienoic acid,[42] several 15-keto-PG, epoxycyclopentenone phospholipids,[43,44] and neuroprostanes.[39] These lipids act through various mechanisms, in some cases not completely understood. Also, the well-studied α,β-unsaturated aldehydes 4-hydroxy-nonenal and acrolein can be formed as products of lipid peroxidation and can react both with thiol groups by Michael addition and with amine groups forming Schiff bases. Several excellent reviews[45,46] have dealt with the biological effects of these reactive aldehydes; therefore, this subject will not be considered here in more detail. In several studies, 4-hydroxy-nonenal has been shown to modify protein targets common to cyPG.[29,30,33] The predominance of protective or toxic effects of these electrophilic compounds appears to be concentration-dependent and cell-type specific. It has been proposed that moderate levels of oxidized lipids elicit an adaptative response to stress that protects cells from subsequent oxidative insults, while high concentrations of electrophilic lipids would induce apoptosis.[47] In addition to endogenous compounds, many natural products, like active principles from medicinal herbs, possess unsaturated carbonyl groups in their structure, which may show reactivity similar to that of cyPG. For instance, parthenolide or the kaurane diterpene kamebakaurin,[48,49] inhibit NF-κB activation through mechanisms analogous to those of cyPG (FIG. 2). In addition, herbimycin A and punaglandins have been reported to inhibit NF-κB through the formation of Michael adducts with critical cysteine residues of components of the NF-κB activation pathway.[50–52] For these reasons, certain natural or synthetic cyclopentenones are being studied in order to assess their potential use as antiproliferative or anti-inflammatory agents.

DERIVATIVES OF cyPG AS TOOLS FOR THE STUDY OF THE INTERACTIONS BETWEEN ELECTROPHILIC EICOSANOIDS AND CELLULAR PROTEINS

Eicosanoids with cyclopentenone structure have been used in various approaches to understand the mechanism of action of electrophilic lipids. The synthesis of cyPG allowed to explore the interactions of these compounds with cellular structures and with proteins.[53] The interaction of cyPG with cells was explored in a series of important studies by Narumiya and co-workers.[54–56] Using radioactively labeled cyPG, these authors showed that these compounds could enter intact cells in significant amounts and that they could be found both in cytosolic and nuclear compartments, with preferential accumulation in cell nuclei, where cyPG were bound to an insoluble fraction. They found that the retained radioactivity could be recovered by treatment with alkali and by proteolytic digestion of the nuclear fraction, which suggested the formation of Michael adducts between cyPG and nuclear proteins.[56] In addition, these authors showed that the extent of the accumulation was dependent on the structure of the cyPG. In the case of PGA_2 both the binding to cellular structures and the inhibitory effect of cell growth could be reversed by extensive washing. In contrast, nuclear accumulation of Δ^{12}-PGJ_2 was only partially reversed by cell washing and the growth inhibitory effect was irreversible.[56] Therefore, this constitutes one of the first evidences of a differential effect of cyPG with diverse structure.

The use of a biotinylated analog of PGA_2 provided one of the first evidences of the binding of cyPG to cellular polypeptides by means of SDS-PAGE and Western blot with streptavidin-alkaline phosphatase. This allowed the visualization of several biotin-positive polypeptides whose labeling correlated with the inhibition of cellular proliferation exerted by PGA_2 in K562 erythroleukemia cells.[57] We used a modification of this procedure to provide the first identification and visualization of a protein that could be modified through the stable binding of a biotinylated analog of 15d-PGJ$_2$.[24] Since then, biotinylated analogs of cyPG have been used in several approaches to get insight into the identity and consequences of protein modification by cyPG. Biotinylated analogs of cyPG enter cells and form adducts with cellular proteins that are resistant to electrophoresis under denaturing conditions. Recently, 15d-PGJ$_2$ modified by incorporation of a fluorescent moiety has been used to visualize the cellular sites of binding of this cyPG.[58] It should be noted, however, that, although modified cyPG possess the reactive cyclopentenone moiety, their properties may not be identical to those of endogenous cyPG due to the presence of the incorporated label. For instance, in a recent work, several lipids containing α,β-unsaturated ketone moieties have been reported to bind to PPARγ covalently. In this model, 15d-PGJ$_2$ has been proposed to occupy the ligand binding site in such a way that the carboxyl group of the PG establishes interactions with the PPAR backbone (formation of a hydrogen bond with Tyr-473 in helix 12 of PPARγ), that are important for binding.[42] This may provide an explanation for the observation that 15d-PGJ$_2$ biotinylated at the carboxyl group did not detectably bind to PPARγ in cells, as explored with a pull-down assay, whereas it still bound to c-Jun or c-Fos.[27] Taken together these evidences suggest that the structural requirements for covalent binding of 15d-PGJ$_2$ to the PPARγ ligand binding domain are different from those involved in PPARγ-independent actions. We recently described that the presence of the endocyclic double bond in 15d-PGJ$_2$ was required for the efficient binding of this cyPG to c-Jun and to various proteins present in a mesangial cell lysate, in an *in vitro* assay,[21] since an analog of 15d-PGJ$_2$ lacking this double bond, namely 9,10-dihydro-15d-PGJ$_2$ failed to mimic the effects of 15d-PGJ$_2$. This indicated that probably, carbon 9 of 15d-PGJ$_2$ was involved in Michael addition with the thiol groups of the proteins studied. However, both 15d-PGJ$_2$ and 9,10-dihydro-15d-PGJ$_2$ bind to and activate PPARγ with similar potency,[21] which indicates that the carbon important for the reaction with the sulfur atom of the cysteine residue in PPARγ is the carbon at position 13, as suggested also by molecular modeling.[42]

Use of Biotinylated Derivatives of Eicosanoids to Confirm the Interaction of cyPG with Potential Targets

The observations described above indicate that electrophilic eicosanoids can potentially bind to additional or different protein targets than their modified counterparts. Therefore, care should be exercised when using these compounds because it would be important to confirm that natural and modified eicosanoids display the same mechanism of action for a particular target. In recent years, a number of reports have used biotinylated analogs of cyPG to assess the binding of these compounds to defined proteins in intact cells. This approach has been used to confirm the binding of cyclopentenone moiety containing compounds to the p50 subunit of NF-κB,[24] the components of AP-1, c-Jun, and c-Fos,[27] the protein thioredoxin,[28] thioredoxin reductase,[29] Keap1,[30] or H-Ras,[32] among others. In some cases, it has been confirmed that the biotinylated cyPG mimics the biological effects of the nonmodified compound.[24,30,59] In addition, some studies infer the modification of the target cysteine residues by the natural electrophilic lipids *in vitro* or in intact cells, by evidencing a reduction in the ability of general cysteine reagents, such as biotinylated iodoacetamide or fluorescent maleimide, to modify the electrophile-treated proteins.[21,30,42] However, it should be taken into account that treatment of cells with cyPG may induce potent oxidative stress,[60–62] and this may be the cause for the occurrence of other modifications of cysteine residues, either reversible, such as thiolation,[63] or irreversible, like the formation of sulfoxides, that could result in decreased accessibility of the cysteine residues for modification by thiol reagents. In addition, we have observed that treatment of human leukemia cells with 15d-PGJ$_2$ induces the formation of intermolecular and intramolecular disulphide bonds (Sánchez-Gómez *et al.*, unpublished observations). Therefore, ideally, the modification of cellular proteins by electrophilic eicosanoids should be confirmed by the direct observation of adducts by methods like mass spectrometry.

Proteomic Studies

Biotinylated analogs of cyPG have been used in several studies to visualize and/or identify protein targets for the modification by cyPG (FIG. 3). We recently observed the binding of biotinylated 15d-PGJ$_2$ to approximately 50 target proteins in mesangial cells by 2D electrophoresis and Western blot with

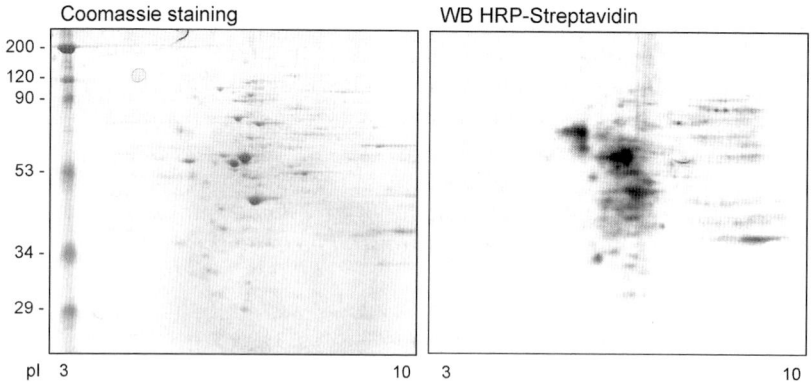

FIGURE 3. Typical pattern of biotinylated 15d-PGJ$_2$-modified proteins in rat mesangial cells. Mesangial cells were incubated with 5 μM biotinylated 15d-PGJ$_2$ for 2 h. Cell lysates were analyzed by 2D electrophoresis and either Coomassie Gold staining or Western blot with horseradish peroxidase streptavidin and ECL.

horseradish peroxidase-conjugated streptavidin.[21] Levonen and co-workers,[30] found nearly 30 proteins modified by biotinylated 15d-PGJ$_2$ in human embryonic kidney 293 cells. For these studies to be mechanistically relevant it is important that the biotinylated analogs of electrophilic eicosanoids mimic the biological effects of the nonmodified compounds. In a recent study we have observed that biotinylated 15d-PGJ$_2$ displays potent anti-inflammatory effects and induces heat shock in mesangial cells, thus mimicking the effects of 15d-PGJ$_2$.[59] In a previous study we showed that the ability of 15d-PGJ$_2$ to modify cellular proteins was required for the anti-inflammatory effects of this PG. Therefore, we reasoned that the biotinylated compound could be a valid tool to identify novel targets for cyPG, whose modification could play a role in the anti-inflammatory effects of these compounds. We have recently reported the identification of some of the modified proteins, which include proteins involved in the heat shock response and in the organization of the cytoskeleton, such as vimentin and tubulin.[59] A recent study using also biotinylated 15d-PGJ$_2$ has identified 17 potential targets of the electrophile responsive proteome in isolated liver mitochondria, thus suggesting that electrophilic lipid oxidation products can target a subproteome in the mitochondria with consequences for cytochrome c release and induction of apoptosis.[64] The elucidation of the role of these targets in the biological effects of electrophilic eicosanoids will grant further developments in this field.

Study of the Selectivity of the Modification: Interactions with GSH

Some cyPG have shown beneficial effects in experimental models of inflammation and tissue injury. Therefore, an important aspect of the study of these compounds is the assessment of their therapeutic potential, which requires a deeper knowledge of their mechanism of action and the identification of novel targets for therapeutic intervention. In this context it is important to ascertain the selectivity of protein modification by cyPG. The studies with biotinylated analogs of cyPG described above have shown that the modification of proteins by these compounds is a selective process, which, in spite of the broad reactivity of cyPG, does not occur randomly but affects a defined subset of cellular proteins. These targets are different from the proteins modified by general cysteine reagents such as biotinylated iodoacetamide.[21] By using biotinylated 15d-PGJ$_2$ and PGA$_1$ we have recently observed that these two cyPG bind to some common targets but also to some selective targets in NIH-3T3 fibroblasts.[65] Therefore, selectivity of protein modification by cyPG depends in part on the structure of the PG. In addition, the selectivity of the modification can be observed in individual proteins. Both p50 and H-Ras possess several accessible cysteine residues, however, only one cysteine residue is modified by 15d-PGJ$_2$ in either protein.[24,32] In H-Ras, cysteine 184, out of the six cysteine residues of the protein, has been found to be the main site for 15d-PGJ$_2$

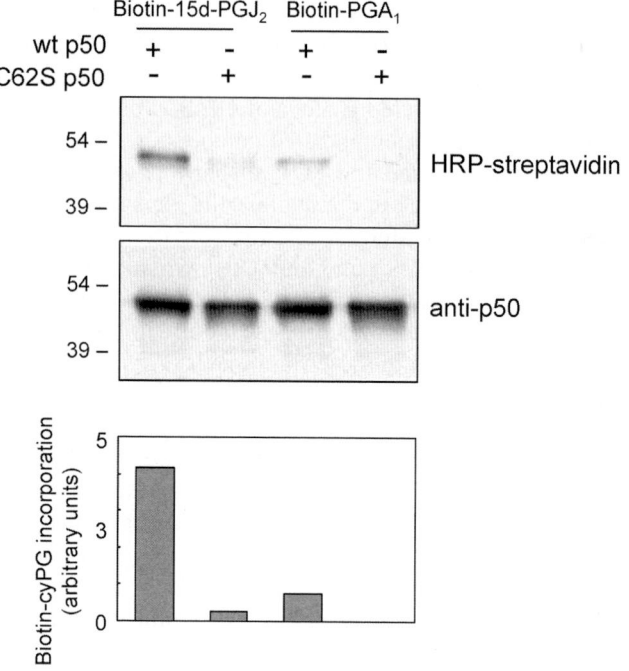

FIGURE 4. Selectivity of the modification of the cysteine residues in p50 by biotinylated 15d-PGJ$_2$ and PGA$_1$. Wild type (wt) or cysteine 62 to serine mutant p50 (C62S) at 0.5 µM was incubated with 1 µM biotinylated 15d-PGJ$_2$ or PGA$_1$ for 1 h at r.t. Biotin incorporation was monitored by Western blot with horseradish peroxidase-conjugated streptavidin and p50 levels were assessed by Western blot with anti-p50 antibody. The intensity of the biotin signal, normalized by the signal given by the anti-p50 antibody, expressed in arbitrary units, is shown in the lower panel. Results are representative of four assays.

binding. In addition, the binding of biotinylated analogs of 15d-PGJ$_2$ and PGA$_1$ to a p50 DNA binding domain construct containing seven cysteine residues is dependent on the presence of cysteine 62, indicating that this is the main site for cyPG addition (FIG. 4). Identification of the selective targets for modification by cyPG could contribute to the elucidation of the mechanism of action of endogenous eicosanoids, the structure of which is very diverse. Also, this knowledge may aid in the identification of compounds that selectively target certain proteins of therapeutic interest.

In addition to PG structural determinants, several factors could influence the selectivity of protein modification by cyPG. These include factors related to the structure of the protein, either through the establishment of additional interactions with the eicosanoid,[42] or due to steric factors. In addition, cysteines with low pKa would be more susceptible to modification, and an acidic pH in the vicinity of the adduct would contribute to its stability. It is interesting to note that interactions of cyPG with proteins can also be influenced by soluble thiols.[66] It has been reported that the formation of GSH-CyPG adducts is a reversible process while cyPG addition to proteins appears to be virtually irreversible under physiological conditions. It is well known that GSH is an important factor in the biological effects of cyPG. Depletion of GSH has been associated in several studies with a potentiation of the effects of cyPG.[67,68] In contrast, supplementation of cells with N-acetyl-cysteine, as a thiol-protective agent, reduces the effects of cyPG.[69] Therefore, there seems to exist an inverse correlation between intracellular levels of GSH and the intensity of the biological effects of cyPG. This can be due to several mechanisms, including the ability of GSH to form adducts with cyPG. This can occur both enzymatically, through the action of glutathione-S-transferase (GST) enzymes, as well as nonenzymatically. GSH-cyPG adducts are substrates for the multidrug transporters, which play an important role in the detoxification of cyPG. We have observed recently that agents that modulate intracellular GSH levels differentially affect the intensity of the binding of biotinylated cyPG, 15d-PGJ$_2$ and PGA$_1$ to cellular proteins.[65] Biotinylated 15d-PGJ$_2$ was more effective than PGA$_1$ at binding cellular proteins. Depletion of GSH with buthionine sulfoximine increased protein modification. Conversely, supplementation with N-acetyl-cysteine reduced the extent of cyPG addition. These changes were more marked in the case of biotinylated PGA$_1$. In addition, the modulation of the binding of cyPG by GSH was not uniform for all polypeptides, thus leading to alterations in the pattern of labeled proteins. These results suggest that GSH levels may contribute not only to the intensity but also to the selectivity of protein modification by biotinylated cyPG. These effects may be related to the differential reactivity of two cyPG toward soluble thiols. The 15d-PGJ$_2$ presents a dienone structure, while PGA$_1$ is a single enone. Adducts of single enone cyPG and GSH have been reported to be more stable than dienone cyPG-GSH adducts.[66] This may be the reason why GSH levels affect the binding of biotinylated-PGA$_1$ to proteins more intensely. However, as outlined above, the effect of GSH does not affect uniformly all proteins. We have addressed this issue by exploring the effect of GSH on the incorporation of biotinylated cyPG to several purified proteins *in vitro*. We have observed that GSH reduces the modification of GST, but it can improve the modification of p50 (Gayarre *et al.*, unpublished observations). These opposite effects could be due to several mechanisms. GSH or GSH conjugates can alter the conformational status of GST modifying the orientation of critical cysteine residues.[70] This could preclude binding of cyPG. On the other hand, our results indicate that GSH may favor the incorporation of cyPG into certain proteins by contributing to the maintenance of critical cysteine residues in a reduced state, and therefore, accessible for modification. Extensive protein oxidation due to oxidative stress or GSH depletion would then lead to an inhibition of the incorporation of cyPG and probably to an attenuation of cyPG-specific effects. It is interesting to consider that under situations of oxidative stress or increased

NO generation, the GSSG or GSNO formed can act as thiolating agents and induce reversible protein glutathiolation. In several proteins, for instance, in c-Jun and p50, the cysteine residues that are targets for modification by cyPG are also targets for glutathiolation. Therefore, these two types of agents, oxidant species derived from GSH and cyPG, may compete for binding to the same residue(s). In fact, glutathiolation under these conditions, could act as a defense mechanism protecting critical cysteine residues in signaling proteins from the irreversible addition of cyPG, as it has been proposed previously in the context of oxidative or nitrosative stress.[71]

The effects of GSH levels on the biological actions of electrophilic eicosanoids could be mediated by an additional mechanism. It has been recently reported that certain synthetic cysteine adducts of cyclopentenones can display potent biological activity.[72] In addition, it is well known that GSH adducts of several endogenous oxyeicosanoids retain biological activity or can even display novel actions.[73] Therefore, the formation of cyPG-GSH adducts could give rise to the appearance of biochemical species with effects and mechanisms of action different from those of the parent cyPG.

Subcellular Localization of Cellular Targets for Electrophilic Eicosanoids

The early studies with radioactive cyPG already detected important differences in the subcellular distribution of cyPG depending on their structure, with single enone cyPG being present as soluble species in a higher proportion than dienone cyPG.[56] The subcellular fate of cyPG has also been explored by means of biotinylated or fluorescent analogs of cyPG. These studies have shown cell type-dependent differences in the subcellular distribution of cyPG. In HeLa cells incubated with biotinylated 15d-PGJ$_2$, a predominantly nuclear staining has been observed,[74] while in mesangial cells both nuclear and cytosolic compartments show biotin staining.[21] These observations may correlate with some of the biological effects of cyPG described in both cell types. 15d-PGJ$_2$ inhibits NF-κB activation in both systems. However, the mechanism of inhibition appears to involve mainly nuclear events in HeLa cells,[24,26] and both cytoplasmic and nuclear processes in macrophages[26] and in mesangial cells.[75] Interestingly, GSH levels and/or GST activity have been proposed to play an important role in the predominance of either cytosolic or nuclear effects of cyPG.[26,53] It is tempting to speculate that these factors could also influence the cellular compartment in which cyPG would preferentially bind to their targets. A recent study using a fluorescent derivative of 15d-PGJ$_2$ indicates that this compound localizes significantly to the mitochondria in endothelial cells.[58] It has been proposed that interaction with mitochondrial proteins contributes to the oxidative stress induced by this compound.[58]

Induction of Protein Cross-Links

One of the mechanisms by which electrophilic lipids can damage proteins is by inducing protein cross-links. This appears particularly relevant in the case of electrophilic aldehydes, such as 4-hydroxy-nonenal, isoketals, and levuglandins.[16] The consequence of this modification can be the accumulation of the modified proteins that cannot be degraded by the proteasome, and the formation of anomalous aggregates that may damage cells and contribute to the pathophysiology of several processes, including inflammatory diseases and neurodegeneration.[16] It is interesting to note that certain cyPG possess a dienone structure, and therefore they could form Michael adducts with two cysteine residues. In fact, 9-deoxy-Δ^9,Δ^{12}(E)-PGD$_2$, which possesses two electrophilic carbons, has been described to form a conjugate with two molecules of GSH.[76] In this context, we have recently described that 15d-PGJ$_2$, a dienone, can induce cross-linking of the protein c-Jun,[27] thus providing the first example of protein cross-linking induced by cyPG. This effect could also be observed in mutant c-Jun constructs which only possess one cysteine residue, like the C320S c-Jun mutant, which only presents the cysteine residue located in the DNA binding domain, cysteine 269 (FIG. 5). The dimer formed by treatment with 15d-PGJ$_2$ was resistant to electrophoresis under denaturing and reducing conditions. In contrast, treatment with H_2O_2 induced the dimerization of C320S c-Jun, which was reversed by incubation with dithiothreitol (DTT). In addition, in assays performed with biotinylated 15d-PGJ$_2$ we could observe the presence of biotin signal in the c-Jun dimer. Taken together, these observations are strongly indicative of the participation of 15d-PGJ$_2$ as the cross-linking bridge in the oligomer. Therefore, it would be important to assess whether other electrophilic lipids with dienone structure can induce a similar phenomenon in c-Jun or other proteins and what could be its functional relevance.

CONCLUSIONS AND PERSPECTIVES

Electrophilic eicosanoids are gaining importance both as signaling molecules and as potential pharmacological agents. In recent years, the work of numerous research groups has greatly broadened our knowledge of their generation, mechanisms of action, and pathophysiological and/or biochemical importance. The use of prostanoids with cyclopentenone structure in diverse experimental approaches has allowed exploring the subcellular localization, target identity, structural implications and functional consequences of the modification of cellular proteins by electrophilic eicosanoids (FIG. 6). In spite of this there are many aspects of the biology of these lipids that remain to be elucidated and will probably be the subject of intense research in the future. Much needs to be learned about the potential selectivity of protein modification by eicosanoids and its importance in their biological actions. It would be also

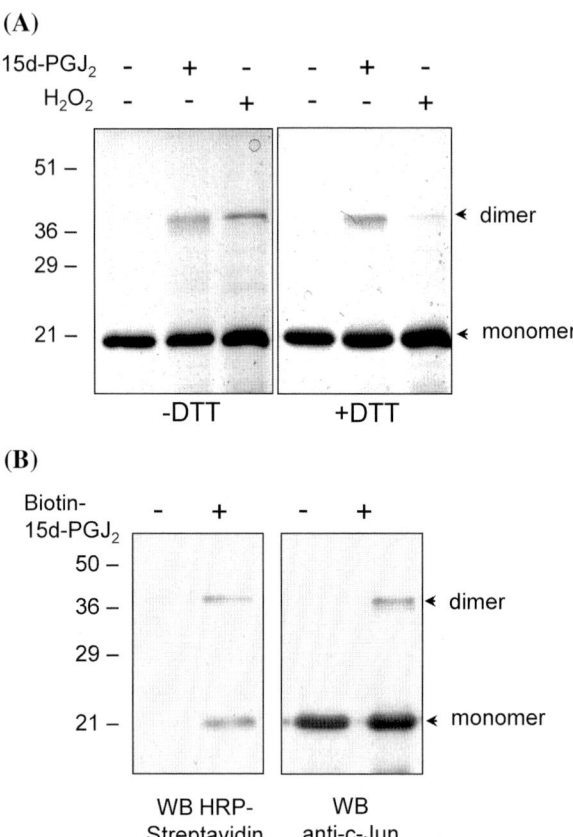

FIGURE 5. Involvement of 15d-PGJ$_2$ in c-Jun cross-linking. (**A**) A C320S mutant c-Jun construct was incubated the presence of 15d-PGJ$_2$ (10 μM) or H$_2$O$_2$ (1 mM) for 30 min at 37°C. Incubation mixtures were treated with 50 mM iodoacetamide before of after treatment with 10 mM DTT, as indicated, and analyzed by nonreducing SDS-PAGE and Coomassie staining. The position of c-Jun monomer and dimer is indicated. (**B**) c-Jun C320S was incubated in the absence or presence of biotinylated 15d-PGJ$_2$, subjected to SDS-PAGE under reducing conditions and analyzed by Western blot with horseradish peroxidase-conjugated streptavidin to reveal biotinylated 15d-PGJ$_2$ incorporation, and with anti-c-Jun antibody to show the position of c-Jun monomer and dimer.

interesting to explore the potential reversibility of these modifications in some proteins, or their effects on protein turnover. The elucidation of the physiological or pathophysiological roles of some electrophilic eicosanoids, such as cyPG, will require a detailed study of their generation in diverse situations and the use of sensitive methods for isolation and quantitation. In this context, a significant challenge will be the detection and identification of proteins modified by endogenous eicosanoids. In parallel, it would be important

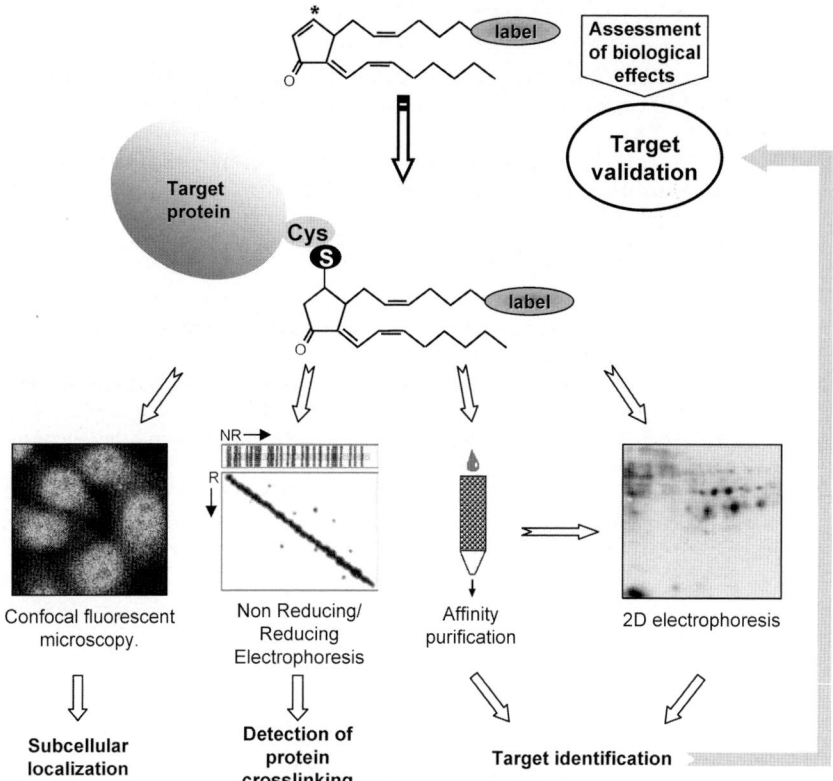

FIGURE 6. Diverse experimental approaches employed in the characterization of the electrophilic eicosanoid-protein interactome through the use of cyPG derivatives.

to continue the assessment of the therapeutic potential of cyPG-based active species.

ACKNOWLEDGMENTS

We wish to acknowledge the work of collaborators who have contributed to the research on this topic, F. J. Sánchez-Gómez, J. Gayarre, D. Sánchez, M. I. Avellano, Drs. E. Cernuda-Morollón, E. Pineda-Molina, F. J. Cañada, L. Boscá, and J.M. Rojas. The technical assistance of M. J. Carrasco is gratefully acknowledged. Our work is supported by Grants SAF2003-03713 from Ministerio de Educación y Ciencia and 0179 from Fundación La Caixa. K.S. is the recipient of an I3P fellowship from CSIS.

REFERENCES

1. MARNETT, L.J., J.N. RIGGINS & J.D. WEST. 2003. Endogenous generation of reactive oxidants and electrophiles and their reactions with DNA and proteins. J. Clin. Invest. **111:** 583–593.
2. FRATELLI, M. *et al.* 2002. Identification by redox proteomics of glutathionylated proteins in oxidatively stressed human T lymphocytes. Proc. Natl. Acad. Sci. USA **99:** 3505–3510.
3. JAFFREY, S.R. *et al.* 2001. Protein S-nitrosylation: a physiological signal for neuronal nitric oxide. Nat. Cell Biol. **3:** 193–197.
4. SCOZZAFAVA, A., A. CASINI & C.T. SUPURAN. 2002. Targeting cysteine residues of biomolecules: new approaches for the design of antiviral and anticancer drugs. Curr. Med. Chem. **9:** 1167–1185.
5. FUNK, C.D. 2001. Prostaglandins and leukotrienes: advances in eicosanoid biology. Science **294:** 1871–1875.
6. KUHN, H. 2005. Biologic relevance of lipoxygenase isoforms in atherogenesis. Expert Rev. Cardiovasc. Ther. **3:** 1099–1110.
7. ROBERTS, L.J., 2nd, J.P. FESSEL & S.S. DAVIES. 2005. The biochemistry of the isoprostane, neuroprostane, and isofuran pathways of lipid peroxidation. Brain Pathol. **15:** 143–148.
8. GAO, L. *et al.* 2003. Formation of prostaglandins E_2 and D_2 via the isoprostane pathway: a mechanism for the generation of bioactive prostaglandins independent of cyclooxygenase. J. Biol. Chem. **278:** 28479–28489.
9. STRAUS, D.S. & C.K. GLASS. 2001. Cyclopentenone prostaglandins: new insights on biological activities and cellular targets. Med. Res. Rev. **21:** 185–210.
10. PÉREZ-SALA, D. & S. LAMAS. 2001. Regulation of cyclooxygenase-2 expression by nitric oxide in cells. Antioxid. Redox Signal. **3:** 231–248.
11. PÉREZ-SALA, D. *et al.* 2001. Posttranscriptional regulation of human iNOS by the NO/cGMP pathway. Am. J. Physiol. Renal Physiol. **280:** F466-F473.
12. SERHAN, C.N. & J. SAVILL. 2005. Resolution of inflammation: the beginning programs the end. Nat. Immunol. **6:** 1191–1197.
13. DÍAZ-CAZORLA, M., D. PÉREZ-SALA & S. LAMAS. 1999. Dual effect of nitric oxide donors on cyclooxygenase-2 expression in human mesangial cells. J. Am. Soc. Nephrol. **10:** 943–952.
14. LEVY, B.D. 2001. Lipid mediator class switching during acute inflammation: signals in resolution. Nat. Immunol. **2:** 612–619.
15. ROBERTS, L.J., 2nd & J.D. MORROW. 2002. Products of the isoprostane pathway: unique bioactive compounds and markers of lipid peroxidation. Cell. Mol. Life Sci. **59:** 808–820.
16. DAVIES, S.S. *et al.* 2002. Effects of reactive gamma-ketoaldehydes formed by the isoprostane pathway (isoketals) and cyclooxygenase pathway (levuglandins) on proteasome function. FASEB J. **16:** 715–717.
17. GILROY, D.W. *et al.* 2004. Inflammatory resolution: new opportunities for drug discovery. Nat. Rev. Drug Discov. **3:** 401–416.
18. GILROY, D.W. *et al.* 1999. Inducible cyclooxygenase may have anti-inflammatory properties. Nat. Med. **5:** 698–701.
19. FORMAN, B.M. *et al.* 1995. 15-Deoxy-$\Delta^{12,14}$-prostaglandin J_2 is a ligand for the adipocyte determination factor PPAR gamma. Cell **83:** 803–812.

20. REILLY, C.M. et al. 2001. Prostaglandin J_2 inhibition of mesangial cell iNOS expression. Clin. Immunol. **98:** 337–345.
21. SÁNCHEZ-GÓMEZ, F.J. et al. 2004. Protein thiol modification by 15-deoxy-$\Delta^{12,14}$-prostaglandin J_2 addition in mesangial cells: role in the inhibition of proinflammatory genes. Mol. Pharmacol. **66:** 1349–1358.
22. CASTRILLO, A. et al. 2000. Inhibition of IkappaB kinase and IkappaB phosphorylation by 15-deoxy- $\Delta^{12,14}$-prostaglandin J_2 in activated murine macrophages. Mol. Cell. Biol. **20:** 1692–1698.
23. ROSSI, A. et al. 2000. Anti-inflammatory cyclopentenone prostaglandins are direct inhibitors of IκB kinase. Nature **403:** 103–108.
24. CERNUDA-MOROLLÓN, E. et al. 2001. 15-Deoxy-$\Delta^{12,14}$-prostaglandin J_2 inhibition of NF-κB DNA binding through covalent modification of the p50 subunit. J. Biol. Chem. **276:** 35530–35536.
25. SHIBATA, T. et al. 2003. An endogenous electrophile that modulates the regulatory mechanism of protein turnover: inhibitory effects of 15-deoxy-Δ12,14-prostaglandin J_2 on proteasome. Biochemistry **42:** 13960–13968.
26. STRAUS, D.S. et al. 2000. 15-deoxy-$\Delta^{12,14}$-prostaglandin J_2 inhibits multiple steps in the NF-κB signaling pathway. Proc. Natl. Acad. Sci. USA **97:** 4844–4849.
27. PÉREZ-SALA, D., E. CERNUDA-MOROLLÓN & F.J. CAÑADA. 2003. Molecular basis for the inhibition of AP-1 DNA binding by 15-deoxy-$\Delta^{12,14}$-prostaglandin J_2. J. Biol. Chem. **278:** 51251–51260.
28. SHIBATA, T. et al. 2003. Thioredoxin as a molecular target of cyclopentenone prostaglandins. J. Biol. Chem. **278:** 26046–26054.
29. MOOS, P.J. et al. 2003. Electrophilic prostaglandins and lipid aldehydes repress redox-sensitive transcription factors p53 and hypoxia-inducible factor by impairing the selenoprotein thioredoxin reductase. J. Biol. Chem. **278:** 745–750.
30. LEVONEN, A.L. et al. 2004. Cellular mechanisms of redox cell signaling: the role of cysteine modification in controlling antioxidant defenses in response to electrophilic lipid oxidation products. Biochem. J. **378:** 373–382.
31. ITOH, K. et al. 2004. Transcription factor Nrf2 regulates inflammation by mediating the effect of 15-deoxy-$\Delta^{12,14}$-prostaglandin J_2. Mol. Cell. Biol. **24:** 36–45.
32. OLIVA, J.L. et al. 2003. The cyclopentenone 15-deoxy-$\Delta^{12,14}$-prostaglandin J_2 binds to and activates H-Ras. Proc. Natl. Acad. Sci. USA **100:** 4772–4777.
33. WAGNER, T.M., J.E. MULLALLY & F.A. FITZPATRICK. 2006. Reactive lipid species from cyclooxygenase-2 inactivate tumor suppressor LKB1 / STK11: cyclopentenone prostaglandins and 4-hydroxy-2-nonenal covalently modify and inhibit the AMP-kinase kinase that modulates cellular energy homeostasis and protein translation. J. Biol. Chem. **281:** 2598–2604.
34. CHATTERJEE, P.K. et al. 2004. The cyclopentenone prostaglandin 15-deoxy-$\Delta^{12,14}$-prostaglandin J_2 ameliorates ischemic acute renal failure. Cardiovasc. Res. **61:** 630–643.
35. KAWAHITO, Y. et al. 2000. 15-deoxy-$\Delta^{12,14}$-PGJ_2 induces synoviocyte apoptosis and suppresses adjuvant-induced arthritis in rats. J. Clin. Invest. **106:** 189–197.
36. DIAB, A. et al. 2002. Peroxisome proliferator-activated receptor-gamma agonist 15-deoxy-$\Delta^{12,14}$-prostaglandin J_2 ameliorates experimental autoimmune encephalomyelitis. J. Immunol. **168:** 2508–2515.
37. IANARO, A. et al. 2003. 2-Cyclopenten-1-one and prostaglandin J_2 reduce restenosis after balloon angioplasty in rats: role of NF-κB. FEBS Lett. **553:** 21–27.

38. ANDO, M. *et al.* 2003. Retrovirally introduced prostaglandin D_2 synthase suppresses lung injury induced by bleomycin. Am. J. Respir. Cell. Mol. Biol. **28:** 582–591.
39. MUSIEK, E.S. *et al.* 2005. Cyclopentenone eicosanoids as mediators of neurodegeneration: a pathogenic mechanism of oxidative stress-mediated and cyclooxygenase-mediated neurotoxicity. Brain Pathol. **15:** 149–158.
40. CHEN, Y., J.D. MORROW & L.J. ROBERTS, 2nd. 1999. Formation of reactive cyclopentenone compounds in vivo as products of the isoprostane pathway. J. Biol. Chem. **274:** 10863–10868.
41. MUSIEK, E. *et al.* 2005. Cyclopentenone isoprostanes inhibit the inflammatory response in macrophages. J. Biol. Chem. **280:** 35562–35570.
42. SHIRAKI, T. *et al.* 2005. a,ß-Unsaturated ketone is a core moiety of natural ligands for covalent binding to peroxisome proliferator-activated receptor. J. Biol. Chem. **280:** 14145–14153.
43. SUBBANAGOUNDER, G. *et al.* 2002. Epoxyisoprostane and epoxycyclopentenone phospholipids regulate monocyte chemotactic protein-1 and interleukin-8 synthesis. Formation of these oxidized phospholipids in response to interleukin-1beta. J. Biol. Chem. **277:** 7271–7281.
44. BIRUKOV, K.G. *et al.* 2004. Epoxycyclopentenone-containing oxidized phospholipids restore endothelial barrier function via Cdc42 and Rac. Circ. Res. **95:** 892–901.
45. DIANZANI, M.U. 2003. 4-hydroxynonenal from pathology to physiology. Mol. Aspects Med. **24:** 263–272.
46. KEHRER, J.P. & S.S. BISWAL. 2000. The molecular effects of acrolein. Toxicol. Sci. **57:** 6–15.
47. CEASER, E.K. *et al.* 2004. Mechanisms of signal transduction mediated by oxidized lipids: the role of the electrophile-responsive proteome. Biochem. Soc. Trans. **32:** 151–155.
48. KWOK, B.H. *et al.* 2001. The anti-inflammatory natural product parthenolide from the medicinal herb Feverfew directly binds to and inhibits IκB kinase. Chem. Biol. **8:** 759–766.
49. LEE, J.H. *et al.* 2002. Kaurane diterpene, kamebakaurin, inhibits NF-κB by directly targeting the DNA-binding activity of p50 and blocks the expression of antiapoptotic NF-κB target genes. J. Biol. Chem. **277:** 18411–18420.
50. MAHON, T.M. & L.A. O'NEILL. 1995. Studies into the effect of the tyrosine kinase inhibitor herbimycin A on NF-κB activation in T lymphocytes. Evidence for covalent modification of the p50 subunit. J. Biol. Chem. **270:** 28557–28564.
51. OGINO, S. *et al.* 2004. Herbimycin A abrogates Nuclear Factor-kappaB activation by interacting preferentially with the IkappaB Kinase beta subunit. Mol. Pharmacol. **65:** 1344–1351.
52. VERBITSKI, S.M. *et al.* 2004. Punaglandins, chlorinated prostaglandins, function as potent Michael receptors to inhibit ubiquitin isopeptidase activity. J. Med. Chem. **47:** 2062–2070.
53. NOYORI, R. & M. SUZUKI. 1993. Organic synthesis of prostaglandins: advancing biology. Science **259:** 44–45.
54. NARUMIYA, S. & M. FUKUSHIMA. 1986. Site and mechanism of growth inhibition by prostaglandins. I. Active transport and intracellular accumulation of cyclopentenone prostaglandins, a reaction leading to growth inhibition. J. Pharmacol. Exp. Ther. **239:** 500–505.

55. NARUMIYA, S. et al. 1986. Site and mechanism of growth inhibition by prostaglandins. II. Temperature-dependent transfer of a cyclopentenone prostaglandin to nuclei. J. Pharmacol. Exp. Ther. **239:** 506–511.
56. NARUMIYA, S. et al. 1987. Site and mechanism of growth inhibition by prostaglandins. III. Distribution and binding of prostaglandin A_2 and Δ^{12}-prostaglandin J_2 in nuclei. J. Pharmacol. Exp. Ther. **242:** 306–311.
57. PARKER, J. 1995. Prostaglandin A_2 protein interactions and inhibition of cellular proliferation. Prostaglandins **50:** 359–375.
58. LANDAR, A. et al. 2006. The interaction of electrophilic lipid oxidation products with mitochondria in endothelial cells and the formation of reactive oxygen species. Am. J. Physiol. Heart Circ. Physiol. **290:** H1777–H1787.
59. STAMATAKIS, K., F.J. SÁNCHEZ-GÓMEZ & D. PÉREZ-SALA. 2006. Identification of novel protein targets for modification by 15-deoxy-$\Delta^{12,14}$-prostaglandin J_2 in mesangial cells reveals multiple interactions with the cytoskeleton. J. Am. Soc. Nephrol. **17:** 89–98.
60. HORTELANO, S. et al. 2000. Contribution of cyclopentenone prostaglandins to the resolution of inflammation through the potentiation of apoptosis in activated macrophages. J. Immunol. **165:** 6525–6531.
61. KONDO, M. et al. 2001. Cyclopentenone prostaglandins as potential inducers of intracellular oxidative stress. J. Biol. Chem. **276:** 12076–12083.
62. LI, L. et al. 2001. 15-deoxy-$\Delta^{12,14}$-prostaglandin J_2 induces apoptosis of human hepatic myofibroblasts. A pathway involving oxidative stress independently of peroxisome-proliferator-activated receptors. J. Biol. Chem. **276:** 38152–38158.
63. ISHII, T. & K. UCHIDA. 2004. Induction of reversible cysteine-targeted protein oxidation by an endogenous electrophile 15-deoxy-$\Delta^{12,14}$-prostaglandin J_2. Chem. Res. Toxicol. **17:** 1313–1322.
64. LANDAR, A. et al. 2006. Induction of the permeability transition and cytochrome c release by 15-deoxy prostaglandin J_2 in mitochondria. Biochem. J. **394:** 185–195.
65. GAYARRE, J. et al. 2005. Differential selectivity of protein modification by the cyclopentenone prostaglandins PGA_1 and 15-deoxy-$\Delta^{12,14}$-PGJ_2: role of glutathione. FEBS Lett. **579:** 5803–5808.
66. SUZUKI, M. et al. 1997. Chemical implications for antitumor and antiviral prostaglandins: reaction of Δ^7-prostaglandin A_1 and prostaglandin A_1 methyl esters with thiols. J. Am. Chem. Soc. **119:** 2376–2385.
67. ATSMON, J. et al. 1990. Conjugation of 9-deoxy-Δ^9, Δ^{12}(E)-prostaglandin D_2 with intracellular glutathione and enhancement of its antiproliferative activity by glutathione depletion. Cancer Res. **50:** 1879–1885.
68. LEVONEN, A.L. et al. 2001. Biphasic effects of 15-deoxy-$\Delta^{12,14}$-prostaglandin J_2 on glutathione induction and apoptosis in human endothelial cells. Arterioscler. Thromb. Vasc. Biol. **21:** 1846–1851.
69. ZHANG, X. et al. 2004. Stress protein activation by the cyclopentenone prostaglandin 15-deoxy-$\Delta^{12,14}$-prostaglandin J_2 in human mesangial cells. Kidney Int. **65:** 798–810.
70. VEGA, M.C. et al. 1998. The three-dimensional structure of Cys-47-modified mouse liver glutathione S-transferase P1-1. Carboxymethylation dramatically decreases the affinity for glutathione and is associated with a loss of electron density in the alphaB-310B region. J. Biol. Chem. **273:** 2844–2850.
71. KLATT, P. & S. LAMAS. 2000. Regulation of protein function by S-glutathiolation in response to oxidative and nitrosative stress. Eur. J. Biochem. **267:** 4928–4944.

72. BICKLEY, J.F. *et al.* 2004. Reactions of some cyclopentenones with selected cysteine derivatives and biological activities of the product thioethers. Bioorg. Med. Chem. **12:** 3221–3227.
73. MURPHY, R.C. & S. ZARINI. 2002. Glutathione adducts of oxyeicosanoids. Prostaglandins Other Lipid Mediat. **68-69:** 471–482.
74. STAMATAKIS, K., F.J. SÁNCHEZ-GÓMEZ & D. PÉREZ-SALA. 2004. Protein modification by cyclopentenone prostaglandin addition: biological actions and therapeutic implications. Gene Ther. Mol. Biol. **8:** 241–258.
75. CERNUDA-MOROLLÓN, E. & D. PÉREZ-SALA. 2006. Regulation of proinflammatory transcription factors by direct modification with cyclopentenone prostaglandins. *In* Trends in DNA Research. C.R. Woods, Ed. 1–31. Nova Science Publishers, Inc. Hauppauge, NY.
76. ATSMON, J. *et al.* 1990. Formation of thiol conjugates of 9-deoxy-$\Delta^9,\Delta^{12}(E)$-prostaglandin D_2 and $\Delta^{12}(E)$-prostaglandin D_2. Biochemistry **29:** 3760–3765.

Index of Contributors

Abdi-Oskouei, F., 52–64
Abdollahi, M., 110–122, 142–150
Accorsi, A., 10–16
Aggarwal, B.B., 151–169
Ahn, K.S., 151–169
Akchurin, R.S., 205–217
Albertini, M.C., 10–16
Algás, R., 412–424
Aliño, S.F., 412–424
Andreeva, E., 490–500
Angel, P., 310–318
Arenz, A., 170–183, 191–204
Attar, F., 52–64
Bantseev, V., 10–16, 17–33
Barišić, I., 225–232
Bartolomé, N., 282–295
Basini, G., 34–40, 17–33
Baumstark-Khan, C., 170–183, 191–204
Beabealashvilli, R.S., 205–217, 319–335
Bergamaschi, A., 1–9, 10–16
Bianchini, E., 509–516
Biro, J.C., 399–411
Bogner, S., 191–204
Borgatti, M., 509–516
Borutinskaite, V.V., 346–355, 368–384
Botella, R., 412–424
Bottone, M.G., 94–101
Breveglieri, G., 184–190
Britareva, V.V., 319–335
Bulkina, O.S., 319–335
Busnadiego, I., 408–418, 419–428
Buza, V.V., 319–335

Cadepond, F., 296–309
Callebaut, I., 531–539
Carlo, P., 455–465
Carretero, O.A., 336–345
Cerella, C., 1–9, 10–16
Chernolovskaya, E.L., 425–436
Cobeño, L., 41–51
Cogolludo, A., 41–51
Concin, N., 270–281
Cristofanon, S., 1–9
Čulić, V. 225–232

D'Alessio, M., 1–9, 10–16
Daryani, N.E., 110–122
Dasí, F., 412–424
De Nicola, M., 1–9, 10–16
Dehghan, G., 110–122, 142–150
Diederich, M., xxvii, 1–9
Dittmer, U., 540–547
Djiane, J., 531–539
Doe, I., 445–459
Dorcaratto, A., 332–343
Dzemeshkevich, S.L., 205–217

Ermonval, M., 123–141
Eshghtork, A., 110–122
Fallot, G., 296–309
Feriotto, G., 184–190
Finotti, A., 184–190
Franceschetti, L., 445–454
Frazziano, G., 41–51

Gambari, R., 184–190, 509–516
Genova, P., 470–489
Gertler, A., 501–508, 531–539
Ghamsari, L., 65–75
Ghibelli, L., 1–9, 10–16
Ghorbani, F., 110–122
Girard, C., 296–309
Goebel, G., 270–281
Golitsyn, S.P., 205–217
Gonen-Berger, D., 531–539
Goryunova, L.E., 205–217, 319–335
Grasselli, F., 34–40, 17–33
Grotheer, H.-H., 170–183
Groyer, A., 296–309
Gussakowsky, E.E., 501–508

Höfer, T., 517–530
Ha, J., 102–109
Hadizadeh, M., 65–75
Hegedűs, C., 344–354
Hellweg, C.E., 170–183, 191–204
Herrero, M.J., 412–424
Heshmat, R., 142–150
Hofstetter, G., 270–281

Hrabia, A., 501–508
Hwa Lee, J., 76–82
Hwang, J.-T., 102–109

Ichikawa, H., 151–169
Ishiki, T., 445–459

Kabilova, T.O., 425–436
Kaledin, V., 490–500
Kang, G.H., 218–224
Kang, S.B., 218–224
Karpov, Y.A., 319–335
Kawabata, A., 445–459
Kawai, K., 445–459
Kellermann, O., 123–141
Kertesz, A., 437–444
Keyhani, E., 52–64, 65–75
Keyhani, J., 52–64, 65–75
Kharlap, M.S., 205–217
Khaspekov, G.L., 205–217, 319–335
Kim, H., 76–82
Kim, J.W., 218–224
Kim, K.H., 76–82
Kim, Y.M., 102–109
Knežević, J., 225–232
Kob, A., 460–469
Konstantinov, S., 470–489
Koutsilieri, E., 540–547
Kovalevskii, D.A., 319–335
Kubo, S., 445–459
Kumei, Y., 311–317
Kunnumakkara, A.B., 151–169

Lambertini, E., 509–516
Lamla, S., 540–547
Lampronti, I., 509–516
Larijani, B., 142–150
Launay, J.-M., 123–141
Lee, H.P., 218–224
Lee, T.S., 218–224
Li, X.C., 336–345
Lodi, F., 41–51

Magnusson, K.-E., 346–355, 356–367, 368–384
Magri, E., 509–516
Magrini, A., 1–9, 10–16
Martínez, M.J., 282–295

Marth, C., 270–281
Matijević, T., 225–232
Matlhagela, K., 233–243
Matsui, H., 445–459
Merzvinskyte, R., 356–367
Mestres, P., 460–469
Mihailova, M., 470–489
Min, G., 244–257, 258–269
Mironova, N., 490–500
Mobasheri, A., 83–93
Mohammadirad, A., 110–122, 142–150
Mondello, C., 94–101
Moreno, L., 41–51
Morguet, A., 460–469
Mouillet-Richard, S., 123–141
Mutel, V., 123–141

Navakauskiene, R., 346–355, 356–367, 368–384
Nikfar, S., 110–122
Nikolin, V., 490–500
Nishikawa, H., 445–459
Niv-Spector, L., 531–539
Noriki, S., 94–101
Nuccitelli, S., 1–9

Ochoa, B., 282–295

Pérez-Sala, D., 548–570
Paczoska-Eliasiewicz, H.E., 501–508
Pandey, M.K., 151–169
Park, I.-J., 102–109
Park, N.H., 218–224
Park, O.J., 102–109
Paternoster, L., 10–16
Pavelič, J., 225–232
Pellicciari, C., 94–101
Penolazzi, L., 509–516
Perez-Vizcaino, F., 41–51
Pietri, M., 123–141
Piva, R., 509–516
Platt, N., 83–93
Popova, N., 490–500

Radogna, F., 1–9, 10–16
Rahimi, R., 142–150
Rešić, B., 225–232
Reicher, S., 501–508

INDEX OF CONTRIBUTORS

Reimer, D., 270–281
Rethwilm, A., 540–547
Rezaie, A., 110–122
Řezáěová, M., 385–398
Richter, K.H., 310–318
Rodríguez, L., 282–295
Ruskin, V.V., 205–217

Sánchez, M., 412–424
Sadr, S., 270–281
Salazar, C., 517–530
Salomon, G., 501–508
Sandur, S.K., 151–169
Santini, S.E., 34–40, 17–33
Sarmay, G., 437–444
Sator-Schmitt, M., 310–318
Savickiene, J., 356–367, 368–384
Scamrov, A.V., 319–335
Scheller, C., 540–547
Schmidt, W., 460–469
Schmitz, C., 191–204
Schneider, B., 123–141
Schorpp-Kistner, M., 310–318
Schweizer-Groyer, G., 296–309
Scovassi, A.I., 94–101
Sekiguchi, F., 445–459
Sethi, G., 151–169
Shakibaei, M., 83–93
Shklyaeva, O., 490–500
Soldani, C., 94–101
Solomon, G., 531–539
Song, Y.S., 218–224
Stamatakis, K., 548–570
Stojicic, N., 170–183
Sung, B., 151–169

Taghavi, B., 110–122
Takacs, B., 437–444
Talitskii, K.A., 319–335

Tamargo, J., 41–51
Taub, M., 233–243
Tavanti, E., 509–516
Textor, B., 310–318
Thedinga, E., 460–469
Thorpe, C., 83–93
Tichý, A., 385–398
Timofeeva, A.V., 205–217, 319–335
Toth, G.K., 437–444
Treigyte, G., 356–367, 368–384
Trenev, V., 470–489
Treves, S., 184–190

Uguccioni, F., 10–16
Uli, V., 225–232
Ullrich, A., 540–547

Vávrová, J., 385–398
Varadi, G., 437–444
Villamor, E., 41–51
Violani, E., 455–465
Vladimirova, A.V., 425–436
Vlassov, V.V., 425–436, 490–500
Vokurková, D., 385–398

Wiedemair, A., 270–281

Youn, H.-Y., 10–16, 17–33
Yousefzadeh, G., 142–150
Yu, J.H., 76–82

Záškodová, D., 385–398
Zamani, M.J., 110–122
Zeimet, A.G., 270–281
Zenkova, M., 490–500
Zennaro, M., 509–516
Zhuo, J.L., 336–345
Zorzato, F., 184–190